Environmental Change and Challenge

Third Edition

Environmental Change and Challenge

A Canadian Perspective

Philip Dearden

Bruce Mitchell

	Trees Saved	Air Emissions Eliminated	Water Saved	Solid Waste Eliminated
Planet Friendly Publishing ✓ Made in the United States ✓ Printed on Recycled Paper GREEN EDITION — Learn more at www.greenedition.org	54	5,417 lb	22,814 gal.	2,114 lb[1]

Manufacturing books in North America ensures compliance with strict environmental laws and eliminates the need for international freight shipping, a major contributor to global air pollution. Printing on recycled paper helps minimize our consumption of trees, water, and fossil fuels.

[1] *Estimates were made using the Environmental Defense Paper Calculator.*

OXFORD
UNIVERSITY PRESS

OXFORD

UNIVERSITY PRESS

8 Sampson Mews, Suite 204, Don Mills, Ontario M3C 0H5
www.oupcanada.com

Oxford University Press is a department of the University of Oxford.
It furthers the University's objective of excellence in research, scholarship,
and education by publishing worldwide in

Oxford   New York

Auckland Cape Town Dar es Salaam Hong Kong Karachi
Kuala Lumpur Madrid Melbourne Mexico City Nairobi
New Delhi Shanghai Taipei Toronto

With offices in

Argentina Austria Brazil Chile Czech Republic France Greece
Guatemala Hungary Italy Japan Poland Portugal Singapore
South Korea Switzerland Thailand Turkey Ukraine Vietnam

Oxford is a trade mark of Oxford University Press
in the UK and in certain other countries

Published in Canada by Oxford University Press

Copyright © Oxford University Press Canada 2009

Library and Archives Canada Cataloguing in Publication

Dearden, Philip
Environmental change and challenge : a Canadian perspective / Philip
Dearden and Bruce Mitchell. -- 3rd ed.

Includes bibliographical references and index.
ISBN 978-0-19-542865-0

1. Environmental management--Canada--Textbooks. 2. Human
ecology--Canada--Textbooks. 3. Nature--Effect of human beings
on--Canada--Textbooks. 4. Global environmental change--Textbooks.
I. Mitchell, Bruce, 1944- II. Title.

GF511.D42 2009   333.70971   C2009-901065-8

Cover image: Gary Braasch/Braasch Environmental Photography

Part- and chapter-opening photo credits: pp. 2–3: Torleif Svensson/Corbis; p. 5: iStockphoto.com/Peeter Viisimaa;
pp. 48–49: iStockphoto.com/Michael Fernahl; p. 53: iStockphoto.com/John Anderson; p. 83: Gary Irving/
PhotoDisc/Panoramic Landscapes/Getty Images; p. 120: iStockphoto.com/Greg Cooksey; pp. 158–59: Hans
Peter Merten/zefa/Corbis; p. 162: iStockphoto.com/Dan Barnes; p. 178: iStockphoto.com/Steve Jacobs; pp. 206–07:
iStockphoto.com/Shaun Lowe; p. 209: iStockphoto.com/Mehmet Can; p. 246: iStockphoto.com/Stanley Creek
Media; p. 292: iStockphoto.com/AVTG; p. 346: iStockphoto.com/Mark Kuipers; p. 393: iStockphoto.com/Jason
Verschoor; p. 442: iStockphoto.com/Rob Fox; p. 481: iStockphoto.com/Peeter Viisimaa; p. 514: iStockphoto.com/
Kim Bunker; pp. 566–67: iStockphoto.com/Andrew Penner; p. 569: iStockphoto.com/Michael Westhoff

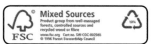

Oxford University Press is committed to our environment. This book is printed on Forest
Stewardship Council certified paper, harvested from a responsibly managed forest, which
contains a minimum of 10% post-consumer waste.

Printed and bound in the United States of America.

3 4 5 – 12 11 10

Contents Overview

Detailed Contents

Chapter 9   Forests   292

Chapter 10   Agriculture   346

Chapter 11   Water   393

Chapter 12   Minerals and Energy   442

Preface

When we wrote the first edition of *Environmental Change and Challenge* more than a decade ago, it was already becoming very obvious that the two themes of 'change' and 'challenge' were going to be major defining characteristics of the twenty-first century. However, the speed with which change has occurred was a little unanticipated. And with that rapid change has come massive challenges.

Scientists in the mid-1990s were well aware of global climate change, but the speed of change was expected to be a concern for the next rather than this generation. The Arctic Ocean was predicted to be ice-free in 50 to 100 years. However, following the colossal ice losses over the last couple of years as positive feedback loops kicked in, that prediction has been revised to within the next five years.

The challenges created by these and other changes will be profound and global. Sea levels will rise, communities will be flooded, ocean currents will change, rainfall patterns will alter, crops will fail, and billions of lives will be affected.

The reality of these changes is difficult for many people to believe, since it counters many of our most deep-seated beliefs. Treaties were signed with First Nations for 'as long as the rains fall, as long as the rivers flow, as long as the winds blow' because these were the immutable constructs of nature that were reliable. The Earth was also conceived as being so large that the impact of humans was trifling in comparison. Photos of our lonely planet floating through space taken from spacecraft helped to dispel this myth.

A fundamental change has taken place over the past couple of decades in the relationship between humans and our fragile planet. No longer is the planet a vast and wild place where change occurs on a geological time scale driven by natural forces; it has, in fact, become the 'greenhouse' of the greenhouse gas analogy in which wild nature is replaced by human constructs, and even the vast atmosphere and oceans reflect human desires as they become increasingly choked by the industrial wastes of a consumer society.

There has never been a more critical time when humans should know how the planet works and especially about the processes that drive our life support system. But environmental management is not about managing natural systems; it is about managing humans and our impacts on these systems. This book was written with these twin goals in mind: that students should gain a basic appreciation of how the planet works and also understand the impacts of humanity on these systems, the challenges created, and potential solutions.

The book is also focused on Canada, and creating such a resource when most texts were American or European was, and still is, our main driving force. Canada is a huge and beautiful country, one of the most magnificent places on Earth. Our geography, people, history, and political culture are different from those of the US and Europe. We believe that Canada warrants its own texts for students rather than a spin-off from texts written for other countries. Canadians can also play a major role in what happens globally in terms of the environment. We are the world's second largest country in terms of area. We are also a rich country. In general, our citizens have a high quality of life and value the environment, but we also create some of the highest per capita impacts in the world in terms of carbon dioxide emissions, water use, and waste production. Changes need to take place. And those changes need to take place far more quickly than is currently the case. Our so-called leaders have only been willing to make those changes if they perceive support for them. That support hinges on having a well-informed and active populace.

We believe it is critical that university students leave our universities when they graduate with a greater understanding of the planetary ecosystems that support life and of their impacts on ecosystems, as well as an awareness of what society and individuals can do to help improve the situation. If all university graduates came out thus informed and acted on this knowledge to create change in their own lifestyles and society, the prognosis for the future would be a little more optimistic.

This book was written to help contribute towards these changes. It was written for students taking a first course in environment to impart an understanding of the biosphere's function and to link basic environmental management principles to environmental and resource problems in a Canadian context. The book provides both a basic background for those who will go on to specialize in fields other than environment and a broad platform upon which more detailed courses on environment can build later.

Part A (Chapter 1) provides an overall introduction to environment, resources, and society and the role of science, both social and natural, in helping us to understand the relationship among them. This relationship is illustrated in more detail by a well-known Canadian case study, the Sydney Tar Ponds. We also provide a global and national context for environmental management and describe some approaches for assessing current progress in dealing with environmental challenges. If we do not know how we are doing, we can hardly judge with any degree of accuracy the severity of the problem or map out suitable strategies to address the problem.

Part B (Chapters 2–4) provides a basic primer on the environmental processes that constitute the Earth's life support system. Primary emphasis is on energy flows, biogeochemical cycles, and biotic responses, with reference to Canadian examples wherever possible. A strong emphasis is placed both here and in subsequent sections on making explicit links between these principles and examples illustrating the principles in action.

Part C (Chapters 5 and 6) reviews what are referred to as different philosophies, processes, and products that should characterize high-quality resource and environmental planning and management. Some refer to such attributes as elements of 'best practice'. Our hope is that by the end of these two chapters, you will be able to develop a mental checklist of the attributes you would expect to see used in planning and management and that you would advocate either as a team member addressing resource and environmental issues or as a member of civil society.

Part D (Chapters 7–14) takes the basic science of Part B and the management approaches of Part C and puts them together by focusing on environmental and resource management themes: climate change, oceans and fisheries, forests, agriculture, water, minerals and energy, urban environmental management, and endangered species and protected areas. In each chapter, we provide an overview of the current situation in Canada and the main management challenges. Selected international examples also are provided. Text boxes highlight particular case studies of interest and also illustrate the connectivity among the different themes.

The final section (Chapter 15) concludes the book with views from three perspectives—global, national, and personal. Here we emphasize solutions and the actions that individuals can take in moving towards a more sustainable society and introduce the 'Law of Everybody', suggesting that if everyone took a few conservation actions, they would add up to a massive contribution to the overall changes required. We question the ways in which values are taken into account in much environmental decision-making and also the way in which development progress is measured. The Happy Planet Index, for example, is one international indicator that has been suggested as an alternative to measure progress that takes into account not only level of human well-being but also the costs of achieving that well-being.

Change and challenge are main themes of this book, and fundamental changes are required in the way by which society manages itself to meet the challenges that lie ahead. We hope that this book will help in some small way to contribute to producing the more sustainable future that must evolve over the next few years and encourage you, the reader, to become part of making this future a reality.

Philip Dearden
Bruce Mitchell

Features

Scientific Content

Thoroughly expanded coverage of basic scientific concepts ensures that students gain a comprehensive understanding of the Earth's ecosystems and processes.

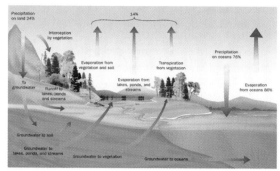

Figure 4.8 The hydrological cycle.

the order of nine to 12 days. These figures have special relevance with regard to the effects of pollution. Although many major rivers have suffered from critical pollution incidents, the flushing action of rivers, combined with the short residence time of the water, means that a relatively rapid recovery is often possible. This is not the case, however, with groundwater pollution, especially deep groundwater pollution.

The hydrological cycle involves the transport of water from the oceans to the atmosphere, through terrestrial and subterranean systems, and back to the ocean, all fuelled by energy from the sun. Eighty-four per cent of the water in the atmosphere is evaporated directly from the ocean surface. The remainder comes from evaporation from smaller water bodies, from the leaves of plants (**transpiration**), or from the soil and plants (**evapotranspiration**). As it evaporates, water leaves behind accumulated impurities. The most common dissolved substance in the ocean is sodium chloride, or table salt, which also contains many other elements in trace amounts. Evaporation acts as a giant purification plant until further pollutants are encountered in the atmosphere.

Once in the atmosphere, the water vapour cools, condenses around tiny particles called condensation nuclei, forms clouds, and is precipitated to earth as either rain, snow, or hail. The warmer the air is, the more water it can hold. Moisture content can be expressed in terms of relative humidity, the amount of moisture held compared to how much could be held if fully saturated at a particular temperature. At a relative humidity

of 100 per cent, the air is saturated, and cloud, fog, and mist form. Clouds are moved around by winds and continue to grow until precipitation (Box 4.7) occurs and the water is returned to earth.

About 77 per cent of precipitation falls into the ocean. The remainder joins the terrestrial part of the cycle in ice caps, lakes, rivers, groundwater, and transport between these compartments. Gravity moves water down through the soil until it reaches the water table, where all the spaces between the soil particles are full of water. This is the **groundwater**

Water in the solid phase of ice may be in the Lowell glacier, located in Kluane National Park, Yukon, for thousands of years before melting and flowing to the ocean.

Guest Statement

Feng Shui Landscapes in the Context of Urban Environments in Hong Kong and South China
Lawal Marafa

Patches of forest landscapes located in urban areas play important ecological roles in the overall context of city structures. As urban areas experience numerous problems such as deterioration of air quality, rising temperature, and psychological stress, urban forests and green spaces filter air and water, trap sunlight, and provide shade and recreation areas for the resident population. In fact, such forest areas can help to reduce the urban heat island effect that characterizes most urban areas. Although many urban forests are remnants of wilderness that either purposely or accidentally escaped development processes, globally their existence reflects value attributed to them regarding both cultural purposes and ecological functions.

In past civilizations, forests attracted people for various reasons that include shelter, wood for homes and fires, meditation and worship, or even physical support. Forests and culture have therefore indeed been intertwined throughout human history. But these relationships have often been altered by the growth of urbanized landscapes and the extension of related values. In Hong Kong and South China, the forested landscapes, often at the periphery of urban areas, have been modified as a result of adherence to the *feng shui*, or the science of spatial placement and traditional geomancy. While geomancy in the Eastern tradition involves the interpretation of topography of space and place, it is also seen as a system of human propensity for pattern recognition. Consequently, the *feng shui* has shaped much of the human landscape of Hong Kong and South China through constituting an established system of land cover and landmarks in the region. Characteristically, the *feng shui* woodland is preceded in the front by agricultural fields, followed by the residential area at the bottom of the hills adjacent to hilly woodland areas. But as agriculture has been reduced or has ceased to exist in most urban areas, these open fields at the bottom of the *feng shui* landscapes are assuming other functions,

such as becoming flat open spaces for recreation or other urban infrastructure.

The *feng shui* woods are a common feature associated with most settlements of South China. These wood groves primarily consist of broad-leaved trees that are often juxtaposed along the residential areas. In fact in historical terms, *feng shui* explicitly states that human beings should live in harmony with nature, leading to retaining trees in settled areas. Reflecting this historical notion, the art and science of *feng shui* are gradually becoming incorporated into various aspects of city planning that is now called an 'ecocity' approach. As societies develop and urbanization escalates in South China, patches of these forests have often been retained in urban areas and in some cases at the periphery of the urban environment. One of the main reasons for this retention is the cultural belief of the East that there is a connection between trees and good fortune. It is partly as a result of this that the *feng shui* woods and forests have been preserved, despite the influence of urbanization and human activities that continuously encroach or exert pressure on the forest woodlands.

Although the *feng shui* woodlands are limited in size, ranging in Hong Kong from 600 square metres to about

Luk Keng is a village in the North East New Territories of Hong Kong. The classical layout includes an agricultural field in the lower topography, a village at the foot of the hill, and an area of *feng shui* woodland on the elevated area adjacent to the village.

National and Global Perspectives

Guest Statement boxes provide perspectives and field experience from Canadian and—new to this edition—international geographers working all over the globe.

Robust Ancillary Suite

An exciting DVD-ROM provided with the text presents video segments and audio podcasts chosen specifically to complement content in each of the chapters, enhancing students' experience of critical concepts and issues. In addition, a full slate of online ancillaries offer support for both students and instructors. A student study guide of review material and instructor tools such as an instructor's manual, test bank, and PowerPoint slides are available on the companion website: www.oupcanada.com/environmentalchangechallenge.

Chapter 13

Urban Environmental Management

Learning Objectives

- To understand the nature and significance of urbanization.
- To understand the quality of environmental conditions in Canadian cities.
- To become aware of the impacts of urban areas on the environment.
- To understand the vulnerability of urban areas to natural and human-induced events.
- To become aware of 'best practices' related to urban environmental management.
- To identify strategies for governments, the private sector, civil societies, and individuals to reduce the impact of urban areas on the environment.

Introduction

According to the United Nations Population Fund (2007a, 1), the world passed a significant milestone during 2008, with more than half the global population, some 3.3 billion people, living in urban areas for the first time in the history of the planet. Looking forward to 2030, it is estimated that almost 5 billion people will be urban residents. The most dramatic changes will occur in Africa and Asia, where urban populations are expected to double between 2000 and 2030.

Regarding such dramatic change in the growth of urban areas, the United Nations Population Fund (2007a, 1) observed:

Cities also embody the environmental damage done by modern civilization; yet experts and policy-makers increasingly recognize the potential value of cities to long-term sustainability. If cities create environmental problems, they also contain the solutions. The potential benefits of urbanization far outweigh the disadvantages: The challenge is in learning how to exploit its possibilities.

Canada is a highly urbanized society, as the box on page 482 illustrates, with four of five Canadians living in cities. As a result, both the environmental challenges and the opportunities identified by the UN Population Fund apply to Canada,

making it important to be aware of the issues and alternative ways to address them.

In this chapter, we examine the nature of **sustainable urban development** at global and Canadian scales, determine what impacts cities have had on the environment, examine how urban areas become vulnerable to environmental variability, and consider strategies to enable cities to become part of the solution rather than the problem regarding environmental quality and sustainable development. In other words, we want to learn not only 'what is the right thing to do' but also 'how to do the thing right' with regard to urban environmental management.

Sustainable Urban Development

The National Round Table on the Environment and the Economy (NRTEE 2003c, 3) observed that it is difficult to define a 'sustainable city' or 'urban sustainability'. The NRTEE noted, however, that preserving or protecting the environment for future generations, and giving equal consideration to economic, social, and environmental issues, are normally accepted as key features of 'sustainability'. Building on those ideas, the NRTEE (2003c, 3) defined urban sustainability as 'The enhanced well-being of cities or urban regions, including integrated economic, ecological, and social components, which will maintain the quality of life for future generations'. You may wish

Comprehensive Coverage

The text engages a thorough discussion of natural science and the processes required for understanding environmental issues, planning and management, the interrelationship between human actions and environmental change, and includes a new chapter on Urban Environmental Management.

Box 7.4   What You Can Do

1. Much of global climate change is influenced by our society's love affair with the automobile. Find something else to love, such as a bicycle or a bus. Simply aim to reduce your consumption of fossil fuels.
2. Use energy more efficiently in your home and in all other aspects of life. Make sure your appliances, lighting, and heating are energy-efficient.
3. Encourage power utilities to invest more effort in renewable energy resources, as discussed in Chapter 12.
4. Consider getting by without air conditioning in your car and at home, or if/when you use it, do so

sparingly. Plant shade trees to reduce the summer heat levels in your house.
5. Consider purchasing carbon offsets related to transportation or energy use (Box 7.5).
6. Plant trees to help absorb atmospheric carbon dioxide.
7. Let your political representatives know that you are in favour of mandatory measures to curb emissions, even if it costs more money in the short term.

Box 7.5   Carbon Offsets

Carbon offsets involve a financial transaction to compensate for the generation of greenhouse gas emissions through our activities, such as travel or use of electricity. Offsets are measured in metric tons of CO_2 equivalent, with one carbon offset equalling one ton of CO_2 or equivalent greenhouse gas.

There are two categories of carbon offsets. One involves governments and companies purchasing carbon offsets in order to comply with caps placed on their GHG emissions. The Kyoto Protocol approved offsets as one option for governments and companies to meet their emission targets. In that context, companies providing energy from

wind, solar, small hydro, geothermal, and biomass all create offsets by substituting or displacing fossil fuels and their associated GHGs.

The other category is voluntary and involves individuals and companies who purchase carbon offsets to mitigate the GHG emissions generated by activities such as air travel. The next time you purchase an airline ticket, check to see whether you can purchase carbon offsets. The funds generated from such purchases are often used to plant trees that absorb GHGs.

For further information, see the David Suzuki Foundation (2007).

Special Topics
In-depth explorations of special topics in each chapter help students understand how concepts, approaches, and theories manifest in the world around them and offer tips for concrete actions they can take to address environmental change.

Currency
Inclusion of breaking events and issues illustrates the effects of environmental change here and now.

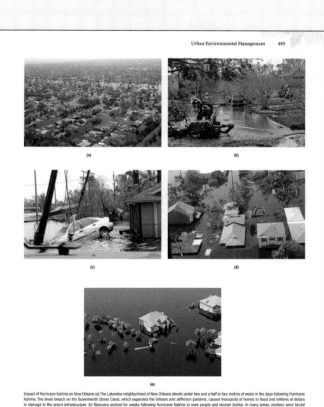

Impact of Hurricane Katrina on New Orleans (a) The Lakeview neighborhood of New Orleans stands under two and a half to four metres of water in the days following Hurricane Katrina. The levee breach on the Seventeenth Street Canal, which separates the Orleans and Jefferson parishes, caused thousands of homes to flood and millions of dollars in damage to the area's infrastructure. (b) Rescuers worked for weeks following Hurricane Katrina to save people and recover bodies. In many areas, workers were forced to wait until flood waters receded before they could access homes in outlying regions. (c) The many levels of destruction caused by Hurricane Katrina may be seen on this vehicle and home. St Bernard Parish, where this picture was taken, lies just to the east of New Orleans and was hit the hardest by the storm. No property, public or private, was spared from destruction in this area. (d) A traditional camel-back home sits in flood waters in the days following Hurricane Katrina. A combination of the storm's surge and poor design and upkeep of local levees led to flooding across more than 80 per cent of New Orleans. (e) Many conspiracy theorists say that New Orleans officials blew the levees to protect the richer communities from flood damage, however, this Lakefront mansion is evidence that even affluent neighbornoods were not spared when the levees broke. (All photos courtesy of Cpl Frans E. Labranche.)

Acknowledgements

Bruce Mitchell gratefully acknowledges the research assistance provided by Ryan Bullock and Chenxiang Li for research related to the third edition. Ryan was also very helpful in identifying images to illustrate many of the chapters. Elizabeth Dawson also assisted in checking page proofs.

The advice of Michael Stone, School of Planning, University of Waterloo, is very much appreciated. Roland Hall, Department of Biology, University of Waterloo, and Mike English, Department of Geography and Environmental Studies, Wilfrid Laurier University, provided photographs related to the Athabasca oil sands development.

Meggan Desmond, Cape Breton Tar Ponds Agency, provided information about the ongoing remediation work as well as photographs related to the tar ponds.

Thanks to Jonathan Price and Paul Parker in the Department of Geography and Environmental Management, University of Waterloo, who provided photographs of wetlands and an 'energy-efficient' residential home, respectively.

Bob and Donna Blakely, who farm near Melfort, provided guidance for a day in northeastern Saskatchewan to view diamond exploration work in that area.

Frans Labranche, a US Marine Corps photojournalist, generously provided aerial and ground photographs taken on 4 and 11 September 2005 illustrating the damage created by Hurricane Katrina in New Orleans and surrounding area. He took these photographs while operating with the 4th Marine Air Wing in the days immediately following the storm and with the 4th Assault Amphibian Battalion about 10 days after the storm.

Graeme Hayward, Manitoba Floodway Authority, arranged for the use of images of the construction on the Red River Floodway.

Frances Hannigan worked diligently to prepare the final versions of chapters and to compile the manuscript before it was submitted to Oxford University Press. She also provided ongoing technical advice.

As always, Joan Mitchell provided support, which allowed Bruce Mitchell the time needed to work on preparing the third edition. This has been a partnership that has continued since 1965 for many projects. Without her encouragement and constructive comments, this work would never have been completed.

Philip Dearden would like to acknowledge the assistance of Jennifer Hewlett in finding relevant materials and Dennis Jelinski for his comments on Chapter 3. Udom Charatcharungkiat provided much appreciated hospitality in Bangkok for writing between field sessions in Cambodia. The sabbatical leave granted by the University of Victoria was also crucial to getting the book completed on time.

We are both indebted to Peter Chambers of Oxford University Press and copyeditor Dorothy Turnbull, who provided editorial and other assistance in completing the text.

Thanks also to the many reviewers whose comments have proved invaluable over the years. In addition to those who provided anonymous feedback on this third edition, the authors and publisher would like to thank the following reviewers, whose thoughtful comments and suggestions have helped to shape all three editions of this text:

Darren Bardati
University of Prince Edward Island

Michael Bardecki
Ryerson University

Stephen Doyle
Okanagan College

Tim Elkin
Camosun College

Andrea Freeman
University of Calgary

Leslie Goodman
University of Manitoba

Johanne Kristjanson
University of Manitoba

Barbara Jean McNicol
Mount Royal College

T. Meredith
McGill University

Brian S. Osborne
Queen's University

Tom Waldichuk
Thompson Rivers University

Barry Weaver
Camosun College

About the Authors

Philip Dearden

I grew up in Britain. Even though home was in one of the wilder parts of Britain, I was always struck with the biological impoverishment of my homeland and dreamed of living in a country where wild nature still existed. My dream was realized when I first came to Newfoundland as a graduate student in the early 1970s. Since that time I have travelled all over Canada, and most of the rest of the world, and have a strong appreciation of the beauty and grandeur of the Canadian landscape.

My main interest is in conservation, and I have taught courses and undertaken research on this topic, based at the University of Victoria, for almost 30 years. Throughout this period, I have taught large introductory classes in society and environment and loved every minute of it. I have a strong belief that the power of individual actions can help to make a better environmental future and that we need to support NGOs working in this area. I have held many positions in the Canadian Parks and Wilderness Society, including chair of the British Columbia chapter, and am currently a trustee emeritus.

My main field of research is protected areas, and I maintain active research programs in Canada and in Asia on this topic. I am a member of the IUCN's World Commission on Protected Areas and have advised many international bodies on protected area management. Author of more than 180 articles and nine books and monographs, including (with Rick Rollins) *Parks and Protected Areas in Canada: Planning and Management* (Oxford, 2008), I have also been recognized for excellence in teaching with an Alumni Outstanding Teacher Award and Maclean's Popular Professor recognition at the University of Victoria. I am Leader of the Marine Protected Area Research Group at the University of Victoria and like nothing better than to be out on the water sailing with my wife Jittiya and children Kate and Theresa.

Bruce Mitchell

I was raised in Prince Rupert, a small community on the northwestern coast of British Columbia whose economy was strongly based on natural resources, especially forests and fish. Thus, from an early age, I recognized the importance of 'resources' and the 'environment'. As a graduate student, I focused on water resources, having become more and more aware of how critical this resource is for natural systems and humankind. Now, in the early twenty-first century, it is apparent that water is a key resource at a global scale.

As a high-school student and an undergraduate, I worked summers as a shoreworker in a fish-processing plant and then as a deckhand on a troller, fishing for salmon. That provided first-hand experience with a resource-harvesting industry. But more importantly, it made me aware of how knowledgeable people who had not finished their formal education in the school system could be. This showed me that what would later be called 'traditional ecological knowledge' is a remarkable source of understanding and insight. This became a lifelong lesson: experiential learning deserves respect, and those pursuing science should continuously look to such knowledge to enhance what they think they know based on science.

Studying geography provided me with a solid background to understand natural systems, as well as the way in which humans interact with or use them. As time passed, I became more and more convinced that many 'resource problems' were often 'people problems'.

As a result, I oriented my work towards planning and management, while always remaining mindful of the need to draw on scientific research. I have published about 130 articles and 25 books, have served as president of the Canadian Water Resources Association, have been a visiting professor at nine universities in various countries, and received the Award for Scholarly Distinction from the Canadian Association of Geographers as well as a Distinguished Teacher Award from the University of Waterloo. I am a Fellow of the Royal Society of Canada and a Fellow of the International Water Resources Association, and in 2008 received the Massey Medal from the Royal Canadian Geographical Society.

As a faculty member, I have had the opportunity to conduct research and consult both in Canada and in developing countries. The work in countries such as Indonesia has made me aware of how fortunate most Canadians are and of how our system of governance and civil society allows us all to become engaged in societal and environmental issues. Barriers and challenges exist, but compared to the people of many nations, Canadians have considerable latitude to participate in and shape our shared future.

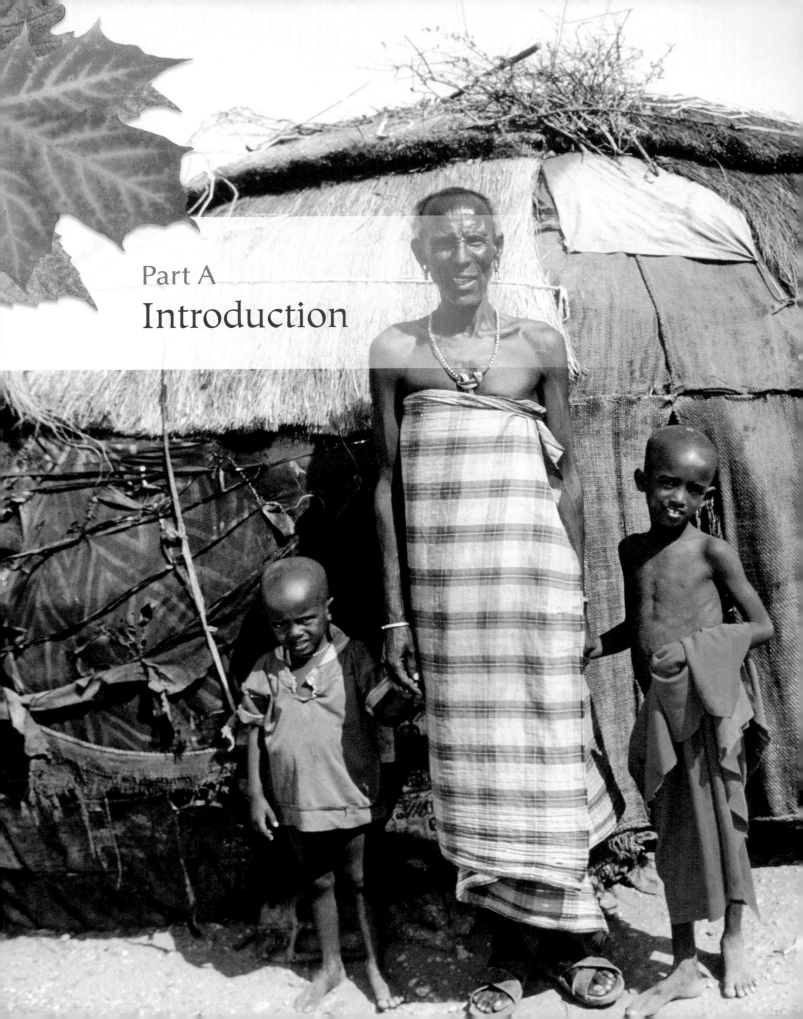

Part A
Introduction

The relationship among environment, resources, and society is one of the most important challenges, if not the most important challenge, currently facing humans on Earth. For many of Earth's human inhabitants, this relationship is an ongoing reality as they try to meet their everyday needs for food, water, and shelter. They do not need to be reminded of how important it is. To ignore the relationship is to perish.

For others, usually urban dwellers in developed countries, this reality seems distant. Food comes from the supermarket, water is piped into homes, even work and home environments have controlled temperature through heating and air conditioning. It is not until disruptions occur in these delivery systems—caused by floods, tsunamis, droughts, ice storms, earthquakes, hurricanes, insect infestations, or similar forces of nature—that many people realize that they too are dependent on the environment for survival, as has been true since before the dawn of human civilization.

This first section introduces some basic concepts regarding the relationship among environment, resources, and society and the ways by which we try to understand complex natural and socio-economic systems. There are many ways of knowing about environment. Here we concentrate mainly on the contribution of the natural and social sciences.

Science, especially environmental science, is also becoming an increasingly collaborative undertaking. This involves not only workers from one discipline but those from many disciplines coming together to contribute their understanding of a particular phenomenon. An understanding of acid precipitation, for example, necessarily requires the input of chemists, biochemists, climatologists, geologists, hydrologists, geographers, biologists, health specialists, economists, and political and legal experts, to name a few. Each discipline has its own expertise and methods of approach, and they can be combined in different ways to yield more effective answers to environmental problems.

The need for a collaborative approach to environment is clearly illustrated by a Canadian case study, the Sydney Tar Ponds in Nova Scotia. Many of the problems resulting from steelmaking and the subsequent pollution of the Muggah Creek estuary came about through failure to effectively use scientific information that was known at the time. Even with the use of science, there can be high levels of uncertainty. How we can try to deal with uncertainty and change is one of the main themes of this book.

The Sydney Tar Ponds case also helps to illustrate some of the complexity entailed in trying to address just one environmental problem in one small place. The tar ponds should also make us pause and consider what desirable future we should be aiming for and striving to create. In that context, Chapter 1 considers two concepts relating to a vision for the future: sustainable development and sustainable livelihoods. Sustainable development, popularized in 1987 with the publication of *Our Common Future*, the report of the World Commission on Environment and Development, has provoked much debate and disagreement, since different groups interpret it in ways that favour their values and interests. Despite conflicting views about what sustainable development means, it frequently appears in policies related to the environment and natural resources. Thus, it is important to have a critical appreciation of its strengths and limitations. More recently, attention is turning towards the concept of sustainable livelihoods as an alternative possible direction for the future. Our intent here is to ensure that you understand what these concepts mean, what they offer, and what their weaknesses are.

Needless to say, the global situation is infinitely more complex than sustainability alone. The next section of Chapter 1 provides an overview of the global situation with regard to environment and society. What are some of the main trends pointing to future directions? Although there are disagreements about the rate and severity of environmental change, few claim that overall conditions are improving. One indicator, the Living Planet Index, suggests that the overall ecological health of the Earth has declined by 35 per cent since 1970.

However, there is some good news as well. Predictions of global population growth, for example, are slightly lower—9 billion people by the year 2050. One important dimension that has only shown growth, however, is resource use, fuelled mainly by the demands of consumers in developed countries—we are reminded of the old comic strip 'Pogo' in which the title

character, a possum living in a swamp, famously proclaimed that 'We have found the enemy, and it is us.' If there is one fundamental message that we would like to convey, it is that the power of individuals to make decisions on a daily basis can reduce these pressures. Canadians have much to contribute in this regard, since we are among the most profligate consumers of energy and water in the world and are also among the most prolific producers of solid waste. Our society has developed into one of the most wasteful on the planet. Only we can turn that around.

The jurisdictional and institutional arrangements for environmental management in Canada constitute one critical factor influencing our relationship with the environment and resources. Such arrangements are often little taken into account by scientists and environmentalists, but they can be the most important factor when considering how and when a particular problem is going to be addressed. Canada is a large country, and the various levels of government are complex. For political reasons rather than taking into account the good of all Canadians, they often work poorly together or even work against each other.

Whether the context is global, national, or regional, we are interested in measuring our progress in addressing environmental change. As noted earlier, however, the situation is very complex, with far more variables and changes than we can possibly measure. The following section of the first chapter provides some background on how we try to go about measuring progress through the use of indicators and outlines the various kinds of indicators and their strengths and weaknesses.

The chapter ends with the presentation of a simple framework that summarizes the process of environmental management. Throughout the book, we return to this framework to illustrate deficiencies in understanding or lack of connection between different elements of the framework.

Part A provides an overall introduction to environmental change and challenge with reference to the global, national, and regional levels. Most of the remainder of the book is concerned with Canada, although we return to a global perspective in the final chapter. Part B provides an overview of the main environmental processes we need to be familiar with to understand many environmental problems. Part C discusses some of the philosophical dimensions and best practices of various aspects of resource management. This is followed by Part D, in which we discuss various thematic aspects of resource management, such as fisheries, water, and climate change. The final section, Part E, draws together some of these themes, returns to global and national summaries of current trends, and points out some of the things that individuals can do to effect change for the better in the environment of tomorrow.

Chapter 1

Environment, Resources, and Society

Learning Objectives

- To appreciate different perspectives related to environment and resources.
- To understand different approaches to analyzing complex environmental and socio-economic systems.
- To understand the implications for change, complexity, uncertainty, and conflict relative to environmental issues and problems.
- To appreciate various aspects that must be addressed to bring 'science' to bear on environmental and resource problems.
- To understand the significance of sustainable development and sustainable livelihoods in relation to resource and environmental management.
- To understand the nature of human population growth.
- To appreciate the impacts of overconsumption on global ecosystems.
- To understand institutional and governance arrangements related to environmental management in Canada.
- To understand that Canada's natural environment and society are part of a global system.
- To describe different ways of looking at progress among nations on environmental matters over time.

Introduction: Change and Challenge

On one thing virtually all Canadians will agree—the weather isn't what it used to be! Whether it's Victoria experiencing record cold temperatures in June 2008, new Canadian records set for humidity on the Prairies in 2007 with a humidex of 53, Ontario experiencing its driest summer in 50 years in 2007, or a green Christmas in such well-known white-spots as Quebec City, Timmins, and Thunder Bay in 2006, weather records are being broken on an increasingly frequent basis.

Natural systems change. They have always changed and will always change. There is strong evidence, discussed later in this book, that human activities have become a main driving force behind environmental change. Whatever the reason, it seems that changes are happening more abruptly and with greater magnitude than previously. They threaten societal well-being, and society must respond—and respond quickly.

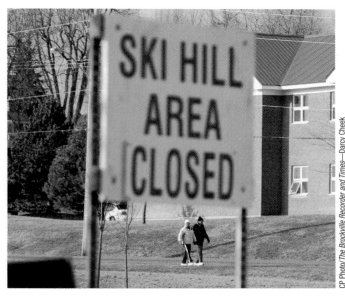

A man and a woman walk down Laurier Boulevard next to a green Memorial Park in Brockville, ON, on 20 December 2006.

Changes also occur in societies as a result of shifts in human values, expectations, perceptions, and attitudes, which may have implications for future interactions between those societies and natural systems. The value of the world economy has increased more than sixfold in the past 25 years. This increase was not merely the result of population growth; the chief cause has been increased consumption. Expectations have changed. Things that were seen as luxuries 50 years ago, such as TVs and automobiles, can now be found in some of the most remote societies on Earth. They have become necessities to many.

Changes in natural and human systems generate challenges. If we wish to protect the integrity of biophysical systems yet also ensure that human needs are satisfied, questions arise about how to determine ecosystem integrity and how to define basic human needs. Such questions force us to think about conditions both *today* and in the *future*. Such questions also remind us that an understanding of environmental and resource systems requires both natural and social sciences. Neither alone provides sufficient understanding and insight to guide decisions. Finally, such questions pose fundamental challenges as to whether we can realistically expect to manage or control natural systems or whether we should focus on trying to manage human interactions with natural systems.

In this chapter, we begin by explaining what we mean by 'environment', 'resources', and 'society' and then consider alternative ways to understand systems, issues, and problems. This is followed by a case study of the Sydney Tar Ponds in Nova Scotia to illustrate opportunities and challenges regarding resource and environmental systems as well as the importance of using both science and social science to inform public decisions and policy-making. The Sydney Tar Ponds case study vividly demonstrates that decisions are often made in the context of changing conditions, incomplete knowledge and understanding, conflicting interests and values, trade-offs, and uncertainty.

These conditions apply not only to the Sydney Tar Ponds or, indeed, to Canada generally but also to the global stage. In this context, we provide an overview of some major environmental trends and the main issues that arise. There is no doubt that human population growth is a stress on this planet, but so are the consumption patterns engaged in by the more affluent sectors of society. These factors are leading to unprecedented changes in global systems. Of particular concern are the challenges posed by global climate change. It is important that we appreciate the role that Canada plays in global environmental change—both as the second largest country on Earth and as a source of major carbon resources, such as our forests and oil resources. The decisions made by Canada regarding these resources have global implications. Therefore, you should understand the institutional aspects of environmental management in Canada to appreciate the ways by which decisions are made and how stakeholders such as you can become more involved.

The changes taking place are very complex. It is important that we grasp the essence of major changes and act accordingly. One way of doing this is to use the various indicators that measure environmental change and response, and we discuss the ways in which indicators, such as ecological footprints, are used in this regard. We conclude the chapter by identifying some key considerations regarding how scientific understanding and insight can be used to inform resource and environmental management and decisions.

Defining Environment and Resources

The **environment** is the combination of the atmosphere, hydrosphere, cryosphere, lithosphere, and biosphere in which humans, other living species, and non-animate phenomena exist. As an analogy, the environment is the habitat or home on which humans and others depend to survive. In contrast, **resources** are more specific and are normally thought of as such things as forests, wildlife, oceans, rivers and lakes, and minerals and petroleum.

Some consider resources to be only those components of the environment with utility for humans. From this perspective, coal and copper were part of the environment but were not resources until humans had the understanding to recognize their existence, the insight to appreciate how they could be used, and the skills or technology to access and apply them. In other words, in this perspective, elements of the environment do not become resources until they have value for humans. This is considered to be an **anthropocentric view** in the sense that value is defined relative to human interests, wants, and needs.

In contrast, another perspective sees resources as existing independently of human wants and needs. On that basis, components of the environment, such as temperate rain forests and grizzly bears, have value regardless of their immediate value for people. This perspective is labelled an **ecocentric** or

A cyclist uses the foot-bridge to cross the waterless Twenty Mile Creek at Balls Falls Park west of St Catharines, ON, 1 August 2007.

biocentric view because it values aspects of the environment simply because they exist and accepts that they have the right to exist.

In this book, we are interested in resources both as they have the potential to meet human needs and with regard to their own intrinsic value. Whichever category is emphasized, we often encounter change, complexity, uncertainty, and conflict. For example, different attitudes at different times may lead to an area being logged or used for mining or designated as a protected area. People and other living beings drink water to live, and at the same time urban areas may compete with farmers for access to water. An area considered of significant value for its biodiversity or ecological integrity may be designated as a national or provincial park—but then might change into an ecosystem of less intrinsic value to humans as a result of ecological processes. For instance, if a fire sweeps through an old-growth forest, the question arises as to whether the fire should be allowed to burn because it is a natural part of ecosystem processes or whether humans should intervene to put it out.

Thus, recognizing anthropocentric and biocentric perspectives does not automatically resolve all the problems that scientists and managers face when dealing with the environment and resources. However, being aware of such viewpoints should help us to understand the positions that individuals or groups take with regard to what is appropriate action. You may wish to consider whether your perspective is oriented more in an anthropocentric or a biocentric direction. Does understanding your fundamental perspective help you to appreciate why your view about the right thing to do regarding a resource situation sometimes conflicts with what others would like to see happen?

Alternative Approaches to Understanding Complex Natural and Socio-economic Systems

Systems have environmental, economic, and social components. The environmental component alone can be subdivided into aspects requiring expertise in disciplines such as biology, zoology, chemistry, geology, and geography. However, while humans have organized knowledge into disciplines for convenience and manageability, the 'real world' is not organized in that way, nor does it recognize disciplinary boundaries. Thus, it is important to understand alternative ways of creating and applying knowledge. These alternative ways include at least the following:

The person who sees a manta ray underwater will never forget such a sight or dispute the right of such species to survive. However, a purely anthropocentric resource view will tend to look more at their value in terms of making money as material for bags.

1. *Disciplinary*. Disciplinary understanding is organized around the concepts, theories, assumptions, and methods associated with an academic discipline. Disciplines have emerged and evolved in the belief that specialization will result in more in-depth understanding, and this is correct. However, since systems of interest to environmental scientists and managers have many components, the danger of a disciplinary approach is that important connections with parts of the system not considered by any one disciplinary specialist will not be taken into account.

2. *Multidisciplinary*. In order to get the in-depth insight of the disciplinary specialist but also gain the benefits of a broader view by drawing on specialists from various disciplines, the multidisciplinary approach emerged. In this approach, different specialists examine an issue, such as biodiversity, from their disciplinary perspectives, such as ecology, economics, and law. The specialists work in isolation, or only with others from the same discipline or profession, and provide separate reports, which are submitted to one person or group, who then draws upon them to

iStockphoto.com/Len Tillim

Philip Dearden

synthesize the findings and insights. In this manner, both depth and breadth are achieved through synthesis of the findings of different specialists *after* they have completed their analyses.

3. *Cross-disciplinary*. While specialists in a multidisciplinary team work in isolation from one another, in cross-disciplinary research a disciplinary specialist 'crosses' the boundaries of other disciplines and borrows concepts, theories, methods, and empirical findings to enhance his or her disciplinary perspective. However, while in this approach the specialist deliberately and systematically crosses disciplinary boundaries to borrow from other disciplines, he or she does not actively engage with specialists from the other disciplines but simply draws on their ideas, approaches, and findings. This approach allows the investigator to make connections throughout an investigation that would not occur in a disciplinary or multidisciplinary approach, and this can be very positive. At the same time, it can also involve misunderstanding of the borrowed material, using theories, concepts, and methods out of context, and overlooking contradictory evidence, tests, or explanations in the discipline from which the borrowing is done.

4. *Interdisciplinary*. To overcome the limitations of the previous three approaches, interdisciplinary investigations involve disciplinary specialists crossing other disciplinary boundaries and engaging with other specialists from the very beginning of a research project. The objective is to achieve the benefits of both depth and breadth from the outset, as well as synthesis or integration, rather than at the end of the process, as occurs in the multidisciplinary approach.

   This approach requires more time than the second and third approaches, because a team of disciplinary specialists must meet at the start and then regularly throughout a study. In addition, the approach requires a high degree of respect, trust, and mutual understanding among the disciplinary specialists, since it is common for one disciplinary specialist to question basic beliefs or assumptions that another specialist takes for granted. The approach also requires patience, because disciplinary specialists have to be prepared to make the effort to learn the 'jargon' of other specialists so that clear communication can occur. Finally, an interdisciplinary approach requires team members to have considerable self-confidence and a willingness to acknowledge the strengths and weaknesses of their disciplines, since their disciplinary views will inevitably be challenged by others.

   It is not uncommon for research projects initiated as interdisciplinary ventures to fail and consequently to evolve into multidisciplinary or cross-disciplinary projects because participants are unwilling to allocate the necessary extra time or do not have the patience to learn and receive critiques from other specialists. However, when a group of disciplinary specialists overcomes these obstacles and achieves mutual respect and trust, then enhanced understanding through depth, breadth, and synthesis is attainable.

5. *Transdisciplinary*. Transdisciplinary research is similar to interdisciplinary research, except that the term is used when the issue or problem under investigation is generally recognized not to be in the domain or the 'property' of any one discipline. A good example is the concept of sustainable development (discussed later in this chapter), which did not emerge from any one discipline yet reflects ideas from environmental, economic, and social fields of inquiry.

In this book, we examine many complex systems, and their understanding is often based on the knowledge derived from more than one discipline. As an individual, you should be aware of what insight you can contribute from a disciplinary or interdisciplinary/transdisciplinary base and what knowledge you should have to be a valued member of an interdisciplinary or transdisciplinary team. As an individual, you can approach research from a disciplinary or cross-disciplinary perspective. To participate in a multidisciplinary, interdisciplinary, or transdisciplinary project, you will need to be part of a group of people from different disciplines or professions.

Science-Based Management of Resources and Environment

In this book, we consider how understanding and insight from science can be used to inform management and decision-making. The nature of science is discussed in more detail in the introduction to Part B. In that context, it is helpful to consider the view of Mills, Quigley, and Everest (2001, 11) regarding the nature of science-based natural resource decisions. In their words, a science-based decision 'must be made with full consideration and correct interpretation of all relevant science information, and the scientific understanding must be revealed to all interested parties. The decision, however, must be left to the appropriate decision-maker(s) with authority to make the decision.' Based on this interpretation, they provide five guidelines for contributions by scientists to effective management of resources and the environment.

Guidelines for Science-Based Management of Resources and the Environment

1. *Focus the science on key issues, and communicate it in a policy-relevant form.* If science is to have value for management decisions, it must address pertinent management issues, and research must be conceived in a

manner relevant to such issues. This does not preclude scientific research from addressing basic or fundamental questions. However, to be perceived as relevant to the needs of managers, scientific work must be focused on and be timely to the needs of managers. In that regard, while scientists can provide important input into establishing management goals, this task is properly in the domain of the value-laden process of decision-making and is not part of scientific research per se.

In their research, scientists should be describing and explaining key relationships within and between systems being managed. Improved understanding of environmental, economic, social, political, and legal systems affected by management is essential to appreciating basic scientific information that might be presented.

2. *Use scientific information to clarify issues, identify potential management options, and estimate consequences of decisions.* A basic challenge for management is to determine whether a problem has been defined in an appropriate manner. Sometimes, because of complexity and uncertainty, managers may be unaware of questions that should be asked. In that regard, science, by helping to clarify relations and trends in systems, can contribute to clarifying known issues and to identifying issues previously overlooked or unknown. Science can also help to calculate the implications of different options related to an issue or problem.

3. *Clearly and simply communicate key scientific findings to all participants.* While it is important for scientists to publish their results in peer-reviewed journals—and that is to be encouraged—if their work is to be relevant for management, then scientists must go beyond that and share their findings in forums and formats accessible and understandable to a wide variety of readers.

Mills et al. suggest that scientific findings can be broadly shared in one or more of the following ways: (1) publishing a summary of key scientific findings in lay terms within an engaging format; (2) condensing the most policy-relevant scientific findings into a few clear bullets for electronic or print media; and (3) organizing conferences intended for mixed audiences and ensuring that presentations are geared to a wide diversity of participants.

4. *Evaluate whether or not the final decision is consistent with scientific information.* Making relevant scientific information available and accessible is necessary but not sufficient. It must be considered and incorporated into decision-making. One way of ascertaining whether that is happening, and of putting pressure on decision-makers to do so, is to conduct systematic and formal evaluations of decisions to determine to what extent they have relied on science.

Many benefits can emerge as a result of evaluations. First, an evaluation is a formal mechanism to demonstrate whether decision-makers have understood and used scientific information appropriately. Second, evaluation encourages scientists to provide their results in a way understandable to non-scientists. Third, as a review is being conducted, decision-makers have an opportunity to make sure that they have understood the scientific information and, if not, to improve their understanding before taking decisions that are practically irreversible.

Mills et al. recommend that a first step in an evaluation should be to identify the principal elements of a decision for which scientific information was available and then, for each element, determine (1) whether all of the relevant scientific information has been addressed and acknowledged in the decision, (2) whether the scientific information has been correctly interpreted, based on current understanding, and then accurately presented, and (3) whether projected outputs, outcomes and impacts, and risks associated with the decision have been considered and shared. In their view, if any element does not conform to these three considerations, then the decision cannot be said to have been consistent with science.

5. *Avoid advocacy of any particular solution.* There is much debate regarding this final guideline, since some scientists believe that they should be advocates for solutions when their knowledge leads them to a preferred conclusion. Others maintain that if scientists and their evidence and interpretations are to be credible, they should not be seen to favour any particular position or solution. For example, if a scientist is a known advocate of the use of herbicides and pesticides to enhance agricultural production, would

Professional Judgment

Since there seldom is time to conduct new research in the middle of a major policy debate, there always will be holes, sometimes big ones, in the science information. The scientists will be asked to at least hypothesize relations that might fill those holes and that will require significant personal judgement. Often, tight time frames will not permit the sort of multiple rounds of peer review that are desirable and typical in the science arena. In these circumstances, faith in the objectivity and independence of the scientists is particularly important.

—Mills, Quigley, and Everest 2001, 14

that person's evidence supporting the use of herbicides and pesticides be viewed as credible? Even if scientists can separate their basic values from their scientific understanding, there is a danger that they will be perceived to favour a particular viewpoint, leaving doubt in some people's minds as to whether data, interpretations, conclusions, and recommendations from such a scientist have been 'contaminated' by those values.

The dilemma, of course, is that nobody is value-free or value-neutral, so to suggest that scientists can or should be value-free or 'objective' is difficult to sustain. However, there is a difference between being perceived to be open-minded in defining a problem or identifying alternative solutions and being known to uphold a particular view or position and consistently producing findings that support only that one particular view or position. A further complication occurs when there is insufficient evidence to support a conclusion and scientists are asked to provide a professional opinion. In this type of situation, a strong argument can be made that the scientist will be viewed as more credible if he or she has no record of advocating a particular perspective, as highlighted in the quote in the previous box.

The following case study of the Sydney Tar Ponds illustrates the different kinds of knowledge required to understand resource and environmental problems. It also shows that the ideal of having 'science' inform policy decisions is often fraught with challenges when both the natural environment and human perspectives are changing.

The Sydney Tar Ponds

Background

Sydney, the third largest city in Nova Scotia, is located on the northern part of Cape Breton Island (Figure 1.1). Extensive deposits of coal and iron ore in the Sydney area resulted in a long history of coal mining and steel production. These resources, in addition to a coastal fishery, forestry, and a striking natural landscape, have been the economic base for what one commentator has characterized as 'a culturally rich but economically poor and government-dependent community' (Rainham 2002, 26).

In the late nineteenth century, the Industrial Revolution was being powered by coal. Geologists were aware that Cape Breton Island had substantial coal deposits in the areas of Inverness, Glace Bay, North Sydney, Sydney Mines, and Dominion. Shafts were sunk, and men and boys began the hard work of digging out the coal. Production steadily grew, and shortly after 1891, when a rail line connected Sydney to Halifax, annual coal production had grown to 1.5 million tonnes. By 1893, the numerous small coal mines joined together as the Dominion Coal Company (DOMCO), and output continued to grow—to more than 6 million tonnes annually by 1913.

Steel Production in Sydney

By 1912 Cape Breton was the source of nearly half of all the steel produced in Canada, and Sydney's Dominion Iron and Steel Company had the largest share of the pie. For half a century business boomed.

—Lahey 1998, 38

The steady expansion of rail lines in North America led to a high demand for steel, prompting the American owner of DOMCO to form a partnership to create a new corporation, the Dominion Iron and Steel Company (DISCO). This new company received land along the Sydney waterfront, and in 1899 construction started on the new steel plant located along Muggah Creek. The aim was to build 400 coke ovens, four 250-tonne-capacity blast furnaces, and 10 open-hearth furnaces. A supply of coal was readily available, but more iron ore was needed, and it was procured from the iron ore mines at Wabana, Newfoundland.

Making high-quality steel requires high-quality ingredients. If coal deposits contain too much sulphur, then inferior coke is the result. Low-grade iron ore leads to the need for larger quantities of limestone to remove the impurities. To assess the quality of the basic inputs to the steelmaking process, science is essential. However, such science was not drawn upon when arrangements were being made for the basic raw materials. This was the beginning of what would become a pattern. As Barlow and May (2000, 11–12) observed:

> . . . neither Arthur Moxham, now vice-president and general manager of DISCO, nor Henry Whitney [the owner of DISCO] stopped to test the quality of their ingredients. In a rush to begin full operation, they failed to run the most basic tests on their coals and ores. DISCO tapped its first steel on New Year's Eve 1901, and right from the start there were problems.
>
> The coal from Cape Breton seams was very high in sulphur, so far more coal had to be baked to produce usable coke. The iron ore from Wabana was full of impurities, such as silica and rock, so far more limestone was required to pull out the impurities as slag. The unusually large amounts of limestone required in the blast furnaces caused the furnace linings to deteriorate rapidly. . . .

The poor quality of the basic ingredients led to higher costs, less marketable and inferior products, and far more waste. In what had been Muggah Creek, the slag would

Figure 1.1 Location of Sydney, NS, and the tar ponds. Source: Rainham 2002, 27, from Joint Action Group. 1999. *Sydney, Cape Breton.*

eventually create a mountain range of waste, stretching hundreds of feet high and reducing the mouth of the estuary by nearly a mile. With more coal being baked to make less coke, the mills were producing larger quantities of tarry sludge, benzenes (volatile liquid used as solvent), ammonium sulphates, naphthalene (a white crystalline substance used to make mothballs), and PAHs [polycyclic aromatic hydrocarbons, the largest group of cancer-causing chemicals in the world]. These by-products and wastes were simply dumped into Muggah Creek.

The coal and steel operations continued for almost a century under various owners. The coke ovens, in which coal was baked at high temperatures to produce a higher-quality fuel source for the steel mill's open-hearth blast furnaces, were closed in 1988 (see photo). The steel plant stopped operating in 2001. The closure of these operations was a serious blow to the economy of Sydney. However, even though the coke ovens closed in 1988, the legacy of the uncontrolled dumping of millions of tonnes of toxic sludge remained a threat to adjacent residents and the environment, and significant industrial pollution in the Muggah Creek watershed remained.

For decades, air pollution originating from the coke ovens and open-hearth furnaces had been clearly evident. Nevertheless, the implications of the air pollution for land, waterways, and the Muggah Creek watershed were in some cases unknown and in other cases only poorly understood.

The area labelled as the 'tar ponds' was created from the deposit of chemical by-products from the coking process, runoff of water used for cooling in the coke ovens and the steel mill, leaching from contaminated soil at the coke and steelmaking

plants, an adjacent garbage dump site in the upper part of the watershed, and discharge of raw sewage from residential and commercial areas of Sydney. The outcome was a chemical and bacteria-laden river system, including the **estuary** full of contaminated sediments, the latter referred to as the tar ponds, 'a two-kilometre-long stretch of contaminated water and sludge which federal officials refer to as the largest chemical waste site in Canada' (Lahey 1998, 37). Various government surveys of this ecosystem identified polycyclic aromatic hydrocarbons (PAHs), polychlorinated biphenyls (PCBs), and other chemicals

Sydney, Nova Scotia, coke ovens, 1987. Aerial view of the coke ovens, showing the quenching plant with steam at left, coal pocket and batteries between exhaust stacks, centre, and conveyor leading from pocket to blending plant, right. By-product building is large brick structure, foreground. Smaller buildings are pump house, carpenter shop, and oil house (*Coke Ovens, SYSCO, 1987, 90–221–19653, Beaton Institute, University College of Cape Breton, Sydney, NS*).

Box 1.1   Health Challenges Related to the Tar Ponds

Cancer rates are far above the national average and higher than the rates in nearby Glace Bay or New Waterford. Cervical cancer in women in Sydney is 134 per cent higher than the provincial rate, which itself exceeds the national rate. All cancers in men and women are higher in Sydney, including brain cancer, breast cancer, stomach cancer, lung cancer, and salivary gland cancer. Alzheimer's, multiple sclerosis, heart disease, and birth defects are all much higher in Sydney.

Source: May and Barlow 2001, 7.

and metals among the pollutants. Muggah Creek and the estuary empty into Sydney harbour, and discovery of PAHs in those waters led to the closure of the lobster fishery in the harbour and its approaches. Beyond the contamination of the aquatic system, the studies determined that the soil on the coke ovens site was highly contaminated. The site also contained storage tanks holding chemical waste.

The tar ponds are in the lower part of the Muggah Creek watershed, and since Dominion Iron and Steel built its plant in 1899, what was once an active, navigable waterway and a habitat for fish and birds became a highly contaminated narrow tidal outlet. A causeway and bridge divide what used to be an estuary into a north pond and a south pond. Surveys by governments and consultants have revealed that the two ponds contain about 700,000 tonnes of sediment contaminated with PAHs, including some 45,000 tonnes also containing PCBs at concentrations above 50 parts per million (ppm).

Challenges for Epidemiological Studies

Various studies began revealing that citizens of Sydney were encountering serious health problems. Unreleased government studies pointed to significantly higher cancer rates in Sydney compared to the rest of Canada. For example, between 1977 and 1980, Sydney experienced 347 cancer deaths per 100,000 people, while the national average was 192. Health and Welfare Canada expressed concerns about the possible link to the toxic wastes from the coke ovens. More recent studies provide evidence that health issues in Sydney deserve attention. One study in the late 1990s found that the life expectancy for male and female residents in Sydney was as much as five years less than for the Canadian population as a whole. The primary contributions to reduced life expectancy were significantly higher

levels of cancer and cardiovascular disease. Another study in the late 1990s showed increased incidence of cancer in the Sydney population over a 45-year period. Yet another study indicated that rates of major birth anomalies are significantly higher among Sydney residents relative to the rest of Nova Scotia.

High Levels of Uncertainty

The interactions among claims and interpretations of evidence provided by empirical study and scientific assessment indicate there indeed exists an inherent complexity and uncertainty surrounding the understanding of health effects from chemical exposures from environmental contamination. Epidemiology, historically viewed as a parsimonious social science, seems to lose credibility and power when questions of causality arise, and its ability to generate meaningful change in Sydney is questionable.

—Rainham 2002, 31

In the mid-1980s, Health and Welfare Canada alerted the Environment Canada Atlantic regional office about health concerns related to the coking process, and the regional official alerted his provincial counterpart. In contrast, the epidemiologist for Nova Scotia took the position that the hazard depended on long-term exposure and that balanced against social and economic benefits of the coke ovens, it was reasonable to allow the coke operations to continue. The province then conducted its own investigation and concluded that unhealthy lifestyles (smoking, alcohol, poor diet such as fatty foods and high salt intake) were more likely to be causes of the higher incidence of poor health.

As Rainham (2002, 30) noted, 'The Sydney issue is characterized by two core groups: stakeholders who believe that the community contamination poses serious risks to human health and stakeholders who believe that the community contamination is both contained (a necessary result of supposed economic prosperity) and negligible (in terms of risk to human well-being).' An important challenge is that evidence based on environmental epidemiology research is often not conclusive because many variables (smoking and other lifestyle factors) can affect human health and there is uncertainty related to estimation of risks. Even though a large body of literature indicates a link between human morbidity and steelmaking, mining, and related industries, others can always argue that other variables are as likely to be the cause of morbidity.

Other factors can confound our understanding. For example, Goodarzi and Mukhopadhyay (2000, 369–70) reviewed the possible source and level of concentrations of PAHs and metals from a sample of bore holes and lakes providing water to the Sydney Regional Municipality. They concluded that 'the origin of the PAHs and the priority metals in the drinking water of the Sydney basin is not known.' They identified at least three possible origins for the PAHs and metals: (1) from contamination through leaching of the combustion residuals from the Sydney coke oven site (the tar ponds) or from the disposal site for the two power plants of the Nova Scotia Power Authority; (2) from contamination through leakage or flow of paleowater from the sedimentary strata that contain hydrocarbon reservoirs and source rocks; and (3) from leaching of underground coal-bearing seams, which are ubiquitous in the Sydney Basin. These three possible explanations illustrate that it is not uncommon during scientific investigations for alternative explanations to emerge. It is not always possible to state definitively which is correct.

Problems When Science Is Not Used to Inform Decisions

As noted above, basic science was not used when arrangements were made for the raw materials (coal, iron ore) for the steel mill. This pattern continued when efforts began to address the industrial pollution in the tar ponds. Once it was acknowledged that the estuary was severely polluted and posed a health threat, alternative actions were identified, including incinerating, burying, and/or removing the waste.

Serious thinking about remediation followed a 1980 federal survey of lobsters in Sydney harbour. The survey revealed that the lobsters were contaminated with the cancer-causing PAH chemicals, as well as with mercury, cadmium, and lead. This led to closure of the lobster fishery in 1982 in the south arm of the harbour. Testing by Environment Canada indicated the obvious source—the steelmaking operations of what was then SYSCO, or the Sydney Steel Corporation, an agency of the Nova Scotia government that had taken control of the failing private-sector operations in late 1967.

In April 1984, the consulting firm Acres International received a contract to determine the scope of the pollution and recommend options. Initial testing indicated that the tar ponds contained the equivalent of 540,000 tonnes (dry weight) of toxic waste, including 4.4 to 8.8 million pounds of PAHs. The sludge on the bottom of the estuary was judged to be between one and four metres deep. The report by Acres focused on the challenges represented by the PAHs.

Although PCBs had been identified in the earlier study of lobsters, they were not considered a problem because they were not a by-product of making coke or steel. The source of the PCBs was waste transformer fluid used in the steel mill. Another source of PCBs was the railway yard operated by Canadian National on the opposite side of Muggah Creek.

The consulting firm conducted random sampling by drilling bore holes in the estuary and on that basis discovered only small quantities of PCBs. From those results and its analysis, Acres identified three options: (1) leave the polluted sludge in place and cover it; (2) remove the sludge and store it somewhere else; and (3) remove the sludge and incinerate it. Acres estimated that incineration would destroy 99.99 per cent of the PAHs. However, PCBs are virtually indestructible at extreme temperatures and would only have been changed into airborne dioxins and other poisons if incinerated. Given the three options, along with the estimate of the high proportion of PAHs that would be destroyed by incineration and the almost 1,500 person-years of work to be generated by incineration, the provincial government selected the incineration option.

In November 1987, the federal minister of the environment and the Nova Scotia ministers of environment and development met in Sydney to announce a $34.3-million package to cover excavation of the toxic waste in the tar ponds and then its incineration. The workers in the coke ovens would have the first opportunity for employment in the cleanup project. At the press conference, the ministers stated that the tar ponds were the worst toxic waste site in Canada and the second worst in North America. Nevertheless, Environment Canada decided after a preliminary screening that a full environmental impact assessment would not be required for the excavation and incineration.

The incinerator was supposed to be operational by 1990. However, as 1992 began, the project was behind schedule and over budget, testing of the systems was still ongoing, and repairs were being made. It also had been decided to conduct further testing of the tar ponds to better understand the contaminants. Then in October 1992, a surprise was encountered. The additional testing by Acres identified a 'hot spot' of PCBs in the south tar pond. It was estimated that this hot spot had 4,000 tonnes of sludge contaminated with PCBs. Canadian law requires PCBs over 50 ppm to be incinerated at a minimum temperature of 1,200°C, and the sample from the south tar pond revealed concentrations up to 633 ppm. The incinerator, designed to destroy only PAHs, had a capacity up to 900°C. At the same time, serious problems were being encountered with the pumping and dredging systems for the sludge and with operating the incinerator.

By the fall of 1994, the problem created by the PCBs was not resolved, although the incinerator and dredging equipment were working (see photo). The originally announced budget of $34 million had climbed to more than $55 million. In December 1994, the province decided that the incinerator option was not going to do the job. It called for tenders for new approaches to deal with the contaminated sludge, and all the bids were above $100 million. The province rejected all the proposals as too expensive. It subsequently invited one Nova Scotia consulting firm, Jacques Whitford, to determine what could be done for $20 million or less. In January 1996, the Nova Scotia minister of supply and services announced that he had accepted the Jacques Whitford plan to use the slag piled up next to the tar ponds to fill in the ponds. Once that work had been done, grass and trees would be planted to create a park. This proposal was greeted with surprise and anger by the people of Sydney, including the mayor, none of whom had participated in the development of this 'solution'.

Notwithstanding the opposition, Jacques Whitford started the first phase of its work, at a cost of $5 million, which was a thorough sampling of the tar ponds to determine the extent of the PCBs. By this time, 10 years had passed since the federal and provincial governments had announced the cleanup, and under federal law, PCBs cannot be buried. The intent was to identify the PCB-contaminated sludge and remove it to a disposal site in Quebec. Throughout the spring, the sampling continued, and by midsummer the estimate was that 45,000

Incinerator plant, November 2003.

tonnes of PCB-contaminated sludge existed, leaving Jacques Whitford to express reservations about its proposal. The outcome was that the 'encapsulation' option was rejected. To this point, $60 million had been spent, and a viable solution had not emerged.

The two levels of government announced that they would pursue a more open and participatory approach and would establish a community-government committee to develop a cleanup plan. The committee was named the Joint Action Group (JAG). In September 1998, an agreement was reached to clarify the relationship of JAG with the three levels of government, and $62 million was committed to complete studies, designs, and other preparations for the cleanup. In 2000, another consulting firm, Conestoga Rovers and Associates, was hired to manage the agreement intended to lead to the cleanup of the tar ponds as well as the coke ovens site. Over the next few years, the following actions were taken: (1) a sewer system was built to divert tonnes of raw sewage flowing daily into the tar ponds; (2) the derelict structures on the coke oven site were demolished and removed; and (3) the old Sydney landfill was closed and capped.

Next Steps

In 2007, a $400-million cleanup of the tar ponds was announced by the federal and provincial governments. This was the fourth major initiative to deal with the legacy of pollutants in the tar ponds and was characterized by Angus MacIsaac, minister of transportation and public works for Nova Scotia, as providing a 'long-lasting, sustainable, and environmentally responsible solution' for the people of Cape Breton. The press release from the governments of Canada and Nova Scotia on 28 January 2007 stated that the solution would involve solidification, containment, and capping of contaminated soils, to be followed by site development and long-term and ongoing monitoring and maintenance. Incineration would not be used.

In announcing the cleanup initiative, the governments emphasized that the remediation proposal had been reviewed through an environmental assessment process that started during 2005. The independent Environmental Assessment Panel strongly recommended that remediation should focus on containment and capping of all materials in the tar ponds rather than incinerating some materials and containment and capping of the remainder. The key milestones of the remediation work were identified as:

- Early spring 2007: The Sydney Tar Ponds Agency issues tenders for the remediation work.
- Late spring 2007: Contractors start building more than two kilometres of channels through the tar ponds to allow clean water from the Coke Oven Brook and Wash Brook to flow through the ponds without becoming contaminated.
- Spring and summer 2008: Sediments from the old channel of the Coke Oven Brook are removed and moved

elsewhere on the site to be stabilized and solidified. High-density polyethylene liner is then placed on the bed of the brook and covered with gravel to allow water to drain off the site without becoming contaminated.

- Summer 2008 to early 2012: Sediments in the tar ponds are stabilized, solidified, and capped. The stabilization and solidification is accomplished by mixing the sediment with cement powder and other drying agents. Once solidified, the material is capped with high-density polyethylene liner or clay and subsequently by layers of either gravel or soil. The finished surface is planted with grass and other vegetation.
- Summer 2008 to 2012: A layered cap is placed over the contaminated areas of the coke ovens.
- Summer 2009 to summer 2012: The tar cell is stabilized, solidified, and capped.
- Post-2012: The provincial government begins a long-term monitoring program regarding the water quality of Sydney harbour, and governments consult with the community regarding a plan for future uses on the remediated sites.

The cost is to be shared by the federal government ($280 million) and the provincial government ($120 million) over the anticipated eight years scheduled for the cleanup. During the construction season, it is expected that up to 150 people will be employed. Full opportunity for participation in the remediation will be available to local companies and to First Nations businesses.

The federal and provincial governments clearly hope that the remedial work described above will resolve the legacy of industrial pollution at the tar ponds. Not all stakeholders share that belief. On 28 January 2007 when the remediation project was announced, John Morgan, mayor of Cape Breton, was reported as stating that he was concerned about the technology to be used and believed that more study was required. As reported by Canadian Press, he said, 'What the province of Nova Scotia has done is steer the cleanup to the very cheapest mechanism possible with no plan for future use of the site and really tenuous science.' Elizabeth May, the leader of the federal Green Party, also expressed doubts to Canadian Press. She commented that the technical experts she had talked to about the 'solidification, containment, and capping' solution were sceptical that it would work. She went on to state that 'It's a Mickey Mouse solution. It's not high tech.'

Engineering design work occurred during 2007, and in December 2007 construction started on what was referred to as the cooling pond project. This project was completed in the spring of 2008. The cooling pond, built in 1912 to contain water used in the steelmaking process, measured about 122 metres in diameter, was about 4.3 metres deep, and held an estimated 16,300 cubic metres of sludge. The sludge contained petroleum hydrocarbons, dioxins/furans, and metals (arsenic, copper, lead, selenium, and zinc), all above the guidelines specified by the Canadian Council of Ministers of Environment.

The former cooling pond. It is now a small, circular piece of property adjacent to the south tar pond. As the grass grows, the former cooling pond will disappear into the landscape.

Warning signs surrounding the tar ponds in the Whitney Pierarea of Sydney.

And as the Jacques Whitford (2006, 3) consulting firm had stated, although the cooling pond had not been used for years, it had to be decommissioned because 'it continues to be an environmental hazard, especially for migratory birds which in some cases have become oiled and suffered mortality after landing in the cooling pond.'

The estimated cost for the work was $4.1 million. The approach involved draining the cooling pond, stabilizing and solidifying the sludge with a mix of cement and other additives, capping it, and then hydroseeding the capped area. This project was the first 'Aboriginal set-aside' project. Three local Aboriginal companies participated in the project, and the goal was to create an opportunity to develop new skills and experience in order to be able to compete for future work.

Completion of the cooling pond project created what is considered to be a 'staging area' for the work to follow regarding the cleanup of the main tar ponds and adjacent lands. Field tests were conducted in the summer of 2008 in the south pond to test different combinations of cement to solidify the sludge,

Perspective on the Tar Ponds
Tim Babcock

I moved to the outskirts of Sydney (on the *other* side of town from the tar ponds) in the fall of 2003 after two decades of working in Indonesia on issues related to resource management. What contrasts between the two places and yet what similarities!

One of the first things that visitors to Sydney would likely notice at that time, if they arrived by car along the Trans-Canada Highway from the mainland, was a sign on the outskirts (since removed) proclaiming Sydney to be the 'Steel Center of Eastern Canada' (see photo). But this is only a small example of the types of disjunctions one soon came across in the area. There was no sign, perhaps on the other side of the highway, proclaiming the existence of the largest polluted industrial site in Canada.

Sydney: Steel Center of Eastern Canada, March 2004.

I very soon learned that the tar ponds cleanup (for which a whole government agency had been set up) continued to be a major topic of discussion—and a source of frustration and anger—among the residents of the area. It was embedded, of course, in a much larger issue—the dismantling of the steel plant itself, the appropriate disposition and use of the land on which it

stood, and, far more serious, the devastating effects of the plant's closure on the economy and on the very fabric of society of Cape Breton Island. Cape Breton, the recently amalgamated Regional Municipality declared, was in a 'state of crisis'. Much uncertainty, at least in the public mind, still existed concerning the choice of cleanup measures, the amount of money available, and even when the program would get into high gear. Argument, not all of high quality, raged on concerning the appropriate means to clean up the site among the Sierra Club, the still extant though no longer funded Joint Action Group, members of the public, and the cleanup agency. We are no further ahead than we were in 1996, opined some informed individuals; others, with intimate knowledge, claimed that progress was now 'rapid'.

At this point, many tens of millions of dollars had been spent to obtain what one presumes are the best scientific analyses and solutions currently available, and many more dollars had been spent on various forms of 'citizen participation' to discuss and propose a range of solutions. Yet the studies continued to be debated, and confusion, and much unhappiness, reigned. The question remains: were the processes flawed, or were they the 'state-of-the-art', the best that available Canadian money could buy?

Over the past several decades, Canada has spent hundreds of millions of other dollars providing what is, one hopes, good advice to developing countries on how to better manage their environments and natural

Tim Babcock

resources. But are we following 'best practice' here in Canada? Would we wish such a lengthy and complex process as the tar ponds case on a poor country with equally serious pollution problems, probably affecting far more people but with far fewer scientific, educational, and financial resources to deal with them? And

we have so far left politics out of our analysis: this adds a further, almost unlimited layer of complexity and uncertainty to the situation, where individuals and groups will inevitably, and selectively, use the results of 'science' to further their own sincerely, or cynically, held positions. There are clearly no simple answers.

Tim Babcock has a PhD in anthropology and has worked on various development projects in Indonesia and elsewhere for more than 25 years.

with contracts to be tendered in 2009 for the solidification and stabilization of the tar ponds system.

What has been learned from this experience so far? At least three lessons stand out. First, when basic science is not used from the outset to inform policy decisions related to environmental issues, the probability is high that funds will not be allocated to effective solutions. Second, even when science is used, understanding can be incomplete, and decisions will have to be taken in the face of considerable uncertainty. Third, when local stakeholders are not included in the process, challenges can be expected to proposed solutions. In this book, our goal is to help you enhance your understanding of scientific aspects of ecosystems and how that can be combined with 'best practices' (discussed in Chapters 5 and 6) related to management of resources and the environment.

Lessons

It would appear that the residents of Sydney . . . have suffered and fought in isolation and perhaps in vain. We appear not to have learned one thing from their ordeal. We continue to talk about the 'trade-off' between jobs and the environment; jobs and health. The tar ponds saga should have taught us that such trade-offs are wrong economically, environmentally, and morally.

—Barlow and May 2000, 200

The Sydney Tar Ponds situation is a vivid example of the challenges we can face from industrial legacies of pollution, but it is not unique. Other communities with similar challenges

exist: the Lubicon First Nation in Alberta, surrounded by sour gas wells; the White Dog Reserve downstream from Dryden, ON, experiencing mercury contamination from a paper mill in Dryden; Deline, NWT, and its abandoned uranium mine; Yellowknife, NWT, with an abandoned gold mine; Port Hope, ON, with waste from a refinery for radium and uranium; Newcastle, NB, and Transcona, MB, with abandoned wood preservative plants; and Squamish, BC, with heavy-metal pollution from an old chloralkali plant and the nearby abandoned Britannia mine site on Howe Sound. All of these communities face formidable challenges, and the development and implementation of solutions will be major and often expensive undertakings. These community-based problems also illustrate that resource and environmental problems can be located within or adjacent to settlements and indeed in highly urbanized landscapes. Later, in Chapters 7, 12, and 13, other examples of environmental challenges in more urbanized areas will be presented.

Sustainable Development and Sustainable Livelihoods

The previous sections have highlighted that we have choices but that deciding what is the right thing to do is often difficult because of changing conditions, uncertainty, complexity, and conflicting views. Many believe that a key element required in making choices is a clear sense of where we want to go. Or, more fundamentally, what is the desirable future that we want to create? Without a clear sense of direction, it is challenging to know what matters deserve priority and how to resolve trade-offs when conflicts occur. The concept of 'vision' often comes up in regard to establishing a clear direction, and this concept is discussed in Chapter 5.

In this section, however, we introduce two concepts—sustainable development and sustainable livelihoods—that are often pointed to as representing the kind of future to which we should aspire.

Sustainable Development

Sustainable Development

Sustainable development is development that meets the needs of the present without compromising the ability of future generations to meet their own needs. It contains within it two key concepts:

- *the concept of 'needs', in particular the essential needs of the world's poor, to which overriding priority should be given; and*
- *the idea of limitations imposed by the state of technology and social organization on the environment's ability to meet present and future needs.*

—WCED 1987, 43

Sustainable development emerged in the late 1980s. In directing us to pursue development that meets the needs of the present without compromising the ability of future generations to meet their own needs, sustainable development directs us to consider both intra- and intergenerational equity. However, sustainable development offers a major challenge to Canada and societies everywhere, since it requires a re-examination of and shift in current values, policies, processes, and practices.

Sustainable development entails three strategic aspects. At one level, it represents a *philosophy* in that it presents a vision or direction regarding the nature of future societies. In sustainable societies, attention is given to meeting basic human needs, achieving equity and justice for present and future generations, realizing self-empowerment, protecting the integrity of biophysical systems, integrating environmental and economic considerations, and keeping future options open. The comments by Mary Ellen Turpel and Ovide Mercredi in Box 1.2 highlight that many of the ideals associated with sustainable development were recognized well before the concept was popularized in the 1980s.

As a *process*, sustainable development emphasizes a system of governance and management characterized by openness, transparency, decentralization, and accessibility. It accepts the legitimacy of local or indigenous knowledge and seeks to incorporate such understanding with science-based knowledge when developing strategies and plans. It also recognizes that conditions change and much uncertainty exists. Thus, it is necessary to be flexible and adaptable, thereby allowing for policies and practices to be modified as experience accumulates.

As a *product* related to specific places or resource sectors, sustainable development seeks to ensure that economic, environmental, and social aspects are considered together and that trade-offs are made in a way that is visible and transparent to those affected.

The concept of sustainable development has generated both enthusiasm and frustration. The enthusiasm has come from those who believe that it provides a compelling vision for the twenty-first century, one in which more attention will be given to longer-term implications of development and to balancing economic, social, and environmental considerations. The phrase 'think globally and act locally' reminds us that while ultimately the planet is a single system in which actions in one part often have implications for other parts, resolution of problems also requires significant action at the local level, thereby stimulating self-empowerment, partnerships, and co-operative approaches to management and development (see Chapter 6).

The frustration has come from those who believe that 'sustainable development' is so vague that it can be defined in ways to suit different and often conflicting interests. It has been argued that sustainable development has been generally accepted only because different interests can interpret or use it to justify their often incompatible positions. Thus, developers like the concept because they can argue that growth must continue if basic human needs are to be met and if standards of living are to continue to rise. In contrast, environmentalists support the concept because they can use it to argue that environmental integrity must be given priority if there is to be long-term and equitable development.

How do you think such incompatibilities and tensions can be reconciled?

Sustainable Livelihoods

As explained in the previous section, sustainable development is often greeted with scepticism today. Many observers conclude that the concept is too often used to justify whatever someone wants to achieve, whether economic development or environmental protection. Thus, for many, the concept is seen as incapable of providing practical guidance. As a result, in some countries and regions, interest has shifted to the concept of **sustainable livelihoods**, which is viewed as more realistic and focused. Sustainable livelihoods emphasize the conditions necessary to ensure that basic human needs (e.g., food, shelter) are satisfied. However, the concept has been criticized by those who view it as too anthropocentric. Critics argue that other living creatures or inanimate components of ecosystems may be sacrificed or degraded to meet human needs. Such concern is legitimate, and ongoing debate is needed.

Nonetheless, the concept of sustainable livelihoods does respond to the concerns of the World Commission on Environment and Development (WCED 1987, 43) that 'Poverty is

Box 1.2　Aboriginal Perspectives on Environment and Resources

First Nations peoples use the expression Turtle Island to refer to North America, which is thought of as a shell of a turtle surrounded by oceans. The images of the protective shell jutting out and the living creature within make a powerful metaphor for the connection with and respect for the land that all First Nations cultures share.

— Mary Ellen Turpel

We have always been here on this land we call Turtle Island, on our homelands given to us by the Creator, and we have a responsibility to care for and live in harmony with all of her creations. . . . There is harmony in the universe, among our relatives, among the animals, and among all creatures. . . . we do not believe in competition, in the survival of the fittest. We believe all should be cared for in our Nations, that caring and sharing, not self-interest, must be our overriding aims. . . .

There are many resources in this country, and Canadians respond to this with massive resource development.

We believe that people have forgotten the importance of protecting the environment for future generations. The peoples that I represent, those with an indigenous philosophy, have a world view that is different from that of the corporate mainstream. We have a view of the environment that does not stop all forms of development but allows it to proceed in a way that respects the environment and ensures that it is protected for future generations. . . . But to respect Mother Earth as a living entity is not easy, particularly when the preoccupation of economic development may well be to exploit natural resources rather than to preserve or sustain them. That endangers our common survival and the survival of future generations who are relying on us to preserve the planet for them.

—Ovide Mercredi

Source: Mercredi and Turpel 1993, 14, 16, 44, 45, 155–6

a major cause and effect of global environmental problems', that 'many present development trends leave increasing numbers of people poor and vulnerable', and that sustainable development contains within it two key concepts, one of which is 'the concept of "needs", in particular the essential needs of the world's poor, to which overriding priority should be given'.

In the sustainable livelihoods literature, attention is given to ways for local people to meet basic needs (food, housing) as well as other needs related to security and dignity through meaningful work, at the same time minimizing environmental degradation, rehabilitating damaged environments, and addressing concerns about social justice. Strategies for sustainable livelihoods usually aim to create diverse opportunities, efficiency, and sufficiency relative to basic needs while also achieving social equity and sensitivity regarding environmental integrity.

Do you think 'sustainable livelihoods' provides a real alternative to 'sustainable development'? If you conclude that both concepts contain major flaws, what alternative 'vision' would you propose to help guide us towards a desirable future condition?

The question of sustainable development and Canada's progress will be reviewed in a later section. The next section provides a quick snapshot of the global context for sustainable development.

The Global Picture

Our home, planet Earth, is different from all the other planets we know. As it hurtles through space at 107,200 kilometres per hour, an apparently infinite supply of energy from the sun fuels a life support system that should provide perpetual

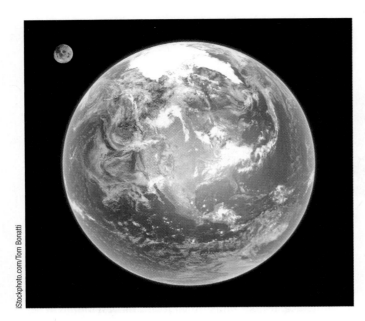

Our home, planet Earth.

sustenance for Earth's passengers. Unfortunately, this seems not to be the case. Organisms are becoming extinct at rates unsurpassed for at least 65 million years. These extinctions cover all life forms and probably represent the largest orgy of extinction ever in the 4.5-billion-year history of the planet. Our seas are no longer the infinite sources of fish we thought they were. Our forests are dwindling at unprecedented rates. Even the atmosphere is changing in composition and making the spectre of significant climatic change a reality. Every raindrop that falls on this planet bears the indelible stamp of the one organism bringing about these changes—you and us.

Concern over this situation led to the request by the UN secretary-general to undertake an assessment of the relationship between planetary ecosystems and the demands placed on them by human activity. Between 2001 and 2005, the **Millennium Ecosystem Assessment** was carried out to assess the consequences of ecosystem change for human well-being and to establish the scientific basis for actions needed to enhance the conservation and sustainable use of ecosystems and their contributions to human well-being. Some 1,360 experts from 95 countries were involved in the assessment (Box 1.3). The experts concluded that there is good evidence that many of the changes are non-linear and once they start, the processes of degradation will increase rapidly. These positive feedback loops are discussed more extensively in Chapter 4.

Population

One main variable that affects our impact on the planetary life support system is the number of passengers being supported. Although countless billions of passengers—from insects and fish to the great blue whale—are on board planet Earth, we are mainly concerned with those who seem to be having the greatest impact on the system—humans, or *Homo sapiens*. This

species, along with a few others such as rats and cockroaches, has experienced a staggering increase in population numbers over the past century.

The steep curve of population increase, shown in Box 1.4, coincides with the time that humans learned how to exploit the vast energy supplies of past *photosynthetic* activity laid down as coal and oil in the Earth's crust. Until then, energy supplies had been limited by daily inputs from the sun. The discovery of this new treasure house of energy allowed humans to increase food supplies dramatically and improve and greatly speed up the processing and transportation of materials (see Chapters 2 and 12 for more discussion on this). Global energy consumption increased more than twelvefold between 1850 and 1970, while population tripled. Energy use has increased another 73 per cent since 1970. There are now almost 7 billion humans drawing upon the planetary life support system for sustenance; before the Industrial Revolution, there were fewer than a billion. Another result of increased energy consumption is the pollution that now chokes this life support system and is causing unprecedented human-induced changes in global climate.

An estimated 4.3 people are born every second around the world. By the end of 2007, there were 6.7 billion people on Earth. The United Nations forecasts an increase to 9.1 billion people by 2050 (United Nations Population Division 2007b). This is lower than earlier estimates of 9.3 billion but still represents more than 80 million additional people per year to feed.

Much of the increase will occur in less developed countries, where populations are predicted to grow by 33 per cent between 2005 and 2050, compared to only 2.4 per cent in developed countries (Figure 1.3). By 2050, the United Nations has forecast that the populations of the world's 50 least developed countries will increase by 56 per cent. China's massive population would continue to grow until 2025, when economic growth would trigger reductions in fertility, and level out at around 1.4 billion. India is predicted to overtake China as the most populous country on Earth by 2025. Nigeria would also experience rapid growth, with the population tripling to 288 million, while that of Bangladesh would double to 254 million.

The different trajectories of the developed and developing countries are epitomized by Nigeria and Japan. In Japan, the birth rate works out to slightly more than one child per woman. Fourteen per cent of the Japanese population is younger than 15, and 21 per cent is older than 65. In contrast, 44 per cent of Nigeria's population is younger than 15, with only 3 per cent over 65. A typical Nigerian woman gives birth to six children over her lifetime.

In all these predictions, there are great uncertainties. Much depends on trends in infant mortality rates and attempts to curb the growth of AIDS in the developing world. In 2007, the Joint United Nations Program on HIV/AIDS (UNAIDS) and the World Health Organization released updated estimates on the global HIV/AIDS epidemic showing some significant reductions from previous estimates. For example, in 2007 the estimate of the number of people with HIV/AIDS was reduced

Box 1.3   Some Findings from the Millennium Ecosystem Assessment

- More land was converted to cropland in the 30 years after 1950 than in the 150 years between 1700 and 1850. Cultivated systems (areas where at least 30 per cent of the landscape is in croplands, shifting cultivation, confined livestock production, or freshwater aquaculture) now cover one quarter of Earth's terrestrial surface.

- Approximately 20 per cent of the world's coral reefs were lost and an additional 20 per cent degraded in the last several decades of the twentieth century, and approximately 35 per cent of mangrove area was lost during this time.

- The amount of water impounded behind dams has quadrupled since 1960, and three to six times as much water is held in reservoirs as in natural rivers. Water withdrawals from rivers and lakes has doubled since 1960; most water use (70 per cent worldwide) is for agriculture.

- Since 1960, flows of reactive (biologically available) nitrogen in terrestrial ecosystems have doubled, and flows of phosphorus have tripled. More than half of all the synthetic nitrogen fertilizer (which was first manufactured in 1913) ever used on the planet has been used since 1985.

- Since 1750, the atmospheric concentration of carbon dioxide has increased by about 32 per cent (from about 280 to 376 ppm in 2003), primarily because of the combustion of fossil fuels and land-use changes. Approximately 60 per cent of that increase (60 ppm) has taken place since 1959.

- More than two-thirds of the area of two of the world's 14 major terrestrial biomes and more than half of the area of four other biomes had been converted by 1990, primarily to agriculture.

- Across a range of taxonomic groups, either the population size or range, or both, of the majority of species is currently declining.

- The distribution of species on Earth is becoming more homogenous; in other words, the set of species in any one region of the world is becoming more similar to the set in other regions, primarily as a result of introductions of species, both intentionally and inadvertently in association with increased travel and shipping.

- The number of species on the planet is declining. Over the past few hundred years, humans have increased the species extinction rate by as much as 1,000 times over background rates typical over the planet's history. Some 10 to 30 per cent of mammal, bird, and amphibian species are currently threatened with extinction. Freshwater ecosystems tend to have the highest proportion of species threatened with extinction.

- Genetic diversity has declined globally, particularly among cultivated species.

- Most changes to ecosystems have been made to meet a dramatic growth in the demand for food, water, timber, fibre, and fuel. Between 1960 and 2000, the demand for ecosystem services grew significantly as world population doubled to 6 billion people and the global economy increased more than sixfold.

- Food production increased by roughly two and a half times, water use doubled, wood harvests for pulp and paper production tripled, installed hydropower capacity doubled, and timber production increased by more than half.

Source: Millennium Ecosystem Assessment 2005

---

to 33.2 million from 39.5 million in 2006, percentage rates dropped from 1 per cent to 0.8 per cent, the number of new infections fell from 4.3 to 2.5 million, and deaths declined from 2.9 to 2.1 million (UNAIDS 2007). In part, these reductions were the result of the availability of better data, but they also suggest that from a global perspective, HIV/AIDS may not

be having a significant impact on global populations, although it certainly will for some countries.

Population age structure is also important. If two countries have similar populations but differing age structures, that can have a dramatic impact on future population growth. Population age structure is usually represented as a

Box 1.4   Population and Exponential Growth

Population change is a result of the interaction between births and deaths. The **crude birth rate** (**CBR**) minus the **crude death rate** (**CDR**) will yield the **crude growth rate** (**CGR**), all usually expressed as per thousand of the population per year. In this way, populations of different countries, regardless of their size, can be compared. The figures are known as 'crude' because they give no insights into factors such as age and sex ratios, figures that are very important for understanding future potential growth. If CBR and CDR are equal, a zero population growth will result, if the effects of migration are excluded.

In 1798, a British clergyman, Thomas Malthus, pointed out that population growth was geometric in nature (i.e., 2, 4, 8, 16, 32, 64, and so on), whereas the growth in food supply was arithmetic (i.e., 1, 2, 3, 4, 5, and so on). This phenomenon would inevitably, said Malthus, lead to famine, disease, and war. Such a viewpoint was not popular in his day when population growth was considered very beneficial. For many years, the Malthusian view was ignored. The opening up of new lands for cultivation in North America and the southern hemisphere and later the development of Green Revolution techniques (Chapter 10) allowed food supplies to increase rapidly.

Increasing numbers of experts, watching the decline in food supplies per capita over the past few years (see Chapter 10) and the increase in population, particularly in less developed countries, now feel that the Malthusian spectre is quite real. Currently, 80 million people are being added every year to the population in less developed countries, compared to about 1.6 million in more developed countries. Figure 1.2 illustrates how global population has grown over the centuries and millennia.

On the other hand, some pundits, particularly economists, feel that more population simply furnishes more resources—human resources—upon which to build increases in wealth for the future. Indeed, there are concerns that some developed countries will start losing population in the future and that this will have a negative impact on their economies. For example, Japan is predicted to lose 20 per cent of its population by 2050, with declines also expected to take place in Germany, Russia, and Italy. In both the US and Canada, the trend is predicted to move in the opposite direction, largely as a result of immigration. By 2050, it is estimated that Canada's population will have increased by 14 per cent to 37 million through immigration despite a fertility rate of 1.5 children per couple.

Political leaders in some of the less developed countries experiencing the most rapid population growth rates have argued that population growth per se is not a problem and

Figure 1.2 The growth of human population over time.

that the main problem is overconsumption in the more developed countries. This distributive concern is echoed by women's groups—also wary of coercive birth control programs—who think that most progress can be achieved by improving the status of women. Women with more education have smaller, healthier families, and their children have a better chance of making it out of poverty. Yet two-thirds of the world's 876 million people who can neither read nor write are women, and a majority of the 115 million children not attending school are girls. Women who have the choice of delaying marriage and childbearing past their teens also have fewer children than teen brides. Yet more than 100 million girls will be married before their eighteenth birthday during the next decade.

However, starting with the landmark 1994 International Conference on Population and Development (ICPD) at which 179 governments adopted a forward-looking, 20-year Program of Action, remarkable progress has been made on achieving consensus on approaches to population control. The ICPD Program of Action, sometimes referred to as the Cairo Consensus, recognized that reproductive health and rights, as well as women's empowerment and gender equality, are cornerstones of population and development programs. Furthermore, at the 2005 World Summit, the largest-ever gathering of world leaders reaffirmed the need to keep gender equality, HIV/AIDS, and reproductive health at the top of the development agenda.

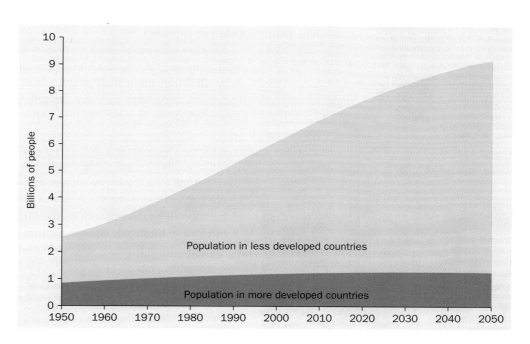

Figure 1.3 World population growth and projections, 1950–2050. Source: UN Population Division 2007b.

population pyramid (Figure 1.4), and the shape gives information about birth and death rates as well as life expectancy. It also reveals the number of dependants. Two groups are recognized, young dependants (aged below 15) and elderly dependants (aged over 65). Dependants rely on the economically active for economic support. Many less developed countries have a high number of young dependants, while many developed countries have a growing number of elderly dependants.

A population with a high number of young dependants and a low life expectancy has a pyramid that is very triangular.

A population with a falling birth rate and rising life expectancy is reflected in a population pyramid that has fairly straight sides. Generally, the shape of a country's population pyramid changes from a triangular shape to a barrel-like shape with straighter edges as the country develops and passes through the **demographic transition**, discussed below. In fact, places experiencing an aging population and a very low birth rate may have a population structure that looks a little like an upside-down pyramid.

Human reproductive age is usually identified as between 15 and 44 years of age. The larger the number of people in

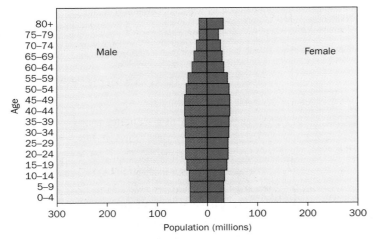

Figure 1.4 Population pyramids for (1) developing and (2) developed countries. Source: UN Population Division 2007b.

What the future holds in terms of population growth depends on the reproductive decisions taken by today's children as they grow up. Children in the central highlands of Vietnam.

this age bracket, then the greater the potential for population increase in the future. **Total fertility rates** represent the average number of children each woman has over her lifetime. If the fertility rate is 2.0, then theoretically this will lead to stable populations as children replace their parents. If the rate is higher than 2.0, it will lead to population growth; if less than that, the population declines. However, because of infant mortality, particularly in less developed countries, the **replacement level fertility** is calculated as higher than 2.0.

The largest generation in history—1.2 billion people—is now between the ages of 10 and 19, and the childbearing decisions made by this generation will be critical. If the world's women have on average just half a child more than predicted, then the population would swell to 10.6 billion by mid-century. And although the average annual global population growth rate has fallen from more than 2 per cent, as it was from the 1950s to the 1990s, to less than 1.3 per cent today, that rate is being applied to a much larger and still increasing population. Furthermore, after early declines in fertility rates, the rates in many countries have now reached a plateau (Figure 1.5).

Another important consideration is the speed of the so-called demographic transition in each country. Demographers, those who study population structure and growth, have noted a relationship between economic growth and population that occurs in four main phases (Figure 1.6):

1. *High equilibrium.* Both death rates and birth rates are high, resulting in very little population growth. This situation usually occurs in pre-industrial societies.
2. *High expanding.* Advances in health care result in declining mortality rates but with no concomitant decrease in birth rates, leading to high population growth. This situation occurs in the early stages of industrialization when some of the benefits of technology and industrial society are starting to be felt but are insufficient to outweigh the desire to have large families. Large families are an advantage in underdeveloped countries, since it means more

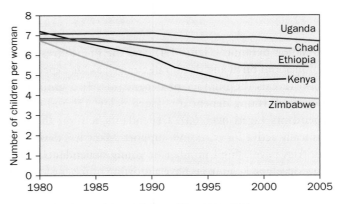

Figure 1.5 Fertility trends in sub-Saharan Africa, 1980–2005.
Source: UN Population Division and national demographic and health surveys.

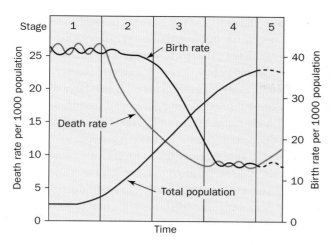

Figure 1.6 The demographic transition.

Life Expectancy Increases

The average life expectancy at birth in less developed countries rose from 41 years in 1950 to 66 years in 2007. The Middle East and North Africa region has experienced the largest increase in life expectancy since the late 1950s: from 43 years to 70 years. Since 1950, the greatest gains in life expectancy at birth occurred among women. In more developed countries, average life expectancy for women rose from 69 years in 1950 to 80 years in 2007, while the average for men rose from 64 years to 73 years.

—Population Reference Bureau 2007

labour to generate family income. Lacking the pension systems of more advanced societies, parents need to have someone to look after them as they age. Large families also compensate for the high rate of child mortality in pre-industrial societies.

3. *Low expanding.* Birth rates start to fall as the benefits of increased income start to erode the advantages of having large families. In Western societies, where the cost of raising children is high, having large families is no longer an overall benefit.

4. *Low equilibrium.* Birth rates and death rates are in balance as a result of the decline in birth rates.

As with most simplified models, the model can be criticized because of its generality, largely based on European experience, and because it does not take a full range of cultural factors into account. Moreover, a fifth phase to the model is emerging in some nations, illustrated in Figure 1.6, as total populations fall.

Historically, the decline in death rates (the **epidemiological transition**) that occurred in most developed countries was relatively slow. Discoveries about the causes of disease and how it could be countered developed in conjunction with growing interest and investments in science. For example, in the late 1800s, Louis Pasteur and others discovered the main infectious agents and the means by which they were transmitted. Vaccines were developed, and whole populations became immune to diseases such as typhoid and smallpox. By the 1930s, antibiotics such as penicillin were being developed. These medications led to cures for many other ailments. Sanitation improved, as did nutrition. However, these developments took time, and the decline in crude death rates was gradual in the developed world. In contrast, these innovations were made available in many less developed countries all at once, often leading to a precipitous decline in death rates without a corresponding decrease in birth rates.

Some countries travel through this sequence more quickly than others, with rapid economic development followed by corresponding adjustments in birth rates. Thailand is a prime example, where a fertility rate of 6.4 in 1960 fell to 1.5 in 2008, with economists voicing fears of further reductions, forecasting a future labour shortage. However, not all countries adjust as rapidly as Thailand did. In Thailand, a Buddhist country, there are no religious obstacles to reducing family size, women play a major role in household decision-making, education is valued for both sexes, and there has been a latent demand for effective contraception and effective means to distribute contraceptive devices. These conditions often do not exist in many countries.

Recognition of the importance of the demographic transition in stabilizing population growth and the role of economic development as a main driving force behind the transition was also a factor in the drive to industrialize the world promoted by many global organizations, such as the World Bank and the United Nations. Furthermore, a strong relationship was seen between some indicators of environmental degradation and economic growth. As economic growth increases, so does environmental degradation—until a threshold is reached. After that point, the wealth generated by increased industrial activity is sufficient to pay for environmental services (e.g., pollution control) that poorer countries cannot afford. Thus, the goal of development planning was to help countries reach that threshold so that they could enjoy the benefits of increased wealth while not succumbing to environmental degradation.

Unfortunately, the relationship among society, stages of development, and economic growth is more complex than these models allow. In particular, they ignore one of the most (and many scientists would argue the single most) important factor—the impact of consumption on the capability of planetary ecosystems to continue to provide life support services.

Although much is being done throughout the world to curb population growth, as seen in this signboard in Vietnam encouraging couples to have only one child, consumption knows no bounds, and we are constantly exhorted to buy more.

Consumption

The Earth's passengers do not all have the same impact on the life support system. Some passengers—those in first class—get special meals, three times a day, wine included; those in the economy section are lucky if they get one meal and must buy their own water, if it is even available. The richest 20 per cent of the world's population are responsible for more than 75 per cent of world consumption, while the poorest 20 per cent consume less than 2 per cent (World Bank 2008). In terms of metal use, for example, the 15 per cent of the global population living in the US, Canada, Japan, Australia, and western Europe account for 61 per cent of aluminum use, 60 per cent of lead, 59 per cent of copper, and 49 per cent of steel. The average North American uses 22 kilograms of aluminum a year, the average African less than one kilogram. Rural populations are significantly poorer than urban populations, and one of the main drivers of future consumption will be the increased urbanization of global populations (Chapter 13).

Urbanization

In 2008, the world reached an invisible but momentous milestone: for the first time in history, more than half its human population, 3.3 billion people, live in urban areas. By 2030, this number is expected to swell to almost 5 billion. Many of the new urbanites will be poor. Their future, the future of cities in developing countries, the future of humanity itself all depend very much on decisions made now.

—United Nations Population Fund 2007b

Energy consumption is also very unequally distributed, with the people in the wealthiest countries using 25 times more per capita than the world's poorest people. More than a third of the global population does not have access to electricity, but demands are growing. There are also large differences in energy consumption among developed countries, with the average Canadian and American consuming 2.4 times as much energy as the average person in western Europe.

The world's auto manufacturers produced a record 67 million vehicles in 2006, putting more cars on the road than ever before (Worldwatch Institute 2007). Global production grew by 4 per cent, but China increased its production by nearly 30 per cent, overtaking Germany to become the third largest producer. China's rise represents the most dramatic change in the world auto industry, with production there more than quintupling over the past decade. Sales within China surpassed the 3-million mark in 2005, with nearly 9 million passenger cars on the country's roads. While this is still a comparatively small fleet, it is likely to grow rapidly in coming years, and China is expected to become a major exporter within the next four years.

Traffic congestion, road accidents, air pollution, climate change, and peak oil are all challenges the world faces from car-centred transportation. On average, urban car travel consumes nearly twice as much energy as urban bus travel.

Canadians are among the top per capita consumers of energy in the world, with an even larger consumption rate than Americans (Figure 1.7). Each Canadian consumes as much energy as 60 Cambodians. Government policies encourage us to be wasteful by subsidizing energy production, and we as individuals normally do not resist. However, precipitous increases in gas prices in 2008 are finally having some impact on consumers, with people cutting down on their driving and looking for smaller cars. Energy is a good index of our planetary impact, reflecting our ability to process materials and disrupt the environment through pollution such as *acid precipitation* (Chapter 4) and the production of *greenhouse*

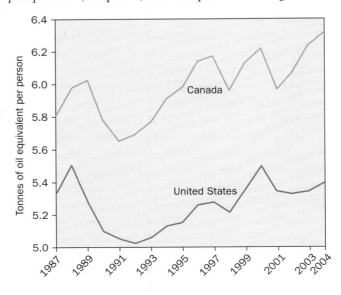

Figure 1.7 Per capita energy consumption of Canada and the US, 1987–2004.
Source: GEO data portal, compiled from IEA 2007 and UNPD 2007.

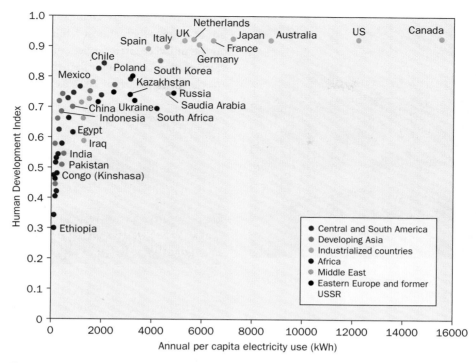

Figure 1.8 The relationship between energy use and the Human Development Index. Source: UNDP 2005.

gases (Chapter 7). However, as shown in Figure 1.8, there is no direct relationship between electricity consumption and human development. In other words, it is possible to have high standards of living without excessive energy consumption, as exemplified by many European countries. Canada has yet to make this transition.

Obviously, very different kinds of passengers share our planet, and the differences among them have grown rather than diminished as a result of increased wealth over the past 20 years. **Gross national product** (GNP) is an index that economists use to compare the market value of all goods and services produced for final consumption in an economy during one year. Over the past two decades, the planetary GNP has risen by $20 trillion, but only 15 per cent of this increase has trickled down to the 80 per cent of the passengers in the economy section of the spaceship. The rest has made the rich even richer.

Global poverty is a major challenge to planetary survival. Living in extreme poverty (less than $1 a day) means not being able to afford the most basic necessities to ensure survival. An estimated 8 million people a year die from absolute poverty. Moderate poverty, defined as earning about $1 to $2 a day, enables households to just barely meet their basic needs, but they still must forgo many of the things—education, health care—that many others take for granted. The smallest misfortune (e.g., a health issue, job loss) threatens survival.

However, progress is being made. Extreme poverty in developing countries fell from 28 per cent in 1990 to 19 per cent in 2002. Over the same period, however, the number of people in developing countries grew by 20 per cent to more than 5 billion, which means that 1 billion people live in extreme

poverty. If economic growth rates in developing countries are sustained, global poverty will fall to 10 per cent by 2015, led by a sustained period of economic growth in China and increasing growth in India. However, such success is not universal. In sub-Saharan Africa, the number of poor people has increased by a third.

Implications

The stresses on the planetary life support system are a consequence of overconsumption and the resulting pollution as well as overpopulation and the resulting poverty. Together, they create pressure on the **carrying capacity** of the planet at all different scales. Although in the past many cultures violated the carrying capacities of their local environments with dire results, never before have we approached these limits at a global scale.

Jared Diamond, a geographer at UCLA, has written a fascinating book on why past societies collapsed and what we can learn from their experiences. Appropriately, the book is titled *Collapse* (2005), and we strongly recommend it as a readable but science-based analysis of the history of collapsed human societies. Diamond suggests that four main reasons explain why societies fail to make the corrections necessary to prevent societal collapse. First, they may not anticipate the problem; second, they may fail to appreciate the severity of the problem even though they are aware of it; third, they may appreciate the problem but neglect to address it; and lastly, they may perceive the problem as a serious threat, try to solve it, and fail. Diamond explores examples of all these situations through scientific analysis. However, his main message is to alert us to how we can forestall such a collapse in modern society. At the

Families in Lesotho are still large, but the planetary impact of this entire family will be a fraction of that of one Canadian child.

heart of Diamond's analysis is the role of environmental degradation in causing societal collapse. One of his most sobering conclusions is that many of the societies that collapsed were very successful societies for which collapse seemed impossible, yet it happened with frightening rapidity.

There have been many warnings about the impact of environmental degradation on society in the future, spanning back to the famous *Limits to Growth* study of the early 1970s (Meadows et al. 1972) through to the report of the World Commission on Environment and Development in 1987 (WCED 1987). Some of the trends were brought to the public stage at the Earth Summit in Rio de Janeiro in 1992 and at the follow-up meeting, the World Summit on Sustainable Development in Johannesburg 10 years later.

In 2000, at the United Nations Millennium Summit, world leaders agreed to a set of time-bound, measurable goals and targets for combating poverty, hunger, disease, illiteracy, environmental degradation, and discrimination against women. Placed at the heart of the global agenda, they are now called the Millennium Development Goals (MDGs). The eight MDGs aim to improve human well-being by reducing poverty, hunger, and child and maternal mortality; by ensuring education for all; by controlling and managing diseases; by tackling gender disparity; by ensuring environmental sustainability; and by pursuing global partnerships. Under each of the MDGs, countries have agreed to targets to be achieved by 2015. Many of the regions facing the greatest challenges in achieving these targets coincide with regions facing the greatest problems of ecosystem degradation. At the 62nd General Assembly of the United Nations in 2007, the goals, targets, and indicators were revised (Table 1.1). The use of indicators is discussed in greater detail later in the chapter.

There is widespread consensus that **global climate change** will present very serious challenges to the ongoing stability of society. Although vast changes have taken place in planetary

ecosystems in the past, change has never occurred so quickly, nor at a time when almost 7 billion humans were dependent on the uninterrupted services provided by ecosystems. So severe are these challenges that the UNDP devoted its *Human Development Report* for 2007/8 exclusively to them.

The Imperative for Action on Global Change

What we do today about climate change has consequences that will last a century or more. The part of that change that is due to greenhouse gas emissions is not reversible in the foreseeable future. The heat-trapping gases we send into the atmosphere in 2008 will stay there until 2108 and beyond. We are therefore making choices today that will affect our own lives but even more so the lives of our children and grandchildren. This makes climate change different and more difficult than other policy challenges.

. . . Even if we were living in a world where all people had the same standard of living and were impacted by climate change in the same way, we would still have to act. If the world were a single country, with its citizens all enjoying similar income levels and all exposed more or less to the same effects of climate change, the threat of global warming could still lead to substantial damage to human well-being and prosperity by the end of this century.

In reality, the world is a heterogeneous place: people have unequal incomes and wealth, and climate change will affect regions very differently. This is, for us, the most compelling reason to act rapidly. Climate change is already starting to affect some of the poorest and most vulnerable communities around the world. A worldwide average 3° centigrade increase (compared to pre-industrial temperatures) over the coming decades would result in a range of localized increases that could reach twice as high in some locations. The effect that increased droughts, extreme weather events, tropical storms, and sea level rises will have on large parts of Africa, on many small island states and coastal zones will be inflicted in our lifetimes. In terms of aggregate world GDP, these short-term effects may not be large. But for some of the world's poorest people, the consequences could be apocalyptic. In the long run, climate change is a massive threat to human development, and in some places it is already undermining the international community's efforts to reduce extreme poverty.

—UNDP 2007

Table 1.1   The Revised Millennium Development Goals and Indicators

Goals and targets	Indicators
Goal 1 Eradicate extreme poverty and hunger	
Halve, between 1990 and 2015, the proportion of people whose income is less than $1 a day	• Proportion of population below $1 (PPP) per day • Poverty gap ratio (*incidence x depth of poverty*) • Share of poorest quintile in national income or consumption
Achieve full and productive employment and decent work for all, including women and young people	• Growth rate of GDP per person employed • Employment-to-population ratio • Proportion of employed people living below $1 (PPP) per day • Proportion of own-account and contributing family workers in total employment
Halve, between 1990 and 2015, the proportion of people who suffer from hunger	• Prevalence of underweight in children under five years of age • Proportion of population below minimum level of dietary energy consumption
Goal 2 Achieve universal primary education	
Ensure that, by 2015, children everywhere, boys and girls alike, will be able to complete a full course of primary schooling	• Net enrolment ratio in primary education • Proportion of pupils starting grade 1 who reach last grade of primary school • Literacy rate of 15- to 24-year-olds, women and men
Goal 3 Promote gender equality and empower women	
Eliminate gender disparity in primary and secondary education preferably by 2005 and in all levels of education no later than 2015	• Ratio of girls to boys in primary, secondary, and tertiary education • Share of women in wage employment in the non-agricultural sector • Proportion of seats held by women in national parliament
Goal 4 Reduce child mortality	
Reduce by two-thirds, between 1990 and 2015, the under-five mortality rate	• Under-five mortality rate • Infant mortality rate • Proportion of one-year-old children immunized against measles
Goal 5 Improve maternal health	
Reduce by three-quarters, between 1990 and 2015, the maternal mortality ratio	• Maternal mortality ratio • Proportion of births attended by skilled health personnel
Achieve, by 2015, universal access to reproductive health	• Contraceptive prevalence rate • Adolescent birth rate • Antenatal care coverage (at least four visits) • Unmet need for family planning
Goal 6 Combat HIV/AIDS, malaria, and other diseases	
Have halted by 2015 and begun to reverse the spread of HIV/AIDS	• HIV prevalence among population aged 15–24 years • Condom use at last high-risk sex • Proportion of population aged 15–24 years with comprehensive correct knowledge of HIV/AIDS • Ratio of school attendance of orphans to school attendance of non-orphans aged 10–14 years
Achieve, by 2010, universal access to treatment for HIV/AIDS for all those who need it	• Proportion of population with advanced HIV infection with access to antiretroviral drugs

(Continued)

Have halted by 2015 and begun to reverse the incidence of malaria and other major diseases	• Incidence and death rates associated with malaria • Proportion of children under five sleeping under insecticide-treated bed-nets • Proportion of children under five with fever who are treated with appropriate anti-malarial drugs • Incidence, prevalence, and death rates associated with tuberculosis • Proportion of tuberculosis cases detected and cured under directly observed treatment short course

Goal 7   Ensure environmental sustainability

Integrate the principles of sustainable development into country policies and programs and reverse the loss of environmental resources	• Proportion of land area covered by forest • Carbon dioxide emissions, total, per capita and per $1 GDP (PPP) • Consumption of ozone-depleting substances • Proportion of fish stocks within safe biological limits • Proportion of total water resources used
Reduce biodiversity loss, achieving, by 2010, a significant reduction in the rate of loss	• Proportion of terrestrial and marine areas protected • Proportion of species threatened with extinction
Halve, by 2015, the proportion of people without sustainable access to safe drinking water and basic sanitation	• Proportion of population using an improved drinking water source • Proportion of population using an improved sanitation facility
By 2020, to have achieved a significant improvement in the lives of at least 100 million slum dwellers	• Proportion of urban population living in slums

Goal 8   Develop a global partnership for development[1]

NOTE: The first four targets in Goal 8 share the first 12 indicators.

Develop further an open, rule-based, predictable, non-discriminatory trading and financial system (includes a commitment to good governance, development, and poverty reduction—both nationally and internationally) Address the special needs of the least developed countries (includes tariff-and quota-free access for exports, enhanced program of debt relief for heavily indebted poor countries (HIPC) and cancellation of official bilateral debt, and more generous ODA for countries committed to poverty reduction) Address the special needs of landlocked developing countries and small island developing states (through the Programme of Action for the Sustainable Development of Small Island Developing States and the outcome of the 22nd special session of the General Assembly)	*Official development assistance* • Net ODA, total and to the least developed countries, as percentage of OECD/DAC donors' gross national income • Proportion of total bilateral, sector-allocable ODA of OECD/DAC donors to basic social services (basic education, primary health care, nutrition, safe water, and sanitation) • Proportion of bilateral official development assistance of OECD/DAC donors that is untied • ODA received in landlocked developing countries as a proportion of their gross national incomes • ODA received in small island developing states as a proportion of their gross national incomes *Market access* • Proportion of total developed country imports (by value and excluding arms) from developing countries and least developed countries admitted free of duty • Average tariffs imposed by developed countries on agricultural products and textiles and clothing from developing countries
Deal comprehensively with the debt problems of developing countries through national and international measures in order to make debt sustainable in the long term	• Agricultural support estimate for OECD countries as a percentage of their gross domestic product • Proportion of ODA provided to help build trade capacity *Debt sustainability* • Total number of countries that have reached their HIPC decision points and number that have reached their HIPC completion points (cumulative) • Debt relief committed under HIPC and MDRI Initiative • Debt service as a percentage of exports of goods and services
In cooperation with pharmaceutical companies, provide access to affordable, essential drugs in developing countries	• Proportion of population with access to affordable, essential drugs on a sustainable basis
In cooperation with the private sector, make available the benefits of new technologies, especially information and communications	• Telephone lines per 100 people • Personal computers per 100 people • Internet users per 100 people

[1] *Some of the indicators will be monitored separately for the least developed countries, Africa, landlocked countries, and small island developing states.*

Source: UN Millennium Development Goals

The evidence for and causes of global climate change are discussed in more detail in Chapter 7, along with Canada's response. Canada, although a signatory to the legally binding Kyoto Protocol, has the second highest per capita emissions of greenhouse gases in the world. Canada agreed under the Kyoto Protocol to target a 6 per cent cut in emissions. Instead, emissions have increased by 27 per cent, and the country is now around 35 per cent above its Kyoto target range. While greenhouse gas intensity has fallen, efficiency gains have been swamped by an increase in emissions from an expansion in oil and gas production. Net emissions associated with oil and gas exports have more than doubled since 1990. This compares with Britain's reduction of 12 per cent over the same time period and Germany's reduction of 17 per cent.

Scientists have suggested that an overall increase in temperatures in excess of 2°C will cause runaway environmental damage and that emissions need to be kept under 450 ppm Ce (carbon dioxide emission equivalents) to avoid this consequence. More recent modelling suggests that 2°C was an optimistic threshold, with 1.5°C being more realistic. Nonetheless, to put emissions in perspective, of the overall planetary carrying capacity and using the suggested annual allowable emission ceiling of 14.5 Gt CO_2, if emissions were frozen at the current level of 29 Gt CO_2 to stay below the threshold, we would need two planets. However, emissions are not equally distributed. In Ethiopia, for example, the average per capita carbon footprint is 0.1 tonnes, compared to 20 tonnes in Canada. The per capita increase in emissions since 1990 for the United States (1.6 tonnes) is higher than the total per capita emissions for India in 2004 alone (1.2 tonnes). The overall increase in emissions from the United States exceeds sub-Saharan Africa's total emissions. The per capita increase for Canada since 1990 (5 tonnes) is higher than per capita emissions for China in 2004 (3.8 tonnes).

Rich countries use 90 per cent of the allowable capacity of the planet with only 15 per cent of the world's population (Table 1.2). If every person living in the developing world had the same carbon footprint as the average for high-income countries, global CO_2 emissions would rise to 85 Gt CO_2, a level that would require six planets. With a global per capita footprint at Australian levels, we would need seven planets, rising to nine for a world with Canadian and US levels of per capita emissions. As a global community, we are running up a large and unsustainable carbon debt, but the bulk of that debt has been accumulated by the world's richest countries.

In terms of addressing the situation, a global perspective is obviously necessary, although responsibilities clearly differ. For example, a 4 per cent cut would be generated in global emissions if a 50 per cent cut were initiated in CO_2 emissions for South Asia and sub-Saharan Africa. A similar cut in high-income countries would reduce emissions by 20 per cent. Equity must also play a role. An average air-conditioning unit in Florida emits more CO_2 in a year than a person in Afghanistan or Cambodia during his or her lifetime. A European dishwasher emits as much CO_2 in a year as three Ethiopians. Some countries obviously have a much greater impact and much longer histories of polluting the global commons that is the atmosphere. What kind of role do you think Canada should assume to shoulder its share of global responsibility?

There are other important global perspectives on Canada that you should bear in mind. Our land is vast, about 13 million square kilometres, and our population small, about 32 million people (Statistics Canada 2006). Between 1980 and 2005, Canada's population grew by 32 per cent, from 24.5 million to 32.3 million (Figure 1.9). Population density is 0.04 people per hectare, compared to Bangladesh at eight people per hectare. Canada would have to have a population of more than 8 billion to equal this density. Immigration, rather than natural increase, is the most important factor in population growth in Canada. Between 1995 and 2000, around 2.6 million migrants per year moved internationally from less developed to more developed regions. More than one-half of these migrants settled in the United States and Canada. Are more people good for Canada when we consider that there is more to Canada than just the economy? Does Canada have a moral obligation to accept migrants from overcrowded countries elsewhere or from countries generating 'environmental refugees' as a result of environmental stress in their home countries? These are some of the important questions that policymakers—and you—must consider.

In terms of numbers alone, Canada is not overpopulated compared to virtually any other country on Earth. Canada is the second largest country in the world in terms of area and includes 20 per cent of the world's wilderness, 24 per cent of its wetlands, 10 per cent of its forests, and 9 per cent

Satellite image of Canada (Processed at the Canada Centre for Remote Sensing, Earth Sciences Sector, Natural Resources Canada).

Toutin, 1997

Table 1.2 Measuring the Global Carbon Footprint—Selected Countries and Regions

	Carbon dioxide emissions[a]								
	Total emissions (Mt CO$_2$)		Growth rate (%)	Share of world total (%)		Population share (%)	CO$_2$ emissions per capita (t CO$_2$)		CO$_2$ emissions or sequestration from forests[b] (Mt CO$_2$/year)
Top 30 CO$_2$ emitters	1990	2004	1990–2004	1990	2004	2004	1990	2004	1990–2005
1. United States	4,818	6,046	25	21.2	20.9	4.6	19.3	20.6	−500
2. China [c]	2,399	5,007	109	10.6	17.3	20.0	2.1	3.8	−335
3. Russian Federation	1.984[d]	1,524	−23[d]	8.7[d]	5.3	2.2	13.4[d]	10.6	72
4. India	682	1,342	97	3.0	4.6	17.1	0.8	1.2	−41
5. Japan	1,071	1,257	17	4.7	4.3	2.0	8.7	9.9	−118
6. Germany	980	808	−18	4.3	2.8	1.3	12.3	9.8	−75
7. Canada	416	639	54	1.8	2.2	0.5	15.0	20.0	..
8. United Kingdom	579	587	1	2.6	2.0	0.9	10.0	9.8	−4
9. Korea (Republic of)	241	465	93	1.1	1.6	0.7	5.6	9.7	−32
10. Italy	390	450	15	1.7	1.6	0.9	6.9	7.8	−52
11. Mexico	413	438	6	1.8	1.5	1.6	5.0	4.2	..
12. South Africa	332	437	32	1.5	1.5	0.7	9.1	9.8	(.)
13. Iran (Islamic Republic of)	218	433	99	1.0	1.5	1.1	4.0	6.4	−2
14. Indonesia	214	378	77	0.9	1.3	3.4	1.2	1.7	2.271
15. France	364	373	3	1.6	1.3	0.9	6.4	6.0	−44
16. Brazil	210	332	58	0.9	1.1	2.8	1.4	1.8	1,111
17. Spain	212	330	56	0.9	1.1	0.7	5.5	7.6	−28
18. Ukraine	600[d]	330	−45[d]	2.6[d]	1.1	0.7	11.5[d]	7.0	−60
19. Australia	278	327	17	1.2	1.1	0.3	16.3	16.2	..
20. Saudi Arabia	255	308	21	1.1	1.1	0.4	15.9	13.6	(.)
21. Poland	348	307	−12	1.5	1.1	0.6	9.1	8.0	−44
22. Thailand	96	268	180	0.4	0.9	1.0	1.7	4.2	18
23. Turkey	146	226	55	0.6	0.8	1.1	2.6	3.2	−18
24. Kazakhstan	259[d]	200	−23[d]	1.1[d]	0.7	0.2	15.7[d]	13.3	(.)
25. Algeria	77	194	152	0.3	0.7	0.5	3.0	5.5	−6
26. Malaysia	55	177	221	0.2	0.6	0.4	3.0	7.5	3
27. Venezuela (Bolivarian Republic of)	117	173	47	0.5	0.6	0.4	6.0	6.6	..
28. Egypt	75	158	110	0.3	0.5	1.1	1.5	2.3	−1
29. United Arab Emirates	55	149	173	0.2	0.5	0.1	27.2	34.1	−1
30. Netherlands	141	142	1	0.6	0.5	0.2	9.4	8.7	−1
Word aggregates									
OECD [e]	11,205	13,319	19	49	46	18	10.8	11.5	−1,000
Central & Eastern Europe & CIS	4,182	3,168	−24	18	11	6	10.3	7.9	−166
Developing countries	6,833	12,303	80	30	42	79	1.7	2.4	−5,092
East Asia and the Pacific	3,414	6,682	96	15	23	30	2.1	3.5	2,294
South Asia	991	1,955	97	4	7	24	0.8	1.3	−49
Latin America & the Caribbean	1,088	1,423	31	5	5	8	2.5	2.6	1,667
Arab states	734	1,348	84	3	5	5	3.3	4.5	44
Sub-Saharan Africa	456	663	45	2	2	11	1.0	1.0	1,154
Least developed countries	74	146	97	(.)	1	11	0.2	0.2	1,098
High human development	14,495	16,616	15	64	57	25	9.8	10.1	90
Medium human development	5,946	10,215	72	26	35	64	1.8	2.5	3,027
Low human development	78	162	108	(.)	1	8	0.3	0.3	858
High income	10,572	12,975	23	47	45	15	12.1	13.3	−937
Middle income	8,971	12,163	36	40	42	42	3.4	4.0	3,693
Low income	1,325	2,084	57	6	7	37	0.8	0.9	1,275
World	22,703[f]	28,983[f]	28	100[f]	100[f]	100	4.3	4.5	4,038

NOTES

a Data refer to carbon dioxide emissions stemming from the consumption of solid, liquid, and gaseous fossil fuels and from gas flaring and production of cement.

b Data refer only to living biomass—above and below ground, carbon in deadwood, soil, and litter are not included. Refer to annual average net emissions or sequestration due to changes in carbon stock of forest biomass. A positive number suggests carbon emissions while a negative number suggests carbon sequestration.

c CO$_2$ emissions for China do not include emissions for Taiwan, Province of China, which were 124 Mt CO$_2$ in 1990 and 241 Mt CO$_2$ in 2004.

d Data refer to 1992 and growth rate values refer to the 1992–2004 period.

e OECD as a region includes the following countries that are also included in other subregions listed here: Czech Republic, Hungary, Mexico, Poland, Republic of Korea, and Slovakia. Therefore, in some instances, the sum of individual regions may be greater than the world total.

f The world total includes carbon dioxide emissions not included in national totals, such as those from bunker fuels and oxidation of non-fuel hydrocarbon products (e.g., asphalt), and emissions by countries not shown in the main indicator tables. These emissions amount to approximately 5% of the world total.

Source: UNDP 2007.

Figure 1.9 Population growth in Canada, 1980–2005. Source: Statistics Canada. CANSIM database http://cansim2.statcan.gc.ca, Table 051-0001.

of its fresh water and has the longest coastline in the world. However, as discussed above, it is not simply numbers of people but rather the impact of those people that is critical. Canadians are among the world's top producers per capita of industrial and household garbage, hazardous wastes, and greenhouse gases (Box 1.5). Some point to the size of the country, the cold in winter, and the heat in summer as the reasons behind our remarkable energy consumption, but it is clear that Canadians can contribute substantially to reducing impacts on the planetary life support system.

We offer suggestions on some of the ways that you can help contribute to the needed changes throughout the text, and we return to this theme in Chapter 15.

Changes in environmental directions for Canada require changes in policies and legislation and the strict implementation of those changes. It is therefore important for you to appreciate the jurisdictional arrangements for environmental management in Canada, the topic of the next section.

Jurisdictional Arrangements for Environmental Management in Canada

As noted in Mitchell (2004), no one government has total authority or responsibility for natural resources and the environment in Canada. Instead, under the Canadian Constitution, authority is divided between the federal and provincial governments, with territorial and municipal governments increasingly having a role. First Nations are also increasing their role commensurate with their recognition as another level of government. In addition, Canada is involved in bilateral arrangements with the United States to address environmental problems such as air pollution and to deal with shared water bodies such as the Great Lakes and the Columbia River, as well as in multilateral arrangements with other nations or international organizations regarding resources

Box 1.5   Canada Facts

- We generate about 360 kilograms of solid waste per capita per year, ranking seventh in the world.
- We generate almost six tonnes of hazardous waste for each US $1 million of goods and services produced; Japan generates less than a quarter of a tonne.
- We have one of the highest per capita uses of water in the world, about 15 cubic metres per day, roughly three times that of Sweden and Japan.

- We are among the highest per capita energy consumers in the world.
- We use our cars nearly 10 per cent more than residents of other industrialized countries.
- We emit 2 per cent of the world's greenhouse gases with 0.5 per cent of the world's population and rank second in global production of greenhouse gases per capita.

such as fisheries, migratory birds and animals, and minerals on or under the ocean floor.

Federal, Provincial, and Municipal Roles

Canada is a federated state, with power and authority shared between federal and provincial governments and with municipal governments receiving their power and authority from provincial legislatures. The Constitution Act, 1982, differentiates between proprietary rights and legislative authority relative to natural resources and the environment. Ownership of all Crown lands and natural resources not specifically in private ownership is given to the provinces, except for the Canadian North (north of 60 degrees latitude), where the federal government has proprietary rights to land and resources until the territories receive such power, and for resources found on or under seabeds off the coasts of Canada (some provinces, however, are challenging this right—see Chapter 12).

Legislative authority is mixed between the federal and provincial governments and often becomes a significant source of conflict. The federal government has jurisdiction over trade and commerce, giving it substantial authority over both interprovincial and export trading of resources (oil and natural gas, water). Alberta, Saskatchewan, and British Columbia, which have oil and natural gas, often object to the federal

government becoming involved in setting prices and determining buyers, arguing that such matters are within provincial authority because of their responsibility for property and civil rights, as well as because these resources are a matter of local interest. The federal government has used its legislative authority for navigation and shipping and for fisheries to create water pollution regulations—even though water within provinces falls within the jurisdiction of the provinces. Thus, there are ambiguities and inconsistencies regarding jurisdiction over resources and the environment. One consequence is that it has been difficult to establish *national approaches* (combined federal and provincial) to deal with resource and environmental issues (Box 1.6).

In the early to mid-1990s, many provincial governments began to download selected responsibilities, which they had traditionally held, to municipalities. The provinces argued that downloading was consistent with the principle of **subsidiarity**, which stipulates that decisions should be taken at the level closest to where consequences are most noticeable. While such an argument is rational, others have suggested that the primary motive for downloading was the desire of provincial governments to shift the cost of many responsibilities to lower levels of government as part of a strategy to reduce provincial debts and deficits. Whatever the motivation, the outcome has been that municipalities have become much

Box 1.6   Achieving Sustainable Development

Because of the multidisciplinary nature of sustainable development and environmental issues, the responsibility for environmental protection cannot be left to only one government department; all departments must co-ordinate their activities within a central framework. An integrated strategy for sustainable development and environmental protection (SDEP) should be adopted and secured by government-wide enforcement. The integration of SDEP considerations requires both vertical and horizontal linkages. *Vertical linkage* means that the country has an integrated SDEP mechanism, which involves federal, state, and local governments united by common themes—ecosystem protection, pollution control, health, equity, respect for nature, concern for the needs of future generations,

natural resources, historical and cultural properties, and so on. Such an integrated, sustainable development system, providing for full accountability for the allocation and expenditure of resources, may not be easy to operationalize. Instead, it may be desirable, therefore, first to institute a well-functioning, horizontal co-ordinating mechanism within the central government and only then attempt to develop an integrated approach.

Such an approach...requires (1) an appointment of an interministerial committee for sustainable development and environmental protection, (2) a dependable monitoring mechanism, and (3) an establishment of a central office to oversee the overall operation of the strategy.

—Dwivedi and Khator 2006, 128–9

more significant players in natural resource and environmental management, since in many instances provinces have withdrawn from related management activities.

There are instances of effective partnerships between provincial and municipal governments. One of the best and most enduring examples is the Ontario conservation authorities, watershed-based organizations established by statute in 1946 to manage many renewable resources within river basins. Individual authorities were established when two or more municipalities in a watershed petitioned the provincial government to establish one. When a majority of the municipalities in a watershed agreed that they would work collaboratively, the province established a conservation authority. Today, 36 authorities exist, primarily in the more settled parts of the province.

While the provincial government would not impose a conservation authority, it initially provided a strong incentive for local governments to form one by offering funds not available to municipalities on their own but available after a conservation authority was established. This cost-sharing arrangement was a powerful stimulus for municipalities to agree to establish authorities, and for many years the cost-sharing was 50/50 between the province and the municipalities. The arrangement meant that, as Lord (1974, x) reported, 'an authority can flourish only when the local people have enough enthusiasm and conviction to support it financially. It has also meant that the authority does not exceed the financial resources of its jurisdiction.' However, in the mid-1990s, the Ontario government significantly reduced its proportion of the funding to the conservation authorities as part of a drive to reduce government activities.

Another challenge in the reallocation of responsibility to municipal governments is the variable degree of competence to deal with resource and environmental matters. As Kreutzwiser and de Loe (2004, 188) have observed, there is significant variation among municipalities and other local-level governments in technical expertise and required data, funding capability, leadership, community awareness and engagement, and ability to implement, monitor, and enforce solutions related to resource and environmental management. Often, it seems, the provincial governments have downloaded responsibility to local governments without having determined whether they had the necessary competence to take it on (de Loe, Di Giantomasso, and Kreutzwiser 2002; Ivey, de Loe, and Kreutzwiser 2002).

Monitoring Progress towards Sustainable Development

Beginning in 1997, the Office of the Auditor General began reporting on progress by 24 federal government departments and agencies regarding sustainable development. The new commissioner of the environment and sustainable development observed in his first report in 1997 that Canadians expected governments to provide strong leadership and a clear vision and to lead by example through fostering a culture of environmental protection and sustainable development within federal organizations. The commissioner also stated that few quick solutions were available for environmental problems and that progress would require persistence, patience, and focused effort. The challenge was characterized as a 'long journey', one requiring systematic change if a real difference were to be realized for present and future generations.

Regarding sustainable development, the commissioner noted that much needed to be done. For example, he commented that in comparison to many other nations, Canada was a large per capita consumer of energy and other natural resources as well as a large generator of pollution and waste. He stated that recent research indicated that Canada was likely to fall 'far short' of its target for reducing greenhouse gas emissions.

Looking to the future, the commissioner wrote that many environmental problems and sustainable development issues were 'difficult to manage' as well as 'scientifically complex' and involved long time frames. Furthermore, rarely did they fit tidily within one department's or government's mandate or jurisdiction. He noted three major aspects where improvement was necessary:

- Federal government agencies' performance often fell well short of stated objectives. As a result, an *implementation gap* existed, since policy direction too often was not translated into effective action.
- Many pressing issues transcended departmental mandates and governmental jurisdiction. Consequently, *lack of co-ordination and integration* was frequent. The need was to manage 'horizontal issues', or those involving shared responsibility.
- There was often inadequate information about the benefits being realized from environmental programs. Therefore, a strong need existed to resolve *inadequate performance review* processes so that both senior managers and parliamentarians could know what was being accomplished.

The commissioner's report in 2007 was not encouraging. He reported that although the 24 federal departments and agencies had been creating sustainable development strategies every three years, it was 'clear that the strategies are not helping or encouraging departments to take environmental issues into account'. It was the view of the commissioner that 'the preparation and tabling of the strategies have become little more than a mechanical exercise, required to fulfill a statutory obligation' and that for the most part, 'senior managers have not demonstrated that they take the strategies seriously.' To resolve these inadequacies, the commissioner stated that the federal government needed to clearly articulate its sustainable development goals and to specify how the government as a whole and individual departments should contribute to realizing them.

A year later, in his 2008 annual report, the commissioner indicated that some progress was taking place but that in many more areas, little or no progress had been made. He explained that where progress was not occurring in a satisfactory manner, it was due to lack of commitment by senior managers and often insufficient funding. Examples of poor progress were found in such areas as protection of species at risk along with their habitat, control of aquatic invasive species, and restoration of heavily polluted areas in the Great Lakes. The commissioner stated that a major concern was poor performance in the conduct of strategic environmental assessments at the time when policy and program proposals were prepared. In the words of the commissioner, 'Strategic environmental assessments and sustainable development strategies are management tools put in place to get departments and agencies to do this [mitigate challenges]. Unfortunately, both tools are broken, and they need to be fixed.'

All of the commissioner's reports since the first one in 1997 can be found on the website http://www.oag-bvg.ca/internet/English/oag-bvg_e_920.html.

Measuring Progress

From the foregoing discussions, you will appreciate that the environmental situation is very complex. In fact, it is so complex that many, including many decision-makers, give up on trying to make sense of it. However, if we do not understand the problem and whether it is getting better or worse, we cannot implement effective management strategies. One of the goals of science is to provide understanding of complex problems. And one way of doing this is through the use of **indicators**.

Ecological footprints enable us to make comparisons among various countries (Box 1.7). Some countries have ecological demands greatly in excess of their capabilities (Figure 1.10), and they import ecological capital from elsewhere to make up for this deficit. Although trade between nations is to be expected, this excess of ecological footprint over capacity allows some nations to live beyond their ecological means. Canada has one of the largest available ecological capacities; we also have one of the largest ecological footprints per capita.

Indicators are not new. For many years, doctors have used body temperature, measured easily by thermometer, as one indicator of the health of the human body. Gross domestic product has been used as an indicator of economic performance, as has the Dow Jones industrial average. These indicators tell us something of the current state of a particular system, but they do not help us to understand why the system is in that state. Over the past 15 years, there has been growing

When you have your morning cup of coffee or tea with sugar, your ecological footprint is reaching out to the tropics where both drinks and sugar are grown. Tea plantation, Sri Lanka (top). Sugarcane fields, South Africa (bottom).

awareness of the need to develop indicators that would gauge the health of other aspects of societal well-being, including the environment. Indicators are often used to provide information on environmental problems that enables policy-makers to evaluate their seriousness, to support policy development and the setting of priorities by identifying key factors that cause pressure on the environment, to monitor the effects of policy responses, and to raise public awareness and generate support for government actions. Box 1.8 describes one framework that helps to develop causal linkages between indicators—the Drivers-Pressures-State-Impact-Response framework, or DPSIR.

One useful example of the DPSIR approach is the joint Environment Canada–US Environmental Protection Agency (Environment Canada 2003b) series of indicators on the state of the Great Lakes, which is broken down into pressure, state, and response indicators. It is interesting to note that two of the richest nations in the world found that there was insufficient data on many of the 80 desired variables. On the basis of the 43 indicators used, the conclusion is that the overall trend is 'mixed'—i.e., some indicators show improving conditions, others deterioration.

Box 1.7   Your Ecological Footprint

Your 'ecological footprint' is the land and sea base required to provide your needs, including all energy and material requirements, and also to dispose of your wastes. In fact, most of a person's footprint comprises the space needed to absorb waste from energy consumption, especially carbon dioxide. Can nature provide enough of these services on an ongoing basis to meet the needs of an individual, community, or nation?

A 2004 study undertaken for the Federation of Canadian Municipalities (see www.fcm.ca/english/communications/eco.pdf) examined the ecological footprints of 20 municipalities—that is, the area of land and sea throughout the world required to produce the amount of food, energy, and other materials the citizens use. The Canadian average was 7.25 hectares per capita, third highest in world rankings. The lowest and highest footprints in Canada were both in Ontario. Sudbury was the lowest at 6.87 hectares, with York-Durham the highest at 10.33. One reason for Sudbury's success is the efforts of the public works engineer in charge of heating and sewage. He has introduced many innovative programs, especially programs related to using local energy sources, such as the

power of the wind, the sun, and the Earth's heat, that have reduced both the cost of heating and resource consumption. It is a striking example of how the initiative and energy of one person can make a significant improvement in planetary resource use.

York-Durham, on the other hand, uses 43 per cent more resources per capita than the average Canadian. This is mainly a result of the fuel consumption generated by the sprawling suburban developments near Toronto largely dependent on car transportation. Chapter 13 provides more detail about Canadian cities and their footprints.

On a global scale, only 1.6 hectares are available for each person, and this amount is shrinking every year, largely as a result of population growth. Yet the collective average global footprint is 2.2 hectares per person, with a North American average double that for Europeans and seven times greater than it is in Asia or Africa. To provide for everyone at Canadian standards would require three Earths, not just one. Estimates suggest that by 2030, only 0.9 hectare will be available per person at a global scale.

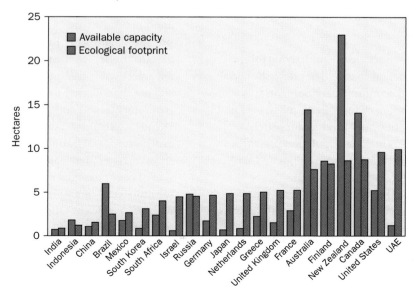

Figure 1.10 Ecological footprint per person in selected nations, 1999. Source: Worldwatch Institute, *State of the World 2004*, www.worldwatch.org.

Box 1.8   The DPSIR Indicator Framework

The most widespread framework for classifying environmental indicators is the Drivers-Pressures-State-Impact-Response (DPSIR) framework developed by the Organisation for Economic Co-operation and Development (OECD) and adopted by all European Union countries, the US, Canada, Australia, Japan, and many developing countries (e.g., Malaysia) and international organizations (such as the Commission for Sustainable Development of the UN, the United Nations Environment Programme, and the World Bank). The framework, as shown in the figure below, is popular because of its organization around key causal mechanisms of environmental problems.

The DPSIR Indicator's Framework

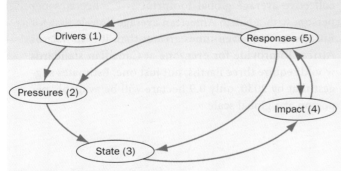

Figure 1.11 The DPSIR Indicator's framework

Driving forces (1). Drivers are the underlying forces causing environmental change. They describe social, demographic, and economic developments in societies and corresponding changes in lifestyles, overall levels of consumption, and production patterns. Examples of drivers include population pressures and the demand for various consumer goods and services (e.g., cars, red meat, increasing travel).

Pressure indicators (2). These indicators are the pressures on the environment resulting from the drivers. Examples include emission of pollutants, use of resources, use of land for roads, water withdrawals, deforestation, and fisheries catch. Initial interest in these indicators focused further down the causal chain in the state indicators, described below. However, there

is now widespread realization that state indicators are mere reflections of changes further up the chain, prompting much greater interest in drivers and pressures. One example of the link between drivers and pressures can be found in a driver such as the number of cars. Not only can we address the driver (by seeking to limit the number of cars through better public transportation, raising the price of gas, or imposing special taxes, licensing fees, or tolls on people who insist on driving their cars into downtown urban areas during the daytime), but we can also try to reduce the pressure by making vehicles more fuel-efficient and less polluting.

State indicators (3). State indicators describe the quantity and quality of physical phenomena (e.g., temperature), biological phenomena (e.g., fish stocks, extinctions), and chemical phenomena (e.g., CO_2 concentrations, phosphorus loading) and tell us the current state of a particular environmental system. They are often tracked over time to produce a trend.

Impact indicators (4). The changes in the state of the environment described by the state indicators result in societal impacts. For example, a rise in global temperatures (a state indicator) has an impact on crop productivity, fisheries value, water availability, flooding, and so on.

Response indicators (5). Response indicators measure the effectiveness of attempts to prevent, compensate for, ameliorate, or adapt to environmental changes and may be collective or individual efforts, both governmental and non-governmental. Responses may include regulatory action, environmental or research expenditures, public opinion and consumer preferences, changes in management strategies, and provision of environmental information. Examples include the number of cars with pollution control or houses with water-efficient utilities, the percentage of waste that communities and households recycle, use of public transport, and passage of legislation. Response indicators are critical to assessing the effectiveness of policy interventions but are often the most difficult to develop and interpret.

In 2003, Environment Canada created a set of national environmental indicators that were a good source of information for monitoring Canada's progress. Unfortunately, this comprehensive array of indicators on everything from agriculture, forestry, fisheries, and biodiversity to pollution was discontinued. The government now publishes a report, *Canadian Environmental Sustainability Indicators* (Environment Canada 2007b), that provides information on just three environmental stresses: air quality indicators that track ground-level ozone and fine particulate matter; a greenhouse gas emissions indicator that tracks the annual releases of the six greenhouse gases that are the major contributors to climate change; and a freshwater quality indicator that uses the Water Quality Index to summarize the status of surface freshwater quality. The first two indicators are further discussed in Chapter 13.

One issue difficult to resolve is the degree of aggregation of information included in an indicator. An almost infinite amount of information could be collected on environmental systems (Figure 1.12). Much of this information might be useful for understanding the basic nature of the system, but it is not necessary for decision-making. Research scientists and line agencies may be involved in the routine collection of such data, and without such data, meaningful indicators cannot be constructed. At a higher level of sophistication, these raw data may form an *integrated database*, such as the integration of social and biophysical data as a basis for integrated watershed management planning. However, synthesis of these data into indicators is often most useful to decision-makers, and indicators themselves may show greater or lesser degrees of aggregation, especially in a spatial sense. The Canadian Environmental Sustainability Indicators mentioned above are examples of the kind of indicators in which there is some spatial aggregation for the whole country.

Higher levels of thematic aggregation produce *indices*. Simple indices are composed mainly of indicators that are all similar. The well-known Dow Jones industrial average, for example, combines changes in market processes for 30 blue chip stocks listed on the New York Stock Exchange.

The **Living Planet Index**, created by the World Wildlife Fund (WWF) is an example of a widely used index (www .panda.org/news_facts/publications/general/livingplanet/ index.cfm) that quantifies the overall state of planetary ecosystems. It tracks populations of 1,313 vertebrate species— fish, amphibians, reptiles, birds, mammals—from around the world. Separate indices are produced for terrestrial, marine, and freshwater species, and the three trends are then averaged to create an aggregated index. Although vertebrates represent only a fraction of known species, it is assumed that trends in their populations are typical of biodiversity overall.

The index shows a 35 per cent reduction overall in the planet's ecological health since 1970. Figure 1.13 shows the temperate and tropical terrestrial Living Planet Indices for 1970 to 2003. The terrestrial index is a spatial aggregation of the temperate and tropical indices. The challenge with aggregation can be seen by comparing the two. The temperate index shows a slight improvement since 1970, which is one of the reasons why inhabitants of the temperate nations of the world have too optimistic a view of what environmental conditions are on the planet. On the other hand, the tropical index shows a clear drop, as is evident to anyone who has spent a lot of time in the tropics over the past 30 years. These differences are hidden when the two are aggregated.

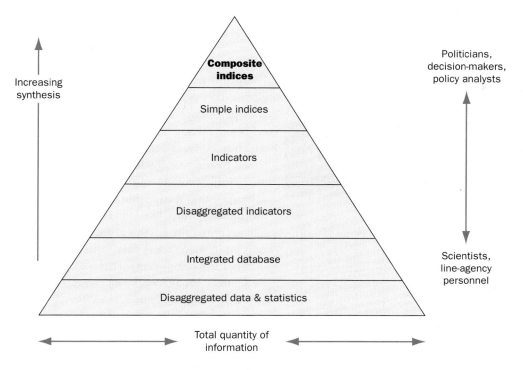

Figure 1.12 Relationship among data, indicators, indices, and users.

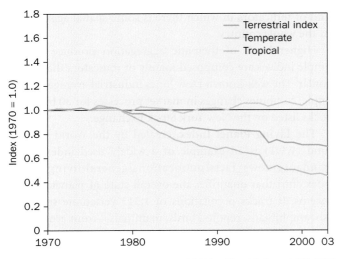

Figure 1.13 Temperate and tropical terrestrial Living Planet Indices, 1970–2003. Source: WWF International, *Living Planet Report 2006.*

Figure 1.14 Humanity's ecological footprint, 1961–2003. Source: WWF International, *Living Planet Report 2006.*

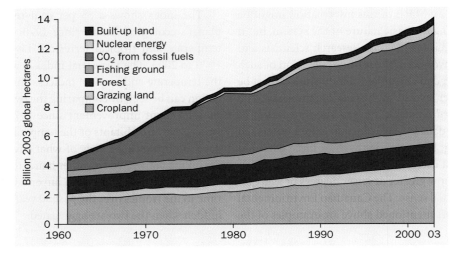

Figure 1.15 Ecological footprint by component. Source: WWF International, *Living Planet Report 2006.*

Humanity's Ecological Footprint

The Earth's regenerative capacity can no longer keep up with demand—people are turning resources into waste faster than nature can turn waste back into resources.

Humanity is no longer living off nature's interest but drawing down its capital. This growing pressure on ecosystems is causing habitat destruction or degradation and permanent loss of productivity, threatening both biodiversity and human well-being.

—World Wildlife Fund 2006, iv

WWF produces another global index, the Ecological Footprint, which shows the extent of human demand on global ecosystems. The index measures the area of biologically productive land and water needed to provide ecological resources and services—food, fibre, and timber, land on which to build, and land to absorb carbon dioxide (CO_2) released by burning fossil fuels (Figure 1.14). The amount of biologically productive area—cropland, pasture, forest, and fisheries—that is available to meet humanity's needs is the **biocapacity**. Since the late 1980s, the Ecological Footprint had exceeded the Earth's biocapacity by as much as 25 per cent by 2003, the latest data available.

The main components of the ecological footprint are shown in Figure 1.15, and again the aggregated index masks large global differences (Figure 1.16). When the ecological footprint is compared with biocapacity, the unsustainable direction on which we are embarked becomes clear (Figure 1.17). The challenge is to devise effective strategies to reduce the footprint below biocapacity before the ecological debt becomes insurmountable. Many scientists feel that point is fast approaching.

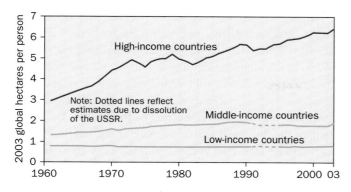

Figure 1.16 Footprint by national average per person income, 1961–2003. Source: WWF International, *Living Planet Report 2006*.

Even if we could determine the value of insect pollinators to Canada's agriculture adequately, money could never buy the services they provide.

Composite indices, such as the ecological footprint, are often the most useful for decision-makers and represent the highest level of aggregation. They are few in number and incorporate many, often very different sub-variables. The Human Development Index created by the United Nations Development Programme, the Environmental Sustainability Index of the United Nations, and GNP are other examples of aggregate indices.

In Canada, the National Round Table on the Environment and Economy (NRTEE) was charged by the minister of finance in 2002 with developing indicators to supplement the main economic indicators with information related to natural and human capital (NRTEE 2003a). It developed indicators related to water quality, forest cover, air quality, greenhouse gas emissions, wetlands coverage, and educational attainment but stopped short of the intended step of synthesizing these indicators into a composite index to represent the total capital of Canada (economic, social, and environmental) that could be tracked on an ongoing basis to assess sustainability. Part of the rationale for not proceeding was the unease with aggregating such different dimensions of capital and assuming that they were substitutable. For example, economists have derived ways of assessing the monetary value of ecosystem services, such as that provided by forests

absorbing rain to reduce the damage caused by flooding or the value of wild insects for pollinating agricultural crops. However, because we think we can measure the value of these services in money does not mean that money can buy these services. They are not substitutable.

Composite indices are highly attractive because they convey a lot of information and are useful for making macro-level policy decisions. However, these highly aggregated indicators also carry risks. They often tell us what is happening at the macro level but add little in terms of explaining why. They may mask the complex detail that decision-makers require to make informed decisions. Composite indices must be highly transparent and capable of being disaggregated to facilitate understanding of why change is occurring.

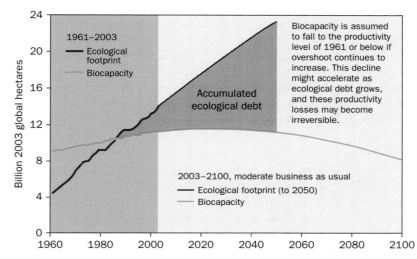

Figure 1.17 Business-as-usual scenario and ecological debt. Source: WWF International, *Living Planet Report 2006*.

Indicators provide some basis for assessing change and comparison among countries, but they raise questions about the role of science in environmental decision-making. This is discussed in more detail in the next section.

Implications

We are violating global thresholds related to the carrying capacity of the life support system of the planet. We have ceased to live off the interest and are consuming the capital at such a rate that it threatens the future viability of the system. Many species reach such carrying capacity limits with their environment, overshoot them, and have their numbers drastically reduced by environmental factors, as discussed in Chapter 3. So far, we have been able to avoid this process because of human technological ability, which has increased carrying capacities. But can we continue to increase our numbers and our habits of consumption indefinitely? Or must even humans accept some limits to activities and numbers?

If the answer to the latter question is yes, then identification of the changes that need to be made in general terms is not that difficult. We need to balance birth and death rates, restore climatic stability, protect our atmosphere and waters from excessive pollution, curb deforestation and replant trees, protect the remaining natural habitats, and stabilize soils. The challenge, however, is charting the course necessary to fulfill these objectives. Before its demise, the Soviet Union had possibly the most stringent and comprehensive environmental protection regulations in existence and yet still ended up as one of the most polluted environments on Earth. The regulations were simply not enforced. The secret is charting a course that not only addresses the goals mentioned above but is actually able to achieve these goals through strategies that can be implemented.

In this book, we aim to provide some background as to how this can be achieved, with particular reference to the Canadian situation. In this chapter, we started by discussing the characteristics of the 'environment' and 'resources'. We also examined different approaches to understanding complex systems and considered issues related to the use of 'science' in decision- and policy-making. The case study of the Sydney Tar Ponds dramatically illustrates the complexity of the environmental challenges we face, even at the local level. They are characterized by uncertainty, rapid change and conflict, and the need to appreciate both the scientific and technical aspects of a problem and the social dimensions. This book attempts to provide an introduction to both of these aspects. Part B outlines some of the main processes of the *ecosphere*, the basic functionings of the planetary life support system, and the ways in which we are disrupting them. Part C details some of the main planning and management approaches that have evolved within the Canadian context to address environmental challenges. Part D provides a thematic assessment of the challenges associated with particular activities such as fisheries, forestry, agriculture, wildlife use, water, energy production, and mineral extraction.

Figure 1.18 illustrates the relationship among these aspects. Natural systems form the basis of all human activity. A system is a recurring process of cause-and-effect pathways (Box 1.9). They range in scale from the giant atmospheric and oceanic circulation systems to the processes underway in a single living cell. The pollination of a flower by a bee, the melting of a glacier, and the biological fixation of nitrogen from the atmosphere are all parts of such systems. These systems are infinitely complex, and there is a great deal of uncertainty as to how they function. One of the goals of natural science is to try to understand this complexity. We do this by constructing simplified models of how we think they work (Figure 1.18, box 1). The models presented in Chapters 2 and 4 on how

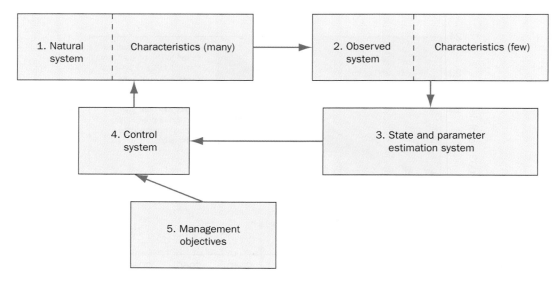

Figure 1.18 Simplified model of interaction of biophysical and social systems in resource management.

Box 1.9   Systems

You are in the educational system. You use the transportation system to go to your college or university, which is warmed by a heating system. What is a system?

Systems are composed of sets of things—e.g., educational institutions, buses, heating components—that are all related and linked together in some sort of functional way. Between these different components there is a flow of material, such as students, passengers, or heat, subject to some driving force—a thirst for knowledge, the need to get somewhere, or the need to get or stay warm. Systems are generalized ways of looking at these processes.

energy flows through the biosphere and on the nature of biogeochemical cycles are examples of these kinds of simplified representations of natural systems.

We do not know all the facts relating to these systems. We do not know all the components, let alone their functional relationships (Figure 1.18, box 2). Many species, especially insects, still await discovery, even in well-explored temperate countries such as Canada. Of those we do know about, we have to select which ones we think are important and worth representing in our simplified models. Only recently, for example, have we become aware of the critical role played by various lichens in the circulation and retention of nitrogen in temperate rain forests (Chapter 4). Furthermore, not all characteristics are measurable, even if we are aware of their existence. Thus, the simplified models that we use to understand natural systems are fraught with uncertainty.

On the basis of these models, we try to estimate the status of a given system (Figure 1.18, box 3). How many fish spawn in a certain river? What proportion of the landscape supports commercial tree growth? What soil characteristics are suitable to support a given crop? If we understand the current status of the system, we can also start to ask what the result will be if certain parameters are changed. What would happen to the system, for example, if we took a certain number of fish from the river before they spawned, if we removed tree growth from a portion of the landscape, or if we grew a given crop in the same soil for a particular time period? In other words, we try to assess the impact of various changes to the system. Quite formal processes of impact assessment have arisen in many jurisdictions. They assess not only the impact on natural systems but the impact on social systems as well, as described in Chapter 6.

On the basis of this understanding, we try to replace natural systems with control systems in which the main decision

regulators are humans rather than nature (Figure 1.18, box 4). Instead of natural forces determining the number of fish that reach the spawning grounds, or the age of trees before they are replaced by other trees, or what species will grow in a particular location, people make these decisions as we modify the environment to our own advantage. These control systems are considered under topics such as forestry, water, energy, and agriculture in Part D of the book.

The moral I labor toward is that a landscape as splendid as that of the Colorado Plateau can best be understood and given human significance by poets who have their feet planted in concrete—concrete data—and by scientists whose heads and hearts have not lost the capacity for wonder. Any good poet, in our age at least, must begin with the scientific view of the world; and any scientist worth listening to must be something of a poet, must possess the ability to communicate to the rest of us his sense of love and wonder at what his work discovers.

—Abbey 1977, 87

Control systems are implemented on the basis of the social, economic, technological, and management constraints of a society (Figure 1.18, box 5). These factors influence the demands for various outputs from the system and the speed of extraction. The environmental management strategies discussed in Part C of the book outline some of the main approaches to mediating between the social and economic demands of the society and the productive capacity of the system. As with the natural system, these strategies are characterized not only by complexity and uncertainty but also by *conflict* among different societal groups regarding the rate of outputs and distribution of benefits. Simply deciding which groups in society have a legitimate interest in a particular environmental issue is quite complex, as described in Chapter 5. Various dispute resolution mechanisms (Chapter 6) have emerged to address the conflicts arising from resource allocation decisions.

As we can already see, the challenges faced in environmental management are complex indeed, and consequently it is necessary to employ an integrated approach, such as the ecosystem approach described in Chapter 5, to understand this complexity. Both natural and social systems are fraught with uncertainty, making an adaptive approach (Chapter 6) a necessity, with strong adherence to the *precautionary principle*, as discussed in Chapter 5. Furthermore, our present predicament is largely the result of modification of natural systems before we had invested the time and effort—or had sufficient data or scientific expertise—to understand the consequences of our actions. The fisheries on both the Pacific and Atlantic coasts of Canada are in trouble (see Chapter 8) because our simplified system models were inadequate as a basis for decision-making. However, in some instances, even when the long-term implications of an activity on the future viability of a resource are understood, the activity continues for political and economic reasons. The overharvesting of timber (Chapter 9) is a good example.

Perhaps the most important message underlying the environmental challenges we face is the need for fundamental changes in the way we view our relationship with nature, as discussed in Chapter 15. These changes must take place at all levels, from international agencies such as the World Bank, through national and regional governments, to household and individual initiatives. Part of the goal of this book is to motivate the reader to become more involved in making these changes happen, both locally and globally.

Summary

1. The *environment* is the combination of the atmosphere, hydrosphere, cryosphere, lithosphere, and biosphere in which humans, other living species, and non-animate phenomena exist.

2. Some consider resources to be only those components of the environment of utility to humans. This view is considered to be *anthropocentric*; value is defined relative to human interests, wants, and needs. Another, contrasting perspective is that resources exist independently of human wants and needs. This view is *ecocentric* or *biocentric*; aspects of the environment are valued simply because they exist, and they have the right to exist.

3. Environmental and resource issues can be approached from disciplinary, multidisciplinary, cross-disciplinary, interdisciplinary, and transdisciplinary perspectives. Each provides a different basis or model for viewing the world. We should strive for interdisciplinary and transdisciplinary approaches in order to understand complex systems.

4. For a science-based approach to management of resources and the environment, we should (1) focus the science on key issues and communicate it in a policy-relevant form; (2) use scientific information to clarify issues, identify potential management options, and estimate consequences of decisions; (3) clearly and simply communicate key scientific findings to all participants; (4) evaluate whether or not the final decision is consistent with scientific information; and (5) be aware of the balance between scientists providing technical information and interpretation and being advocates for particular approaches or solutions.

5. The Sydney Tar Ponds case illustrates how attitudes and approaches towards environmental issues evolve and how what may be viewed as appropriate behaviour at one time can be viewed as inappropriate later.

6. There are at least three lessons from the Sydney Tar Ponds experience: (1) when basic science is not used from the outset to inform policy decisions related to environmental issues, the probability is high that effective solutions will not be found; (2) even when science is used, understanding can be incomplete, and decisions will have to be taken in the face of considerable uncertainty; and (3) when local stakeholders are not included in the process, challenges can be expected to proposed solutions. The third point reminds us that while science is important, it is often not sufficient by itself.

7. Having a vision of or sense of direction towards a desirable future condition is essential for planning and management. In that regard, *sustainable development* and *sustainable livelihoods* have been proposed as appropriate ideals to characterize what societies should aspire towards.

8. On the global scale, there is undeniable evidence of unprecedented environmental degradation as a result of human activities. Growing global population is a continuing challenge, as are the consumer demands of people in the wealthier countries.

9. An estimated 4.3 people are born every second around the world. By the end of 2007, there were 6.7 billion people on Earth. The United Nations forecasts an increase to 9.1 billion people by the year 2050. This is lower than earlier estimates of 9.3 billion but still represents more than 80 million additional people per year to feed.

10. Conditions continue to deteriorate in many poorer countries. The richest 20 per cent of the world's population is responsible for more than 75 per cent of world consumption, while the poorest 20 per cent consume less than 2 per cent.

11. The eight Millennium Development Goals adopted by the United Nations in 2000 aim to improve human well-being by reducing poverty, hunger, and child and maternal mortality; by ensuring education for all; by controlling and managing diseases; by tackling gender disparity; by ensuring environmental sustainability; and by pursuing global partnerships. Under each of the MDGs, countries have agreed to targets to be achieved by 2015.

12. In order to stay below the threshold level of what scientists predict would be runaway change and if emissions were frozen at the current level, we would need the absorptive capacity of two planets.

13. Emissions are not equally distributed. In Ethiopia, the average per capita carbon footprint is 0.1 tonnes, compared to 20 tonnes in Canada. The per capita increase for Canada since 1990 (5 tonnes) is higher than per capita emissions for China in 2004 (3.8 tonnes). If every person living in the developing world had the same carbon footprint as the average person in high-income countries, global CO_2 emissions would require six planets.

14. Canada is one of the most privileged countries, covering some 13 million square kilometres and with a population of some 32 million people. However, our environmental impacts are considerable. Our per capita consumption of water and energy is among the highest in the world. We also have some of the highest production per capita of waste products, including greenhouse gases.

15. Canada, although a signatory to the legally binding Kyoto Protocol, has the second highest per capita emissions of greenhouse gases in the world. Canada agreed under the Kyoto Protocol to target a 6 per cent cut in emissions. Instead, emissions have increased by 27 per cent, and the country is now around 35 per cent above its Kyoto target range.

16. Responsibility for the environment and natural resources is divided between the federal and provincial governments, with the territories and municipalities taking on increasingly important roles. First Nations also are much more involved. The shared responsibility often requires collaboration and partnerships, which can create tensions because of differing interests and perspectives.

17. Indicators are one way by which science can assist decision-makers to appreciate current trends. There are different kinds of indicators with different levels of complexity. Decision-makers often want few indicators that contain the most information to be able to understand the situation. Indicators that combine many different elements are called composite indices.

18. One such index is the Living Planet Index. The index shows a 35 per cent reduction overall in the planet's ecological health since 1970. Ecological Footprints are another index that show the extent of human demand on global ecosystems. The index measures the area of biologically productive land and water needed to provide ecological resources and services—food, fibre, and timber, land on which to build, and land to absorb carbon dioxide (CO_2) released by burning fossil fuels.

19. A comparison of humanity's ecological footprint with biocapacity (the amount of biologically productive area—cropland, pasture, forest, and fisheries—available to meet humanity's needs) shows that since the late 1980s, the ecological footprint exceeds the Earth's biocapacity by more than 25 per cent.

20. The main challenge for humanity is to implement effective strategies to ensure that the ecological footprint falls below biocapacity as soon as possible.

Key Terms

anthropocentric view
biocapacity
carrying capacity
crude birth rate (CBR)
crude death rate (CDR)
crude growth rate (CGR)
demographic transition

ecocentric (biocentric) view
ecological footprint
environment
epidemiological transition
estuary
exponential growth
global climate change

gross national product
indicators
Living Planet Index
Millennium Ecosystem
    Assessment
population age structure
replacement level fertility

resources
subsidiarity
sustainable development
sustainable livelihoods
total fertility rates

Questions for Review and Critical Thinking

1. What information is available in your municipality or province regarding environmental hazards? Could a Sydney Tar Ponds occur in your community?

2. Who should be responsible for dealing with environmental hazards resulting from earlier resource use and environmental standards no longer acceptable today?

3. If you were hired to provide recommendations related to the Sydney Tar Ponds, what information would you need to make a decision about the potential risk to health for people living in the community? How would you place a monetary value on any potential risk?

4. What is the distinction between sustainable development and sustainable livelihoods? What other concepts could provide a foundation for characterizing an ideal future condition?

5. Outline the main arguments for considering population growth a threat to global carrying capacity or as a building block for future economic growth.

6. What moral obligations, if any, do Canadians have to assist people in the developing world whose standards of living do not meet basic human needs?

7. Is population growth or environmental degradation the major problem in the less developed countries? Which is cause, and which is effect?

8. What are indicators used for? What are the main types of indicators? Are there any indicators used by your province that give insight into environmental changes?

9. What are some of the main initiatives of Canadian governments to address environmental problems?

10. What is a system? Outline the components of a system that you use on a regular basis.

Related Websites

Community-based indicators: http://www.sustainable.org

Fraser River Basin indicators:
http://www.fraserbasin.bc.ca/indicators.html

Living Planet Index:
http://www.pandaorg/news_facts/publications/general/livingplanet/index.cfm

Selection of indicators for Great Lakes:
http://www.binational.net

National Roundtable on Environment and Economy (NRTEE): http://www.nrtee-trnee.ca

Office of the Auditor General, Commissioner of the Environment and Sustainable Development. *Reports on the Environment and Sustainable Development*:
http://www.oag-bvg.gc.ca/internet/english/oag-bvg_e_920.html

United Nations Environment Programme:
http://www.unep.org/geo/geo4/media/index.asp

United Nations Population Fund: http://www.unfpa.org

Worldwatch Institute: http://www.worldwatch.org

Further Readings

Note: This list comprises works relevant to the subject of the chapter but not cited in the text. All cited works are listed in the section 'References' at the end of the book.

Balmford, A., et al. 2002. 'Economic reasons for conserving wild nature'. *Science* 297 (August): 950–3.

Furimsky, E. 2002. 'Sydney Tar Ponds: Some problems in quantifying toxic waste'. *Environmental Management* 30 (6): 872–9.

Guernsy, J.R., et al. 2000. 'Incidence of cancer in Sydney and Cape Breton County, Nova Scotia, 1979–1997'. *Canadian Journal of Public Health* 91 (4): 285–92.

Joffres, M.R., et al. 2001. 'Environmental sensitivities: Prevalence of major symptoms in a referral center: The Nova Scotia Environmental Sensitivities Research Center Study'. *Environmental Health Perspectives* 109 (2): 161–5.

O'Leary, J., and K. Covell. 2002. 'The tar pond kids: Toxic environments and adolescent well-being'. *Canadian Journal of Behavioural Science* 34 (1): 34–43.

Vitousek, P.M., et al. 1997. 'Human alteration of the global nitrogen cycle: Causes and consequences'. *Issues in Ecology* 1: 2–16.

Wackernagel, M., et al. 1993. *How Big Is Our Ecological Footprint? A Handbook for Estimating a Community's Appropriate Carrying Capacity*. Vancouver: University of British Columbia, Department of Family Practice.

Worldwatch Institute. 2007. *Vital signs: The trends that are shaping our future*. New York: W. W. Norton.

Part B
The Ecosphere

On Spaceship Earth, there are no passengers;
we are all members of the crew.

—Marshall McLuhan

The first section of the book provided an overview of the global and Canadian situations regarding the relationship between humans and the Earth. As the Canadian communications theorist Marshall McLuhan noted, we are not passive bystanders in terms of this interaction. We are members of the crew, helping to direct what happens to our planet. As crew members, we need to have some idea of the workings of our 'Spaceship Earth' and particularly of the nature of its life support system. Imparting such an overview is the prime purpose of Part B. Here we describe the natural systems through simplified models, following the conceptual framework outlined in Chapter 1 (Figure 1.10). This will provide the background you need to understand many of the environmental challenges facing society today.

Most of what follows is derived from just one form of environmental knowledge—natural science (see Box B.1). Natural science tries to find order in and to understand nature so that accurate predictions can be made regarding the outcome of given changes that may occur. An important underlying assumption is that patterns in nature can be discerned if we approach things in the right manner with the right tools. The *scientific method* lays the foundation for how scientists approach this task. Another important assumption of the scientific method is that the same results will be obtained by different scientists if they repeat the experiment or observations in the same manner as the original observations. This is one reason why scientists must give detailed descriptions of their methodology and attempt to be as objective as possible. What use would a thermometer be, for example, if its readings of temperature varied according to who took the measurement?

Box B.1   Traditional Ecological Knowledge

Most of the concepts presented in this book are the result of the 'scientific approach' to understanding different phenomena. There are other approaches, however, and one gaining increasing attention is **traditional ecological knowledge (TEK)**. Scientists around the world have found that indigenous peoples often have a detailed knowledge of their local environments, which is not entirely unexpected for peoples gaining their sustenance directly from that environment. They have also found that indigenous peoples tend to undertake the same kinds of tasks as Western scientists, such as classification and naming of different organisms and studies of population dynamics, geographical distributions, and optimal management strategies. Unlike Western science, however, this knowledge is rarely recorded in written form but is handed down orally from generation to generation.

Only now are we becoming more aware of this body of knowledge, as highlighted in the guest statement by Elise Ho in Chapter 7. In Canada, this awareness has come about particularly as a result of increasing industrial interest in northern regions and the potential impact of resource extraction on Native communities. Inevitably, this has given rise to discussions about which form of ecological knowledge, Western or traditional, is the 'best'. Both have their advantages and disadvantages. Modern science is informed by developments around the world but is limited in terms of its knowledge of changes over time in a particular place, an area where traditional knowledge is particularly rich. Scientists also tend to concentrate on information that can be tested by replication and to ignore idiosyncratic and individual behaviour, which is given substantial weight by indigenous hunters.

A graphic example of these differences came to light in March 2008. Scientists from the Department of Fisheries and Oceans (DFO) claimed that the bowhead whale populations in the eastern Arctic were so low—about 5,000 animals—that the species was listed as threatened under the Species at Risk Act, discussed in Chapter 14. However, for many years the Inuit had claimed that there were far more animals and that the hunting quota of one whale every two years should be increased. As a result of new scientific information, this latter claim was found to be correct, and DFO scientists expanded their estimates to 14,400. That is nearly 300 per cent higher and roughly equal to the 11,000 whales that are thought to have frequented waters such

as the Davis Strait and Lancaster Sound during the nineteenth century, when whales were vigorously hunted for their blubber as the main source of lamp oil. However, this example is not meant to imply that traditional knowledge is always superior; it just happened to be more accurate in this particular situation. A similar dispute is now occurring over polar bear numbers, and it will be interesting to see how the issue is resolved over the next few years.

Management systems also differ between indigenous and scientific approaches. The traditional system is self-regulating, based on communal property arrangements. Conservation practices, such as rotation of hunting areas, were commonly practised. However, the system is not infallible, especially with the onslaught of outside influences and commercialization. There are many examples throughout the world in which indigenous peoples have hunted species to extirpation within their homelands. Similarly, the modern system of private property rights and state allocation of harvesting rights does not always work. This was recognized all too clearly with the North Atlantic cod fishery: if scientists and policy-makers had given more credence to the local ecological knowledge of inshore fishers in Newfoundland outport communities, the destruction of the fishery might have been averted. Scientists and indigenous peoples are now realizing the benefits offered by the two systems of knowledge and management approaches and are trying to use both through co-management arrangements.

There is considerable dispute between scientists and indigenous peoples about the numbers of polar bears remaining.

iStockphoto.com/David T. Gomez

Although the scientific method is designed to minimize the effects of scientists on the phenomenon they are studying, it is a myth to believe that science is 'objective'. Scientists work in a social environment and are greatly influenced by their peers and by society. The research topics selected and the ways in which research questions are framed are key components of the scientific method and yet are heavily influenced by the individual scientist. However, the fact that such social biases and values influence science should not be an excuse for not trying to maintain as value-free an approach as possible and always making sure that biases are explicit and documented.

Scientists collect data or *facts* (observations that are widely accepted as truthful) about the environment and then try to make some order out of those facts. This order is called *theory*. The theory of natural selection, first outlined by Charles Darwin and Alfred Russel Wallace in 1858, is a good example. When there is universal acceptance of the theory, a *scientific law* may be established. A scientific law represents the most stringent form of understanding and lays down a universal truth that describes in all cases what happens in certain circumstances. In the next two chapters, you will be introduced to some of these laws. They are useful because they provide firm building blocks upon which we can build our scientific understanding.

The scientific method links, with minimum error, the testing of *hypotheses* with the existing body of knowledge. It prescribes a series of steps that, over time, scientists have found are most likely to provide understanding about phenomena under study, irrespective of the observer undertaking the observations or experiment. The process starts with a question often derived either from observations of the environment or from existing literature. A hypothesis is generated to explain the phenomenon of interest, and experiments or observations are designed to disprove the hypothesis. If the hypothesis is not rejected, then additional experiments may be designed in further efforts to disprove it. If all efforts fail, the hypothesis becomes incorporated into theory and is accepted as a valid answer to the question. However, natural scientists do not have to follow one scientific method. Many important scientific insights have come about through quite irregular approaches. Nonetheless, it is wise to learn from previous experience and to know the kinds of procedures usually followed in any given area of inquiry.

That said, we should not feel that any of the hypotheses, theories, or even laws advanced by science are beyond question. The whole purpose of science is to ask questions, and science advances by continually changing and modifying previous knowledge. 'Be kind to scientists (and teachers!) but ruthless in your questions' is not a bad dictum. Debate and disagreement in science are normal. In the following chapters, for example, ideas on some topics are changing rapidly, and even some concepts (e.g., keystone species, climax vegetation) are being questioned by some scientists. This active debate is often misunderstood by those outside the academic community as implying that scientists do not know anything. It can also be used for political purposes to support inaction on measures that might be unpopular, such as limits on industrial emissions to reduce acidic precipitation. Debate and dispute, however, are signs of the vitality and strength of scientific thinking—and of the urgent need to get closer to truth—rather than the converse.

When we think of science, we often tend to think of it within the context of natural and physical sciences, and most of the next three chapters focuses on this understanding of 'science'. However, social science is also critical to understanding and addressing environmental management. Just as natural scientists hope to provide greater understanding of the natural world, social scientists address the same need for social dimensions. How can we understand how individuals think, how to change their behaviour, how governments and the economy operate, and how legislation is formed and enforced? What can we learn from societies elsewhere and how they have interacted with their environmental surroundings in the past? These are some of the kinds of questions that we look to disciplines such as geography, anthropology, history, psychology, sociology, political science, and economics to address.

The methods of social science can also have much in common with the natural sciences, with the testing of hypotheses as a means of building theory and establishing laws that will help to provide generic understanding of seemingly disparate events. Yet the great variability in social systems, the difficulties with isolating and controlling many of the variables under study in a social context (as opposed to natural phenomena or laboratory work), and the challenges of repeatability (it is often impossible to duplicate the circumstances for repeated experiments) make adherence to strict scientific procedures impossible. These difficulties mean that social scientists must apply a very broad range of approaches to gaining social understanding. Thus, although theory formation is a critical part of the social sciences, the formulation of universal laws is very rare.

Furthermore, although both natural and social scientists strive for *quantitative* (numerical) data because of the precision that numbers provide, *qualitative* (non-numerical) data are sometimes all that can be obtained or even all that can or should be sought, depending on the particular research question. A case in point: Over many years, federal fisheries scientists tended to dismiss the qualitative data from inshore Newfoundland fishers as 'anecdotal' and chose to rely largely on their own estimates of stock biomass, which were based on offshore data from the trawler fishery and from their own survey boats. As it turned out, of course, the anecdotal observations and warnings based on generations of knowledge and experience were far more accurate than the quantitative data the scientists chose to believe. Both kinds of data are important. For example, if someone tells you that the weather was 'cool', that does not convey as much information as telling you that the temperature was −25°C. Data on the beauty of a certain landscape is less amenable to quantitative assessment, although some scientists have developed numerical ways of approaching such problems!

The three chapters in Part B provide a basic overview of the main processes that maintain the planetary life support system. A simple model of the planet would look like the layers of an onion (Figure B.1). We are most concerned with the outer layer, the **ecosphere**, which consists of three main layers:

- The **lithosphere**, which is the outer layer of the Earth's mantle and the crust and contains the rocks, minerals, and soils that provide the nutrients necessary for life.
- The **hydrosphere**, which contains all the water on Earth. Water, in a frozen state, is referred to as the **cryosphere**, a very important component for much of Canada.
- The **atmosphere**, which contains the gases surrounding the lithosphere and hydrosphere. It can be further divided into four main sub-layers. The innermost layer, or **troposphere**, contains 99 per cent of the water vapour and up to 90 per cent of the air and is responsible for our weather. Two gases, nitrogen (78 per cent) and oxygen (21 per cent), account for 99 per cent of the gaseous volume. This layer extends

on average to about 17 kilometres before it gives way to the second layer, the **stratosphere**, wherein lies the main body of ozone that blocks out most of the ultraviolet radiation from the sun. At about 50 kilometres from the Earth's surface is the **mesosphere** and above that the **thermosphere**. As distance from Earth increases, the pressure and density of the atmosphere decreases as it melds into space.

These three layers combine to produce the conditions necessary for life in the **ecosphere**, stretching from the depths of the ocean trenches up to the highest peaks, a layer some 20 kilometres in width, no larger in scale than the peel of an apple, that contains some 30 million different organisms. The following chapters impart some idea of the main environmental processes of the ecosphere and describe how human activities interrupt these processes.

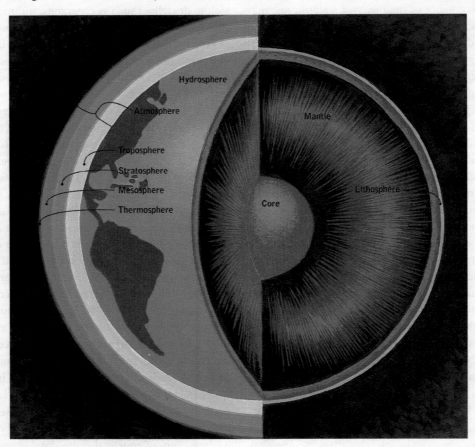

Figure B.1 A simplified model of the Earth showing the ecosphere.

Key Terms

atmosphere	hydrosphere	stratosphere	traditional ecological
cryosphere	lithosphere	thermosphere	knowledge (TEK)
ecosphere	mesosphere		troposphere

Chapter 2

Energy Flows and Ecosystems

Learning Objectives

- To know the nature of energy and the laws that govern its transformation.
- To understand the way energy flows through the ecosphere and links ecosystem components.
- To outline the main influences on the structure and nature of ecosystems.
- To appreciate some of the implications of global climate change on species distributions and abundance.
- To learn the importance of reducing energy use in society.

Introduction

The annual arrival of the capelin to the beaches of Newfoundland to spawn on the first full moon of June had long been a bounty—not only for many animal species but also for the settlers who collected the fish for consumption and application to their gardens as fertilizer. The capelin is a small fish of the North Atlantic and an important food supply for many other species, including cod, salmon, halibut, mackerel, seals, various whale species, and many species of seabirds such as puffins and murres. In the late 1970s and early 1980s, the numbers of fish coming to spawn declined markedly.

Normally, 80 to 90 per cent of the Atlantic puffin's diet consisted of capelin. In the early 1980s, scientists noticed a large decline in this proportion, down to 13 per cent, resulting in severe malnutrition of puffin chicks and subsequent declines in the population owing in part to starvation. Their numbers fell as a direct result of the removal of their food base, the capelin. The energy flow between the species had been interrupted by the opening of an offshore capelin fishery that had removed the capelin from the food chains that nourish so many other marine species. The puffins were one noticeable victim of this appropriation, but other species feeding on the capelin

The colourful bill is the most striking feature of the Atlantic puffin, which breeds among the rocks of sea islands.

Spawning capelin on a beach in Newfoundland.

suffered the same consequence. These species in turn would affect the abundance of other species at all levels in the food web, since the numbers of some species are controlled mainly by their predators. This example illustrates the importance of understanding how energy links species and flows through ecosystems. Changing the energy available at one part of the food chain will have repercussions throughout the ecosystem.

Reading this book, taking notes in class, even snoozing at home all require energy. That energy comes ultimately from the radiant energy of the sun and is transformed into chemical energy in the form of food supplies before being converted to mechanical energy in terms of physical exertion and activity. In this chapter, you will gain an appreciation of energy in relation to such transformations, how energy flows through ecosystems, and the ecosystem consequences that result.

Traditional societies in many lesser developed countries, such as the Lisu people in the highlands of northern Thailand, still rely on wood as the main source of energy. Although this can lead to deforestation if demands are greater than growth, it also limits the amount of energy that can be transformed by these societies.

Energy

Energy is the capacity to do work and is measured in calories. A **calorie** is the amount of heat necessary to raise one gram or one millilitre of water one degree Celsius (C), starting at 15 degrees. Energy comes in many forms: radiant energy (from the sun), chemical energy (stored in the chemical bonds of molecules), as well as heat, mechanical, and electrical energy. Energy differs from matter in that it has no mass and does not occupy space. It affects matter by making it *do* things—work. Energy derived from an object's motion and mass is known as **kinetic energy**, whereas **potential energy** is stored energy that is available for later use. The water stored behind a dam is potential energy that becomes kinetic energy as it pours over the dam. The gas in a car is potential energy before it is poured into the engine to create mechanical energy for propulsion.

Most of the energy available for use is termed **low-quality energy**, which is diffuse, dispersed at low temperatures, and difficult to gather. The total energy of all moving atoms is referred to as **heat**, whereas temperature is a measure at a particular time of the average speed of motion of the atoms or molecules in a substance. The oceans, for example, contain an enormous amount of heat, but it is very costly to harness this energy for use. They have high heat content but low temperature. On the other hand, **high-quality energy**, such as a hot fire or coal or gasoline, is easy to use, but the energy disperses quickly. Much of our economy and technology is now built around the transformation of low-quality energy into high-quality energy for human use. It is important that we match the quality of the energy supply to the task at hand. In other words, the aim is not to use high-quality energy for tasks that can be undertaken by low-quality supplies. Heating space, such as your house, for example, requires only low-temperature heat, yet many homes are heated through the conversion of high-quality energy sources that entail significant energy losses in generation, transport, and application. Nuclear energy, involving high-quality heat at several thousand degrees, which is then converted to high-quality electricity transmitted

to homes and used in resistance heating, is very inefficient. The most efficient way to provide space heating is through having super-insulated houses and passive solar heating.

All organisms, including plants, require energy for growth, tissue replacement, movement, and reproduction. To gain a comprehensive perspective on life, we must understand energy and how it is transformed and used. Box 2.1 provides an introductory definition of life.

Laws of Thermodynamics

Two laws of physics (or physical laws) describe the way in which trillions of energy transformations per second take place all over the globe. They are known as the *laws of thermodynamics*. The first one, the **law of conservation of energy**, tells us that energy can neither be created nor destroyed; it is merely changed from one form into another (nuclear is a form of potential energy—the energy is simply held in the nucleus of an atom). Organisms do not create energy; rather, they obtain it from the surrounding environment. When an organism dies, the energy of that organism is not 'lost'. It flows back into the environment and is transformed into different types of energy, the total sum of which adds up to the original amount. Similarly, we all know that cars obtain their energy from gasoline. As the fuel gauge goes from full to empty, this does not indicate that energy has been consumed; it has merely been transformed from chemical energy into other forms of energy, including the mechanical energy to move the car.

The **second law of thermodynamics** tells us that when energy is transformed from one form into another, there is always a decrease in the quality of usable energy. In any transformation, some energy is lost as lower-quality, dispersed energy that is dissipated into the surrounding environment, often as heat. The amount of energy lost varies depending on the nature of the transformation. In a coal-fired generating station, for example, 35 per cent of the coal's energy at most is

Box 2.1   What Is Life?

We have said that energy is essential for all life, but what is life? Living organisms have a number of common characteristics, including:

- They use energy to maintain internal order.
- They increase in size and complexity over time.
- They can reproduce.
- They react to their environment.
- They regulate and maintain a constant internal environment.
- They fit the biotic and abiotic requirements of a specific habitat.

We think we have a fairly good idea of what constitutes life, but there is still a lot of debate as to how life developed on Earth. More than 80 years ago, two scientists proposed a theory, called the Big Bang Theory, explaining the origin of the universe as the result of a massive explosion that occurred some 15 billion years ago. The solar system came from the resulting matter. As the chunks of matter grew in size, they heated up. As the Earth cooled, warm seas formed, and precipitation occurred that helped to create a nutrient-rich environment. Over time, the constant bombardment of this nutrient-rich soup by high energy levels from the sun created chemical reactions producing simple organic compounds, such as amino acids. Scientists have managed to recreate several organic compounds necessary for life from inorganic molecules by bombarding them with energy.

Over billions of years, larger organic molecules came to be synthesized until the first living cells, probably bacteria, developed between 3.6 and 3.8 billion years ago. These cells passed through several stages over billions of years, with increasingly complex development. This activity took place in the ocean environment, protected from ultraviolet (UV) radiation. Between 2.3 and 2.5 billion years ago, a major change occurred when photosynthetic bacteria developed that emitted oxygen into the atmosphere as they manufactured carbohydrates from the carbon dioxide in the atmosphere. Over time, the oxygen reacted with the abundant and poisonous methane in the atmosphere, reducing levels of this gas and leading to the

atmosphere we know today. Some oxygen was also converted to ozone in the lower stratosphere that protected evolving life from UV radiation and allowed the emergence of life from deeper to shallower waters and eventually onto land itself.

To simplify the vast array of life on Earth, biologists recognize five main kingdoms. The most simple kingdom (#1) consists of *monerans* that include bacteria and the photosynthetic blue-green bacteria and are single-celled micro-organisms. The genetic material of monerans is not contained within a nuclear envelope, and they are known as *prokaryotic* and were the first to evolve. The other four kingdoms (#2 to #5) all have nuclei and a high degree of internal structure and are known as *eukaryotic*. The *protists* (#2) comprise a large variety of unicellular and multi-cellular species such as algae, protozoans, slime moulds, and foraminifera. Kelp species, found in abundance around much of Canada's coastline, are multi-cellular algae from this kingdom. The kingdom consists of 14 phyla with more than 14,000 species described. *Fungi* have their own kingdom (#3), one that evolved relatively recently, some 400 million years ago. The kingdom includes both fungi, which are multi-cellular, and yeasts, which are unicellular. All fungi are heterotrophs and mostly digest dead organic matter or act as parasites. They are key components of the biogeochemical cycles described in Chapter 4. Many are asexual. Mushrooms are a good example.

The remaining two kingdoms (#4, #5) will be most familiar to you. The *plantae* (#4) are mostly photosynthetic, although there are some exceptions as described in Box 2.4. Unlike algae, plants are always multi-cellular. They dominate terrestrial ecosystems and contain two main groups, the *bryophtyes* and *vascular* plants. Bryophytes, such as mosses and liverworts, are restricted to moist environments because of the lack of a waxy cuticle to cover their foliage. They also lack vascular tissues. Vascular plants are very complex and have vascular tissues in their stems to convey water and nutrients. There are a further nine divisions within the vascular plants, including ferns, conifers, and flowering plants. It is estimated that there are more than 235,000 species of flowering plants.

Other divisions, such as the gingko, have only one surviving species.

The final kingdom (#5) is the *animalia*, which are heterotrophic, multi-cellular organisms that have the ability to move. They ingest their food and digest it within their bodies. Most reproduce sexually. This is the largest kingdom, largely because of the vast numbers of insects. Insects are examples of the *invertebrates*, having no backbone. *Vertebrates*, on the other hand have backbones and include amphibians, fish, birds, reptiles, and mammals. There are about 4,500 species of mammals. Mammals feed their young with milk, are homeotherms (i.e., they can regulate their body temperatures at a constant level), are hairy, and have a four-chambered heart. This book is mainly about the impacts of one mammal, humans, on the rest of life on this planet.

converted into electricity. The rest is given off as waste heat to the environment.

In a car, only about 10 per cent of the chemical energy of the gasoline is actually converted into mechanical energy to turn the wheels. The remainder is dispersed into the environment. Put your hand onto the hood of a car that has just stopped running. The heat you feel is a result of this second law of energy or the *law of entropy*. **Entropy** is a measure of the disorder or randomness of a system. High-quality, useful energy has low entropy. As energy becomes dispersed through transformation, the entropy increases.

For organisms, the second law is particularly important because they must continuously expend energy to maintain themselves. Whenever energy is used, some of that energy is lost to the organism, creating a need for an ongoing supply that must exceed these losses if the organism is to survive. If losses exceed gains for an extended period of time, then the organism dies.

There are many other important ramifications of this law. It tells us, for example, that energy cannot be recycled. It flows through systems in a constantly degrading manner. We think of 'advanced' societies as being energy consumers. Large dams and nuclear power stations, for example, are visible signs of a modern economy. As we become more economically developed, we find new ways to transform energy. Cars, telephones, electric can openers, blenders, microwaves, hot tubs, computers, and compact disk players are all energy transformers. Yet as more energy is transformed, more is dispersed into the atmosphere because entropy increases. This dispersion can be likened to a bar of soap in a bowl of water. As the soap is used over time, it dissolves into the water, making it less and less useful. Energy may be thought of in a similar fashion as use gradually disperses useful and concentrated forms into the atmosphere.

Some of the principal transformations that have to take place to achieve a sustainable society are: to view high energy consumption as undesirable; to reduce energy waste; and to switch from the non-renewable sources of energy that now dominate (coal and oil particularly) to renewable sources, such as those discussed in Chapter 12. Until the Industrial Revolution, the speed of processing raw materials was limited by the energy available, supplied largely by human and animal labour combined with wood, wind, and water power. These sources were in turn limited by the input of solar energy over a relatively short time period. The use of coal, and later oil, to fuel steam engines removed these limitations and made accessible a vast storehouse of potential energy created by the sun over millions of years through the remains of compressed plants. Acid rain, climatic change, and many of our current environmental problems result directly from this transformation of the energy base of society from a renewable to a non-renewable one. In geological terms, we have released the energy input of millions of years in the blink of an eye—the past 250 years. Many problems are a result of this increase in entropy.

Energy Flows in Ecological Systems

Energy is the basis for all life. The source of virtually all this energy is the sun. More than 150 million kilometres away, the sun, a giant fireball of hydrogen and helium, constantly bombards the Earth with *radiant energy*. This energy, although it is only about 1/50 millionth of the sun's output, fuels our life support system, creates our climate, and powers the cycles of matter discussed in Chapter 4. About a third of the energy received is reflected by the atmosphere back into space (Figure 2.1). Of the remainder, about 42 per cent provides heat to the Earth's surface, 23 per cent causes evaporation of water, and less than 1 per cent forms the basis for our ecological systems. When we think of solar energy, it is important to remember not just the direct heat from the sun but also these indirect forms of energy created by heat input.

Figure 2.1 illustrates the law of conservation of energy. The total amount of energy received by the Earth is equal to the total amount that is lost. One of the changes caused by human

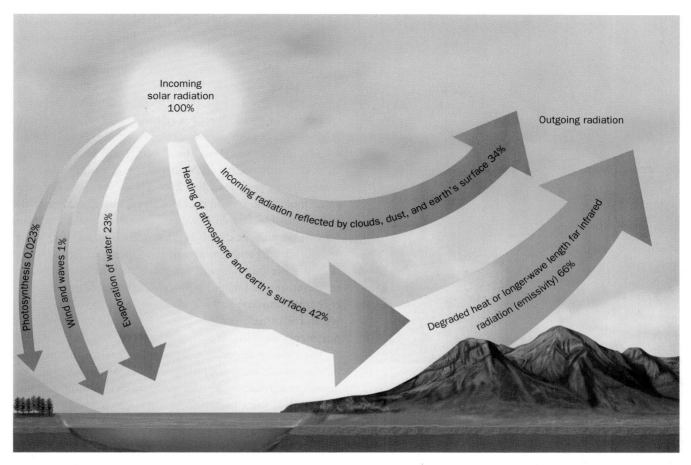

Figure 2.1 The Earth's energy input and output, a good example of the first law of thermodynamics.

activity is delaying the loss of heat to space by trapping it in the atmosphere through increased levels of heat-trapping gases such as carbon dioxide and methane. The implications of this are discussed more fully in Chapter 7.

Producers and Consumers

The sun's energy is transformed into matter by plants through the process of **photosynthesis** (photo = light, synthesis = to put together). Through this process, plants combine carbon dioxide and water, using energy from the sun, into high-energy carbohydrates such as starches, cellulose, and sugars (Figure 2.2). Green pigments in the plants, called **chlorophylls**, absorb light energy from the sun. Photosynthesis also produces oxygen, some of which is used by plants in various metabolic processes. The rest goes into the atmosphere. Hundreds of millions of years of evolution have served to produce the oxygen in the atmosphere that we depend on for life.

Organisms with the ability to capture energy and manufacture matter are known as **autotrophs** (auto = self, trophos = feeding) or **producers**. All other organisms obtain their energy supply through eating other organisms and are known as **heterotrophs** (heter = different) or **consumers**. There are two kinds of autotrophs, **phototrophs** and **chemoautotrophs**. Phototrophs obtain their energy from light; chemoautotrophs gain their energy from chemicals available in the environment. Although most of us are aware of the critical role played

Plants that grow on the forest floor have differing strategies to obtain enough light to survive. Most, such as many ferns, can survive on relatively low light levels. Some, such as the devil's club shown here, grow very large leaves (over 40 cm wide for the devil's club) in order to expose as much photosynthetic surface as possible to the low light levels. The devil's club is a member of the ginseng family and well known among indigenous peoples in western North America for its medicinal properties.

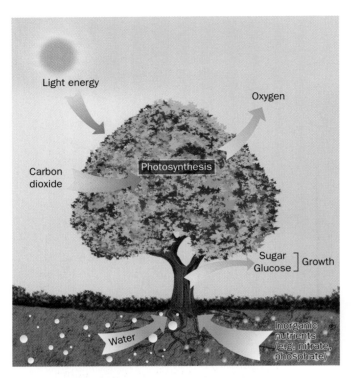

Figure 2.2 The process of photosynthesis.

by phototrophs (plants) in our life support system, the chemoautotrophs play an equally critical yet not so visible role (Box 2.2). Most of them are bacteria and play a fundamental role in the **biogeochemical cycles**, discussed in more detail in Chapter 4.

Phototrophs convert the light energy of the sun into chemical energy, using carbon dioxide and water to produce carbohydrates. The second law of thermodynamics instructs us that some energy will be lost in this transformation; indeed, the efficiency rate is only between 1 and 3 per cent. In other words,

97 to 99 per cent of the energy will be lost. Nonetheless, this conversion is sufficient to produce billions of tons of living matter, or **biomass**, throughout the globe.

Besides photosynthesis, *cellular respiration* is another essential energy pathway in organisms. In both plants and animals, this involves a kind of reversal of the photosynthesis process in which energy is released rather than captured. High-energy organic carbohydrates are broken down through a series of steps to release the stored chemical bond energy. In other words, the potential energy is now realized as kinetic energy in the way described above. This produces the inorganic molecules, carbon dioxide, water, heat (because of the law of entropy), and energy that can be used by the organism for various purposes, such as growth, feeding, seeking shelter, communicating with one another, producing seeds, and maintaining basic physiological functions such as constant body temperature and breathing. Since we are unable to obtain energy from photosynthesis or through chemotaxis, this is how we, and all other organisms unable to fix their own energy, get our energy supplies.

For cellular respiration to occur, most organisms must have access to oxygen or they will die. Such organisms are known as **aerobic** organisms. Some species, anaerobic organisms, such as some bacteria, can survive even without oxygen. This makes them useful in the breakdown of organic wastes, such as sewage.

Food Chains

Some of the energy captured by autotrophs is subsequently passed on to other organisms, the consumers, by means of a **food chain** (Figure 2.3). **Herbivores** eat the producers and are in turn the source of energy for higher-level consumers, or **carnivores** (Box 2.3). Decomposers will feed on all these organisms after they die. Each level of the food chain is known as a **trophic level**. A giant Douglas fir tree on the Pacific coast and

Animals as different as the grasshopper and the elephant are on the same trophic level.

Box 2.2   Deep-Sea Ecosystems

We think of the deep-sea floor as a biological desert. In the 1970s, however, scientists discovered that rich biological communities were supported at hydrothermal vents on the sea floor, mainly bacteria that derive their energy from sulphide emissions. Similar kinds of chemoautotrophic-based communities were discovered on whale skeletons found at depth, nourished by sulphides produced as the carcasses decay. Discoveries of fossils suggest that dead whales may have provided dispersal stepping stones for these communities for more than 30 million years. The question then becomes—what was the impact on these communities when whales were virtually eliminated from the oceans by whalers? Scientists do not yet have the answer to this question.

However, we do know that the biodiversity of the sea floor is much greater than imagined and may equal that of shallow-water tropical reefs. In addition to the vents, the sea floor contains many other habitats, including cold seeps, seamounts, submarine canyons, abyssal plains, oceanic trenches, and asphalt volcanoes, sure to contain a large number of endemic species, with total numbers perhaps as high as 10 million. Although seemingly far removed from human activities, deep seabeds are threatened by many activities, including pollution, shipping, military activities, and climate change. However, fishing is probably the greatest threat as new technologies penetrate the depths and new markets develop.

Deep-sea bottom trawling is a major concern and very damaging to seamounts and the coldwater corals they sustain. These habitats are home to several commercial bottom-dwelling fish species. Seamounts are also important spawning and feeding grounds for species such as marine mammals, sharks, and tuna, which makes them very attractive fishing grounds. Deep-sea fish are particularly vulnerable to large-scale fishing activities because of their long life cycles and slow sexual maturation. Lack of information on deep-sea environments and their species makes it difficult to establish whether sustainable fisheries can take place.

One way to protect such environments is by establishing marine protected areas (MPAs), as discussed in Chapters 8 and 14. Canada has designated Sable Gully, the largest underwater canyon in eastern Canada, as an MPA. The gully is located approximately 200 kilometres off the coast of Nova Scotia at the edge of the Scotian Shelf, where the sea floor suddenly drops by more than two kilometres. More than 70 kilometres long and 20 kilometres wide, this area is home to many interesting and unusual species. The gully is a productive ecosystem that supports a diversity of marine organisms. The world's deepest-diving whale, the bottlenose whale, is a 'vulnerable' species that lives in the gully year-round. Fin whales and northwest Atlantic blue whales, both also classified as 'vulnerable' by the Committee on the Status of Endangered Wildlife in Canada (Chapter 14), make use of the gully throughout the year. Deep-sea corals are a significant feature of the benthic fauna in the area, and nine species are confirmed to live in the gully. MPA regulations prohibit disturbing, damaging, destroying, or removing any living marine organism or habitat within the gully. The MPA contains three management zones, providing varying levels of protection based on conservation objectives and ecological sensitivities. The regulations also control human activities in areas around the gully that could cause harm within the MPA boundary.

a minute Arctic flower on Baffin Island are on the same trophic level—autotrophs. Herbivores, on the second level, range in size from elephants to locusts. The role in energy transformation, rather than the size of the organism, is the important factor in determining trophic level.

Some organisms, such as humans, raccoons, sea anemones, and cockroaches, are **omnivores** and can obtain their energy from different trophic levels. When we eat vegetables, we are acting as **primary consumers**; when we eat beef, we are at the second trophic level, acting as **secondary consumers**; and

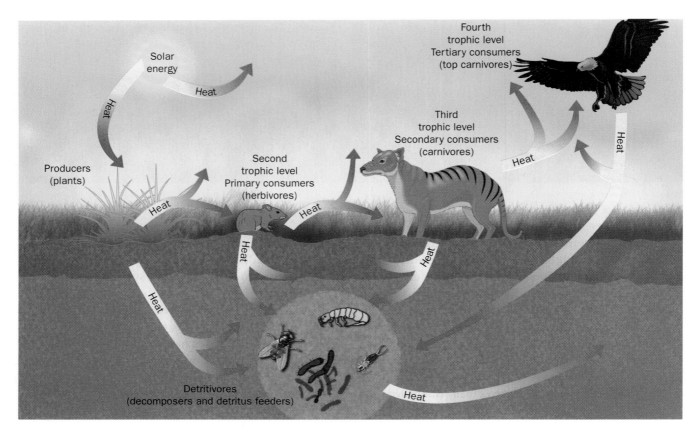

Figure 2.3 A food chain.

Box 2.3   Carnivorous Plants

Not all plants are autotrophs. Carnivorous plants, such as the pitcher plant, the floral emblem of Newfoundland and Labrador, gain their energy from ingesting the bodies of insects that become trapped in their funnel-shaped leaves. The plant, which grows in boggy areas across Canada, has no photosynthetic surfaces. Instead, the leaves act as 'pitchers' to hold a soapy liquid from which a hapless insect cannot escape. It is now thought that the plant may be aided in the decomposition of dead insects by other insects that have developed immunities to the decomposing enzymes produced by the plant. In fact, the plant plays host to several insects that seem to thrive on the environment it provides. This is an example of mutualism in which both species benefit from a relationship.

The carnivorous pitcher plant, the provincial flower of New foundland and Labrador, grows in abundance.

when we eat fish that have derived their energy from eating smaller organisms, we may be **tertiary consumers** at the top of the food chain. The level at which food energy is obtained has some important implications, to be discussed later.

We tend to concentrate on these grazing food chains, but equally important are the decomposer food chains (Figure 2.4). These chains are based on dead organic material or **detritus**, which is high in potential energy but difficult to digest for the consumer organisms described above. However, various species of micro-organisms, bacteria, and fungi are able to digest this material as their source of energy. Indeed, many large grazing animals such as cows and moose have such bacteria in their stomachs to help break down the cellulose in plant material. These decomposers (or saprotrophs) derive their energy from dead matter (sapro = putrid). They are joined by consumers such as earthworms and marsh crabs, known as detritivores, which may consume both plant and animal remains.

These **decomposer food chains** play an integral role in breaking down plant and animal material into products such as carbon dioxide, water, and inorganic forms of phosphorus and nitrogen and other elements. For example, fungi that consume simple carbohydrates, such as glucose, first break down dead wood. Following this phase, other fungi, bacteria, and

organisms such as termites break down the cellulose that is the main constituent of the wood. Were it not for these organisms, wood and other dead organisms would accumulate indefinitely on the forest floor.

The relative importance of grazing and detrital food chains varies. The latter often dominate in forest ecosystems, where less than 10 per cent of the tree leaves may be eaten by herbivores. The remainder dies and becomes the basis for the detritus food chain. In the coastal forests of British Columbia, for example, there are some 140 different species of birds, mammals, and reptiles through which energy can flow. By way of contrast, there are more than 8,000 known, as well as many unknown, species involved in breaking down the soil litter. The same is often true in freshwater aquatic systems, where there may be relatively little plant growth but abundant detritus from overhanging leaves and dead insects. However, the converse is true in marine ecosystems (see Box 2.4), where 90 per cent of the photosynthetic **phytoplankton** (phyto = plant, plankton = floating) may be grazed by the primary consumers, the **zooplankton**.

In general, ecological theory suggests that the more species in the ecosystem, the more alternative pathways are available for energy flow and the better able the ecosystem is to withstand stress. In the Arctic, for example, a simple food chain

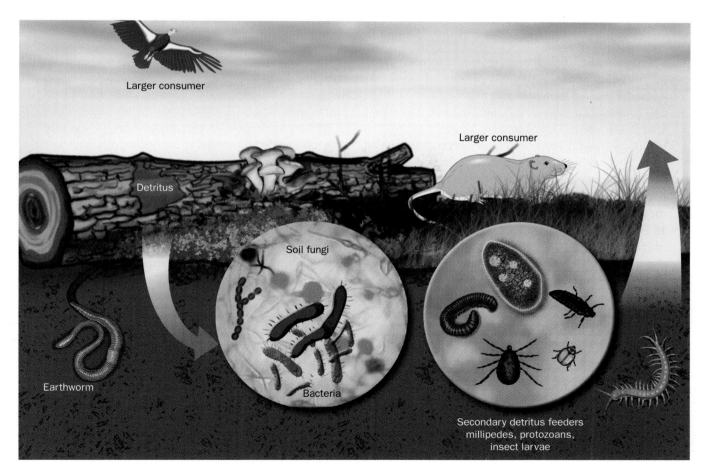

Larger consumer

Larger consumer

Detritus

Soil fungi

Earthworm

Bacteria

Secondary detritus feeders
millipedes, protozoans,
insect larvae

Figure 2.4 Detritus-based food chain.

might be phytoplankton to zooplankton to cod to ringed seal to polar bear. All these species are heavily dependent on the species at the preceding trophic level. Were one of these species to be drastically reduced in number or made extinct, the chances of the role of that species being compensated for by other species is quite low, and the whole food chain might well collapse. This situation can be compared to that at the other extreme, such as a tropical forest, where there are many times more species and the chance of other species combining to fulfill the ecological role of a depleted one is much higher. In practice, however, many other factors are involved, such as the relative degree of specialization of the various organisms, and examples of relatively unstable complex systems abound, so care must be exercised before generalizing the theory to all ecological systems.

Rarely are food chains organized in the simple manner shown in Figure 2.3. Usually, there are many competing organisms and energy paths. Thus, scientists tend to think of food chains more as complex **food webs** in which numerous alternative routes exist for energy to flow through the ecosystem (Figure 2.6). Ecologist Bristol Foster has described an unusual example of this breadth—on the BC coast, he witnessed a garter snake foraging in a tidal pool and swallowing a small fish while at the same time being held by a green sea anemone.

The number of species increases—from the poles to the tropics—as conditions become more amenable for life (Figure 2.7). In the Arctic, for example, there are relatively few species and therefore relatively few alternative pathways for energy flow. If a prey species, such as the Arctic hare, decreases in number, then so will the organism dependent on it higher

Box 2.4   Oceanic Ecosystems

From space, the Earth appears to be a blue, not a green, planet, reflecting the fact that 71 per cent of the Earth's surface is covered by oceans. Life originated in this blueness, perhaps 3.5 billion years ago, and only came onto land some 450 million years ago. Hence, much of our biological ancestry lies within these waters. Although we know about more different species on land than in the oceans, the number of phyla, distinguished by differences in fundamental body characteristics, is higher in the oceans. Of the 33 different animal phyla, for example, 15 exist exclusively in the ocean, and only one is exclusively land-based. We share the same phyla as the fishes, the chordata, characterized by a flexible spinal cord and complex nervous system.

Through their photosynthetic activity, the early bacteria that started in the oceans helped to create the conditions under which the rest of life evolved. Current photosynthetic activity is no less important to our survival.

Scientists estimate that the phytoplankton in the **euphotic zone** of the oceans produce between one-third and one-half of the global oxygen supply. In doing so, they also extract carbon dioxide from the atmosphere. Some 90 per cent of this is recycled through marine food webs, but some also falls into the deep ocean as the detritus of decaying organisms and is stored as dissolved carbon dioxide in deep ocean currents that may take more than 1,000 years to reappear at the surface. The oceans contain at least 50 times as much gas as the atmosphere and are playing a critical role in helping to delay the so-called greenhouse effect, discussed in more detail in Chapters 7 and 8.

These phytoplankton, so important to atmospheric regulation, are also the main autotrophic base for the marine food web. From tiny zooplankton through to the great whales, almost every marine animal has phytoplankton to thank for its existence. Phytoplankton flourish best in areas where ocean currents return nutrients from the deep ocean back to the euphotic zone. This occurs in shallow areas, such as the Grand Banks, where deep ocean currents meet the coast or where two deep ocean currents meet head on. Such areas are the most productive in what is generally an unproductive ocean, and they are the best sites for fisheries. Ninety per cent of the marine fish catch comes from these fertile near-shore waters. Unfortunately, these waters are also the sites of greatest pollution. The blueness of most of the rest of the ocean is a visible symbol of the low density of phytoplankton. That is why the sea is blue, not green.

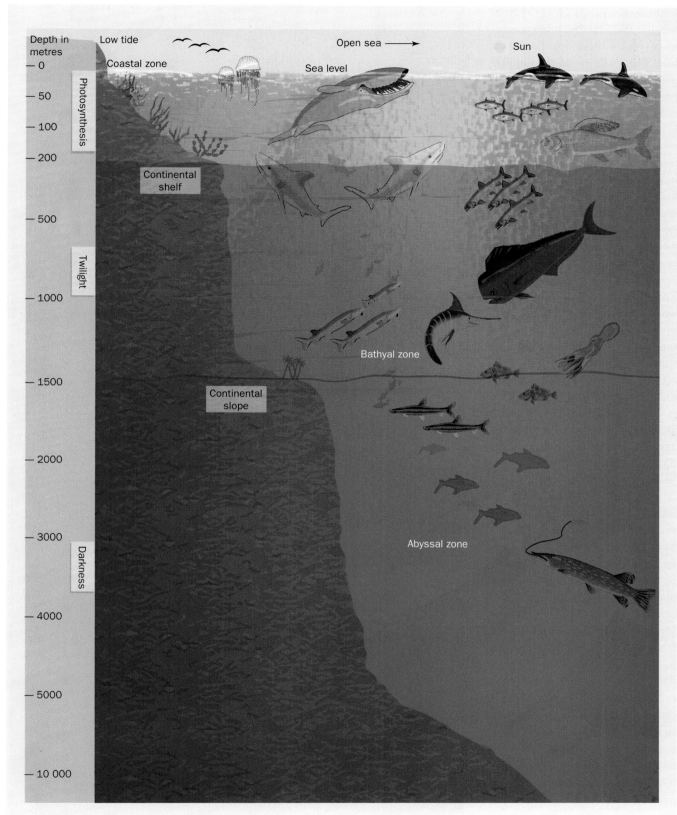

Figure 2.5 Key elements of marine ecosystems.

in the food chain, such as the lynx, because there are few other species upon which these organisms can feed. This gives rise to the familiar population cycles in the North as predator numbers closely reflect the availability of dominant prey species (Figure 2.8).

However, as discussed in Box 2.5, this relationship might not be quite so simple, and ecologists have long debated whether ecosystems are mostly controlled by predators (*top-down control*) or by prey populations (*bottom-up control*). In the former, predators restrict the size of the prey population. This seems to occur, for example, when wolves control deer, elk, or moose populations or when sea otters control urchin populations. Conversely, in some systems, the quality of available forage limits the number of herbivores, which in turn limits the number of predators. The predator numbers are essentially limited by the energy flow through the previous trophic levels. If herbivore populations fall as a result of disease or lack of forage, this will result in a drop in predator populations because of a lack of food. Ecosystems in which controls are dominantly bottom-up tend to have marked limits on plant productivity through abiotic factors, such as low nutrient supply, lack of water, and similar factors, or very close relationships between a specific plant, a herbivore, and a carnivore. Ecosystems reflecting top-down control typically lack these features. However, as with most ecological phenomena, these are general guidelines—most ecosystems contain elements of both top-down and bottom-up control.

Biotic Pyramids

The second law of thermodynamics describes how energy flows from trophic level to trophic level,

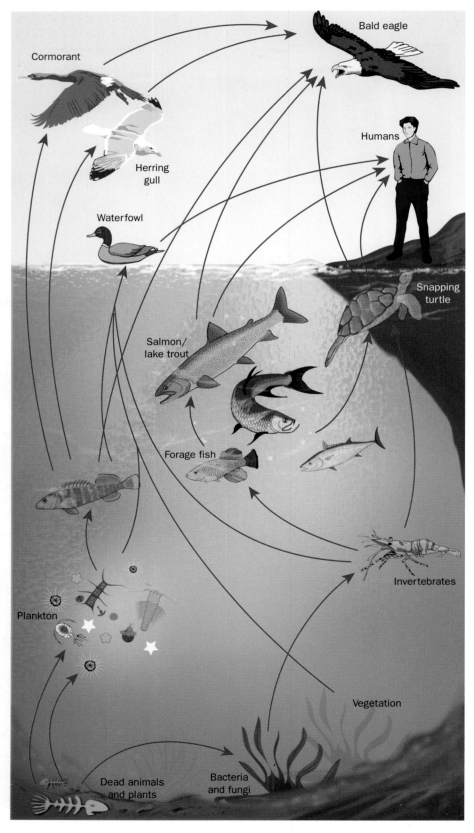

Figure 2.6 A simplified Great Lakes food web. Source: Environment Canada. 1991. *Toxic Chemicals in the Great Lakes and Associated Effects*. Toronto: Department of Fisheries and Oceans; Ottawa: Health and Welfare Canada.

Figure 2.7 The number of mammal species per latitude. Source: After Simpson 1964.

Figure 2.8 Lynx and hare cycles. Data based on pelts received by Hudson's Bay Company. Source: D.A. MacLulich, University of Toronto Studies, Biological Series No. 43, 1937.

ecosystem in order to be supported by the energy base below. The case of the Atlantic puffins, described in the introduction to this chapter, provides a good example. The biomass of carnivores (puffins) could no longer be supported by the energy from the preceding trophic level, the capelin.

Some ecosystems, however, may display an inverted biomass pyramid. In natural grasslands such as those found in southern Saskatchewan, the dominant species, such as grasshoppers, are small-bodied and do not have a large biomass. In contrast, many herbivores in this system, such as antelope and mule deer, are large-bodied and long-lived, with a large total biomass. The same situation exists in the oceans, discussed in more detail in Chapter 8. However, in both cases, the productivity of the plant base is much greater than that of the herbivores.

There are several reasons for the low energy efficiencies of natural food chains. First, not all the biomass at each trophic level is converted into food for the next trophic level. Many organisms have developed characteristics to avoid getting eaten by something else. For example, many plant species have thorns or produce secondary chemicals to deter herbivores. Others have low nutritive levels. Generally, only between 10 and 20 per cent of the biomass of one trophic level is harvested by the next level. Furthermore, of that which is consumed, not all is digested. Humans, for example, are not well equipped to break down and consume the bones or fur of animals, nor are they equipped, compared to moose and other members of the deer family, to break down woody tissue. The proportion of ingested energy actually absorbed by an organism is the *assimilated food energy*. Finally, as cellular respiration occurs to liberate energy for the growth, maintenance, and reproduction of the organism, energy is further released as heat.

The longer the food chain, the more inefficient it is in terms of energy transformation, reflecting the second law of thermodynamics. An Arctic marine food chain that starts from the producers (phytoplankton) to primary consumers (zooplankton) that are subsequently grazed by the largest animals ever to exist on earth (whales) is very efficient because it is so short, with only three energy transformations in which energy is lost. Longer food chains involve a proportionately larger loss of energy because of the greater number of energy transformations. Entropy dictates that long food chains with five or six trophic levels, such as that supporting a killer whale, are very scarce.

The *energy pyramid* also has important implications for humans. For example, it takes between eight and 16 kilograms of grain to produce one kilogram of beef. This means that more land must be cultivated to provide people with a diet high in meat as opposed to a diet based on grains. Since humans are one of the species that can access food energy at several different trophic levels, in terms of energy efficiency it would be better to operate as low on the food chain as possible—that is,

with a loss of usable energy at each succeeding transformation. In natural food chains, the *energy efficiency,* or amount of total energy input of a system that is transformed into work or some other usable form of energy, may be as low as 1 per cent. In general, we expect about 90 per cent of the energy to be lost at each level (Figure 2.9). Similar losses may be experienced in biomass and numbers of organisms at each trophic level. This explains why there are fewer secondary than primary consumers and fewer tertiary than secondary. Carnivores must always have the lowest numbers in an

Box 2.5   Arctic Population Fluctuations

Populations of many animals in the North show distinctive fluctuations in numbers over regular time periods. A three- to four-year cycle is recognized for smaller animals, such as lemmings and meadow voles, and a longer one of nine to 10 years for larger animals, such as snowshoe hares and several species of grouse and ptarmigans. The predators of these herbivores show similar fluctuations in numbers as their energy source becomes critically depleted.

One of the themes that this book illustrates is the uncertainty surrounding many aspects of environmental management. In this case, there is a fair amount of certainty about the dates of fluctuations in these populations, but there is considerable uncertainty as to why these fluctuations occur. Four main ideas have been advanced:

1. The seasonality of the Arctic environment means that plants grow for only a few months every year, yet herbivores must eat throughout the year. Thus, when population levels rise, overexploitation of the food supply can occur rapidly and for a relatively long period, but the plants will take a long time to recover. The relationship between the herbivores and their food supply is the determining factor, and the predator numbers simply reflect those of the prey.
2. A second hypothesis builds on this idea but suggests that as populations of small animals such as

lemmings increase, then more nutrients vital to plant productivity become tied up in this higher trophic level. This lack of nutrients causes reductions in plant productivity and quality, leading to starvation for the herbivores.

3. Other ideas postulate more of an interaction between the predators and prey. Keith et al. (1984) studied the ecology of snowshoe hares in northern Alberta to test a food supply–predation hypothesis. It suggested that food supply shortages halt populations of herbivores that are subsequently caught by predators until numbers fall enough to permit plant recovery. Keith et al. reported that malnutrition of hares was evident but that predators caused 80 to 90 per cent of deaths, thereby supporting the hypothesis.
4. The final idea suggests that food supplies play a negligible role in population cycles and that these cycles would not occur in the absence of predators (Trostel et al. 1987). Scientists have found that sudden drops in hare populations occur despite supplementary feeding programs, a finding that would appear to support this hypothesis (Smith et al. 1988).

In short, ecological systems are not easy to understand. In many cases, several factors may contribute to changes such as in population cycles.

as primary consumers or vegetarians. This topic is discussed in more detail in Chapter 10.

Productivity

Productivity in ecosystems is measured by the rate at which energy is transformed into biomass, or living matter, and is usually expressed in terms of kilocalories per square metre per year. **Gross primary productivity (GPP)** is the overall rate of biomass production, but there is an energy cost to capturing this energy. This cost, *cellular respiration* (R), must be subtracted from the GPP to reveal the **net primary productivity (NPP)**. This is the amount of energy available to heterotrophs.

All ecosystems are not the same in terms of their ability to fix biomass. Light levels, nutrient availability, temperature, and moisture, among other factors, regulate the rates of photosynthesis. The most productive ecosystems per unit area are estuaries, swamps and marshes, and tropical rain forests (Figure 2.10). Recent data indicate that the temperate rain forests, such as those that grow in the Pacific Maritime ecozone, are just as productive as the tropical forests. Other ecosystems are more limited because of deficiencies in one or more of the characteristics noted above. A desert, for example, lacks water, the Arctic lacks heat, and the ocean lacks nutrients.

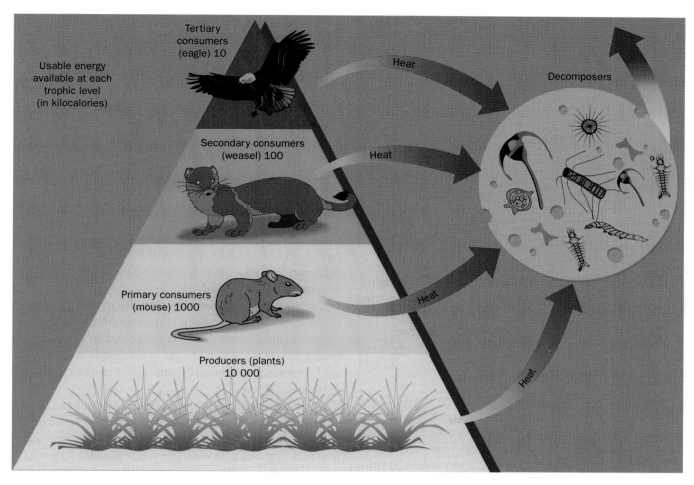

Figure 2.9 Generalized pyramid of energy flow.

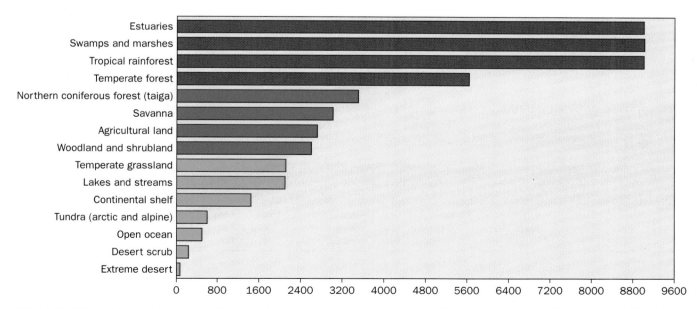

Figure 2.10 Estimated annual average net productivity of producers per unit of area in principal types of life zones and ecosystems. Values are given in kilocalories of energy produced per square metre per year.

Unfortunately, in Canada some of our most highly industrialized and polluted lands are adjacent to estuaries. There is not one sizable estuary on the east coast of Vancouver Island, for example, that is not used by the logging industry either as a mill site or for log storage. The estuary of the Fraser River has been extensively dyked, industrialized, and polluted. Paradoxically, relative little data exist on the effects of these intrusions on our most productive ecosystems. However, one

Estuaries are among the most productive ecosystems. Unfortunately, they are also very convenient sites for industrial activity, such as the log boom storage shown here, which inhibits productivity.

estuary, the Musquash in the Bay of Fundy, should experience little further degradation. It has been declared a marine protected area under Canada's Oceans Act (Chapter 8), and one of the main characteristics noted for its designation was its high productivity. It is one of the last ecologically intact estuaries in a region that has seen extensive modification of its salt marshes.

Humans already take about 40 per cent of terrestrial NPP for their own use. The remainder supports all the other organisms on Earth, which in turn maintain the environmental conditions that keep us alive. The human population is projected to increase by about 40 per cent over the next 50 years. It is highly doubtful that the Earth's systems could withstand a concomitant increase in the amount of NPP being appropriated for human use—another illustration of the carrying capacity challenge we face.

In addition to primary productivity, measurements also can be made of *net community productivity* (NCP), including heterotrophic and autotrophic respiration. Measurements indicate that as communities mature, although GPP and NPP rise, an increasing proportion of the energy of the community is devoted to heterotrophic respiration (Table 2.1). In mature communities, the amount of respiration may be sufficient to account for all the energy being fixed by photosynthesis. There is thus no net gain, leading to the characterization of such communities by some foresters as 'decadent' because they are mainly interested in the productivity of the autotrophs.

Over time, natural systems mature towards maximization of NCP. On the other hand, humans are often concerned with maximizing NPP. This is an example of the 'decision regulators' discussed in Chapter 1. Natural forest system decision regulators may allow trees to achieve ages of several hundred to more than a thousand years before they die. The control system exerted by forest management determines that the life of the trees will be that which maximizes NPP, before considerable amounts of energy become devoted to heterotrophic respiration (Figure 2.11). The age of the trees in systems managed for forestry will hence be much younger than is the case in natural systems.

Auxiliary energy flows allow some ecosystems and sites to be especially productive. For example, tidal energy in an estuary is a form of auxiliary energy flow that helps to bring in nutrients and dissipate wastes so that organisms do not have to expend energy on these tasks and can devote more energy to growth. Agriculture, as discussed in Chapter 10, relies extensively on the inputs of auxiliary energy in the form of

Table 2.1   Production and Respiration As (kcal/m²/yr) in Growing and Climax Ecosystems

	Alfalfa field (US)	Young pine plantation (England)	Medium-aged oak–pine forest (NY)	Large flowering spring (Silver Springs, FL)	Mature rain forest (Puerto Rico)	Coastal sound (Long Island, NY)
Gross primary production	24,400	12,200	11,500	20,800	45,000	5,700
Autotrophic respiration	9,200	4,700	6,400	12,000	32,000	3,200
Net primary production	15,200	7,500	5,000	8,800	2,500	2,500
Heterotrophic respiration	800	4,600	3,000	6,800	13,000	2,500
Net community production	14,400	2,900	2,000	2,000	little or none	little or none

Source: Adapted from E.P. Odum. 1971, Fundamentals of Ecology. 3rd edn, p. 46. Toronto: Holt, Rinehart and Winston, CBS College Publishing. Copyright © 1971 by Saunders College Publishing, reproduced by permission of the publisher.

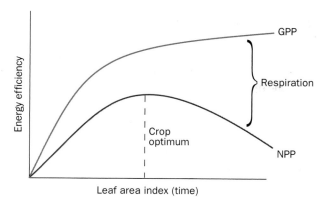

Figure 2.11 Diagram to illustrate general relationship between productivity and time as a forest matures. Foresters might consider the optimal stage of the forest to be at maximum NPP, even though GPP continues to increase over time.

pesticides, fertilizer, tractor fuel, and the like to supplement the natural energy from the sun to augment crop growth. In many cases, this subsidy, mostly derived from fossil fuels, is in excess of the amount of energy input from the sun. Without this subsidy, productivity would be much reduced. There is a cost to the subsidy, however, in terms of high energy costs and the environmental externalities created as the subsidy disperses into the environment in the form of pollution.

Ecosystem Structure

The energy flows described above are all part of the ecosphere. The ecosphere can be broken down in size to smaller units. At the smallest level is the individual *organism*. A group of individuals of the same species is a **population**. The populations in a particular environment are known as a *community*. The **ecosystem** is a collection of communities interacting with the physical environment. However, ecosystems represent a somewhat abstract conceptualization of the environment that can range greatly in scale. Because of the highly interactive nature of the relationship between organisms and their environment, it is often difficult to define precisely the boundary of an ecosystem. Furthermore, ecosystems are *open systems* in that they exchange material and organisms with other ecosystems. Ecosystems and communities thus provide useful abstractions for the study of the environment but should not be taken as very precise categories that will be agreed upon by all scientists.

Many ecosystems taken together and classified according to their dominant vegetation and animal communities form a **biome**. Canada, because of its size, has as many biomes as any country in the world (Figure 2.12). The main factors that control biome distribution are water availability and temperature. Figure 2.13 summarizes how these factors influence biomes on the global scale.

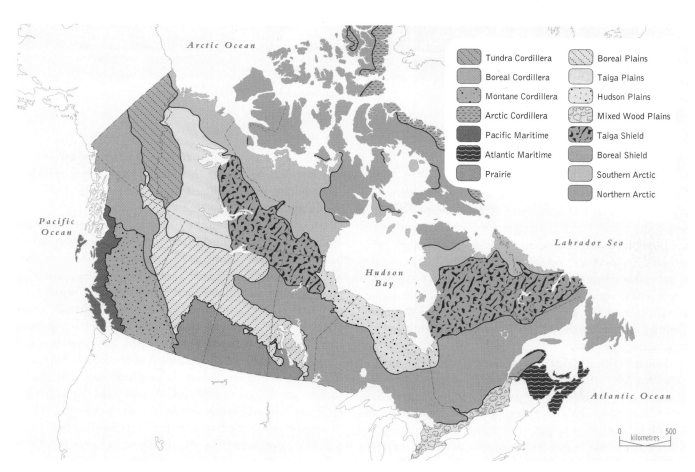

Figure 2.12 Terrestrial ecozones of Canada. Source: E. Wiken. 1986. *Terrestrial Ecozones of Canada*. Ottawa: Environment Canada.

Figure 2.13 Influence of temperature and rainfall on biome.

Third pg

Abiotic Components

The food chains described above constitute the living or **biotic components** of ecosystems. **Abiotic components** play an important role in determining how these biotic components are distributed. Important abiotic factors include light, temperature, wind, water, and soil characteristics such as pH, soil type, and nutrient status. All these factors influence different organisms in various ways. The interaction among these characteristics and the organisms and between the organisms themselves determines where each organism can grow and how well it may grow.

Soils are critical in determining the vegetation growth of an area (Box 2.6). Soil is a mixture of inorganic materials such as sand, clay, and pebbles, decaying organic matter such as leaves, and water and air. This mixture is home to billions of microorganisms that are constantly modifying and developing the soil. In the absence of these organisms, earth would be a sterile rock pile rather than the rich life-supporting environment it now is. Most of these organisms are in the surface layer of the soil, and one teaspoon may contain hundreds of millions of bacteria, algae, and fungi. In addition, many larger species—roundworms, mites, millipedes, and insects—play vital roles in this complex ecology.

Most soils form from the *parent material* where they are found. This may originate from the weathered remains of bedrock or where sediments have been deposited from elsewhere by water, ice, landslides, or the wind. Over centuries, ongoing physical and chemical weathering and organic activities modify this mixture. As the parent material breaks down, inorganic elements such as calcium, iron, manganese, and phosphorus (Chapter 4) are released. The amount of nutrients in the material and the speed of breakdown are major influences on the fertility of the resulting soil. Hard rocks, such as granite, break down slowly and yield few nutrients. Different soils will result, depending on the location. These various processes result in different layers forming in the soil, called **soil horizons**. A view across these horizons is called a *soil profile*. Figure 2.15 shows a generalized profile. However, not all soils have all these different horizons.

Box 2.6 Soils in Canada

Just as we can define ecozones, soil scientists can define soil zones that group together soils that are relatively similar in terms of their measurable characteristics. A glance at the soil map of Canada (below) will reveal a close resemblance to the ecozone map, since at this scale both tend to reflect the gross climatic and geological conditions of the region. The Canadian System of Soil Classification includes nine orders, the largest category of classification:

Brunisols cover 8.6 per cent and are brown soils found mainly under forests.

Chernozems cover 5.1 per cent, occur under grasslands, and are some of the most productive soils.

Cryosols are the dominant soils in Canada, covering some 40 per cent of the country's land mass, and are found in association with permafrost.

Gleysols only cover 1.3 per cent and are found in areas that are often waterlogged.

Luvisols cover 8.8 per cent and occur in a wide variety of wooded ecosystems. They have higher clay content than brunisols.

Organics, which cover 4.1 per cent of Canada, form in wetland ecosystems where decomposition rates are slow.

Podzols are found beneath heathlands and coniferous forests, are relatively nutrient poor, and cover 15.6 per cent of Canada.

Regosols cover less than 1 per cent of Canada and vary little from their parent material.

Solonets are saline soils, covering 0.7 per cent of Canada, and are found mostly in grassland ecosystems.

Dominant Soil Type

- Chernozemic
- Solonetzic
- Luvisolic
- Podzolic
- Brunisolic
- Regosolic
- Gleysolic
- Organic
- Cryosolic
- Unclassified

Figure 2.14 Soil zones of Canada.

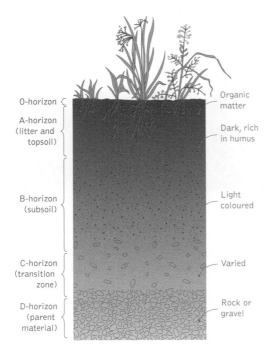

O-horizon

A-horizon
(litter and
topsoil)

B-horizon
(subsoil)

C-horizon
(transition
zone)

D-horizon
(parent
material)

Organic
matter

Dark, rich
in humus

Light
coloured

Varied

Rock or
gravel

Figure 2.15 Generalized soil profile.

Time is also a critical factor in soil development. Soils that have been exposed to millions of years of chemical and physical weathering, such as many tropical soils, have often lost their entire nutrient content. Conversely, where glaciation has scraped all the soil away, as in much of Canada as recently as 10,000 years ago, then the hard rocks, such as the granite of the Canadian Shield, have had little opportunity for weathering. They are also infertile. However, soils can be very fertile where ice sheets, as they retreated, deposited large quantities of clay rich in nutrients, as on the Prairies.

Soils also differ in their *texture*, or sizes of different materials. Clay is the finest, followed by silt, sand, and then gravel, the coarsest. Soils that contain a mixture of all these with decomposed organic material, or *humus*, are called *loams* and often make the best soils for vegetation growth. Texture is a main determinant of *soil permeability*, or the rate at which water can move through the soil. Water moves very slowly through soils composed mainly of the smallest particles, clay, and the soil easily becomes waterlogged. On the other hand, the large spaces between particles of sand or gravel lead to rapid drainage, and the soils may be too dry to support good vegetation growth. Plants obtain their nutrient supply necessary for growth from ions dissolved in the soil water, and so permeability is critical.

Soil has many different chemical characteristics. One of the most important is the pH value (see Chapter 4), measuring the acidity/alkalinity of the soil, which helps to determine which minerals are available and in what form. Different plants have different mineral requirements. Farmers often try to change the acidity of their soils, for example, by adding lime if the soil is too acidic or sulphur if the soil is too alkaline.

Just as the laws of thermodynamics explain energy flows, some useful principles help us to understand how organisms react to different abiotic influences. The first of these is known as the **limiting factor**. This principle tells us that all factors necessary for growth must be available in certain quantities if an organism is to survive. Thus, a surplus of water will not compensate for an absence of an essential nutrient or adequate warmth. In other words, a chain is only as strong as its weakest link. The weakest link is known as the *dominant limiting factor*. A major goal of agriculture is to remove the effect of the various limiting factors. Thus, auxiliary energy flows are employed to ensure that a crop has no competition from other plants (weeding), or that water supply is adequate (irrigation), or that the plant has optimal nutrient supply (fertilizer) (Chapter 10).

The corollary of the above is that all organisms have a range of conditions that they can tolerate and still survive. This is known as the **range of tolerance** for a particular species. This range is bounded on each side by a zone of intolerance for which limiting factors are too severe to permit growth (Figure 2.16). There may, for example, be too much or too little water. As conditions improve for the particular factor, certain individuals within the population can tolerate the conditions, but because the conditions still are not optimal, relatively few individuals can exist. This is known as the **zone of physiological stress**. Still further amelioration creates a range where conditions are ideal for that species, the *optimum range*. Here, in theory, barring other factors, there will be the highest population of the particular organism.

The concepts of limiting factors and range of tolerance can be illustrated by an example from Saskatchewan, where smallmouth bass are introduced every year into lakes for sport fishing. These hatchery-raised fish survive from year to year (if they are not caught), proving that they are within their range of tolerance for survival as individuals. They do not, however, breed successfully, because the hatchlings need a slightly warmer temperature to develop than do the mature fish to survive. Thus, different levels of tolerance exist for species at different life stages. The dominant limiting factor for the smallmouth in this environment is the low temperatures during hatching.

Water availability is often the critical factor that determines differences between communities. Where precipitation is in excess of about 1,000 millimetres per year, for example, trees will usually dominate the landscape if other factors are suitable. Below 750 millimetres, the range of tolerance for trees is exceeded, and grasses will dominate because they have a much lower range of tolerance for water stress, in the order of 100 millimetres per year. Below that level even grasses run into their zone of intolerance, and cacti, sagebrush, and other drought-resistant species dominate.

We must keep in mind that organisms are not reacting to just one abiotic factor, such as water availability, but to all the factors necessary for growth. Sometimes the optimal range for one factor will not overlap with the optimal range for other factors, which would place the organism in the zone of physiological stress for that factor, and it would become the dominant limiting factor. Organisms may also be out-competed for a particular factor in their optimum range by another organism with a greater tolerance to that environmental factor and again be forced into a zone of physiological stress. In other words, the simple single-factor model represented in Figure 2.13 is more complicated because of the numerous abiotic and biotic influences that must also be taken into account. The model does, however, provide a useful conceptual tool to help us understand the spatial distribution of organisms.

Biotic Components

Other species also have an important role in influencing species distributions and abundance. Species interact in several ways, including competition for scarce environmental resources. Each species needs a specific combination of the physical, chemical, and biological conditions for its growth. This is known as the **niche** of that species. Where the species lives is known as the **habitat**.

The **competitive exclusion principle** tells us that no two species can occupy the same niche in the same area. Most species have a *fundamental niche*, representing the potential range of conditions that they can occupy, as well as a narrower *realized niche*, representing the range actually occupied. The physical conditions for growth exist throughout the fundamental niche, but the species may be out-competed in parts of this area through the overlapping requirements of other species. **Specialist** species have relatively narrow niches and are generally more susceptible to population fluctuations as a result of environmental change. The panda is a classic example of such a specialist, with a total concentration on one plant, bamboo, as a source of food. Whenever the bamboo supply falls, as it does after it flowers, this specialist species has few suitable alternative sources of food. Historically, when bamboo was abundant this did not particularly matter, because the pandas simply moved to a new area. However, as the animals have become increasingly restricted to smaller and more isolated reserves, it has become a major problem.

In Canada, specialist species include many of those discussed in later chapters, such as the burrowing owl and the whooping crane. **Generalist species**, on the other hand, like the black bear and coyote, may have a very broad niche, where few things organic are not considered a potential food item. Such generalist species have adapted most successfully to the new environments created by humans.

Competition

Intraspecific competition occurs among members of the same species, whereas **interspecific competition** occurs between different species. Both forms of competition are a result of demands for scarce resources. Intraspecific competition occurs particularly where individual species densities are very high. Interspecific competition occurs where species niches are similar. Competition may be reduced through **resource partitioning** in which the resources are used at different times or in different ways by species with an overlap of fundamental niches. Hawks and owls, for example, both hunt for similar types of prey but at different times, since owls are mainly nocturnal.

Philip Dearden

In northern climates, vegetation structure is very simple as both low temperatures and low rainfall result in growth conditions under which few species can survive. In the short Arctic summer, however, richer areas are ablaze with brightly coloured flowers, as on Herschel Island off the north coast of Yukon.

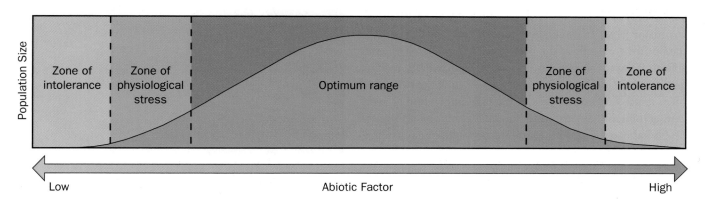

Figure 2.16 Range of tolerance.

Philip Dearden

The panda is a classic example of a specialist species.

iStockphoto.com/Frank Leung

Some species, such as the black bear, are very adaptable and have a relatively broad range of tolerance. On the Pacific coast, black bears are frequent scavengers of the intertidal zone, where many items are considered potential food. The year-round availability of an abundant and varied food source results in very large individuals.

Intraspecific competition may lead to the domination of specific areas by certain individuals; the area is known as a *territory* and may be aggressively defended against intruders. Grizzly bears establish such territories, which may be as large as 1,000 square kilometres for dominant males, although the possibility of defending such a large territory from intruders at all times is remote. During the breeding season, male robins establish and defend nesting territories, the boundaries of which are advertised in song. This kind of behaviour aims to establish sufficient resources for breeding pairs to be successful. Ultimately, intraspecific competition contributes to regulation of population size in areas where favourable habitat is limited, since those individuals unable to defend territories are outcast to less favourable areas where their likelihood of success is limited.

Biotic Relationships

There are other kinds of relationships between species besides competition. In predation, for example, a **predator** species benefits at the expense of a **prey** species. The lynx eating the hare and the osprey eating the fish are familiar examples of this kind of relationship, although in a broader sense we should also consider the herbivore eating the plant. Predation is a major factor in population control and usually results in the immediate death of the prey species. A predator must be able to kill and overwhelm prey on a regular basis without getting hurt. Usually, predators are bigger than their prey and often target weaker members of the prey population to avoid getting injured. They may also hunt as a group to improve the likelihood of a kill and minimize the possibility of a debilitating injury.

One theory that addresses the relationship between the benefit of making a kill and feeding against the cost of the energy expended to make the kill is **optimal foraging theory**. The theory recognizes that there is a point of compensation between the benefit of obtaining the prey and the costs of doing so and that the predator's behaviour adjusts to optimize the benefits. It may be more worthwhile, for example, to hunt a smaller prey more often, even though it will result in less food intake, if the smaller prey can be dispatched with little fear of injury and eaten quickly so that another predator cannot steal it. Optimal foraging theory also suggests that as one type of prey becomes scarce, most predators will switch prey if they can. Several examples of this kind of behaviour are discussed within the marine context in Chapter 8.

Prey species have evolved many strategies to avoid being transferred along the food chain. Some plants develop physical defences such as thorns, while others may evolve chemical defences such as poisons to deter their predators. The chemicals manufactured by plants provide the raw material for many of our modern medicines, such as aspirin, which comes from willows. Animal species employ a wide variety of predator

Philip Dearden

This rather stunned-looking red fox has just escaped from the clutches of a large female golden eagle that managed to lift it some three metres from the ground before one of the authors happened on the scene and unwittingly rescued one carnivore from becoming the prey of another one at the next trophic level.

avoidance strategies ranging from camouflage, alarm calls, and grouping to flight (Box 2.7).

A special kind of predator-prey relationship is **parasitism**, where the predator lives on or in its prey (or host). In this case, the predator is often smaller than the prey and gains its nourishment from the prey over a more extended time period that may lead to the eventual death of the host. This may cause the death of the parasite too, although some parasites, such as dog fleas and mosquitoes, can readily switch hosts. Tapeworms, ticks, lamprey, and mistletoe are all examples of parasites.

Not all relationships between species are necessarily detrimental to one of the species. **Mutualism** is the term used

iStockphoto.com/George Clerk

Rubber is a natural product made by rubber tree to help deter herbivores. Depicted here is tree-tapping on Koh Lanta, Western Thailand.

Philip Dearden

These Maldivian anemone fish live in a mutualistic relationship with the anemone, which provides protection from potential predators with its stinging tentacles while benefiting from the food scraps brought by the fish.

Guest Statement

Landscape Ecology
Dennis Jelinski

The problem of relating phenomena across scales is the central problem in biology and in all of science.

— S. Levin, 1992 Presidential Address to the Ecological Society of America

'Space' has been declared the 'final frontier' in ecology. Geographers have known this for years, and over the past two decades, ecology and geography have combined to spawn the new sub-discipline of landscape ecology, which deals explicitly with the effect of space on ecological processes and patterning. A landscape is simply a heterogeneous land area, such as a river valley in Waterton Lakes National Park, AB, a patchwork of logged areas in western Ontario, or the intertidal zone in Clayoquot Sound, BC. Landscape ecology seeks to understand the relationship between patterns and processes, explaining the spatial variation in landscapes at multiple scales as affected by natural causes and human society. In other words, landscape ecologists are interested in understanding how heterogeneity in the structure of the landscape might affect processes such as the movement of pollinators, the location of food supplies, refugia from predators, denning and calving areas, spawning beds, and shelter from inclement weather.

Landscape ecology, both as a science and especially as a way of examining the natural world, was instrumental in creating a major paradigm shift in ecology from the equilibrium view of ecosystem function to the non-equilibrium view. The former view embraced the notion of a 'balance of nature', a theory that had been front and centre in ecology for most of the twentieth century. The paradigm had at its core a belief that ecosystems were closed, integrated, and self-regulated and able to respond to positive feedbacks in accordance with the mechanistic principles of cause and effect, all directed towards achieving equilibrium or 'balance'. By zooming out in scale and examining systems at the coarser landscape scale, students of ecology discovered that in fact ecosystems were subject to a great degree of openness wherein energy, materials, and organisms were frequently exchanged across system boundaries. They were also subject to frequent disturbance and a great deal of chance (more formally termed stochastic events). This flux in nature is most apparent at landscape scales.

Over the past 15 years, my research and that of my students has focused on landscape ecology in a number of different environments. For example, the effect of habitat fragmentation on birds has been a major focus of research in avian conservation. Much research has focused on identifying several area-sensitive species. On wet meadow grasslands flanking the Platte River, Nebraska, Chris Helzer and I studied a combination of habitat area and shape and found that a perimeter-area ratio of habitat patches was a more effective determinant of the presence and richness of grassland bird species than patch area alone (Helzer and Jelinski 1999). In other words, some grassland bird species avoid habitat edges and habitat patches with high perimeter-to-area ratios. This means that ecosystem restoration efforts must try to ensure that habitat patches are as compact as possible so that brood parasites and generalist predators cannot gain ready access to all portions of the habitat, as is possible when patches are elongated.

In a very different way, Lindsay Mulkins, myself, and others show how landscape boundaries are very permeable. We (Mulkins et al. 2001) studied the flow of dissolved carbon from coastal old-growth forests in

Dennis Jelinski

Clayoquot Sound, BC, and found that in nearby marine waters small zooplankton called mysids, which are an important component of the diet in grey whales, were seasonally affected by this importation, or spatial subsidy, of carbon. Thus, grey whales that in turn feed on these mysids are fuelled in part by carbon originating in terrestrial forests. This is a profound illustration of how open ecosystems really are!

In a more recent study in Barkley Sound, BC, we examined the effect of wave-tossed marine algae (such as kelps) on the nitrogen budget of beach plant communities. Nitrogen is in short supply in most ecosystems but especially on these soil-poor beaches. The algae plus marine carrion, diatoms, and driftwood that is also washed ashore are collectively termed wrack. This spatial subsidy of nutrients on otherwise low-nutrient cobble beaches elevates the level of nitrogen in a species of beach grass. The wrack is also important in the diets of terrestrial isopods (you can find them under most decaying logs), which migrate from nearby forest and forage on the dried algae and beach grass. Thus, both plant and animal members of beach and nearby forest communities benefit from materials that originate in the ocean.

Combined, these studies show that plants, animals, their ecological communities, and indeed entire ecosystems are strongly affected by the landscape in which they are embedded. Processes in the landscape are affected by spatial patterning at a great many scales, from those in a small patch of a square metre of beach, to patches of grasslands, to the juxtaposition of marine and terrestrial ecosystems. Landscape ecology as a management tool holds tremendous potential in addressing a great many environmental problems that are played out at landscape and regional scales, including threats to biodiversity, pollution, habitat fragmentation, parks management, and sustainable resource development.

Dennis Jelinski is an associate professor in the Department of Geography, University of Victoria.

Box 2.7   Canada's Olympic Champion: The Pronghorn Antelope

The pronghorn is the sole survivor of a family of North American antelopes, unrelated to the African antelopes, that at one time contained more than two dozen different species. Pronghorn are migratory, moving north in summer and south in winter, at which times they may gather together in large numbers. Their main predators (other than humans)—the plains wolf and grizzly—have become extirpated from their range, leaving the coyote and, farther south, the bobcat as predators. Unlike all other members of the bovid family (sheep, oxen, and goats), pronghorns shed their horns after the rut. Populations of pronghorn are now protected in Grasslands National Park in Saskatchewan.

Pronghorn are reckoned to be the fastest middle-distance runners on Earth, having been clocked at speeds of up to 98 kilometres per hour for more than 24 kilometres. They are superbly adapted for speed, with small stomachs and big lungs, long strides, small feet, flexible spines like the cheetah, and additional ligaments and joints, all designed to make this animal the fastest on the prairie. Speed is the logical defence mechanism for a vulnerable animal easily seen by predators on the open prairies. Confidence in their escape mechanism made them rather curious, and early settlers found them easy to attract and shoot. Initial herds estimated at between 20 and 40 million animals in North America were soon reduced to fewer than 30,000 animals as a result of hunting pressure and land-use change. There are now some half a million pronghorn remaining.

to describe situations in which the relationship benefits both species. These benefits may relate to enhanced food supplies, protection, or transport to other locations. The relationship between the nitrogen-fixing bacteria and their host plants, described in Chapter 4, is an example of such a relationship that results in enhanced nutrition for both species. Other examples include the relationship between flowering plants and their pollinators, which results in the transport of pollen to other plants, and the protection offered by ants to aphids in return for the food extracted from plants by the aphids. Interactions that appear to benefit only one partner but do not harm the other are examples of **commensalism**. The growth of *epiphytes*, plants that use others for support but not nourishment, is one example.

Keystone Species

Species with a strong influence on the entire community are known as **keystone species**. They are named after the final wedge-shaped stone laid in an arch. Without the keystone, all the other stones in the arch will collapse. In Canada, our national symbol, the beaver, is a good example of such a species (Box 2.8). Beavers can have a profound impact on their environments through the dams they build that raise and lower water levels. This, in turn, affects the limits of tolerance of other species in the community that may suddenly find themselves submerged under a beaver

dam or facing lower water levels downstream. Different species will have different reactions to this change, depending on, for one thing, their range of tolerance relating to water. However, when a keystone species is removed, there is generally a cascading effect throughout the ecosystem as other species are affected. Interestingly, the same species may be a keystone is some communities and not in others, depending on the community composition in that particular locale.

It is especially significant when a keystone species is removed from an area, or **extirpated**, by human activity. Such changes may take some time before they become noticeably manifest. Changes to soil characteristics created by the extermination of major herbivores, such as bison from the prairie, may take centuries before they become noticeable and are generally not reversible. The same is true for the other large grazers, such as the great whales (discussed in Chapter 8), which have been decimated over the last couple of centuries.

Impacts of Global Change

Global **climate change** will have profound impacts on the numbers and distributions of species in the world. Overall, climate is the main determinant of the patterns of life. As temperatures and precipitation change (discussed in more detail

Box 2.8   Canada's National Symbol—The Beaver

The beaver is found all the way from Mexico to the Arctic and from Vancouver Island to Newfoundland. However, the beaver is mostly associated with the northern woods and their waterways, where it is well-known for its capability for water engineering. Many different species of beaver could once be found throughout the northern hemisphere. A Eurasian counterpart remains in small populations, but it is the North American beaver that has flourished and become one of the continent's most successful mammals. It also played a critical role, as did the sea otter on the west coast, in attracting the colonial gaze of Europeans to the resources of North America.

The beaver is a rodent. In fact, it is the second largest rodent in the world. All rodents are distinguished by their sharp incisor teeth, designed to gnaw though bark, crack nuts, or attack any other edible vegetable matter in a similarly efficient manner. The success of this strategy is attested to by the proliferation of rodents, which comprise nearly 40 per cent of all mammal species. The specialty of the beaver, of course, is its ability to fell trees (some as large as a metre in diameter), which can then be used as a food supply and as building material for its familiar dams and lodges. Trees, particularly hardwoods such as poplars, are felled close to the water's edge so that they can be dragged into the water, which is the beaver's preferred medium. With their broad flat tails, sleek coats, powerful and webbed hind feet, beavers are well equipped to undertake their aquatic construction activities. Their dams impede the flow of water, giving them greater access to trees and making them less vulnerable to terrestrial predators. They can use the pond so created as a low-energy way of transporting food to their lodge, which, surrounded by water, is virtually impregnable to predators.

Probably no other animal, except humans, has the ability to cause such a radical and deliberate change to the environment. Beaver dams benefit not only beavers but also other water-oriented organisms, such as waterfowl, otters, muskrats, and frogs and other amphibians. With an estimated population of more than 60 million animals prior to European settlement in North America, their ecological impact on the landscape would have been substantial. Although they were trapped out of large areas, they are now starting to recolonize as a result of conservation activities. Industrious, clever, effective, resilient, but not spectacular—is this the Canadian character?

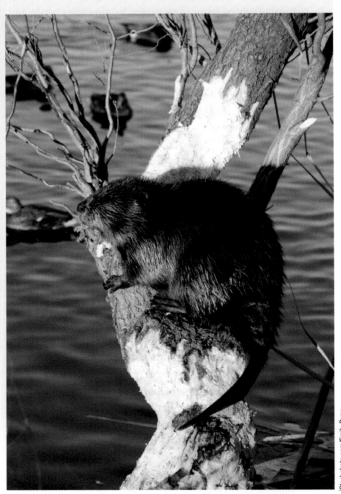

The beaver, the national animal of Canada, can have a very significant impact on other species through its dam-building activites. Here, a beaver perches in a gnawed tree.

in Chapter 7), the changes will have a profound effect on these patterns. In general, there will be a poleward shift of life zones. The east to west orientation of the main terrestrial ecozones of Canada, shown in Figure 2.12, is likely to be replaced by a dominantly north-south pattern. Prairie ecozones will grow and forested ecozones will contract as precipitation levels fall. Species dependent on grasslands will increase their range; those dependent on forests will contract.

Already, there are many documented examples of species range changes in response to climate change (Parmesan 2006). Drought in the Prairie Potholes region (southeastern Alberta and northeastern Montana to southern Manitoba and western Minnesota) will lead to significant reductions in the populations of 14 species of migratory waterfowl; 30 to 50 per cent fewer prairie ponds will hold water in spring by 2060, with an associated 40 to 50 per cent decline in the numbers of ducks settling to breed in the area. In addition, egg-laying, flowering, and spawning are occurring earlier for many species, in some cases disrupting delicate cycles that ensure that insects and other food are available for young animals. For example, tree swallows across North America advanced egg-laying by as much as nine days from 1959 to 1991. Unfortunately, the hatchlings are now emerging before major insect hatches, and as a result populations of tree swallows are declining because of chick starvation.

Other changes are also taking place. Spring migration is occurring earlier and fall migration later for many species. For example, 25 migratory bird species are arriving in Manitoba earlier than they did some 60 years ago; only two are arriving later. However, as usual, there are complications. Researchers have found that short-distance migrants are migrating earlier because they can read the local cues that conditions are right for them in their destination. However, a bird migrant in Costa Rica has few cues to tell it that conditions in Canada are right for its return. Consequently, it arrives at the normal time, only to find that its food supply has already waned (Wiley-Blackwell 2008).

Extinctions will take place because some species are incapable of adjusting at such rapid rates. Already, the US has listed the polar bear as threatened under the Endangered Species Act as a result of global warming, citing rapid melting of the sea ice that the bears depend on, and Canadian scientists are reporting similar findings (Stirling et al. 2008). There are some interesting human dimensions to this move. One implication is that the US will now block the import of polar bear skins into the US, effectively removing the incentive for US-based hunters to hunt the bear. However, many Inuvialuit people in Arctic coastal communities depend on the hunt for their livelihood. Hunters pay from $25,000 to $40,000 for the chance to shoot a polar bear, and the Northwest Territories estimates the total loss to the territorial economy at around $3 million annually.

There are also some interesting political dimensions to the endangered species listing. The listing was long delayed by the US government and came about as a result of a lawsuit by an environmental group. During the delay, several additional exploratory licences for oil and gas were issued by the government.

Because of the strict provisions of the Endangered Species Act, it is unlikely that these licences would have been issued after the listing. In announcing the listing, the government bemoaned the 'inflexible nature' of the Act that forced it to determine the listings on purely ecological grounds, irrespective of economic considerations (*The Globe and Mail* 15 May 2008, A6). On the other hand, environmentalists consider this 'inflexibility' one of the strongest aspects of the Act. There are no such inflexibilities in the Canadian Act. In Chapter 14, you will see that this feature of the US Act is in direct contrast to our own Species at Risk Act under which decisions on listing endangered species are made by politicians, not biologists, and can take into account economic and political considerations.

Scientists predict that 9 to 52 per cent of all terrestrial species (up to 1 million plants and animals) will be on an irreversible path to extinction by 2050. These figures appear to be supported by past temperature changes. Mayhew, Jenkins, and Benton (2008) analyzed the fossil record for the past 520 million years against estimates of low-latitude sea surface temperature for the same period. They found that global biodiversity (the richness of families and genera) is related to temperature and has been relatively low during warm 'greenhouse' phases, while during the same phases, the extinction and origination rates of taxonomic lineages have been relatively high. Calculations suggest that the world's biodiversity hotspots, the main engines of evolution, are particularly vulnerable to extinctions from climate change because of the high level of change predicted in these biomes and the low levels of out-migration (Malcolm et al. 2006).

Climate change will also influence the functioning of ecosystems, the characteristic ways in which energy and chemicals flow through the plants, herbivores, carnivores, and soil organisms that comprise the living components of ecosystems, as described earlier in this chapter. Productivity will change; in some places it will increase, in others decline. Food webs will be disrupted as predators and prey react differently to the changing conditions, as illustrated above with the example of the tree swallow. These changes will affect marine ecosystems as much as, if not more than, terrestrial ecosystems. Migratory whales, for example, will face shrinking crucial Antarctic foraging zones, which will contain less food and will be farther away. Levels of global warming predicted over the next 40 years will lead to winter sea-ice coverage of the southern ocean declining by up to 30 per cent in some key areas. Migratory whales may need to travel 200 to 500 kilometres farther south to find the 'frontal' zones that are their crucial foraging areas. The migratory whale species that will be affected include the blue whale, the Earth's largest living creature, and the humpback whale, which is only now coming back from the brink of extinction after populations were decimated by commercial whaling, mainly during the first half of the twentieth century.

Both species build up the reserves that sustain them throughout the year in the frontal zones, which host large populations of their primary food source, krill. Shrinking ice-covered areas affect krill production in two ways: sea ice is a refuge for krill larvae in winter and an area of intense algal blooms on which the krill feed in summer. Krill is so fundamental to the southern ocean ecosystem that the impact will not be confined to whales but will also affect seals, seabirds, and penguins as well as fisheries productivity. 'Frontal zones' are areas where water masses of different temperatures meet. They are associated with upwelling of nutrients supporting large plankton populations on which species such as Antarctic krill feed. As frontal zones move southward, they also move closer together, reducing the overall area of foraging habitat available. Since the krill is dependent on sea ice, less sea ice is also expected to reduce the abundance of food for whales in the feeding areas (Tynan and Russell 2008).

Implications

The above discussion points to important implications for society and species distributions:

- All of the Earth's inhabitants are interlocked in environmental systems that depend on one another for survival. Perturbations in part of the system will have impacts on other parts of the system.
- The basic scientific laws that govern the transformation of matter and energy dictate that sooner or later, society must transform itself from a throwaway society built on processing ever-increasing matter and energy flows to one in which energy efficiencies are improved and matter flows are reduced.
- A species may have a wide range of tolerance to some factors but a very narrow range for others.

- Species with the largest ranges of tolerance for all factors tend to be the most widely distributed. Cockroaches and rats, for example, enjoy virtually global distribution.
- Many weed and pest species are successful because of their large range of tolerance. Eurasian water milfoil, a significant nuisance in many waterways in Canada, is an alien that can grow in conditions from Canada to Bangladesh.
- Response to growth factors is not independent. Grass, for example, is much more susceptible to drought when nitrogen intake is low.
- Tolerance for different factors may vary through the life cycle. Critical phases often occur when organisms are juveniles and during the time of reproduction.
- Some species can adapt to gradually changing conditions for some factors, up to a point. However, after this **threshold** of change is reached, the population will collapse.
- Global climate change will place considerable stress on many species in terms of their limits of tolerance. This will lead to changes in range and abundance, and some species will go extinct.

From this discussion, it should be apparent that ecosystems are very complicated. A complex set of interrelationships exists among organisms and between organisms and their environment. A change in part of this matrix will often result in corresponding changes throughout. Humans are now such a dominant influence on global environmental conditions at all scales that significant changes are underway as a result of human activities. There is considerable uncertainty as to how ecosystems and the entire life support system of this planet will react to these changes. Yet even under natural conditions, ecosystems are not static. The next chapter will focus on how ecosystems change over time.

Summary

1. Energy is the capacity to do work. Energy comes in many forms, including radiant energy (from the sun), chemical energy (stored in chemical bonds of molecules), and heat, mechanical, and electrical energy. Energy differs from matter in that it has no mass and does not occupy space.
2. Understanding energy flows is critical to an understanding of the ecosphere and environmental problems. The laws of thermodynamics explain how energy moves through systems. The first law states that energy can be neither created nor destroyed but merely changed from one form to another. The second law informs us that at

each energy transformation, some energy is converted to a lower-quality, less useful form.
3. Energy is the basis for all life. Through the process of photosynthesis, certain organisms transform carbon dioxide and nutrients in the presence of radiant energy from the sun into organic matter. This matter forms the basis of the food chains by which energy is passed from trophic level to trophic level. At each transference, the second law of thermodynamics dictates that some energy is lost, typically as much as 90 per cent.
4. Productivity is a measure of the abilities of different communities to transform energy into biomass. The

most productive communities are found in estuaries, wetlands, and rain forests.

5. The ecosphere is the thin, life-supporting layer of the Earth characterized by interactions between the biotic and abiotic components. It can be further subdivided into communities, ecosystems, and biomes.

6. The concepts of limiting factors and range of tolerance help us to understand the interaction between the biotic and abiotic components of the ecosphere.

7. Each species needs a specific combination of physical, chemical, and biological conditions for its growth. This is the niche of that species.

8. The principle of competitive exclusion tells us that no two species can occupy the same niche in the same area at the same time.

9. Species compete for scarce resources in any given habitat. However, there are many other forms of relationship between species, such as predation, parasitism, mutualism, and commensalism.

10. Species with a strong influence on the entire community are known as keystone species.

11. Global climate change will have a very significant impact on the distribution and abundance of species. Some will flourish. Others will decline, and some will become extinct.

Key Terms

abiotic components
aerobic
autotrophs
biogeochemical
    cycles
biomass
biome
biotic components
calorie
carnivores
chemoautotrophs
chlorophylls
climate change
commensalism
competitive exclusion
    principle
consumers
decomposer food
    chains

detritus
ecosystem
energy
entropy
euphotic zone
extirpated
food chain
food web
generalist
    species
gross primary
    productivity (GPP)
habitat
heat
herbivores
heterotrophs
high-quality energy
interspecific
    competition

intraspecific
    competition
keystone species
kinetic energy
law of conservation
    of energy
limiting factor
low-quality energy
mutualism
net primary
    productivity (NPP)
niche
omnivores
optimal foraging
    theory
parasitism
photosynthesis
phototrophs
phytoplankton

population
potential energy
predator
prey
primary consumers
producers
range of tolerance
resource partitioning
second law of
    thermodynamics
secondary consumers
soil horizons
specialist
tertiary consumers
threshold
trophic level
zone of physiological
    stress
zooplankton

Questions for Review and Critical Thinking

1. What are the main biotic and abiotic components of ecosystems?

2. How do the laws of thermodynamics apply to living organisms?

3. How do the laws of thermodynamics apply to environmental management?

4. What are chemoautotrophs, and what role do they play in ecosystem dynamics?

5. On what trophic level is a pitcher plant? Why? Are there plants on the same trophic level in your area? What are they, and where do they grow?

6. In what kinds of ecosystems do detritus food chains dominate?

7. What roles do phytoplankton play in maintaining ecospheric processes?

8. What are the management implications of recognizing concepts such as specialist, generalist, and keystone species? Can you think of any examples in your area?

9. What is optimal foraging theory?

10. What do you think the dominant limiting factors are for plant communities in your area?

11. Draw a cross section across (E–W) and down (N–S) your province or territory, and show the main environmental gradients and the vegetational response.

12. What are the implications of global climate change on species distributions and abundance?

13. What are some of the main transformations that have to take place in society to reflect the implications of the laws of thermodynamics and law of conservation of matter?

Related Website

Environment Canada, ecosystem information:
http://ecoinfo.ec.gc.ca/index_e.cfm

Further Readings

Note: This list comprises works relevant to the subject of the chapter but not cited in the text. All cited works are listed in the section 'References' at the end of the book.

Allan, J.D., et al. 2003. 'Influence of streamside vegetation on inputs of terrestrial invertebrates to salmonid food webs'. *Canadian Journal of Fisheries and Aquatic Sciences* 60 (3): 309–20.

Gende, S.M., and T.P. Quinn. 2004. 'The relative importance of prey density and social dominance in determining energy intake by bears feeding on Pacific salmon'. *Canadian Journal of Zoology* 82: 75–85.

Hodges, K.E., and A.R.E. Sinclair. 2003. 'Does predation risk cause snowshoe hares to modify their diets?' *Canadian Journal of Zoology* 81: 1,973–85.

Krebs, C.J., et al. 2003. 'Terrestrial trophic dynamics in the Canadian Arctic'. *Canadian Journal of Zoology* 81: 827–43.

Methratta, E.T. 2003. 'Top-down and bottom-up factors in tidepool communities'. *Journal of Experimental Marine Biology and Ecology* 299 (1): 77–96.

Mills, E.L., et al. 2003. 'Lake Ontario: Food web dynamics in a changing ecosystem (1970–2000)'. *Canadian Journal of Fisheries and Aquatic Sciences* 60: 471–90.

Scott, G.A.J. 1995. *Canada's Vegetation: A World Perspective*. Montreal and Kingston: McGill-Queen's University Press.

Chapter 3

Ecosystems Are Dynamic

Learning Objectives

- To understand the nature of ecosystem change and its implications for society and environmental management.
- To understand the process of primary and secondary succession and the ways in which humans alter these processes.
- To realize what is meant by dynamic equilibrium in ecosystems.
- To appreciate the role of disturbance such as fires, insect infestations, and major storms as often being an integral and natural part of healthy ecosystem function.
- To explore the impact and management of invasive species.
- To recognize the main factors affecting species population growth.
- To appreciate the nature of evolution and extinction.
- To understand the range of values in maintaining high levels of natural biodiversity.
- To understand the ecological implications associated with the loss of biodiversity.

Introduction

Communities and ecosystems change over time. The rate of change depends on factors driving change, the response of individual species, how species interact with one another, and how they respond collectively and individually to their abiotic environment from an ecosystem perspective. Part of the response of the species comprising these systems is related to their range of tolerance, to such factors as the amount of light, nutrients, and soil type in the case of plants, discussed in Chapter 2. For animals, the response may be related to the type, distribution, and availability of food resources or potential for predation. Some changes are very rapid, such as a forest fire. Others, such as climate change under natural conditions, occur over long time periods and allow communities to adjust slowly to the new environment. Unfortunately, the speed of change now occurring as a result of greenhouse gas emissions is faster than any previously experienced, and many species will be unable to adapt at this speed. As vegetation communities change, so do the heterotrophic components dependent on plants for food. Similarly, if the components of the food web change, it may well cause a change in vegetation.

In this chapter, we examine aspects of change in ecosystems, starting with the process of ecological succession, and then discuss the concept of ecosystem function and its dynamic characteristics. Next we examine the role of a species population growth and how and why it varies. Last, we look at the role of longer-term change in the processes of evolution and extinction on biodiversity and ecosystem function. The impact of humans on these processes is often to alter their natural function relative to a time when humans were far less populous on Earth.

Ecological Succession

Ecological succession is a relatively slow process. It involves the gradual replacement of one assemblage of species by another as environmental conditions change over time. Some of these changes are created by the species themselves and others more indirectly. We divide succession into two basic types with

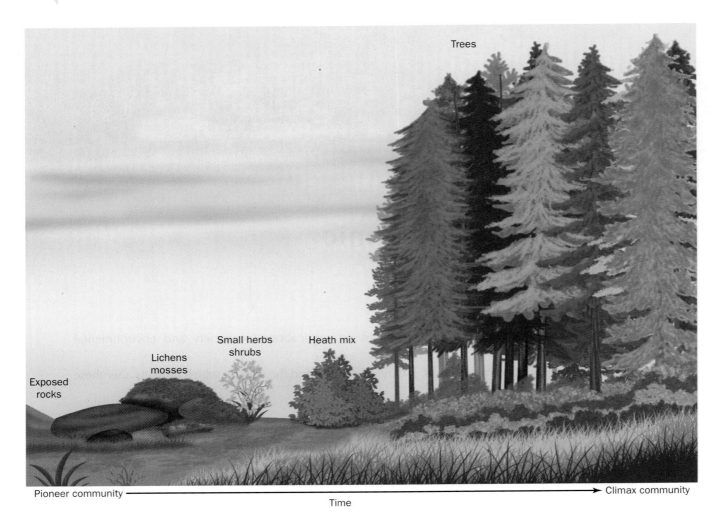

Trees

Small herbs
shrubs

Heath mix

Lichens
mosses

Exposed
rocks

Pioneer community ⟶ Climax community

Time

Figure 3.1 A general model of primary succession over time, from a bare rock surface to a forest community.

some additional variants. **Primary succession** is the colonization of a previously unvegetated surface, such as when a glacier retreats or a landslide removes all traces of the vegetation of the previous ecosystem (Figure 3.1). Little or no soil exists, and the first species to occupy the area, known as *primary colonizers*, must be able to withstand high variability in temperatures and water availability and highly limited nutrients. Few species can tolerate such conditions.

Lichens are typically the first colonizers because they can establish on bare rock surfaces that are virtually devoid of nutrients and have the ability to hold water (Box 3.1). Over time, lichens, in combination with other physical and chemical processes, break down rocks. Their biomass traps water and nutrients. Over centuries, their accumulating biomass and alteration of the environment make it possible for other species to colonize; mosses most often follow. Mosses grow faster than lichens, resulting in yet greater accumulation of biomass and the beginnings of soil. The lichens are eventually out-competed by the faster-growing mosses.

The next stage in successional advance is typically invasion by herbaceous plants such as grasses and species that we often think of as 'weeds'. Most of these species are annuals or biannuals. Such species are able to colonize a wide range of habitats

and have reproductive strategies to disperse widely. Dandelions and fireweed are good examples. While some plant species physically disperse into the habitat patch, others are already present in the form of seeds lying dormant in the soil, sometimes for decades! These seeds germinate when environmental conditions, such as the availability of light, become favourable

Only 10,000 years ago, most of Canada was covered in a thick layer of ice.

Philip Dearden

Box 3.1   Lichens

Some environments have such challenging growing conditions that virtually nothing can survive. However, lichens are one of the few types of organism that can be found in such places. Lichens are partnerships, part of an evolutionary mutualistic relationship between fungi and photosynthetic algae such that each benefits from the presence of the other (see Box 2.1). The fungi are able to cling to rocks or trees with their filaments and to retain water. In turn, the algae produce food for both groups of species through photosynthesis. This may include the fixing of nitrogen from the atmosphere by cyanobacteria, as discussed in the next chapter. This combination is able to survive intense cold and drought and has been evolving for more than a billion years, making lichens one of the most primitive of living organisms. Individual lichens may be more than 4,000 years old. Over centuries, sufficient growth of lichens may occur so that the thinnest of soils is produced, allowing other species able to tolerate harsh conditions to colonize. Lichens are therefore very important primary colonizers.

More than 18,000 species of lichens have been described throughout the world, and undoubtedly many others have yet to be discovered. They have different life forms—dust, crust, scale, leaf, club, shrub, and hair. The best known in Canada include the leafy variety found growing on trees, an encrusting variety that grows on rocks and sometimes trees, and the so-called (and misnamed) reindeer mosses found throughout northern Canada.

Lichens are not only important agents of succession, but some species rely on them for food. There are a couple of species found not in early successional environments but in old-growth forests that constitute the main winter food supply for BC mountain caribou, which are species precariously on the brink of extinction.

Besides providing an essential food supply for caribou, lichens have also been used by humans as flour (when dried and ground up) and as a dye for wool and other fabrics. They are now finding other uses as well. Lichens, because of their adaptive ability to absorb mineral requirements directly from the air, are very efficient accumulators of pollution. Unlike many other plant species, they concentrate pollutants to exceed their own tolerance levels and hence are excellent indicator species for air pollution, since lichens will be absent from heavily polluted areas. Wong and Brodo (1992) documented that of 465 species of lichen in Ontario, 52 are believed to be regionally extinct owing to pollution.

Lichen.

for growth. The seeds that lie 'in wait' are said be part of the soil **seed bank**.

Over time, these early herbaceous species create an environment conducive for the next successional stage to establish, which includes hardy shrubs and light-tolerant trees that in turn further ameliorate conditions until shade-tolerant tree species become established. Examples of light-tolerant ('sun-loving') trees are birch, oak, and trembling aspen. Species that can establish in the shade include western hemlock and western red cedar, which we typically find in old-growth forests. In areas where precipitation and temperature are adequate, trees typically dominate the final stage of this successional process, with fewer species in the understory. Each stage along the way is known as a **seral** stage.

In the first half of the past century, it was believed that vegetation would ultimately reach a well-defined, stable stage known as the **climax community** and that this final successional stage was in equilibrium with the environment.

Philip Dearden

Fireweed, seen here growing alongside the Tatshenshini River in northern British Columbia, is a common herb in early successional sites throughout Canada.

Philip Dearden

The term 'tree line' is used to describe areas where vegetation communities dominated by trees give way to those dominated by other types of vegetation, such as herbs and grasses. Rarely, however, is there a sharp line but rather an ecotone where patches of both tree- and grass-dominated communities exist together.

Beginning in the second part of the twentieth century, it became increasingly obvious that equilibrium conditions are rare and that **disturbance** (such as fires, insect infestations, flooding, ice storms) are so common that most ecological systems never reach a dynamically stable climax stage even in large landscapes such as Banff National Park. Many agents of disturbance are a natural and integral part of the healthy functioning of ecosystems. This is contrary to our intuitive sense that phenomena like fires, flooding, and windstorms are more or less quite harmful. More formally, disturbance can be defined as a relatively discrete event in time and space (such as a flood) that alters the structure and function of populations, communities, and ecosystems. This change occurs because the amount, kind, and distribution of resources is altered or the physical environment (including substrate availability) is significantly changed.

An example of a major disturbance currently unfolding relates to the mountain pine beetle invasion in more than 9 million hectares of British Columbia forest. The beetles have killed and are killing millions of trees and thus are preventing these forests from achieving or maintaining a state of climax. As such, they are 'setting back the successional clock' to an environment represented by early successional states. The pattern of recovery following this disturbance will depend on the features of the species themselves, the nature of interactions among species, and many unpredictable factors. Thus, ecosystems and landscapes are dynamic, interacting in complex ways, often unpredictable, over large space-time scales. Although it is not possible to predict which species will occupy different areas undergoing succession, it is often possible to predict the type of plant that will dominate.

The foregoing is not an inevitable linear progression. It is a guideline to help us understand the changes that may take place in ecological communities. In some instances—in recently glaciated terrain, for example—very hardy species of trees, such as willows and alders, may become established in favoured sites with little previous colonization having

Figure 3.2 Number of forest fires in Canada (thousands) and area burned (millions of hectares). Source: Environment Canada. 2003, 46. www.nrcan-rncan.gc.ca/cts-sct/science/prodserv/firereport.

occurred. *Cyclic succession* may also occur where a community progresses through several seral stages but is then returned to earlier stages by natural phenomena such as fire (Box 3.2) or intense insect attack. The different seral stages are not discrete but may blend from one into another. These blending zones tend to be the areas with the highest species diversity, since they contain species from more than one community. They are known as **ecotones** and occur as relatively richer zones between communities.

Box 3.2   Fire, Management, and Ecosystem Change

In many areas, fire is a natural occurrence that has a profound impact on plant and animal communities. In some communities, it may be the dominant agent of disturbance, and if suppressed by human interference, those communities may significantly change in species composition. Fire has been used as a tool by humans since earliest times to manipulate ecosystems to produce desired effects, such as removing forests to facilitate agriculture, burning grasslands to generate new grass growth, and herding animals so that they can be more readily hunted. Fire is used in forest management: hazard reduction for silviculture, insect and disease control, wildlife habitat enhancement, and range burning.

Fire has several important ecological and social implications:

- It favours the growth of certain species over others. Some species are fire-resistant (such as the Douglas fir), while the heat from fire may aid in the germination of other species. For example, lodgepole pine seeds can only be released from their cones when sufficiently high temperatures melt the resin that once held the cone tight shut. The phenomenon is termed **serotiny**. Fire may result in the death of other species.
- At moderate levels of intensity and frequency, it tends to increase the diversity of species in a community. Fire releases nutrients from the biomass into the soil and atmosphere; some may be lost from the site, while the remainder help to stimulate growth of some species— for example, the pine seedlings mentioned above.
- It stimulates the growth of various grasses and herbs that provide fodder for herbivores, which may in turn increase carnivore populations.
- Soil temperatures are increased not only during the fire but also afterwards—the site has a lower **albedo** and is more open to the sun. This also influences chemical and biological properties of the soil, stimulating microbial activities and enhancing decomposition.
- Fires that are highly intense or very frequent may cause sufficient nutrient impoverishment of a site to preclude further growth of trees, and the vegetation may become dominated by grasses and low shrubs. This is known as a plagioclimax. Many of the heathlands of northern Europe were created in this manner, and clear-cutting and fire in nutrient-poor black spruce forests in Canada can have the same effect.

Fire is also a highly emotive topic for management. Early concepts of forestry and conservation encouraged policies of total fire suppression, with little attention given to the role of fire in various ecosystems. One can see evidence of this in the message of Smoky the Bear: forest fires are 'bad'. This mindset led to unanticipated changes in some ecosystems. For example, in the absence of fires as a result of fire suppression, lodgepole pine cannot establish (as explained above) and thus establish what otherwise would be a lodgepole pine–dominated forest. Instead, these ecosystems with an altered fire regime may be dominated by species such as trembling aspen. Further, fire suppression results in the accumulation of organic debris such as dead trees. If and when fire occurs, this debris will help fuel a fire such that the fire will jump from the forest floor to the canopy. Often, these fires are so intense and spread over such a large area that they cannot be controlled. Managers of protected areas such as parks now realize that if fire is a natural part of an ecosystem, fire suppression policies are altering the ecosystem in unnatural ways. This has led to **prescribed burning** programs in many parks, such as Banff National Park (Chapter 14). A decision on whether fire should be suppressed or not should reflect knowledge of an ecosystem's natural fire regime. The regime includes factors such as the frequency, intensity, and spatial magnitude of this agent of disturbance. Such knowledge enables managers to mimic the regime, thus maintaining the natural state of the ecosystem. Some fires may be ecologically appropriate. Others may result from human carelessness or lack of ecological understanding. Furthermore, we cannot ignore the potentially destructive effects of fires on human livelihoods.

It is widely believed that global warming (see Chapter 7) will result in more frequent and intense fires. The burning of millions of tons of carbon that is biologically fixed in the biomass of the trees releases carbon dioxide, further exacerbating the build-up of greenhouse gases in the atmosphere. This is an example of a **positive feedback loop**. The hotter it gets, the drier it gets, the more fires we have, the more carbon dioxide is released, and the warmer it gets. Scientists predict that temperature rise associated with global warming will be in the order 4 to 6°C within 40 years for the boreal forest biome. They also predict lower rainfalls. This will lead to greater drying of the land surface and, again, increased frequency and area of fire. Overall, forest ecosystems will show a high degree of disturbance not typical of that which they experience as part of their natural fire regime. Species dependent on old-growth ecosystems, such as woodland caribou, will be put under increasing pressure. Caribou are highly dependent on the forest for lichens, which form the major part of their winter diet and only grow in forests more than 150 years old.

Guest Statement

Controls on High-Elevation Treeline in Westernmost Canada
Ze'ev Gedalof and Dan J. Smith

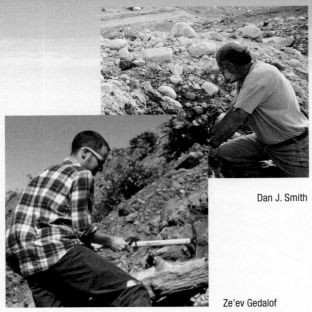

Dan J. Smith

Ze'ev Gedalof

An important challenge for scientists and resource managers is predicting precisely how forests will respond to global environmental change. Trees are typically long-lived, and they reproduce relatively slowly. Unlike most animals, they cannot change their location in response to changing environmental pressures. As the Earth's climatic and chemical properties change, forests are being exposed to different stressors than existed at the time of their establishment, and the rate and magnitude of change is so rapid that trees will not be able to adapt. As trees become increasingly stressed, they will become more vulnerable to wildfire, insect attack, fungal pathogens, drought, and disease. Understanding how these vulnerabilities will change over time is necessary for resource managers to seek ways of preserving forest biodiversity and ecosystem function.

One method of predicting how the distribution of trees might change in response to the changing climate is to evaluate how individual trees have responded to climatic variability in the past. In temperate regions such as Canada, trees produce growth rings each year. The width of these rings often varies, depending on the availability of heat, moisture, nutrients, light, or other limiting factors. At particularly severe locations, such as high elevation, there will often be a single limiting factor (for example, summer temperature). Variability in the width of tree rings at these locations can therefore be used to understand how the species might respond to future changes in this limiting factor and also to reconstruct how the factor varied over the lifetime of the tree—which can be considerably longer than the interval over which people have kept records. In Canada, the oldest known living tree is a yellow cedar (*Chamaecyparis nootkatensis*) on Vancouver Island, which is more than 1,200 years old (Laroque and Smith 1999).

In westernmost Canada, mountain hemlock (*Tsuga mertensiana*) is the most common treeline species. By collecting core samples from mountain hemlock trees growing in Strathcona Provincial Park on Vancouver Island, we have been able to develop a tree-ring chronology that extends back to 1412 (see Figure 3.3), thus predating Columbus's discovery of North America. Annual radial growth at this site is limited by low summer temperatures and enhanced by low winter precipitation. These results are consistent with the hypothesis that the length of the growing season

Figure 3.3 An example of a tree-ring sample as seen by a high-resolution scanning/analysis system.

is the most critical factor influencing growth at treeline environments. Deep, persistent snowpacks keep soil temperatures low and tend to delay the initiation and rates of metabolic processes, reducing growth. In contrast, warm temperatures during the growing season melt snow more quickly, increase root and shoot temperatures, enhance metabolic rates, and increase photosynthesis, thereby increasing radial growth rates.

It is possible to quantify the relationship between temperature, precipitation, and radial growth and then forecast the expected changes to growth under various global warming scenarios. However, the relationship is complicated. Using a response surface model to visualize the relationship between these parameters in three dimensions reveals that the relationship is complex and highly nonlinear (Figure 3.4). From this figure it is apparent that

Figure 3.4 A response surface graph showing the relationship between summer temperature, the depth of the April 1st snowpack, and annual radial growth of mountain hemlock.

the species' response to summer temperature is highly dependent on the amount of snow that has accumulated. In years with a deep spring snowpack, radial growth is highly sensitive to variability in summer temperatures. In contrast, when the snowpack is shallow, good growth occurs regardless of summer temperature.

The future health of mountain hemlock forests therefore depends not only on changes to temperature but also on changes to precipitation—changes that are much less certain in the current generation of climate models. If the future turns out to be warmer and drier, mountain hemlock may be well pre-adapted and thus benefit from climate change. However, mountain hemlock's competitors, western hemlock (*Tsuga heterophylla*) and Pacific silver fir (*Abies amabilis*) are probably better adapted to a warmer, less snowy climate and may ultimately out-compete mountain hemlock. Different tree species exhibit distinct responses to climatic variability, and only by studying these individual responses can scientists hope to understand the potential impacts of global environmental change on forest ecosystems.

Ze'ev Gedalof is an assistant professor in the Department of Geography at the University of Guelph. Dan Smith is head of the Tree Ring Laboratory in the Department of Geography at the University of Victoria.

Sand dune succession is another common form of primary succession in which the primary colonizers are not lichens but grasses that have the ability to withstand not only the high variability in temperature and water but also the constantly shifting sand. The grasses help to stabilize the sand until mat-forming shrubs invade. Later, conditions may become suitable for hardy trees such as pines that may in turn be replaced by other tree species such as oaks.

Climax should be thought of as a relative rather than an absolute stage. Communities do not change up to the climax and then cease to change. However, the nature of the species assemblage is more constant over time once a climax stage is reached. Some ecologists prefer the term *mature community* to describe this stage rather than climax, since the latter has stronger implications of mature communities being static over time. Even in mature communities, future changes in pathogens, predation, and climate will generate ongoing changes.

The climax vegetation for most areas is strongly influenced by the prevailing climate and is therefore known as a *climatic climax*. In some areas, other factors such as soil conditions may be more important than climate in determining community composition and structure. These are known as *edaphic climaxes* (Box 3.3).

In addition to the primary succession described above, successional processes also occur on previously vegetated surfaces such as abandoned fields or avalanche tracks or following a fire, where soil is already present. This process is known as **secondary succession**. The earlier soil-forming stages of primary succession are not repeated, so the process is much shorter, with the dispersal characteristics of invading species being a main factor in community composition. Annual weeds again dominate the community until perennial weeds, such as goldenrod, start to become established. Where conditions are suitable, the community will eventually be invaded by shrub and ultimately tree species. A major challenge for agriculture and forest managers is to prevent this natural recolonization taking place by species that may not yield the required products. As a result, chemical herbicides, as discussed in greater detail in Chapters 9 and 10, are often used to arrest secondary succession.

Similar kinds of processes also occur in aquatic environments. Here, the natural aging is called **eutrophication** (eu = well, trophos = feeding) as nutrient supplies increase over time with inflow and the growth and decay of communities. The process can be relatively rapid in shallow lakes, because the nutrients (one of the auxiliary energy flows discussed in Chapter 2) promote increased plant growth that leads to more biomass and nutrient accumulation. The lake becomes shallower over time, with less surface area of water, and the aquatic communities may eventually be out-competed by marsh and ultimately terrestrial plants. This process is another example of a positive feedback loop (the shallower the lake gets, the stronger the forces become to make it shallower), which is

Sand dunes are a good place to observe the successional changes over time, as shown here. With increasing distance from the sea, the communities change to those in later seral stages representing the build-up and colonization of the sand.

Box 3.3   Edaphic Climax: Table Mountain, Newfoundland

The west coast of Newfoundland (like most of the rest of the island) is dominated by the boreal forest (Chapter 9). In Gros Morne National Park (Figure 3.5), however, and at other locations on the west coast, this greenery (white spruce, paper birch, balsam fir) is punctuated by practically treeless orange-coloured outcrops that bear little if any similarity to the surrounding vegetation. These outcrops result from the distinctive chemical composition of the bedrock, known as 'serpentine'. Along with three other serpentine outcrops in western Newfoundland, the Table Mountain massif in Gros Morne was formed on the floor of the Atlantic Ocean millions of years ago and rafted up to its present position through the process of continental drift.

This geological history has given the serpentines a composition characterized by high levels of nickel, chromium, and magnesium and low levels of calcium. Most of the species of the surrounding forests cannot tolerate these conditions; if they grow at all, they are stunted. Instead, the serpentines are host to relict communities of tough arctic-alpine species that have survived since the retreat of the glaciers and have not been displaced through the process of succession like the arctic-alpines on the surrounding bedrock. These serpentine communities are edaphically driven where the underlying geology has been more important in determining plant cover than climate has.

Figure 3.5 Location of Table Mountain.

The difference between the dominant vegetation of the edaphic climax of Table Mountain and the surrounding boreal forest can be clearly seen along the geological boundary.

The inhospitable soil chemistry has allowed rare species, such as this *Lychnis alpina*, to continue to grow in the area as relicts from the ice age.

discussed in more detail in the next section. Eutrophication may also constitute a significant management problem, since the species being replaced often have higher values to humans than the species replacing them. This problem is discussed in more detail in Chapter 4.

Indicators of Immature and Mature Ecosystems

As successional changes take place in communities, several trends emerge. For example, annual net primary productivity declines as the slower-growing species establish, and diversity increases as more specialized species come to dominate the community and more finely subdivide the resources of the particular habitat. However, the increase in diversity will not continue indefinitely, according to the intermediate disturbance hypothesis (Figure 3.6). This hypothesis suggests that ecosystems subject to moderate disturbance generally maintain high levels of diversity compared to ecosystems that experience low levels of disturbance or those that have high levels of disturbance. Under low levels, competitive exclusion by the dominant species reduces diversity. With high disturbance, only those species tolerant of the stress can persist. Disturbance occurs at different scales, from small scale such as that associated with a gap created in a forest when a tree falls over from death or windthrow to large scale associated with widespread fire.

Despite the flexibility of the successional concept, certain differences between mature and immature systems are generic (Table 3.1). In general, mature ecosystems tend to have a high level of community organization among a large number of larger plants and have a well-developed trophic structure. Decomposers dominate food chains, with a high efficiency of nutrient cycling and energy use. Net productivity is low. Immature ecosystems tend to have the opposite of these characteristics.

Effects of Human Activities

Humans have a profound influence on ecological succession. Many activities are directed towards keeping certain communities in early seral stages. In other words, humans seek to maintain the characteristics of the immature ecosystems shown in Table 3.1 as opposed to those of the mature ecosystems that would result if natural processes were allowed to proceed. Agriculture, for example, usually involves large inputs of auxiliary energy flows to ensure that succession does not take place as weeds try to colonize the same areas being used to grow crops. The same can be said for commercial forestry. Maintaining ecosystems in early successional stages has several implications:

- The productivity of early successional phases is often higher than later phases.

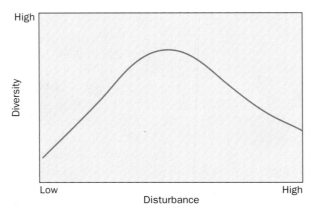

Figure 3.6 The intermediate disturbance hypothesis.

Table 3.1   Characteristics of Immature and Mature Ecosystems

Characteristic	Immature Ecosystem	Mature Ecosystem
Food chains	Linear, predominantly grazer	Weblike, predominantly detritus
Net productivity	High	Low
Species diversity	Low	High
Niche specialization	Broad	Narrow
Nutrient cycles	Open	Closed
Nutrient conservation	Poor	Good
Stability	Low	Higher

Source: Modified from E. Odum. 1969. 'The strategy of ecosystem development'. Science 164: 262–70. Copyright © 1969 by the American Association for the Advancement of Science.

- Nutrient cycling, discussed in more detail in the next chapter, is often more rapid in early stages. Trees, for example, not only hold nutrients in their mass for a longer time than herbaceous plants, but they also maintain relatively low temperatures in soils. High temperatures result in more rapid breakdown of organic material and release of nutrients to the environment. Water uptake and storage by plants is also much reduced. Consequently, disturbance may result in a significant loss of nutrient capital from a site through losses in soil water to streams.
- Overall biodiversity tends to be reduced.
- The species most adversely affected are often highly specialized ones at higher trophic levels.
- The species that benefit most are usually pioneer species (weeds and pests) that have broad ranges of tolerance and efficient reproductive strategies for wide dispersal.

Changing Ecosystems

In the early 1970s, residents of the Okanagan Valley in British Columbia began to complain to the government about excessive weed growth in some of the lakes in the valley. Several popular beaches were becoming virtually unusable because of the weeds, and the extent of the invasion appeared to be growing rapidly. This was of considerable concern to the residents, not only because of the impact on their recreational activities but also because of the impact on the economy of this tourist area for which water-based recreation was the main attraction.

The culprit was Eurasian water milfoil, which arrived in the area in the 1970s and over the next couple of decades would spread not only to all the lakes in the Okanagan but also to many other lakes in southern BC and other provinces. The government spent significant amounts of money trying to control the spread of the species but to no avail. Originating in Eurasia, the milfoil had reached the eastern shores of this continent probably a century ago and since that time had spread across the continent, replacing native aquatic plants in many water bodies.

This ecological event, the spread of a Eurasian plant into North America, also illustrates the dynamic relationship among the biophysical, socio-economic, and management systems that is the main focus of this book. In BC, for example, the dependence of local economies such as that of the Okanagan Valley on water-based tourism triggered a strong response to milfoil that involved the use of the chemical 2,4-D. This created considerable conflict between different stakeholders regarding the relative impact of the plant versus that of the control mechanism. Critics claimed that management had failed to consider the broader perspectives that would have been included had they adopted an ecosystem-based approach to the problem and that management failed to adapt to the changing parameters of the situation. Chapter 6 discusses various approaches to these kinds of resource management issues in greater detail.

Situations such as this are not uncommon. We tend to think of ecosystems as having relatively constant characteristics, of being in a balance in which internal processes adjust for changes in external conditions. It is not a static state but one of **dynamic equilibrium**. James Lovelock (1988) postulated the *Gaia hypothesis*, which claims that the ecosphere itself is a self-regulating homeostatic system in which the biotic and abiotic components interact to produce a balanced, constant state (Box 3.4). This is an example of a highly integrated system in which there is a strong interaction between the different parts of the system. Other systems may not be so highly dependent on one another. Cells in a colony of single-celled organisms, for example, may be removed and have little effect on the remainder because of the low integration of the system.

Not all ecosystems are equal in their ability to withstand perturbations. **Inertia** is the ability of an ecosystem to withstand

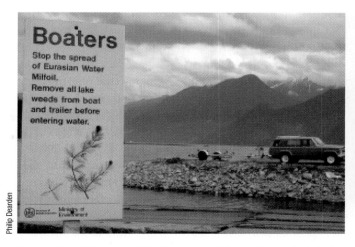

Signs warning of the spread of Eurasian water milfoil were placed at boat-loading ramps throughout BC but did little to stem the colonization.

In the tourist economy of the Okanagan Valley where resorts rely on water-based activities to attract clientele, there was considerable conflict among different stakeholders regarding the most appropriate means of controlling the spread of milfoil.

Box 3.4   The *Gaia Hypothesis*

As humans became capable of probing deeper and deeper into space and thus able to view Earth from this unique perspective, it became increasingly clear that our planet was significantly different from all the millions of others. It seems a happy coincidence that of the vast range of temperatures that could be experienced, those on Earth are just right for life, between the freezing and boiling points for water, even though the energy output of the sun has increased by more than 30 per cent during the past 3.6 billion years. The gaseous composition of the atmosphere is also just what we need to breathe.

One hypothesis, the *Gaia hypothesis*, named for the ancient Greek goddess of the Earth by the originator of the idea, James Lovelock, postulates that the Earth acts as one giant self-regulating super-organism to help maintain these conditions necessary for life. Organisms act like cells in a body, with integrated functions to promote the health of that body, or in this case, optimum conditions for life. Active, automatic feedback processes among the atmosphere, lithosphere, hydrosphere, and ecosphere maintain this equilibrium.

There is no doubt that since the beginning of life on Earth, organisms have adapted to existing conditions and have also modified these conditions in ways that are beneficial to life. However, few scientists believe that Earth is a super-organism with all living things interacting to maintain an equilibrial environment. The historical record clearly shows that organisms do not regulate natural cycles. Furthermore, it is mere wishful thinking to assume that Earth will adapt and compensate for all of the changes that are now being initiated by human activity in ways that are beneficial to humans.

change, whereas **resilience** refers to the ability to recover to the original state following disturbance. Ecosystems can have low inertia and high resilience or any combination thereof. In terms of human usage, it is best to work with systems that have both high inertia and high resilience. This means that they are relatively difficult to disturb and even when disturbed will recover quickly. Such systems are relatively stable. The best growth sites for forestry—alluvial sites in nutrient-rich areas at low elevations—would fit into this category. Many tropical and Arctic sites would fit into the opposite combination in which sites are readily disturbed and recover only very slowly, if at all.

Ecosystems have various attributes, such as feedback, species interactions, and population dynamics, that tend to maintain this equilibrium. However, most scientists feel that there is no such thing as a fixed equilibrium, that ecosystems are constantly changing, and that only dynamic equilibriums exist. In some cases, as with the milfoil described above, this is obviously true. The milfoil invasion involved the replacement of a variety of native aquatic species with a mono-specific stand of the alien species. Similar effects are common with other non-native invaders, such as Scottish broom, purple loosestrife (Box 3.5), sea lamprey, and zebra mussels.

Invasive Alien Species

Organisms found in an area outside their normal range, such as Eurasian water milfoil and purple loosestrife, are considered **alien species**. The UN Convention on Biological Diversity defines alien species as a species introduced outside its normal past or present habitat. It includes any part, seeds, or eggs that might survive and subsequently reproduce. Many species transported to a new environment do not survive. However, many others multiply rapidly, out-compete native species, change native habitats, and are now considered as **invasive** alien species. Their impacts are often irreversible. Invasive species are second only to habitat destruction as a leading cause of biodiversity loss. On islands, they are often the main cause of extinctions, since there is little opportunity for the indigenous species to escape. Twenty-two per cent of species listed as endangered in Canada are in such a perilous state because of the effects of invasive species in their respective habitats (Venter et al. 2007).

In Canada, we have felt the impact of thousands of alien species invasions (Figure 3.7), and their numbers are increasing. Alien species total 27 per cent of all vascular plant species, and there are 181 insect species that feed on woody plants, as well as 24 bird, 26 mammal, two reptile, four amphibian, and 55 freshwater fish species. Several of the invasive alien species on the World Conservation Union's list of the 100 worst are

Box 3.5   Purple Loosestrife: Alien Invader

Purple loosestrife was inadvertently introduced to North America from Europe more than a century ago. Ocean-going ships typically carry 'ballast water'—that is, water to balance their cargo load in heavy seas—and it is taken on in the originating port. When the ship reaches calm water near its destination, this ballast and everything in it, including biological organisms, is discharged. An aggressive invader of aquatic systems, the purple loosestrife has spread through thousands of hectares of wetlands in Quebec and Ontario. In Manitoba, it is thought to be the most serious noxious weed. It is estimated that 190,000 hectares of wetland habitat in North America is invaded by purple loosestrife each year. After its woody root systems have become established, native plants and the animals that depend on them for food are forced out.

At the University of Guelph, experiments with the *Galerucella pusilla* beetle have showed promising results in controlling this invader plant. The beetle has a voracious appetite for purple loosestrife. They eat the metre-high plant at such a rate that the plant's capacity to produce seed (about 2.5 million per plant per year) is reduced by 99 per cent. Thus, use of the beetles to control purple loosestrife is promising, since previous control efforts that relied on physical removal, burning, mowing, and spraying produced negligible results. However, the beetles do forage on native plant species. Manitoba has initiated a biological control program using the highly host-specific weevil *Nanophyes marmorates* that is showing promising signs for controlling loosestrife populations.

Sources: http://www.purpleloosestrife.org; www.ducks.ca/purple

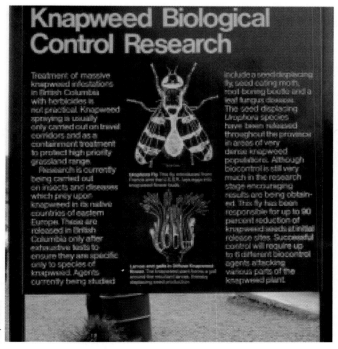

Philip Dearden

A roadside sign describes the knapweed problem.

Figure 3.7 Alien agricultural and forest pests introduced in Canada, 1870–2000. Source: Auditor General of Canada 2002, 4:5. Reproduced with the permission of the Minister of Public Works and Government Services, 2004.

established in Canada: Dutch elm disease, purple loosestrife, leafy spurge, Japanese knotweed, green crabs, spiny water fleas, gypsy moths, carp, rainbow trout, starlings, domestic (feral) cats, and rats (http://www.issg.org/database/species/search.asp?st=100ss). Canada's annual timber loss due to invasive alien species is estimated at 61 million cubic metres, which is equivalent to $720 million in financial losses to stumpage, royalties, and rent revenues. The cost of the damage caused by invasive alien species affecting forestry and agriculture has been estimated at $7.5 billion annually (Canadian Food Inspection Agency 2004). More than 500 species of alien plants in Canada have developed into agricultural weeds. They cost farmers millions of dollars every year to control, costs that we all pay when purchasing grains, vegetables, and fruit grown in Canada (Table 3.2).

One example is the various species of knapweed introduced into Canada and the US from the Balkan states, probably in shipments of alfalfa. The diffuse knapweed causes the most problems; it has a wide range of tolerance and a very effective seed dispersal system that it has used to colonize vast areas of rangeland in western Canada. It is also thought that this species may be *allelopathic*—that is, it can directly inhibit the growth of surrounding species through production of chemicals in the soil. The species displaces native species and considerably reduces the carrying capacity of the rangelands. Cattle will eat it only as a last resort, and the nutritive content is less than 10 per cent of that of the displaced native species. Initial control efforts relied on chemical sprays. A more integrated approach is now being taken, using biological control and attempting to limit its spread through stricter controls on vehicular access to rangelands, one of the main means of seed distribution as seeds get distributed by vehicle tires.

Besides plants, many other species have also proved troublesome. Two fungi, chestnut blight and Dutch elm disease, for example, have had significant impact on the landscape of central and eastern Canada. Both attack native trees that at one time were conspicuous parts of the deciduous forests. The American chestnut was attacked by an Asian pathogenic fungus that was introduced on stocks of Japanese chestnuts during the past century and the elm by a European fungus

Table 3.2   Some Introduced Agricultural and Forestry Pests and Diseases in Canada

Species	Place and Date of Discovery	Current Area	Impacts/Notes
Asian long-horned beetle *Anaplophora glabripennis*	Brooklyn and Amityville, NY: 1966; Chicago, IL: 1998; no known established population in Canada	Intercepted in warehouses in both Ontario and British Columbia	Poses serious threat to Canadian hardwood forests (maple being the preferred host); expensive eradication effort (more than $5 million to date in New York alone) undertaken in Brooklyn, Amityville, and Chicago after detection of established populations
Brown spruce longhorn beetle *Tetropium fuscum*	Likely arrived in Port of Halifax, NS; found in Point Pleasant Park around 1990	Halifax Regional Municipality, NS	No chemical pesticide currently registered to control the beetle; 5,309 trees are confirmed positive for removal in the Halifax Regional Municipality, 2,645 of these in Point Pleasant Park; native to Europe and Asia
Colorado potato beetle *Lepinotarsa decemlineata*	Extended its range during 1850s	Across Canada	Significant pest of potato crops; native to the western United States but fed on another species prior to the arrival of the potato in the 1850s; after discovering new host, quickly followed the potato path back across North America; good example of a species spread by habitat change
Dutch elm disease *Ophiostoma ulumi* and *O. novo-ulmi*	Quebec: 1944 (first discovery; introduction was probably before 1940)	Present in all provinces except Newfoundland and Labrador, Alberta, and British Columbia	A fungus spread by the native elm bark beetle (*Hylurogopinus rufipes*) and the European elm bark beetle (*Scolytus multistriatus*); controlled through public education campaigns, surveillance, insecticides, and sanitation (removal of diseased elms)

From Statistics Canada, *Human Activity and the Environment: Annual Statistics 2003*. Cat. no. 16-201, 3. Dec. 2003. Ottawa: Statistics Canada, 75.

transmitted between trees by beetles. More than 600,000 elm trees were killed in Quebec alone, and 80 per cent of Toronto's elms died within one year in the 1970s.

In Winnipeg, the boulevard elm trees are worth $307 million and contribute an estimated $160 million in property value. Annual budgets for control in Winnipeg have been as high as $2.25 million, with more than $30 million in control and research costs. Researchers at the universities of Manitoba and Toronto are using a natural toxin called *mansonone* that occurs in some elms to breed seedlings that are more resistant to Dutch elm disease.

Another fungus, the white pine blister rust, illustrates the complexity of the impact of invasive species. The fungus, originating in Eurasia, attacks five-needled pines and causes extensive mortality. Whitebark pine is a key component of the subalpine ecosystems of the Canadian Rockies. It has a mutualistic relationship (Chapter 2) with Clark's nutcracker, a type of crow-like bird that caches the seeds for forage during the winter. Unlike those of many pines, the whitebark's cones are not opened by fire but by animal activity. The seeds cannot be carried by wind and rely on the nutcracker for dispersal. The bird caches the seeds in forest openings for easy retrieval, creating perfect conditions for germination of the seed. However, the birds, while remarkable in their ability to remember hundreds of cache sites, invariably 'forget' some. These seeds may then germinate, resulting in the establishment of seedlings. Beyond the mutualistic relationship between these species in that both benefit, it is important to note that the seeds of the pine are too heavy to disperse very far, which means that the nutcracker is a keystone species, as discussed in Chapter 2. When keystone species are lost in an ecosystem, that system is subject to significant change. Stuart-Smith et al. (2002) measured mortality rates of the pine in excess of 20 per cent in some areas of the national parks as a result of fungus attack. There is concern that if mortality rates increase, it will lead to population declines in the Clark's nutcracker.

Often, invasive species have been deliberately introduced by humans for one reason or another and can have much the same impact as species that are introduced accidentally. One example is the introduction of Sitka black-tailed deer into Haida Gwaii as a food source for local people in the late nineteenth century. In the absence of predators such as wolves and cougars, the deer populations and distribution expanded rapidly. However, because of the nature of the archipelago, some islands were colonized early, others later, and others not at all. This created ideal conditions for scientists to study the impact of the deer over different time periods. Stockton et al. (2005) found that vegetation cover exceeded 80 per cent in the lower vegetation layers on islands without deer. This contrasted with 10 per cent for islands that had supported deer for longer than 50 years. Overall plant species richness was similar, but at the plot level it was reduced from 20 to 50 per cent on islands that had deer for more than 50 years. In general, these results show the potential of seemingly innocuous species to greatly simplify ecosystems.

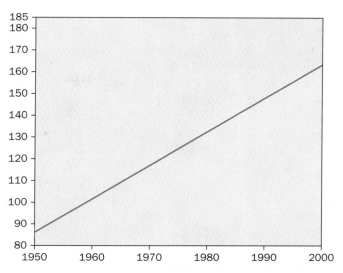

Figure 3.8 Cumulative number of aquatic invasive species in the Great Lakes, 1950–2000. Source: Auditor General of Canada 2002, 4:14; 2008. Reproduced with the permission of the Minister of Public Works and Government Services, 2004, 2008.

Many of the most serious invasions occur in aquatic ecosystems. One notable recent invasion is the spread of the zebra mussel into the waters of the Great Lakes. The mussel, named for its striped shell, joins a long line of aliens in the Great Lakes, including the sea lamprey, alewife, and rainbow smelt (Figure 3.8). It is thought that the mussel, a native of the Black and Caspian seas, was introduced from the ballast of freighters in the mid-1980s. It was first found in 1988 in a sample of aquatic worms collected from the bottom of Lake St Clair at Windsor-Detroit, which connects lakes Erie and Huron. Evidence from Europe indicated that the species was an aggressive colonizer, able to displace most native species. In a short period of time, it displaced 13 species from Lake St Clair and caused the near-extinction of 10 species in western Lake Erie.

The mussels start their three- to six-year life as free-swimming larvae before attaching to a hard surface, usually in the top three to four metres of the water, although they can live as deep as 30 metres. By the end of 1988, the mussels had colonized half of Lake St Clair and two-thirds of Lake Erie at densities as high as 30,000 per square metre. Densities as high as 600,000 per square metre have been recorded as testament to the mussel's ability to grow on itself.

The mussel has now spread extensively through the Great Lakes, where it appears capable of colonizing any hard surface. It has encrusted water intakes and discharges, severely reducing their efficiency and necessitating significant expense to remove them. Water intakes may be reduced by as much as 50 per cent. Many different approaches are being undertaken to screen out the mussels, but they appear to be able to pass through most physical barriers. At the moment, chlorination is the most common measure, but this raises problems related to the potential for formation of toxic organochlorines. Ontario Power Generation has spent more than $20 million on

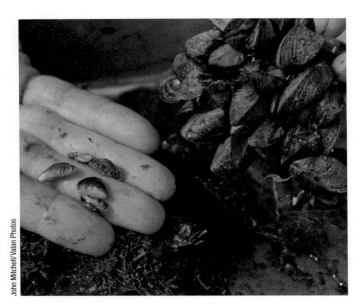

Zebra mussel infestation.

John Mitchell/Valan Photos

installing and maintaining chlorine applicators at its Great Lakes facilities and inland facilities and another $13 million on research to reduce chlorine use. Estimates of the damage to all Great Lakes utilities range from $200 to $500 million per year. The mussels also colonize spawning sites for other fish, with as yet undetermined impact on their populations or the $4.5-billion fishing and tourism industry in the region.

— Impacts on the population levels of other species are likely to come about more indirectly through food-chain effects. The mussels are filter-feeders that remove phytoplankton from the water, thereby affecting all the species higher in the food chain, such as walleye, bass, trout, and perch. In some European locations, invasion by the mussels has led to clearer water as a result of the removal of phytoplankton. These changes may benefit some species, even fish species. Bottom-feeders, such as carp and whitefish, and invertebrates such as crayfish may benefit as more nutrients are returned to the lake bottoms, either in the form of mussels themselves or mussel feces.

However, the mussels do not remove all species of phytoplankton equally. This is creating problems with blooms of blue-green algae, such as the toxic *Microcystis aeruginosa*, that are not ingested by the mussels. Some scientists believe that the algae may be primarily responsible for Lake Erie's 500- to 1,000-square-kilometre dead zone, which had mostly been attributed to chemical pollutants.

It remains to be seen whether species higher in the food chain, such as scaup and other waterfowl, can help to control the spread of the mussel. Already, numbers of some of these species, which stop over to feed during their migration, appear to have risen considerably. Realistically, it appears that the ducks may have some impact, as they have in Europe, but that the infestation will be too large and the number of ducks too small for the problem to be controlled in this manner.

Furthermore, once a species becomes established, it is difficult to prevent further spread. Despite major efforts in the US to stop the spread of zebra mussels there, they were found for the first time in early 2008 in Lake Mead in the desert near Las Vegas, about 2,000 kilometres from the Great Lakes.

In 2005 another deadly invader, perhaps the worst to date, suddenly appeared: viral haemorrhagic septicaemia, or VHS, dubbed the Ebola for fish. Great Lakes fish have little immunity to it, and this has led to massive die-offs as they become infected. The virus is one of the world's most dreaded fish diseases, normally found only in salt water, and one of the first foreign pathogenic microbes to become established in the Great Lakes. The virus has been identified in 19 species, and in the St Lawrence River hundreds of thousands of round gobies have succumbed to the disease. Gizzard shad die-offs from VHS in Lake Ontario west of Rochester and in Dunkirk Harbor on Lake Erie also have been reported.

Another threat has appeared on the horizon, the Asian big head carp. The carp has a voracious appetite, eating half its body weight every day and reaching almost a metre in length. The carp escaped from fish farms in the southern US in the 1990s and invaded the Illinois and Mississippi River systems. Only a canal in Chicago that connects to Lake Michigan, protected by an electric fence, prevents the fish from entering the Great Lakes. Biologists feel that it is only a matter of time before the carp enter the lakes, which would lead to the demise of the entire fishery. One possible means of entrance is through the live fish trade. Carp are brought live from fish farms in the US to markets and restaurants in Toronto. The water is subsequently discarded into the drainage system, along with any fingerlings.

Many invasive aquatic species, including the zebra mussel, arrive in their new habitat courtesy of ocean freighters, which take on water for ballast in one part of the world and release it in another. More than 3,000 species are being transported around the world every day through this process. Scientists have known for a long time that ballast water is a major threat to aquatic communities. It also transports human pathogens. In 1998, a study in the Great Lakes found a wide range of pathogens in ballast water, including giardia, salmonella, fecal coliforms, and *Vibrio cholerae*. Given the magnitude of these introductions, it is inevitable that from time to time one or two of these species will not only find a tolerable home in their new location but also explode into great numbers. The US requires freighters to discharge ballast water offshore. Canada had voluntary discharge with little or no monitoring for many years.

In 2004, a new international convention to prevent the potentially devastating effects of the spread of harmful aquatic organisms carried by ballast water was adopted by the International Maritime Organization, the United Nations agency responsible for the safety and security of shipping and the prevention of marine pollution from ships. The convention requires all new ships to implement a Ballast Water and Sediments Management Plan. They will have to carry a Ballast Water Record Book and will be required to carry out ballast water management procedures to a given standard. Existing ships are required to do the same but after a phase-in period.

Canada has legislation and programs that ostensibly deal with the problem of invasive species, especially those that may damage agricultural and forest crops or pose a danger to human health. Under the terms of the United Nations **Convention on Biological Diversity**, discussed in Chapter 14, Canada is also committed to containing invasives that threaten biodiversity. However, an audit undertaken by the Auditor General (2002) discerned no progress in addressing the problem. The report recommended that Fisheries and Oceans Canada develop and implement a means of identifying and assessing the risks posed by aquatic invasive species and use it as a tool for setting priorities and objectives for prevention, eradication, or control of those risks. A monitoring and reporting system to track the effectiveness of actions taken towards aquatic invasive species objectives was also recommended, as well as a requirement to report annually on progress. The department agreed with the recommendations and provided an expected completion date of fall 2003 for addressing them.

Transport Canada was asked to define best management practices for ship ballast water and establish regulations requiring the application of those practices. A means of monitoring, maintaining records, and reporting annually on compliance was also recommended. The department accepted the Auditor General's recommendations and committed to regulating and monitoring ballast water accordingly. A follow-up audit was undertaken in 2007, and the results are summarized in Box 3.6. They are not encouraging.

The draft plan to address the threat of invasive alien species, *An Invasive Alien Species Strategy for Canada,* was delivered in 2004, two years behind schedule (Canadian Food Inspection Agency 2004). The strategy contained no specific steps to achieve its key objectives: prevention, detection, rapid response, and management of invasive alien species. A phase two has now been developed, consisting of preliminary action plans, but as yet little action seems to be forthcoming, other than regarding ballast water control (http://www.inspection. gc.ca/english/plaveg/invenv/action/phase2e.shtml).

In June 2006, the previously voluntary ballast measures became mandatory under the Ballast Water Control and Management Regulations. All ships arriving from beyond the exclusive economic zone (EEZ) and entering waters under Canadian jurisdiction must now undertake one of the following: exchange their ballast water, treat their ballast water, discharge their ballast water to a reception facility, or retain their ballast water on board. In 2007, the government announced $4.5 million in funding over the following five years to enforce the regulations. The funding will be used to increase the number of marine inspectors enforcing ballast water regulations, to support the development of technologies to better address ballast water issues, and to equip marine inspectors with the proper tools to enforce the ballast water regulations. Increased aerial surveillance to combat marine pollution has also been undertaken. For example, during fiscal year 2006–7, Transport Canada conducted 1,649 hours of dedicated productive pollution patrol. This resulted in the detection of 98 marine pollution incidents from the 10,063 vessels that were flown over.

Figure 3.9 Areas for ballast water control on the east coast.

Box 3.6   Summary of Findings of the Commissioner of the Environment and Sustainable Development on Control of Aquatic Invasive Species

- Fisheries and Oceans Canada has made unsatisfactory progress in implementing our 2002 recommendations. While the Department has identified some aquatic invasive species that pose major risks to Canada and put in place biological risk assessment guidelines, the rate at which new alien species are becoming established is exceeding the rate at which the Department is assessing risks. The Department has failed to assess economic and social risks, and priorities and objectives for prevention, control, or eradication of risks posed by aquatic invasive species have not been set.

- In addition, Fisheries and Oceans Canada does not have plans or mechanisms in place for early detection of, or rapid response to, aquatic invasive species and is therefore unprepared to prevent, control, or eradicate potential new aquatic invasive species. It has not monitored or reported how effective its efforts have been at preventing, controlling, or eradicating the aquatic invasive species it has identified.

- Transport Canada has made satisfactory progress in implementing one of our 2002 recommendations. Mandatory regulations for the control and management of ballast water came into force through the Canada Shipping Act in 2006. These regulations constitute a major step forward in addressing the issue of aquatic invasive species. Ships entering Canadian waters are now required by law to take steps to manage ballast water to reduce the likelihood of introducing aquatic invasive species.

- Transport Canada has made unsatisfactory progress on our 2002 recommendation related to monitoring and reporting on compliance. While Transport Canada has developed the tools it needs and is gathering information on compliance with its regulations for ships entering the Great Lakes, gaps remain in the Department's compliance monitoring and reporting at the national level.

- We found that risks posed by aquatic invasive species have not been adequately assessed or effectively managed. The federal government is not yet in a position to prevent, control, or eradicate invasive species that pose the greatest threat to Canada's aquatic ecosystems and economy. Much remains to be done to meet commitments made in the federal government's 1995 Canadian Biodiversity Strategy.

Source: Auditor General of Canada 2008.

Although regulations have slowly been tightened up, the days of allowing ocean freighters into the Great Lakes might be numbered in any case. If the approximately 500 ocean ships that now enter the Great Lakes each year were allowed no closer than the Port of Montreal and their cargo transferred to railways, lake freighters, or barges, the cost would be about $55 million (US) a year in extra freight charges. Only about 7 per cent of tonnage on the Great Lakes is carried by ocean freighters, but they are the main source of ecosystem change. Lake freighters, because they do not leave the Great Lakes area, are not responsible for bringing in foreign species. The cost of ending ocean shipping would be relatively low because it moves only a marginal amount of freight. Estimates suggest that as few as four trains could accommodate all the cargo carried by ocean freighters, while the transportation costs for companies now relying on the ocean vessels would rise by about 6 per cent. This seems a small price to pay to eliminate the current ecosystem damage.

Species Removal

Just as the introduction of species to new habitats can disturb ecosystem function, so can the removal of species from food webs. The reduction of some species, the so-called keystone species discussed in the previous chapter, may be particularly disruptive. One well-known example relates to the extirpation of the sea otter from the Pacific coast.

When James Cook anchored at Nootka Sound on the west coast of Vancouver Island in 1778, he reported that the fur of the sea otter 'is softer and finer than that of any others we know of; and, therefore, the discovery of this part of the continent of North America, where so valuable an article of commerce may be met with, cannot be a matter of indifference.' Indeed, it was not. The

British, seeking trading goods to barter with the Chinese in exchange for tea, discovered that sea otter pelts were in great demand in China and thus made every effort to ensure that the west coast became British (rather than Spanish or Russian!) Columbia.

The sea otter is a large sea-going weasel of the outer coasts, flourishing in the giant kelp beds. They lack a protective layer of blubber but have a very fine fur that traps air and insulates them from the cold Pacific waters. They also need a lot of food (up to nine kilograms per day) to fuel the fast metabolism that counteracts energy loss to the environment. Favourite prey are sea urchins, crabs, shellfish, and slow-moving fish.

The otters were easy to catch, and Russian, American, and Spanish hunters, aided by local Native populations, finished off what the British had begun. Within 40 years, populations were reduced from more than half a million to 1,000 to 2,000. On the coast of British Columbia, it is likely that they were completely extirpated. However, relict populations remained both to the south, around Monterey in northern California, and to the north in the Aleutian Islands. Individuals from this latter population have now been reintroduced to the coast of British Columbia, where small but vibrant populations thrive again.

Scientists discovered the otters' key role in maintaining ecosystems after studying two groups of islands off Alaska. They noticed that although the two groups were very similar in terms of location and physical conditions, one group had much more life—bald eagles, seals, kelp beds, and otters—than the other. Otters play a critical role in controlling sea urchin populations (Estes, Duggins, and Rathbun 1989). Sea urchins are voracious eaters of kelp (large, brown seaweed), which may be the world's fastest-growing plant, with increments of up to 60 centimetres per day. Given the support of the ocean, kelp does not need to invest much energy in heavy support structures, leaving more energy for growth (another example of the auxiliary energy flow discussed in Chapter 2).

Sea otters are making a slow comeback in British Columbia's coastal areas.

Kelp plays a major role in coastal communities. It provides food and habitat for many other species. Diatoms, algae, and microbes grow on the fronds of the kelp, along with colonies of filter-feeding bryozoans and hydroids. Predators abound. Fish come to feed off the colonists or to seek protection from open-water predators such as seals, sea lions, and killer whales. When overgrazed by sea urchins, this productive habitat disappears. The urchins eat through the holdfasts that anchor the kelp to the ocean floor, and the kelp is soon washed away into the open ocean or onto land. As the kelp disappears, so do the species dependent on it. On the two islands in Alaska, one island had managed to escape the fur rampage that eliminated otters elsewhere, and this one displayed the rich coastal community that should extend all along the outer coast of the North Pacific. Otter populations, through their control of the urchin population, are therefore critical to maintaining the productivity of the entire community, right up to bald eagle populations. The fact that the fashion tastes of Chinese mandarins 200 years ago, met by traders from the other side of the world who wanted to enjoy afternoon tea, is still reflected in bald eagle populations 7,000 kilometres away on the BC coast indicates the complex interactions between biophysical and human systems.

Feedback

Feedback is an important aspect of maintaining stability in ecosystems whereby information is fed back into a system as a result of change. Feedback initiates responses that may exacerbate (**positive feedback**) or moderate (**negative feedback**) the change. There is, for example, considerable debate regarding the role of feedback loops in global climate change, as discussed in Chapter 7. One positive feedback loop that may have a strong influence in Canada is the effect of increased temperatures in the North. It would increase the area of snow-free land in summer and is known as polar amplification. Snow has a high albedo; in other words, it reflects rather than absorbs much of the incoming radiation. As temperatures rise, the area covered in snow will be replaced by areas free of snow, uncovering rocks and vegetation with lower albedo values. This will cause more heat to be absorbed, which in turn will contribute to global warming. A similar situation with regard to forest fires was noted in Box 3.2.

On the other hand, negative feedback loops may also be in operation and serve to counteract such positive feedback loops. One of them has to do with the possible role of phytoplankton in global warming. Phytoplankton produce a gas called dimethyl sulphide. When seawater interacts with the gas, sulphur particles formed in the atmosphere serve as condensation nuclei for cloud droplets. As the planet heats up, the productivity of the phytoplankton should increase, leading to an increase in the amount of gas and cloud droplets produced. This will have the effect of increasing cloud cover and reflecting away solar radiation, which could lead to cooling of the Earth. However, scientists feel that this cooling will be offset by the overall impact of global warming, as discussed in Chapter 7.

Almost all the examples in this chapter can be used to illustrate some aspect of feedback mechanisms. The allelopathic qualities of the diffuse knapweed, for example, show a positive feedback loop that promotes the spread of the species. The more the species spreads, the more conditions are created into which only *it* can spread. The sea otters produce a negative feedback loop on the sea urchin–kelp relationship. If the urchins become too numerous and overgraze the kelp beds, increases in otter populations will help to reverse this imbalance. When this negative feedback loop was removed from the system, there was nothing to maintain the dynamic balance of the system.

Similar examples of feedback loops occur at all scales, even down to the regulation of temperatures in individual organisms. Sometimes these feedback messages can be rapid, as in the case of organism thermoregulation. In other cases, there can be considerable delay between the stimulus for change and the resulting feedback response. Unfortunately, as the example of the positive feedback loop and snowmelt described above indicates, sometimes the delay between the stimulus and the response may be so long that we are not conscious of it. By the time we are aware, it may be too late to try to moderate the stimulus, and a powerful positive feedback loop may already, albeit slowly, have been set in motion. This is one reason why many scientists support immediate actions to cut down on the emission of greenhouse gases (see Chapter 7), even though we do not yet have a clear understanding of all the relationships involved.

We are also becoming more aware of the chaotic nature of many systems in which a slight perturbation becomes greatly enhanced by positive feedback. The so-called *butterfly effect*, for example, traces how the turbulence of a butterfly flapping its wings in South America might, through cascading effects on airflows, influence the weather in North America (see www.imho.com/grae/chaos/chaos.html). Further research has revealed the existence of similar phenomena in many different systems in which very small changes can have a great influence on outcomes. Chaos theory tries to discern pattern and regularity in such systems and allow for greater predictability.

Synergism

Synergism is another important characteristic that may influence change in ecosystems. A synergistic relationship occurs when the effect of two or more separate entities together is greater than the sum of the individual entities. One example is the problem of acid deposition, discussed in Chapter 4. The effects of acid deposition are often exacerbated by the presence of other pollutants, such as ground-level ozone. Individually, both these forms of pollution may cause a certain amount of damage to an ecosystem. In combination, however, their effects are magnified.

On an individual scale, we can see the same effects for smokers. The potential for developing lung cancer is increased both by smoking and by the inhalation of radon gas. However, if someone who smokes also inhales radon gas, the chances of getting lung cancer are greater than the sum of the individual exposures to each hazard. This is because the particles of tobacco ash that accumulate in the lungs as a result of smoking also act as anchors for radon particles. Together, the two exposures are far more dangerous than the sum of the individual exposures.

Ecological Restoration

From the foregoing, it should be very clear that ecosystems change. Many changes result from human activities, and these changes have a negative impact on ecosystem components and function. However, this does not mean that ecosystems so damaged have to stay that way, and restoration ecology has developed as a field of study and practice to help repair environmental damage. This book contains many examples, ranging from remediation of the Sydney Tar Ponds in Chapter 1, reclamation of the areas around Sudbury rendered treeless by acid rain (Chapters 4 and 13), to efforts to reintroduce endangered species into national parks (Chapter 14). The goals of these efforts vary enormously—from merely stabilizing an area with a self-maintaining cover of vegetation, such as the reclamation of many industrial sites, to efforts to restore areas to their pre-disturbance condition. One of the common difficulties in the latter situation is ascertaining the nature of the original ecosystem. Restoration ecologists are now concentrating more on trying to restore natural processes in an area rather than reintroducing components. However, **ecological restoration** is very challenging and very costly, and there is widespread agreement that it is better to avoid degrading ecosystems in the first place rather than trying to restore them afterwards.

Population Growth

The number of individuals in a species is known as the **population**. When calculated on the basis of a certain area, such as the number of sea otters per hectare, it becomes *population density*. The number of organisms in a population is important, because low numbers will make a species more vulnerable to extinction. Changes in population characteristics are known as *population dynamics*.

Populations change as a result of the balance among the factors promoting population growth and those promoting reduction. The most common response is through adjustments in the birth and/or death rates to the factors shown in Figure 3.10, although emigration and immigration can be important factors in some species. Population change is calculated by the formula $I = (b - d) N$, where I is the rate of change in the number of individuals in the population, b is the average birth rate, d is the average death rate, and N is the number of individuals in the population at the present time. As long as births are more numerous than deaths, then a population will increase exponentially over time (Figure 3.11) until the environmental resistance of the factors shown in Figure 3.10 begin to have an inhibiting effect that will serve to flatten out the curve.

The **carrying capacity** of an environment is the number of individuals of a given species that can be sustained in a

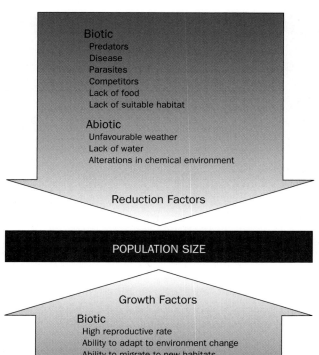

Figure 3.10 Factors affecting population growth.

given area indefinitely, given a constancy of resource supply and demand. Most species will grow rapidly in numbers up to this point and then fluctuate around the carrying capacity in a dynamic equilibrium (Figure 3.12). The carrying capacity is not one fixed figure, however, but will vary along with changes in the other abiotic and biotic parts of the ecosystem. In the puffin-capelin example in Chapter 2, the carrying capacity of the North Atlantic waters to support the puffin population was severely reduced as a result of a reduction in their food supply caused by competition from another organism, humans. Management inputs, such as provision of supplementary feeding or other habitat requirements, are often used to change the capacity of an area to meet human demands. The concept of carrying capacity is now often applied to park management, with park managers establishing the number of visitors a park can sustain without either suffering ecological damage (known as *ecological carrying capacity*) or seeming too crowded (known as *social carrying capacity*).

Organisms that demonstrate the kind of S-shaped growth curve of Figure 3.12 are *density-dependent*, and as the population density increases, the rate of growth decreases. In other words, the larger the population, the lower the growth rate. This view is in accord with the equilibrium view of ecosystems discussed earlier. Some organisms, however, are *density-independent*,

and the population operates with a positive feedback loop—the more individuals there are in the population, the more that are born, and the population grows at an increasing rate to demonstrate a J-shaped curve. At some point, this population meets environmental resistance, causing the population to crash back to or below the carrying capacity. The algae blooms that you notice on ponds in the late spring or early summer are a result of this kind of growth. In reaction to the increased nutrient availability after winter, spectacular growth can occur until this food supply is exhausted and the population crashes. The spectacular and economically very damaging increase in the numbers of the mountain pine beetle in BC is another example. Ecologists now accept that given the absence of cold winters because of climate change, the population will only reach its limit when it has exhausted its food supply (Chapter 9).

In some locations in Europe, this is what has happened with the zebra mussels discussed earlier. Mussel populations soared to a peak and stayed there for a few years before exhausting

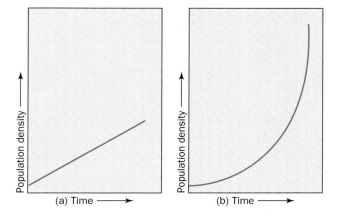

Figure 3.11 Geometric (a) and arithmetic (b) growth patterns.

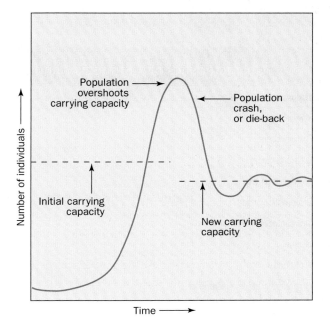

Figure 3.12 Carrying capacity and population growth rates.

Jellyfish and the Pacific white-sided dolphin are good examples of marine r and K species respectively.

the food supply and crashing to between 10 and 40 per cent of the original numbers. However, in other locations, such as Sweden, the expected crash has yet to occur. Given the enormous food supplies of the Great Lakes and the low numbers of predators such as waterfowl, it may be a very long time before any natural population crash happens there.

The capacity of species to increase in number is known as their **biotic potential**, the maximum rate at which a species may increase if there is no environmental resistance. Different species, however, have different reproductive strategies. Some species, such as zebra mussels, are known as *r-strategists*, which produce large numbers of young early in life and over a short time period but invest little parental energy in their upbringing. Most of their energy is spent on reproduction, and

they have few resources left to devote to maintaining a longer lifespan. Such species are usually small and short-lived and can respond to favourable conditions through rapid reproduction. They are opportunists, and their reproductive strategy is essentially based on quantity. Such species tend to dominate the early seral stages of the successional process.

K-strategists, on the other hand, produce few offspring but devote considerable effort to ensuring that these offspring reach maturity. Their strategy is based on quality. Individuals live longer and are usually larger. Populations of K-strategists often reach the carrying capacity of an environment and are relatively stable compared to r-strategists, which may experience large variations in population size. Table 3.3 summarizes the characteristics of these different strategists.

Table 3.3   Characteristics of K-Strategists and r-Strategists

K-Strategists	r-Strategists
Late reproductive age	Early reproductive age
Few, larger young	Many, small young
More care of young	Little care of young
Slower development	Rapid development
Later reproductive age	Early reproductive age
Greater competitive ability	Limited competitive ability
Longer life	Short life
Larger adults	Small adults
Live in generally stable environments	Live in variable or unpredictable environments
Emphasis on efficiency	Emphasis on productivity
Stable populations usually close to carrying capacity	Large population fluctuations usually far below carrying capacity

Typical examples of r-strategists include insects, rodents, algae, annual plants, and fish. A mature female codfish, for example, may produce more than 9 million eggs in one season. However, fewer than 5 per cent of these offspring may mature and last the first year. Most K-strategists are larger organisms, such as the larger mammals (including humans). Their lower biotic potential and ability to disperse often means that they are more restricted to the later seral stages of succession. Many endangered species (see Chapter 14) are K-strategists. The great whales (Chapter 8), with perhaps only one offspring every three years, are a good example. When the conditions to which they have become accustomed, and under which they

evolved their reproductive strategy, change dramatically, such as with the introduction of new predators (humans), they have little capacity to respond in terms of increasing their reproductive rate. Muskox are another example (Box 3.7).

In addition to the factors outlined above, chance also plays an important role in determining population size. Severe winters, disease outbreaks, fires, droughts, and similar factors often have a major impact on populations that continually find themselves recovering from one natural catastrophe or another. Peary caribou, for example, exist north of the 74th parallel by digging under the snow to feed on vegetation. In 1974–5, heavy snows and freezing rains led to high mortality as the herd starved to death, unable to reach their food source.

Box 3.7   Muskox

Perhaps no animal epitomizes the Arctic more than the prehistoric-looking muskox. With its great shaggy head and flowing mane facing into an Arctic storm, it seems to symbolize the determination of life to survive in even the most inhospitable environments. Muskox are in fact superbly adapted to the rigours of the North. They have a long, coarse outer coat covering a fine underlayer of soft wool; all of their extremities (e.g., tail and ears) are covered by this coat. Their keen sense of smell enables them to find food in the winter darkness, they have low metabolic rates, and they minimize energy loss by remaining relatively inactive.

Muskox are not widely distributed throughout the Arctic—they concentrate in areas where vegetative conditions are best, where they can graze and browse on many Arctic herbs, grasses, and shrubs. In these areas, they stay in bands of females with a single bull that takes the leadership role in finding routes and repelling predators, mainly Arctic wolves. It is against such predators that they evolved their well-known defensive strategy of forming a circle around the calves with their massively horned heads pointing outward. Although effective against wolves, this strategy proved much less effective against humans with guns, who could stand at a distance and shoot this rather large target at will. More than 16,000 hides were shipped out of Canada between 1864 and 1916. In addition, Arctic explorers used them as a main source of meat. Consequently, with their low birth

Muskox.

rate (one calf every second year) and a few hard winters, numbers fell drastically. They are now protected but may be hunted by licence and by indigenous hunters. About 8,000 muskox are left in the Canadian Arctic, ranging from the Arctic islands south to the Thelon game sanctuary and the coast of Hudson Bay.

Unfortunately, these are the very same weather conditions predicted to become more common as a result of global climate change. In 1993, there were more than 3,000 caribou on Bathurst Island in the High Arctic. By 1997, as a result of repeated bad winters, the number was down to 75.

Evolution, Speciation, and Extinction

Over the long term, populations adapt to changing conditions through **evolution**, a change in the genetic makeup of the population with time. This can be achieved through mutations that are passed on to subsequent generations, eventually creating a new species. Within any population, there will also be some variation in the genetic composition that may predispose a certain segment of the population to adapt to certain conditions. If change occurs to favour those conditions, then the success of the part of the population that is genetically better adapted to the new conditions will be enhanced. In this way, over time, *natural selection* can lead to changes in the characteristics of a population.

Phyletic evolution is the process in which a population has undergone so much change that it is no longer able to interbreed with the original population and a new species is formed. This is the process of **speciation**. It can happen as a result of geographical isolation, when a single population becomes fragmented by a geographical feature, such as a mountain range or a water body, and the populations evolve separately from one another. If conditions differ in the respective environments of the different breeding groups, then natural selection will favour those individuals best suited to those conditions. Over time, they become so different that they can no longer interbreed. Another example occurs when part of a population adapts better to a new food source. It is better for individuals so adapted to mate with similar individuals to enhance the ability to exploit that food source, and over time this process might create sufficient differences that they become a different species. Sometimes the effects of these influences may combine.

The evolution of the polar bear from the grizzly bear is one example. It is thought that the polar bear evolved as a separate species from the grizzly bear some 10,000 years ago. Bears with characteristics that helped them hunt seals on ice flows, such as lighter-coloured fur and greater strength, would be relatively more successful in the Far North rather than in the rest of the range, where a brown pelt and greater mobility might be an advantage. In this way, a single bear species became two bear species through adaptation to different environments and the process of natural selection. This process of local adaptation and speciation is known as *adaptive radiation*. However, instances of interbreeding between the two species have recently come to light.

Genetic diversity helps to protect species from extinction. The resilience of a species depends partly on the magnitude of the environmental change, how rapidly it takes place, and the capacity of the gene pool of the species to respond to these changes. In general, the broader the gene pool, the greater the capacity to adapt to change.

Changes in the abiotic environment are not alone in promoting evolutionary change. Species may also change through **coevolution**, whereby changes in one species cause changes in another. Each species may become an evolutionary force affecting the other. A typical case is a prey species evolving to be more effective in avoiding a predator. In turn, the predator may evolve more efficient hunting techniques to detect the prey. Many such relationships have evolved in the tropical forests, especially between specific plants and animals, because of the long period of evolutionary change that has taken place in such environments. Canada has many examples as well, particularly relating to pollination in which various insects, birds, and bats have evolved to pollinate flowering plants and, in turn, a great diversity of plant shapes, sizes, and colours have developed as a direct response to the activities of the pollinators.

The processes of evolution described above have always been thought to take place very slowly. Recently, however, biologists have identified processes occurring much more quickly as a result of human activities in a process known as **contemporary evolution** (Stockwell, Hendry, and Kinnison 2003). One of the main pressures for contemporary evolution is human harvesting of prey populations. In established fisheries, once fish enter targeted age classes, predation by humans occurs at rates two to three times higher than that of natural predators, often exceeding 50 per cent (Stokes and Law 2000). Our predation of terrestrial vertebrates, especially ungulates, is also well developed. Fa, Peres, and Meeuwig (2002) estimated that human predators kill more than 5 million tonnes of wild mammals in neotropical and Afrotropical forests per year. Gauthier et al. (2001) reported that half of the adult

Many tropical orchids are products of coevolution. The flower has evolved to imitate the female wasp of the species that pollinates the flower. The male is deceived into thinking that it is a female, flies into the flower and in so doing picks up pollen that is subsequently taken to the next imitator, and pollination occurs.

Philip Dearden

mortality of North American snow geese results from hunting, and Festa-Bianchet (2003) asserted that hunting accounts for most of the adult mortality of many of Europe and North America's large herbivores. Human predators induce microevolutionary change because we are now replacing natural predators as dominant agents of selection.

Traditional harvesting strategies have often concentrated on taking the oldest and largest members of a population. For example, we shoot the largest ram and catch the largest fish, letting the others go so that they too can grow larger and be harvested later. However, research is now indicating that all individuals do not have the same capacity to become that large and that by eliminating the largest, we are systematically selecting for smaller individuals in the future. In general, individuals' size and growth rates decline, while reproductive investment increases and individuals become reproductively mature at smaller sizes and earlier ages. This is happening with many different species as average sizes continue to decline. Jachmann, Berry, and Imae (1995) reported an increase in tusklessness among African elephants. Tuskless males increased in the population from approximately 1 per cent in the early 1970s to about 10 per cent in 1993 and tusklessness

among females from 10.5 per cent in 1969 to roughly 38 per cent in 1989 following intense poaching that targeted individuals with ivory-bearing tusks.

For fish, scientists have formulated the Big Old Fat Female Fish (BOFFF) hypothesis. These old fish are irreplaceable in that they not only contain more eggs but also proportionately more viable eggs. As BOFFFS are harvested, fish populations have increasingly greater difficulty in achieving replacement population, and a negative feedback loop sets in.

Extinction is the opposite of evolution and represents the elimination of a species that can no longer survive under new conditions. The fossil record suggests that perhaps close to 99 per cent of the species that have lived on Earth are extinct. The fact that we may still have up to 50 million species, more than have ever existed before, indicates that speciation has exceeded the extinction level. However, speciation takes time. Even with r-strategists, it may take hundreds and thousands of years; with K-strategists, it may take tens of thousands of years. Evidence suggests that in recent times, the activities of humanity have tipped the scale in favour of extinction over speciation (Box 3.8), as discussed in Chapter 14. Table 3.4 gives some examples of species that at one time existed in Canada but are now extinct.

Box 3.8   Humans and Extinction

Extinction, as pointed out in the text, is a natural process. Scientists can try to estimate the average rate of species extinction by examining the fossil record, which suggests that extinctions among mammals might be expected to occur at the rate of about one every 400 years and among birds one every 200 years. Current extinction rates are difficult to estimate, because we do not have a full inventory of species and so we do not know what we are losing. Based on current rates of habitat destruction for tropical forests, estimates range as high as 100,000 extinct species per year. Many of these extinct species are likely to be undescribed arthropods, since they comprise the majority of species in tropical forests. More conservative estimates suggest that somewhere between 2 and 8 per cent of the planet's species will become extinct over the next 25 years, many more than what would have occurred in pre-human times.

Although there may be disagreements over the rate of extinction, there is widespread consensus that humans have vastly increased this rate, and there are now more than 16,000 species on the Red List of the World Conservation Union (IUCN). Fewer than 10 per cent of the world's species have undergone status assessments, yet more than 30 per cent of amphibians, 23 per cent of mammals, and 12 per cent of birds are threatened, according to IUCN. Species at risk of extinction are discussed in more detail in Chapter 14.

Evidence suggests that humans have had a major impact on biodiversity for quite some time. Paul Martin (1967) was one of the first to suggest that humans may have been a major factor in causing the extirpation of several species of large mammals from North America at the end of the last ice age, some 10,000 years ago. During that time, at least 27 genera comprising 56 species of large mammals, two genera and 21 species of smaller mammals, and several large bird species became extinct. The extinctions included 10 species of horse, four species of camel, two species of bison, a native cow, four elephant species, the sabre-toothed tiger, and the American lion. Although this period was also a time of global climatic change, no such extirpations were associated with the same period in Eurasia. The period also saw a substantial in-migration of humans from the Asian continent, who began to prey on animals unfamiliar with and therefore not adapted to human hunting. This hunting, combined with the environmental stresses experienced through habitat alteration and repercussions through the food chain, was sufficient to extirpate the species.

Charles Kay (1994) has conducted research into the subsequent impact of Native Americans on ungulate populations before the onset of European influences. He concludes that even then, humans were the main limiting factor on ungulate populations in the intermountain West and that elk in particular were overexploited. The people had no effective conservation strategies and hunted to maximize their individual needs, irrespective of environmental impacts. Thus, the image of North America as a vast wilderness unaffected by human activities before the coming of the Europeans appears to be a

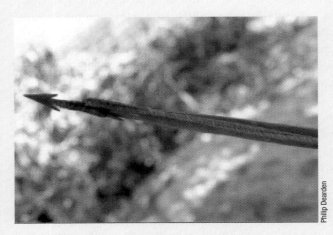

It seems difficult to believe that what we now consider primitive weapons, such as this fishing spear from the Warao people of the Orinoco delta, may have enabled humans to hunt many other species to extinction.

myth. Even that mightiest symbol of the wild, the grizzly bear, was apparently under pressure from Aboriginal hunters in Alaska (Birkedal 1993).

Recent reviews agree that human predation was the central cause among a suite of forces responsible for Pleistocene extinctions (Barnosky et al. 2004; Brook and Bowman 2004), and it is obvious that humans are super-predators who excel all other species.

Table 3.4 Some Canadian Vertebrate Species That Are Now Extinct

Species	Distribution	Last Recorded	Probable Causes
Great auk (*Alca impennis*)	Canada, Iceland, UK, Greenland, Russia	1844	Hunting
Labrador duck (*Camptorhynchus labradorius*)	Canada, US	1878	Hunting, habitat alteration
Passenger pigeon (*Ectopistes migratorius*)	Canada, US	1914	Hunting, habitat alteration
Deepwater cisco (*Coregonusjohannae*)	Canada, US	1955	Commercial fishing, introduced predators
Longjaw cisco (*Coregonus alpenae*)	Canada, US (Great Lakes)	1978	Commercial fishing, introduced predators

Extinction, like speciation, should not be considered a smooth, constant process but one punctuated by relatively sudden and catastrophic changes. It appears that multicellular life, for example, has experienced five major and many minor mass extinctions. Scientists think that the age of the dinosaurs, a remarkably successful dynasty that effectively relegated mammals to minor ecological roles for more than 140 million years, was brought to an end 65 million years ago by the impact of a large extraterrestrial object. And then the mammals took over. Perhaps the dinosaurs, through the processes of evolution and speciation, managed to out-compete the mammals for a long period of time and, were it not for the chance impact of the asteroid, might still be the dominant animal life. However, this chance occurrence not only led to the demise of the cold-blooded dinosaurs but also favoured the survival of the rodent-like mammals with their smaller body size, less specialization, and greater numbers. Small body size was likely a sign of the mammals' inability to challenge the dinosaurs during the normal evolutionary process; however, small body size became a positive feature favouring survival under the new conditions.

There are other examples of the non-random impact of mass extinction on life. The features that make some species successful during ordinary times may be completely unrelated to the new conditions, making life's pathway somewhat chaotic and unpredictable rather than the smooth path that evolutionary theory might suggest.

Finally, as emphasized in Box 3.9, we should not think that evolution is fundamentally a story that demonstrates the benefits of complexity over simplicity. Humans—the most complex

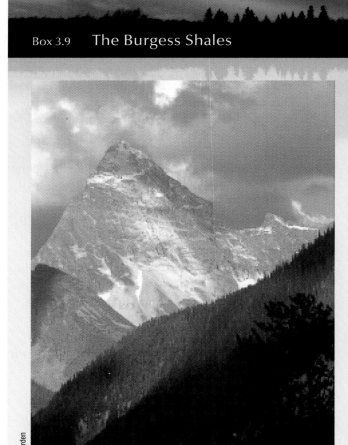

Box 3.9   The Burgess Shales

Stephen Jay Gould, the well-known Harvard paleontologist, called the Burgess Shales in Yoho National Park in British Columbia the single most important scientific site in the world. The reason for this superlative is the extensive bed of fossils that can be found high on the flanks of Mount Wapta. They are fossils from the Cambrian era, some 530 million years ago, when there was a great flourishing of diverse life forms. The special feature of the site is that the fossils from this era are preserved in great detail, even down to the soft body parts, such as stomach contents.

The story revealed by this detail is one of great diversity at a time when all but one phylum of animal life made a first appearance in the geological record. The site also contains many body patterns for which there are no current counterparts. Thus, it seems as though life could be characterized as three billion years of unicellularity, followed by this enormously diverse Cambrian flowering in a brief five-million-year period and a further 500 million years of variations on the basic anatomical patterns set in the Cambrian period. Why, or how, this flowering took place is uncertain. It would seem to require a combination of explanations. First, there was literally an open field available for colonization—an environment ripe to support life but with little life in it. Therefore, species did not have to

Philip Dearden

The site of the Burgess Shales World Heritage Site on the slopes of Mount Wapta in Yoho National Park.

be particularly good competitors to survive. Virtually anything could survive. Since this time, even after mass extinctions, sufficient species have remained to make it pretty tough competition for any newcomers. Second, it seems as though the early multi-cellular animals must have maintained flexibility for genetic change and adaptability that declined as greater specialization arose and organisms concentrated on refining the successful designs that had already evolved. Furthermore, we have little idea why most of these early experiments in life died out and yet others remained. There seem to be no common traits shared by the survivors to indicate that they were the victors of Darwinian strife. Perhaps just the lucky ones survived.

Gould, in his fascinating book *Wonderful Life*, suggests that this challenges our established view of evolution as an inevitable progression over time from the primitive and few to the sophisticated and many. It also radically challenges our view of ourselves as being the logical endpoint of evolutionary change, the rightful inheritors of the Earth. In Gould's words: 'If humanity arose just yesterday as a small twig on one branch of a flourishing tree [of evolution], then life may not, in any genuine sense, exist for us or because of us. Perhaps we are only an afterthought, a kind of cosmic accident, just one bauble on the Christmas tree of evolution' (Gould 1989, 44). In other words, we should be humble!

Archeologists work at a dig site of a fossilized dinosaur at Alberta's Dinosaur Provincial Park. More species of extinct dinosaurs have been found and identified at this World Heritage Site than anywhere else in the world.

organisms—are not the pinnacle of life's achievement or the most successful. For that accolade, we have to travel to the other end of the complexity spectrum. From the beginning of the fossil record until present times, bacteria have demonstrated the most stable presence. According to Stephen Jay Gould (1994, 87):

> Bacteria represent the great success story of life's pathway. They occupy a wider domain of environments and span a greater range of bio-chemistries than any other group. They are adaptable, indestructible, and astoundingly diverse. We cannot even imagine how anthropogenic intervention might threaten their extinction, although we worry about our impact on nearly every other form of life. The number of *Escherichia coli* cells in the gut of each human being exceeds the number of humans that has ever lived on this planet.

Biodiversity

Over billions of years, interaction between the abiotic and biotic factors through the process of evolution has produced many different life forms. The main classification is the **species**, life forms that resemble one another and can interbreed successfully. Two species are created from one as a response of natural selection to changes in environmental conditions, as explained in the previous section. **Biodiversity** is the sum of all these interactions and is usually recognized at three different levels:

- *genetic diversity*, the variability in genetic makeup among individuals of the same species. Individuals generally vary in the genetic information encoded in their DNA. However, some individuals do not. Genetically uniform *clones* may exist in separate individuals. One example is the trembling aspen: through vegetative propagation, genetically identical individuals can cover an area of up to 50 hectares. In general, genetic diversity in a population increases the ability to avoid inbreeding and withstand stress.
- *species diversity*, the total number of species in an area. The number of different species in an area is often known as *species richness*.
- *ecosystem diversity*, the variety of ecosystems in an area. Some ecosystems are more vulnerable to human interference than others (Box 3.10). Estuaries and wetlands, for example, are highly productive but are often used for industrial and agricultural activities. As these ecosystems are replaced by human-controlled ecosystems, natural diversity at the landscape level is reduced.

Box 3.10 Carolinian Canada

Carolinian Canada is the wedge of land stretching from Toronto west to Windsor that contains 25 per cent of the country's population. It also contains the highest number of tree species in the country as the mixing zone between the eastern deciduous forests to the south and mixed coniferous-deciduous forests to the north. Its location in the southernmost part of the country, with the mediating effects on climate of the southern Great Lakes, allows semi-tropical tree species such as the cucumber and sassafras to spread up into this land of ice and snow. After Vancouver and Victoria, Windsor ranks as the third warmest city in Canada, and it is the most humid city in the country. It is little wonder that in summer the humidity and southern vegetation can give the appearance of a much more southern location.

Besides the distinctive vegetation, the Carolinian zone also supports a noteworthy bird population. Point Pelee is noted as one of the top birding spots in North America. Not only do birds (exhausted from crossing Lake Erie on their migrations north in winter) like to rest here, but it is also part of the Carolinian forest and the nesting habitat for many species that are unusual for Canada, particularly warblers. Of the 360 bird species seen, about 90 stay to nest, and in spring there may be 25 to 30 different warblers spotted on a good day. In the fall, the birds are joined by thousands of monarch butterflies as they too pause here before heading south on their 3,600-kilometre journey to the Gulf of Mexico for the winter.

Point Pelee is a national park. Most of the rest of the Carolinian forest is not so well protected and is heavily fragmented by agriculture and urban development. As a result, it is estimated that close to 40 per cent of Canada's rare, threatened, and endangered species are primarily Carolinian. The Carolinian Canada Program, started in 1984, has co-ordinated the efforts of government agencies and private landowners to try to protect the remaining forest. About half of the 38 targeted sites now have some degree of protection, but biologists still worry that these fragments are too small and isolated to be capable of protecting this most diverse area of Canada.

One of the threats to the Carolinian zone comes from invasive plants. A recent arrival is the so-called Tree-of-Heaven (*Ailanthus altissima*), which has long been recognized as a serious problem in other areas of North America and Europe. Currently, the tree is established only in the Rondeau Provincial Park area in the western basin of Lake Erie, and efforts are underway to control the species before it spreads extensively (Meloche and Murphy 2006). Past experience shows that all too often, we ignore small infestations of invasive species and by the time they become significant problems, it is too late to do anything about it.

Scientific knowledge of biodiversity is very primitive (Box 3.11). There may be up to 100 million species, although most scientific estimates suggest between 5 million and 20 million, of which we have identified some 1.8 million (Figure 3.13). Some 56 per cent of these species are insects, 14 per cent are plants, and just 3 per cent are vertebrates such as mammals, birds, and fish. Even new mammals are still being discovered, such as the giant muntjac and saola discovered recently on the borders of Vietnam and Laos. However, most species awaiting discovery are probably invertebrates, bacteria, and fungi in the tropics. We also know relatively little about the ocean (Chapter 8). Only about 15 per cent of described species are from the oceans. Most biologists agree that there are fewer species to be found there than on land. On the other hand, there are 32 phyla in the oceans, compared with only 12 on land.

Species identification is only the first building block in biodiversity. We also need to understand the differences in genetic diversity within species and how species interact in ecosystems to really understand how the life support system of the planet works (Box 3.12).

Box 3.11   Counting Critters

Since actually discovering and describing new species is a very slow process, scientists have devised several means to help estimate just how many species we might have on Earth.

Species-area curves is one of the most popular approaches. Generally, the larger the area, the larger the number of species found there, but the new species gradually level off as the same species are encountered repeatedly. This relationship has enabled scientists to construct species-area curves that show the number of species likely to be found in areas of different sizes and hence predict how many species might be found in larger unsampled areas.

Rain forest insect samples work from the idea that most of the species we have yet to describe are probably rain forest insects. Studies indicate that there are up to 1,200 beetle species in the canopy of a single tree in Peru. Since we know that 40 per cent of insects are beetles, this suggests that there may be more than 3,000 insect species in the canopy alone. About half this number will be found lower down and on the trunk, leading to estimates of 4,500 species on one tree alone. Given that there are more than 50,000 species of tropical tree, the numbers that may be found in total will be vast. Some researchers predict that there are more than 30 million insect species.

Ecological ratios can be used to predict the populations of little-known groups from their relationships with better-known groups. For example, the ratio of fungus to plant species in Europe is 6:1. If this holds worldwide, then there should be more than 600,000 species of fungus. At the moment, fewer than 70,000 have been described.

None of these approaches (and these are just three of the more popular ones) can deliver an absolute estimate of the numbers of species on the planet. Each one, however, is consistent in indicating that we have a long way to go before we can claim that we truly know the nature of life on this planet.

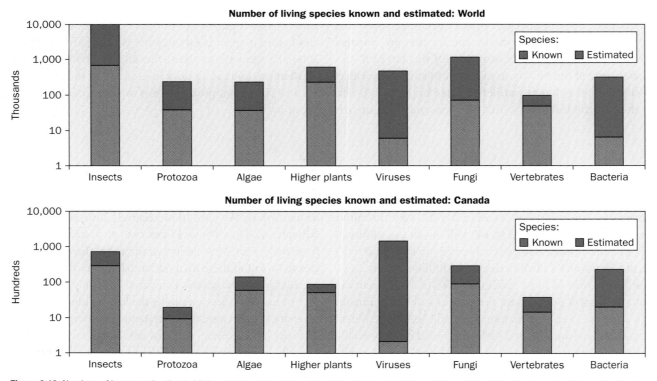

Figure 3.13 Numbers of known and estimated living species in the world and in Canada. Source: B. Groombridge. 1992. *Global Biodiversity: Status of the World's Living Resources*, p. 17. London: Chapman and Hall. Reprinted with kind permission of Springer Science and Business Media.

Box 3.12   Biological Uncertainty

Attention has been drawn to the high degree of biological uncertainty regarding the numbers of species in Canada and elsewhere, not to mention the ecological functions of each species. Few things in the natural world are absolutes. Even gender differences blur. We have known this for some time for more primitive species, such as slugs and earthworms, which are hermaphrodites—that is, one individual has both male and female sexual functions. Still, it was somewhat of a surprise to Charles Francis, a biologist with the Canadian Wildlife Service undertaking field research in Malaysia, to find the first free-ranging wild male mammals that lactate. He was collecting Dayak fruit bats when he found that several males had breasts and milk. He speculates that this characteristic may have evolved among males that are monogamous. Another theory suggests that something in the bats' diet contains steroids that mimic female hormones.

There is also uncertainty regarding many of the basic ecological principles described in these last two chapters. Concepts such as succession, ecosystems, and communities have been found unsatisfactory for addressing many real-life ecological problems. Ecosystems often exhibit many different states that can be reached by various paths and that may produce no identifiable local climax. Furthermore, species distributions along environmental gradients may overlap broadly and be subject to many natural and human-induced disturbances. Heterogeneity is the norm rather than the exception. These observations have changed the way that ecologists look at ecological systems, recognizing the openness of systems, the importance of episodic events, and the numerous possibilities for intervention in ecological processes. Nonetheless, the classical concepts provide a useful background against which to organize this new, more flexible, 'non-equilibrium' approach.

Biodiversity in Canada

Biodiversity is not evenly distributed around the world. Some biomes, mainly tropical forests, are extremely diverse (Box 3.13), but in temperate latitudes, there is much less diversity. Overall, as discussed in Chapter 2, species numbers decline in a gradient from the tropics to the poles. Latin America, for example, is home to more than 85,000 plant species. North America has 17,000, of which only 4,000 occur in Canada. A similar gradient for birds is shown in Table 3.5. Several reasons have been advanced to account for the latitudinal gradient in species richness. Rohde (1992), following a review of the various reasons, concludes that many factors contribute at different scales but the primary cause appears to be the effect of solar radiation (i.e., temperature) that increases evolutionary speed at lower latitudes.

Estimates suggest that Canada has more than 140,000 different species, of which about half have been named. The taxonomic groups containing the most numbers of species are shown in Figure 3.14. Mosquin (1994) points out that although the groups represented in this graph are not as well known as other groups, such as birds and mammals, they undertake key functions in ecosystems, often functions that we are only just becoming aware of and that support the more familiar and larger organisms. Beneficial insects, for example, fertilize

Philip Dearden

Most of the world's species are insects, and many more await discovery.

flowers and control pests; crustaceans provide food for fish; bacteria recycle nutrients; and bread, beer, and penicillin all come from fungi. The Canadian Endangered Species Conservation Council provides five-yearly assessments of the status of more than 7,000 species in Canada (Chapter 14).

Box 3.13   The Tropical Forests

Charles Darwin, who described the mechanisms of evolution in *On the Origin of Species* (1859), originated most of his ideas in the tropics. It was in the tropics—where life is speeded up through high energy inputs and abundant moisture, where adaptation is at its most complex and intricate, and where the struggle for survival is most dramatic—that evolution could most readily be appreciated.

The diversity of the tropical forests is astounding—estimates suggest that at least half of the world's species are within the 7 per cent of the globe's surface that the tropics cover. For example: in 100 square metres in Costa Rica, researchers found 233 tree species; one tree in Venezuela was home to at least 47 different species of orchids; there are 978 different species of beetles that live on sloths; and more than 1,750 different species of fish live in the Amazon basin. In general, the rain forests of South America are the richest in species, followed by Southeast Asia and then Africa. Several factors account for this abundance.

1. Tropical rain forests have been around for more than 200 million years, since the time of the dinosaurs and before the evolution of the flowering plants. It is thought that at that time, there was just one gigantic landmass, before continental drift started to form the continents as we now know them. The vegetation of many areas was subsequently wiped out by succeeding glacial periods, which had minimal impact on the rain forests. Hence, evolutionary forces and speciation have had a long time to operate in the tropics.
2. Over the long period of evolution, there is a kind of positive feedback loop. As more species have developed and adapted, it has caused further adaptations as more species seek to protect themselves from being eaten and also to improve their harvesting of available food supplies. It is thought, in particular, that plant diversity has been partly the result of the need to adapt defences against the myriad of insects that graze on them. As the plants develop their defences, insects constantly adapt to the new challenge. The very high biodiversity of these groups is due to the speed of these evolutionary processes. In a system where most plants are immune to most insects but highly susceptible to a few, it pays to be a long way from a member of your own species.

Successful trees are hence widely distributed, which allows more opportunity for speciation to occur.
3. The tropics receive a higher input of energy from the sun than other areas of the globe. Not only are they closer to the sun, but they also have little or no winter. The flux in solar input at the equator between the seasons is 13 per cent, but at latitude 50 degrees, the variation is 400 per cent.
4. Tropical rain forests receive a minimum of 2,000 millimetres of precipitation that is evenly distributed throughout the year. Moisture is therefore not a limiting factor, allowing for constant growth. There is a strong correlation between diversity and rainfall.
5. Tropical rain forests are the most diverse ecosystems that have evolved on Earth. They are also characterized by examples of coevolution and mutualism in which two species are absolutely co-dependent on one another. More than 900 species of wasp, for example, have evolved to pollinate the same number of fig trees. Each wasp has adapted to just one species of fig. Should anything destroy the food supply in such a finely tuned system, then the other species will also meet its demise.

While the evolutionary process has benefited from most of these characteristics, the soils have suffered. They have been exposed to weathering processes for a very long time, with no renewal and remixing from glaciation. The

Species that evolved among the complexity of tropical forests have developed many adaptations to protect themselves. The camouflage of the leaf insect in this photo gives it some protection from predators.

warm temperatures and abundant moisture are perfect for chemical weathering to great depths, and most tropical soils have long since had their nutrients washed out. A fundamental difference between tropical and temperate ecosystems is that in the tropics, unlike more temperate climes, most of the nutrients are stored in the biomass and not in the soils. When tropical vegetation is removed—by logging, for example—this removes most of the nutrients.

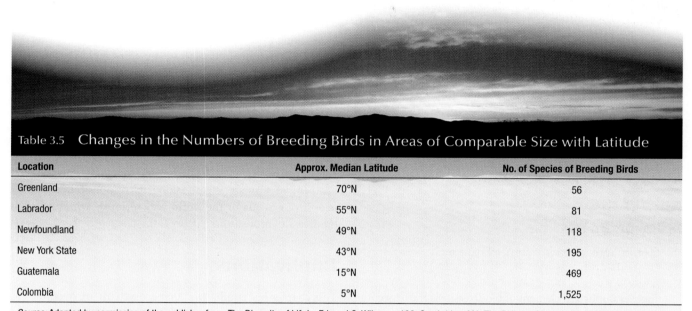

Table 3.5   Changes in the Numbers of Breeding Birds in Areas of Comparable Size with Latitude

Location	Approx. Median Latitude	No. of Species of Breeding Birds
Greenland	70°N	56
Labrador	55°N	81
Newfoundland	49°N	118
New York State	43°N	195
Guatemala	15°N	469
Colombia	5°N	1,525

Source: Adapted by permission of the publisher from *The Diversity of Life* by Edward O. Wilson, p.196, Cambridge, MA: The Belknap Press of Harvard University Press. Copyright © 1992 by Edward O. Wilson.

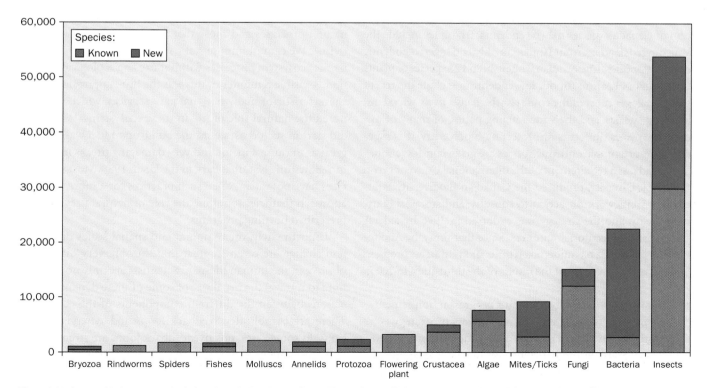

Figure 3.14 Groups with the most species in Canada (excluding viruses). Source: Figure, 'Group with the most species in Canada', from Mosquin et al., *Canadian Biodiversity: The Variety of Life, Its Status, Economic Benefits, Conservation Costs and Unmet Needs* (Ottawa: Canadian Museum of Nature, 1995), 58. Reprinted by permission of the publisher.

Jared Hobbs/Getty Images

The Vancouver Island marmot is Canada's only endangered endemic mammal species.

Another important element of biodiversity is the concept of endemism. **Endemic species** are ones that are found nowhere else on Earth. In Canada, we have relatively few endemic species compared, for example, to southern Africa, where some 80 per cent of the plants are endemic, or southwest Australia, where 68 per cent are endemic. It is estimated that between 1 and 5 per cent of Canada's species may be endemic. Examples include the Vancouver Island marmot (Canada's only endangered endemic mammal species), the Acadian whitefish, and 28 species of plants in the Yukon. Reasons for our low endemism include the recent glaciation over most of the country, which effectively wiped out localized species, and the wide-ranging nature of many of our existing species. In terms of protecting biodiversity, it is especially important that endemic species are given consideration.

Canada, as a requirement of the international Convention on Biological Diversity, produced the Canadian Biodiversity Strategy as a framework for action to conserve and sustainably use biodiversity, and the strategy was endorsed in 1996 by federal, provincial, and territorial ministers. Since that time, the Auditor General of Canada has undertaken three audits to assess progress. Following is a summary of the results of the third audit in 2005:

- Overall, the federal government's progress in implementing key aspects of the Canadian Biodiversity Strategy is *unsatisfactory*. Nearly 10 years after endorsement of the strategy, momentum has stalled on several fronts. Problems identified in our previous two audits of the strategy persist, and several commitments to deliver on biodiversity priority areas have not been met.

- The federal government still *lacks a coherent plan* for implementing the strategy, and while it has made implementation efforts, the government does not know how well these are meeting the strategy's goals.
- Commitments to improve Canada's capacity to understand its biodiversity and manage biodiversity information have *not been fulfilled*, despite numerous calls to improve these areas.
- There is no consolidated, comprehensive report on the state of biodiversity in Canada. While the government is working toward such reporting and reporting is taking place in various sectors, this is the third audit in which we have pointed out that *there is no overall picture of the state of biodiversity* in Canada or how it is changing over time. This lack of information limits the federal government's ability to help conserve biodiversity, use it sustainably, and reduce its loss (Auditor General of Canada 2005b).

Unfortunately, it is clear from the foregoing that the government of Canada is not taking its responsibilities to the people of Canada or the world very seriously with respect to biodiversity protection.

Implications

This chapter has emphasized that ecosystems are dynamic entities that change over time. Without such change, we would not have evolved, and the dinosaurs would not have become extinct. The main implication of this is that we should accept and try to understand the nature of these changes and be able to distinguish between those that are essentially the result of natural processes and those that are the result of human activities. We cannot impose static management regimes on dynamic ecosystem processes without causing ecological disruption. A visible reminder of this was the fire-suppression policy characteristic of many national park services, which often ignored the natural role of fire in these ecosystems. When fires did start in such ecosystems, the build-up of fuel was often so great as to cause a major and very damaging fire, as happened in Yellowstone National Park in the US in 1988. Most park services have now abandoned such practices for a more dynamic approach that tries to mimic the role of natural fires through prescribed burning programs.

Unfortunately, the temporal and spatial scales of ecosystem change are often so great that they are very difficult to observe in the human lifespan. Scientists are only now beginning to unravel the mysteries of some of these dynamic interactions between the different components of the ecosphere. There are complicated feedback loops and synergistic relations. In some cases, positive feedback loops are strengthened and accelerate undesirable changes that underlie some of the most serious environmental challenges facing humanity, such as global warming, a topic discussed in more detail in Chapter 7. When faced with such dynamic ecosystem changes, we

Box 3.14   What You Can Do

It may seem that with all the complexities of ecosystem change, there is little that an individual can do to influence the situation. However, this is far from the case, since many changes are brought about by individuals and it is the sum total of their actions that adds up to the tremendous changes described in this chapter. Specific actions that you can take include:

- Try to minimize your contribution to global warming. The many ways of doing this are outlined in Chapter 7.
- Try not to help speed up eutrophication processes by polluting waterways, using excessive fertilizer on your garden, and using phosphate-based detergents and other nutrient additives.

- Patronize organic farmers who do not use chemical additives.
- Take care not to introduce new species into the environment either deliberately by releasing them or inadvertently by, for example, transporting their seeds or fragments.
- Join a campaign to help eradicate an alien invasive species.
- Start your own campaign to eradicate an alien invasive species.
- Get to know the local flora and fauna in your area so that you can recognize alien invasive species.
- All change is not bad. Join a group that is trying to change an ecosystem for the better, sometimes called ecological restoration.

must use equally dynamic thinking to confront the challenges of the future.

The loss of biodiversity also has enormous implications, as discussed in Chapter 14. Before the rise of biodiversity as a concept, the human-induced extinctions of species were looked on as tragic, isolated events. However, biodiversity has helped us to reframe the problem and acknowledge the systematic nature of the process as well as the implications

for ecological processes overall as genetic, species, and landscape impoverishment occurs at ever-increasing rates. Such is the concern over biodiversity loss that an international treaty, the Convention on Biological Diversity, is now mobilizing global responses to the problem. However, the progress of the government of Canada on biodiversity protection has been slow. We discuss this in more detail in Chapter 14.

Summary

1. Ecosystems change over time. The speed of change varies from very slow over evolutionary time scales to rapid, caused by events such as landslides and volcanic eruptions.

2. Ecological succession occurs as a slow adaptive process involving the gradual replacement of one assemblage of species by another as conditions change over time. Primary succession occurs on surfaces that have not been previously vegetated, such as surfaces exposed by glacial retreat; secondary succession occurs on previously vegetated surfaces, such as abandoned fields. Fire is an important element in ecosystem change. Some ecosystems, such as much of the boreal forest, have evolved in conjunction with periodic fires. Fire suppression in such ecosystems can be detrimental to these natural processes.

3. Ecosystems tend towards a state of dynamic equilibrium in which the internal processes of an ecosystem adjust

for changes in external conditions. Not all ecosystems are equal in their ability to withstand perturbations. Inertia is the ability of an ecosystem to withstand change; resilience is the ability to recover to the original state following disturbance. Both contribute to the stability of the system.

4. Important causes of ecosystem change include the introduction of alien species and the removal of native keystone species.

5. Feedback mechanisms exist in ecosystems that may either exacerbate (positive feedback loops) or mitigate (negative feedback loops) change.

6. Population change occurs as a result of the balance between factors promoting growth (e.g., increase in birth rates or reduction in death rates) and those promoting reduction (e.g., declines in birth or survival rates or increase in death rates).

7. Different species have different reproductive strategies. K-strategists produce few offspring but devote considerable effort to ensuring that these offspring reach maturity. In comparison, r-strategists produce large numbers of young starting early in life and over a short time period and devote little or no energy to parental care.

8. Populations adapt to changing conditions over the long term through evolution. Evolution results in the formation of new species as a result of divergent natural selection responding to environmental change. This is speciation. Extinction results in the elimination of species that can no longer survive under new conditions.

9. Although evolution can take thousands of years, scientists now detect evolutionary changes on the scale of tens of years as a result of humans acting as predators on a massive scale.

10. Biodiversity involves the variety of life at three different scales: genetic, species, and landscape. Estimates suggest that Canada has more than 140,000 species, of which about half have been named.

11. The progress of the government of Canada in implementing its biodiversity strategy has been very slow.

Key Terms

albedo
alien or invader species
biodiversity
biotic potential
carrying capacity
climax community
coevolution
contemporary evolution
Convention on Biological
    Diversity

disturbance
dynamic equilibrium
ecological restoration
ecological succession
ecotones
endemic species
eutrophication
evolution
extinction
inertia

invasive
negative feedback
population
positive feedback loop
prescribed burning
primary succession
resilience
secondary succession
seed bank
seral

serotiny
speciation
species
species-area curves
synergism

Questions for Review and Critical Thinking

1. What are the different kinds of succession? Can you identify different seral stages in your area?

2. What is an edaphic climax? Can you find some local examples and identify the dominant limiting factor?

3. How does the concept of succession relate to environmental management?

4. How important was fire in the development of vegetation patterns in your region? Is there a fire management plan in your region? If so, what are its management goals?

5. Identify the main plant and animal species in your region that are not native. What effect are they having on the local ecosystems? What are the implications for management?

6. Can you think of any other examples of negative and positive feedback loops in the ecosphere besides those mentioned in the text?

7. Are K-strategists or r-strategists most vulnerable to environmental change?

8. How does genetic diversity help to protect a species from extinction?

9. What is coevolution? Can you think of any examples of coevolution among species in Canada?

10. What place in Canada has been called the most important scientific site in the world, and why?

11. What is endemism, and why does Canada have relatively few endemic species?

12. What progress is Canada making on implementing its biodiversity strategy?

Related Websites

Invasive species: www.invadingspecies.com; http://www.oag-bvg.gc.ca/internet/English/aud_ch_cesd_200803_06_e_30132.html#ch6hd3a; http://www.imo.org/ome.asp?topic_id=548; http://globallast.imo.org; http://cws-scf.ec.gc.ca/habitat/ramsar; http://www.iucn.org/biodiversityday/100booklet.pdf; http://www.invasivespecies.gov/geog/canada.shtml

Biodiversity Convention Office: http://www.bco.ec.gc.ca/en/default.cfm

Biodiversity in Canada: http://www.cbin.ec.gc.ca/default_e.cfm; http://canadianbiodiversity.mcgill.ca/english/species

Canadian Biodiversity Strategy: http://www.bco.ec.gc.ca/documents/CBS_E.pdf; http://www.oag-bvg.gc.ca/internet/English/aud_ch_cesd_200509_3_e_14950.html

IUCN Red List of Threatened Species: http://www.redlist.org

Wildlife and nature, Atlantic region: http://www.atl.ec.gc.ca/wildlife/index.html

Further Readings

Note: This list comprises works relevant to the subject of the chapter but not cited in the text. All cited works are listed in the section 'References' at the end of the book.

Balmford, A., R.E. Green, and M. Jenkins. 2003. 'Measuring the changing state of nature'. *Trends in Ecology and Evolution* 18 (7): 326–30.

Biodiversity Science Assessment Team. 1994. *Biodiversity in Canada: A Science Assessment for Environment Canada*. Ottawa: Environment Canada.

Dalpe, Y. 2003. 'Mycorrhizal fungi biodiversity in Canadian soils'. *Canadian Journal of Soil Science* 83 (3): 321–30.

Dearden, P. 1979. 'Some factors influencing the composition and location of plant communities on a serpentine bedrock in western Newfoundland'. *Journal of Biogeography* 6: 93–104.

———. 1983. 'An anatomy of a biological hazard: *Myriophyllum spicatum L.* in the Okanagan Valley, British Columbia'. *Journal of Environmental Management* 17: 47–61.

Donlan, C.J., and P.S. Martin. 2004. 'Role of ecological history in invasive species management and conservation'. *Conservation Biology* 18 (1): 267–9.

Gedalof, Z., and D.J. Smith. 2001. 'Dendroclimatic response of mountain hemlock (*Tsuga mertensiana*) in Pacific North America'. *Canadian Journal of Forest Research* 31: 322–32.

Gould, S.J. 2002. *I Have Landed: The End of the Beginning in Natural History*. New York: Harmony Books.

Grigorovich, I.A., et al. 2003. 'Ballast-mediated animal introductions in the Laurentian Great Lakes: Retrospective and prospective analyses'. *Canadian Journal of Fisheries and Aquatic Sciences* 60 (6): 740–56.

Jablonski, D. 2004. 'Extinction: Past and present'. *Nature* 427 (6,975): 589.

Jones, G.A., and G.H.R. Henry. 2003. 'Primary plant succession on recently deglaciated terrain in the Canadian High Arctic'. *Journal of Biogeography* 30 (2): 277–96.

Kerr, R.A. 2003. 'Mass extinction: Extinction by a whoosh, not a bang?' *Science* 302 (5,649): 1,315.

McNeely, J.A. 2001. *The Great Reshuffling: Human Dimensions of Invasive Alien Species*. Gland, Switzerland: World Conservation Union.

Mosquin, T., and P.G. Whiting. 1992. *Canada Country Study of Biodiversity*. Ottawa: Canadian Museum of Nature.

Neave, D., et al. 2002. *Canada's Forest Biodiversity: A Decade of Progress in Sustainable Management*. Ottawa: Canadian Forest Service, Natural Resources Canada.

Noonburg, E.G., B.J. Shuter, and P.A. Abrams. 2003. 'Indirect effects of zebra mussels (*Dreissena polymorpha*) on the planktonic food web'. *Canadian Journal of Fisheries and Aquatic Sciences* 60 (11): 1,353–68.

Perrings, C. 2002. 'Biological invasions in aquatic systems: The economic problem'. *Bulletin of Marine Science* 70 (2): 541–52.

Quammen, D. 1988. *The Flight of the Iguana: A Sidelong View of Science and Nature*. New York: Touchstone Books.

Chapter 4

Ecosystems and Matter Cycling

Learning Objectives

- To understand the nature of matter.
- To be able to describe why human intervention in biogeochemical cycles is a fundamental factor behind many environmental issues.
- To learn the main components and pathways of the phosphorus, nitrogen, sulphur, and carbon cycles.
- To be able to identify the main components of the hydrological cycle and the nature of human intervention in the cycle.
- To understand the causes, effects, and management approaches to eutrophication and acid deposition.

Introduction

The collapse in the Atlantic puffin population, described in Chapter 2, was a result of human interference with energy flow through the ecosystem. There are implications, however, for other aspects of ecosystem functioning. Puffins and most other seabirds play an important role in recycling nutrients, particularly phosphorus, from marine to terrestrial ecosystems. If these systems are disturbed, then the efficiency of the recycling mechanisms can be greatly reduced. Since the phosphorus cycle has very limited recycling capabilities from aquatic to terrestrial systems, the impact of interfering with it in this way could be substantial. This chapter explains how matter, such as phosphorus, cycles in the ecosphere and some of the implications of disturbing these cycles. The most critical environmental challenges facing the Earth, such as global warming, acid deposition, and the spread of dead zones in the ocean, result from cycle disturbance. Consequently, it is critical that you understand the nature of biogeochemical cycles if you are to fully appreciate the nature of these problems and their potential solutions.

Matter

Everything is either matter or energy. However, in contrast to the supply of energy, which is virtually infinite, the supply of matter on Earth is limited to that which we now have. **Matter,** unlike energy, has mass and takes up space. Matter is what things are made of and is composed of the 92 natural and 17 synthesized chemical elements such as carbon, oxygen, hydrogen, and calcium. *Atoms* are the smallest particles that still exhibit the characteristics of the element. Subatomic particles include *protons, neutrons,* and *electrons* that have different electrical charges. At a larger scale, the same kinds of atoms can join together to form molecules. When two different atoms come together, they are known as a **compound**. Water

Water is the only substance that occurs in all three phases of matter at the ambient temperatures of the Earth's surface.

(H_2O), for example, is a compound made up of two hydrogen atoms (H) and one oxygen atom (O). Four major kinds of organic compounds—carbohydrates, fats, proteins, and nucleic acids—make up living organisms.

Matter also exists in three different states (solid, liquid, and gas) and can be transformed from one to another by changes in heat and/or pressure. At the existing temperatures at the surface of the Earth, we have only one representative of the liquid state of matter, water. We can also readily see water in its other two states as ice (solid) or vapour (clouds).

Just as the laws of thermodynamics explain energy flow, the **law of conservation of matter** helps us to understand how matter is transformed. This law tells us that matter can neither be created nor destroyed but merely transformed from one form into another. Thus, matter cannot be consumed so that it no longer exists; it will always exist but in a changed form. When we throw something away, it is still with us, on this planet, as matter somewhere. There *is* no 'away'. All pollution stems from this law. The huge super-stacks on large smelters such as at Inco in Sudbury do not dispose of waste (see Chapter 12); they just disperse those wastes over a much larger area. The matter dispersed is the same and ultimately falls as acid deposition somewhere else. The same is true for all the wastes that we wash down our sinks. They do not disappear but collect in larger water bodies and create pollution problems.

Biogeochemical Cycles

For millions of years, matter has been moving among different components of the ecosphere. These cycles are as essential to life as the energy flow described in Chapter 2. About 30 of the naturally occurring elements are a necessary part of living things. These are known as **nutrients** and may be further classified into **macronutrients**, which are needed in relatively large amounts by all organisms, and **micronutrients**, required in lesser amounts by most species (Table 4.1). About 97 per cent of organic mass is composed of six nutrients: carbon, oxygen, hydrogen, nitrogen, phosphorus, and sulphur. These nutrients are cycled continuously among different components of the ecosphere in characteristic paths known as **biogeochemical cycles**.

We know from studies of chemistry that our bodies are reorganized star-dust, recycled again and again, so that, truly, our bones are of corals made.

—Rowe 1993

Figure 4.1 shows a generalized model of such a cycle. Like all the subsequent diagrams of cycles in this chapter, it exemplifies the types of simplifying models that scientists construct to try to represent the vast complexity of Earth processes, as described in Chapter 1. Nutrients can be stored in the different

According to the law of matter, emissions from stacks such as these do not simply disappear but end up somewhere else, often with undesirable consequences, such as acid deposition or global warming.

Slash-and-burn agriculture is a common way to transfer nutrients from the biomass to the soil to increase agricultural productivity. This photograph shows several swiddens (fields cut in the forest) by the hill tribe people of northern Thailand. The soils in the swiddens rapidly lose fertility caused by the burning of the biomass, and they are then abandoned for secondary succession to occur.

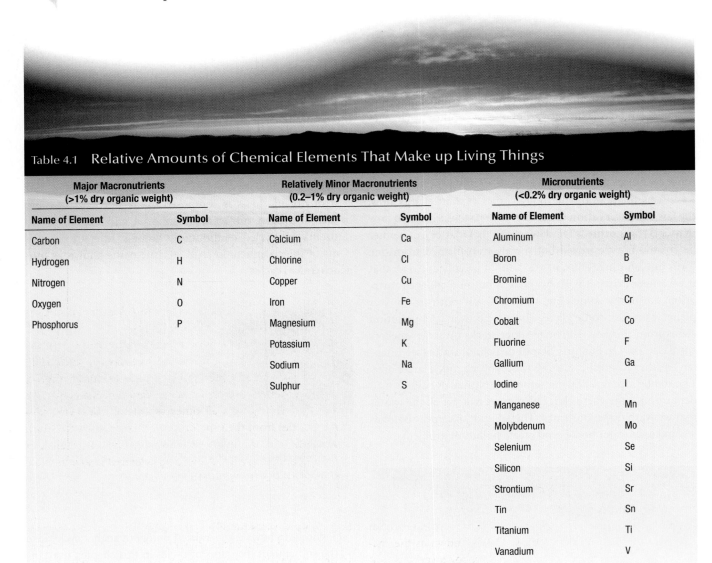

Table 4.1   Relative Amounts of Chemical Elements That Make up Living Things

Major Macronutrients (>1% dry organic weight)		Relatively Minor Macronutrients (0.2–1% dry organic weight)		Micronutrients (<0.2% dry organic weight)	
Name of Element	Symbol	Name of Element	Symbol	Name of Element	Symbol
Carbon	C	Calcium	Ca	Aluminum	Al
Hydrogen	H	Chlorine	Cl	Boron	B
Nitrogen	N	Copper	Cu	Bromine	Br
Oxygen	O	Iron	Fe	Chromium	Cr
Phosphorus	P	Magnesium	Mg	Cobalt	Co
		Potassium	K	Fluorine	F
		Sodium	Na	Gallium	Ga
		Sulphur	S	Iodine	I
				Manganese	Mn
				Molybdenum	Mo
				Selenium	Se
				Silicon	Si
				Strontium	Sr
				Tin	Sn
				Titanium	Ti
				Vanadium	V
				Zinc	Zn

Source: Kupchella and Hyland 1989.

compartments shown in Figure 4.1 for varying amounts of time. In general, there is a large, relatively slow-moving abiotic pool that may be in the atmosphere or lithosphere and is chemically unusable by the biotic part of the ecosystem or is physically remote. There is a more rapidly interacting exchange pool between the biotic and abiotic components. Nutrients move at various speeds from the biotic to the abiotic pools. For example, very rapid exchange takes place through respiration as carbon and oxygen move rapidly between the biotic and atmospheric components. The elements that now make up your body have undergone millions of years of recycling through these various compartments. You are a product of recycling!

Ecosystems also vary substantially in terms of the speed of cycling and the relative proportion of nutrients in each compartment. Some systems have nutrient-poor soils, for example, and have developed different mechanisms to store nutrients in other compartments. Tropical forest ecosystems are classic examples. Most of the nutrients are stored in the biomass as

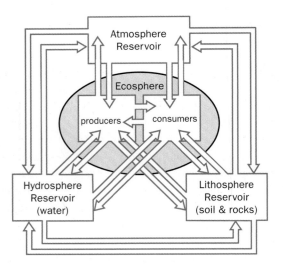

Figure 4.1 Generalized model of the biogeochemical cycle.

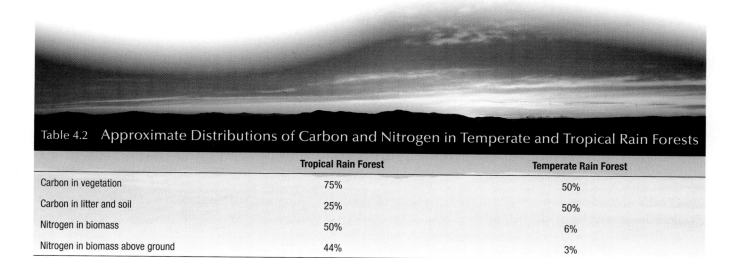

Table 4.2	Approximate Distributions of Carbon and Nitrogen in Temperate and Tropical Rain Forests	
	Tropical Rain Forest	**Temperate Rain Forest**
Carbon in vegetation	75%	50%
Carbon in litter and soil	25%	50%
Nitrogen in biomass	50%	6%
Nitrogen in biomass above ground	44%	3%

opposed to the soil system (Table 4.2). When leaves fall to the ground, they are rapidly mined for nutrients by plant roots before those nutrients have a chance to be leached out of the system. In contrast, many temperate forests have soils of high fertility. Removal of the nutrients in the biomass, through logging for example, does not remove as high a proportion of the site nutrient capital as removal in tropical ecosystems. This is discussed in more detail in Chapter 9.

Speed of cycling may also change within a cycle, depending on season. For the carbon cycle, for example, there is greater uptake of CO_2 in spring and summer as deciduous trees grow leaves. In fall, there is a correspondingly greater release as the leaves fall off and decompose (see Box 4.1). On average, a carbon dioxide molecule stays in the atmospheric component of the cycle from five to seven years. This is known as the *residence time*. It takes, on average, 300 years for a carbon molecule to pass through the lithosphere, cryosphere, atmosphere, hydrosphere, and biotic components of the carbon cycle. By way of contrast, it may take a water molecule two million years to make a complete cycle. The speed of cycling is influenced by such factors as the chemical reactivity of the substance. Carbon, for example, participates in many chemical reactions. It also occurs as a gas. In general, a gaseous phase allows for a speeding up of a cycle, because gas molecules move more quickly than molecules in the other states of matter.

Cycles can be classified according to the main source of their matter. **Gaseous cycles**, as the name would suggest, have most of their matter in the atmosphere. The nitrogen cycle is a good example. **Sedimentary cycles**, such as the phosphorus and sulphur cycles, hold most of their matter in the lithosphere. In general, elements in sedimentary cycles tend to cycle more slowly than those in gaseous cycles, and the elements may be locked into geological formations for millions of years.

Under natural conditions, recycling rates between components achieve a balance over time in which inputs and outputs are equal. Human activity serves to change the speed of transference between the different components of the cycles. Many of our pollution problems result from a human-induced build-up in one or more components of a cycle that cannot be effectively dissipated by natural processes.

In addition to the biogeochemical cycles, some attention will be given in this chapter to the hydrological cycle. This cycle is critical to all other cycles, since water plays a major role in the mobilization and transportation of materials. The energy for this, as with all other aspects of the cycles, ultimately comes from the sun. Photosynthesis powers the biotic aspects of the cycles, and atmospheric circulation, fuelled by the sun's energy, controls the water power that is so important for weathering and erosional processes.

Sedimentary Cycles

Sedimentary cycles mobilize materials from the lithosphere to the hydrosphere and back to the lithosphere. Some, such as sulphur, may involve a gaseous phase, while others, such as phosphorus, do not. These cycles rely essentially on geological uplift over long periods to complete the cycle. Lack of a gaseous phase means that the cycle is missing one potential route for more rapid recycling, which can lead to problems when mobilization rates are increased through human activity. Phosphorus and sulphur will be discussed here, but other elements, such as calcium, magnesium, and potassium, follow similar pathways.

Phosphorus (P)

Phosphorus, a macronutrient incorporated into many organic molecules, is essential for metabolic energy use. It is relatively rare on the Earth's surface in relation to biological demand, so it is essential that phosphorus cycles efficiently between components. Many organisms have devised means of storing this element preferentially in their tissues, and phosphorus moves very readily within plants from older tissues to more active growth sites. Deciduous trees may recirculate up to 30 per cent of their phosphorus back to their more permanent components before the leaves fall in an effort to preserve the nutrient.

Under natural circumstances, phosphorus is a prime example of a nutrient held in a tight circulation pattern between the biotic and abiotic components. Replenishment rates through

Box 4.1   The Decomposers

In Chapter 2, attention was drawn to the importance of decomposer organisms and detritus food chains. These are the main means by which nutrients in the biotic component of the ecosphere are returned to the abiotic so that plants can once again use them. Photosynthesis has been described as the process of making a complicated product out of simple components; decomposition is the reverse process of making simple components out of that complicated product.

Decomposer organisms such as fungi may attack leaves while still on the plant; the fungi release products such as sugars, which are then washed to the ground by rainfall. Once leaves fall to the ground, they are broken down progressively by various groups of organisms. Larger organisms such as earthworms, slugs, snails, beetles, ants, and termites help to break up the leaf material initially. Many gardeners are fully aware of the ability of slugs, for example, to devour green leaves in great quantities.

Fungi and heterotrophic bacteria further break down the organic matter, releasing more resistant carbohydrates, followed by cellulose and lignin. The humus—the organic layer in the soil—is composed mainly of products that can resist rapid breakdown. A chemical process, oxidation, is mainly responsible for the decay of this material.

As everyone who has witnessed leaf decay in autumn knows, the process can occur quite rapidly. The speed varies depending on the environment. Warm environments tend to promote more rapid microbial activity. Leaf decay in the tropics takes place in a matter of weeks. In the boreal forest, however, where conditions are cold and the leaves, such as spruce and pine needles, quite resistant, recycling of the nutrients held in the leaves may take decades. Overall, the average recycling time for organic material in the wet tropics is five months; in the boreal

forest, it is 350 years. The high amounts of lignin found in leafs of needle-leafed trees help to protect the trees against freezing conditions but offer little food value to decomposers, so decay is slow. In comparison, deciduous trees, such as maple, have a high reward for decomposers: high nitrogen levels and little protective lignin, so they decay very quickly. However, researchers are now discovering that the slow breakdown of organic matter in the boreal forest also plays a major role in global carbon storage. As the forests are removed through logging and other activities, not only does the supply of leaves disappear, but the built-up carbon is released either through burning associated with these activities or the higher ground temperatures resulting from the reduced amount of shade.

Slugs play an important role in breaking down vegetable matter. In the wet west coast forests the biomass of slugs is greater than that of any other animal in the ecosystem.

weathering and soil availability are limited; thus, the amount retained by the biomass is quite critical. The residence time of phosphorus in terrestrial systems can be up to 100 years before it is leached into the hydrosphere. Phosphorus is often the dominant limiting factor (Chapter 2) in freshwater aquatic systems and for plant growth in terrestrial soils. Agricultural productivity relies heavily on augmenting this supply (auxiliary energy flow) through fertilizer application. Gruber and Galloway (2008) suggest that both the nitrogen and carbon cycles in the ocean are ultimately controlled by phosphorus and, since these cycles are key to global warming response, the phosphorus cycle will be a main determinant of global futures.

Box 4.2 outlines the impact humans have on the phosphorus cycle.

Box 4.2   Human Impacts on the Phosphorus Cycle

Humans intervene in the phosphorus cycle in several ways that serve to accelerate the mobilization rate:

- mining of phosphate-rich rocks for fertilizer and detergent production, creating excessive runoff into aquatic environments;
- biomass removal, leading to accelerated erosion of sediment and solutes into streams;
- concentration of large numbers of organisms such as humans, cattle, and pigs, creating heavy burdens of phosphate-rich waste materials;

- removal of phosphorus from oceanic ecosystems through fishing, with the phosphorus returned to fresh water again and ultimately the marine system through the dissolution of wastes.

The major implication of all these interventions is for excessive phosphorus accumulation in freshwater systems, resulting in eutrophication. Human activity is now estimated to account for about two-thirds of the phosphorus reaching the oceans. The environmental impact of this nutrient enrichment will be discussed in more detail later.

The availability of phosphorus in the soil is influenced by soil acidity. Acidity is measured on the pH scale, which is discussed in more detail later in this chapter. Below pH 5.5, for example, phosphorus reacts with aluminum and iron to form insoluble compounds. Above pH 7, the same thing happens in combination with calcium. Obviously, things that change soil pH, such as acid precipitation, can have a critical effect on phosphorus availability. This is an example of the kind of synergistic reaction discussed in Chapter 3 in which the combination of either high or low pH values with low phosphorus availability, as a result of chemical reactions, can have a stronger effect than the sum of the two individually.

Philip Dearden

The Head-Smashed-In World Heritage Site in southern Alberta is rich in phosphorus as Native peoples used the cliff to kill stampeding bison.

Rocks in the Earth's crust are the main reservoir of phosphorus (Figure 4.2). Geological uplift and subsequent weathering (Box 4.3) make phosphorus available in the soil, where it is taken up by plant roots. Phosphate ions are the main source of phosphorus for plants and are released from slowly dissolving minerals such as iron, calcium, and magnesium phosphates. Many higher plants have a mutualistic relationship with soil fungi, or mycorrhizae, which helps them to gain improved access to phosphorus in the soil. Once incorporated into plant material, the phosphorus may be passed on to other organisms at higher trophic levels.

Animal wastes are a significant source of phosphorus return to the soil. All organisms eventually die, and the organic material is broken down by the decomposer food chains. This may take some time, since a considerable amount of the phosphorus is within animal bones. In the past, farmers have used concentrated sources of animal bones, such as the bison jumps used by indigenous peoples on the prairies, as a source of phosphate fertilizer.

Following breakdown in the soil, the phosphorus is then either taken up again by plants or removed by water transport. Bacteria mineralize the returned organic phosphorus into inorganic forms so that plants can take it up once more. Most of the water transport occurs in particulate form by streams, which is one reason to be concerned about excessive sedimentation occurring through land-use activities such as logging and agriculture, as described in Chapters 9 and 10.

Stream transport ultimately ends up in the ocean. Estuaries have such high productivities, as discussed in Chapter 2, in part because of this nutrient input from upstream. The circulation patterns within estuaries tend to trap nutrients, but some

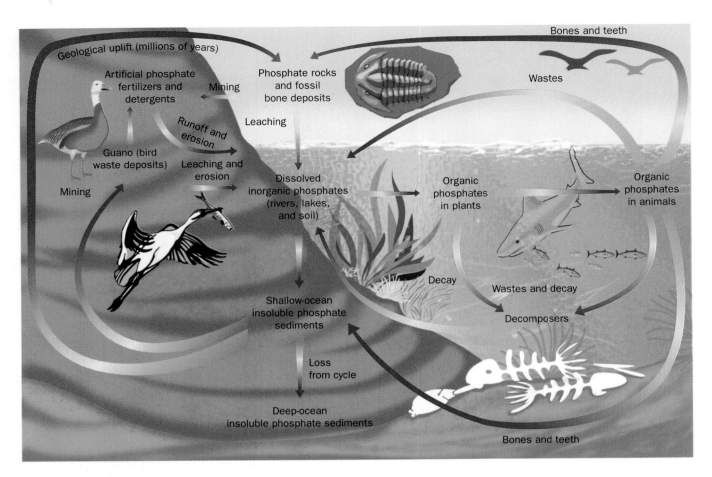

Figure 4.2 The phosphorus cycle.

phosphorus finds its way into the shallow ocean areas of the coastal zone. It may be fixed in biomass by phytoplankton or other aquatic plants in the euphotic (eu = well, photos = light) zone and once again incorporated into the food chain. The coastal zones, with this plentiful supply of nutrients and photosynthetic energy from the sun, cover less than 10 per cent of the ocean's surface but account for more than 90 per cent of all ocean species.

Beyond the coastal zone and the continental shelves, water depth increases into the open ocean. Phosphorus and other nutrients that have not been incorporated into food chains, plus elements from the death of oceanic organisms, filter through to the bathyal and ultimately the abyssal zones (see Figure 2.5). Here, uptake by organisms is extremely limited, and the nutrients must either be moved back to the euphotic zone by upwelling currents or wait to be geologically uplifted over millions of years to move into another component of the cycle. Where such upwellings occur, such as off the west coasts of Africa and South America, plentiful fisheries are found because of the combination of high nutrient and energy levels. Some fish species, such as salmon, are **anadromous**, spending part of their lives in salt water and part in fresh water, where they die after spawning. When they die, the nutrients they have collected during their ocean phase are returned to the freshwater system, resulting in a significant input of nutrients, including phosphates (Chapter 8).

Two other recycling mechanisms also occur. One is the biotic one described earlier as marine birds, such as puffins, cormorants, and other fish-eating birds, return phosphorus to land in the form of their droppings, representing the phosphorus that has concentrated through the marine food chain. This phosphorus, known as guano, constitutes the largest source of phosphorus for human use and is heavily mined for fertilizer production. A small amount of phosphorus is also returned to land through the atmosphere as sea spray.

The nutrients that have sustained this salmon are now being recycled.

Box 4.3   Weathering, the Rock Cycle, and Plant Uptake

The weathering of the rocks of the Earth's crust plays an important role in supplying long-term inputs to biogeochemical cycles. Weathering is part of the **rock cycle** whereby rocks that have been uplifted are eroded into different constituents. The rock cycle involves the transformation of rocks from one type to another, such as when volcanic rocks are eroded and washed into the ocean. Over millions of years, the resulting sediments are turned into sedimentary rocks. In turn, these sedimentary rocks may be compressed within the Earth's crust and altered by heat and pressure before once more being uplifted through the process of continental drift.

Weathering involves numerous different processes. In Canada, mechanical weathering involves the physical break-up of rocks as a result of changing temperatures. The action of water is important. Chemical processes, such as hydration and carbonation, further the process by removing elements in solution. Secondary clay minerals are produced from primary rock minerals by hydrolysis and oxidation. These clays are very important in terms of holding the nutrients in the soil. The soil can be thought of as a giant filter bed in which each particle is chemically active. As water percolates through, containing many different nutrients in solution, some of these nutrients are held by the clays and become available for plant uptake.

Plants constantly lose moisture from their leaves. This creates a moisture gradient within the plant that serves to draw water up to replace what has been lost. Water moves from the roots, and more nutrient-laden water is taken in by the roots. It is the job of the roots to keep the plant supplied with water. As nutrients are removed from the soil water around the plants, new nutrients move within the soil water to replace them.

These sedimentary rocks have been compressed and folded as part of the rock cycle.

David Parker/Science Photo Library

Sulphur (S)

Like phosphorus, sulphur is a sedimentary cycle, but it differs from phosphorus in two important ways. First, it has an atmospheric component and therefore better recycling potential. Sulphur is not often a limiting factor for growth in aquatic or terrestrial ecosystems. Second, like most of the other cycles but unlike phosphorus, it has strong dependencies on microbial activity. Sulphur is a necessary component for all life and a building component of proteins.

Sulphur is not available in the lithosphere and must be transformed into sulphates to be absorbed by plants. Bacteria are critical here, changing sulphur into various forms in the soil (Figure 4.3). The exact form depends on factors such as the presence (**aerobic**) or absence (**anaerobic**) of oxygen, which is usually a reflection of the relationship of the particular site of transformation to the water table and

the presence of other elements such as iron. From these microbial transformations by chemoautotrophs (discussed in Chapter 2), gases such as hydrogen sulphide (H_2S) may be released directly into the atmosphere (giving the familiar 'rotten egg' smell we associate with marshlands), or sulphate salts (SO_4) are produced. Through their roots, plants can then absorb the sulphates, sulphur enters the food chain, and the same processes occur as in the biotic components of the other cycles.

The complexity of these cycles is illustrated further by some of the interactions that occur between cycles. For example, the phosphorus cycle benefits when iron sulphides are formed in sediments and phosphorus is converted from insoluble to soluble forms, where it becomes available for uptake.

The upward movement of the gaseous phase of the sulphur cycle is also important, since significant quantities of sulphur

Box 4.4   Human Impacts on the Sulphur Cycle

Humans intervene in the sulphur cycle mainly through:

- the burning of sulphur-containing coal, largely to produce electricity;
- the smelting of metal ores that contain sulphates.

Almost 99 per cent of the sulphur dioxide and about one-third of the sulphur compounds reaching the atmosphere come from these activities. These sulphur compounds react with oxygen and water vapour to produce sulphuric acid (H_2SO_4), a main component of acid deposition, as discussed later in this chapter.

---

are returned to the atmosphere, thereby shortening the long sediment uplift time that characterizes the phosphorus cycle. This is fortunate, because average ocean residence times are quite long and sulphur is continually lost to the ocean floor. From the upper reaches of the oceans, sulphur can be returned to the atmosphere by phytoplankton or photochemical reactions. However, unlike phosphorus, a relatively small proportion of sulphur is fixed in organic matter, and availability is not usually a problem. As with phosphorus, human intervention in the sulphur cycle (Box 4.4) is significant.

Gaseous Cycles

Nitrogen (N)

Nitrogen is a colourless, tasteless, odourless gas required by all organisms for life. It is an essential component of chlorophyll, proteins, and amino acids. The atmosphere is more than 78 per cent nitrogen gas (N_2) and also contains other forms of gaseous nitrogen such as ammonia (NH_3), nitrogen dioxide (NO_2), nitrous oxide (N_2O), and nitric oxide (NO). Excess quantities of these other forms are involved in many of our most challenging environmental problems, such as acid deposition, ozone depletion, and global climate change (Box 4.5).

Most organisms cannot gain access to the necessary nitrogen from the atmosphere. The nitrogen is instead obtained from the soil as nitrates. The main way in which the atmospheric reservoir is linked to the biotic components of the food chain is through **nitrogen fixation** and **denitrification**, both mediated through microbial activity (Figure 4.4). The historical record shows a close coupling between the speed of these processes and atmospheric CO_2 levels.

Biological nitrogen fixation occurs as bacteria transform atmospheric nitrogen (N_2) into ammonia (NH_3). The most important fixers are bacteria of the *Rhizobium* family that grow on the root nodules of certain plants, such as members of the pea

or legume family (e.g., peas, beans, clover, alfalfa). The bacteria and roots of the plant communicate through chemical stimuli that result in the bacteria infecting root cells. Once infected, the cells swell into the nodules that you can see on the roots of the peas or beans in your garden. In a remarkable example of coevolution, the plant supplies the products of photosynthesis to the relationship, and the bacteria transform the atmospheric nitrogen into nitrates. It is one of the few known examples of two organisms cooperating to make one molecule.

The resulting enrichment explains why farmers grow crops such as alfalfa and clover as part of a crop rotation to help build up nitrates in the soil. About one-half of the nitrogen circulating in agricultural ecosystems comes from this source. The increasing cost of fertilizer worldwide has focused more attention on biological nitrogen fixation (BNF) as a part of meeting the global food challenges of the future (see Chapter 10). Through genetic engineering, for example, it may be possible to inject other crops, such as cereals, with similar symbiotic unions between plants and nitrogen-fixing bacteria. However, it may not be that simple. For example, research indicates that species involved in dinitrogen fixation may also be particularly susceptible to phosphorus deficiencies, given their high P and energy requirements.

Some wild species such as alder, lupines, and vetch have similar bacteria and hence play a valuable role when they act as primary colonizers in the successional process or when they help to recolonize sites that have been logged (Chapter 9) or otherwise disturbed. These relationships are mutualistic in that both organisms gain. The plant receives enhanced nutrient supply, and the organisms find a home in which the plant supplies them with various sugars. One of the most celebrated examples of such a partnership is that between the fungi and algae in lichens, discussed in more detail in Chapter 2.

Other bacteria and algae that fix nitrogen are not attached to specific plants. In the Arctic, where nitrogen is a limiting

Box 4.4   Human Impacts on the Sulphur Cycle

Humans intervene in the sulphur cycle mainly through:

- the burning of sulphur-containing coal, largely to produce electricity;
- the smelting of metal ores that contain sulphates.

Almost 99 per cent of the sulphur dioxide and about one-third of the sulphur compounds reaching the atmosphere come from these activities. These sulphur compounds react with oxygen and water vapour to produce sulphuric acid (H_2SO_4), a main component of acid deposition, as discussed later in this chapter.

---

are returned to the atmosphere, thereby shortening the long sediment uplift time that characterizes the phosphorus cycle. This is fortunate, because average ocean residence times are quite long and sulphur is continually lost to the ocean floor. From the upper reaches of the oceans, sulphur can be returned to the atmosphere by phytoplankton or photochemical reactions. However, unlike phosphorus, a relatively small proportion of sulphur is fixed in organic matter, and availability is not usually a problem. As with phosphorus, human intervention in the sulphur cycle (Box 4.4) is significant.

Gaseous Cycles

Nitrogen (N)

Nitrogen is a colourless, tasteless, odourless gas required by all organisms for life. It is an essential component of chlorophyll, proteins, and amino acids. The atmosphere is more than 78 per cent nitrogen gas (N_2) and also contains other forms of gaseous nitrogen such as ammonia (NH_3), nitrogen dioxide (NO_2), nitrous oxide (N_2O), and nitric oxide (NO). Excess quantities of these other forms are involved in many of our most challenging environmental problems, such as acid deposition, ozone depletion, and global climate change (Box 4.5).

Most organisms cannot gain access to the necessary nitrogen from the atmosphere. The nitrogen is instead obtained from the soil as nitrates. The main way in which the atmospheric reservoir is linked to the biotic components of the food chain is through **nitrogen fixation** and **denitrification**, both mediated through microbial activity (Figure 4.4). The historical record shows a close coupling between the speed of these processes and atmospheric CO_2 levels.

Biological nitrogen fixation occurs as bacteria transform atmospheric nitrogen (N_2) into ammonia (NH_3). The most important fixers are bacteria of the *Rhizobium* family that grow on the root nodules of certain plants, such as members of the pea or legume family (e.g., peas, beans, clover, alfalfa). The bacteria and roots of the plant communicate through chemical stimuli that result in the bacteria infecting root cells. Once infected, the cells swell into the nodules that you can see on the roots of the peas or beans in your garden. In a remarkable example of coevolution, the plant supplies the products of photosynthesis to the relationship, and the bacteria transform the atmospheric nitrogen into nitrates. It is one of the few known examples of two organisms cooperating to make one molecule.

The resulting enrichment explains why farmers grow crops such as alfalfa and clover as part of a crop rotation to help build up nitrates in the soil. About one-half of the nitrogen circulating in agricultural ecosystems comes from this source. The increasing cost of fertilizer worldwide has focused more attention on biological nitrogen fixation (BNF) as a part of meeting the global food challenges of the future (see Chapter 10). Through genetic engineering, for example, it may be possible to inject other crops, such as cereals, with similar symbiotic unions between plants and nitrogen-fixing bacteria. However, it may not be that simple. For example, research indicates that species involved in dinitrogen fixation may also be particularly susceptible to phosphorus deficiencies, given their high P and energy requirements.

Some wild species such as alder, lupines, and vetch have similar bacteria and hence play a valuable role when they act as primary colonizers in the successional process or when they help to recolonize sites that have been logged (Chapter 9) or otherwise disturbed. These relationships are mutualistic in that both organisms gain. The plant receives enhanced nutrient supply, and the organisms find a home in which the plant supplies them with various sugars. One of the most celebrated examples of such a partnership is that between the fungi and algae in lichens, discussed in more detail in Chapter 2.

Other bacteria and algae that fix nitrogen are not attached to specific plants. In the Arctic, where nitrogen is a limiting

Humans disrupt the nitrogen cycle in many ways:

- Chemical fixation (the Haber-Bosch process) to supply nitrates and ammonia as fertilizer. The amount fixed currently is currently estimated to be about equal to that produced by natural processes. However, the Millennium Ecosystem Assessment (2003) predicts large increases in the future (Figure 4.5), and Gruber and Galloway (2008) predict that humans will double the turnover rates of the nitrogen cycle of the entire Earth. The main impacts are through runoff of excess fertilizer (contributing to eutrophication) and denitrification (contributing to climatic change). Agricultural fertilizers are implicated in both. Eutrophication will be discussed in more detail in the next section. Denitrification transforms nitrogen fertilizers into nitrous oxide, a greenhouse gas (Chapter 7), and is also involved in the catalytic destruction of the ozone layer. There are also health concerns related to excessive nitrate levels from fertilizers running into water supplies (Box 4.8).
- Removal of nitrate and ammonium ions from agricultural soils through the harvesting of nitrogen-rich crops.
- High-temperature combustion, which produces nitric oxides (NO) that combine with oxygen to produce nitrogen dioxide (NO_2), which reacts with water vapour to form nitric acid (HNO_3), a main component of acid deposition.

factor on plant growth, an important source of nitrogen fixation is the cyanobacteria, or blue-green algae, such as *Nostoc commune*. Sometimes, combinations of various cyanobacteria and other organisms form crusts over the soil surface, known as cryptogamic crusts, which are important in facilitating successional processes. Other important free-living nitrogen-fixing micro-organisms are found in the ocean. Estimates suggest that in terrestrial ecosystems, about twice the amount of nitrogen is fixed by mutualistic relationships as by these free-floating relationships. The major supply, however, comes from the breakdown of existing biomass by decomposer food chains. Nitrogen is tightly circulated in most ecosystems between the dead and living biomass.

In addition to these biological mechanisms, some nitrogen is also fixed through atmospheric fixation that occurs largely during thunderstorms. Lightning causes extremely high temperatures that unite oxygen and nitrogen to form nitric acid (HNO_3), which is subsequently carried to earth as precipitation and converted into nitrates (NO_3). These nitrates can then be taken up by plant roots. Estimates on the importance of atmospheric fixation vary, but 10 per cent of total fixation would be a maximum figure, and most estimates place it at about half this percentage.

Once fixed in the soil, ammonia and nitrates are incorporated into plant matter and then through the food chain. When the plants die, the biomass is converted back to ammonia (NH_3) and ammonium salts (NH_4) by bacterial action and returned to the soil. This is known as **mineralization**. Some ammonia and ammonium is changed to nitrites and then to nitrates by chemotrophic bacteria such as *Nitrosomonas* and *Nitrobacter* in a process known as nitrification. Other bacteria—anaerobic bacteria—through the denitrification process convert nitrates into nitrogen gas and release it into the atmosphere (Figure 4.4). Denitrification occurs in anaerobic conditions, especially where there are large amounts of nitrates available, such as on flooded agricultural fields.

Nitrates are highly soluble in water and if not held tightly may be lost to the ecosystem by surface runoff. Ammonia is also susceptible to loss by soil erosion, since it tends to adhere to soil particles. Like phosphorus, nitrogen is often a limiting factor for growth. When excessive concentrations occur in water, it is a major contributor to the process of eutrophication. Unlike phosphorus, however, nitrogen is not immobilized in deep-ocean sediments but has an effective feedback mechanism to the atmosphere from the ocean through microbial denitrification.

Scientists are trying to understand the relationship between the nitrogen cycle and the major elements of global climate change, such as the carbon cycle. They know for sure, based on historical records, that the cycles are closely linked but are still unsure as to the direction and magnitude of changes that might be expected in the future and whether the nitrogen cycle will form a positive or negative feedback loop with rising atmospheric carbon levels (Figure 4.6). On the one hand, ocean acidification resulting from the ocean's taking up anthropogenic

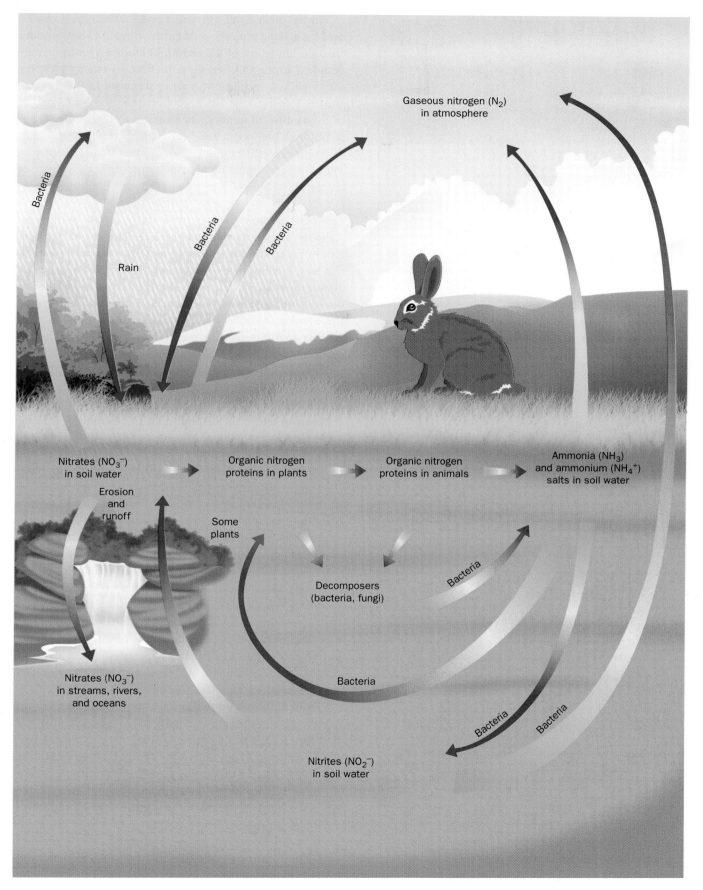

Figure 4.4 The nitrogen cycle.

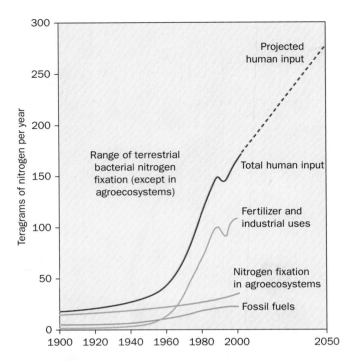

Figure 4.5 Global trends in the creation of reactive nitrogen on Earth by human activities, with projections to 2050. Source: Millennium Ecosystem Assessment 2005.

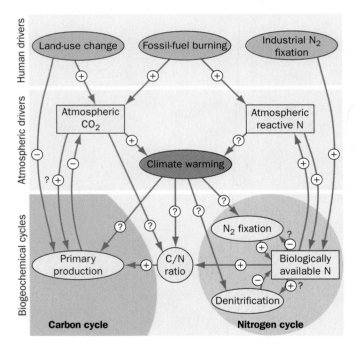

Figure 4.6 Nitrogen-carbon-climate interactions. Source: Reprinted by permission from Macmillan Publishers, Ltd: N. Gruber and J.N. Galloway. 2008. 'An Earth-system perspective of the global nitrogen cycle.' *Nature* 451: 293–6. © 2008 Nature Publishing Group.

CO_2 might lead to an increase in the C/N uptake ratio of marine phytoplankton and enhanced nitrogen fixation. If this happens, the marine biosphere would act as a negative feedback for climate change, since the resulting enhanced fixation of carbon would draw additional carbon from the atmosphere, thus reducing the accumulation of anthropogenic CO_2 in the atmosphere.

On the other hand, current climate–carbon cycle models used for making projections of Earth's climate do not consider nitrogen limitation of the terrestrial biosphere but generally assume a strong CO_2 fertilization effect. In other words, the additional CO_2 available will act to stimulate further organic growth. However, the latter would require the availability of large quantities of nitrates, since nitrates are a necessary ingredient for all life, as discussed earlier. If they are not available, then nitrogen will become a major limiting factor (Chapter 3). Thus, nitrogen limitation will significantly affect the ability of the terrestrial biosphere to act as a CO_2 sink in the future.

Carbon (C)

Although carbon dioxide gas (CO_2) constitutes only 0.03 per cent of the atmosphere, it is the main reservoir for the carbon that is the building block for all necessary fats, proteins, and carbohydrates that constitute life. Plants take up carbon dioxide directly from the atmosphere through the process of photosynthesis and at the same time emit oxygen. The carbon becomes incorporated into the biomass and is passed along the food chain. Residence times can vary greatly, but older forests constitute a significant repository for carbon for centuries. Respiration by organisms transforms some of this carbon back into carbon dioxide (Figure 4.7), and the cellular respiration of decomposers helps to return the carbon from dead organisms into the atmosphere. Most of this is in the form of CO_2 but also methane (CH_4) in anaerobic conditions. Thus, the cycling of carbon and the flow of energy through food chains are intimately related.

Besides this relatively rapid exchange, some carbon can also be stored in the lithosphere for extended periods as organisms become buried before they decompose. This is particularly true under relatively inefficient anaerobic decay conditions such as in peat bogs. Through geological time, millions of years of photosynthetic energy have been transformed into 'fossil fuels' by this process as a result of heat and compression. The highly productive forests and marine environments of the distant past have become the coal, oil, and natural gas that fuel the world's economy today (see Box 4.6). Scientists predict that there will be a positive feedback loop between increased atmospheric CO_2 and the terrestrial elements of the cycle that will serve to further augment atmospheric CO_2 (Heimann and Reichstein 2008).

Some of the carbon dioxide is dissolved into the shallower ocean before re-entering the atmosphere. Residence time is in the order of six years in these shallower waters but much longer (up to 350 years) when mixed with deeper waters. These residence times are now of considerable scientific interest because of the rising levels of carbon dioxide in the atmosphere and the potential for the oceans to absorb these increases (see Chapter 8).

Large amounts of carbon are stored for much longer periods in the ocean. When marine organisms die, their shells of calcium carbonate ($CaCO_3$) become cemented together to form rocks such as limestone. Over millions of years, the limestone may be uplifted to become land and then is slowly weathered to release the carbon back into the carbon cycle.

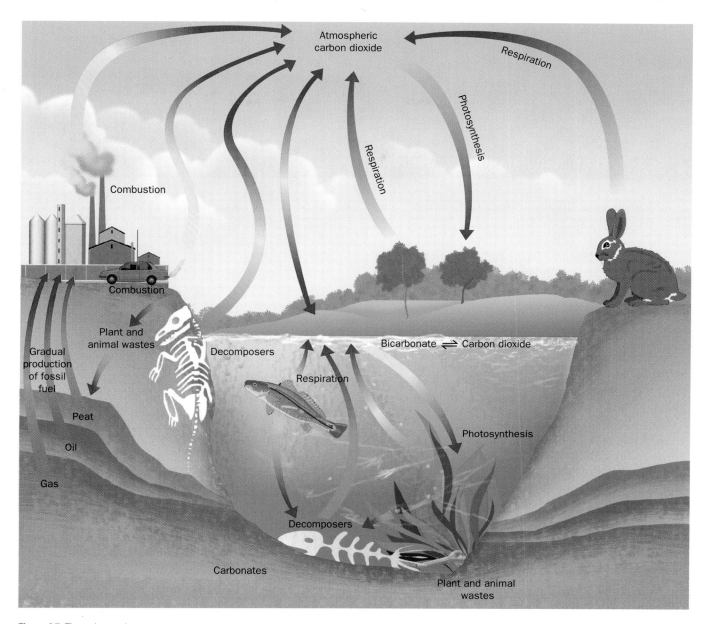

Figure 4.7 The carbon cycle.

Large amounts of carbon are stored in the lithosphere, such as in these coal-beds, the product of millions of years of photosynthetic activity.

Coral reefs, such as here in the Maldives, store large amounts of carbon from the remains of thousands of years of coral growth.

Box 4.6   Human Impacts on the Carbon Cycle

As human populations have increased, two major changes to the carbon cycle have occurred:

- Natural vegetation, usually dominated by tree growth, has been replaced by land uses, such as urban and agricultural systems, that have reduced capacity to uptake and store carbon.
- For the past 200 years or so, human activity, particularly industrial activity, has mobilized large amounts of fossil fuels from the lithospheric component of

the cycle to the atmospheric component. Estimates suggest that this mobilization represents the release of one million years of photosynthetic activity every year. The concentration of carbon dioxide in Earth's atmosphere now exceeds 380 parts per million (ppm), which is more than 80 ppm above the maximum values of the past 740,000 years, if not 20 million years. Furthermore, increasing CO_2 has driven an increase in the global oceans' average temperature by 0.74°C and sea level by 17 centimetres.

The Hydrological Cycle

Water, like the nutrients discussed above, is necessary for all life. You are 70 per cent water. Although other planets such as Venus and Mars have water, only on Earth does it occur as a liquid. It also occurs in a fixed supply that cycles between various reservoirs driven by energy from the sun. By far the largest reservoir is the ocean, containing more than 97 per cent of the water on Earth. Most of the rest is tied up in the polar ice caps,

with only a small amount readily available as the fresh water that sustains terrestrial life (Table 4.3). Water travels ceaselessly between these various reservoirs through the main processes of evaporation and precipitation known as the **hydrological cycle** (Figure 4.8).

Average residence times in the reservoirs vary greatly (Table 4.3). In the deep ocean, it may take 37,000 years before water is recycled through evaporation into the atmosphere, whereas once in the atmosphere, average residence time is in

Table 4.3   Global Water Storage

Reservoir	Average Renewal Rate	Per Cent of Global Total
World oceans	3,100 years	97.2
Ice sheets and glaciers	16,000 years	2.15
Groundwater	300–4,600 years	0.62
Lakes (freshwater)	10–100 years	0.009
Inland seas, saline lakes	10–100 years	0.008
Soil moisture	280 days	0.005
Atmosphere	9–12 days	0.001
Rivers and streams	12–20 days	0.0001

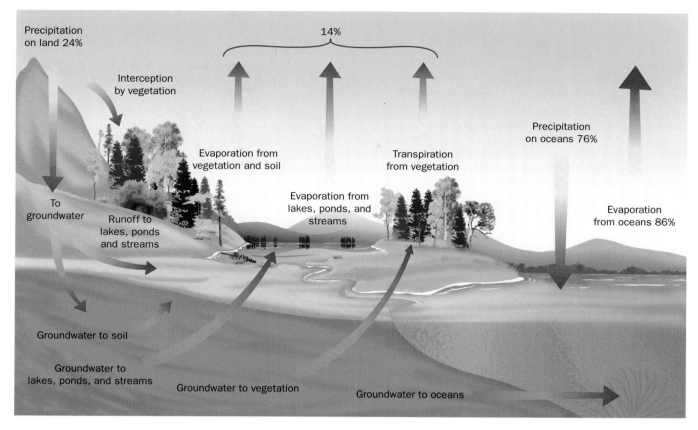

Figure 4.8 The hydrological cycle.

the order of nine to 12 days. These figures have special relevance with regard to the effects of pollution. Although many major rivers have suffered from critical pollution incidents, the flushing action of rivers, combined with the short residence time of the water, means that a relatively rapid recovery is often possible. This is not the case, however, with groundwater pollution, especially deep groundwater pollution.

The hydrological cycle involves the transport of water from the oceans to the atmosphere, through terrestrial and subterranean systems, and back to the ocean, all fuelled by energy from the sun. Eighty-four per cent of the water in the atmosphere is evaporated directly from the ocean surface. The remainder comes from evaporation from smaller water bodies, from the leaves of plants (**transpiration**), or from the soil and plants (**evapotranspiration**). As it evaporates, water leaves behind accumulated impurities. The most common dissolved substance in the ocean is sodium chloride, or table salt, which also contains many other elements in trace amounts. Evaporation acts as a giant purification plant until further pollutants are encountered in the atmosphere.

Once in the atmosphere, the water vapour cools, condenses around tiny particles called condensation nuclei, forms clouds, and is precipitated to earth as either rain, snow, or hail. The warmer the air is, the more water it can hold. Moisture content can be expressed in terms of relative humidity, the amount of moisture held compared to how much could be held if fully saturated at a particular temperature. At a relative humidity

of 100 per cent, the air is saturated, and cloud, fog, and mist form. Clouds are moved around by winds and continue to grow until precipitation (Box 4.7) occurs and the water is returned to earth.

About 77 per cent of precipitation falls into the ocean. The remainder joins the terrestrial part of the cycle in ice caps, lakes, rivers, groundwater, and transport between these compartments. Gravity moves water down through the soil until it reaches the water table, where all the spaces between the soil particles are full of water. This is the **groundwater**

Water in the solid phase of ice may be in the Lowell glacier, located in Kluane National Park, Yukon, for thousands of years before melting and flowing to the ocean.

Box 4.7   Precipitation

Precipitation occurs in several forms: rain, snow, hail, dew, fog, and rime ice (frost). It occurs when the accumulated particles of condensed water or ice in clouds become large enough that they overcome the uplifting air currents and fall to earth as a result of gravity. Some of this precipitation may never reach the ground. Lower air layers may be warmer and drier, and re-evaporation may occur as the precipitates pass through these layers—an excellent example of the speed of some of these mini-cycles that occur as parts of the larger Earth cycles.

Distribution of precipitation is one of the main factors influencing the nature and location of global biomes and the ecozones of Canada. In Canada, precipitation varies from almost nothing in the Arctic to more than 3,000 millimetres annually on the west coast (Figure 4.10).

Differences in precipitation occur for various reasons. At the global scale, heating of equatorial regions causes air to rise. As it rises, it cools and condenses, clouds form, and precipitation occurs. As a result, equatorial regions tend to have consistent high rainfall. Where this air falls as it cools, over subequatorial regions, it tends to be dry, such as in the Sahara Desert.

In Canada, much of the precipitation comes from low-pressure systems, large cells of rising air that form along the boundary between warm and cold air masses. The main factor influencing relative precipitation levels in Canada is moisture-laden winds crossing the ocean and being forced to rise as a result of mountain barriers. The most extreme example occurs with westerlies coming across the Pacific and meeting the western Cordillera, thus creating the highest precipitation levels in the country. As the air warms up in its descent from the mountains, it can hold more moisture, and precipitation levels fall considerably to produce a **rainshadow effect** that accounts for the small amounts of precipitation across the Prairies, with as little annual precipitation as 300 to 400 millimetres. Precipitation levels rise again in central Canada because of disturbances bringing moisture from the Gulf of Mexico and Atlantic Ocean. In southern Ontario, precipitation increases to 800 millimetres and more than 1,000 in the lee of the Great Lakes, a major source of moisture for downwind localities. In Atlantic Canada, exposure to maritime influences increases once more, with up to 1,500 millimetres of precipitation falling annually on the south coast of Newfoundland.

The Arctic is very dry because the prevailing winds from the north are very cold and therefore have little capacity for carrying moisture, they pass over terrain that has relatively few sources of water evaporation (sources often remain frozen for a proportion of the year), and there is an absence of low-pressure systems in winter. Topography is also an important determinant of whether the precipitation falls as snow or rain.

More localized precipitation may be produced by convection as warmer and lighter air rises and cooler, heavier air sinks. This mechanism is important for localized storms in summer when the air is heated by the warm ground and may send columns of moist, warm air to great elevations, resulting in thunderstorm activity.

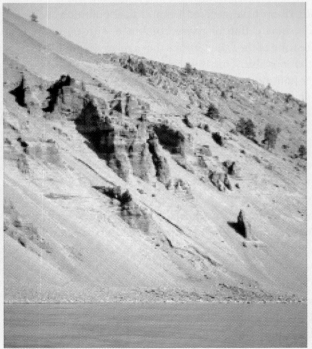

Philip Dearden

Desert-like conditions occur in some areas of Canada largely due to rainshadow effects. Although we associate the Fraser River with the high rainfalls of the west coast, the interior of British Columbia is quite dry.

(Box 4.8). Lakes, streams, and other evidence of surface water occur where the land surface is below the water table. Surface water is a major factor in sculpting the shape of the surface of the Earth. At greater depths, the groundwater may penetrate to occupy various geological formations, known as aquifers.

As mentioned in Box 4.9, water is unique in that at the ambient temperatures and pressure of the Earth's surface, it is the only substance that exists in all three phases of matter (solid, liquid, vapour). Water is stored in all three forms within the hydrological cycle and moves among these forms by the processes shown in Figure 4.11. *Sublimation* is the process for direct transfer between the solid and vapour phases of matter, regardless of direction. This explains why on bright sunny winter days when the air is dry, snowbanks may lose size without any visible melting. About 75 per cent of the world's fresh

Relative humidity is high most of the year in Atlantic Canada and produces some beautiful atmospheric effects.

Box 4.8   Groundwater

Groundwater is found within spaces between soil and rock particles and in crevices and cracks in the rocks below the surface of the Earth. Above the water table is the unsaturated zone where the spaces contain both water and air. In this zone, water is called soil moisture. Groundwater moves the same way as surface waters, downhill, but rarely as quickly and not at all through impermeable materials such as clay. Permeable materials allow the passage of water, usually through cracks and spaces between particles. An **aquifer** is a formation of permeable rocks or loose materials that contains usable sources of groundwater. They vary greatly in size and composition. Porous media aquifers consist of materials such as sand and gravel through which the water moves through the spaces between particles. Fractured aquifers occur where the water moves through joints and cracks in solid rock. If an aquifer lies between layers of impermeable material, it is called a confined aquifer, which may be punctured by an artesian well, releasing the pressurized water to the surface. If the pressure is sufficient to bring water to the surface, the well is known as a flowing artesian well.

Areas where water enters aquifers are known as recharge areas; discharge areas are where the water once more appears above ground. These discharge areas can contribute significantly to surface water flow, especially in periods of low precipitation. Groundwater, of course,

is a very significant part of the Canadian water supply. Dependencies range from 100 per cent in Prince Edward Island to 17 per cent in Quebec, with Ontario having the largest total consumption.

One example of the interaction between groundwater and biogeochemical cycle disruption is contamination of the groundwater of the Abbotsford Aquifer in the Lower Mainland of BC. Agriculture is the main land use. Changes in agricultural products over the past 25 years have led to high concentrations of nitrogen. These changes include a shift in animal production from locally fed dairy/beef cattle to poultry production based on outside feed. Crops have changed from grass and hay to raspberry production, which requires less nitrogen. Poultry manure is used to fertilize the raspberries. Nitrogen use is less than one-half of that produced. The remainder infiltrates to the groundwater. Of 2,297 groundwater samples taken over the years, 71 per cent have exceeded the Guideline for Canadian Drinking Water Quality of 10 milligrams per litre (mg/L), with some as high as 91.9 mg/L. This is cause for concern. These nitrates may be transformed to nitrites in the digestive systems of babies, which in turn may lead to an oxygen deficiency in the blood, known as methemoglobinemia, or blue-baby syndrome. High nitrate levels have also been linked to cancer. More information on this can be found at Environment Canada's 'ecoinfo' website, listed at the end of this chapter.

Figure 4.9 Groundwater flow.

Box 4.9   Some Important Properties of Water

Water has several properties that make it unique:

- Water is a molecule (H$_2$O) that can exist in a liquid, gaseous, or solid state.
- These molecules have a strong mutual attraction, promoting high surface tension and high capacity to adhere to other surfaces; these properties allow water to move upward through plants.
- Water has a high heat capacity, meaning that it can store a great deal of heat without an equivalent rise in temperature; this is the reason why the oceans have such a moderating influence on climate.
- It takes a lot of heat to change water from liquid to gaseous form; this is why evaporation results in a cooling effect.
- Few solids do not undergo some dissolution in water; this allows water to carry dissolved nutrients to plants, but it also means that water is easily polluted.

- Unlike other substances, when water passes from a liquid to a solid state, it becomes less rather than more dense; this is why ice floats on top of water and permits aquatic life to exist in cold climates.

Most of these properties spring from the fact that although the water molecule is electrically neutral, the charges are distributed in a bipolar manner. In other words, a positive charge is at one end of the molecule, and a negative charge is at the other. This means that water molecules have a strong attraction for each other and also explains why water is such a good solvent, since the charges increase the chemical reactivity of other substances.

Figure 4.10 Average annual rain and snow for Canada (cm). Source: After D. Phillips. 1990. *The Climate of Canada*, p. 210. Ottawa: Minister of Supply and Services Canada.

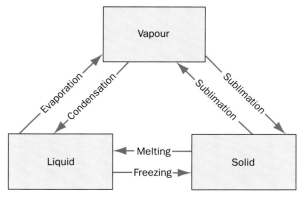

Figure 4.11 Changes of phase in the hydrological cycle.

water is stored in the solid phase, and it may stay in this phase for a long time. Measurements in the Antarctic, for example, indicate that some of the ice is more than 100,000 years old.

Although over the short term there are relatively constant amounts of water in the different storage compartments, over the long term these amounts can change markedly. For example, large amounts of water are evaporated from the oceans and precipitated on land as snow during glacial periods. Over time, the snow accumulates and builds ice fields that may be more than a kilometre thick. This effectively removes water from the oceanic component, causing the sea level to fall. During glacial times, then, the area of land will increase relative to the ocean, and the area covered by ice may increase by up to 300 per cent.

Box 4.10   Human Impacts on the Hydrological Cycle

Human activities have also affected the hydrological cycle. Changes include:

- the storage and redistribution of runoff to augment water supplies for domestic, agricultural, and industrial uses;
- the building of storage structures to control floods;
- the drainage of wetlands;
- the pumping of groundwater;
- cloud seeding;
- land-use changes such as deforestation, urbanization, and agriculture that affect runoff and evapotranspiration patterns;
- climatic change caused by interference with biogeochemical cycles.

In Canada, the solid phase of water is particularly important. Canada is estimated to have up to one-third of the world's fresh water, but most of it is held in a solid state. Compared to the surface supply in Canada, the country's 100,000 glaciers contain more than 1.5 times the volume of fresh water. Furthermore, more than 95 per cent of the country is snow-covered for part of the winter. Spring melt is hence a critical part of the hydrological cycle in Canada as water moves from the solid to liquid phase. This creates a runoff regime for many Canadian rivers, characterized by low late-winter flows and a high spring melt flow that slowly diminishes over the summer into the winter lows as water becomes stored in the solid phase once more. This marked seasonality is one of the reasons why Canada has developed considerable expertise in the construction of water storage facilities. At the same time, human impacts on the hydrological cycle are significant (Box 4.10).

Canada also has abundant storage of fresh water in lake systems, covering almost 8 per cent of the area of the country. These lakes are replenished by river flow that contains approximately 9 per cent of the total river discharge in the world (Table 4.4). About 60 per cent of the discharge in Canada drains north to the Arctic Ocean (Figure 4.12), whereas 90 per cent of the Canadian

Table 4.4   Mean Annual Stream Discharge to the Oceans for Selected Canadian Rivers

River	Area (km²)	Discharge (m³s⁻¹)
Saguenay	90,100	1,820
St Lawrence	1,026,000	9,860
Churchill	281,300	1,200
Nelson	722,600	2,370
Albany	133,900	1,400
Koksoak	133,400	2,550
Yukon (at Alaska border)	297,300	2,320
Fraser	219,600	3,540
Columbia (at Washington border)	154,600	2,800
Mackenzie	984,195	10,800

Source: D. Briggs et al. 1993.

Ocean Basin Region	River Basin Region		Area in 000s km²
Pacific	1	Pacific Coastal	352
	2	Fraser-Lower Mainland	234
	3	Okanagan-Similkameen[a]	14
	4	Columbia[a]	90
	5	Yukon	328
Arctic	6	Peace-Athabasca	487
	7	Lower Mackenzie	1300
	8	Arctic Coast–Islands	2025
Gulf of Mexico	9	Missouri[a]	26
Hudson Bay	10	North Saskatchewan	146
	11	South Saskatchewan[a]	170
	12	Assiniboine-Red[a]	190
	13	Winnipeg[a]	107
	14	Lower Saskatchewan–Nelson	363
	15	Churchill	298
	16	Keewatin	689
	17	Northern Ontario	694
	18	Northern Quebec	950
Atlantic	19	Great Lakes[a]	319
	20	Ottawa	146
	21	St Lawrence[a]	116
	22	North Shore–Gaspé	403
	23	St John–St Croix[a]	37
	24	Maritime Coastal	114
	25	Newfoundland-Labrador	376
CANADA			9974

Figure 4.12 Drainage regions of Canada. [a]Canadian portion only; area on US side of international basin regions excluded from total. Source: Environment Canada. 1985. *Currents of Change: Final Report, Inquiry on Federal Water Policy,* p. 35. Ottawa: Environment Canada. Reproduced with the permission of the Minister of Public Works and Government Services Canada, 1997.

population lives within 300 kilometres of the US border, creating the potential for water deficits in this water-rich country (see Chapter 11).

As water demands grow, we have become increasingly dependent on groundwater sources. Once they become polluted with agricultural biocides or industrial wastes, however, they may be unsuitable for human use for centuries. The importance of this is underscored by the fact that Canada is estimated to have 37 times the amount of water in underground sources as in surface sources and one-quarter of the Canadian population relies on groundwater for domestic use, while some communities, such as Fredericton, are almost totally dependent on groundwater.

Biogeochemical Cycles and Human Activity

Despite the apparent sophistication of human society, the humble fact remains that society could not exist without biogeochemical cycles and those unpretentious bacteria that make them work. Yet all the cycles are susceptible to perturbations by human activity. Such is the scale of human actions that the major transfers taking place between some of the reservoirs in the cycles are human-induced. Some of the most notable and difficult environmental challenges now faced by society spring from these transfers. The purpose of this section is to discuss two of them in more detail: eutrophication and acid deposition. A third example, global climatic change—largely resulting from disruptions in the carbon and nitrogen cycles—is of such significance that Chapter 7 is devoted to it.

Eutrophication

Eutrophication is a natural process of nutrient enrichment of water bodies that leads to greater productivity.

Phosphorus and nitrogen are often the two main limiting factors for plant growth in aquatic ecosystems. Systems with relatively low nutrient levels, **oligotrophic** ecosystems, have quite different characteristics from those with high levels (**eutrophic**), as summarized in Table 4.5. Natural terrestrial ecosystems are relatively efficient in terms of holding nutrient capital. The progression from an oligotrophic to eutrophic condition, through the process of succession discussed in Chapter 3, may take place over thousands of years. This rate is influenced by the geological makeup of the catchment area and the depth of the receiving waters. Catchments with fertile soils will progress more quickly than those with soils lacking in nutrients. Depth is important, because shallower lakes tend to recycle nutrients more efficiently.

How Is Eutrophication Caused?
Cultural eutrophication (eutrophication caused by human activity) may speed up the natural eutrophication process by several decades, mainly through the addition of phosphates and nitrates to the water body. As lakes become shallower as a result of this input, nutrients are used more efficiently, productivity increases, and eutrophication progresses. This is a classic example of a positive feedback loop, with change in the system promoting even more change in the same direction. Additional phosphates and nitrates come from many different sources (Table 4.6), and in accordance with the law of conservation of matter discussed earlier, they do not simply disappear but accumulate in aquatic ecosystems.

What Are the Effects?
This enrichment promotes increased growth of aquatic plants, particularly favouring the growth of floating phytoplankton over **benthic** plants rooted in the substrate. As the benthic plants become out-competed for light by the phytoplankton, they produce less oxygen at depth. Oxygen is critical for the

Table 4.5 Characteristics of Oligotrophic and Eutrophic Water Bodies

Characteristic	Oligotrophic	Eutrophic
Nutrient cycling	low	high
Productivity (total biomass)	low	high
Species diversity	high*	low
Relative numbers of 'undesirable' species	low	high
Water quality	high	low

*Lakes that are extremely non-productive (e.g., high mountain lakes) will have low species diversity.

Table 4.6 Main Nutrient Sources Contributing to Cultural Eutrophication

Runoff from	fertilizers (N and P)
	feedlots (N and P)
	land-use change, such as cultivation, construction, mining natural sources
Discharge of	detergents (P)
	untreated sewage (N and P)
	primary and secondary treated sewage (N and P)
Dissolved nitrogen oxides	(from internal combustion engines)

maintenance of more diverse, oxygen-demanding fish species such as trout and other members of the salmonid family, which also start to decline in number. The oxygen produced by photosynthesis by the phytoplankton tends to stay in the shallower water, escaping back to the atmosphere rather than replenishing supplies at greater depths.

Oxygen depletion is further exacerbated by the decay of the large mass of phytoplankton produced. Dead matter filters to the bottom of the lake where it is consumed by oxygen-demanding decomposers. Once broken down, nutrients may be returned to the surface through convection currents and provide more food for more phytoplankton and algae. Blue-green algae replace green algae in eutrophic lakes, which further exacerbates the problem, since most blue-green algae are not consumed by the next trophic level, the zooplankton.

These effects of oxygen depletion in a water body also occur whenever excess organic matter is added. Under natural conditions, water is able to absorb and break down small amounts of organic matter, with the amount depending on the size and flow and the lower temperature of the receiving water body. The greater the size and flow are and the lower the temperature, the greater the ability to absorb organic materials and retain oxygen levels.

When organic wastes are added to a body of water, the oxygen levels fall as the number of bacteria rises to help break down the waste. This is known as the **oxygen sag curve** (Figure 4.13) and is measured by the **biological oxygen demand** (BOD), the amount of dissolved oxygen needed by aerobic decomposers to break down the organic material in a given volume of water at a certain temperature over a given period. At the discharge source, the oxygen sag curve starts to fall, and there is a corresponding rise in the BOD. As distance from the input source increases and the bacteria digest the wastes, then the oxygen content returns to normal, and the BOD falls.

Major sources of nitrates and phosphates, such as runoff from feedlots and sewage discharge, also contain large amounts of oxygen-demanding wastes. Heat is another source of oxygen stress. The overall result is a progression to a less useful and less healthy water body. The composition of the fish species changes to those less dependent on high oxygen levels, species that are generally less desirable for human purposes. Populations of waterfowl may fall as aquatic plants die off. The water becomes infested with algae, aquatic weeds, and phytoplankton, making swimming and boating unpleasant and giving off unpleasant odours. Water treatment for domestic or industrial purposes becomes more expensive.

What Can We Do about It?

The main way to control eutrophication is through limiting the input of nutrients into the water body (Table 4.6). Domestic and animal wastes must be treated to remove phosphates. Advanced treatment can remove up to 90 per cent of these wastes. More difficult problems occur with diffuse, **non-point sources**, such as runoff from urban areas and agricultural land. Such flows really have to be controlled at the source, since they enter the water body, by definition, in so many different locations. In the past, measures to control water pollution have been largely directed towards **point sources** of pollution, or single discharge points, such as effluent discharges from sewage plants or industrial processes.

iStockphoto.com/Jason Lugo

Animal feedlots are a major source of nutrients, such as phosphates and nitrates, which speed up eutrophication.

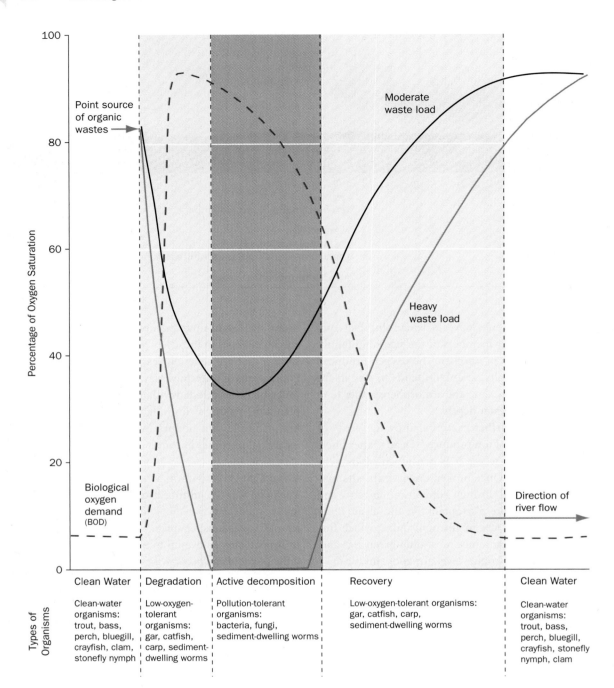

Figure 4.13 Oxygen sag curve and biological oxygen demand (BOD).

By and large, because of the high visibility of such sources, they are easy to identify and monitor, and pollution from such sources has fallen as a result. Increasing attention is now being directed towards the non-point sources (see Chapter 11).

Water quality is one of the three environmental sustainability indicators still measured by Environment Canada. Phosphorus levels in southern Canada exceeded limits set under the water quality guidelines for aquatic life more than half the time at 127 of 344 monitoring sites (Canada 2007a).

Eutrophication used to be considered a problem of smaller water bodies, but now entire areas of the world's largest water bodies, the oceans, are becoming so eutrophic they are being described as 'dead zones'. More than 200 such zones have been recognized, some as large as 70,000 square kilometres in size. One of the largest and best-known areas occurs in the Gulf of Mexico, which receives all the excess fertilizers brought down from the Mississippi watershed. However, similar areas are now found on every continent, and researchers expect that their number and size will increase as global climatic change generates greater rainfall and greater runoff in many areas. The next section explores one of Canada's most notable eutrophication challenges, Lake Erie, and points to some of the ways to address the problem.

Lake Erie: An Example of Eutrophication Control

In all parts of the country, there are many examples of eutrophic lakes. Lake Erie is a particularly well-known case because of its size and importance. Lake Erie is the second smallest and also the shallowest of the Great Lakes, which contain almost 20 per cent of the world's fresh water. Erie has experienced considerable changes in fish species composition since the early explorers described a highly diverse community. Gone are the lake sturgeon, cisco, blue pike, and lake whitefish as the human population in the basin increased and water quality declined. Some 11 million people live in the Erie drainage basin, and 39 per cent of the Canadian and 44 per cent of the American shore is taken up by urban development. There is also considerable industrial use around the shore and intensive agricultural use throughout the basin.

In the past, up to 90 per cent of the bottom layer of the central zone of the lake became oxygen-deficient in the summer. Huge algae mats more than 20 metres in length and a metre deep became common. Beaches were closed. The natural eutrophication that might have taken thousands of years was superseded by cultural eutrophication within the space of 50 years.

In 1972, Canada and the US signed the Great Lakes Water Quality Agreement to try to come to grips with this problem. The signing of the 1985 Great Lakes Charter, in which the two countries agreed to take a cooperative and ecosystem-based approach to the lakes, further strengthened international efforts. Since the 1970s, phosphorus controls, implemented under the Canada Water Act, have led to significant reductions in the phosphorus concentration of the water (Figure 4.14). Phosphate-based detergents were banned and municipal waste treatment plants upgraded.

These measures have led to improvement of water quality, but significant problems still remain. The controls are largely on point-source pollution, discharges that have a readily identifiable source, such as waste treatment plants and industrial complexes. However, much of the remaining nutrient load comes from non-point sources, such as runoff from agricultural fields, lawn fertilizer, and construction sites that are much more difficult to regulate. Figure 4.14 shows that phosphorus levels are once more rising in the lake but remain below the target. The levels of nitrite

and nitrate concentrations have also been rising, leading to increased eutrophication as a result of other nutrients.

Although significant reductions in nutrient loadings have been achieved, phosphorus concentrations in Lake Erie appear to be increasing again.
—Environment Canada and US Environmental Protection Agency 2003, 17

Acid Deposition

In 1966, fisheries researcher Harold Harvey was puzzled to find that the 4,000 pink salmon he had introduced to Lumsden Lake in the La Cloche Mountains southwest of Sudbury, ON, the previous year had all disappeared. Their passage upstream and downstream of the lake had been blocked. To unravel the mystery, he began to take more measurements of the lake and look into its past history. The results were startling—not only had the salmon disappeared but many other species of fish indigenous to the lake had gone missing as well (Table 4.7). The reason soon became apparent. Between 1961 and 1971, Lumsden Lake had experienced a hundredfold increase in the acidity of its waters, as had many other lakes in the same region (Table 4.8). The changes had shifted the lakes outside the limits of tolerance of the species, as discussed in Chapter 2. The indigenous fish species, and many of the species upon which they depended for food,

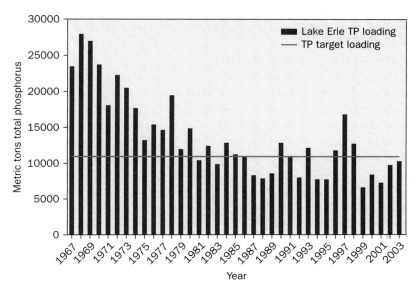

Figure 4.14 Lake Erie total phosphorus loadings. Source: Lake Erie Millennium Network 2006. http://www.lemn.org.

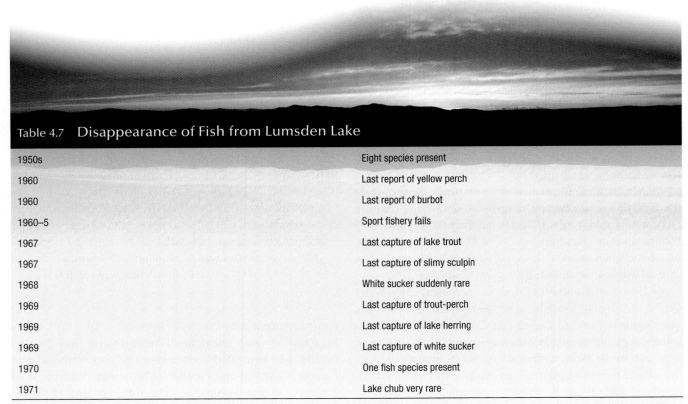

Table 4.7   Disappearance of Fish from Lumsden Lake

1950s	Eight species present
1960	Last report of yellow perch
1960	Last report of burbot
1960–5	Sport fishery fails
1967	Last capture of lake trout
1967	Last capture of slimy sculpin
1968	White sucker suddenly rare
1969	Last capture of trout-perch
1969	Last capture of lake herring
1969	Last capture of white sucker
1970	One fish species present
1971	Lake chub very rare

Source: H. Harvey, unpublished speech, based on R. Beamish and H. Harvey. 1972. 'Acidification of the La Cloche Mountain lakes, Ontario, and resulting fish mortalities'. Journal of the Fisheries Research Board of Canada *29 (8): 1,135. Reprinted with the author's permission.*

Table 4.8   Lake Acidification in the La Cloche Mountains, 1961–71

Lake	pH 1961	pH 1971
Broker	6.8	4.7
David	5.2	4.3
George	6.5	4.7
Johnnie	6.8	4.8
Lumsden	6.8	4.4
Mahzenazing	6.8	5.3
O.A.S.	5.5	4.3
Spoon	6.8	5.6
Sunfish	6.8	5.6
Grey (1959)	5.6	4.1
Tyson (1955)	7.4	4.9

Source: H. Harvey, unpublished speech, based on R. Beamish and H. Harvey. 1972. 'Acidification of the La Cloche Mountain lakes, Ontario, and resulting fish mortalities'. Journal of the Fisheries Research Board of Canada *29 (8): 1,135. Reprinted with the author's permission.*

simply could not tolerate the new conditions and perished. They were victims of the effects of **acid deposition**.

What Is Acid Deposition?

Acids are chemicals that release hydrogen ions (H+) when dissolved in water, whereas a *base* is a chemical that releases hydroxyl ions (OH−). When in contact, acids and bases neutralize each other as they come together to form water (H_2O). Acidity is a measure of the concentration of hydrogen ions in a solution and is measured on a scale, known as the pH scale, that goes from 0 to 14 (Figure 4.15). The midpoint of the scale, pH 7, represents a neutral balance between the presence of acidic hydrogen ions and basic hydroxyl ions. The pH scale is logarithmic. A decrease in value from pH 6 to pH 5 means that the solution has become 10 times more acidic. If the number drops to pH 4 from pH 6, then the solution is 100 times more acidic.

Precipitation, either as snow or rain, tends to be slightly acidic, even without human interference, because of the chemical reaction as carbon dioxide in the atmosphere combines with water to form carbonic acid. Generally, a pH value of 5.6 is given to 'clean' rain. Acid rain is defined as deposition that is more acidic than this, and in Canada rainfall has been recorded with pH levels much lower. Acidic deposition is a more generic term that includes not only rainfall but also snow, fog, and dry deposition from dust.

How Is Acid Deposition Caused?

The increases in acidity reflected in the pH levels of the lakes in Table 4.8 are due to human interference in the sulphur and nitrogen cycles. The largest sources are the smelting of sulphur-rich metal ores and the burning of fossil fuels for energy. These processes change the distribution of the elements between the various sources shown in Figures 4.3 and 4.4, and consequently, natural processes are inadequate to deal with the build-up of matter. Increased amounts of sulphur and various forms of nitrogen are ejected into the atmosphere, where they may travel thousands of kilometres before being returned to the lithosphere as a result of depositional processes. Human activities account for more than 90 per cent of the sulphur dioxide and nitrogen oxide emissions in North America.

Excessive sulphur is produced when ore bodies, such as copper and nickel, are roasted (smelted) at high temperatures to release the metal. Unfortunately, such ores often contain more sulphur than metal, and the sulphur is released into the atmosphere as a waste product of the process. A similar effect is created when sulphur-containing coal is burned as the energy source in power plants. In Canada, the smelting of metal ores accounts for 69 per cent of SO_2 emissions, with power generation and other sources accounting for 30 per cent (Figure 4.16). In the US, electrical utilities are the largest source, accounting for 70 per cent of emissions. The effects of these emissions became obvious fairly early on around smelting plants such as those at Inco in Sudbury and in Trail, BC. Trees were destroyed over large areas. Now there are very encouraging signs of rehabilitation, as discussed in Chapter 13.

These obvious signs of ecological damage were ignored for many years. However, as the ecological implications became better known, governments responded by encouraging industry to build higher stacks to eject the waste further into the atmosphere. Inco, for example, built a 381-metre super-stack in 1972. This improved matters somewhat at the local scale but merely served to create problems elsewhere, especially in Quebec, as entire air masses became acidified and dropped their acid burdens over a larger area. Weather patterns are not random, and so acidified air masses tend to travel in the same kinds of patterns. In central and eastern Canada, as air masses travel from southwest to northeast, they bring a heavy pollution burden from the heavily industrialized Ohio Valley in the US, which falls mostly in Canada. It is estimated that approximately half of the sulphate falling in Canada originates in the US.

These point sources of pollution are, however, easier to monitor and control than the other main source of acids—nitrogen emissions—of which some 22 per cent comes from various means of transport as a result of high-temperature combustion (Figure 4.16). The remainder is split between emissions from thermoelectric generating stations and other industrial, commercial, and residential combustion processes.

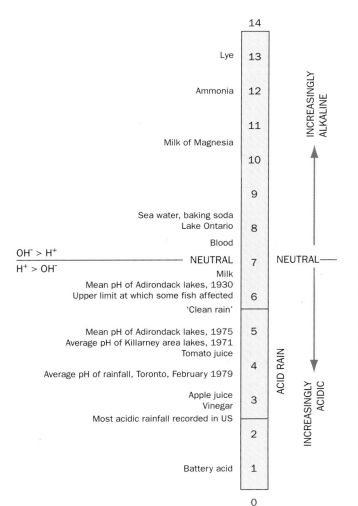

Figure 4.15 The acid (pH) scale.

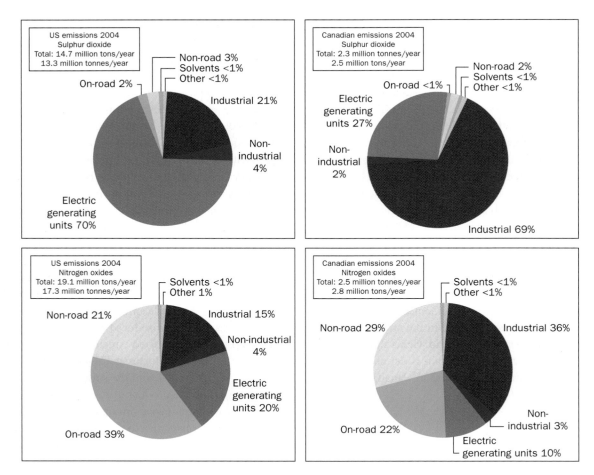

Figure 4.16 Main sources of sulphur dioxide and nitrogen oxide emissions in Canada and US in 2004. Source: Environment Canada 2006a.

What Are the Effects of Acid Deposition?

Aquatic Effects

The effects of acid deposition on the fish of Lumsden Lake and other aquatic ecosystems are one visible sign of some of the impacts of acid deposition. Other species are also affected as the pH of the water body declines. Indeed, as can be seen in Figure 4.17, fish are often not the most sensitive species and are really more of an indicator of the damage that has already occurred. As insects such as mayflies are eliminated, species higher in the food chain that feed on them become affected through food depletion. The same is true of fish-eating birds such as loons, whose young have been shown to have a lower chance of survival on acidified lakes because of starvation.

Unfortunately, some of these impacts are permanent. Examination of historical angling records in Nova Scotia, for example, indicates that of 60 main salmon rivers, 13 runs of salmon are extinct and a further 18 are virtually extinct. It is estimated that more than half of the total salmon stock has been lost as a result of declining pH levels. Atlantic salmon populations in rivers of the Southern Upland region of Nova Scotia continue to be severely negatively affected and will likely become extinct if adult survival rates remain at current low levels and pH recovery continues to be delayed (Environment Canada 2004). It has also been suggested that the endangered Acadian whitefish is threatened by declining pH levels.

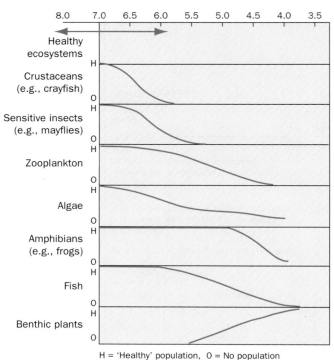

Figure 4.17 Sensitivity of various aquatic organisms to pH level. Source: Environment Canada 1991.

Even where fish themselves manage to survive, they may be grossly disfigured, with twisted backbones and flattened heads, because their bones have been deprived of the necessary nutrients for strength. Reproductive capacities may be sufficiently impaired that it leads to eventual population declines. Generally, the time of reproduction is the most sensitive part of the life cycle. The critical factor is often the lower pH level of the water as the snow melts. At this time, the build-up of acids over the winter can result in even higher acidity than experienced through the rest of the year. This pulse of acidity is called **acid shock**, and it may also be one of the causes of stress on amphibious creatures, such as frogs, that often use small temporary pools of water for breeding in the spring following runoff. At Lumsden Lake, for example, spring runoff produced a pH as low as 3.3, more than 100 times more acidic than the 1961 levels.

More acidic water and the subsequent food chain effects are not the only concern. Other chemical changes also occur. The increased acidity, for example, releases large amounts of aluminum from the terrestrial ecosystem into the rivers and lakes. Here it forms a toxic scum, lethal to many forms of aquatic life.

Terrestrial Effects

Terrestrial effects of acid deposition first became visible around emission sources, such as at Trail, BC, and Sudbury, ON, as trees began to die. Before joining the soil, the acids eat away at the sensitive photosynthetic surfaces of the leaves. Broad-leaved trees, such as sugar maples, the source of Quebec's $40-million annual maple syrup industry, are thought to be particularly susceptible because of the large surface area of their leaves.

Once in the soil, the acids leach away the nutrients required for plant growth, leading to nutrient deficiencies. The high levels of aluminum released by the acids also help to inhibit the uptake of nutrients. The bacteria so critical to the workings of many biogeochemical cycles, as described earlier, are also adversely affected and cause changes in natural soil processes. Decomposition and humus formation are retarded. Soil contains many organisms, including mites, insects, worms, bacteria, fungi, and many others involved in the critical ecological functions of biogeochemical cycling and energy flow. There may be more than 100 million bacteria and several kilometres of fungal hyphae in a single gram of healthy soil. Mycorrhizal fungi are very important for the growth of many plants because they help to transport nutrients from the soil water into the roots. Research on Jack pine, undertaken to try to understand the potential impact of increased acidity on these mutualistic relationships, found that changes in calcium-to-aluminum ratios caused by increased acidity influenced the succession of mycorrhizal fungi on tree root systems. The actual physical contact of the acids with the plant roots can also inhibit growth and lower resistance to disease. The long-term effects of these kinds of changes on tree growth, the ecological health of the community, and forest yields are still highly uncertain, but they are not going to be beneficial.

Eastern Canadian watersheds are now exhibiting releases of sulphur from soils in excess of deposition. Two internal catchment sources, sulphate desorption and release via decomposition of organic matter, are the likely causes of the budget imbalance. The release of this extra sulphur acts as an additional acid load for soils and downstream waters and may be partly retarding the recovery of surface waters in eastern Canadian forested watersheds (Environment Canada 2004). However, increased nitrogen deposits are not a problem, since most areas are nitrogen-deficient and the increased inputs act as fertilizer.

Damage from acid deposition is now visible over much wider areas than those surrounding sources of high emissions. Extensive areas of damage have been recorded in Europe and in the eastern United States. At these larger scales, it is often difficult, however, to single out one cause, such as acid deposition. It is likely that other factors—climate change and other pollutants, such as high levels of ozone brought about by excessive nitrogen oxides—are also important in placing stress on these communities in synergistic reactions.

In central and eastern Canada, 89 per cent of the high-capability forest land receives in excess of 20 kilograms per hectare of acid deposition annually, the figure originally defined by the Canadian government as being an 'acceptable' level of deposition. Severe impacts have been noticed. A survey of white birch around the Bay of Fundy, for example, found 10 per cent of the trees already dead and virtually all others showing signs of damage.

The impact on plant life is not restricted to natural ecosystems. Significant changes can also occur on agricultural lands as direct damage to crops or more indirect changes through changes to soil chemistry. The growth of crops such as beets, radishes, tomatoes, beans, and lettuce is inhibited, and biological nitrogen fixation is diminished at a pH of 4. In central and eastern Canada, 84 per cent of the most productive agricultural lands receive more than the 20 kilograms per hectare. In some areas, the application of lime to the soil has become a routine agricultural procedure in an attempt to neutralize the acids by adding more basic ions.

The smelter at Trail caused extensive damage to the surrounding vegetation.

CP Photo/Jeff Bassett

Heterotrophs can also be affected. As the forest cover diminishes, so does the habitat for many species. Toxic metals such as cadmium, zinc, and mercury, released from the soil by the acids, may be concentrated by certain species of plants and lichens and accumulated in the livers of species eating them, such as moose and caribou.

Sensitivity

Not all ecosystems are equally sensitive to the effects of acid deposition. Some areas have a high capacity to neutralize the excess acids because of the high base capacity of the bedrock and soils. For example, the Prairies are not as sensitive to the effects of acid deposition, because underlying carbonate-rich rocks, deep soils, and other factors combine to provide high *buffering* capacity. On the other hand, areas with difficult-to-weather rocks with low nutrient content (e.g., granite) and with thin soils following glaciation often have very low buffering capacities. Much of central and eastern Canada and coastal British Columbia fit within this category (Figure 4.18). The provinces most vulnerable to acid deposition in terms of amount of area ranked as highly sensitive are Quebec (82 per cent), Newfoundland (56 per cent), and Nova Scotia (45 per cent). The spatial coincidence of low buffering capacities and high deposition rates explains why most attention in Canada has centred on the central and eastern parts of the country.

Socio-economic Effects

The environmental effects described above have socio-economic implications, and there are also direct effects of acid deposition on human health. In aquatic ecosystems, for example, declines in fish populations obviously have implications for those involved in fishing, whether commercially or for sport. The impacts of acid deposition on tree growth and the forest industry are also of considerable concern. Some studies indicate that tree growth reductions of up to 20 per cent might be experienced. Preliminary estimates of the market value of lost wood production resulting from acid rain are in the hundreds of millions of dollars in Nova Scotia and New Brunswick alone (Canadian Council of Ministers of the Environment 2006). The full effects, however, are likely to be much greater than this. European research indicates that a time lag of some 20 to 30 years is likely before the effects of acid rain are reflected in reduced tree growth.

Many values, however, are difficult to express in monetary terms. Thousands of Canadians, for example, maintain lakeside cottages. It is challenging, and some would say impossible, to ascribe a value to the changes that might occur as the lakes become devoid of life. Common loons, for example, are for many the quintessential symbol of the Canadian wilderness, yet populations have been found to be quite sensitive to lake acidification. Below pH levels of 4.5, loons do not seem able to find enough food to feed their young. However, as a result of emissions regulations in Canada and the US, overall improvements in the capacity of many lakes to support aquatic biota are being observed. For instance, a general increase in the number of breeding fish-eating water birds (such as loons) has been observed in lakes in

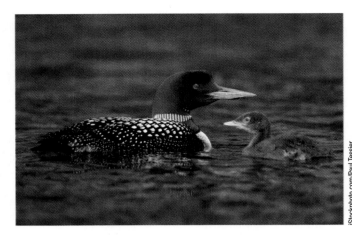

Loons, with their beautiful coloration and haunting cries, are a wilderness symbol for many Canadians, but their breeding success has been reduced by the effects of acid rain.

iStockphoto.com/Paul Tessier

Ontario, Quebec, and Newfoundland, particularly those in close proximity to reduced emission sources. At the same time, algae, invertebrates, and water-bird food chains in many lakes in this region continue to show acidification impacts (i.e., direct effects of acidification, metal toxicity, loss of prey species, and reduced nutritional value of remaining prey), particularly in lakes and rivers where fish communities have been affected.

It is not only the natural environment that is damaged as a result of acid deposition but also the human-built environment. The acids eat away at certain building materials. The effects have been most damaging in Europe where there are many old monuments, but such effects can also be seen on the Houses of Parliament and nearby statues in Ottawa. It has been estimated that acid rain causes $1 billion worth of damage in Canada every year (Environment Canada 2003a).

The most direct impact on human health may arise from inhalation of airborne acidified particles, which can impair respiratory processes and lead to lung damage. There appear to be significant relationships between air pollution levels in southwestern Ontario, for example, and hospital admissions for respiratory illnesses. Comparative studies between heavily polluted areas in Ontario and less polluted areas in Manitoba found diminished lung capacity in about 2 per cent of the children in the more polluted areas. Further studies have confirmed this relationship, although whether high sulphate or ozone levels are responsible has yet to be conclusively determined. Calculations by the US government suggest that southern Ontario will reap health benefits of more than $1 billion per year by the year 2010 as a result of current efforts to reduce emissions in the US. Benefits of acid rain control programs in the US have turned out to be considerably higher than predicted and costs considerably less, resulting in a benefit/cost ratio of 40 to 1 (Chestnut and Mills 2005).

Humans can also be affected by ingesting some of the products of acid rain. Regarding drinking water, for example, in some areas where older delivery systems are still in place, the increased acidity of water can corrode pipes and fittings and result in elevated levels of lead in the water supply. This is

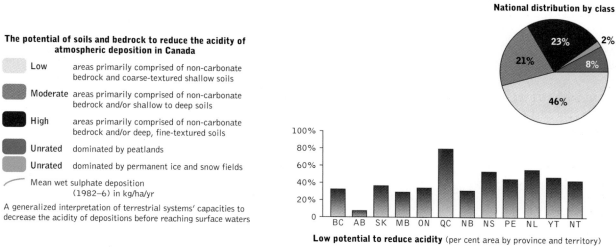

The potential of soils and bedrock to reduce the acidity of atmospheric deposition in Canada

Low — areas primarily comprised of non-carbonate bedrock and coarse-textured shallow soils

Moderate — areas primarily comprised of non-carbonate bedrock and/or shallow to deep soils

High — areas primarily comprised of non-carbonate bedrock and/or deep, fine-textured soils

Unrated — dominated by peatlands

Unrated — dominated by permanent ice and snow fields

Mean wet sulphate deposition (1982–6) in kg/ha/yr

A generalized interpretation of terrestrial systems' capacities to decrease the acidity of depositions before reaching surface waters

National distribution by class

23% 2% 8% 46% 21%

Low potential to reduce acidity (per cent area by province and territory)

Figure 4.18 The potential of soil and bedrock to reduce the acidity of atmospheric deposition in Canada. Source: Environment Canada 1988.

one reason why in certain areas, such as Victoria, it is recommended that schools flush their water fountains early in the morning to spill the water that has been held overnight and that might show elevated lead levels.

Acidified water may also hold other substances in solution that are deleterious to human health. Excess levels of aluminum have already been noted. When metals such as mercury, chromium, and nickel are leached from the substrate into water, they may be taken up and concentrated along the food chain and eventually cause a human health problem.

What Can We Do about It?

Had emission levels continued at 1980 levels, it was predicted that 600,000 lakes in Canada would eventually be virtually lifeless because of excess acidity. Thankfully, this has not happened because of various measures undertaken to combat the problem. Critical early measures were to conduct more scientific research to ascertain the scale and dimensions of the problem and then to raise public and political awareness to gain support for the necessary changes.

One of the main challenges associated with acid deposition is that it is not limited to the areas generating the emissions and causing the problem. In Canada, estimates suggest that more than half of the acid deposition originates in the US. Furthermore, the acidic fog that forms in the Canadian Arctic in spring is largely a result of emissions from Europe and Russia. To address such problems, international efforts are required, as is a greater concern on the part of individuals (see Box 4.11).

Canada has addressed the problem through national, bilateral, and multilateral efforts. In 1983, the Canadian Council of Resource and Environment Ministers agreed on an annual target deposition of 20 kilograms per hectare as an acceptable goal, taking political and economic costs into account. This is an important qualification. The critical load represents a **policy target value** (PTV) set by politicians. However scientists estimate that a further 75 per cent reduction in SO_2 emissions would be needed to address the situation adequately. This is a **scientific target value** (STV). Many of the environmental problems you will read about in this text represent instances in which PTVs were established that conflicted with STVs. Establishment of *total allowable catch* in fisheries (Chapter 8) and *annual allowable cut* in forestry (Chapter 9), as well as designation of endangered species (Chapter 14), are good examples of this conflict.

In 1985, an agreement was reached among the provinces east of Saskatchewan (the area defined as 'eastern' Canada within the context of acid rain) that set specific emission reductions to reach this target. Progress was rapid. The area of eastern Canada receiving 20 kilograms per hectare or more of wet sulphate per year declined by nearly 59 per cent from 0.71 million square kilometres in 1980 to about 0.29 million in 1993.

In 1998, the provinces, territories, and federal government signed the Canada-Wide Acid Rain Strategy for Post-2000, committing them to further actions to deal with acid rain.

The targets for eastern Canada emissions are shown in Table 4.9. Under this scenario, and assuming a further 50 per cent reduction in US emissions over and above those already committed to, the area receiving more than the critical load would be reduced to 247,000 square kilometres. By 2004, SO_2 emissions were down to 2.3 million tonnes, a 50 per cent reduction from the 1980 level of 4.6 million tonnes (Environment Canada 2006a).

Box 4.11   What You Can Do

Many of the challenges and problems discussed in this book are international in scope and require the co-ordinated efforts of different levels of government, industry, and individuals. The following are some of the ways by which individuals can try to have a positive influence.

1. Acid deposition is profoundly influenced by the personal decisions we all make regarding our use of fossil fuels in transport and electricity consumption. Think about your decisions and how you can minimize use.

2. Many consumer items, such as TVs, cellphones, computers, and other electronic products, contain materials such as lead and nickel that contain sulphur. As you buy and dispose of these items, you are helping to increase acid deposition.

3. Use chemical fertilizers sparingly on your gardens.

4. Eat less meat.

5. Let your political representatives know that you are in favour of mandatory measures to curb sulphur emissions and treat livestock wastes, even if this costs you more money.

Table 4.9   Sulphur Dioxide Reduction Targets for Ontario, Quebec, New Brunswick, and Nova Scotia

	Former Eastern Canada Acid Rain Program Cap	New Target under the Canada-Wide Acid Rain Strategy	Timeline for New Target
Ontario	885 kt	• 442.5 kt (50% reduction)	2015[a]
Quebec	500 kt	• 300 kt (40% reduction) • 250 kt (50% reduction)	2002 2010
New Brunswick	175 kt	• 122.5 kt (30% reduction) • 87.5 kt (50% reduction)	2005 2010
Nova Scotia	189 kt	• 142 kt (25% reduction) • 94.5 kt (50% cumulative reduction goal)[b]	2005 2010

[a]Ontario has proposed and is consulting on the proposal to advance this timeline to 2010.
[b]Ninety-four and a half kilotonnes is a reduction target and not a cap. Nova Scotia's commitment is to reduce SO_2 emissions by 25 per cent from the existing cap by 2005 and to further reduce emissions to achieve a cumulative reduction goal of 50 per cent by 2010 from existing sources.
Source: Canadian Council of Ministers of the Environment 2005.

Advances in science have allowed more comprehensive assessments of critical loads to be defined that combine sulphates and NOx for both aquatic and terrestrial ecosystems. Sulphur and nitrogen have different atomic weights, so the combined critical load cannot be expressed as mass units (e.g., kilograms per hectare). They are now expressed as ionic charge balance in terms of equivalent/hectare/year, and the old load of 20 kilograms/hectare/year is represented by 416 equivalent/hectares/year. Current critical loads are shown in Figure 4.19. Calculations show that between 21 and 75 per cent of eastern Canada is still receiving deposits in excess of the critical load. The very wide range represents the uncertainty regarding long-term effects of nitrogen and degree of future absorption in soils (Environment Canada 2006a).

Bilaterally, early efforts to convince decision-makers in the US to control emissions met with little success. However, in 1990, intense lobbying efforts came to fruition. The Canada-US Air Quality Agreement was signed to create a flexible and dynamic mechanism for bilateral environmental cooperation in reducing trans-boundary air pollution. The initial focus of the agreement was on reducing emissions of sulphur dioxide and nitrogen oxides, the major contributors to acid rain. Both Canada and the US have surpassed the emission reduction requirements in the agreement.

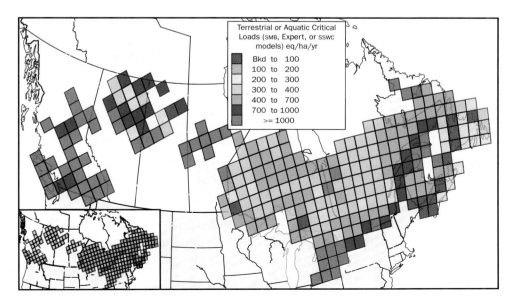

Figure 4.19 Critical loads for acid deposition in Canada. Source: Environment Canada 2006a.

Figures 4.20 and 4.21 Estimated emissions of sulphur dioxide in the US and Canada and estimated emissions of nitrogen oxides in the US and Canada. Source: Environment Canada 2006a.

The US Clean Air Act was revised to cut by half the 1980 sulphur emission levels by the turn of the century, and significant progress has been made (Figure 4.20). Total emissions of SO_2 declined by approximately 40 per cent between 1980 and 2000 and are predicted to decline by approximately 38 per cent from 2000 levels by 2020. Overall, the amount of acid rain falling on eastern Canada originating in both countries is down an estimated 33 per cent since controls were introduced. More than half the eastern Canadian emission reductions have occurred at the smelters at Sudbury, ON, and Rouyn-Noranda, QC. However, since that time, reductions in Canada have been minimal, with western Canada showing increasing emissions (Figure 4.20), largely due to emissions from the Alberta oil industry.

Efforts to control the emissions of nitrogen oxides have not been as successful. Between 1980 and 1990, both Canadian and US emissions remained fairly constant, and Canadian emissions show an increase to 2000 (Figure 4.21). Canada was committed under the terms of the Canada-US Air Quality Agreement to a 10 per cent reduction from stationary sources by the year 2000. Annual emissions in the US should have been reduced by 1.8 million tonnes below the 1980 level by the year 2000. Mobile emission reductions are being sought by introducing more stringent performance standards on exhaust emissions from new vehicles and are predicted to decline by approximately 17 per cent between 2000 and 2020. In eastern Canada, emissions of NOx decreased by 17 per cent between 1985 and 2000 and are predicted to decrease by approximately 39 per cent between 2000 and 2020. These declines have been largely the result of power plant emissions reductions. Emissions from power plants in the Ontario portion of the Pollutant Emission Management Area were approximately 78 kilotonnes in 1990 and had decreased by almost half by 2004. In 2005, the Lakeview Generating Station closed, eliminating annual emissions of approximately 4,000 tonnes of NOx and 15,000 tonnes of SO_2 upwind of the Greater Toronto Area. Ontario achieved a 12 per cent reduction in electricity use by the government in 2007 and is now aiming for a 20 per cent reduction by 2010. Further action in the province to achieve reductions includes agreements to purchase power from 19 new renewable energy

projects, including three water power projects, three landfill gas and biogas projects, and 13 wind farms.

However, in the West, NOx emissions increased by 29 per cent between 1985 and 2000 and are predicted to increase by approximately 5 per cent between 2000 and 2020. Emissions of NOx in western Canada surpassed emissions in eastern Canada by the year 2000 and will continue to increase, driven by the oil sands developments in northern Alberta (see also Chapter 12). Greater reductions are predicted for the US, where NOx emissions declined slightly between 1985 and 2000 and are predicted to decline by approximately 47 per cent from 2000 levels by 2020 (Figure 4.20).

An analysis of the US Acid Rain Program estimated annual benefits of the program to both Canada and the United States at $122 billion in 2010 and costs for that year at $3 billion (in 2000 dollars)—a 40 to 1 benefit/cost ratio. These benefits flow from such factors as improved air quality prolonging lives, reducing heart attacks and other cardiovascular and respiratory problems, and improving visibility.

Multilaterally, Canada also has been active, signing the 1979 Convention on Long-Range Trans-Boundary Air Pollution and then taking international leadership in the signing of the 1985 Helsinki Protocol that committed member countries to a 30 per cent reduction in emissions by 1994, a target that was met. Canada is also a signatory to the UN Economic Commission for Europe Protocol on Acidification, Eutrophication and Ground Level Ozone. However, to ratify the protocol, Canada will have to commit to emission ceilings for SO_2, NOx, and volatile organic compounds (VOC), which it has yet to do (Environment Canada 2005).

Canada has met all its goals and commitments, some with considerable time to spare, in terms of acid deposition reductions. There has been a decreasing trend in lake sulphate levels in southeastern Canada in response to reductions in SO_2 emissions; however, many of these lakes are still acidified, and many do not meet a pH condition of ≥ 6, a key threshold for the sustenance of fish and other aquatic biota. There is no doubt that acid deposition remains a significant problem in Canada and will remain so for years to come. To some extent, the issue has been overshadowed by public and political interest in global

Automobile emissions reduce the air quality in many urban centres.

climate change. However, there are still significant causes for concern. Unlike the US, Canada does not have a Clean Air Act that can be enforced by law, creating the possibility of further non-compliance with targets. Emission controls have focused largely on point-source control of sulphate emissions. It is more difficult to address the more diffuse nitrogen derivatives coming mainly from the transportation sector. The pattern of wet nitrate deposition has changed little over the last decade.

Concern has focused on eastern Canada, but pockets of acidity exist all across the country. The presence of acid-sensitive geology and increasing emissions of SO_2 and NOx suggests that monitoring should expand into the western provinces to ensure that acid deposition does not damage ecosystems in that region. Large increases in emissions of SO_2 and NOx from oil sands operations in northern Alberta are raising concerns that

acid deposition could potentially negatively impact the West. Between 2000 and 2020, emissions of SO_2 are predicted to decline by 21 per cent in eastern Canada and increase by 15 per cent in western Canada. Emissions of SO_2 in western Canada are predicted to exceed emissions in eastern Canada by 2010.

Furthermore, although some lake pH levels are showing signs of recovery, it is still too early to assess the biotic implications in terms of community health.

Emission targets have been consistently set below levels required to ensure biological recovery. The pH levels are still low enough in many areas to have damaging effects on sensitive species.

Implications

Understanding the nature of matter and the way that nutrients cycle in the ecosphere is fundamental to appreciating many of the more challenging environmental issues that society currently faces. Acid deposition, eutrophication, and global climate change all have their roots in disruption of biogeochemical cycles. Science is only just starting to unravel some of the secrets of these cycles, but we know enough to understand their significance. Research also clearly shows their complexity and their interconnectedness. We add nitrogen to soils to try to boost productivity and produce more food. However, this also results in eutrophication, depletion of the ozone layer (Chapter 7), and other problems. Clearly, we have to understand the basic science of this interconnectivity. However, this understanding has to be linked to our ability to manage the situation. The chapters in Part C discuss some of the main approaches in environmental planning and management.

Summary

1. Matter has mass and takes up space. It is composed of 92 natural and 17 synthesized chemical elements. The law of conservation of matter states that matter can neither be created nor destroyed but merely transformed from one form into another. Matter cannot be consumed.
2. Elements necessary for life are known as nutrients. They cycle between the different components of the ecosphere in characteristic paths known as biogeochemical cycles.
3. Humans disturb these cycles through various activities, resulting in environmental problems such as acid rain and global warming.
4. Cycles can be classified into gaseous or sedimentary, depending on the location of their major reserves.
5. Phosphorus is an example of a sedimentary cycle. The main reservoir of phosphorus is the Earth's crust. Phosphates are made available in the soil water through erosional processes and are taken up by plant roots and

passed along the food chain. There is no atmospheric component to the cycle, making it especially vulnerable to disruption. The main human use for phosphorus is as fertilizer. It is a main cause of eutrophication.
6. Sulphur is also a sedimentary cycle but differs from phosphorus in that it has an atmospheric component. Like phosphorus, it is an essential component for all life. Bacteria enable plants to gain access to elemental sulphur by transforming it to sulphates in the soil. Sulphur is a main component of acid deposition.
7. Nitrogen is a gaseous cycle. Almost 80 per cent of the atmosphere is composed of nitrogen gas, yet most organisms cannot use it as a source of nitrates. Instead, various bacteria help to transform nitrogen into a form that can be used by plants. As with the other cycles, these nitrates are then passed along the food chain. Nitrates are used as fertilizers and contribute to

eutrophication. Various nitrous oxides also contribute to acid deposition and the catalytic destruction of ozone.

8. Carbon dioxide constitutes only 0.03 per cent of the atmosphere, but it is the main source of carbon—the basis for life—through the process of photosynthesis. Carbon becomes incorporated into the biomass and is passed along the food chain. Respiration by organisms transforms some of this carbon back into carbon dioxide, and the cellular respiration of decomposers helps to return the carbon from dead organisms into the atmosphere. Carbon dioxide emissions from burning fossil fuels are a main contributor to global climatic change.

9. Water travels between the different components of the ecosphere by means of the hydrological cycle fuelled by energy from the sun. Ninety-seven per cent of water is in the oceans. Less than 1 per cent is readily available for human use.

10. Canada has up to one-third of the world's fresh water. Most of this is held in the solid phase as ice. Canada also has high storage capacities for liquid water, with lakes covering an estimated 8 per cent of the country. These lakes are replenished by a river flow containing approximately 9 per cent of the total river discharge in the world.

11. Major pollution problems, such as eutrophication, acid deposition, and global warming, are the result of human disruption of biogeochemical cycles.

12. Eutrophication is the nutrient enrichment of water bodies over time. Although it is a natural process, human disruption of the phosphorus and nitrogen cycles has caused a marked acceleration in the rate of eutrophication. This promotes excessive plant growth that leads to oxygen depletion when the plants die and start to decay. Over time, this leads to changes in the composition of fish species and makes water treatment more expensive. The rate of eutrophication can be slowed down by limiting the inputs of nutrients into the water body.

13. Lake Erie is a classic example of eutrophication. Although phosphorus loadings have been reduced considerably, they are still above mandated guidelines and appear to be increasing.

14. Precipitation tends to be naturally acidic. However, as a result of disturbances in the sulphur and nitrogen cycles, the acidity has increased dramatically over much of Canada during the past few decades. The largest impacts are caused by the burning of sulphur-rich fossil fuels or the smelting of sulphur-rich metal ores. The resulting sulphur dioxide mixes with water in the atmosphere to produce sulphuric acid. Emissions of various nitrogen oxides as by-products of high-temperature combustion account for most of the remainder. The increase in acidity has a damaging effect on aquatic and terrestrial ecosystems as well as on human health.

15. Emission controls have been agreed upon to try to limit these impacts, and targets have been exceeded. However, concomitant improvements in reducing acid rain and the recovery of aquatic systems have not been seen. Many scientists believe that more stringent measures are called for.

Key Terms

acid deposition	compound	matter	rainshadow effect
acid shock	denitrification	micronutrients	rock cycle
aerobic	eutrophic	mineralization	scientific target value
anadromous	eutrophication	nitrogen fixation	sedimentary cycles
anaerobic	evapotranspiration	non-point sources	transpiration
aquifer	gaseous cycles	nutrients	
benthic	groundwater	oligotrophic	
biogeochemical cycles	hydrological cycle	oxygen sag curve	
biological oxygen demand (BOD)	law of conservation of matter	point sources	
	macronutrients	policy target value	

Questions for Review and Critical Thinking

1. Summarize some of the key differences and similarities between energy and matter.
2. Why is life dependent on biogeochemical cycles?
3. Explain why decomposer organisms are important in biogeochemical cycling.
4. What are some of the important implications of biogeochemical cycling for forestry and agricultural activities?
5. Outline the main characteristics of the hydrological cycle in Canada.
6. Which biogeochemical cycles are most responsible for eutrophication, and in what ways are they changed by human activities?
7. Are there any eutrophic lakes in your region? If so, what are the main inputs causing eutrophication, and where do they come from?

8. How can eutrophication be controlled? Discuss one example.
9. What is the pH scale, and what is it used for?
10. Which biogeochemical cycle is most responsible for acid deposition, and how do disruptions occur as a result of human activities?
11. What are the effects of eutrophication and acid deposition on aquatic ecosystems? What do you think might be the impacts of both of them together?

12. Are all areas equally sensitive to the impact of acid deposition? If not, what influences the relative vulnerability of different areas?
13. What are the main socio-economic impacts of acid deposition likely to be?

Related Websites

Environment Canada, acid rain:
http://www.ec.gc.ca/acidrain

Environment Canada, air pollutant emissions:
http://www.ec.gc.ca/pdb/ape/cape_home_e.cfm

Environment Canada, clean air:
http://www.ec.gc.ca/air/introduction_e.cfm

Environment Canada, sustaining the environment and resources for Canadians:
http://www.environmentandresources.gc.ca/default.asp?lang_En

Environment Canada, water: http://www.ec.gc/water_e.html

Nitrate levels: http://www.ecoinfo.ec.gc.ca/env_ind/region/nitrate/nitrate_e.cfn

Further Readings

Note: This list comprises works relevant to the subject of the chapter but not cited in the text. All cited works are listed in the section 'References' at the end of the book.

Browning, M.H.R., and T.C. Hutchinson. 1991. 'The effects of aluminum and calcium on the growth and nutrition of selected ectomycorrizal fungi of jack pine'. *Canadian Journal of Botany* 69: 1,691–9.

Dalton, H. 2003. 'Nitrogen: The essential public enemy'. *Journal of Applied Ecology* 40(5): 771–81.

Galloway, J.N., and E.B. Cowling. 2002. 'Reactive nitrogen and the world: 200 years of change'. *Ambio* 31(2): 64–71.

Hendrickson, O. 2003. 'Influences of global change on carbon sequestration by agricultural and forest soils'. *Environmental Reviews* 11(3): 161–92.

Janzen, H.H., et al. 2003. 'The fate of nitrogen in agroecosystems: An illustration using Canadian estimates'. *Nutrient Cycling in Agroecosystems* 67(1): 85–102.

Jeffries, D.S., et al. 2003. 'Assessing the recovery of lakes in southeastern Canada from the effects of acidic deposition'. *Ambio* 32: 176–83.

———, et al. 2003. 'Monitoring the results of Canada/USA acid rain control programs: Some lake responses'. *Environmental Monitoring and Assessment* 88(1–3): 3–19.

Keller, W., J.H. Heneberry, and S.S. Dixit. 2003. 'Decreased acid deposit and the chemical recovery of Killarney, Ontario, lakes'. *Ambio* 32: 183–7.

Magee, J.A., et al. 2003. 'Effects of episodic acidification on Atlantic salmon (*Salmo salar*) smolts'. *Canadian Journal of Fisheries and Aquatic Sciences* 60(2): 214–21.

Pilgram, W., et al. 2003. 'Changes in acid precipitation–related water chemistry of lakes from southwestern New Brunswick, Canada, 1986–2001'. *Environmental Monitoring and Assessment* 88(1–3): 39–52.

Rabalais, N.N. 2002. 'Nitrogen in aquatic ecosystems'. *Ambio* 31(2): 102–12.

Tao, F., et al. 'Terrestrial water cycle and the impact of climate change'. *Ambio* 32(4): 295–301.

Vile, M.A., et al. 2003. 'Atmospheric sulfur deposition alters pathways of gaseous carbon production in peatlands'. *Global Biogeochemical Cycles* 17(2): article no. 1,058.

Vitousek, P.M., et al. 2003. 'Nitrogen and nature'. *Ambio* 31(2): 97–101.

Winter, J.G., et al. 2002. 'Total phosphorous budgets and nitrogen loads: Lake Simcoe, Ontario (1990 to 1998)'. *Journal of Great Lakes* 28(3): 301–14.

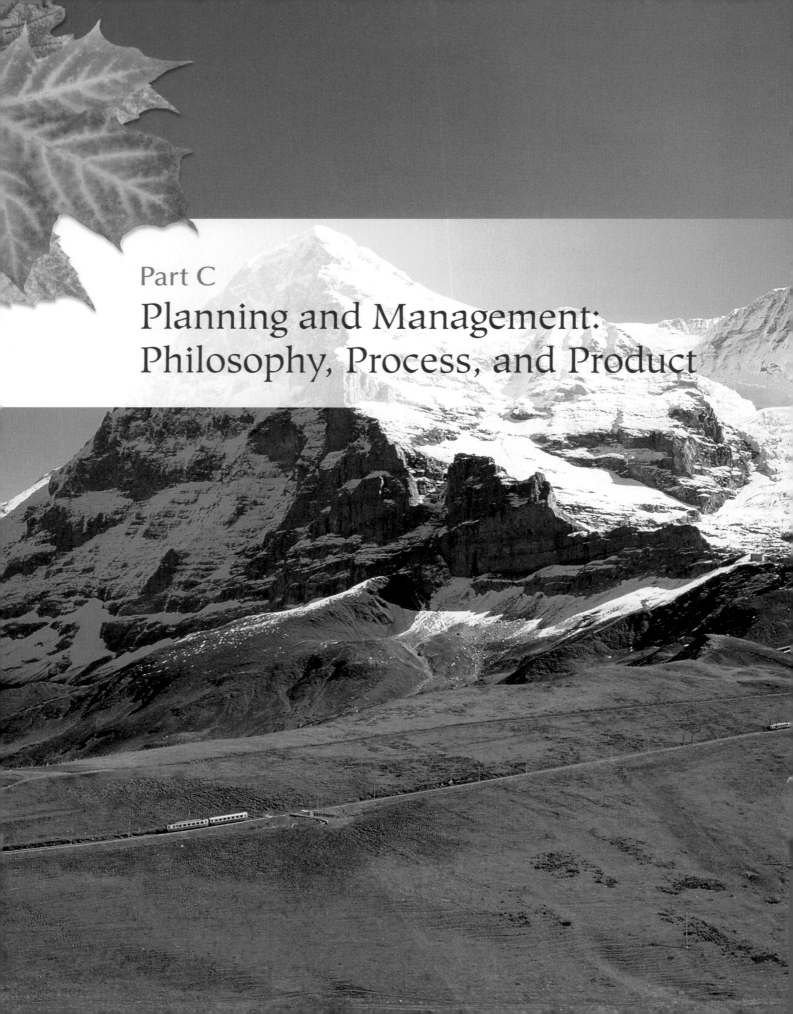

Part C
Planning and Management: Philosophy, Process, and Product

The chapters in Part B focused mainly on the natural science pertinent to resources and the environment in Canada. In the following two chapters, attention is given to planning and management related to philosophies, processes, and products that can be applied in resource and environmental management. However, a key point needs to be emphasized. Often, it is inappropriate to think that humans 'manage' the environment or natural resources. Instead, we usually attempt to manage the *interaction* between humans and the environment. This is why resource and environmental management involves more than application of 'science' or 'technical expertise'. It also requires sensitivity to various, and often different, values, interests, needs, and wants. Indeed, these are often the most important variables affecting resource management outcomes.

Chapter 5 focuses on the 'philosophy' of planning and management relating to resources and environments, while Chapter 6 considers 'processes and products'. Together, they provide an overview of concepts, approaches, and methods that can be drawn upon to inform the subsequent discussions in Part D on management of various resources.

Chapter 5 begins by considering what is meant by the concept of *best practice*. It then examines the importance of *context* for a problem-solving situation and the need to be able to design solutions to fit specific situations. In this regard, we face a basic challenge. The unique characteristics of a given place and time suggest that it is best to develop approaches and solutions specific to a situation. However, when this is done, a potential problem is that some people may perceive other people or regions receiving preferential treatment, since different arrangements are being applied to them. As a result, there often is pressure, in the spirit of being equitable, to use the same or a similar approach in all places, regardless of differences among them. Thus, debate arises over the merits of using a standard approach versus allowing the development of different approaches, depending on the circumstances prevailing in a particular place.

While appreciating the importance of context is necessary, best practice stipulates that we also need to understand the distinction among most probable, desirable, and feasible futures and why a *vision* or *direction* needs to be established to help in making choices about what is the right thing to do. Often, those concerned with environmental problems give most of their attention to what is the most likely or probable future and how we should prepare to deal with it. Understanding what is likely to occur in the future is very important, but it is also important to recognize that choices can often be made and that the most probable future is not necessarily the most desirable future. Thus, best practice indicates that we should have a clear sense of what kind of desirable future we aspire to so that we then can judge, knowing what is likely to occur, whether by intervening it is possible to move closer towards what is desired. A fundamental challenge, of course, is that societies are not homogeneous, and therefore at any given time there may be competing visions about what kind of future is desirable. Thus, a major task is to identify, and achieve, a shared vision.

Various possibilities for what could constitute a desired future have been identified, and we focused in Chapter 1 on two of them: *sustainable development* and *sustainable livelihoods*. In determining what might be a desirable future, it is important to appreciate that basic values shape perspectives. In that spirit, Chapter 5 examines the difference between *biocentric* and *anthropocentric* views. As part of this discussion, the guest statement by Dan Shrubsole draws attention to different interpretations about what constitutes a 'resource' and how that can influence thinking about what are appropriate decisions to take.

A basic concept associated with best practice is a *systems perspective* and how that can be applied as an ecosystem approach. In Chapter 5, discussion focuses on how the ecosystem approach can be translated from concept to practice, which is not always easy, especially since most often, administrative or political boundaries do not reflect or respect ecosystem boundaries. Another best practice element is the need for thinking simultaneously in the short, medium, and long terms. Too often, our society expects instant results or gratification, which places emphasis on the short term. In addition, elections at municipal, provincial, and federal levels usually occur within five years and sometimes as frequently as every two or three years, which further drives attention towards the immediate and short term. The point is not that we should think only in the long term but that we should be thinking simultaneously at several time scales and also have the patience and understanding to recognize that results may take some time to emerge.

Aspects of planning and management continue to be the focus in Chapter 6, but here the emphasis shifts to processes and products. For example, if a systems perspective, noted in Chapter 5, is to be used, it requires considerable *collaboration* and *co-ordination* among governments and First Nations, public agencies, the private sector, and non-government organizations. Both are needed to overcome what is often termed the *silo effect*, or the propensity of organizations or individuals to focus only on their own interests and responsibilities and not to consider those of others. One means of achieving collaboration and co-ordination is through *stakeholder* and *participatory approaches* and by continuously seeking to enhance *communication* among participants. In Chapter 6, alternative approaches for facilitating participation and communication are examined. None is perfect, and each has different strengths and weaknesses. The challenge is to determine which mix of approaches to use so that the strengths of one can offset the limitations of another.

Best practice increasingly includes *adaptive management*, *impact and risk assessment*, and *dispute resolution*. These processes all explicitly recognize that there are high levels of change, complexity, and uncertainty in resource and environmental management and conflicts often occur. Therefore, the challenge is not to eliminate or avoid change, complexity, uncertainty, and conflict but instead to manage within the reality of their presence. *Adaptive management* involves monitoring experience in order to make systematic adjustments as a result of that experience. *Impact and risk assessment* encourages us to be proactive—that is, to look ahead and anticipate positive and negative results from management actions so that we can enhance the positive and mitigate the negative. *Dispute resolution* methods and processes have been created to enable us to deal with the presence of fundamentally different values, needs, interests, and behaviour, which need to be reconciled or at least managed.

A challenge in resource and environmental management is that strategies, programs, and plans have to gain credibility or legitimacy. This is normally realized by having a statutory or legislative base, political commitment, or administrative endorsement. The more of these factors in place, the more visibility and authority there will be for strategies, programs, or plans. One way to enhance their credibility is to link them

Toronto Harbourfront Parks and Open Spaces public consultation process.

explicitly to other management tools that have a statutory base, such as *regional and land-use plans*. When connections are made with such tools or methods, the probability of initiatives being sustained and implemented usually goes up markedly.

The discussion of philosophies, processes, and products in Part C provides an overview of elements of best practice relating to resource and environmental management. The purpose of presenting these elements is threefold. First, being aware of them will enable you to examine ongoing initiatives to determine whether they reflect best practice. If they do not, then you must begin to consider what would have to be added or changed for best practice to be achieved. Second, understanding the elements of best practice will allow you to incorporate them into solutions you want to develop relative to emerging or new issues. Third, your awareness will enable you to consider and explore how science can be used to provide a more solid foundation for best practices in planning and management.

Chapter 5

Planning and Management: Philosophy

Learning Objectives

- To appreciate the significance of different planning and management approaches.
- To understand the importance of context for a problem-solving situation and the need to be able to design solutions to fit particular contexts.
- To appreciate the distinction among government, governance, and management.
- To distinguish among most probable, desirable, and feasible futures and why it is important to identify a vision or direction to help in making choices about what is the right thing to do.
- To know the difference between ecocentric and technocentric perspectives.
- To understand the significance of a systems perspective and how that can be applied as an ecosystem approach.
- To realize the importance of thinking simultaneously in the short, medium, and long terms.
- To understand the significance of 'social learning' as a foundation for resource and environmental management.

Introduction

Improved resource and environmental management is likely to be achieved if two principles are taken into consideration. First, and as already discussed in Chapter 1, using science to inform decision-making is desirable. Second, management and decisions should reflect what are accepted as the best planning and management approaches in terms of concepts, processes, and methods. In this chapter, we focus on what might be termed *philosophy* and consider basic concepts widely accepted around the world when addressing resource and environmental problems. In Chapter 6, attention will turn to *processes*, *methods*, and *products*, or those processes (such as public participation and community-based approaches), methods (impact and risk assessment), and outcomes (effective, efficient, equitable) generally recognized as leading to more effective environmental management.

Awareness of different approaches enables us to assess what is being done in any given situation and to judge whether the approach is likely to be appropriate. In that sense, the ideas

reviewed in Chapters 5 and 6 serve as a framework or template through which you can see whether the way a problem is being addressed reflects what many professionals around the world consider the most suitable approaches.

Planning and Management Components

Context

The **context**—i.e., specific characteristics of a time and place—needs to be systematically considered when developing a strategy, plan, or approach for a resource or environmental management problem. Biophysical, economic, social, and political conditions frequently differ from place to place and from time to time, demonstrating that it is usually inappropriate to proceed as though one model or approach were sufficient in every situation. Instead, it is often necessary to custom-design solutions depending on the conditions in different places and at different times. In other words, 'one size does not fit all situations'

Management Challenges

Resource management is at a crossroads. Problems are complex, values are in dispute, facts are uncertain, and predictions are possible only in a limited sense. The scientific system that underlies resource management is facing a crisis of confidence in legitimacy and power. Top-down resource management does not work for a multitude of reasons, and the era of expert-knows-best decision-making is all but over.

Some of the new directions that have been proposed include adopting learning-based approaches in place of set management prescriptions . . . , using a broader range of knowledge . . . , dealing with resilience and complexity . . . , and sharing management power and responsibility. . . . As Pahl-Wostl and Hare (2004, 193) put it, 'management is not a search for the optimal solution to one problem but an ongoing learning and negotiation process where a high priority is given to questions of communication, perspective-sharing, and the development of adaptive group strategies for problem solving.'

—Berkes, Armitage, and Doubleday 2007, 308

(Box 5.1). Considering context is especially important in a large country like Canada where there is a wide variety of ecosystems and a rich tapestry of differing cultural perspectives.

The importance of context reaffirms that it is advisable to include local people when developing a strategy or implementing initiatives, since they often have special insight into the conditions of a region or place (discussed further in Chapter 6). This is especially critical in situations involving First Nations populations. Understanding the significance of context also means understanding that context can change as conditions change, requiring the capacity and willingness to modify strategies and initiatives to ensure that they remain relevant. This point is discussed further in Chapter 6 with regard to the concept of adaptive management.

While context is important, it is not unusual in practice for public agencies to prefer a standardized approach to problem-solving. The rationale is that a standardized approach is easiest to defend or justify. If every area or region is treated the same way, then no region can be perceived as receiving special or preferential treatment. But if a specific approach is designed for one region or place, people in other regions might charge that the region was favoured over theirs.

We suggest that despite possible criticism about **custom-designed solutions**, managers should recognize the specific conditions of a place and time and design accordingly. It is better to develop an approach specially designed for the needs of a place and accept criticism about perceived favouritism than to use a standardized, cookie-cutter approach that forestalls criticism but does not really suit the specific conditions or needs.

Context in the Big Picture

Experience since the Rio Summit in June 1992 and the Summit in Johannesburg in 2002 has confirmed that management of natural resources and the environment involves many organizations and jurisdictions that frequently have overlapping and/or conflicting legal mandates and responsibilities, numerous and often conflicting interests regarding access and rights to environmental systems, and growing scepticism about the formal mechanisms of **governments**—local, state, national, regional, international—to deliver services effectively, efficiently, and equitably.

Behind such challenges is the reality that **governance** of resources and the environment takes place in situations defined by high levels of *complexity* and *uncertainty*. Furthermore, managers function in situations characterized by rapid *change* and at the same time often need to become agents of positive change. Finally, managers increasingly deal with *conflict* because of the many different interests related to resources and the environment. As Homer-Dixon (2000, 1) has observed, 'the complexity, unpredictability, and pace of events in our world, and the severity of global environmental stress, are soaring.' In his view, such complexity, unpredictability, and rapid change create an **ingenuity gap**, or growing disparities between those who adapt well and those who do not. Homer-Dixon (2000, 3) defines ingenuity as 'ideas for better institutions and social arrangements, like efficient governments and competent governments'.

Governance and Management

The difference between the two is subtle but important. Simply put, '. . . governance' refers to the decision-making process we follow, whereas '. . . management' refers to the operational procedures we adopt. Governance refers to how we make decisions and who gets to decide; management refers to the models, principles, and information we use to make those decisions. Obviously the two are interrelated; however, management is often the focus of debate, whereas governance is often overlooked.

—Bakker 2007, 16

Box 5.1   Custom-Designing to Deal with Agricultural Diffuse Pollution

Stonehouse (1994) reviewed Canadian experience regarding adoption and use of soil conservation practices, with implications for non-point-source pollution control. He concluded that both cultural and biophysical factors influence conservation decisions by farmers and that each set of factors operates at a localized, individual farm or land-unit level and at a broader regional or national scale. As a result, he recommended that a targeted approach is the best way to apply public policies to help solve degradation problems, in contrast to the dominant universal application approach normally used in Canada. The latter approach assumes that farmers and landowners are homogeneous and that they contribute to degradation in more or less the same way. In contrast, he argued that farmers are extremely heterogeneous with respect to their personal characteristics, economic circumstances, and managerial abilities. Furthermore, their farms vary in size, natural resource characteristics, and enterprise specializations. Such differences result in wide variations in conservation effort, leading to the need for differential treatment of farmers regarding remedial or preventive public policy measures.

Stonehouse suggested that farmers can be identified as innovators, early adopters, mainstream adopters, laggards, or apathetic types, in descending order of conservation effort. In a targeted approach, the type of farmer is matched with different mixes of policy instruments (ranging from

voluntary to enforced compliance). For example, for innovators (or those most likely to adopt new ideas), some form of public recognition of their stewardship and public leadership would be appropriate, along with, where necessary, information and extension assistance for newly emerging conservation practices. Early adopters, in contrast, would most likely be better assisted by information and extension assistance. Mainstream adopters would need some positive financial incentives, particularly for practices for which profitability was unclear or risk-loading was significant. For laggards, people with a much lower inclination to adopt conservation measures, much stronger incentives would be required because of their lack of problem awareness, low management skills, economic problems, or other reasons. Thus, for laggards, the most effective policy approach would be positive financial incentives and necessary information and extension support. Members of the final group, the apathetic, are often aware of degradation problems but are unwilling to take remedial action. For them, strong measures must be used, and the possibilities include regulation and control of farming activities associated with resource depletion, supported where needed by litigation, taxes, and penalties.

Key messages from Stonehouse's evaluation of Canadian experience are the need to custom-design solutions, to apply a mix of policy instruments, and to use a targeted rather than a universal approach.

Governance, Management, and Monitoring

. . . governance is the process of resolving trade-offs and providing a vision and direction for sustainability, management is the realization of this vision, and monitoring provides feedback and synthesizes the observations to a narrative of how the situation has emerged and might unfold in the future.

—Olsson 2007, 269

Experience since the early 1990s suggests that four other contextual aspects are also important for understanding progress related to managing resources and the environment wisely. One is the preoccupation of many national governments with debt and deficit reduction. Since Rio in 1992, many governments have significantly reduced their expenditures on environmental infrastructure and services, which usually has had a serious negative impact on agencies responsible for resources and the environment.

For example, in Canada during the mid-1990s, the federal agency responsible for water management had its budget reduced by 35 per cent. In Ontario, the Ministry of Environment had its operating budget cut by 44 per cent, and the Ministry of Natural Resources lost 42 per cent of its budget between 1995–6 and 1997–8—an approach that Kreutzwiser (1998, 138) characterized as a 'scorched earth retreat' by the provincial government with regard to its normal responsibilities for resources and the environment. Similar decisions were taken in Alberta and British Columbia. Events such as the contamination of the water supply in Walkerton, ON, during May 2000 (Chapter 11) were partly the result of a lack of funding to government resource agencies and led to rethinking about such cuts. Yet significant cuts to environmental budgets continue to occur.

In September 2007, the CBC reported that budget problems at Environment Canada were threatening wildlife programs and services, freezing allotted money for some and reducing funds to zero for others. The Canadian Wildlife Service had its service budget frozen for the rest of the fiscal year, meaning that all of its scientific field and survey work was halted. The budget for national wildlife areas, which protect bird and other animal habitats, was cut to zero. The Environmental Monitoring and Assessment Network, which observes changes in ecosystems, lost 80 per cent of its budget. The Migratory Bird Program, which monitors the health of bird populations, had its budget cut by 50 per cent. Although the environment minister, John Baird, was not available for comment, his department reported that it was prioritizing spending for climate change initiatives (Chapter 7). This incident illustrates the way that governments sometimes proudly announce new programs to address environmental challenges but are a little less forthcoming in detailing what programs might have been cut in order to support the new programs. In many cases, no new money is being allocated to the environment; it is merely shifted from other programs.

Second, and emerging from the concern about debt and deficit reduction, many national and state governments have been (1) downloading responsibilities for environmental services to lower levels of government, which usually do not have the human or financial resources to maintain levels of service; (2) commercializing such services; or (3) privatizing these services. With regard to the Ontario government's decision to close provincial water testing laboratories in 1996, Kreutzwiser (1998, 139) concluded that the government made such cutbacks 'without any consideration of the capacity in the private sector, or among municipalities, to test municipal water supplies'.

Such initiatives are usually justified by the principle of *subsidiarity* (allocating responsibilities to levels of government closest to where the services are used or received) or *efficiency* (providing services at least cost). However, some believe that these decisions are at least as much ideological as they are rational, reflecting a belief in market-based economies as the most effective way to allocate scarce societal resources and that less government involvement is desirable. Thus, while the rationale for the subsidiarity principle is sensible because it allows people closest to the outcome to participate directly in decisions that affect them, it also requires willingness to maintain,

create, or enhance human and financial capacity at local levels to deal with resource and environmental management issues. When senior levels of government download their traditional management responsibilities to local levels without regard for the existing capacity to handle those responsibilities or without consideration as to how the necessary capacity can be created, they may cite subsidiarity as the basis for decisions that in fact may have been driven more by financial cost-cutting considerations. Various ideological perspectives present quite different views of the environment and resources, and these differing views can affect planning and management dramatically.

Third, while many governments favour less government intervention, more reliance on the private sector and market forces to deliver products and services efficiently, acceptance of the value of globalization, and modified management processes to reflect a 'business model' that emphasizes efficiency, results-based management, and tangible products, they demonstrate less interest in using systematic and thorough consultation processes regarding development and implementation of policy. Indeed, in some jurisdictions, consultation that is transparent and accessible to the general public has almost disappeared. In such cases, while the senior levels of government embrace subsidiarity as an appropriate principle to guide decisions, in practice their actions do not reflect adequate consideration of all the preconditions required for subsidiarity to function effectively.

Fourth, since the late 1990s, many governments have been steadily backing away from concern for or commitment to environmental issues and instead have been emphasizing strategies for economic growth. An often cited example is the decision by president George W. Bush to have the United States withdraw from the Kyoto Protocol for reducing global warming (see also Chapter 7) and to refocus energy policies in the US on supply solutions with little or no attention to demand management strategies. There is some irony in this situation, given that many polls indicate that the general public has become increasingly concerned about environmental issues over the past five years.

As Mascarenhas (2007) has explained, the above characteristics are usually associated with the concept of **neo-liberalism**, a view of political theory based on the belief that humans' well-being is best achieved by encouraging and facilitating individual freedom and minimizing the role of government. Defining features of neo-liberalism are strong private property rights, free markets, and free trade.

Winfield and Jenish (1998) reviewed the implications of neo-liberalism for resource and environmental management in Ontario during the mid-1990s. They concluded that neo-liberalism prompted changes that limited or constrained nearly all provincial legislation and regulations related to protection of the environment or management of natural resources, major reductions in the key agencies responsible for the environment and natural resources, a scaling back of opportunities for public participation in decision-making, and a rearranging of responsibilities for the environment among provincial and municipal governments and the private sector. Regarding the role of the private sector, much more scope was allowed for voluntary compliance and self-monitoring.

Thus, much of what is (or is not) happening in resource and environmental management around the world may be attributed to the shifting political power of differing ideologies. To be an informed citizen, planner, or manager, individuals should aim to learn about the basic values and assumptions of various ideologies so that they can determine whether their arguments realistically present the rationale for policies and actions. In Box 5.2, the environmental commissioner of Ontario identifies some basic challenges and poses fundamental questions that deserve our attention. While he refers to Ontario, his comments can be generally applied across the entire country.

Vision

Ends need to be distinguished from means. In other words, before deciding how to deal with resource and environmental management problems or opportunities, managers should determine what ends or desirable future conditions are sought, as highlighted in Box 5.2. This consideration is often referred to as a need to have a clear vision or sense of direction.

According to Nanus (1992, 8–17), a **vision** represents a realistic, credible, and attractive future for a region, community, or group. It is ideal to have a shared vision, one to which many people are committed. Achieving a shared vision is challenging, however, since many interests exist in a society, some of which are mutually exclusive. If a shared vision about a desirable future is to be achieved, it is important to involve many stakeholders in the management process, a matter discussed further in Chapter 6.

Box 5.2   Thinking beyond the Near and Now

When did we become so focused on the present moment and our immediate situation? It seems to me there was a time not so long ago when we were focused as a society on building a better future for our children and our grandchildren. We had a broader sense of the connections between the landscape, the communities of Ontario, and the economy. . . . And we valued that. We worked for a better tomorrow, but at the same time we didn't forget about the past, about those who had made sacrifices and worked hard to create the opportunities we enjoyed.

Somehow the awareness of the past and concern about what the future might become seem to be missing from current public discourse and decision-making. In the same way, there is little concern about how other areas of the province are impacted by decisions made in the urban centres. We accept as inevitable that the population of the GTA [Greater Toronto Area] will continue to swell by 100,000 new residents every year. Yet have we really come to grips with what that will mean for our remaining forests, watercourses, air quality, and our most productive agricultural land? Do we care about our countryside?

Alas, life moves too quickly to linger on such considerations. We must grow at an expanding rate. We can't worry about those things now. Approve those subdivisions—

they're needed now. Build more highways—otherwise we can't get to work. Just get the garbage to Michigan—and make sure it's nowhere near us. Do something now about that new crisis in this morning's newspaper.

It's rather like speeding down a dark northern Ontario highway on a moonless June night with only your low beams on. You have that confident, comfortable feeling because there is no one on the road and you're making good time. But if you just click on your high beams, you'll see the moose standing there only a few seconds in front of you. . . .

The analogy of a speeding car works quite well in a discussion about the 'near and now', because as most drivers know, we have a tendency to focus on objects and surroundings just in front of the car. It takes training and discipline to bring your eyes up to the distant horizon where you become aware of events and objects far ahead—in both time and space. But only then can you acquire the capacity to anticipate, plan, and react to future hazards. And so it is with public decision-making that affects the environment. We all must keep our eyes on the horizon.

Source: Gord Miller, Environmental Commissioner of Ontario, 2003, 4.

In Chapter 1, sustainable development and sustainable livelihoods were examined as possible visions for the future. The examples below illustrate what different organizations or groups have developed in terms of vision.

• *Bay of Fundy Ecosystem Partnership*
Promoting the ecological integrity, vitality, biodiversity, and productivity of the Bay of Fundy ecosystem, in support of the social well-being and economic sustainability of its coastal communities.

Facilitating and enhancing communication and co-operation among all citizens interested in understanding, sustainably using, and conserving the resource, habitats, and ecological resources of the Bay of Fundy.
—http://www.bofep.org/vision.htm

• *Fraser River Basin Charter*
The Fraser Basin is a place where social well-being is supported by a vibrant economy and a healthy environment.
—Fraser Basin Council 1997

• *Vision for Nova Scotia's Drinking Water Strategy*
To conserve and protect Nova Scotia's drinking water supplies for the environmental, health, economic, and social benefits of current and future generations.
—Nova Scotia Department of Environment and Labour 2002, 6

• *Toronto and Region Conservation Authority*
The quality of life on Earth is being determined in rapidly expanding city regions. Our vision is for a new kind of community. The Living City, where human settlement can flourish forever as part of nature's beauty and diversity.
—Toronto and Region Conservation Authority 2006

• *Nova Scotia's Vision for Sustainable Coastal Tourism Development*
Nova Scotia's tourism industry will be recognized globally as a leader in adopting sustainable practices for coastal tourism, including an adaptation of Canada's Code of Ethics and Guidelines.

Tourism operators will display stewardship for the protection and enhancement of natural, cultural, and community tourism assets and will be recognized and rewarded for these efforts.

There will be increased integration between the tourism industry and other users of Nova Scotia's coastal resources.

Government planning and policy development will be implemented to protect and enhance Nova Scotia's coastline in recognition that it is an essential component of our tourism product in the long term.

There will be an unparalleled and authentic experiential tourism industry based on both the cultural and natural components of our coastline, which satisfies the body, mind, and soul and encourages return visits.

These experiences will also draw visitors to inland areas to experience all that Nova Scotia has to offer.

Tourism will be planned at a local level to integrate coastal tourism into the broader economy and lifestyle of Nova Scotia's coastal communities and to foster an appreciation of the coastal tourism assets and the economic benefits of tourism by local residents.

A more vibrant, viable industry will develop associated with coastal experiences.

Tourism will be developed with an understanding of global tourism market trends and knowledge of Nova Scotia's key current and potential markets.

Resources and product development programs will be in place to overcome the most significant infrastructure gaps affecting Nova Scotia's coastal tourism product.

There will be tourism businesses open on a year-round basis along our entire coastline.
—Nova Scotia Tourism Partnership and Nova Scotia Tourism, Culture and Heritage 2007, 9

A Vision for Your Area or Community

What would be the key features of a vision for resource and environmental management for your college or university, or community, region, or province? Does one exist now?

Visions can be developed in various ways. One is to ask three questions: (1) What is likely to happen? (2) What ought to happen? (3) What can happen? In much work in resource and environmental management, the focus is on the first question. An answer to this question helps to establish the likelihood of some future state, assuming continuation of current conditions or estimating changes in some current conditions. Such answers are useful because they help to establish the most probable future. However, this question does not to help determine whether the most probable or likely future is also the most desirable. To deal with the issue of desirability, we also need to ask what ought to happen and therefore consider what would be desirable future conditions for a society or place. These two questions (what is most likely, what is most desirable) reflect the difference between forecasting and backcasting, as explained by Tinker in the box on page 168. Finally, the third question imposes discipline by requiring us to consider what is feasible or practical. We need to address all three of these questions instead of relying only on the first, which is the common practice.

Forecasting versus Backcasting

Forecasting takes the trends of yesterday and today and projects mechanistically forward as if humankind were not an intelligent species with the capacity for individual and societal choice. Backcasting sets itself against such predestination and insists on free will, dreaming what tomorrow might be and determining how to get there from today. Forecasting is driving down the freeway and, from one's speed and direction, working out where one will be by nightfall. Backcasting is deciding first where one wants to sleep that night and then planning a day's drive that will get one there.

—Tinker 1996, xi

Ethics and Values

To ensure that a shared vision is endorsed by a group or society, it must be consistent with and reflect basic ethics and values. Alternatively, a vision may outline a desirable future significantly different from the present situation, and if it is to be achieved, there will have to be a shift in fundamental values. The vision can help to clarify what different values will have to prevail for the desirable future to be realized.

As Matthews, Gibson, and Mitchell (2007, 337) have observed, an ethic is 'a set or system of moral principles or values that guides the actions or decisions of an individual or group'. Such principles or values help us to determine right from wrong and how to behave appropriately. At the same time, no set of ethics gives every direction that is needed, and sometimes ethical principles conflict with each other. Furthermore, we normally do not understand or cannot predict all the consequences of our decisions or actions, and even when we can, many anticipated consequences contain 'shades of grey', making it difficult to know how much of a good or bad feature will be present.

Notwithstanding the challenges outlined above, it is desirable to have a clearly articulated foundation, based on ethical principles, from which we can make decisions. Ethical principles direct us regarding appropriate behaviour (e.g., always treating others as you like to be treated yourself) and also regarding process issues (such as striving for accountability, transparency, and equity). At a minimum, articulating such principles should help us in deciding among different options, assuming that we intend to act in accordance with them. In Chapter 11, we outline current thinking regarding a set of universal principles to guide decision-making for water.

Victoria Park, Kitchener: 1870s (above) and present (below).

Orchard, Creston, BC.

Ecocentric satire of technocratic values. © 1999 P.S. Mueller

We also need to be able to appreciate that the values in different societies usually reflect a mix of explicit principles as well as implicit and unstated principles, with the latter simply being understood as common knowledge among all members of the society.

Values normally do not appear in a constitution, charter of rights, or other legal documents. The presence of both explicit and implicit values and principles, of course, often raises challenges in cross-cultural situations, when someone from outside a society may not recognize or understand the significance of implicit values unless they have been exposed to that society for a significant length of time.

Two sets of values that are significant for resource and environmental management can be identified. At one end of a continuum are **ecocentric values**, which include a belief that there is a natural order governing relationships between living things and that there is a harmony and balance reflecting a natural order that humankind tends to disrupt through ignorance and presumption. Other key values include reverence for, humility and responsibility towards, and stewardship of nature. Those with ecocentric values are not against technology per se but favour application of low-impact technology, oppose bigness and impersonality in all forms, and advocate behaviour consistent with ecological principles of diversity and change (O'Riordan 1976).

In contrast, a **technocentric perspective** is based on the assumption that humankind is able to understand, control, and manipulate nature to suit its purposes and that nature and other living and non-living things exist to meet human needs and wants. While ecocentrics are concerned about choosing appropriate ends and using consistent means, technocentrics focus more on means because of their confidence in human ingenuity and rights, which makes them less concerned about the moral aspects of activities or consequences. Technocentrics admire the capacity and power of technology and believe that technology and human inventiveness will be able to overcome possible resource shortages as well as remediate or rehabilitate environmental degradation. Most but not all technocentrics live in major urban centres.

While recognizing that a spectrum of values is important if resource and environmental managers are to understand why ideas are supported or opposed, we should not assume that the values held by a group or society can be neatly allocated into these two categories. The boundaries are often blurred and indistinct, and many people will support certain aspects in both categories, depending on the conditions. In that regard, O'Riordan (1976) shares a story about a man who asked a socially conscious friend what he would do if he had two houses. The friend replied that he would keep one and give one to the state. The man then asked his friend what he would do if he had two cows. Again, the friend answered that he would keep one and give one to the state. The man next asked what he would do if he had two chickens. The friend responded that he would keep them both. When asked why, the friend stated, 'Because I have two chickens.' This story highlights that circumstances (referred to earlier as 'context') can be very important in shaping outlooks about what is needed, important, and desirable. Thus, it is important for resource and environmental managers to be aware of dominant and secondary values in a society so that they can determine how these values can be used or may have to be altered if a vision is to be defined and then achieved. In his guest statement, Dan Shrubsole provides further insight into the shades of meaning that can be associated with perspectives reflecting different basic values.

Systems and Ecosystem Perspective

The Saskatchewan Round Table on Environment and Economy, in its *Conservation Strategy* published back in 1992, stated that 'ecosystems consist of communities of plants, animals, and micro-organisms, interacting with each other and the non-living elements of their environment.' In the same year, the final report of the Royal Commission on the Future of the Toronto Waterfront argued that an ecosystem approach emphasizes that human activities are interrelated and that decisions made in one area affect all others. The royal commission concluded that 'dealing effectively with . . . environmental problems . . . requires a holistic or ecosystem approach to managing human activities.' The timing of these two statements reminds us that there has been awareness for some time of the value of an ecosystem approach for planning and management. Indeed, most researchers agree that the concept of ecosystem was formulated in 1935 by Arthur Tansley,

Guest Statement

Five Elements of a Resource Management Philosophy

Dan Shrubsole

Part of the reason why resource management is at a crossroads pertains to the lack of clarity in understanding and/or communicating a consistent management philosophy. A resource management philosophy comprises at least five key elements—world view, methodology, politics, ethics, and aesthetics. The following commentary provides a brief and limited insight into each element.

First, we should enunciate a world view—that is, some generalized view of the world and the relationship between people and the world. Ecocentrism and technocentrism, which are mentioned in this chapter, are two environmental world views. Another aspect of world view is related to the universality of problems and solutions. For instance, should we be able to custom-design solutions to deal with diffuse source pollution, source water protection, or wetland management, or are more consistent/uniform approaches within Canada, province/territory, or region most appropriate?

Second, an environmental philosophy would consider issues related to methodology, which refers to general ways of structuring and organizing our thinking. Deductive and inductive thinking represent two alternative forms of logic. The former is sometimes referred to as 'top down', because we identify a theory (e.g., resiliency) and establish a scientific experiment (i.e., use a null hypothesis) to test the theory. In other words, deduction tests a general theory on the basis of specific cases. Inductive thinking, also termed 'bottom up' thinking, works from the specific case to the general. By making specific observations and measurements over time and space, we may detect patterns and regularities. On this basis, generalizations or theories may emerge. Resource and environmental management research involves both types of thinking.

Dan Shrubsole

Third, a resource management philosophy should consider politics. This does not refer to whether the Conservative, Green, Liberal, New Democratic, or other political party should be elected. At a philosophical level, attention is paid to how an ordered society is achieved, specifically to the role of government and citizens. 'Bottom up' and 'top down' decision-making is one way to consider how governments might make decisions. Another level considers the breadth and depth of government involvement. For instance, must water supply be owned and operated by government (and which level), or is there a role for the private sector, and if so, what is it? Should natural resources and their industries be nationalized or protected from foreign takeovers? What role, if any, should the market play in guiding management activities?

Ethics is the fourth element. It guides our sense of right and wrong. For instance, the need for and means of achieving effective stakeholder involvement is governed not only by laws but by a sense that public participation represents a 'fair and required' element in decision-making.

The fifth aspect pertains to the consideration of aesthetics or beauty. When you close your eyes, what do you imagine as the 'ideal landscape'? Does it involve any people, or is it strictly 'natural'? If it is the former, is it an urban setting with highly manicured parks with paved walkways, or is it at the urban-rural fringe? While identifying these five elements of an environmental philosophy is easy, articulating them is much harder. For many of us, our philosophy changes over time relative to our experience. As well, many of us inherit rather than consciously select our philosophy. Inheriting a philosophy can happen quite subtly. For instance, we have traditionally labelled resource managers according to their knowledge about resource stocks, such as 'water managers', 'foresters', 'agronomists', 'wildlife managers', or 'land managers'. These labels imply an orientation that overemphasizes functional and biophysical processes relative to the interrelationships between natural and human systems and within human systems. These types of labels may also present obstacles to more ecosystem-based or interdisciplinary-based management approaches. Consciously reflecting on the elements of our philosophy may assist you in better determining where differences of opinion exist and identifying effective solutions.

Dan Shrubsole is a professor and chair, Department of Geography, University of Western Ontario. His research focuses on sustainable water management, flood hazards, and risk management.

who argued that organisms were affected by more than other organisms and that other key factors included variables such as soils, water, weather, and climate.

Such interpretations of ecosystems and an ecosystem approach reflect the ideas presented in Chapters 2, 3, and 4 in which emphasis was placed on systems, interrelationships or linkages, energy flows, and ongoing change. In this section, we turn our attention to the characteristics, opportunities, and

Aerial view of Grasslands National Park, Saskatchewan.

Toronto waterfront.

challenges presented by the systems or ecosystem approach for resource and environmental management.

Characteristics of the Ecosystem Approach

Slocombe (2004, 421) has suggested that the ecosystem approach has a set of core characteristics, including systems concepts and analysis, ethical perspectives, stakeholder and public participation, a bioregional place-based focus, efforts to identify and develop common goals, and gaining a systematic understanding of the ecosystem of interest. Slocombe (2004, 420) also argued that the ecosystem approach was developed in order to address a mix of problems frequently encountered in resource and environmental management. In his view, these problems include:

- viewing people and their activities as separate from nature;
- adopting single-discipline perspectives;
- emphasizing single resource uses or economic sectors and avoiding or ignoring conflicts over possible alternative uses;
- failing to recognize the many ways in which ecological and socio-economic systems are interconnected;
- ignoring the propensity of biophysical and socio-economic systems to change, sometimes rapidly and unexpectedly;
- reacting to rather than anticipating change and problems, leading to attempts to eliminate uncertainty by controlling complex, dynamic systems instead of adapting to them.

Viewed in the context of the characteristics and problems identified by Slocombe, it becomes apparent why the Royal Commission on the Future of the Toronto Waterfront stated that an ecosystem approach:

- includes the whole system, not just parts of it;
- focuses on the interrelationships among the elements;
- recognizes the dynamic nature of the ecosystem, presenting a moving picture rather than a still photograph;
- incorporates the concepts of carrying capacity, resilience, and sustainability—suggesting that there are limits to human activity;

- uses a broad definition of environments—natural, physical, economic, social, and cultural;
- encompasses both urban and rural activities;
- is based on natural geographic units such as watersheds rather than on political boundaries;
- embraces all levels of activity—local, regional, national, and international;
- understands that humans are part of nature, not separate from it;
- emphasizes the importance of species other than humans and of generations other than the present;
- is based on an ethic in which progress is measured by the quality, well-being, integrity, and dignity it accords natural, social, and economic systems.

Opportunities through the Ecosystem Approach

If the above attributes are accepted as defining an ecosystem approach, they present some basic challenges to contemporary environmental and resource management. First, in a Western, industrialized society such as Canada, many humans believe that they have a dominant role relative to nature and that the environment and natural resources exist to satisfy human needs and wants. This is an anthropocentric perspective, in contrast to the ecocentric world view described earlier. By emphasizing that humans are part of nature rather than separate from it and by recognizing the inherent value of non-human species and things, the ecosystem approach questions such a belief.

Second, by taking a holistic perspective focusing on interrelationships, the ecosystem approach reminds us of the need to consider management problems and solutions in the context of linked 'systems'. It forces us to appreciate that decisions made about one system, such as land, can have consequences for other systems, such as water or wildlife, and vice versa. For example, many flooding problems are often associated with land-use practices such as removing vegetation, which accelerates runoff, or with allowing development in areas subject to regular inundation. Concentrating only on the aquatic system to reduce flood damages is unlikely to be effective, since many land-based activities exert some influence on this system.

In contrast, the conventional approach to environmental or resource management has often focused on systems in isolation from one another, as reflected by having one government agency responsible for forestry, another for wildlife, another for water, another for agriculture, and another for urban development. All of these agencies have a shared interest in resources such as wetlands, but if they each focus only on their own goals or interests, they may undermine the activities or values of other agencies.

Third, the ecosystem approach demands that the links between natural and economic or social systems be considered. This focus is also one of the basic thrusts behind sustainable development. When such linkages are recognized, it becomes apparent that certain thresholds normally exist in natural systems and that exceeding these thresholds leads to deterioration and degradation. For example, agricultural production can be increased by adding chemical inputs (fertilizers, pesticides, herbicides). However, the cumulative effects of agrochemicals may eventually make the product grown (e.g., fruit or vegetables) unsafe for human consumption. Other concerns revolve around introducing chemicals into adjacent environments, resulting in eutrophication and pollution, as discussed in Chapters 4 and 10. This reminds us that while sustainable development accepts the need for development to meet basic human needs, some kinds of growth are not sustainable because they lead to degradation of natural, economic, or social systems.

Fourth, the holistic perspective reminds us that decisions made at one place or scale can have implications for other places or scales. If a community deposits untreated sewage into an adjacent river, people and communities downstream will bear some costs related to that action. Or if a community, state, or nation is unwilling to impose emission reductions on factories, some of the costs will be borne by people, states, or nations downwind, since air pollutants are often carried well beyond the borders of the community or state in which the pollutants are generated, as is the case with acid deposition, discussed in Chapter 4.

Fifth, given the impact of decisions in one place for people and activities in other places, the ecosystem approach raises questions regarding what is the most appropriate areal or spatial unit for planning and management. The conventional management unit has usually been based on political or administrative boundaries, such as municipal, regional, provincial, or national boundaries. In contrast, the ecosystem approach suggests that areas identified on the basis of other units, such as watersheds or airsheds, may have more functional value. For example, in managing for migratory birds that travel from the Gulf of Mexico to the Canadian Arctic, national boundaries have little relevance. The management area in this situation comprises at least three nations (Canada, the United States, Mexico). If managers in each nation develop plans and strategies without regard to what is being done in other jurisdictions, the effectiveness of their separate initiatives is likely to be modest, as discussed in Chapter 11.

Sixth, an ecosystem approach highlights that systems are dynamic or continuously changing. An ecosystem, whether a local wetland, prairie grassland, boreal forest, or urbanizing area, is not static. In addition to daily, seasonal, and annual variations, ongoing longer-term changes occur, as illustrated by the transition of natural grasslands to cultivated cropland or of farmland to urban land use. Climate change, discussed in Chapter 7, exemplifies how ecosystems may change and 'migrate' as result of different patterns of temperature, precipitation, and other climate variables. That is why managers and management strategies must be capable of adapting or adjusting to evolving situations. This imperative has led to an interest in *adaptive management*, discussed further in Chapter 6.

Spatial Units for Ecosystem-Based Management

At the most basic level, problems derive from the definition of the units for which planning and management are undertaken. Management units often bear no relationship to the realities of ecological problems (even the home range of the species for which protection is sought), their connections to economic and social processes, or local people's cultural and political identity. Instead, they are arbitrary units defined by lines drawn on the map—lines often drawn by someone who has never been to the region and who, for example, decides a river would make a good boundary.

—Slocombe 1993, 616

In summary, the ecosystem approach incorporates the key ideas that humans are part of nature rather than separate from it, that a perspective emphasizing interrelationships is needed, and that critical thresholds exist. When these aspects are combined, it can be appreciated why almost two decades ago, the Royal Commission on the Future of the Toronto Waterfront (1992, 31–2) concluded that

> the ecosystem approach is both a way of doing things and a way of thinking, a renewal of values and philosophy. It is not really a new concept: since time immemorial, aboriginal peoples around the world have understood their connectedness to the rest of the ecosystem—to the land, water, air, and other life forms. But, under many influences, and over many centuries, our society has lost its awareness of our place in ecosystems and, with it, our understanding of how they function.

The need and rationale for adopting an ecosystem approach was spelled out some time ago by the Conservation Authorities of Ontario (1993, 2):

> The fundamental problem that exists in resource management . . . is not financial constraint. It is that the current body of legislation, agency structures, and mandates do not recognize the concept of ecosystem-based management. The overlapping of mandates between the Ministries of Natural Resources, Environment and Energy, Agriculture and Food, and Municipal Affairs, and Conservation Authorities and Municipalities is evident to everyone.

The situation has evolved over time as public agencies react to specific problems with specific solutions. This issue-by-issue approach results in a situation that, when viewed from an ecosystem perspective, borders on the ludicrous. Thus, many reasons can be identified as to why an ecosystem approach should be used more frequently in Canada for environmental management. However, as we have already pointed out in earlier chapters, implementing an ecosystem approach requires adjustments to arrangements for governance and management systems, aspects considered in more detail in later chapters.

Long-Term View

In resource and environmental management, it is important to have a long-term view (more than 15 years) while also being able to identify actions that should be taken in the short term (less than five years) and middle term (five to 15 years). The rationale is that systems often change slowly and that a significant period of time may be required to shift values, attitudes, and behaviour.

Furthermore, many of our environmental problems have emerged after many decades, or even centuries, so it is

Old-growth forest.

Plantation forest in Finland.

unrealistic to assume that they can be reversed or 'fixed' in a few years. For example, changes in the aquatic ecology of the Great Lakes (see Chapters 3, 4, and 13) reflect decades of different kinds of land use (agriculture, industry, settlement) on lands adjacent to the lakes.

Unfortunately, our society usually wants 'instant results' and does not often show the patience required to deal with problems that have been created over a long period.

There are many reasons for the usual short-term perspective. One is the relatively short time between elections (up to five years for federal and provincial elections and two to three years for municipal elections). Politicians normally want to produce tangible results so they can show what has been accomplished during their term of office and why they should be re-elected. In contrast, those running for office for the first time emphasize how little has been accomplished by those previously elected and why they should be given an opportunity to demonstrate what they could do instead. This mindset drives decisions focusing on short-term, tangible results and usually results in low priority for long-term strategies involving intangible outcomes.

In many instances, changes are required in basic values, attitudes, and behaviour, such as Canadians modifying their basic patterns of activity so that the greenhouse gas emission targets identified for Canada in the Kyoto Protocol can be achieved (see Chapter 7). There was much outcry from 2002 to 2008 against the targets the federal government committed to reach under the Kyoto agreement, based on fear that they would cause economic hardship or disadvantage. The former premier of Alberta, Ralph Klein, was one of the most aggressive in criticizing the targets, since they could eventually lead to less reliance on a 'petro-economy', the backbone of the Alberta economy. Thus, while focused on the possible negative short-term consequences for the Alberta economy, Klein's criticism reflected the reality that when changes occur, some gain and some lose. Those who might lose are understandably not usually very enthusiastic about the changes.

The focus on tangible results also explains why one of the first 'casualties' of budget cutbacks in the public education system usually is outdoor education and field trips, which are viewed as a luxury. And yet, as supporters of outdoor education and field trips argue, if young students are to appreciate the role of natural systems and change their attitude towards them, what better way to accomplish that than by exposing the students directly to these systems through such programs? Unfortunately, the 'product' or result of these programs—changed values and attitudes—is difficult to document in a tangible way, and it often takes many years before behaviours are sufficiently influenced and changed that they have significant positive consequences for the environment.

Thus, while a long-term perspective is necessary, accompanied by patience, perseverance, determination, and commitment, most of our society is often more preoccupied with shorter-term and visible results. This creates many difficulties for those who believe that solutions require a commitment of funds and human resources over an extended and sustained period of time.

Social Learning

As Diduck (2004, 503) has observed, 'Social learning is an emerging perspective in resource and environmental management. At its heart is the suggestion that learning is an idea that applies not only to individuals but also to social collectives, such as organizations and communities.' He and others argue that it is not enough to create opportunities for stakeholders to become engaged in resource and environmental management processes. Such processes should be designed so that both individuals and organizations are able to learn from their experience and thereby become more knowledgeable and effective in the future.

Based on the 'theory of action', those interested in the concept of **social learning** differentiate between 'single-loop' and 'double-loop' learning. The emphasis in **single-loop learning** is to ensure a match between intent and outcome. A metaphor often used is the thermostat. Reflecting single-loop learning, a thermostat is designed to monitor the temperature, and when the temperature becomes either too cold or too hot, it sends a signal to turn the heat on or off. Thus, the thermostat receives information (about temperature in a room), takes corrective action (turns heat on or off), and in that way ensures an outcome consistent with what is intended or desired.

Double-loop learning, in contrast, addresses a different condition—when there is a mismatch between intention and outcome. Double-loop learning happens when such a mismatch exists and is addressed first by challenging underlying values and behaviour rather than assuming the prevailing values and behaviour are appropriate. Returning to the thermostat metaphor, if double-loop learning were to occur, the thermostat would question the basic underlying value (why is the desired temperature set for 22°C? why not for 20°C and adapt to a cooler temperature?) or question the prescribed behaviour (rather than requiring more heat, put on a sweater and/or close an open window).

In resource and environmental management, we should be drawing on both single- and double-loop learning—but especially the latter, because it encourages 'out of the box' thinking. Another way of putting it is to state that single-loop learning emphasizes the right way to get something done, whereas double-loop learning focuses on the right thing to do.

In the next chapter, we will examine specific approaches to facilitating public participation, and when considering choices in that regard, you should recall the distinction between single- and double-loop learning and why we should strive to ensure that conditions are created that encourage both.

Implications

The concepts, or 'philosophy', that reflect best practice for management of natural resources and the environment are becoming more clearly identified. Important concepts include (1) recognizing the contextual factors that characterize a problem situation and being willing to design solutions that address specific attributes of the problem; (2) establishing a vision that identifies a desirable future so that appropriate means can be identified to achieve the desired end; (3) appreciating the strengths and limitations of potential desirable futures; (4) clarifying underlying values that influence attitudes and behaviour and developing ethical principles or guidelines consistent with the desired future; (5) adapting a systems perspective to ensure that the interactions of various environmental and human subsystems are considered; (6) looking beyond the present and immediate future to consider the longer term; (7) appreciating the significance of 'social learning'; and (8) recognizing the importance of governance issues. In the spirit of transforming these ideals into action, Box 5.3 outlines initiatives that you can undertake as an individual or as part of a group.

Box 5.3   What You Can Do

1. Ask elected officials and designated experts to explain how solutions to problems have been designed with regard to the particular conditions in the area of concern.
2. Identify a desirable future consistent with ideas presented in this chapter and book, and then ensure that your behaviour is consistent with this vision.
3. Urge your college or university to include ideas and practices consistent with ideas in this chapter in its curriculum and practices.
4. Strive to incorporate more ecocentric values into your personal outlook on living.
5. Expect public agencies to use a 'systems approach' when dealing with resource and environmental problems, and challenge them when that does not seem to be the case.
6. Advocate both short- and long-term thinking and action related to environmental problems.
7. Actively support an NGO that reflects your own values.

Summary

1. Context refers to the specific conditions related to a time and place. Since context can vary significantly from place to place and time to time, it is desirable to design solutions to fit specific environmental problems. A challenge in doing so is that some people may conclude that a region or group is receiving special consideration.

2. Attention to context reinforces the importance of incorporating the experiential knowledge of local people with scientific knowledge when seeking to understand problems and develop solutions. Because they have lived in an area for many years, local people often have insight and understanding that scientists do not.

3. Important aspects of context include rapid change, high complexity and uncertainty, and significant conflict. We should not be surprised to encounter such aspects when dealing with environmental issues.

4. Neo-liberalism has become an increasingly important ideology shaping the context for environmental problem-solving. Characteristics include reducing the scope of government, relying on market mechanisms to allocate scarce resources, and accepting globalization. This ideology has driven many governments to a preoccupation with deficit and debt reduction and to commercialize or privatize management functions previously taken on by governments.

5. The subsidiarity principle, which stipulates that people closest to problems should participate directly in decisions affecting them, encourages involving local people in decisions. It also is used to justify approaches based in a neo-liberal ideology: decisions to download or privatize responsibilities are defended using the subsidiarity principle.

6. A vision represents a realistic, credible, and attractive future for a region, community, or group. It is ideal to have a shared vision, one to which many people are committed. Achieving a shared vision is challenging, because many interests exist in a society, some of which are mutually exclusive. Without a vision or a well-defined endpoint, managers, their political masters, and society itself have difficulty in determining the most appropriate means for achieving the desired ends.

7. Explicit ethical principles should be identified in order to provide guidance when choosing among options.

8. Ecocentric and technocentric world views reflect different basic values and interests. Their existence is one explanation for the conflicts that arise when decisions must be made regarding the environment or resources.

9. Ecosystems consist of communities of plants, humans, animals, and micro-organisms interacting with each other and the non-living elements of their environment. Their management requires a systems or holistic perspective.

10. A long-term view (15 years or more) needs to be maintained, but at the same time, there must be a capability to deal with immediate or short-term issues. Too often, only a short-term perspective is taken.

11. Social learning emphasizes the importance of appreciating that both individuals and organizations can and should learn from engagement in resource and environmental management decision-making.

12. Single- and double-loop learning models remind us of the difference between doing the thing right and doing the right thing.

Key Terms

context
custom-designed solutions
double-loop learning
ecocentric values

ingenuity gap
neo-liberalism
single-loop learning

social learning
technocentric perspective
vision

Questions for Review and Critical Thinking

1. Can science be 'objective' and provide unbiased input to inform management decisions related to natural resources and the environment? Before responding, consider what is meant by the word 'objective'. What are the reasons for your position on objectivity and unbiased evidence?

2. Why do many believe that it is desirable to custom-design solutions to the specific conditions of a problem rather than design approaches to apply uniformly throughout a region?

3. What is the 'ingenuity gap'?

4. Explain why the principle of subsidiarity generates debate.

5. What are the implications of neo-liberalism for resource and environmental management?

6. Why do many consider that establishing a vision is important for resource and environmental management?

7. What might be a vision for resource and environmental management for your community?

8. Do you believe that you are primarily ecocentric or technocentric in your values and behaviour? What are the implications of this view for the way you interact with the natural environment?

9. What are the major obstacles to effective implementation of an ecosystem-based approach? What things must first be changed if implementation is to improve?

10. If there were a model of an ideal ecosystem approach, what characteristics would it have?

11. Why is there a predisposition to favour short-term rather than long-term strategies regarding resources and the environment? What is the best example of a long-term strategy in your community or province?

12. What should be done to facilitate social learning by both individuals and organizations?

13. What is the distinction between 'single-loop' and 'double-loop' learning, and what is its significance for resource and environmental management?

Related Websites

Environmental Commissioner of Ontario:
http://www.eco.on.ca

Environmental justice:
http://www.epa.gov/compliance/environmentaljustice

Envirolink: The Online Environmental Community:
http://www.envirolink.org

International Union for Conservation of Nature:
http://cms.iucn.org/index.cfm

National Round Table on Environment and Economy:
http://www.nrtee-trnee.ca

Sustainable Business: http://www.sustainablebusiness.com;
http://www.sustainable.org.nz/cms

Sustainable Livelihoods:
http://www.livelihoods.org/index.html; http://www.undp.org

World Summit on Sustainable Development, Johannesburg,
2002: http://www.un.org/esa/sustdev/index

Further Readings

Note: This list comprises works relevant to the subject of the chapter but not cited in the text. All cited works are listed in the section 'References' at the end of the book.

Bernstein, S., and B. Cashore. 2000. 'Globalization, fourth paths of internationalization and domestic policy change: The case of ecoforestry in British Columbia, Canada'. *Canadian Journal of Political Science* 33: 67–99.

Brown, L.M. 2000. 'Scientific uncertainty and learning in European Union environmental policy-making'. *Policy Studies Journal* 28: 576–97.

Carney, D., et al. 2000. 'Livelihood approaches compared'. Draft prepared November 1999, revised February 2000. www.undp.org/sl/documents.

Coward, H., R. Ommer, and T. Pitcher. 2000. *Just Fish: Ethics and Canadian Marine Fisheries*. Social and Economic Papers no. 23. St John's: ISER Books.

Diduck, A., J. Moyer, and E. Briscoe. 2005. 'A social learning analysis of recent flood management initiatives in the Red River basin, Canada'. In D. Shrubsole and N. Watson, Eds, *Sustaining Our Futures: Reflections on Environment, Economy and Society*, 126–64. University of Waterloo Department of Geography Publication Series no. 60. Waterloo, ON: University of Waterloo.

Dwivedi, O.P., et al., Eds. 2001. *Sustainable Development and Canada: National and International Perspectives*. Peterborough, ON: Broadview Press.

Fiorino, D.J. 2001. 'Environmental policy as learning: A new view of an old landscape'. *Public Administration Review* 61: 322–34.

Keen, M., V.A. Brown, and R. Dyball, Eds. 2005. *Social Learning in Environmental Management*. London: Earthscan.

Krajnc, A. 2000. 'Wither Ontario's environment? Neo-conservatism and the decline of the Environment Ministry'. *Canadian Public Policy* 26: 111–27.

Leeuis, C., and R. Pyburn. 2002. 'Social learning in rural resource management'. In C. Leeuwis and R. Pyburn, Eds, *Wheelbarrows Full of Frogs*, 85–105. Assen, Netherlands: Koninklijke Van Gorcum.

McKenzie, J.I. 2002. *Environmental Politics in Canada: Managing the Commons into the Twenty-first Century*. Toronto: Oxford University Press.

Ommer, R. 2000. *Just Fish: Ethics and Canadian Marine Fisheries*. St John's: Institute of Social and Economic Research, Memorial University of Newfoundland.

Pahl-Wostl, C. 2002. 'Towards sustainability in the water sector: The importance of human actors and the processes of social learning'. *Aquatic Sciences* 64: 394–411.

Peterson, B. 2007. 'Using scenario planning to enable an adaptive co-management process in the Northern Highlands Lake District of Wisconsin'. In D. Armitage, F. Berkes, and N. Doubleday, Eds, *Adaptive Co-management: Collaboration, Learning, and Multi-level Governance*, 286–307. Vancouver: University of British Columbia Press.

Reed, M.G. 2003. *Taking Stands: Gender and the Sustainability of Rural Communities*. Vancouver: University of British Columbia Press.

Schusler, T.M., D.J. Decker, and M.J. Pfeffer. 2003. 'Social learning for collaborative natural resource management'. *Society and Natural Resources* 15: 309–26.

Stewart, J.M. 1993. 'Future state visioning—A powerful leadership process'. *Long Range Planning* 26: 89–98.

Zimmermann, E. 1964 [1933]. *Introduction to World Resources*. New York: Harper and Row.

Chapter 6

Planning and Management: Process, Method, and Product

Learning Objectives

- To understand the strengths and limitations of key methods and processes related to resource and environmental management.
- To know the distinction between collaboration and co-ordination.
- To appreciate the benefits and limitations of participatory approaches.
- To understand the obstacles inhibiting communication of scientific results and conclusions to policy-makers and the lay public.
- To examine why adaptive management has been designed to expect uncertainty and surprise.
- To appreciate the concept of adaptive co-management.
- To gain an understanding of the connection between risk and impact assessment.
- To understand the role of strategic environmental assessment and sustainability assessment.
- To identify alternative approaches to resolving conflicts.
- To realize the relationship between regional and land-use planning and resource and environmental management.

Introduction

In Chapter 5, we argued that planning and management can reflect basic philosophies, processes, methods, and products. In that chapter, the emphasis was on concepts reflecting philosophy. In this chapter, attention turns to various processes, methods, and products with regard to resource and environmental management. This chapter provides a lot of information. To make it less 'abstract', we suggest that after reading each section you pause and consider the implications of what you have just read for the Sydney Tar Ponds case study in Chapter 1, as well as what this reading might mean relative to an issue or problem in your own community.

Collaboration and Co-ordination

In Chapter 5, we noted that the systems or ecosystem approach emphasizes that different components of resource and environmental systems are interconnected and that therefore a holistic perspective is required. However, for practical reasons, public agencies often focus on a subset of resources or the environment. Hence, we have departments of agriculture, environment, forestry, water, and so on. If a holistic approach is to be taken, there must be collaboration among the various agencies and with other stakeholders.

Collaboration and Co-ordination

Collaboration: working together
Co-ordination: harmonious adjustment

For Himmelman (1996, 29), **collaboration** involves exchanging information, modifying activities in light of others' needs, sharing resources, and enhancing the capacity of others to achieve mutual benefit and to realize common goals or purposes. Selin and Chavez (1995) argue that collaboration also involves joint decision-making to resolve problems, with power being shared and stakeholders accepting collective responsibility for their actions and the outcomes.

Rationale for Collaboration

Many of our problems appear insoluble because our . . . conceptions about how to manage in an increasingly interconnected world are limited. Our current problem-solving models frequently position participants as adversaries, pit them against one another, and leave them to operate with an incomplete appreciation of the problem and a restricted vision of what is possible.

The impasses are typically characterized as technical or economic. I argue . . . that the roadblocks are as much conceptual and organizational as they are technical and economic. Many of our problems are unsolvable because our systems of organizing are not geared for a highly interdependent environment.

—Gray 1989, xvii–xviii

Collaboration is needed within, between, and among organizations. Once collaboration is agreed to, then **co-ordination**—the effective or harmonious working together of different departments, groups, and individuals—can

be sought through various arrangements. For example, interdepartmental committees or task forces often provide a way to co-ordinate the activities of different agencies and stakeholders. Such interdepartmental mechanisms provide the means through which effective collaboration can be achieved. Different public participation processes, addressed in the next section, also can be used to facilitate collaboration.

Collaboration is increasingly accepted as desirable, because the complexity and uncertainty associated with resource and environmental issues create a challenge for any individual or organization in terms of having sufficient knowledge or authority to deal with them. Indeed, this is why multi-, inter-, and transdisciplinary research teams are often used, as explained in Chapter 1. Furthermore, differing values and interests contribute to conflict, which is another reason for various stakeholders to come together to determine how they can meet their respective needs. Done well, collaboration involves sharing information and insight to achieve multiple goals. When this is accomplished, collaboration can contribute to more open, participatory processes and to solutions to which different stakeholders feel committed. Thus, there is greater likelihood that acceptable solutions leading to effective implementation will be found.

However, collaboration is not always accepted or endorsed by everyone. Some individuals or organizations may decide not to reach out to other groups because they are determined to satisfy their own interests and meet their own goals, regardless of what others want. And when such stakeholders are powerful, they often choose to ignore other interests and concerns as they single-mindedly pursue their own interests. Another potential disadvantage of collaboration is that all the parties may compromise their principles or interests to reach a 'common denominator' that may not always represent a good, long-term decision. Therefore, we should always remember that collaboration and co-ordination are means to an end, not an end in themselves. We should seek to apply collaborative approaches aimed at achieving an accepted vision for a desirable future. Box 6.1 outlines a notable success in that regard.

Stakeholders and Participatory Approaches

The Manitoba Round Table on Environment and Economy (1992) argued some time ago that one of 10 basic principles for sustainable development should be *shared responsibility* and that one of six guidelines to achieve sustainable development should be *public participation*. The Manitoba Round Table concluded that it was essential to 'encourage and provide opportunity for consultation and meaningful participation in decision-making processes by all Manitobans' as well as to ensure 'due process, prior notification, and appropriate and

Box 6.1   Boundary Waters Treaty, 1909

Canada and the United States share many lakes, rivers, aquifers, and wetlands along their 8,840-kilometre border. Indeed, more than half of the border passes through water bodies. Examples include the Saint John and Saint Croix rivers in New Brunswick, the Great Lakes–St Lawrence River system in Ontario and Quebec, the Souris River in Saskatchewan and Manitoba, the Red River in Manitoba, the St Mary and Milk rivers in Alberta, and the Columbia River in BC.

The two countries signed the Boundary Waters Treaty in 1909 to provide a framework for a collaborative approach and resolution of disputes over water resources shared by the two countries.

The International Joint Commission, or IJC, has six commissioners, three from Canada and three from the United States. The IJC has power to adjudicate regarding issues associated with water development proposals in one country that could affect water levels or flows in the other country and to serve as a commission of inquiry when matters are referred to it by the two federal governments.

The six commissioners are expected to consider what would be best for both countries rather than being advocates for their own country, and the record indicates that they have effectively done exactly that.

timely redress for those affected by policies, programs, decisions, and developments'.

More recently, Manitoba Water Stewardship (2006) declared that *public participation* was one of 13 principles guiding its programs and activities. More specifically, it explained that 'Watershed management and planning will provide forums that will encourage and provide opportunity for consultation and meaningful participation in decision-making processes by Manitobans. Transparent processes and opportunity for information sharing and open communication will be provided.'

As another example, the Nova Scotia Department of Environment and Labour (2002, 7) identified the following principle related to the provincial drinking water strategy: '. . . the province will foster partnerships among individuals, organizations, governments, communities, and businesses for the purpose of developing water management programs. Partnerships enhance communication and support and will make the strategy more effective and more affordable.' And regarding its *Water for Life* strategy, the government of Alberta stated that one of three core areas would be partnerships and that 'citizens and stakeholders will have opportunities to actively participate in watershed management

on a provincial, regional, and community basis' (Alberta Environment 2005).

Here we focus on the nature of public participation as a means of reallocating *power* among participants and on alternative ways of facilitating *empowerment* of people relative to the environmental management process. The idea of sharing power has introduced a significant new dimension to public participation.

Degrees of Sharing in Decision-Making

In Chapter 1, the basic characteristics of sustainable development and sustainable livelihoods were outlined. Several characteristics—achieving inter- and intragenerational equity, increasing self-determination—provide the rationale for sharing power in environmental management. However, redistribution of power for environmental management from government agencies and private firms to First Nations people and other members of the public can challenge vested interests and/or undermine regulatory authority allocated through legislation. As a result, such redistribution is not always easily or readily accepted. Furthermore, power-sharing raises challenging questions about accountability and responsibility for decisions.

Tensions Exist from Use of Participatory Approaches

Useful and legitimate synthesis of expert knowledge with democratically accountable deliberation and decision-making poses grave challenges, both conceptual and practical, to the design of policy processes and institutions. At a conceptual level, the domains of science and of democratic politics have different goals, standards of merit, norms of participation, and procedures for resolving differences. At a practical level, desired knowledge is often unavailable, and available relevant knowledge is often not adequately used. Knowledge is often inadequate to give high confidence in the consequences of decisions, and decisions sometimes cannot be delayed until high confidence is obtained. Uncertainty is thus unavoidable and pervasive in environmental governance.

—Parson 2000, S128

Arnstein (1969) provided a helpful perspective regarding the issue of power redistribution by identifying what she called 'rungs' on the ladder of citizen participation (Table 6.1). Even though the ladder has been modified in many subsequent writings, the essential steps have remained the same.

As long ago as the late 1960s and early 1970s in Canada, public involvement programs had moved up to Arnstein's rungs of informing, consultation, and placation. In other words, environmental managers often incorporated public participation into environmental management initiatives. However, because of a belief that public agencies were accountable for resource allocation decisions and expenditure of public funds, the position usually taken by public agencies was that information and advice received through public participation was only one of several sources to be considered and that the public agency would retain decision-making authority.

Resource and environmental managers believed that they had the legal mandate, responsibility, and power to decide which trade-offs best reflected societal needs and interests and to make final decisions. This viewpoint was usually reinforced by the conviction that no one public interest existed but rather that many different interests were present and frequent conflicts among them were common. Giving responsibility for decision-making to citizens was often viewed by public agencies as dangerous, since it could too easily evolve into a form

Table 6.1   Rungs on the Ladder of Citizen Participation

Rungs	Nature of Involvement	Degrees of Power
8. Citizen control		
7. Delegation	Citizens are given management responsibility for all or parts of programs.	Degrees of citizen power
6. Partnership	Trade-offs are negotiated.	Degrees of tokenism
5. Placation	Advice is received from citizens but not acted on.	
4. Consultation	Citizens are heard but not necessarily heeded.	
3. Information	Citizens' rights and options are identified.	Non-participation
2. Therapy	Power-holders educate or 'cure' citizens.	
1. Manipulation	Rubberstamp committees	

Source: Adapted from Arnstein 1969.

Stakeholders from various governmental organizations (federal, territorial, and First Nations/Aboriginal) involved in Nunavut take part in a Natural Resources Canada workshop on geomatics products intended to support sustainable development in Canada's North.

of anarchy in which there would be an absence of government and laws and no one would be responsible or accountable for decisions or behaviour.

During the 1980s, dissatisfaction with the process, methods, and products associated with many resource and environmental management decisions began to rise. Growing numbers of Canadians rejected the idea that 'technically correct' answers could always be found. Instead, a prevailing view emerged that such decisions ultimately depended on weighing conflicting goals, aspirations, and values. In these situations, technical or scientific expertise was seen as a legitimate input but only one of several. Furthermore, as decisions became more complex, with more considerations and trade-offs, it seemed unlikely that any one expert or organization could have all of the necessary insight to reach a 'wise' decision.

From these considerations arose the idea that 'stakeholders' had a right to participate in decisions. **Stakeholders** are those who should be included because of their direct interest, including (1) any public agency with prescribed management responsibilities; (2) all interests significantly affected by a decision; and (3) all parties who might intervene in the decision-making process to facilitate, block, or delay it. Because more and more decisions of an increasingly complex nature had to be made, the traditional forum for public participation—the political process with elected representatives consulting and then reflecting constituents' views—no longer seemed adequate. As a result, various individuals and non-governmental organizations (NGOs) began pressing for public involvement to move higher up on Arnstein's ladder. While few politicians and public servants believed that total 'citizen control' was feasible or desirable, expectations were that 'partnerships' and

'delegated power' that gave effective power to the public were desirable and achievable.

As an example, by the early 1990s the BC Ministry of Environment, Lands and Parks (1992, 4) stated that it was committed to 'shared responsibility and partnerships in reaching the goals of environmental protection and sustainability; consultation and discussion; . . . strong representation and advocacy for environmental protection and enhancement'. For its part, the Saskatchewan Round Table (1992) commented that there was 'value [in] a full and open declaration of diverse interests, early in the consultative process', which could be used to develop trust and build consensus. However, the Saskatchewan Round Table also recognized that such an approach can be 'time-consuming, costly, and sometimes frustrating'. Nevertheless, it concluded that unilateral decisions that superficially appear efficient very often end up in 'costly lengthy appeals or court action'.

While the above statements support the use of participatory and stakeholder approaches and endorse the sharing of power with members of the public, it is not always achieved in practice. For example, an environmental impact assessment of a hog-processing plant and associated waste treatment plant in Brandon, MB, did not reflect all of these ideals. Maple Leaf Pork decided to locate a $120-million hog-processing plant in Brandon after a nine-month search that considered proposals from 42 communities in western Canada. The final decision to locate in Brandon involved active participation in the negotiations with Maple Leaf Pork by the premier of Manitoba and the mayor of Brandon, and together the two levels of government offered almost $20 million in subsidies. For the provincial government, the motive was to promote the hog industry in Manitoba as part of a policy of agricultural diversification. For the city, there was the prospect of 2,100 new jobs and an overall population increase of 7,000.

The impact assessment was conducted under provincial legislation. As the process began, the business community, led by the chamber of commerce, supported the project. A citizens' group was formed by those who opposed the project for various reasons, the most important being opposition to public subsidies to private corporations, concern about negative environmental impacts, and unhappiness about lack of transparency in the process. However, the relevant ministry decided not to hold public hearings despite demands for them. Instead, the public was given an opportunity to participate by commenting on a public registry, submitting their views during specified comment periods, and having access to appeal procedures. Ten public information meetings were also organized. (See box on page 183)

Despite situations such as this in which participation is very limited, the idea of **partnerships** among governments, private companies, and the general public has become increasingly popular. The partnership concept has been implemented

Empowerment Is Not Always Achieved: Environmental Assessment of A Hog-Processing Plant in Brandon, Manitoba

This was a classic case of 'decide-announce-defend' that saw little or no input from members of the community beyond the political and business leadership. In many respects, the EA [environmental assessment] became a sidelined issue that further marginalized the already disempowered project opponents.

. . . The EA process in this case was at best legitimating and certainly not participatory, empowering, or equitable.

—Diduck and Mitchell 2003, 360

through **co-management** initiatives and other approaches that reflect a genuine reallocation of power to citizens and away from elected officials or technical experts. Co-management arrangements have been developed particularly with First Nations peoples regarding management of forests, fish, and wildlife. In these situations, power is allocated to First Nations or other local people. Some of the challenges associated with power-sharing with First Nations are highlighted in the context of marine resources in Chapter 8.

Co-management in Northern Canada

With diverse institutional structures, integration of traditional knowledge, partnering with local indigenous groups, and successful implementation, northern Canada co-management institutions are building and strengthening cross-scale interaction.

—Trainor et al. 2007, 638

Communication

At an international conference focused on 'climate change communication', the organizers observed that communication has three main purposes: (1) raise awareness; (2) confer understanding; and (3) motivate action (Scott et al. 2000, iii). Furthermore, at the same meeting, Andrey and Mortsch (2000, WP1) argued that 'communication is thought to be effective only when these changes in awareness and understanding result in attitudinal adjustments and/or improve the basis upon which decisions are made.'

Carpenter (1995) identified important aspects to consider regarding how communication of scientific understanding should be conducted. They include:

- Much of the public does not understand science or how scientific research is conducted.
- Other than for weather and gambling, much of the public does not understand 'probability', and the idea of risk as part of life is rejected by many persons. Association and causation are often confused or assumed to be the same.
- The media do not deal well with the 'ebb and flow' of scientific research, which progresses by new research disproving or challenging existing understanding. This creates confusion and doubt about the authority or credibility of scientists when different views exist.
- In the courts, expert witnesses appear for different sides in a case and present conflicting testimony. A divergence of opinion should be understood as a characteristic of scientific 'findings'.

To overcome communication challenges, we first must recognize that a range of target audiences exist, such as other scientists, planners and managers, elected decision-makers, and the general public. Consequently, we should be sure that messages are crafted with regard to who the recipient will be and what their level of understanding can reasonably be assumed to be. To guide the preparation of messages, Carpenter recommended attention to four complementary questions, regardless of the audience:

1. What do we know, with what accuracy, and how confident are we about our data?
2. What don't we know, and why are we uncertain?
3. What could we know, with more time, money, and talent?
4. What should we know in order to act in the face of uncertainty?

With regard to the final question, Carpenter reminds us that risk assessment does not necessarily lead to risk reduction if the uncertainty is not understood and appropriate action is not taken. To illustrate, he tells an amusing tale about a prospective air traveller worried about riding on an airplane in case another passenger might be carrying a bomb. He called the airline office and asked what the probability was of a passenger being on board his flight with a bomb. The answer: a very low probability. He then asked about the probability of two passengers being on the same flight, each carrying a bomb. The answer: completely improbable. So he decided to carry a bomb on the flight himself!

Communicating Uncertainty: Catch-22

If environmental professionals are candid about uncertainties, the client/recipient will likely be disappointed and perhaps berate the professionals for not being helpful and just simply telling them what is going to happen and what to do. If the environmental professionals ignore or obscure the uncertainties and give unambiguous predictions and advice, events may very well show them to be substantially wrong. Then their credibility will be gone, and they will not be consulted in the future.

—Carpenter 1995, 129

The important message in this section is that while it is important to achieve understanding of natural and human systems and their interactions, it also is important to determine how this knowledge and insight can be shared with others who may not have the same scientific background but are still key stakeholders in terms of taking, facilitating, or thwarting action. To test yourself, consider the four questions above in relation to the Sydney Tar Ponds case study in Chapter 1, and determine how you would answer each question and how you would craft your answer as the target audience changed from scientists, to elected decision-makers, to the general public.

Adaptive Management

Surprise, Turbulence, and Change

Trist (1980) observed that there is no such thing as *the future* but instead there are *alternative possible futures.* Which future actually emerges depends very much on choices made and on actions taken to implement those choices. He argued that 'the paradox is that under conditions of uncertainty one has to make choices and then endeavour actively to make these choices happen rather than leave things alone in the hope that they will arrange themselves for the best.'

A challenge in making choices and taking initiatives is what Trist referred to as the *turbulent conditions* that have become increasingly prevalent. For example, energy plans in Canada became obsolete in the early 1970s when OPEC countries rapidly quadrupled oil prices. Such an increase in prices had not been included in the forecasts and assumptions on which energy plans had been based. It is interesting to reflect on whether we have learned much from that experience, considering the rapidly increasing oil prices between 2007 and 2009 (see Chapter 13). Would anyone have been taken seriously in 2005 if they had suggested that crude oil would be selling for more than $100 a barrel in 2008? During the 1980s when decisions were being made about the east coast fishery, there was no expectation that in the early 1990s, the cod fishery would be effectively closed and thousands of people in Atlantic Canada would lose their jobs (see Chapter 8). Few anticipated the terrorist attack on the World Trade Center in New York City in mid-September 2001. Today, to what extent is the possibility of terrorist attacks on basic infrastructure such as oil pipelines or water supply reservoirs incorporated into strategies and plans (see Chapter 13 for more detailed discussion)? These and other events, such as the economic 'meltdown' in the autumn of 2008, came as surprises to most people. They created bewilderment and anxiety and raised doubts about the capability of science, planning, planners, managers, and policy-makers, because decision-makers and their decision process apparently could not anticipate or adapt to rapid change.

Adaptive Environmental Management

Awareness of the need for **adaptive environmental management** is not new, having been popularized by Holling (1978; 1986) and his colleagues. They concluded that policies and approaches should be able to cope with the uncertain, the unexpected, and the unknown. Holling and his co-workers observed that the customary way of handling the unknown has been through trial and error. Errors or mistakes provide new information so that subsequent activity can be modified. Thus, 'failures' generate new information and insight, which in turn lead to new knowledge.

However, effective trial-and-error management has preconditions. The experiment should not destroy the experimenter. Or at a minimum, someone must remain to learn and benefit from the experiment. The experiment should also not cause irreversible, negative changes in the environment. Furthermore, the experimenter should have the will and ability to learn and to begin again.

In Holling's view, a major challenge for the use of adaptive environmental management is that it can be difficult to satisfy the preconditions. For example, the concern about climate change, discussed in Chapter 7, reflects the fear that we may not be able to reverse such change before serious problems have occurred. Moreover, even when errors are not in theory irreversible, the magnitude of the original capital investment, as well as personal and political investment in a particular decision or course of action, often makes reversibility unlikely. Many people in industrialized societies simply do not like to admit to

Commentary: The Nature of Adaptive Management

*Adaptive management is an approach to natural re-
source policy that embodies a simple imperative:
policies are experiments; learn from them. In order to
live we use the resources of the world, but we do not
understand nature well enough to know how to live
harmoniously within environmental limits. Adaptive
management takes that uncertainty seriously, treating human interventions in natural systems as experimental probes. Its practitioners take special care with
information. First, they are explicit about what they
expect so that they can design methods and apparatus
to make measurements. Second, they collect and analyze information so that expectations can be compared
with actuality. Finally, they transform comparison into
learning—they correct errors, improve their imperfect
understanding, and change action and plans. Linking
science and human purpose, adaptive management
serves as a compass for us to use in searching for a
sustainable future.*

—Lee 1993, 9

or pay for past mistakes but prefer to try to correct them. The
outcome of trying to correct an inappropriate initiative often
is increasing investment, increasing costs related to control and
maintenance, and progressive loss of future options.

Adaptive Co-management

Earlier in this chapter, we introduced the concept of co-
management. A recent innovation has been connecting the
concepts of co-management and adaptive management.

Armitage, Berkes, and Doubleday (2007, 1) differentiate
between co-management and adaptive management in the
following way:

- *Co-management* is primarily concerned with user participation in decision-making and with linking communities
  and government managers.
- *Adaptive management* is primarily concerned with learning-
  by-doing in a scientific way to deal with uncertainty.

In their view, when the two concepts are combined, the outcome is a combination of the iterative *learning dimension* from
adaptive management with the *linkage dimension* of collaborative management through which rights and responsibilities are
shared jointly. When this connection is made between the two
concepts, several characteristics become prominent, including:
(1) learning-by-doing; (2) integration of different knowledge systems; (3) collaboration and power-sharing among community,
regional, and national levels; and (4) management flexibility.
Plummer and FitzGibbon (2007) have also argued that **adaptive
co-management** and social learning (discussed in Chapter 5)
complement each other and can be used to mutual benefit.
Box 6.2 outlines key features of adaptive co-management.

Box 6.2   Features of Adaptive Co-management

1. Shared vision, goal, and/or problem definition to
   provide a common focus among stakeholders and
   interests.
2. A high degree of dialogue, interaction, and collaboration among multi-scaled participants.
3. Distributed or joint control across multiple levels,
   with shared responsibility for action and decision-
   making.

4. A degree of autonomy for different stakeholders at
   multiple levels.
5. Commitment to the pluralistic generation and sharing of knowledge.
6. A flexible and negotiated learning orientation with an
   inherent recognition of uncertainty.

Source: Armitage, Berkes, and Doubleday 2007, 6.

The concept of visions and visioning was introduced in Chapter 5, and the first point in Box 6.2 refers to the role of a shared vision in adaptive co-management. Olsson (2007) has explored the significance of having a vision for adaptive co-management. He (2007, 269) explains the link between governance (introduced in Chapter 1) and an adaptive approach as follows:

> Adaptive governance relies on networks that connect individuals, organizations, agencies, and institutions at multiple organizational levels. . . , where governance is the process of resolving trade-offs and providing a vision and direction for sustainability, management is the realization of this vision, and monitoring provides feedback and synthesizes the observations to a narrative of how the situation has emerged and might unfold in the future.

Olsson (2007, 269–70) argues that visioning processes should promote two key qualities: *adaptability*, or the capacity to sustain a system on a desired trajectory or trajectories in the context of changing conditions and disturbances, and *transformability*, or the capacity to create a fundamentally new system when evolving ecological, economic, social, and political conditions make an existing system untenable or unsustainable.

He explains how these two basic ideas underlay the vision for a wetland area in Sweden: 'to preserve and develop the natural and cultural values of the Kristianstads Vattenrike wetland area, while at the same time making careful and sustainable use of these values, and thus set a good example that can help promote the region.' (Olsson 2007, 272). In order to achieve this vision, attention had to be given to building trust among stakeholders, building a sense of the place among the stakeholders, identifying common interests, facilitating learning, encouraging both horizontal and vertical collaboration, and managing conflict.

Among various lessons learned from Kristianstads Vattenrike, Olsson (2007, 280) highlighted that creating a vision does not ensure successful adaptive co-management. There is always a risk that creating a vision may stimulate other stakeholders to articulate a different vision. Those involved in the Swedish wetland understood this challenge. They dealt with it by initially inviting a small number of people to become engaged and then gradually expanded the number as others became aware of the positive aspects of the vision.

Another lesson learned relates to the power of what was labelled 'transformational leadership'. The Kristianstads Vattenrike initiative moved forward effectively because one creative and committed individual helped to define the desired direction, align people to the vision, and motivate and inspire them to become committed and engaged. All of these accomplishments were important for the ultimate goal of generating movement or change in perceptions, attitudes, and values related to the wetland area. Much of the accomplishment was due to the remarkable leader's ability to recognize windows of opportunity as well as to identify and modify constraints or limitations (such as conflicts of interest, values, and opinions). Of course, the downside of reliance on a transformational leader is that the initiative becomes vulnerable if that individual retires or leaves the region.

In summarizing, Olsson (2007, 281–2) provides valuable insights regarding the role of a vision in achieving adaptive co-management. Concerning the experience in Sweden, he observed that it

> . . . helps develop values and builds motivation for ecosystem management among actors by envisioning the future together and developing, communicating, and building support for this vision. Key factors for the success of this process are dialogue, trust-building, and sense-making. The vision defines an arena for collaboration, frames the adaptive co-management processes, and fosters the development of social networks and interactions among actors, including those dealing with conflict resolution. Visioning processes can therefore be as important as implementation of rules and regulations for framing and directing adaptive co-management.

Impact and Risk Assessment

Environmental impacts, intended and unintended, positive and negative, are common to all development initiatives. The use of environmental impact assessment (EIA) has been a response to the increasing size and complexity of projects, greater uncertainty in predicting impacts, and growing demands by the general public and special interest groups to become more involved in planning and decision-making processes (Lawrence 2003).

Many definitions of **environmental impact assessment** have been put forward. For example, some time ago, Beanlands and Duinker (1983, 18) defined it as 'a process or set of activities designed to contribute pertinent environmental information to project or programme decision-making'. In a slightly different way, the Canadian Environmental Assessment Research Council (1988, 1) defined it as a 'process which attempts to identify and predict the impacts of legislative proposals, policies, programs, projects, and operational procedures on the biophysical environment and on human health and well-being. It also interprets and communicates information about those impacts and investigates and proposes means for their management.'

Initially, environmental impact assessment emphasized the physical and biological resources that might be affected by a project, with attention to reducing negative consequences. However, partly in response to public pressure, the focus gradually broadened to incorporate social concerns, leading to the concept of social impact assessment. In addition, more attention has been given to basic policy questions, such as establishing the appropriateness of the objectives a project is designed to meet, considering alternative projects that also could meet the same objectives, and examining how compensation could be provided for impacts or losses that cannot be mitigated.

Risk assessment underlies impact assessment, since it focuses on determining the probability or likelihood of an environmentally or socially negative event of some specified magnitude, such as an oil spill, and the costs of dealing with the consequences. Of course, since risks have to be estimated, our calculations may be incorrect. Thus, at the Earth Summit of 1992, the **precautionary principle** was endorsed. This principle states that in order to protect the environment when there are risks of serious or irreversible damage, lack of full scientific certainty regarding the extent or possibility of risk should not be used as an excuse for postponing cost-effective measures to prevent environmental degradation. In other words, decision-makers and managers must always err on the side of caution.

Challenges in Impact Assessment

People conducting **impact assessment** have to balance technical matters and value judgments. Often, there are no right or wrong answers but rather different answers depending on the starting point for the assessment and the assumptions made.

The Types of Initiatives to Be Assessed

In Canada and most other countries, impact assessments have primarily been conducted for development and waste management *projects*, especially major capital projects such as dams and reservoirs, nuclear or other types of power plants, oil or natural gas drilling or pipelines, waste disposal facilities,

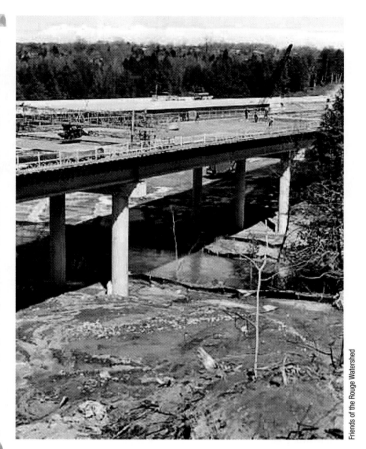

Highway 407 construction, Rouge Valley, Ontario.

Bruce Nuclear Power Plant, Ontario.

major highways, and runway expansions at major airports. The rationale has been that such development usually has the potential for significant environmental and social impact and that there are readily identifiable stakeholders—proponents, people, communities—that would be affected.

It has been argued for some time, however, that impact assessments could and should be completed for *policies* and *programs* as well. The argument in favour of this approach, called *strategic environmental assessment* (SEA), is that projects often are simply the means of implementing policies and programs. Thus, waiting to conduct an impact assessment until a policy or program evolves into a project means that the assessment may come too late. At the policy or program level, decisions may already have been taken to preclude or eliminate possible alternatives, and the project may become virtually irreversible. Partidario and Clark (2000, 4) have defined **strategic environmental assessment** as follows:

SEA is a systematic, ongoing process for evaluating, at the earliest appropriate stage of publicly accountable decision-making, the environmental quality, and consequences of alternative visions and development intentions incorporated in policy, planning, program initiatives, ensuring full integration of relevant biophysical, economic, social, and political considerations.

Guest Statement

From Projects to Priorities: Towards a More Strategic Environmental Assessment Process

Bram Noble

First introduced in the United States by the National Environmental Policy Act of 1969, environmental assessment is now practised in more than 100 countries. The concept of environmental assessment is quite simple—identifying, assessing, and finding ways to mitigate the potential impacts of proposed projects on the human and biophysical environment. Thousands of environmental assessments are completed in Canada each year. At the federal level, the majority of them are 'screening assessments' carried out for relatively routine projects such as highway upgrades or shoreline stabilization. Only about a dozen or so larger-scale 'comprehensive study' and 'review panel' assessments are completed, reserved for those highly complex projects such as mine sites and hydroelectric dams that have the potential for significant environmental effects.

Procedurally, when a project is proposed and the need for an environmental assessment determined, the *most likely* impacts of that project are assessed. A decision is then made as to whether the project might cause significant adverse environmental effects and whether such effects can reasonably be mitigated. In this way, environmental assessment ensures that at least some forethought has been given to the impacts of a project before it becomes a reality. Environmental assessment does not, however, ask whether the project being assessed is the most appropriate development or use of environmental resources; it simply seeks to identify the most likely outcomes of a predetermined development option.

The limitations of environmental assessment became increasingly evident in the face of several mega-project developments throughout the 1990s. In 1995, for example, during public hearings for the Terra Nova offshore oilfield project, Newfoundland, the public's request for a broader, area-wide review of cumulative impacts and consideration of offshore energy policy and planning for future development was at odds with the scope of the

environmental assessment and hardly within the purview of the project proponent. Similarly, in 1997 Environment Canada raised issue with the Alberta Energy Utilities Board concerning development of the Athabasca oil sands. It noted that a project-by-project review was not the most effective approach to assessment and that a much broader and more proactive environmental assessment was required for consideration of the future development of the Athabasca region.

The Inuvialuit Game Council (IGC) echoed this view more recently when Imperial Oil Resources announced its application for development of the Mackenzie Gas Project, Northwest Territories. Based on concerns over the project's potential impacts, including the impacts of future development activity that might result from the project, in 2004 the IGC called on the federal minister of environment for a broader and more strategic assessment to guide the future planning and development of oil and gas resources in the Mackenzie Delta–Beaufort area. In 2007, the first draft of the Beaufort Sea Strategic Regional Plan of Action was released.

Projects are the physical actions of higher-order policy and planning initiatives, or lack thereof, playing out on the ground. Thus, if we are concerned about the real source of environmental impacts, then a more strategic form of environmental assessment is required—one that occurs at the level of policies, plans, and programs (PPPs). Strategic environmental assessment (SEA), the environmental assessment of PPP initiatives and future development scenarios, is gaining widespread recognition as a tool for integrating environmental considerations at the earliest possible stage of decision-making. As a higher-order environmental assessment process, SEA occurs before development decisions are made and when alternative futures and options for development are still open. The focus is on opportunities, regions, and sectors as opposed to projects. The objective is to integrate environmental considerations in the development of PPPs and to identify preferred futures and the means to achieve them, rather than focusing on mitigating the most likely outcomes of an already foreclosed development decision. By focusing on the source of environmental impacts, the benefits of early environmental thinking should cascade downward, resulting in more environmentally sustainable development actions.

Applying environmental assessment before irreversible project decisions are taken seems to make good sense. However, aside from the Canadian Cabinet Directive, 2004, requiring SEA for federal government departments and agencies, there are no formal systems of SEA in Canada. Furthermore, how to do SEA and implement the results remains unclear. Most applications have been informal in nature, adopting SEA principles and perhaps even the SEA nametag but neither following a consistent approach nor occurring within the context of any

Bram Noble

formal system of PPP decision-making. As a result, the outputs of SEA have often been pure rhetoric without actions—divorced from downstream development activities.

That being said, these are only the early days of SEA. In 2008, the Canadian Council of Ministers of Environment's Environmental Assessment Task Group identified SEA as a priority area, particularly the development of meaningful linkages between SEA and project environmental assessment. SEA holds considerable promise. It is not a matter of whether we need SEA but rather how to ensure that it is applied and that when applied, it is done in a meaningful way. Addressing environmental impacts at their source is certainly preferable to mitigating their symptoms.

Bram Noble, PhD, is an associate professor in the Department of Geography and School of Environment and Sustainability at the University of Saskatchewan. His research focuses on environmental assessment, resource planning, and development.

An example of the need to consider policy issues before assessing a project occurred when BC Hydro proposed a pipeline across the Strait of Georgia to provide natural gas to Vancouver Island for generating power. The EIA was to be conducted on the specific impact of the pipeline on the environment. However, environmental groups argued for a much broader EIA that would examine the need for the power and the alternatives available rather than merely assessing mitigation strategies for one option (i.e., a pipeline).

While it is easy to accept the logic of having impact assessments completed for policies and programs as well as for projects, such assessments can be difficult. As Bregha et al. (1990, 3) explained almost 20 years ago, policies can be general or specific, stated or implicit, incremental or radical, independent or linked to other policies. In that regard, Thérival (1993) has suggested that strategic environmental assessment is most logically applied in three types of situations: (1) sectoral policies, plans, and programs (e.g., mineral extraction, energy generation, waste management); (2) regional or area-based policies, plans, and programs (e.g., land-use plans, development plans); and (3) indirect policies, plans, and programs (e.g., fiscal, trade, or science and technology policy).

Other challenges arise regarding which activities should be subject to impact assessments. One issue is particularly difficult. On the one hand, society expects government regulations to be reasonable and efficient. In that regard, it is normal to have some lower limit or threshold below which assessments are not required. For example, it is unlikely one homeowner building an outdoor barbecue in her backyard would adversely affect air quality in the community. However, if every homeowner decided to build an outdoor barbecue, air quality possibly could be affected. The issue, then, is how to balance reasonableness and efficiency against the dilemma that many small developments *in aggregate* might have significant implications. This problem has been described as one of *cumulative effects*.

Stinchcombe and Gibson (2001, 368) provide an excellent summary of the advantages and limitations of strategic environmental assessment as it can be used to help achieve sustainability. Table 6.2 presents these advantages and limitations, and we invite you to consider them and reach a conclusion on the role and value of strategic environmental assessment.

When Impact Assessments Should Be Done

The final report by the federal Environmental Assessment Panel (1991, 2) that reviewed the Rafferty-Alameda Dam and reservoir projects in southeastern Saskatchewan stated that 'environmental impact assessment should be applied early in project planning. That is the intent of both provincial and federal processes.' The panel concluded that the Rafferty-Alameda projects were 'well advanced, however, when both the first and

Table 6.2   Advantages of and Barriers to Strategic Environmental Assessment for Sustainability

Ten Advantages of SEA for Sustainability	Ten Barriers to Effective Implementation
1. Provides a process for integrated pursuit of sustainability objectives in policy-making and planning	1. Limited information and unavoidable uncertainties
2. Operationalizes sustainability principles	2. Boundary-setting complexities
3. Improves the information base for policy-making, planning, and program development	3. Primitive methodologies
4. Is proactive and broad in ways that strengthen consideration of fundamental issues	4. Difficulties in defining the proper role of public participants and ensuring effective involvement
5. Improves analysis of broad public purposes and alternatives	5. Difficult to co-ordinate and integrate strategic assessment with assessment processes at other levels
6. Facilitates proper attention to cumulative effects	6. Institutional resistance
7. Facilitates greater transparency and more effective public participation at the strategic level	7. Conflict between integrated assessment and bureaucratic fragmentation
8. Provides a framework for more effective and efficient project-level assessment	8. Jurisdictional overlap
9. Provides a base for design and implementation of better projects when project-level assessment is not required	9. Limitations of the standard rational planning and policy-making model
10. Facilitates establishment of a more comprehensive overall system of sustainability application at all levels, from the setting of decision objectives to the monitoring of implementation effects	10. Resistance to integration of strategic assessment in core decision-making

Source: Stinchcombe and Gibson 2001, 368.

this Panel became involved. This put some limits on the usefulness of the review.' Unfortunately, too often in Canada, it appears that developments continue to be well advanced before environmental impact assessments are conducted and that little has changed since the review of the Rafferty-Alameda project in the early 1990s.

The creators of environmental impact assessment had intended that EIA would be used jointly with other analyses to determine the appropriateness of development proposals and to help in the design of mitigating measures. However, as with Rafferty-Alameda, environmental assessments have often been conducted after the basic decision has already been taken regarding whether or not the project will be carried out. As a result, the impact assessment has rarely been used to determine whether or not a project should be approved; rather, it has become a tool for establishing which mitigation measures could be used to reduce or soften negative impacts.

Determining the Significance of Impacts and Effects

As already discussed, a difficult challenge is to determine the significance or implications of impacts and effects. This is not solely a scientific or technical issue. For example, scientific data may be able to determine whether water is too polluted to support fish, but whether the absence of fish is significant is influenced by the prevailing values in a society at a specific place and time. What might be considered significant in one place or at one time will not necessarily be viewed in the same way in another place or at another time. Furthermore, people from different cultural backgrounds or of different value systems and ideologies in the same place and at the same time might have quite different perspectives regarding what is important to protect or preserve. Accepting that judgments about significance are not only technical or scientific matters strengthens the rationale for extending partnerships in environmental management and for ensuring that key stakeholders participate in planning and assessments.

One of the major challenges in determining 'significance' is that the issues in a dispute often do not lend themselves to a monetary valuation; rather, they are characterized by intangible features. What is the value of an undisturbed stretch of whitewater in a river that might be inundated by a dam built to generate hydroelectricity? What is the value of a wetland that is to be drained to allow the building of a subdivision or to make it possible for a farmer to increase food production?

Aside from the economic costs of the oil that was lost and the cleanup and the economic value of ecosystem services, what is the value or cost of an oil spill such as the *Exxon Valdez* spill, and how do we measure it? 'After the *Exxon Valdez* spilled oil into the pristine waters of Prince William Sound, a small army was deployed to clean up the mess. One of techniques for washing the oil off the beach rocks was to blast the beaches with high-pressure sea water heated to 160 degrees. The spray created a toxic cloud of oil mist and oil aerosols that enveloped the workers, made many sick, and may have caused permanent damage to thousands of workers.' (http://www .conservationinstitute.org/csi_report/csireport1_05.html)

What is the cost to wildlife of disturbance to their habitat resulting from a mine or a pipeline? What is the ecological value of a stand of old white pine or Douglas fir? Questions such as these pose major challenges for scientists in designing appropriate metrics and are often raised in impact assessments. There is usually no single answer, and people of differing views and with differing vested interests may come to quite different conclusions about their significance and what action is appropriate.

Inadequate Understanding of Ecosystems

The scientific understanding of ecosystems is often incomplete, as discussed in Chapter 2. Furthermore, information about a specific ecosystem is often not sufficiently adequate to permit estimates about what might be the effects of human intervention.

Even the most basic ecological concepts are not without problems. For example, some time ago, McIntosh (1980) commented that ideas such as community, stability, succession, and climax were causing (and still do cause) major disagreements among ecologists. In some instances, scientists disagree over terminology and definitions, even after more than three-quarters of a century of research. Some of these difficulties were considered in Chapters 2, 3, and 4. With uncertainty and disagreement over basic ecological concepts, it is understandable why predictions are difficult to make with confidence or why the same data result in different scientific interpretations, especially when such data are incomplete. This situation partly explains why it is not unusual for proponents and opponents at environmental hearings to have their own scientific experts who have reached opposite conclusions about impacts or risks.

To address in part the problem of incomplete information and understanding, Nakashima (1990) argued that more use should be made of **indigenous knowledge** or traditional ecological knowledge (TEK). He illustrated his argument by discussing the ecological understanding of seabirds on the part of the Inuit in the coastal communities of Inukjuak and Kuujjarapik, QC, and Sanikiluaq, Nunavut, along southeastern Hudson Bay. He noted that Inuit hunters are well positioned to make a contribution to the protection of the environment, since they have accumulated excellent information about the range and behaviour of seabirds as a direct outcome of their hunting lifestyle. Such knowledge can be invaluable in estimating and assessing the potential impact of an oil spill on wildlife. In Chapter 7, the guest statement by Elise Ho provides further insight into the role of indigenous knowledge regarding global change.

Incomplete understanding and inadequate data will remain the reality for those involved in impact assessments in Canada. For this reason, the adaptive environmental management approach is attractive, given its emphasis on learning by trial and error and its acceptance of uncertainty. Such problems also highlight the valuable contribution that indigenous knowledge can make to environmental management.

The Nature of Public Involvement

As discussed earlier in this chapter, several reasons are cited to justify the role of public involvement. Public involvement has at least three functions in impact assessment:

1. It opens up the decision-making process to the public and in so doing contributes to making that process fair. Furthermore, making decision-making accessible to the public enhances the credibility of the process. However, there is always a danger that the public or individuals will be 'co-opted' through such processes.
2. Public involvement helps to broaden the range of issues and alternative solutions considered and allows the public to share in devising mitigation measures. In this manner, citizens share in establishing the conditions under which a proposal will be approved.
3. Public participation can contribute to social change. That is, by participating in the decision-making process, the public in a particular place becomes more aware of conditions in their own environment. Such enhanced awareness can lead to new initiatives within the community to identify and begin to address other problems, as well as a greater likelihood that the government will appreciate the concerns of a community. This point is often referred to as 'social learning', a concept discussed in Chapter 5.

The Development of Monitoring

Chapters 2, 3, and 4 highlighted that many ecological processes unfold over *decades* and that therefore, the effects of some ecosystem changes may emerge only slowly over a lengthy time span. We have also observed that because of incomplete ecological science, it is often difficult to predict which changes may occur in ecosystem structures or processes as a result of development. If we are to improve our knowledge of the

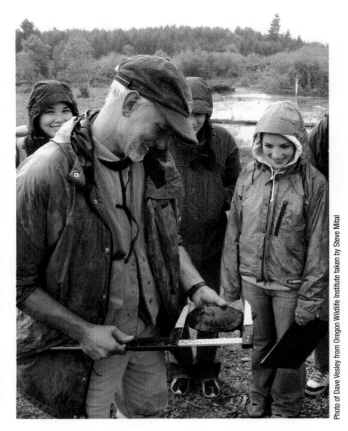

Students participating in the western pond turtle monitoring project, Oregon. It has been estimated that western pond turtle populations have declined by 80 per cent. The intent is to gather information on the quality and quantity of the remaining turtles as a first step to restoration and population recovery.

Photo of Dave Vesley from Oregon Wildlife Institute taken by Steve Mital

resiliency, sensitivity, and recuperative powers of ecosystems, **monitoring** is essential. Monitoring can also help to ensure that mitigating measures recommended in an impact assessment have actually been implemented. It can track public concerns or fears regarding a project and thereby help to ensure that they are recognized and addressed.

Too often in environmental and resource management in Canada and other countries, such monitoring is not conducted. It is usually time-consuming and expensive, and the results may not become useful until many years of monitoring have taken place. When financial and human resources are scarce, managers may be tempted to reduce or eliminate monitoring and redirect resources to new development activity, as happened with the Experimental Lakes Area in northwestern Ontario (Box 6.3). However, by following that route, we curtail the opportunity to learn from past experience and thus make it difficult to apply the adaptive management approach. One way around this problem is to enlist the local community in monitoring activities, an approach that has yielded promising results in some instances.

Sustainability Assessment

Earlier, we noted growing support for strategic environmental assessment. A more recent innovation is a concept known as

Box 6.3   The Experimental Lakes Research Area

Generating enough electricity to fuel society's demands while minimizing environmental impacts is one of the main challenges we face today. Hydroelectric power has generally been regarded as preferable to coal- and oil-generating stations, which are linked to acid deposition, and to nuclear power plants with their associated difficulties in waste disposal. So it was somewhat of a surprise when researchers at the Experimental Lakes Research Area (ELA) in northwestern Ontario discovered that reservoirs created for hydroelectric generation were responsible for releasing large amounts of carbon dioxide and particularly of methane into the atmosphere. The emissions occurred as a result of bacterial decomposition of flooded peat and forest biomass.

This was not the ELA's first finding of global significance. Since its founding in 1968, ELA researchers have also made significant contributions to the understanding of eutrophication and acid deposition. Fifty-seven small lakes and their watersheds are used as experimental sites to help determine the impact of various environmental perturbations. Research on eutrophication at the ELA was instrumental in developing the phosphorus control strategies in the Great Lakes Water Quality Agreement. In 1987, research on acidification was initiated and contributed to new estimates regarding damage to aquatic ecosystems. Again, the results were used as the basis for international accords to limit emissions. The lakes have also provided a valuable function as monitoring sites to track changes in environmental conditions over time, such as global warming. In view of the national and global importance of the research at the ELA, it is unfortunate that government funding has been sharply cut back.

sustainability assessment, an extension of the idea of sustainable development that we reviewed in Chapter 1.

Gibson (2007, 73–4) reminds us that environmental assessment emerged as one way to change decision-making by ensuring that environmental aspects are considered along with economic and technological matters. As environmental assessment has evolved, Gibson suggests that various shifts have occurred, including it becoming

- more mandatory and codified, such as its incorporation through legally based processes, greater specification of requirements, and less opportunity to treat it as discretionary;
- more broadly applied, such as by covering smaller as well as large capital-intensive projects, covering more sectoral and regional initiatives, and addressing policies and programs as well as projects;
- more accessible, transparent, and participatory, such as by making engagement easier for civil society organizations and local residents as well as for stakeholders

from government and the private sector and technical specialists;
- more closely monitored, such as by the courts, government agencies, civil society organizations, communities, and individuals.

All of the above changes are positive. Gibson (2007, 80–7) points out that sustainability assessment is a recent innovation. In his words, its core focus is on 'efforts to apply some form of sustainability analysis, appraisal, or assessment or otherwise to adopt sustainability objectives as core guides for evaluations and decisions' (Gibson 2007, 81). He further notes that the proliferation of sustainability initiatives can be readily surveyed through any reputable Internet search engine, and we encourage you to do that and determine whether any such activities are taking place in your province, region, or community.

To stimulate your thinking about sustainability assessment, Table 6.3 presents what Gibson refers to as the basic sustainability requirements in the form of criteria for decision-making that can guide sustainability assessments.

Table 6.3   Sustainability Requirements As Criteria for Sustainability Assessment

1. Socio-ecological system integrity
Build human-ecological relations to establish and maintain the long-term integrity of socio-biophysical systems and protect the irreplaceable life-support functions on which human as well as ecological well-being depends.

2. Sufficiency and opportunity
Ensure that everyone and every community has enough for a decent life and that everyone has opportunities to seek improvement in ways that do not compromise future generations' possibilities for sufficiency and opportunity.

3. Intragenerational equity
Ensure that sufficiency and effective choices for all are pursued in ways that reduce dangerous gaps in sufficiency and opportunity (and health, security, social recognition, political influence, etc.) between the rich and the poor.

4. Intergenerational equity
Favour present options and actions that are most likely to preserve or enhance the opportunities and capabilities of future generations to live sustainably.

5. Efficiency
Provide a larger base for ensuring sustainable livelihoods for all while reducing threats to the long-term integrity of socio-ecological systems by reducing extractive damage, avoiding waste, and cutting overall material and energy use per unit of benefit.

6. Democracy and civility
Build the capacity, motivation, and habitual inclination of individuals, communities, and other collective decision-making bodies to apply sustainability requirements through more open and better informed deliberations, greater attention to fostering reciprocal awareness and collective responsibility, and more integrated use of administrative, market, customary, and personal decision-making practices.

7. Precaution and adaptation
Respect uncertainty, avoid even poorly understood risks of serious or irreversible damage to the foundation for sustainability, plan to learn, design for surprise, and manage for adaptation.

8. Immediate and long-term integration
Attempt to meet all requirements of sustainability together as a set of interdependent parts, seeking mutually supportive benefits.

Source: Gibson 2007, 84.

If you were to apply the sustainability requirements in Table 6.3 to present or future development proposals in your area, how well would they rate against these criteria, and what changes might you suggest for them? If the same requirements were to be applied in the Sydney Tar Ponds experience, what conclusions would you draw, and what recommendations would you offer?

Dispute Resolution

Conflicts and disputes occur for many reasons. They may emerge because of clashing or incompatible values, interests, needs, or actions. In an environmental context, conflicts may arise as a result of either substantive or procedural issues, or both. At a substantive level, disputes may arise about the effects of resource use or project development, about multiple use of resources and areas, about policies, legislation, and regulations, or over jurisdiction and ownership of resources. At a procedural level, conflicts may occur regarding who should be involved, at what times, and in what ways.

However, conflict is not necessarily bad or undesirable. Conflict can help to highlight aspects of a process or system that hinder effective performance. Conflict can also be positive if it leads to clarification of differences stemming from poor information or misunderstandings. Approached in a constructive manner, conflict can result in creative and practical solutions to problems. On the other hand, conflict can be negative if it breeds mistrust or misunderstanding or reinforces biases. It can be negative as well if it is ignored or set aside, leading to the escalation of a problem or to the creation of stronger obstacles that must be overcome later.

Conflicts over resource management have often become high-profile news issues in Canada. Some of these conflicts are highlighted in subsequent chapters, such as the conflicts over the seal hunt, Native whaling, and the lobster fishery in New Brunswick, discussed in Chapter 8. One of the largest conflicts in Canada over resource management focused on the temperate rain forests of Clayoquot Sound on the west coast of Vancouver Island (Box 6.4).

A more localized but still potentially significant conflict at the national level is the dispute between Six Nations people and developers in Caledonia, a community adjacent to the Grand River in southern Ontario. In 1784, the 'Haldimand Tract' was granted to the Six Nations by the British government in appreciation for their support during the US War of Independence. The original Haldimand Tract extends for about 10 kilometres on both sides of the Grand River from the mouth for a considerable distance upriver and covers a total of 384,450 hectares. It includes much of the cities of Kitchener, Waterloo, Cambridge, and Brantford.

Over time, much of the land in the tract was sold. For example, in 1798, the tract lands in what is now the Regional Municipality of Waterloo were sold by Joseph Brant, the Mohawk chief. However, with regard to situations such as this, Six Nations traditional chiefs have since argued that Brant did not have the right to sell the land and/or that some of the land was purchased from Native leaders who did not appreciate what the sale involved. As a result, the Six Nations traditional government has registered a land claim for the tract lands and maintains that it has jurisdiction over them. The Six Nations traditional chiefs have not claimed ownership over all the land but insist upon being consulted before any new developments occur.

Box 6.4   Clayoquot Sound, Vancouver Island

On 13 January 1993, a full-page advertisement appeared in the *New York Times* with the question 'Will Canada do nothing to save Clayoquot Sound, one of the last great temperate rainforests in the world?' It was paid for by eight major international conservation groups. Six months later, Greenpeace International in London produced a 17-page colour booklet, *British Columbia's Catalogue of Shame*, outlining the background to the Clayoquot decision and demanding 'an end to the all clear-cut logging in Clayoquot Sound, full inventory of all plants and animals to be carried out, and outstanding Native land claim issues to be settled'. Robert Kennedy Jr flew in as a lawyer representing the Natural Resources Defense Council of the US and promised the support of his organization. A resource and environmental management issue that had made nightly headline news in Canada also resulted in the arrest of more than 800 protesters and captured worldwide attention.

At stake was the future of one of the greatest remaining temperate rain forests of the world and the largest remaining tract of old-growth forest on Vancouver Island, which was slated to undergo forest harvesting. The area was also under First Nations land claim negotiations, coveted by the mining industry, and an arena for conflict between the rapidly expanding aquaculture industry and fishers and recreationists, as well as including part of a national park.

The attention drawn to Clayoquot Sound as a result of the protests resulted in several innovations. The area planned for complete protection was expanded. Part of the sound is now included in an international biosphere reserve recognized by UNESCO. The reserve serves in helping to plan a more sustainable future. Some logging operations are now joint operations between logging companies and First Nations and are thus helping to address some of the chronic poverty in some First Nations communities. Tourism has continued to increase in the area, again helping to combat poverty and attracting more support for further preservation.

Figure 6.1 Haldimand Tract.

The conflict erupted during February 2006 when a group of Six Nations protesters occupied the Douglas Creek subdivision, a housing development in Caledonia, a small community south of Hamilton. The Native protesters moved onto the property and insisted that they would not leave until the land claim was resolved. There have been noisy and sometimes violent confrontations between the Natives occupying the development site and residents of Caledonia, and in 2007 the provincial government bought the property from the developer as compensation to him.

At the one-year anniversary of the occupation in February 2007, Janie Jamieson, a spokesperson for the Natives occupying

Caledonia protesters at blockade.

the site, stated that the occupation represented too much to the Six Nations for them to back down. She was quoted as saying that the protesters had made it through a cold winter in 2006 and would remain on the land for several more winters if they had to. In her words, 'It's not close to being over yet. We are fully prepared to stay for as long as it takes.' (Canadian Press 2007, C11).

In terms of jurisdiction, the Ontario provincial government's position has been that the federal government through Indian and Northern Affairs Canada has the responsibility and authority to deal with land claims. The federal minister for Indian affairs, Jim Prentice, maintained in 2007 that the Haldimand Tract land claim was one of the oldest in Canada, involved many challenging issues, and thus required time and patience to resolve. Negotiations have been underway between the Six Nations and the federal government for many years.

There has been a conscious decision by the provincial government and police not to forcibly remove the Six Nations occupiers from the Caledonia site for fear such action would precipitate the type of violence experienced at Oka, QC, in 1990, which left a police officer dead during a conflict between the Mohawks and the Canadian army, or during the Ipperwash Provincial Park dispute during which one Native protester, Dudley George, was shot and killed when Ontario Provincial

Police raided the park to remove members of the Chippewas of Kettle and Stony Point First Nations who had occupied the park.

Tensions escalated in the watershed in September 2007 when the traditional chiefs announced that any person or company planning to build on Haldimand Tract land would need approval from the Six Nations. Approval required obtaining a permit and paying a fee of up to $7,000. The announcement prompted a lead editorial in one of the newspapers in the Grand River basin with the headline 'Six Nations can't dictate new laws'. The editorial began by stating that 'The rule of Canadian law is breaking down in the Grand River valley' (*Kitchener-Waterloo Record* 2007, A16). For its part, the provincial government maintained that landowners were governed by provincial land title system rules, which do not require First Nations' approval for development. This view was shared by local governments in the basin. Nevertheless, by the end of February 2008, 10 developers had paid about $50,000 to the Six Nations planning agency, and the fees had ranged from $3,000 to $7,000 for each approval. The Six Nations also announced that it expected governments to pay fees for three new bridges to be built over the Grand River. By the end of February 2008, the estimated cost of this dispute to provincial taxpayers stood at $54 million, including

Turbine construction.

Shelburne Wind Farm turbines.

Ontario Provincial Police costs of $32 million, $16 million for the province's purchase of subdivision land from developers, nearly $5 million in negotiation costs, and $1 million provided by the provincial government to the Six Nations to cover its negotiation costs.

In part, concerns escalated because a few days before the First Nations announced the planning permit system, a developer was hospitalized after being assaulted by a group of Aboriginal men who maintained that they had authority over

the land on which he was building in Caledonia. The editorial referred to above commented that 'This kind of outrage should have no place in Ontario. The demand for a development fee, backed up by thugs accountable to no one but themselves, is not a reasonable request; it is blackmail and a precursor to anarchy.'

The Caledonia conflict is not unique. In August 2007, a $275-million wind farm project near Shelburne in the northern part of the Grand River basin was stalled because a First Nations group claimed that it owned both the land and the wind above the land (Burt 2007a; 2007b). In this instance, the Six Nations served notice of its land claim rather than occupying the site. The developer of the wind farm, Calgary-based Canadian Hydro, planned a two-phased development, with 45 turbines in the first phase. The second phase, with an additional 88 turbines, was being challenged.

One of the spokespersons for the Six Nations, Kahentinetha Horn, was quoted as saying that 'We've been robbed of our land and now we want it back. It is our land and it is our wind. They need our permission to use it. We are very concerned about the use of the air, land, and water' (Burt 2007a, B1). The wind farm project is also located on Haldimand Tract land and hence subject to the same challenge by the Six Nations as the development at Caledonia. The chief financial officer for Canadian Hydro commented that a delay of one year could increase costs by up to $10 million because of the incremental expenses for storing, handling, and transporting the turbines, along with expected higher prices for materials and services in the future.

Whether conflict is positive or negative, it is often present because people see things differently, want different things, have different beliefs, and live their lives in different ways. The basic differences among people and their values, interests, needs, and activities create conflict. Such differences can be exacerbated by other factors. These factors include lack of understanding of other people or groups; using different kinds and sources of information; differences in culture, experience, or education; and differing values, traditions, principles, assumptions, experiences, perceptions, and biases. In fact, conflicts are a normal part of life, and we need to devise ways to deal with them. We should accept the legitimacy of conflict and recognize that environmental planning and management can often serve as a process for resolving conflicts.

Approaches to Handling Disputes

Disputes usually centre on three main issues: *rights*, *interests*, and *power*. The traditional means of dealing with disputes are political, administrative, and judicial. The latter is the means most familiar to most citizens and involves court action. In litigation, the main issues of concern are *fact*, *precedent*, and *procedure*. Attention focuses on establishing a winner or on punishing an offender.

The strength of the judicial or litigation approach is that it uses a process that has evolved over centuries. Standards for procedure and evidence are well established. Accountability is ensured through appeal mechanisms and the professional certification of lawyers. On the other hand, the judicial process is often viewed as unduly adversarial, time-consuming, and expensive. An adversarial and adjudicative process may also encourage participants to exaggerate their private interests, conceal their 'bottom lines', withhold information, and try to discredit their opponents. In addition, the courts do not always provide a level playing field between, for example, a small group of private citizens or a First Nation opposing a resource project and a large multinational corporation as the proponent. Because of its greater financial resources, a corporation can afford expensive legal expertise and invest in gaining access to and influence within the corridors of political power.

One alternative to the judicial approach is often referred to as alternative dispute or conflict resolution. **Alternative dispute resolution** (ADR) emphasizes the interests and needs of the parties involved. And while the judicial approach ends by declaring a winner and loser or identifying a party to be punished, the focus of alternative dispute resolution is on reparation for harm done and on improving future conduct. Another key distinction between the two is that the judicial approach emphasizes *argument*, while the alternative dispute resolution approach stresses *persuasion*.

Attributes of Alternative Dispute Resolution

At least six strengths or advantages of alternative dispute resolution can be identified (Shaftoe 1993):

1. an emphasis on the issues and interests rather than on the procedures;
2. an outcome that normally results in a greater commitment to the agreement;
3. attainment of a long-lasting settlement;
4. constructive communication and improved understanding;
5. effective use of information and experts;
6. increased flexibility.

These strengths highlight some of the limitations of the judicial or court-based approach. However, we should not conclude that legally based approaches are never appropriate for dealing with environmental issues. No approach is perfect, and alternative dispute resolution is no exception. Thus, while in some circumstances alternative dispute resolution may be more effective than litigation, the key is to recognize the strengths and weaknesses of each approach and to determine which one would be most effective.

Types of Alternative Dispute Resolution

The features of specific types of alternative dispute resolution approaches include public consultation, negotiation, mediation, and arbitration. Various aspects of *public consultation* were reviewed earlier in the context of the ideas of partnerships and stakeholders. Public participation or citizen involvement has been used explicitly since the late 1960s for resource and environmental management in Canada. Initially, such consultation focused on having the public help in identifying key issues and in reviewing possible solutions. However, this public input was simply one of many inputs considered by the managers who ultimately determined which trade-offs were appropriate and then made the final decisions. Members of the public had no real power or authority in the management process.

By the mid-1980s, this approach had been modified as public participation moved towards the concepts of partnership and delegated power. Some co-management initiatives illustrated the shift towards giving real power to the public. However, public consultation is not normally considered one of the emerging types of alternative dispute resolution, because all decisions related to the dispute are the exclusive domain of stakeholders in any case.

Negotiation is one of the two main types of the alternative dispute resolution technique. Its distinctive characteristics include having the parties involved in a dispute come together in a voluntary, joint exploration of issues with the goal of reaching a mutually acceptable agreement. Because participation is voluntary, participants can withdraw at any time. Through joint exploration, the parties strive to identify and define issues of mutual concern and to develop mutually acceptable solutions. The normal procedure is to reach an agreement by consensus.

Mediation is the second main type of the alternative dispute resolution technique. Its distinguishing feature is that it includes a neutral third party (called a mediator) whose task is to help the disputants overcome their differences and reach a settlement. The third party has no power to impose any outcome. The responsibility to accept or reject any solution remains exclusively with the stakeholders in the dispute. In addition, the third party has to be acceptable to all parties in the conflict.

Mediators play a variety of roles. They assist the parties to come together and in this role act as facilitators. They can also help the parties with regard to fact-finding. The mediators do not necessarily have the expertise to provide the needed information, but they can help to identify what information is needed and then assist in finding the necessary data. During mediation, the mediator often first meets separately with the parties and subsequently holds meetings with all parties present. Mediators try to help each stakeholder to understand the interests and objectives of the other stakeholders, find points in common, and settle differences through negotiation and compromise. Key roles for the mediator include

Consensus

Traditional [indigenous] society was based upon the principle of consensus for government. . . . Consensus must be the most perfect form of democracy known because it means that there is no imposition of the rule of the majority. Everyone has input and no one is excluded.

—Ovide Mercredi, in Mercredi and Turpel 1993, 115

maintaining momentum in the negotiations, keeping the parties communicating with each other, and ensuring that proposals are realistic.

Arbitration differs significantly from negotiation and mediation because it normally involves stakeholders accepting a third party with the responsibility to make a decision on the issue(s) in conflict. In mediation, the third party has no power to impose a settlement. In arbitration, the arbitrator's decision is usually binding on the parties. However, in instances known as 'non-binding arbitration', the arbitrator makes a decision on the conflict, but the stakeholders may accept or reject it. In some ways, the judicial or court-based approach is quite similar to binding arbitration, since a judge reaches a decision that is then imposed on those involved in the dispute. The main difference is that in arbitration, the stakeholders usually have a voice in selecting who the arbitrator will be. In the judicial procedure, the participants play no role in determining which judge will hear their case.

As already noted, public consultation or public participation has been used for environmental management in Canada for about four decades, so there has been considerable experience with it. Judicial or court-based approaches have been used ever since the country was settled by people from Europe. Negotiation and mediation are newly emerging approaches, although aspects of negotiation have certainly been used as part of public consultation and judicial approaches. Some of the case studies in Part D provide further details about the way that negotiation and mediation are being used to address conflicts over the environment. At this point, you might consider how ADR methods could be used to deal with the different interests represented in the Sydney Tar Ponds situation or in the

dispute involving the Six Nations, developers, the community of Caledonia, and the provincial and federal governments.

Regional and Land-Use Planning

Regional and land-use planning represents a process and a product, and ideally the product (or plan) reflects a vision of how development should occur in a region. Resource and environmental managers should be aware of arrangements for regional and land-use planning for several reasons. First, work undertaken to create or update a plan is often directly relevant to resource and environmental management because it can provide valuable information and insight. Second, regional and land-use plans frequently have a statutory or legislative basis and thus become part of the official documentation governing activity in an area. In contrast, many resource and environmental management plans, such as a watershed plan, usually do not have a legal basis, and this often creates difficulty for achieving effective implementation. Once a resource management plan has been created, it is common for various agencies to assume responsibility for implementing specific recommendations that fall within their jurisdictions. However, such agencies have other responsibilities as well, and they must determine what priority recommendations from resource management plans should have. As a result, if resource managers can link their plans to official regional land-use plans, there is greater likelihood that their recommendations will be acted upon.

The same argument can be made with regard to environmental impact assessment. As with regional and land-use plans, environmental impact assessment has a legal basis at both the federal and provincial levels. Thus, the likelihood of action being taken on recommendations in a resource management plan will be further enhanced if they are related to associated impact assessment statements and to regional and land-use plans. Put another way, if such links are not created, it becomes too easy for decision-makers to overlook or ignore resource management plans because they usually do not have any legal underpinning.

In the section on system and ecosystem perspectives in Chapter 5, we noted that ecosystems are dynamic and therefore continuously changing. Thus, the context (also discussed in Chapter 5) within which a resource management plan must function can be expected to change. One implication is that resource management plans need to be updated and modified to ensure that they remain relevant. The notion of monitoring and modifying plans in light of changing circumstances, new knowledge, and lessons learned is also embodied in adaptive environmental management, as already discussed. In most jurisdictions, provision is made for reviewing, updating, and modifying regional and land-use plans. This is another reason why it is important to link resource management plans to them, since both sets of plans can be reviewed and updated at the same time.

Implications

The approaches discussed in this chapter represent what many would view as ideals for resource and environmental management regarding processes, methods, and products. Some of them pose fundamental challenges to us as individuals and society. For example, basic values in Western industrialized societies include self-interest and competition. It is often assumed or believed that scarce resources—natural and/or human—are allocated most efficiently and equitably through competition as each agent—individual citizens, interest groups, corporations, resource managers—pursues its own best interests while governments try to mediate the process in the greater public interest. However, the concepts of *collaboration, co-ordination, partnerships,* and *stakeholders* suggest a different perspective, one based on a willingness to recognize the legitimacy of many interests and needs and on a desire to try to satisfy or accommodate them. In a similar manner, *alternative dispute resolution* is predicated on the idea of groups working together for mutual gain rather than determining winners and losers. In a different way, *adaptive management* rejects a belief that humans can completely understand and control natural systems. Instead, this approach accepts that our understanding will always be incomplete and limited, resulting in ongoing surprises that will require us to adapt and modify our policies and practices. *Impact* and *risk assessments* further reflect acceptance of uncertainty with regard to resource and environmental systems and encourage us to strive to anticipate and monitor outcomes—intended and unintended, desirable and undesirable—so that we can determine where and when adjustments are needed.

In Chapter 5, we stressed the importance of a systems approach, and the discussion in this chapter about *regional and land-use planning* highlights that resource and environmental management does not occur in a vacuum. Other management processes occur in parallel, which means that it is important to be constantly watching to determine how connections can be made with these processes, particularly when they have a legal basis.

Finally, *communication* reminds us that it is critically important to ensure that stakeholders share information, insights, needs, and priorities with each other in problem-solving situations. Indeed, it is impossible to allocate too much time to improving communication among participants in a management process.

Given the above points, what can you do as an individual? By being aware of the characteristics of processes, methods, and products, you can first critically examine any initiative related to resource or environmental problems and determine whether these approaches and their inherent values have been included. If one or more is absent, you can draw attention to the need to incorporate them. Second, you can reflect on whether the values you have been acculturated to believe—such as self-interest and competition—are appropriate for guiding behaviour in problem-solving. Alternative approaches exist, and we should not always approach problems without questioning why and how we instinctively try to deal with them. Third, you can pay attention to related approaches, such as regional and land-use planning and risk and impact assessment, to see how they can support or advance resource and environmental management. And fourth, you can do your best to share the information and understanding you have with others in the spirit not only of improving our understanding of structures and processes related to natural and human systems but also of enhancing appreciation of the almost inevitable range of underlying values, assumptions, and attitudes that will shape behaviour. Awareness of all these processes, methods, and products, as well as sensitivity to them, will help to generate diverse ways to define, frame, and solve problems.

Summary

1. If a systems approach is to be used, collaboration and co-ordination are required.
2. Collaboration involves exchanging information, modifying activities in light of others' needs, sharing resources, enhancing the capacity of others in order to achieve mutual benefit and realize common goals or purposes, and joint decision-making to resolve problems during which power is shared and stakeholders accept collective responsibility for their actions and the outcomes.
3. Collaboration is increasingly accepted as desirable because the complexity and uncertainty associated with resource and environmental issues create a challenge for any individual or organization in terms of having sufficient knowledge or authority to deal with them.

4. Participatory approaches aim to incorporate insight from individuals and groups affected by decisions or with responsibility for issues and ideally lead to sharing power and authority. Not all groups welcome power-sharing, since they may view any relinquishment of power as undermining their role in environmental management.

5. Stakeholders are those who should be included in decision-making because of their direct interest, including (1) any public agency with prescribed management responsibilities; (2) all interests significantly affected by a decision; and (3) all parties who might intervene in the decision-making process or might block or delay the process.

6. Co-management is one innovative approach to sharing power by which local people are given responsibility and authority for certain aspects of resource and environmental management. Co-management has been applied in situations involving forestry and wildlife.

7. Effective communication of science and local knowledge is essential, and communications should be prepared in light of who the recipient is and what their level of understanding can reasonably be assumed to be.

8. Effective communication should address four complementary questions: (1) What do we know, with what accuracy, and how confident are we about our data? (2) What do we not know, and why are we uncertain? (3) What could we know, with more time, money, and talent? (4) What should we know in order to act in the face of uncertainty?

9. The concept of adaptive environmental management accepts that (1) surprise, uncertainty, and the unexpected are normal; (2) it is not possible to eliminate them through management initiatives; and (3) management should give allowance for them. Thus, management is viewed as an experiment, requiring systematic monitoring of results so that we can learn from experience.

10. Adaptive co-management includes (1) learning by doing; (2) integration of different knowledge systems; (3) collaboration and power-sharing among community, regional, and national levels; and (4) management flexibility.

11. Environmental impact assessment is used to identify and predict the impact of legislative proposals, policies, programs, projects, and operational procedures on the biophysical environment and on human health and well-being. It also interprets and communicates information about these impacts and investigates and proposes means for their management.

12. Risk assessment focuses on determining the probability or likelihood of an event of some specified magnitude, as well as the likelihood of the associated consequences. It recognizes that since risks have to be estimated, our calculations may be incorrect. For this reason, the precautionary principle is used, which states that in order to protect the environment when there are risks of serious or irreversible damage, lack of full scientific certainty should not be used as an excuse for postponing cost-effective measures to prevent environmental degradation.

13. Sustainability assessment focuses on applying some form of sustainability analysis, appraisal, or assessment or otherwise adopting sustainability objectives as core guides for evaluations and decisions.

14. Disputes usually centre on three main issues: rights, interests, and power. The traditional means of dealing with disputes are political, administrative, and judicial. Increasingly, attention is being drawn to alternative dispute resolution in which information and understanding are shared and efforts are made to find solutions that address the needs of all stakeholders.

15. Regional and land-use plans can be valuable because they often have a statutory or legislative basis and become part of the official documentation governing activity in an area. In contrast, many resource and environmental management plans do not have a legal basis, and this often creates difficulty for effective implementation. By connecting resource and environmental management plans to regional and land-use plans, statutory authority can be obtained.

Key Terms

adaptive co-management	collaboration	indigenous knowledge	risk assessment
adaptive environmental management	co-management	mediation	stakeholders
	co-ordination	monitoring	strategic environmental assessment
alternative dispute resolution	environmental impact assessment	negotiation	
		partnerships	sustainability assessment
arbitration	impact assessment	precautionary principle	

Questions for Review and Critical Thinking

1. Explain the difference between collaboration and co-ordination. Why are both needed in resource and environmental management?

2. What does the word 'stakeholder' imply for resource and environmental management? How would you go about identifying stakeholders in a specific problem-solving situation?

3. What are the underlying principles of 'co-management'?

4. Why is it often difficult to communicate results from scientific research to the lay public? What can be done to improve such communication?

5. Do you believe that scientists are objective when they conduct research?

6. What motivated people to develop the concept of adaptive management? What are the greatest strengths and weaknesses of the approach? How might it be applied to an environmental problem in your community or province?

7. What are the main characteristics of adaptive co-management?

8. Why is the precautionary principle considered important? Has it been effective in practical terms?

9. What is the difference between strategic environmental assessment and sustainability assessment?

10. What are the distinctive features of alternative dispute resolution? In what kinds of situations might it be a better way to deal with conflicts than the judicial approach?

11. What is the benefit of connecting or relating resource and environmental management to regional and land-use plans and to environmental impact assessments? Is that being done in your community or province?

12. To what extent did collaboration and co-ordination occur during the process used to deal with the Sydney Tar Ponds problem discussed in Chapter 1? What initiatives should be taken to achieve a more collaborative and co-ordinated approach in the future?

Related Websites

Adaptive co-management, resilience alliance: http://www.resalliance.org/2448.php

Clayoquot Sound: http://www.forests.org; http://www.focs.ca; http://www.clayoquotbiosphere.org

Co-management: http://www.pcffa.org http://www.indiana.edu/~iascp

Conflict management:

The Conflict Resolution Information Source: http://www.crinfo.org

Public dispute resolution: http://ncinfo.iog.unc.edu/programs/dispute/index.html

Environmental Impact Assessment:

Canadian Environmental Assessment Agency: http://www.ceaa.gc.ca

Environment Canada, Environmental Assessment Program: http://www.ec.gc.ca/ea-ee/home/home_e.asp

Parks Canada, EIA Overview: http://www.pc.gc.ca/nature/eie-eia/index_e.asp

Experimental Lakes Research Area: http://www.dfo-mpo.gc.ca/regions/central/science/enviro/ela-rle_e.htm; http://www.umanitoba.ca/institutes/fisheries

Fraser River Estuary Management Program: http://www.bieapfremp.org/main_fremp.html

International Joint Commission: http://www.ijc.org

Kristianstads Vattenrike Biosphere Reserve: http://www.vattenriket.kristianstad.se/eng/biosphere/index.htm

Manitoba Round Table for Sustainable Development: http://www.gov.mb.ca/conservation/susresmb/mrtsd/index.html

Monitoring:

Ecological Monitoring and Assessment Network: http://www.eman-rese.ca/eman

The Canadian Community Monitoring Network: http://www.ccmn.ca/english

Citizen Science: http://dev.stewardshipcanada.ca/communities/citizenScience/home/csnIndex.asp

Monitoring and Reporting State of Eastern Ontario's Forests: http://sof.eomf.on.ca/Introduction/introduction_e.htm

Further Readings

Note: This list comprises works relevant to the subject of the chapter but not cited in the text. All cited works are listed in the section 'References' at the end of the book.

Armitage, D. 2005. 'Adaptive capacity and community-based natural resource management'. *Environmental Management* 35: 703–15.

———. 2005. 'Collaborative environmental assessment in the Northwest Territories, Canada'. *Environmental Impact Assessment Review* 25: 239–58.

Gibson, R.B. 2000. 'Favouring the higher test: Contribution of sustainability as the central criterion for reviews and decisions under the Canadian Environmental Assessment Act'. *Journal of Environmental Law and Practice* 10 (1): 39–54.

———. 2002. 'From Wreck Cove to Voisey's Bay: The evolution of federal environmental assessment in Canada'. *Impact Assessment and Project Appraisal* 20 (3): 151–9.

Gibson, R.B., et al. 2005. *Sustainability Assessment: Criteria and Processes*. London: Earthscan.

Hahn, T., P. Olsson, C. Folke, and K. Johansson. 2006. 'Trust-building, knowledge generation and organizational innovations: The role of a bridging organization for adaptive co-management of a wetland landscape around Kristianstad, Sweden'. *Human Ecology* 34: 573–92.

Hanna, K.S., Ed. 2005. *Environmental Impact Assessment: Practice and Participation*. Toronto: Oxford University Press.

———. 2005. 'Planning for sustainability: Experiences in two contrasting communities'. *Journal of the American Planning Association* 71: 27–40.

Huxham, C., Ed. 1996. *Creating Collaborative Advantage*. Thousand Oaks, CA: Sage.

Ingles, A.W., A. Musch, and S. Qwist-Hoffman. 1999. *The Participatory Process for Supporting Collaborative Management of Natural Resources: An Overview*. Rome: FAO.

McManus, P.A. 2000. 'Beyond Kyoto? Media representation of an environmental issue'. *Australian Geographical Studies* 38: 306–19.

Noble, B.F. 2000. 'Strategic environmental assessment: What is it and what makes it strategic?' *Journal of Environmental Assessment Policy and Management* 2: 203–24.

———. 2002. 'The Canadian experience with SEA and sustainability'. *Environmental Impact Assessment Review* 22: 3–16.

———. 2002. 'Multi-criteria analysis: Strategic environmental assessment of Canadian energy policy'. *Impact Assessment and Project Appraisal* 20 (3): 177–88.

———. 2003. 'Auditing strategic environmental assessment practice in Canada'. *Journal of Environmental Assessment Policy and Management* 5 (2): 127–47.

———. 2004. 'Applying adaptive environmental management'. In B. Mitchell, Ed., *Resource and Environmental Management in Canada: Addressing Conflict and Uncertainty*, 442–66. Toronto: Oxford University Press.

———. 2004. 'Integrating strategic environmental assessment with industry planning: A case study of the Pasquai-Porcupine Forest Management Plan, Saskatchewan, Canada'. *Environmental Management* 33: 401–11.

———. 2004. 'A state-of-practice survey of policy, plan, and program assessment in Canadian provinces'. *Environmental Impact Assessment Review* 24: 351–61.

———. 2005. 'Integrating human health into environmental impact assessment: Case studies of Canada's northern mining resources sector'. *Arctic* 58: 395–405.

———. 2005. 'Strategic environmental quality assurance: Evaluating and improving the consistency of judgments in assessment panels'. *Environmental Impact Assessment Review* 24: 2–25.

———. 2006. *Introduction to Environmental Impact Assessment: A Guide to Principles and Practice*. Toronto: Oxford University Press.

Noble, B.F, and K. Storey. 2001. 'Towards a structured approach to strategic environmental assessment'. *Journal of Environmental Assessment Policy and Management* 3: 483–508.

———. 2005. 'Towards increasing the utility of follow-up in Canadian EIA'. *Environmental Impact Assessment Review* 25 (2): 163–80.

Olsson, P., C. Folke, and F. Berkes. 2004. 'Adaptive co-management for building resilience in social-ecological systems'. *Environmental Management* 34: 75–90.

Palerm, J.R. 2000. 'An empirical-theoretical analysis framework for public participation in environmental impact assessment'. *Journal of Environmental Planning and Management* 43: 581–600.

Partidario, M.F. 1996. 'Strategic environmental assessment: Key issues emerging from recent practice'. *Environmental Impact Assessment Review* 16: 31–55.

Plummer, R., and J. FitzGibbon. 2004. 'Some observations on the terminology in co-operative environmental management'. *Journal of Environmental Management* 70: 63–72.

Saarikoski, H. 2000. 'Environmental impact assessment (EIA) as collaborative learning process'. *Environmental Impact Assessment Review* 20: 681–700.

Schultz, L., C. Folke, and P. Olsson. 2007. 'Enhancing ecosystem management through social-ecological inventories: Lessons from Kristianstads Vattenrike, Sweden'. *Environmental Conservation* 34: 140–52.

Sinclair, A.J., and A.P. Diduck. 2001. 'Public involvement in EA in Canada: A transformative learning perspective'. *Environmental Impact Assessment Review* 21: 113–36.

Storey, K., and P. Jones. 2003. 'Social impact assessment, impact management and follow-up: A case study of the construction of the Hibernia offshore platform'. *Impact Assessment and Project Appraisal* 21 (2): 99–107.

Trist, E. 1980. 'The environment and system-response capability'. *Futures* 12 (2): 113–27.

Walker, P.A., and P.T. Hurley. 2004. 'Collaboration derailed: The politics of "community-based" resources management in Nevada County'. *Society and Natural Resources* 17: 735–51.

Warner, J., Ed. 2007. *Multi-stakeholder Platforms for Integrated Water Management*. Aldershot, UK: Ashgate.

Wollenberg, E., D. Edmunds, and L. Buck. 2000. 'Using scenarios to make decisions about the future: Anticipatory learning the adaptive co-management of community forests'. *Landscape and Urban Planning* 47: 65–77.

Wood, C. 2003. *Environmental Assessment: A Comparative Review*. 2nd edn. Harlow, UK: Pearson Education.

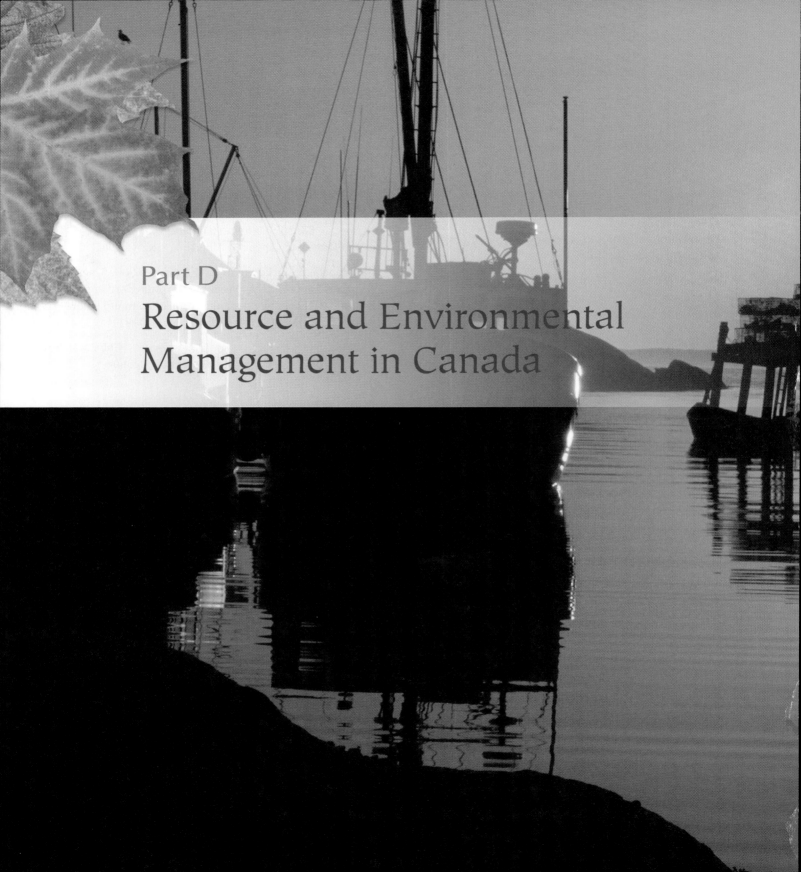

Part D

Resource and Environmental Management in Canada

Parts B and C focused on science and management related to resources and the environment. In Part D, we discover how ideas and methods from science and management can be applied in practical problem-solving situations.

Several comments about the structure of this section should be helpful. It may appear contradictory to argue for a systems or ecosystem approach in Parts B and C and then organize Part D on the basis of specific 'resources' or attributes of the environment, such as agriculture, forestry, or wildlife. Indeed, it would be inappropriate to isolate components of resource or environmental systems and examine them on a sector-by-sector basis, as the chapter headings in Part D might suggest. However, it is appropriate to use one resource or environmental aspect as the starting point for a discussion, as long as we give attention to other elements of the ecosystem and their linkages. This is the approach used in this section.

For example, in Chapter 11 we examine challenges related to flooding and human use of flood plains. Flooding is clearly a problem of too much water at a particular place at a given time. Understanding the process of flooding, however, requires attention not only to aquatic systems but also to connected terrestrial systems, since land-use practices can be a major variable influencing vulnerability to flooding. Similarly, Chapter 7 focuses on climate change and on changes in the atmosphere that can lead to the alteration of climatic conditions. Yet for us to understand what causes climate change, we need to consider other systems as diverse as fossil-fuel resources and transportation industries, as well as examining basic values and attitudes that lead many people to decide that sport utility vehicles are best suited to their transportation needs.

Four types of resources or environments are considered in the following eight chapters. The first, addressed in Chapter 7, is the atmospheric system, with emphasis on climate change. The second, the focus of Chapters 8 to 11, comprises various renewable resources—oceans, forestry, agriculture, and water systems. The third type, considered in Chapter 12, deals with non-renewable resources—minerals and fossil fuels that are created (or renewed) over the course of a geological time span rather than human life spans. The fourth type, place-based, focuses on the environmental opportunities and challenges in urban areas (Chapter 13), in which four of five Canadians now live, and on endangered species and protected areas (Chapter 14).

In all of these chapters, interest centres on examining how science can be used to inform analysis and management and how elements of best practice in management have been or could be applied. For example, after examining the nature of weather and climate, Chapter 7 turns to assessing the scientific evidence related to climate change and to the role of climate models in aiding our understanding of changes. On this basis, it is possible to critically assess the scientific conclusions and explanations offered regarding climate change. We also find that while solid scientific understanding of environmental systems and how they may be changing is of critical importance, it is equally important to communicate that understanding to non-technical specialists. Consequently, this chapter also addresses the challenges encountered by researchers when they seek to communicate scientific insight about climate change to policy-makers and the general public. These challenges are formidable, and too often they are not systematically addressed by scientists, who are frequently more interested in the 'purity' of their scientific work and in communicating their findings to peers. As in other chapters, we attempt to relate scientific understanding to 'real-world' situations. In that context, we present a detailed case study—how the potential rise of sea level caused by climate change would affect Prince Edward Island. In addition, we thoroughly discuss the Kyoto Protocol, along with future possible frameworks for climate change. Besides examining basic issues and approaches represented by this international agreement, we focus on federal, provincial, and territorial government perspectives.

Case studies in other chapters give you a further opportunity to learn about how science and management can be connected. For example, Chapter 8 on oceans looks at the challenges involved in the depletion of the Pacific salmon fishery, the debate over the seal hunt off the east coast, and the closure of the groundfish fishery in Atlantic Canada. In later chapters, case studies cover the environmental impact of the James Bay hydroelectric project in northern Quebec, the development of diamond mines in the Northwest Territories and Nunavut, the rapid growth of the Alberta oil sands, the implications of the Walkerton, ON, experience for water security, the lessons learned about managing natural hazards from the 1997 Red River flood in Manitoba, and the opportunities presented by wind and other renewable sources of energy production.

The chapters in Part D do not always provide detailed scientific or management background, since these aspects were covered in Parts B and C. For example, the case studies on water in Chapter 11 do not examine aspects of the hydrological cycle or contaminant pathways because they were addressed in Chapter 4. Furthermore, other case studies in Parts B and C highlighted the relevance of science and management concepts and methods.

The issues and examples considered here raise fundamental questions about humans' relationships with the environment and resources. We see a range of attitudes towards the environment and other living things, covering the spectrum from humans dominating nature to humans striving to live in harmony with nature. The extinction of species, such as those reported in Chapter 14, reminds us that we have been (and can still be) incredibly arrogant in believing that it is acceptable for us to eliminate some species forever. In contrast, as also shown in Chapter 14, conscious decisions are being taken to protect valued areas, in some instances because we believe it is important to protect examples of different biomes. Notwithstanding some significant and positive accomplishments, current lifestyles in Canada continue to contribute to global climate change in a major way, as discussed in Chapters 7 and 15. The changes associated with climatic warming may confound attempts to identify and protect examples of biospheres if these biospheres change in the future because of different climatic conditions.

Finally, the case studies and examples illustrate how pervasive conflict can be in resource and environmental management. This reinforces our belief that resource and environmental management is not only a scientific or technical exercise. To be effective, it must recognize, identify, and incorporate different values and interests. For this reason, managers often spend significant time trying to resolve conflicts. Given this reality, scientists must also develop a greater appreciation that their work will usually be used in situations in which values and emotions can be as important as, or of greater importance than, theories, models, and quantitative evidence.

The mix of case studies and examples in Part D and throughout this book will help you to appreciate the change, complexity, uncertainty, and conflict that are integral parts of resource and environmental management.

Chapter 7

Climate Change

Learning Objectives

- To understand the difference between weather and climate.
- To know the difference between climate change and global warming.
- To appreciate why the science of climate change is characterized by complexity and uncertainty.
- To understand the nature of scientific evidence regarding climate change.
- To understand the scientific explanation for climate change.
- To realize the implications of climate change for natural and human systems.
- To appreciate the challenges of sharing information and insight related to climate change.
- To comprehend the Kyoto Protocol and different perspectives in Canada related to it.
- To appreciate Canada's role in the global context as a contributor to both climate change challenges and solutions.
- To appreciate the importance of including both mitigation and adaptation in a strategy for reducing vulnerability to climate change.
- To discover what you can do as an individual to minimize the impact of climate change.

Introduction

Climate is naturally variable. It is never exactly the same from one period to another. Sometimes it can shift dramatically within a few hundred or thousand years, as it does when ice ages begin and end. Usually it varies within much narrower limits. For most of the past 1,000 years, for example, the world's average temperature has remained within about half a degree of 14°C.

Over the past 100 years or so, however, the world's climate has changed noticeably. The world's average temperature was approximately 0.6°C warmer at the end of the twentieth century than it was at the beginning, and the 1990s was the hottest decade in 140 years of global climate records. Such changes may seem trifling, but the difference between global temperatures now and at the peak of the last ice age is a mere 5°C. Evidence of earlier climates suggests that global temperatures

have warmed more during the twentieth century than in any other century during the past 1,000 years (Canadian Council of Ministers of the Environment 2003, 5).

As Andrey and Mortsch (2000, WP1) observed, by early in the twenty-first century, climate change had become part of the vocabulary of the North American public. Climate change cartoons and greeting cards had become common (Figure 7.1). In their opinion, perhaps the ultimate proof appeared during the popular TV game show 'Who Wants to Be a Millionaire?' when the following question was posed: 'What gas is primarily responsible for the greenhouse effect? (a) hydrogen, (b) oxygen, (c) nitrogen, or (d) carbon dioxide.' The significance was not that such a question was posed but the $1,000 value for the answer. This relatively low amount highlighted that almost any contestant was expected to know the answer. On the other hand, as the quotation box on page 210 indicates, much complexity and uncertainty is associated with climate change.

Uncertainty Related to Climate Change

Climate change is now a scientifically established fact. The exact impact of greenhouse gas emission is not easy to forecast, and there is a lot of uncertainty in the science when it comes to predictive capability. But we now know enough to recognize that there are large risks, potentially catastrophic ones. . . .

Prudence and care about the future of our children and their children requires that we act now. This is a form of insurance against possibly very large losses. The fact that we do not know the probability of such losses or their likely exact timing is not an argument for not taking insurance. We know the danger exists. We know that the damage caused by greenhouse gas emissions is irreversible for a long time. We know it is growing with each day of inaction.

—Human Development Resource Office 2007, v

Figure 7.1 Kyoto and world governments. Source: Malcom Mayes/artizan.com.

Nature of Climate Change

The condition of the atmosphere at any time or place—that is, the weather—is expressed by a combination of several elements, primarily (a) *temperature* and (b) *precipitation* and *humidity* but to a lesser degree (c) *winds* and (d) *air pressure.* These are called the elements of weather and climate because they are the ingredients out of which various weather and climatic types are compounded. The **weather** of any place is the sum total of its atmospheric conditions (temperature, pressure, winds, moisture, and precipitation) for a *short* period of time. It is the momentary state of the atmosphere. Thus, we speak of the weather, not the climate, for today or of last week.

Climate, on the other hand, is a composite or generalization of the variety of day-to-day weather conditions. It is not just 'average weather', because the variations from the mean, or average, are as important as the mean itself.

Weather and Climate

Canada is perceived to be a cold, northern country, and the average annual temperature of −3.7°C compared to the global average of 15°C supports this view. However, as usual, averages mask a lot of variation, especially in a country as big as Canada. Vancouver, for example, has an average temperature of 10°C, Toronto 8°C, and Halifax 6°C, compared with Alert in the Arctic at −18°C. Precipitation also varies widely, with more than 3,200 millimetres along parts of the west coast and less than 200 millimetres in the Arctic. This variation, spanning more than 40 degrees of latitude between the northern and southern extremities, is reflected in the ecozones described in more detail in Chapter 3.

—After Trewartha 1954, 4

A distinction should also be made between climate change and global warming. Climate represents average day-to-day weather conditions as well as seasonal variations for a particular place or region. In that context, **climate change** is defined as 'a long-term shift or alteration in the climate of a specific location, a region, or the entire planet' (Hengeveld, Bush, and Edwards 2002, 1). A shift is measured for variables associated with average weather conditions, such as temperature, precipitation, and wind patterns (velocity, direction). A change in variability of climate is also included as climate change. In contrast, **global warming**, often mentioned by the media, addresses changes only in average surface *temperatures.* It does not address whether conditions are becoming wetter or drier, for example. A frequent misunderstanding

Philip Dearden

This coral was found near the Arctic Ocean and illustrates how climates have changed in the past.

as heat and moisture are redistributed around the world. The temperature balance of the Earth is maintained through the return of the continually absorbed solar radiation back to space as infrared radiation, consistent with the first law of thermodynamics (Chapter 2). Long-term temperature changes result from shifts in the amount of energy received or absorbed. These shifts may be caused over long cycles (100,000 years) by factors such as the shape of the Earth's orbit around the sun, wobbles of the Earth's axis, and the angle of tilt. Such a 100,000-year cycle of glaciation can be traced over 600,000 years, using evidence from sources such as glacier ice and the chemical characteristics of marine sediments. It is more difficult, however, to explain some of the shorter-term fluctuations that occur.

Natural events, such as the eruption of large volcanoes and changes in ocean currents such as El Niño, are known to have an influence on climate.

Volcanoes, illustrated by the eruption of Mount Pinatubo in 1991, eject large quantities of dust and sulphur particles into the **atmosphere**, which reduce the amount of solar radiation reaching the surface of the Earth. Changes in ocean currents can also be influential (Chapter 8). **El Niño** represents a marked warming of the waters in the eastern and central portions of the tropical Pacific as westerly winds weaken or stop blowing, usually two to three times every decade. In normal years, the trade winds amass warm water in the western Pacific. As the winds slacken, this water spreads back eastward and towards the pole

is that global warming means uniform warming throughout the world. Such an interpretation is not correct. An increase in average global temperatures drives alterations in atmospheric circulation patterns, which can contribute to some areas warming at higher rates, others at lower rates, and others even to become cooler.

What are the 'causes' of climate? The Earth's surface and atmosphere are heated differentially by short-wave radiation from the sun. The differences in heat and pressure between the poles and the tropics fuel the global circulation system

Greenhouse Effect

*The **greenhouse effect** describes the role of the atmosphere in insulating the planet from heat loss, much like a blanket on our bed insulates our bodies from heat loss. The small concentrations of greenhouse gases within the atmosphere that cause this effect allow most of the sunlight to pass through the atmosphere to heat the planet. However, these gases absorb much of the outgoing heat energy radiated by the Earth itself and return much of this energy back towards the surface. This keeps the surface much warmer than if they were absent. This process is referred to as the 'greenhouse effect' because, in some respects, it resembles the role of glass in a greenhouse.*

—Hengeveld, Bush, and Edwards 2002, 2

Bernhard Edmaier/Science Photo Library

Following a volcanic eruption, large amounts of sulphur dioxide (SO_2), hydrochloric acid (HCl), and ash spew into the Earth's stratosphere. In most cases, HCl condenses with water vapour and is rained out of the volcanic cloud. SO_2 from the cloud is transformed into sulphuric acid. The sulphuric acid quickly condenses, producing aerosol particles, which linger in the atmosphere for long periods of time.

into the rest of the Pacific. This triggers weather changes in at least two-thirds of the globe, causing droughts and extreme rainfall in areas along the Pacific and Indian oceans, including Africa, eastern Asia, and North America.

However, it is increasingly apparent that climatic change may occur more rapidly than ever before as a result of human activities. We examine these factors below.

Scientific Evidence Related to Climate Change

In the context of the distinction among weather, climate, global warming, and climate change, the following statements are supported by solid scientific evidence:

1. The world has been warming, with the average global temperature at the Earth's surface having increased by about 0.6°C, with an error range of plus or minus 0.2°C, since the late nineteenth century (Figure 7.2). This increase is a global average, and in some areas, especially over continents, warming has been several times greater than the global average. Between 1995 and 2006, 11 of the 12 years ranked among the 12 warmest years since global surface temperatures have been recorded, starting in 1850 (IPCC 2007c, 1).

2. The increase in the average temperature of the northern hemisphere during the twentieth century was the largest of any century in the past 1,000 years (Figure 7.3).

3. **Greenhouse gas** concentrations, especially those of carbon dioxide, methane, nitrous oxide, and tropospheric ozone, have been rising for several decades (Figure 7.4). Carbon dioxide and methane are at higher concentrations now than at any time over the past 420,000 years (Figure 7.5). There is no scientific dispute about such increases.

Figure 7.3 Variation in northern hemisphere average surface temperature. Based largely on ice core, tree ring, and coral reef data (thin light line). Also shown is the 20-year running mean of the annual paleoclimatic data and the 1000–1900 trend line (thick lines) and the directly observed temperature variation of Figure 7.2 (thin dark line). Source: Miller 2002, A-2, prepared by Professor Danny Harvey, Department of Geography, University of Toronto, using paleoclimatic and historical data from UK website (see Figure 7.2) and data from the US National Oceanographic and Atmospheric Administration paleoclimatology website http://www.ngdc.noaa.gov/paleo.

Figure 7.2 Variation in global average surface temperature between 1856 and 2000. Source: Miller 2002, A-2, prepared by Professor Danny Harvey, Department of Geography, University of Toronto, using data in electronic form available from the UK Meteorological Office website http://www.meo.gov.uk.

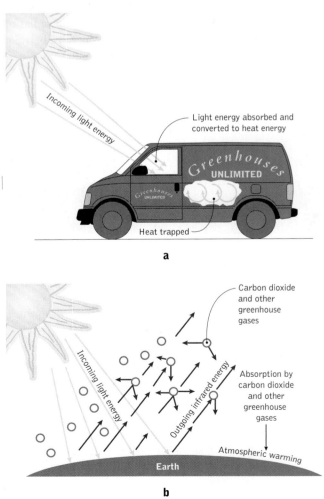

Figure 7.4 The greenhouse effect.

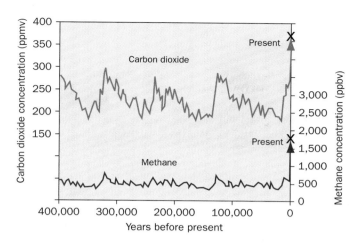

Figure 7.5 Variation in atmospheric concentrations of carbon dioxide and methane to 400,000 years before present. Measured in the Vostok ice core in Antarctica (thin lines) and during the past 200 years (heavy line). Source: Miller 2002, A-1, prepared by Professor Danny Harvey, Department of Geography, University of Toronto, using data in electronic form obtained from the US National Oceanographic and Atmospheric Administration paleoclimatology website http://www.ngdc.noaa.gov/paleo.

4. In most parts of the world since 1980, glaciers have lost more mass than they have gained. Furthermore, most mountain glaciers have been retreating during the past 100 years. Exceptions are found in Norway and New Zealand, where some glaciers have been advancing. The explanation for these exceptions is that they are associated with remarkable increases in precipitation rather than with decreases in temperatures.

5. In many areas of the world, reduced snow cover has been documented, as well as earlier spring melting of ice on rivers and lakes. For example, snow cover in the northern hemisphere has decreased by approximately 10 per cent since 1996. The cold temperatures and large amounts of snow in the winters of 2004 and 2008 across most of Canada were an exception to this general pattern. Based

on satellite data, since 1978 the extent of Arctic sea ice has declined, on average, at a rate of 2.7 (2.1 to 3.1) per cent each decade (IPCC 2007c, 1).

6. Measurements show that permafrost is warming in many regions.

7. The Intergovernmental Panel on Climate Change (IPCC 2001c) reported that the average rate of sea-level rise increased from 0.1 to 0.2 millimetres per year during the past 3,000 years to one to two millimetres per year in the twentieth century, a tenfold increase. Since 1961, global average sea levels have risen at an average rate of 1.8 (1.3 to 2.3) millimetres annually and since 1993 at 3.1 (2.4 to 3.8) millimetres annually (IPCC 2007c, 1). Caution is needed in interpreting such findings, however, since it is recognized that in some areas, land is still rebounding from the weight of the last glaciation. Thus, data from tidal gauges must be interpreted in light of the combination of sea-level rise and land rebound, which could mask the increase in sea-level rise.

Hengeveld (2006, 28) has effectively summarized the above findings with his comment that

> The debate about climate change within the science community is gradually shifting away from whether it is happening and how serious it will be to how to deal with it. The general consensus is that global warming is already occurring, that **fossil fuels** combustion is the primary cause, and that the negative impacts of weather and climate events will get much worse.

One attribute of 'good science' is the use of cross-checking data sources to ensure that findings are not unduly influenced by measurement error or limitations of any single data source. The findings above related to temperature, greenhouse

The forest industry is a major contributor to the rising levels of carbon dioxide in the atmosphere, not only through deforestation but also through emissions from processing plants.

Some countries, such as the Maldives, are so low-lying that they could be mostly flooded as early as 2050 if global sea levels change as predicted.

A view of the Moose River from Moose Factory Island. Traditional environmental knowledge is vital to understanding the complex changes that occur in regions such as Moose Factory and Moosonee. Climate change has not only affected average temperatures but also animal and bird migration, weather patterns, and the freeze/thaw cycles of the Moose River (image from the Cree Village Ecolodge website).

gas concentrations, glaciers, snow cover, river and lake ice break-up, permafrost, sea-level rise, and traditional environmental knowledge all indicate that climate change is occurring. In the next section, we turn our attention to the reasons for this change. Once the pattern and causes of climate change are understood, we have a foundation for considering the implications for designing possible policies and actions.

Glaciers and Climate Change in Canada

Since 1950, the greatest warming in Canada has occurred in the West and Northwest. Most glaciers in these regions are also shrinking rapidly. The 1,300 or so glaciers on the eastern slopes of the Rockies, for example, are now about 25 per cent to 75 per cent smaller than they were in 1850. The area of warming also covers many of the High Arctic islands in Nunavut, where glaciers such as the Melville Island South Ice Cap have been shrinking gradually since at least the late 1950s. In eastern Nunavut, however, the situation is more complex: some glaciers are shrinking, while others are growing.

The melting of glaciers is a concern for Alberta, Saskatchewan, and Manitoba. Farmers depend on glacier-fed rivers like the Saskatchewan and the Bow for irrigation water, and cities like Edmonton, Calgary, and Saskatoon rely on them for municipal water supplies and recreation. At The Pas in Manitoba, reduced flows on the Saskatchewan could interfere with the native fishery and hydroelectric power generation.

—Canadian Council of Ministers of the Environment 2003, 20

Implications of Sea-Level Rise

Rising sea levels threaten familiar shoreline environments. Coastal wetlands, which are important ecosystems and barriers against shoreline erosion, gradually disappear. Bluffs and beaches are more exposed to erosion by waves, groundwater is more likely to become contaminated by salt water, and low-lying coastal areas may be permanently lost. In addition, wharves, buildings, roads, and other valuable seaside property face a greater risk of damage as a result of flooding from storms.

—Canadian Council of Ministers of the Environment 2003, 13

Modelling Climate Change

The uncertainty associated with global climate change is encouraging scientists to explore many different ways of assessing past and future climates (Box 7.1). One approach is **climate modelling**. While concern about climate change due to greenhouse gas (GHG) emissions is relatively recent, climate modelling is not. The earliest global climate models date back to the 1950s, far ahead of when scientists became concerned about carbon dioxide emissions and their effect on the atmosphere. However, more recent concerns about global warming have propelled the science of climate modelling to the forefront.

Climate Models

All climate models consider some or all of five components in order to predict future climates.

- radiation—both incoming (solar) and outgoing (absorbed, reflected);

Guest Statement

Traditional Ecological Knowledge (TEK), Science, and Climate Change
Elise Ho

Climate change will affect many areas of the globe, particularly northern regions. Thus, the western James Bay region in Ontario is expected to experience drastic changes due to climate change. For the predominantly First Nations communities in this area, climate change will affect transportation networks, building infrastructure, recreational opportunities, traditional lifestyles, animal and plant populations, river and sea-ice characteristics, water levels, and weather. Unfortunately, current General Circulation Models (GCMs) (discussed in this section) are not able to predict future changes in climate at this regional scale. Furthermore, Hudson and James bays (which are mostly covered by ice during the winter) create complex feedback loops that confound many regional climate projections. Lastly, these relatively remote northern regions are not monitored to the same extent as southern regions in Canada, and weather stations are sparsely located.

I investigated this problem for my master's thesis. I examined how conventional scientific approaches to understanding climate change could be complemented by traditional environmental knowledge (TEK). Also known as indigenous knowledge, TEK is a broad term used to describe 'various systems of knowledge, practice, and belief gained through experience and culturally transmitted among members and generations of a community' (Huntington et al. 2005, 64). Regrettably, TEK is regarded by some to be anecdotal and less factual than what many

Elise Ho

people consider to be 'science', and it has not been given enough recognition as a valuable and insightful knowledge system in conventional environmental management and assessment practices.

I visited two First Nations communities in the western James Bay region, Fort Albany and Moose Factory. The elders in these communities shared their knowledge of climate change with me, which I was able to compare with the scientific evidence of climate change in the region. I found that TEK was able to identify a large range of climate change variables that contributed to a more complete understanding of how climate change has affected and might continue to affect the western James Bay area. TEK provided in-depth and specific insight regarding aspects such as changing weather characteristics, animal migration patterns and behaviour, and changing seasonal averages. Furthermore, while scientific analyses of aspects such as break-up dates for river ice are informative, TEK was able to complement this information through knowledge of the severity of the break-up events, changes in ice thickness, and other ice characteristics. I concluded that TEK and conventional scientific approaches are complementary in climate studies and that scientific studies of climate change in this region were limited by number and scope.

My study is part of the growing research that seeks to understand how TEK and science can work as complementary knowledge systems, especially in northern climate studies. Hopefully, as we seek to better understand climate change and its effects on our environment, we will become increasingly open to different yet also valuable knowledge systems such as TEK.

Elise Ho completed her PhD at the University of Toronto in 2008. She is employed with an environmental consulting firm, Gartner Lee, in Markham, ON, as an environmental social scientist. Her work focuses on the social aspects of energy projects as well as northern projects.

Box 7.1   Measuring Climate Change

An essential step in attempting to assess climatic change is to see how present variations in climate compare with those of the past. Current data are largely instrument-based weather observations—i.e., the instrumental record. Even here there are difficulties. More modern and accurate data—for example, data on the upper atmosphere gathered by satellite—are available only for the past two and a half decades or so. This period also coincided with the greatest human impact on climate and does not provide any type of control for climate change in the absence of industrialization.

Former climates are reconstructed by scientists using proxy information from many different sources. For example, examination of historical records of climate-influenced factors such as the price of wheat in Europe over the past 800 years, the blooming dates of cherry trees in Kyoto, Japan, since AD 812, the height of the Nile River at Cairo since AD 622, the number of severe winters in China since the sixth century, examination of sailors' and explorers' logs, and other such sources all contribute to building up a picture of past climates.

Scientists also use climate-sensitive natural indicators such as tree rings and glacial ice. Cores obtained from ice in Greenland and Antarctica that go back tens of thousands of years have been analyzed using the ratio of two oxygen isotopes that indicate the air temperature when the original snow accumulated on the glacier surface. Tree rings are also very useful. Outside the tropics where there are noted differences in seasons, the width and density of tree rings reflect growth conditions, including climate. Some species—red cedar in coastal BC, for example—may live for well over 1,000 years and can provide valuable indicators as to past climates. The same kinds of rings also characterize the growth of many long-living corals in the tropics, which may be more than 800 years old, and provide valuable evidence about previous El Niño events.

- dynamics—the horizontal and vertical movement of energy around the globe;
- surface processes—the effects of the Earth's surface (snow cover, vegetation) on climate (i.e., albedo, emissivity);
- chemistry—the chemical composition of the atmosphere and its interactions with other Earth processes (i.e., carbon cycling);
- time step and resolution—the time step of the model (minutes or decades) and the spatial scale of the model (your backyard or the entire globe).

The nature of the Earth's climate and its complexity make comprehensive climate modelling difficult. The many components, interactions, and feedback loops in the global climate cannot be entirely represented by any mathematical model, and therefore all models simplify certain aspects of climate. There are four main types of climate models, each increasing in complexity. They are outlined in Box 7.2.

CO_2 x 2, or CO_2 x 4?

In June 2004, G. Beauchemin, responsible for disaster preparedness for the government of Quebec, commented at the annual conference of the Canadian Water Resources Association in Montreal that the CO_2 x 2 scenario is the future we expect and are preparing for. However, he said, this is very likely to be the 'best-case' scenario. He argued that risk managers should be worrying about the 'worst-case' scenario, such as CO_2 x 4.

Unlike EBMs, GCMs attempt to examine all of the climatic elements and processes, making these models very complex. GCMs model the Earth's atmosphere and oceans under certain climate change scenarios, the most popular being $2 \times CO_2$. In this situation, the Earth's climate is modelled to indicate the changes that would occur if atmospheric concentrations of carbon dioxide were doubled from pre–Industrial Revolution levels, which many scientists believe will occur by 2050.

In a GCM, the Earth's surface is divided into a grid, in which case a larger grid results in a simpler model and a smaller grid requires more calculations. For each grid, a series of fundamental equations are solved at the surface of the grid (sea level) and for several layers of the atmosphere and subsurface layers (the vertical dimension). The fundamental equations deal with:

- conservation of momentum;
- conservation of mass;
- conservation of energy;
- ideal gas law.

Beginning with present-day or known values, the solutions for these equations are solved and repeated at each time step of the simulation, and then the results are interpolated between the grid points to cover the Earth's entire surface. The two key constraints, spatial resolution (grid size) and temporal resolution (length of time step), are compromised to meet computational capacity. Most models operate at spatial resolutions of a few degrees latitude and longitude and at time steps of less than one hour. The vertical dimension is often divided into 10 layers, with two subsurface layers. Because of these simplifications, GCMs are best used for global or overall climate modelling, not for regional representations of climate change.

Other aspects of the global climate are simplified as well, thus limiting the predictive capabilities of many GCMs. For example, known or present-day values are required to run many models, but in some areas of the world, these values are unavailable or scarce for some variables (temperature, sea-ice cover, cloud cover). Therefore, assumptions are made to fill in the missing values, which may not be accurate. In addition, many complicated feedbacks cannot all be accounted for in

Box 7.2   **Four Types of Climate Models**

1. Energy Balance Models (EBMs)
   EBMs can be either non- or one-dimensional. In the first case, the Earth (or any point on the Earth) is treated as a single entity, and only incoming and outgoing radiation are modelled. In one-dimensional EBMs, temperature is modelled as a function of latitude and radiation balance.
2. One-Dimensional Radiative-Convective (RC) Climate Models
   In this model, the one dimension is altitude. One-dimensional RC models take into account incoming and outgoing solar radiation, as well as convective processes that affect the vertical distribution of temperature. These models are useful for examining the vertical distribution of solar radiation and cloud cover and are very useful for examining the effects of volcanic emissions on temperature.
3. Two-Dimensional Statistical-Dynamic (SD) Climate Models

This type of model takes into account either the two horizontal dimensions or one horizontal dimension and the vertical dimension. The latter are most frequently modelled, thus combining the latitudinal EBMs with the vertical RC models. These models can examine wind speed, direction, and other horizontal energy transfers.

4. General Circulation Models (GCMs)

The first three types of climate models described above are still used for various purposes in climate research. However, since the 1980s, **general circulation models (GCMs)** have largely taken over the field of climate modelling, and most model development is devoted to them. It is by far the most complex type of model, since the GCM takes into account the three-dimensional nature of the Earth's atmosphere, oceans, or both.

GCMs, partly because of their complexity and **uncertainty** about how they react under given circumstances. While the relationship between greenhouse gas emissions and temperature is a relatively straightforward positive feedback loop (where a positive change in one variable results in a positive change in the other), the relationship between increased temperature and cloud cover is relatively uncertain and relies on many other variables. Finally, many of the climatic interactions at the Earth's surface are difficult to model and are underrepresented in many GCMs. For example, ocean layers and interactions are difficult to model, but their effect on regional and global climate can be quite significant.

In summary, while GCMs are becoming increasingly sophisticated, many complex aspects and interactions of the global climate need to be understood more fully, and computational facilities need to be better developed to support these new models.

Limitations of GCMs

While GCMs provide overall indications of future climates, their limitations for policy and planning need to be appreciated. Many scientists have recognized that the coarse spatial resolution, poor predictive capacity for precipitation, relatively weak simulation of oceans, lack of baseline data, and many other limitations cause GCM outputs to be highly variable. Some researchers have become discouraged by the dominance of GCMs in climate change sciences and by the scientific disregard of other climate models (i.e., EBM, RC, SD, or combinations). Others caution against the misinterpretation that GCMs are accurate and realistic models of global climate and stress that much more improvement is needed so that GCMs may best represent the complex nature of the Earth's climate.

Intergovernmental Panel on Climate Change

The Intergovernmental Panel on Climate Change (IPCC) was established in 1988 by the World Meteorological Organization and the United Nations Environment Programme.

The IPCC was created to assess scientific, technical, and socio-economic information related to understanding the risks from human-induced change to climate. The IPCC does not conduct original research, nor does it monitor climate data. Its assessments are based on peer-reviewed and published scientific literature.

The IPCC has three working groups and a task force. Its First Assessment Report, published in 1991, had an important role related to the UN Framework Convention on Climate Change, which was adopted at the Earth Summit at Rio de Janeiro in 1992.

Its Second Assessment Report, published in 1995, became a significant input into negotiations that resulted in the Kyoto Protocol in 1997. The Kyoto Protocol is discussed later in the chapter.

The Third Assessment Report, produced in 2001, was the product of the work of more than 2,000 scientists from many disciplines from all around the world.

A Fourth Assessment Report was published in 2007 and is discussed later in this chapter.

Scientific Explanations

In its 2007 report, the Intergovernmental Panel on Climate Change (IPCC) concluded that, as noted earlier in this chapter, worldwide trends in the twentieth century consistently and strongly reveal an increase in global surface temperature. There is now virtually no scientific debate about this finding. Furthermore, as also noted above, greenhouse gas concentrations in the atmosphere have increased significantly. There is strong scientific consensus that the increase in greenhouse gases has been caused by human activities. The basis for this conclusion comes from data showing that (1) the rate of increase of greenhouse gases in the atmosphere over the past century closely matches the rate of human-driven emissions into the atmosphere; (2) atmospheric oxygen has been falling at the same rate that fossil-fuel emissions of carbon dioxide have been increasing; and (3) significantly, the change in proportions of carbon isotopes in the atmosphere is evidence that the atmosphere is being enriched with carbon from fossil-fuel sources rather than from natural sources. Regarding the third point, the Intergovernmental Panel on Climate Change (IPCC 2007b, 2) concluded that 'The global increases in carbon dioxide concentrations are due primarily to fossil fuel use and land-use change, while those of methane and nitrous oxide are primarily due to agriculture' (see Chapter 4).

Natural and human variables both contribute to climate change, but it is difficult to determine their relative contribution, since they usually operate at the same time. Such variables alter the balance of incoming and outgoing energy in the Earth's atmospheric system and are usually referred to as 'radiative forcings'. Positive forcings produce warming; negative forcings lead to cooling. Greenhouse gases, aerosols, variations in solar output, and volcanic eruptions are viewed to be the most influential radiative forcings.

Figure 7.6, from the Intergovernmental Panel on Climate Change, shows the relative contribution of different radiative forcings in 2005. Greenhouse gases, and in particular carbon dioxide, have been the main forcing factor. In comparison, the effect of variation in solar output has been minor. Working in the opposite direction, sulphate aerosols, emissions from volcanic eruptions, biomass aerosols, and depletion of the stratospheric **ozone layer** have contributed to cooling at the Earth's surface. Figure 7.6 is derived from what is viewed as a high level of scientific understanding about most greenhouse gases, based on excellent data regarding their concentrations in the atmosphere and good insight about their radiative forcing characteristics.

Gustoimages/Science Photo Library

Although this rural scene may look bucolic, the rising numbers of cattle on the Earth contribute significantly to two of the problems discussed in this book: eutrophication and increased methane levels leading to global climatic change.

Figure 7.6 Global average radiative forcing (RF) estimates and ranges in 2005 for anthropogenic carbon dioxide (CO_2), methane (CH_4), nitrous oxide (N_2O), and other important agents and mechanisms, together with the typical geographical extent (spatial scale) of the forcing and the assessed level of scientific understanding (LOSU). The net anthropogenic radiative forcing and its range are also shown. They require summing asymmetric uncertainty estimates from the component terms and cannot be obtained by simple addition. Additional forcing factors not included here are considered to have a very low LOSU. Volcanic aerosols contribute an additional natural forcing but are not included in this figure because of their episodic nature. The range for linear contrails does not include other possible effects of aviation on cloudiness. Source: IPCC 2007b, 16.

Box 7.3   Ozone Depletion

Ultraviolet radiation from the sun causes some oxygen molecules to split apart into free oxygen atoms. These may recombine with other oxygen molecules to form **ozone** (O_3) in the outer layer of the atmosphere, known as the **stratosphere**. This layer of ozone helps to filter out ultraviolet (UV) radiation from penetrating to the Earth's surface where it destroys protein and DNA molecules. Without this protective layer, it is doubtful whether life could have evolved on Earth at all.

Although there are natural causes of variation in ozone levels, recent observations have indicated that this layer is being broken down by the emission of various chemicals from the Earth. Since 1979, the amount of stratospheric ozone over the entire globe has fallen by about 4 to 6 per cent per decade in the mid-latitudes and by 10 to 12 per cent in higher latitudes. These decreases have led to average increases in exposure to ultraviolet-b (UV-b) of 6.8 per cent per decade at 55°N and 9.9 per cent in the same latitude in the southern hemisphere. In general, penetration of UV-b radiation increases by 2 per cent for every 1 per cent decrease in the ozone layer. UV-b radiation is responsible for various potentially negative health effects on humans and animals, mainly related to eyes, skin, and immune systems. Human vulnerability to UV-b depends on a person's location (latitude, altitude), duration and timing of outdoor activities, and precautionary behaviour (use of sunscreen, sunglasses, or protective clothing).

Canada hosted an international meeting in 1987 to design a program to eliminate ozone-depleting substances. Despite achievement of the targets set by this **Montreal Protocol** (policy target values: see Chapter 4), ozone depletion continued to increase. In a subsequent follow-up meeting in 2007, hailed by a UN under-secretary general as 'the most important breakthrough in an international negotiation process for at least five or six years', also hosted by Canada, more ambitious targets were set, more in accord with the scientific target values (STVs: see Chapter 4). The new agreement saw developed countries capping production of ozone-depleting **hydrochlorofluorocarbons (HCFCs)** at 2009–10 levels by 2013, replacing the earlier date of 2016. Agreement was also reached to reduce production by 75 per cent by 2010 and by 90 per cent by 2015. Significantly, developing countries, including India and China, agreed to end production by 2020 rather than 2030.

What about the role of humans and their activities regarding climate change? The reports from the Intergovernmental Panel on Climate Change (IPCC 2007a; 2007b; 2007c) are explicit that most warming since the mid-twentieth century is associated with human activities. This view is based on climate modelling showing that when only the influences of variation in solar output and emissions from volcanic eruptions are included, simulated temperature changes do not match the observed changes very well. In contrast, when the models include the forcing factors related to human activities (greenhouse gases, stratospheric ozone depletion, sulphate aerosols), better matches are found between the observed and simulated patterns of temperature change. Not surprisingly, the best match occurs when both natural and human forcing factors are combined.

There is no guarantee that greenhouse gases have caused the temperature increases at the Earth's surface, but the evidence is strong and credible. As the National Research Council in the US (2001) noted, the view that most of the observed increase in temperatures is likely due to an increase in greenhouse gas concentrations 'accurately reflects the current thinking of the scientific community on this issue'.

Implications of Climate Change

The implications of climate change are potentially widespread and significant. In this section, we briefly outline some of the impacts on different resource-based systems. Following this overview, we present a detailed case study focused on the possible impact of sea-level rise driven by climate change on the coast of Prince Edward Island.

Terrestrial Systems

It is conceivable that within your lifetime, many terrestrial systems, along with the associated fauna and flora, will change dramatically. For example, on the Canadian Prairies, boreal forests may shift anywhere from 100 to 700 kilometres to the north, to be replaced by grasslands and more southern forest species. In the Arctic, the southern permafrost border could move 500 kilometres northward, and the tree line could move from 200 to 300 kilometres to the north. These shifts are illustrated in Figure 7.7. Boreal forests in particular would also be affected by increases in insect infestation, disease, and fires.

The consequences of change to terrestrial systems could be dramatic. For example, polar bears may no longer remain to breed in Wapusk National Park in Manitoba, yet the park was created in 1996 to protect the habitat of polar bears. At the other extreme, the hoary marmot in the Rockies and other areas in the Western Cordillera is likely to thrive, since changed climate leads to more avalanches, which will expand its preferred habitat of open meadow. These examples indicate that the rationale for national and provincial parks, which were created to protect representative ecosystems (Chapter 14), may disappear or dramatically change as the distinctive ecosystems currently protected by such parks evolve into something totally different.

Agriculture

One of the major limitations on agricultural activity in most areas of Canada is our cold climate. In southern Canada, the frost-free growing season is about 200 days, and in the Far North the growing season is normally just a few weeks. Furthermore, early frosts or severe winters can damage even dormant vegetation.

Given the above conditions, the scientific consensus is that on balance, Canada would be one of the countries to

Jackie Dawson

Polar bear on the Hudson Bay coast near Wapusk National Park.

benefit from **global warming**, since it would extend the growing season and reduce the damage from severe cold or frosts. For example, some scenarios indicate that by 2050, growing conditions in Whitehorse and Yellowknife would approximate those now found in Edmonton and that in New Brunswick, conditions would become similar to those now experienced in the Niagara Peninsula in Ontario (Hengeveld, Whitewood, and Fergusson 2005, 37). However, as in most situations, challenges may also arise. Many plants are vulnerable to heat stress and drought, and if temperatures increased appreciably or water became limited, crops could be adversely affected. In the next subsection, we look at the effect of climate change on aquatic systems.

Moreover, as shown in the box below, climate change may have significant effects on food production in other regions of the world, and the first people to be hurt would be the poorest farmers. Such conditions and consequences could lead to significant increases in migration, which could cause regional instability, as well as in international migration. The latter would contribute to growing numbers of 'environmental' refugees for whom Canada could be a destination of choice.

Climate Change and Environmental Refugees

Future climate change is expected to have considerable impacts on natural resource systems, and it is well-established that changes in the natural environment can affect human sustenance and livelihoods. This in turn can lead to instability and conflict, often followed by displacements of people and changes in occupancy and migration patterns. Therefore, as hazards and disruptions associated with climate change grow in this century, so, too, may the likelihood of related population displacements.

—McLeman and Smit 2003, 6

Freshwater Systems

As a result of the changes documented so far, every part of Canada except the southern Prairies has become wetter, with precipitation increasing from 5 per cent to 35 per cent since 1950. At the same time, generally higher temperatures cause higher rates of evapotranspiration. What might be the

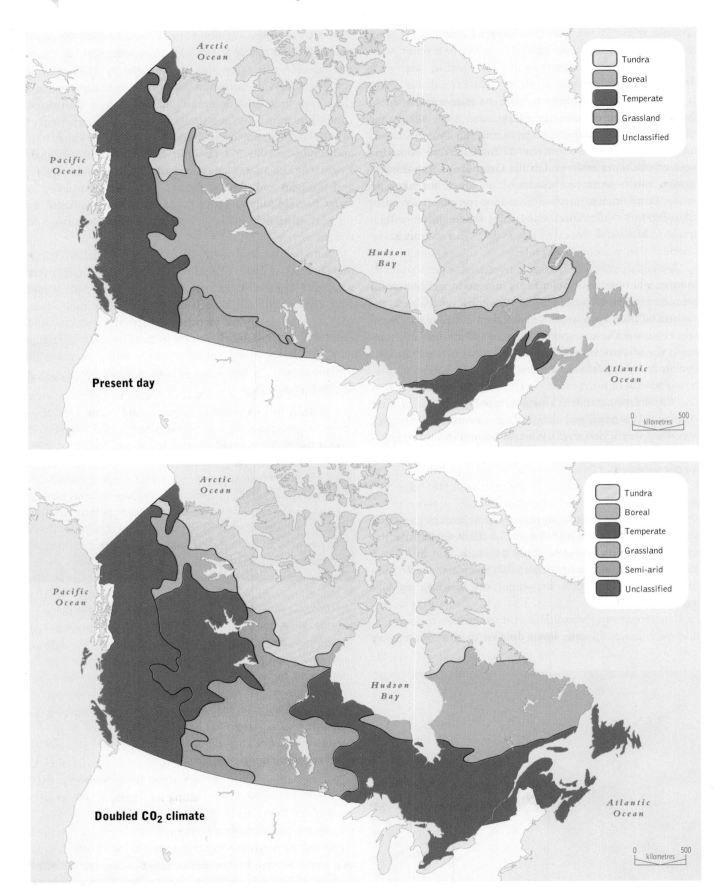

Figure 7.7 Changes in forest and grassland boundaries resulting from a typical doubled CO_2 climate. Source: Hengeveld 1991, 44. Reproduced with the permission of the Minister of Public Works and Government Services, 2005.

outcomes? On the west coast of British Columbia, increased cloud cover and more rain can be expected. As a result, water supplies should be relatively secure, but tourism may be adversely affected if potential tourists avoid an area that is already known for its abundance of rainfall, sometimes referred to as 'liquid sunshine'. In other areas, agriculture operations may become more vulnerable, leading to pressure for expansion of irrigation systems, which may place high pressure on surface and groundwater systems. On the Great Lakes, the shipping season may be extended because of less ice on the lakes, but at the same time, drier conditions may contribute to a drop in lake levels so that lakers must carry less freight in order to navigate locks and other passages with depth constraints.

Fisheries

Marine and freshwater fisheries are important for commercial and recreational activity on the west, east, and Arctic coasts and on inland lakes such as the Great Lakes and Lake Winnipeg. The fishery is a key component of the economy of many small coastal and remote interior communities adjacent to lakes and is also a foundation for many aspects of life of indigenous peoples across the country.

Fish are vulnerable to changes in temperature, precipitation, wind patterns, and chemical conditions in or related to aquatic systems. Hengeveld, Whitewood, and Fergusson (2005, 39) have noted that climate change may be an important cause of the decline in salmon stocks in British Columbia and that water temperature increases in the Atlantic Ocean have most likely contributed to the decrease in flounder in those waters. In fact, changing oceanic conditions as a result of global climate change may well underlie the failure of the Atlantic cod stocks to recover from overfishing, as discussed in Chapter 8. On the other hand, in Arctic waters it has been reported that pink and sockeye salmon are being found well beyond what has been considered their normal range, and this is most likely due to warmer water conditions in northern waters.

In the future, if water levels drop or there are more periods of lower water levels, the mortality of spawning salmon

Greenland ice sheet melt, 1992–2002 (Walsh 2004).

in BC rivers and streams is likely to increase, thereby reducing the number of salmon successfully completing their spawning cycle. A further negative impact could result from more frequent short, intense rainstorms, which could trigger flash floods that in turn could damage the gravel beds in streams used by salmon as spawning beds.

For freshwater systems, warmer water would enhance conditions for warm-water fish such as sturgeon and bass but create additional stress for cold-water fish such as trout and lake salmon. Fish better adapted to warm-water conditions could migrate into waters that have become warmer and compete with and perhaps even prey on species already present. Higher temperatures and increased evapotranspiration could also lower water levels in lakes, and one result would be degradation of shoreline wetlands that provide key habitat for some species of fish. On the positive side, warmer winters could reduce ice cover during the winter months and result in reduced mortality of some fish species during that period.

Cryosphere

Warmer temperatures in higher latitudes are expected to cause melting of ice, such as the Greenland ice sheet.

As ice cover in Arctic latitudes is reduced through melting, various consequences will follow. One of the most obvious would be rises in sea levels because of the incremental water added to the oceans from melting ice sheets, as well as from glaciers that drain to the ocean via rivers and lakes.

Evidence indicates that melting of ice sheets in the Arctic is occurring. For example, in September 2007 scientists examining remote sensing images were surprised to observe that Arctic sea ice had been reduced to about 4 million square kilometres. This number represented a reduction of 23 per cent from the previous low of 5.3 million square kilometres recorded in 2005. The area of ice that had melted in one season was equivalent to

Great Lakes freighter leaving Duluth harbour, Minnesota.

Glaciers Melting at Increasing Rate

In March 2008, the United Nations Environment Pro-gramme reported that data from about 30 glaciers in nine mountain ranges around the world indicated gla-ciers were thinning and melting at a rate that had more than doubled between 2004–5 and 2005–6. This find-ing reflected the pattern of accelerated ice loss from glaciers over the previous two and a half decades.

Achim Steiner, UN under-secretary general and UNEP executive director, commented that 'There are many canaries emerging in the climate change coal mine. The glaciers are perhaps among those making the most noise, and it is absolutely essential that everyone sits up and takes notice.'

—UNEP 2008

Sea Ice in the Arctic and Implications for Cruise Tourism

Records show that surface air temperatures are increas-ing in the Canadian Arctic. The extent of northern hemi-sphere sea ice has also been decreasing since 1979. One conclusion is that these two patterns should lead to an increase in cruises in Arctic waters, which currently extend from about the end of July until mid-September. In 2006, 22 cruise vessels travelled through the Canadian Arctic, the largest number to that point.

Stewart et al. (2007, 377) observed, however, that the 37-year record of sea ice provides 'little to no evi-dence' in support of claims that climate change has affected sea-ice conditions in the Canadian Arctic to the extent that it would allow easier movement of ships through the Northwest Passage. In their words:

While some increases in open water have been recognized, the navigable areas through the Northwest Passage actually have exhibited increases in hazardous ice conditions; naviga-tion choke points remain and are due primarily to the influx of multi-year ice into the channels of the Northwest Passage. Thus, cruise opera-tors working in the Northwest Passage face con-siderable uncertainty. In the future: rather than widespread accessibility, as some have claimed, there is likely to be much more variability of ice conditions across this region.

the area of the province of Ontario. One 'positive' outcome of the reduction in Arctic ice cover was that the Northwest Pas-sage became nearly ice-free and therefore readily navigable for about five weeks in the late summer. The Northwest Passage is some 2,300 kilometres in length, and by the end of the sum-mer of 2007, ice covered only about 20 kilometres rather than the normal 400 kilometres. For some Canadians, however, eas-ier passage raises concerns about Canadian sovereignty in this area, since non-Canadian ships without special ice-breaking capacity could navigate the passage.

Melting of ice has not been confined to the Arctic. In the Rockies, glaciers less than 100 metres thick are vulnerable to melting and could disappear by 2030. Such glaciers are relied upon to provide water to the rivers that drain from the eastern slope of the Rockies across the Prairie provinces, causing con-cern about water security in the mid and long term.

Another consequence will be degradation of permafrost in both alpine and high-latitude regions. One result will be changes in the hydrology of aquatic systems in northern re-gions as well as stability of land. As Hengeveld, Whitewood, and Fergusson (2005, 41) have observed, 'decaying permafrost would also destabilize coastal land areas, increasing the risks of landslides and compounding the coastal erosion caused by rising sea levels and reduced ice cover.'

Another effect of reduced ice cover and degradation of per-mafrost would be a reduction of the expansive and stable ice cover that northerners count on for winter travelling as well as for hunting and other traditional activities. There would be a

reduced season and lower load capacity for 'winter roads' con-structed over snow and ice and relied on for the delivery of goods to and from remote communities.

On a positive note, milder winters would lessen frost damage to roads and railway lines in southern Canada and reduce snow removal costs in urban areas and for provincial highways.

Ocean and Coastal Systems

It appears that both sea temperatures and sea levels will in-crease. The Intergovernmental Panel on Climate Change (IPCC 2007b, 4) concluded that since 1961, the average global water temperature has increased to depths of at least 300 metres and that the ocean has been absorbing more than 80 per cent of the additional heat to the climate system. The IPCC also noted that this warming causes sea water to expand, which contributes

Historical extent of Moving Glacier, Strathcona Provincial Park, Vancouver Island, 1931 to 1981. Milla Lake, on the right side of these aerial photographs, has grown in extent with the melting of the glacial ice up-slope (Smith and Laroque 1996).

A pair of trucks drive the Ice Road in the Northwest Territories. Truckers drive the frozen ice highway on the Mackenzie River in the few short months it is safe for use.

to sea-level rise. The latter result will affect coastal communities, the severity depending on the nature of the coastline and the amount of increase. However, it is also clear that wave action may change, becoming more severe in areas previously covered by ice for part or most of the year. Wave action on shorelines will contribute to erosion and to changes in wetland complexes, both enhancing and damaging them. The case study below outlines the possible impact on Charlottetown and coastal areas of Prince Edward Island.

Infectious Diseases

Given the predictions of the Intergovernmental Panel on Climate Change (IPCC 2007a) about climate change in North America—warmer temperatures, more rainfall, more frequent droughts, and more frequent extreme weather events (hurricanes, tornadoes)—Health Canada (2007) has indicated that Canadians can expect to experience a greater incidence of disease.

Greer, Ng, and Fisman (2008, 716) have noted that most people recognize the close association among climate, environment, and infectious disease in developing countries, but this association is not so obvious in developed nations like Canada because people normally enjoy clean drinking water, lower exposure to insect vectors, and higher-quality housing. All of these advantages reduce the vulnerability of residents. Nevertheless, Greer, Ng, and Fisman conclude that predicted climate changes can be expected to increase both the incidence and burden of infectious diseases.

Examples include:

1. Lyme disease (a tick-borne borreliosis), currently uncommon in Canada, is found mainly in southern Ontario and British Columbia. Temperature is the key factor limiting the northern extent of Lyme disease, and anticipated warmer temperatures may lead to expansion of the ticks 200 kilometres north by 2020. If this were to occur, Lyme disease could appear in Alberta and Saskatchewan.
2. Other diseases, such as dengue fever, could spread northward as the habitat for mosquitoes expands.
3. Earlier springs will also facilitate expansion of West Nile virus, which first appeared in Canada in 2002.
4. More diseases may be imported as a result of Canadians travelling abroad and being exposed to disease vectors in subtropical and tropical countries. Dengue fever and malaria are expected to increase in Latin America, the Caribbean, Asia, and Africa, so it is highly probable that a greater number of returning travellers will carry these diseases into Canada.

Case Study: Impact of a Rise of Sea Level on Prince Edward Island

So far in this chapter, we have concentrated on understanding the nature of climate change, the manner in which climate has been changing, and reasons for the change. In the subsections above, we looked at the general impact of climate change on several resource and environmental systems. Before we turn to possible policies and actions, it is useful to examine the possible consequences of climate change 'on the ground'. For example, what would happen in Prince Edward Island as a result of rising sea levels if anticipated climate change continues?

Al Harvey/www.slidefarm.com

Prince Edward Island coastline.

Prince Edward Island and Rising Sea Level

At a global scale, records document that mean sea level has been rising at a rate of 0.1 to 0.2 metres per century during the past 100 to 200 years. However, with anticipated increased concentrations of greenhouse gases leading to continued climate change, the Intergovernmental Panel on Climate Change (IPCC 2007c, 7) has predicted that by 2100, mean sea level at a world scale may increase by 0.18 to 0.59 metres. Changes in sea level at local scales could be higher or lower. For east-central Prince Edward Island during the past 6,000 years, mean sea-level rise has been about 0.3 metres per century, but the rate has been

slower during the past 200 years. At Charlottetown, records indicate that since 1911, the rate of sea-level change has been 0.32 metres per century.

The coast of Prince Edward Island is one of the areas in Canada most vulnerable to sea-level rise (Figure 7.8). As Shaw and his colleagues (2001, 14) observed, this vulnerability is due to many factors, including the presence of soft sandstone bedrock, a sandy and dynamic shore zone starved of sediment in places, an indented shoreline with many estuaries and marshes, low terrain inland from the shoreline leading to high potential for flooding, high rates of shoreline retreat, and submergence of the coast, which continues today. Against that background, Shaw and others analyzed the physical and socio-economic impact of climate change and accelerated sea-level rise on the coast of PEI.

Sea-Level Rise

Figure 7.9 shows that since the last ice age (during the past 10,000 years), the relative sea level has gone up by 45 metres or more in central and eastern PEI. One result has been inundation of the land once connecting PEI to the Nova Scotia and New Brunswick mainland. The relative rate of sea-level rise (an outcome of both land subsidence and rising sea level) during the past 6,000 years has averaged some 30 centimetres per century.

One reason for the change in relative sea level has been crustal subsidence, resulting from adjustments to different ice and water loads in the post–ice age period. It has been estimated that crustal subsistence has been about 20 centimetres each century in PEI, indicating that the balance of 10 centimetres per century is due to global and regional sea-level rise. As sea levels increase, the impact of storm surges and ocean waves also increases because the water hitting the coastline is carried to higher elevations on the shore.

Ice Cover in the Gulf of St Lawrence

Between 1971 and 2000, annual ice cover in the Gulf of St Lawrence varied considerably, as shown in Figure 7.10. Extent of ice cover is measured by 'total accumulated ice coverage' (TAIC). Figure 7.10 indicates that over this 30-year period, the TAIC varied from a low of 1.1 million square kilometres in 2000 to a high of just under 3 million square kilometres in 1993. The figure also suggests that there may be a 15-year cycle in variability. Climate change is one potential cause of changes in ice cover, but scientists also recognize that other factors have

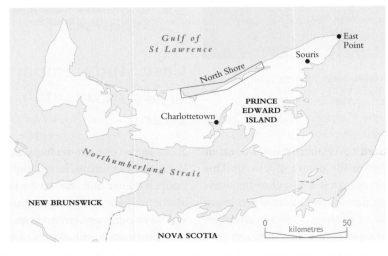

Figure 7.8 The coast of Prince Edward Island. Source: Shaw et al. 2001, 15. Reprinted with permission of the Minister of Public Works and Government Services, Canada, 2005, and courtesy of Natural Resources Canada, Geological Survey of Canada.

Figure 7.9 Relative sea-level increase in coastal and eastern PEI since the last ice age. Note: This represents the combination of sea-level change and vertical land movement. The black broken line indicates the general trend, and the solid line is the best estimate of trend for Charlottetown and the central north shore over the past 6,000 years. Source: Shaw et al. 2001, 35. Reprinted with permission of the Minister of Public Works and Government Services, Canada, 2005, and courtesy of Natural Resources Canada, Geological Survey of Canada.

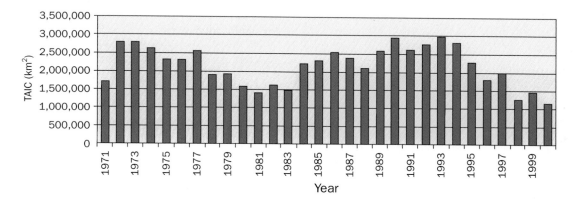

Figure 7.10 TAIC: Total accumulated ice coverage in the Gulf of St Lawrence, 1971–2000. Source: Shaw et al. 2001, 46. Reprinted with permission of the Minister of Public Works and Government Services, Canada, 2005, and courtesy of Natural Resources Canada, Geological Survey of Canada.

an influence, including El Niño, Southern Oscillation, North Atlantic Oscillation, solar flux, and volcanic eruptions.

Calculations based on the Canadian Climate Change Model indicate that by 2050, the extent of sea ice in the northern hemisphere will be reduced (LeDrew 2002). The Gulf of St Lawrence could become free of ice for the entire year. This latter estimate is important, because sea ice in the gulf inhibits development of waves and therefore reduces erosion from winter storms on the coastline. If sea ice decreases or disappears in the gulf, the size of waves can be expected to increase, with an associated increase in erosion. In addition, it is possible that the impact of storm surges could increase, but this is more difficult to determine.

Another consideration is that if wave action increases in frequency or magnitude but there is still some ice in the gulf adjacent to coastlines, damage could result from ice ride-up or pile-up against the coastline. For example, on 21 January 2000, an ice ride-up during a storm pushed a lighthouse in Charlottetown off its foundation and caused major damage to a golf course near the city. In various places along the coast of PEI and other provinces in Atlantic Canada, the ice ride-up and pile-up damaged wharves and other coastal infrastructure as well as some homes. On the other hand, on the north coast of PEI, damage to the shore from waves was minor, mainly because of the protection provided by sea ice along the shoreline, which bore the brunt of the energy from wind-driven waves.

Oscillation

The Southern Oscillation is based on the monthly or seasonal fluctuation in the air pressure differences between Tahiti and Darwin, Australia, in the South Pacific Ocean. Sustained negative values often indicate El Niño episodes, associated with sustained warming of the central and eastern tropical Pacific Ocean. Positive values lead to La Niña episodes, usually generating wetter than normal conditions in eastern and northern Australia. The North Atlantic Oscillation, also referred to by some as the Arctic Oscillation, relates to opposing atmospheric patterns in northern middle and high latitudes. Positive phases involve low pressure over the polar region and high pressure at mid-latitudes. Such conditions result in ocean storms being driven further north, leading to wetter weather in Alaska, Scotland, and Scandinavia and drier conditions in the western United States and the Mediterranean. In negative phases, the pattern is reversed.

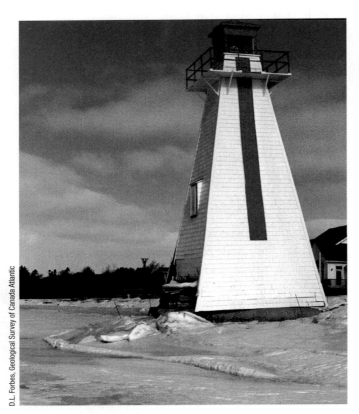

The 21 January 2000 ice ride-up moved this Charlottetown lighthouse and damaged a golf course.

Increased Storminess and Storm Surges

Earlier in this chapter, we noted that climate change relates to more than changes in temperature and precipitation. It can also involve alterations in wind speeds and patterns, as well as new extremes in weather conditions.

Storm surges reflect the meteorological consequences on sea level in an area and are the difference between an observed water level and what would be expected under predicted tidal conditions. A large storm surge combined with an exceptionally high tide creates conditions when flooding and damage are likely to occur. In contrast, a 'negative' storm surge can cause sea levels to be lower than predicted from tidal charts. Between 1960 and 1998, storm surges exceeding 60 centimetres occurred approximately eight times a year in Charlottetown, primarily associated with winter storms. Protective works for the Charlottetown harbour are more than adequate to protect it from surges causing water levels to increase up to 60 centimetres. However, storm surges in the future, accompanied by sea-level increases due to climate change, have the potential to cause damage because the high water levels could exceed the capacity of protective works. Furthermore, not only can the severity of storm surges be expected to increase but so can the frequency of higher storm surges, consistent with more variability as a result of climate change. Finally, if more storm surges to higher levels are accompanied by reduced ice cover in the Gulf of St Lawrence, there will be greater potential for significant damage.

Climate Change Future for Charlottetown and the North Coast

Shaw and his colleagues examined three scenarios regarding sea-level rise affecting Charlottetown: (1) a recreation of the storm of 21 January 2000 to test the accuracy of the model for flooding that occurred with a maximum water level of 4.23 metres above 'chart datum' or CD (the lowest predicted water level averaged over several years); (2) a lower storm surge than occurred in January 2000 but superimposed on anticipated future sea-level rise to give a flood level of 4.70 metres; and (3) a water level associated with the January 2000 storm combined with estimated sea-level rise of 0.7 metres by 2100 to give a flood level of 4.93 metres above CD.

Under scenario 1, about 460 properties would be either flooded or at risk of flooding, and the assessed property value at risk would amount to some $172 million. In scenario 2, the at-risk property value would increase to $190 million, of which one-third would comprise commercial property. In addition, some farmland would be at risk. Scenario 3 would see the value of property at risk climbing to an estimated $202 million, with commercial property again representing one-third.

One complication in estimating damage potential is that Charlottetown has many designated heritage areas and structures, most in the central part of the city. The majority of some 335 municipally designated properties are located in the downtown core, as are about a dozen federally designated sites.

D.L. Forbes, Geological Survey of Canada Atlantic

Many designated sites and areas are within the probable flood plain identified in the modelling study, indicating that potential costs to heritage values would not be trivial. Under scenarios 2 and 3, the value of the municipally designated heritage properties at risk of flooding is estimated at $10.5 million and $11.3 million, respectively.

Damage to properties is not the only risk from flooding exacerbated by sea-level rise and storm surges. Municipal infrastructure, including sewer, water, and storm systems, is also vulnerable. Possible impacts on the storm/sewage system include: (1) a significant rise in water level would lead to a surcharge from outfalls and lift stations into the sewage lines, resulting in backups into residential and commercial areas; (2) water levels high enough to reach the lift station (on the waterfront, pumping sewage to the primary sewage treatment plant) could lead to sea water being pumped along with or in place of sewage to the main treatment plant, causing corrosion and possible lack of capacity to move sewage to the treatment plant; and (3) extended flooding of a lift station or the treatment plant could make them incapable of functioning as a result of temporary or permanent damage. The damage could range from just over $300,000 for a small lift station to more than $25 million for a main sewage treatment plant. Other infrastructure at risk would include sidewalks and streets, with damage estimated at $13 million at least.

Other risks to Charlottetown could occur through waterborne diseases, loss of employment, and disruption to education for students.

Along the north shore of PEI, it is estimated that heightened sea levels, reduced sea ice, and increased wave energy would create severe erosion damage and possibly a reconfiguration of some coastlines. Since the mid-1930s, some coastal areas have already experienced serious erosion, while other areas have experienced shoreline recovery and growth of sand dunes. A long-term perspective indicates that the north coast has generally been retreating at a rate of at least 0.5 metres annually for several thousand years, which has led to property literally disappearing, encroachment on wetlands, and coastal infrastructure being placed at risk.

Adaptation to Climate Change

Five types of adaptation are usually recognized: (1) prevent the loss by adopting measures that reduce vulnerability; (2) tolerate the loss by doing nothing and absorbing the cost of losses when they happen; (3) spread or share the loss by distributing the costs over a larger population, such as through insurance; (4) change the affected activity by ceasing to do certain things or by shifting to other activities; and (5) change the location of the activity by moving to a less vulnerable location. For PEI, Shaw and his colleagues identify three adaptation measures: protection, accommodation, and retreat.

Protection usually involves constructing structural measures to protect property, buildings, and infrastructure. Structural measures can involve individual initiatives or major public works projects, such as sea walls, revetments, and groynes designed to trap sediment. Shaw and his colleagues (2001, 55) noted that 'in general, protection is costly and may have limited long-term effectiveness in exposed locations, though it may be successful as flood protection where wave energy is limited.'

There can also be side effects from some structural measures. For example, groynes are often effective in trapping sediments and therefore in protecting the properties adjacent to the area in which sediment is trapped. However, locations downdrift of the groynes may experience reduced sediment deposition and thus become more vulnerable to erosion, slumping of shorelines, and eventual increased rate of shoreline loss.

Accommodation usually involves a mix of approaches—redesign of structures to reduce their vulnerability, zoning to guide appropriate land use involving low capital investment in vulnerable areas, and other measures such as rehabilitating coastal dune systems, renewing wetlands, nourishing beaches, and replacing causeways with bridges.

However, 'stabilizing' natural systems can undermine the natural functioning of ecosystems and can be counterproductive. For example, the endangered piping plover requires an active beach habitat and prefers environments including washover channels through dune systems. However, cottage and other property owners usually do not favour unstable shorelines, and as a result there can be conflict between various groups, such as bird watchers and cottagers, with different interests and priorities.

Retreat, the third general approach, seeks to avoid vulnerability. It usually involves recognizing the high risk or vulnerability of a place and taking a conscious decision to relocate buildings, other capital works, or infrastructure away from hazardous places. The initial cost of relocating is normally very high, but in the long term the costs are usually much lower than what would have been spent on rebuilding or repairing properties after each damaging natural event.

Piping plover.

Erosion caused by high surf destroyed this home.

Retreat is the easiest approach when development has not already occurred and zoning can be used to keep development away from vulnerable places. However, many Canadians have chosen to live or work close to rivers, lakes, or coastlines, so relocation would often require a very substantial investment.

Designing a Strategy for PEI

An interesting project for you would be to design an appropriate strategy for Charlottetown and the north shore of PEI, using a mix of approaches. What would be your first priority for an overall approach? How might the strategy change if you addressed the challenge for another community vulnerable to sea-level rise? Do you live in a community that might become more vulnerable to flooding as a result of climate change? If so, what strategies might you design?

Complexity and Alternative Explanations

The media have a hard time dealing with the ebb and flow of scientific research, which characteristically attempts to falsify one, or perhaps competing, hypotheses and moves, usually unevenly, towards a most strongly accepted description of reality. Ecosystem problems usually involve multiple causes, so more than one hypothesis may be correct, a source of even more confusion.

—Carpenter 1995, 128

Communicating Global Change

In Chapter 6, we identified communication as one of seven attributes of best practice related to processes, methods, and products in resource and environmental management. In that regard, Andrey and Mortsch (2000) have highlighted several challenges for communicating information or understanding about global change.

1. *Global change is a complex issue.* The global climate system is enormously complex, mainly because of the many linkages and feedback mechanisms in the atmospheric system. Furthermore, the associated socio-economic system is complex and continuously changing.

A related complication is that while many people have heard about 'global warming' or 'global change', the level of in-depth understanding is usually poor. Polls consistently show that many Canadians have a poor understanding of the meaning, causes, or effects of global change. In addition, few see the connections between energy use and deforestation and climate change. Another complication is that the media often provide misleading or incorrect information. In that regard, Hengeveld (2006, 29) identified a particular problem for scientists related to media reporting:

> Scientists need to challenge journalists on their tendency to seek 'balanced reporting' by presenting opposing views of a topic with equal weight, without considering or reporting the credibility or marginality of these views. Such reporting can create a significant bias in communication—a bias that some argue is particularly apparent in high-profile North American media.

A further complication is that scientists and the general public often do not speak the same language when they talk of global change. As Chalecki (2000, A2, 15) observed, 'Scientists often examine small pieces of larger environmental problems in great detail within the limits of their discipline, while most non-scientists have a somewhat fuzzy understanding of the larger issues, often fed by outdated knowledge and half-formed opinions.'

2. *Uncertainties exist regarding almost every aspect of the global change issue, and these uncertainties increase when moving from natural to human systems.* There are four main sources of uncertainty: (1) statistical randomness, or the variability in nature; (2) lack of scientific understanding of the processes involved; (3) lack of or inadequate data; and (4) imprecision in risk assessment methods because of varying protocols for conducting research. All of these are relevant in global change research. They collectively contribute to uncertainty, which encourages a 'wait-and-see' attitude on the part of some policy-makers because they are sceptical about the information and understanding provided by scientists (Fraser Institute 1999). Ehrlich and Ehrlich (1996) refer to such a view as 'brownlash' because the intent is to 'minimize the seriousness of environmental problems' and 'help to fuel a backlash against "green" policies'. The Conservative government of Prime Minister Stephen Harper appeared to fall into

this camp before the publication of the 2007 report by the IPCC. The distinguishing feature of this report was the declaration that the evidence that global change resulted from human activities was 'unequivocal', based on the consensus of more than 2,000 of the world's top climate scientists. Almost immediately after the report was published, the prime minister began making public statements accepting the scientific consensus, something notably lacking beforehand.

3. *The impacts of global change will be disproportionately heavier on people in less developed countries and on future generations.* Human-induced climate change impacts will fall mainly on future generations. Areas at greatest risk are those with limited freshwater, prone to drought, along coasts, and generally in less developed nations. One consequence is that many people in developed nations, confronted by the various issues and problems in their lives, will give less attention to global change challenges. Most give priority to issues with some immediacy or urgency, and global change does not fall into that category. As Andrey and Mortsch (2000) observe, the consequences of climate change are diffuse rather than concentrated, indirect rather than direct, unintended rather than intended and affect statistical or anonymous people rather than identifiable individuals.

4. *The basic causes of global change are embedded in current values and lifestyles.* In the developed world, including Canada, relatively high standards of living and materialistic lifestyles depend on extensive use of energy based on fossil fuels. Much of this use is devoted to residential heating and cooling and personal transportation by car. It is the cumulative effect of millions of people going about their normal lives that contributes to global warming. Thus, it is easy for any one individual to conclude that a change in his or her lifestyle will make virtually no difference at a global scale. This creates a dilemma for any individual, city, province, or country, because the scale of the challenge requires unprecedented collaboration (see Chapter 6). In short, individuals believe that they are helpless to make a difference on their own, while for many people, more immediate issues compete for attention and resources.

This leads us to an important cautionary note. We should not allow ourselves to become so preoccupied with providing information and increasing understanding that we fail to encourage a sense of ownership and empowerment on this issue. As is often argued, the purpose of communication is not only to provide knowledge to raise awareness and increase understanding; it should also lead to changes in attitudes and behaviour.

Kyoto Protocol

During December 1997, representatives from more than 160 countries met in Kyoto, Japan. The outcome was an agreement, popularly referred to as the **Kyoto Protocol**, with targets for 38 developed nations as well as the European Community to ensure that 'their aggregate anthropocentric carbon dioxide equivalent emissions of the greenhouse gases [e.g.,

AP/Getty Images

Delegates from more than 160 countries at the conference in Kyoto, Japan, that led to the Kyoto Protocol, December 1997.

carbon dioxide (CO_2), methane (CH_4), nitrous oxide (N_2O), hydrofluorocarbons (HFCs), perfluorocarbons (PFCs), sulphur hexafluoride (SF_6)] . . . do not exceed their assigned amounts . . . with a view to reducing their overall emissions of such gases by at least 5 per cent below 1990 levels in the commitment period 2008 to 2012' (Kyoto Protocol 1997, Article 3). The protocol would become legally binding when ratified by at least 55 countries accounting for at least 55 per cent of the developed world's 1990 emissions of carbon dioxide. Table 7.1 shows the targets for selected countries. Developing countries, including China and India, were not included in the targets because their per capita emissions were much lower than those of developed countries. Another reason was that their economies were judged to be much less able to absorb the costs of changing to cleaner fuels, since the main source of greenhouse gas emissions is carbon dioxide from use of fossil fuels.

Canada was to reduce greenhouse emissions to 6 per cent below 1990 levels by between 2008 and 2012. After much debate, considered later in this chapter, Canada ratified the protocol in December 2002. However, at the end of March 2001, shortly after taking office, US president George W. Bush stated that he opposed the protocol, the US would not agree to it, and the US would develop its own approach. He argued that it was inappropriate for China and India, countries with the largest populations, not to be included in the Kyoto targets. Bush was quite correct that these two countries have the largest populations, but he ignored the fact that their per capita emissions of greenhouse gases are much lower than those of the United States, which has the worst record. With only 4 to 5 per cent of the world's population, the US accounts for between 20 and 25 per cent of the global emissions of greenhouse gases. Bush's position on the Kyoto Protocol was consistent with his stand on other international matters, ranging from a land mine agreement, an international court of justice, and terrorism—the United States would take unilateral action rather than participate in multilateral agreements.

Bush explained in February 2002 that the US would use a 'voluntary approach' related to greenhouse gas emissions, with

Table 7.1	Greenhouse Gas Emission Reduction Targets under the Kyoto Protocol for Selected Countries

Country	Reduction commitment as percentage of base year
Australia	108
Canada	94
European Community	92
France	92
Germany	92
Iceland	110
Japan	94
Netherlands	92
New Zealand	100
Norway	101
Russian Federation	100
Sweden	92
Ukraine	100
United Kingdom	92
United States	93

Source: Kyoto Protocol 1997, Annex B.

the purpose of reducing 'greenhouse gas intensity' by 18 per cent over 10 years, a general approach later endorsed by Prime Minister Harper. **Greenhouse gas intensity** is the ratio of greenhouse gas emissions to economic output. For the US, this translates into a goal to reduce greenhouse gas emissions from an estimated 183 mega tonnes per million dollars of gross domestic product in 2002 to 151 mega tonnes in 2012. Unlike the Kyoto Protocol, which requires an absolute reduction in greenhouse gas emissions, the American approach would result in emissions continuing to increase as its economy grows but at a slower rate than without this arrangement.

Critics of Bush's position noted that the United States introduced various voluntary measures following the **Rio Declaration** in 1992 in order to meet its pledge to reduce emissions to 1990 levels by 2000. While emission growth in the US was reduced by 2 per cent compared to what would have happened under 'business as usual', overall US emissions increased by 13.6 per cent between 1992 and 2002. In contrast, if the US had accepted its target in the Kyoto Protocol, emissions would have had to be reduced by approximately 30 per cent below 1990 levels by the 2008–12 deadline. Those supporting Bush's approach have argued that accepting the target in the Kyoto Protocol would cost millions of Americans their jobs and discourage long-term investment in clean-energy technology.

The American position and approach are significant because they were based on determination to protect the US economy in the short term, to ensure that jobs are not lost because of the costs associated with reducing emissions, and to maintain its international economic competitiveness. Because the US is such a dominant player in the global economy, its position has been cited by people in other countries, including Prime Minister Harper in Canada, who believe it would be economically foolish to accept the Kyoto Protocol targets when the nation with the largest economy has decided not to do so.

With the United States government choosing not to ratify the Kyoto Protocol, attention turned to Russia, since the other industrialized nations that had ratified the protocol accounted for 44.2 per cent of 1990 emissions and Russia accounted for 17.4 per cent. If Russia ratified, the total would be above the 55 per cent of 1990 emissions agreed to in 1997 in Japan. However, at a conference in Milan, Italy, in early December 2003, Russia announced it would not ratify Kyoto. The main reason was that the pollution cuts required of Russia under the agreement would slow down the economic growth that president Vladimir Putin had made a top priority. His goal was to have Russia double its gross domestic product by 2010. His top adviser commented that it would be unfair to Russia to have to curb emissions and jeopardize its growth while the United States and other nations, which account for the significant portion of global emissions, had refused to ratify. At the same conference, an American representative, Republican Senator James Inhofe, chairman of the US Senate Committee on Environment and Public Works, reiterated the US opposition to Kyoto, stating that the protocol and its policies were 'inconsistent with freedom, prosperity, and environmental policy progress'.

A significant shift occurred at the end of September 2004, however, when Russia's cabinet approved the Kyoto Protocol. The European Union had long lobbied Russia to ratify the pact, and in the spring of 2004, president Putin pledged to have Russia ratify in exchange for support from the EU for Russia to join the World Trade Organization. On 22 October 2004, the lower house of the Russian parliament, the Duma, approved ratification by a vote of 334–73; five days later, Russia's upper house followed suit in a vote of 139–1, with one abstention.

Against this background, we can now turn to some of the specific aspects of the protocol and then to Canada's approach.

Specific Features of the Kyoto Protocol
Legal Basis
Unlike the Framework Convention on Climate Change signed at the Earth Summit in Rio de Janeiro in 1992, which committed countries only to 'aim' to stabilize emissions at 1990 levels by 2000, the Kyoto Protocol commitments are legally binding on nations under international law.

Assigned Amounts
For the commitment period 2008–12, the protocol states that overall average emissions are to be 94.8 per cent relative to 1990 levels. 'Assigned amounts' are identified for each developed nation. While the targets were set for allowed emissions with reference to variables such as population, gross national product, and carbon intensity of economies, the final targets were determined politically. Canada is to reduce its emissions by 6 per cent. Australia is allowed to increase emissions by 8 per cent, and Iceland can increase by 10 per cent.

Greenhouse Gases
The protocol identifies six greenhouse gases. Three are viewed as the main greenhouse gases produced by human activity: carbon dioxide, nitrous oxide, and methane. The other three—hydrofluorocarbons, perfluorocarbons, and sulphur hexafluoride—are released in small quantities but are long-lasting and significant contributors to climatic change.

Exclusion of Most Forest and Soil Sinks
The assigned emission amounts for most nations are a percentage of gross emissions in 1990. Gross emissions are the anthropocentric (human-caused) greenhouse gas emissions from energy, industrial processes, agriculture, and waste. However, they do not include carbon fluxes from forests, soil, and other carbon reservoirs.

When a nation calculates whether it is in compliance with its target emissions, it must count emissions and carbon flux changes due to afforestation, reforestation, and deforestation since 1990. In Canada, the view of the federal government after signing the protocol was that it could interpret the protocol to include loss of carbon from agricultural soil in calculating the balance between emissions and carbon flux removal. Indeed, the target for Australia of 108 per cent of 1990 emissions was partly based on arguments that it had positive net emissions related to land-use change and forestry in 1990. As a result, the 1990 allowable emissions were calculated to reflect the 1990 gross emissions plus the net emissions from land-use change and forestry.

Because of uncertainty and methodological challenges in measuring emissions from land-use change and forestry, some observers are concerned that countries will use forest and soil sinks to claim credits that will be difficult to verify. There are also problems in reaching agreement about the meaning of key terms such as reforestation, afforestation, and deforestation. In Canada, reforestation is normally interpreted to mean replanting and natural regeneration after logging and afforestation to planting trees in areas traditionally forested. In contrast, the Intergovernmental Panel on Climate Change defines reforestation as planting of forests on lands historically forested but later converted to another use and afforestation as planting of new forests on lands historically not forested. These different interpretations are important. If reforestation is interpreted to include planting trees after harvesting, as Canada does, then emissions from harvesting are not counted. As a result, only the credit from planting would be used in calculating the carbon reservoir credit.

At the Milan conference in December 2003, the signatories to the Kyoto Protocol reached agreement on how industrialized countries can earn credit towards their emission targets by preserving or establishing forests. The specific details about the agreed protocol can be found on the home page of the United Nations Framework Convention for Climate Change, Ninth Session of the Conference of the Parties and the Nineteenth Sessions of the Subsidiary Bodies, 1–12 December 2003, Milan, Italy, at http://unfccc.int/cop9. Further insight can be found at http://www.iisd.ca/climate/cop9.

Clean Development Mechanism

Emission reduction commitments can be fulfilled through a clean development mechanism, allowing **emission credits** in countries not given targets through the protocol to be used by countries included in the protocol targets. Initiatives are certified as satisfying the clean development mechanism when they involve voluntary participation by each party; real, measurable, and long-term benefits for mitigation of climate change; and emission reductions in addition to those that would have occurred without the initiative. Such initiatives are also intended to help the host nation. The protocol allows countries with targets to meet their commitments for the 2008–12 period by claiming certified clean development emission reductions achieved between 2000 and 2007.

The major concern about the clean development mechanism is that emission credit may be given for projects that would have occurred without such a mechanism in place. The protocol only requires reductions in emissions that would be in addition to what would have occurred without an initiative. This is referred to as 'emissions additionality'. However, the protocol does not require an initiative to be one that would have occurred only if the clean development mechanism did not exist. This is referred to as 'project additionality'. A concern, therefore, is that credit will be given for initiatives that would have occurred regardless of the protocol. This is a basic challenge in any system for identifying emission credits, since projects for which credit is given may have occurred for other reasons, such as reduction of air pollution to safeguard public health.

Emissions Trading

Under the protocol, a country can meet its emission commitments by acquiring from other countries 'emission reduction units' related to initiatives that cause a reduction in emissions or enhancement of sinks incremental to what would otherwise have occurred without the initiative. When a nation buys some emission reduction units, they are added to its allowable emissions and subtracted from the allowable emissions of the selling country.

However, because developing countries are not given emission targets under the protocol in order to allow them to develop their economies, they cannot sell emission credits to developed countries, even when such sales could benefit them economically. Developing countries also cannot agree to voluntary emission targets, which could benefit some if they could introduce low-cost emission reductions and then sell emission credits.

The theory of **emission trading** is based on the belief that it is more efficient for one country to purchase emission credits from another country that can generate credits in a less costly manner. In such a way, overall compliance goals can be reached at a reduced overall cost. However, as with all approaches, there are problems. For example, because the economies of eastern European countries such as Russia and the Ukraine have collapsed, their emission allowances for 2008 to 2012, based on 1990 levels, will probably be higher than their actual emissions. As a result, they could sell their surplus emission rights (sometimes referred to as 'hot air') to other countries that would then be able to keep increasing their emissions. The outcome would be an overall increased emission of greenhouse gases, whereas if such 'hot air' credits were not transferable, the overall levels would decrease.

Another concern is that the buyer of emission credits from another country does not have to be concerned about whether the selling nation will actually comply with its emission amounts after the sale. This is referred to as a 'seller beware' system, as opposed to the 'buyer beware' warning we are familiar with in economic transactions. The net effect, if the selling country cannot meet its emissions targets after selling some emission credits, will be an overall increase in emissions. Dealing with this issue is a challenge, since a sophisticated international monitoring and enforcement capacity is required.

Canadian Federal Government's Initial Response to the Kyoto Protocol

On 10 December 2002, the House of Commons voted 195 to 77 in favour of a motion to ratify the Kyoto Protocol. The Liberal majority led those supporting the motion, while the Canadian Alliance and Progressive Conservative parties opposed it. Through this vote, Canada committed itself to cut average greenhouse gas emission levels to 6 per cent below 1990 levels by 2008–12.

Prime minister Jean Chrétien signed Canada onto the Kyoto Protocol on 16 December 2002. Canada's ratification brought the total number of countries committed to the Kyoto Protocol to almost 100, accounting for about 40 per cent of 1990 emissions.

Canada's Initial Approach to Implementing the Kyoto Protocol

The Canadian government ratified the Kyoto Protocol without a clear plan on how it would be implemented in Canada. The prime minister argued that details would be worked out. This position was supported by the minister of the environment,

The role of the oceans in helping to mitigate the impacts of global warming through absorption of carbon dioxide is still uncertain, as is the oceanic response to warmer temperatures. Scientists are already detecting larger wave swells in many parts of the world that may be linked to these changes.

David Anderson, who compared the challenge of global warming to World War II. He was often quoted as saying that 'If Winston Churchill had said in 1939 that we are not going to challenge the Nazis until we know exactly how much it will cost and how long the war will last, then we would have never won the war.' In contrast, Stephen Harper, leader of the then Canadian Alliance party, argued that it was inappropriate to ratify the protocol without providing a clear plan regarding how it would be achieved. His view was that 'I don't think we have any idea what the government is going to do' and that the implementation would turn into a great disaster for Canada and Canadians.

The federal government allocated $2 billion in its 2003 budget for Kyoto Protocol initiatives over a five-year period. One specific initiative, the creation of Sustainable Development Technology Canada, was identified. This arm's-length foundation was to receive $250 million to support technologies so new that they were not yet commercially viable. The remaining $1.75 billion was to be allocated to other initiatives once the details were worked out, with the budget indicating only that the money would be 'to support climate change science, environmental technology, and cost-effective climate change measures and partnerships in areas such as renewable energy, energy efficiency, sustainable transportation, and new alternative fuels'.

The components of a plan for implementing the Kyoto Protocol included requiring major industrial emitters to reduce their greenhouse gas emissions, taxes on private vehicles such as sports utility vehicles, minimum requirements for fuel alcohol, subsidies to install energy-efficient windows, and an emissions trading framework.

The federal government compiled a list of large industrial emitters of greenhouse gases, such as oil and gas, mining, and pulp and paper companies, each of which generates large amounts of carbon dioxide per unit of product. Each industry was to be required to reduce emissions extensively. However, the political nature of this approach was illustrated in January 2003 when the gigantic Ontario automotive industry was removed from the list of large industrial emitters as a result of lobbying by the auto assembly industry.

Accomplishments in Reducing GHGs in Canada

In the early fall of 2004, the federal deputy minister of natural resources, speaking at a conference in Australia, stated for the first time that Canada would not meet its commitments under Kyoto and would be unlikely to realize even two-thirds of the reductions of GHGs that Canada had committed to achieve under Kyoto. Indeed, by December 2004, data indicated that emissions of GHGs in Canada had increased by 20 per cent relative to the base year of 1990.

A Change of Federal Government and a Change of Course regarding Kyoto

In January 2006, a federal election led to a minority Conservative government. Stephen Harper, the new prime minister, when previously leader of the opposition, had consistently opposed the Kyoto Protocol, arguing that it would retard economic growth in Canada. In addition, Harper had frequently expressed doubt about the credibility of the scientific conclusions that climate change was a significant issue.

As prime minister, Harper indicated that the commitment Canada had made to Kyoto was unrealistic and unachievable. He also noted that if Canada were to remain economically competitive with its largest trading partner, the United States, it would be unwise to aggressively seek to reduce GHG emissions unless and until the US had also accepted a binding target to reduce GHGs. Furthermore, he argued that it was not reasonable that countries such as China and India were not included in the agreement with binding targets.

The lack of progress in reducing GHG emissions under the Liberal government and the negative view by the Conservative government regarding the Kyoto targets generated considerable criticism of Canada. For example, on the first day of a major two-week-long UN climate change conference in Nairobi in November 2006, Canada, along with Australia, received the 'fossil of the day' award from the Climate Action Network, an environmental non-governmental organization that highlights nations that have made the least progress in reducing GHGs. During the same conference, another group, Germanwatch, based in Bonn, ranked Canada 51st out of 56 countries because of its poor performance on climate change. The only countries on the list that had performed worse than Canada were Kazakhstan, the US, China, Malaysia, and Saudi Arabia. At the other end of the scale, the 10 best nations were led by Sweden, followed by the United Kingdom, Denmark, Malta, Germany, Argentina, Hungary, Brazil, India, and Switzerland.

The Canadian Government Rejects the Kyoto Protocol

In the second week of September 2007, Prime Minister Harper attended an APEC (Asia Pacific Economic Co-operation) summit of 21 Pacific Rim countries in Sydney, Australia. The main outcome was a joint statement regarding climate change. John Howard, then prime minister of Australia, said that the joint statement was a major milestone, since it endorsed a long-term but unspecific target to cut GHG emissions by all participants, including the US and China. This unspecified non-binding target was characterized as 'aspirational'. Those attending the meeting claimed the significance of the agreement was that it represented the first time that the United States and China had agreed to sign a climate change agreement. In addition, other developing countries had agreed to participate, something they had not been required to do under the Kyoto Agreement. Harper was reported as having said at the summit, 'Kyoto divided the world into two groups, those that would have no targets and those that would reach no targets. The reality is that the world is now making efforts toward a new protocol post-2012.' Harper acknowledged that much more work needed to be done but that having China and the United States 'on board' was a noteworthy accomplishment.

Later the same month, Prime Minister Harper attended a session at the United Nations headquarters in New York that had been arranged to create momentum for a new climate change treaty. In his remarks in New York, Harper stated that Canada's intent was to pursue a slow and steady approach to reducing GHG emissions that would allow economic growth to continue while also protecting the environment. He observed that such an approach was needed to achieve international consensus. He also indicated that his government supported a goal of reducing global GHG emissions by 50 per cent by the year 2050. At the same time, he announced that Canada had joined in talks with a climate change group known as the Asia-Pacific Partnership on Clean Development and Climate, or the AP6. This group was created in 2006 by the governments of Australia, China, India, Japan, South Korea, and the United States with a goal of addressing climate change through 'practical, pro-growth, technology-driven efforts'. Harper supported the work of the group, since it emphasized voluntary reductions of GHG by the countries in the AP6, including China and India. Critics of this approach have been sceptical about what would actually be achieved through an **aspirational approach** not binding any nation to action. James Travers (2007), a national media columnist in Canada, made the following comment about the prime minister's position at that time:

> With notable variation but shared purpose, Canada is advancing with the US and Australia a post-Kyoto protocol that will be more inclusive and accommodating by being less demanding.
>
> More than a rubbery response to a concrete threat, that relaxed approach to safeguarding humanity's nest contrasts with the iron fist that these countries are shaking at the far less existential danger of global terrorism.
>
> Pity the former prime minister who has to autobiographically explain why compelling evidence of an environmental death spiral was ignored while the government was otherwise engaged in a war on a tactic that, no matter how vile, has no lasting power to knock confident democracies off course.

UN-Sponsored Climate Change Conference, Bali, December 2007

In December 2007, a major UN-sponsored climate change conference took place in Bali, Indonesia, with 192 countries attending over a two-week period. The purpose of the **Bali Conference** was to start a process for creating a new framework to replace the Kyoto Protocol when it expires in 2012. At the conference, UN Secretary-General Ban Ki-moon remarked that 'The situation is so desperately serious that any delay could push us past the tipping point, beyond which the ecological, financial, and human costs would increase dramatically.' Global attention to climate change had increased after the 2007 Nobel Peace Prize was awarded in mid-October jointly to former US vice-president Al Gore for his film *An Inconvenient Truth* and to the UN Intergovernmental Panel on Climate Change in recognition of the significance of their work and the seriousness of climate change for worldwide stability and peace.

The outcome at Bali was mixed. The participating nations agreed to continue meeting to determine what to do about climate change and in particular how to reduce GHG emissions. The negotiations are to be concluded over two years to allow time for an agreement to be ratified by individual nations before the Kyoto Protocol ends in 2012. There was no agreement, however, on binding targets to reduce GHG emissions. Furthermore, China and India called on the developed countries to help transfer 'green' GHG emission-reducing technologies to developing countries so that they could meet specific targets.

The negotiations are to take place in two sets. One set will involve 38 developed countries, including Canada. The goal will be to set tougher targets for them than those under Kyoto. These discussions will be guided by the view of international scientific experts that GHG emissions need to be reduced by 25 to 40 per cent by 2020 relative to 1990 levels if serious climate change impacts are to be avoided. However, such targets were not accepted as binding by the nations agreeing to participate in the discussions. Indeed, at the Bali Conference, the federal minister of environment John Baird rejected 2020 as a reasonable target year, arguing that it would require Canada to reduce emissions by between 33 and 53 per cent over 12 years, an objective that would be too disruptive to the economy. He argued for 2050 as a more reasonable target year.

In parallel, other discussions will take place among countries that already have targets under Kyoto, as well the 150 others, such as the United States, China, and India, that have not yet committed to specific emission reductions. The agreement that all nations, developed or developing, should commit to reducing GHG emissions is a significant change from the agreement under Kyoto in which no targets were set for developing nations. However, with the targets of reducing GHG emissions by 25 to 40 per cent not binding on any national government, the door was left open for countries like Canada and the United States to continue doing relatively little in terms of actively requiring reductions of GHGs. In contrast, the European Union countries argued that explicit numerical targets were essential.

Throughout the Bali Conference, federal government representatives from Canada and the United States argued that GHG reduction commitments should be required for all countries but that specific numerical targets should not be identified. At the last moment during negotiations, the US government shifted its position regarding explicit targets after a delegate from Papua New Guinea made the following remark directed at the US representatives: 'We seek your leadership. But if for some reason you are not willing to lead, leave it to the rest of us. Please get out of the way.'

Delegates at the Bali Conference.

Canada's position at the Bali Conference was the same as those of the US, Australia, New Zealand, Russia, and China. The environment minister was explicit and emphatic that Canada would not accept any binding targets unless they were applied to all developing countries. The developing countries of special interest were China and India, whose rapidly growing economies generate about half the world's GHG emissions in total even though their per capita emissions are much lower than those of the most developed countries, such as the US and Canada.

Canada's position at the conference was labelled as that of a 'climate hypocrite' by Yvo de Boer, the head of the UN climate change agency, because Canada was calling for binding targets on developing countries but refused to accept them for itself under Kyoto. Rajendra Pachauri, the head of the UN Intergovernmental Panel on Climate Change, suggested that Canada's position made it clear that the Canadian government was not prepared to take significant action on climate change. And once again, Canada received numerous fossil of the day awards.

At the conference, Canada also lobbied hard to change the base year of 1990 established for the Kyoto Protocol. The base year of 1990 has been the accepted international standard since the Kyoto Protocol was created in 1997, and any decision to move that date forward would reward nations that had been slow in curbing GHG emissions. In that regard, it was also noted that the Conservative government's approach to the baseline year has been to refer not to 1990 but rather to 2006, the year that the Conservatives were elected to power.

Representatives of provincial governments also attended the Bali Conference, and during the conference it became clear that

China and Climate Change

Although China is often criticized for emphasizing economic development without regard for accompanying pollution, Gorrie (2008) has argued that China is moving aggressively to cut all types of ghg emissions and is going well beyond what Canada and the US have done in terms of regulations and targets for car emissions, renewable fuels, carbon storage, forest renewal, energy efficiency, and industrial pollution.

In terms of fuel efficiency standards for automobiles, Gorrie notes the following: China: 36 miles per gallon (current), 43 by 2009; Canada: 27 miles per gallon (current average, no standard), an unspecified standard, starting in 2011; US: 25 miles per gallon (current average, with 35 proposed by 2030); California: 25 miles per gallon (current), 36 proposed by 2016); Europe: 40 miles per gallon (current), 48.9 proposed by 2012; Japan: 40 miles per gallon (current), 48.9 proposed by 2015.

not all provinces supported the federal government's position. For example, the environment ministers of Quebec and Ontario, Line Beauchamp and John Gerresten, stated that they disagreed with the position taken by federal environment minister John Baird. At a joint interview involving Beauchamp and Gerresten, the latter expressed concern about the federal government's position, commenting that 'If the US is not willing to sign on, does that mean nobody should sign on? We don't like this attitude.' In contrast, as noted in Chapter 12, the Alberta government has supported a more voluntary approach.

Post-Bali
International

In the first half of June 2008, more than 2,000 delegates from 162 countries assembled in Bonn for a two-week meeting, the next post-Bali step in 18 months of negotiations through eight scheduled meetings in different cities to develop an agreement to succeed the Kyoto Protocol. The goal is to have a new agreement ready by December 2009, which will be signed in Copenhagen.

During the opening session, some speakers emphasized that climate change was already negatively affecting millions of people in poor nations, especially because it aggravated serious food shortages. As a result, and as a representative of the Maldives noted, climate change was a present rather than a distant reality.

In early July 2008, the heads of government of the G8 countries (Britain, Canada, France, Germany, Italy, Japan, Russia, United States) met in Rusutsu in northern Japan. Climate change was on the agenda for discussion. At the end of the meeting, the G8 leaders endorsed a plan to reduce greenhouse gas emissions by half by 2050. Prime Minister Stephen Harper explained that this meeting represented a significant turning point regarding climate change, since it was the first time that both the United States and Russia had agreed on the need for a long-term target to reduce greenhouse gas emissions. Harper further stated that developing nations would have to do much more to cut their greenhouse gas emissions, given that by 2050, the current leading developed nations would probably only account for 20 per cent of emissions. He was reported to have said that it would not be possible to realize a 50 per cent cut from that 20 per cent of emissions.

Critics of the G8 declaration noted that the approach endorsed by the G8 leaders had significant weaknesses, including: (1) the declaration does not bind the G8 countries to halve their emissions by 2050; (2) the base year against which cuts are to be measured is not specified; and (3) no interim targets were identified. One editorial in a Canadian newspaper referred to the announcement as '99 per cent hot air'. Furthermore, the leaders of the countries with the five largest emerging economies (Brazil, China, India, Mexico, South Africa) rejected the declaration, urging the leading developed countries to take more action to reduce their emissions. As a result, a clear consensus on how to address climate change had not been reached on a global scale.

Domestic

On 2 June 2008, the governments of Ontario and Quebec held a historic joint cabinet meeting as one of the many events to mark the 400th anniversary of Quebec City. One outcome was an announcement by premiers Jean Charest and Dalton McGuinty that the two governments had agreed to develop and implement a 'cap-and-trade' plan to reduce carbon emissions because they did not believe that the federal Conservative government's strategy to develop 'intensity-based' emissions targets would be effective.

A joint cap-and-trade strategy means that companies producing fewer GHG emissions than permitted under their caps can sell their unused quota on an open market as credits to other companies exceeding their caps. The intent is to have such a cap-and-trade system in place by 1 January 2010. In the view of Charest and McGuinty, a cap-and-trade system would lead to an absolute reduction in GHG emissions from central Canada relative to 1990 as a base year, the year identified in the Kyoto Protocol. Their view was that the federal intensity-based emissions, in contrast, would lead to a steady increase in GHG emissions and that the Conservative plan to use 2006 as its base year would result in a much higher level of baseline emissions.

The federal minister of environment John Baird quickly criticized the initiative from Ontario and Quebec, stating that the strategy was a 'rogue' and 'flawed' attempt to cut emissions and that there needed to be one national plan rather than various provincial initiatives. In reply, the two premiers stated that the federal intensity-based approach to emissions was flawed and mirrored the US strategy, whereas European countries had embraced a cap-and-trade approach.

Later in June 2008, Stéphane Dion, the leader of the federal Liberal party, proposed a **carbon tax** under which GHG emissions would be taxed initially at $10 per tonne, increasing to $40 per tonne after four years. By the fourth year, $15.4 billion would be collected annually.

The purpose of such a tax is to modify human behaviour in favour of activities that generate lower GHG emissions. And to avoid harmful economic consequences, the carbon tax would not be applied to gasoline for vehicles, while a Liberal government would introduce an equivalent cut in income and business taxes, along with tax breaks for the poor, elderly, and northern and rural residents who would be hardest hit by the consequent increase in prices for necessities such as home heating fuels and electricity.

The proposed carbon tax mirrored a similar one introduced by the BC government earlier in 2008, discussed later in this chapter. Prime Minister Harper responded by saying that the carbon tax was 'crazy economics' and 'crazy environmental policy' and that it would 'shaft' all Canadians. In the federal election held in October 2008, the Conservatives were returned as a minority government, and many commentators suggested that one of the problems for the Liberals during the campaign was a difficulty in explaining the carbon tax—referred to as the 'Green Shift'—in easily understood terms.

The differing positions taken by the federal government, the main opposition party, and the governments of Ontario and Quebec highlight the fact that policy-makers have a range of choices, each of which has strengths and weaknesses. What are your views on the relative merits of 'intensity-based', carbon tax, and 'cap-and-trade' approaches?

Policy and Action Options

Regarding climate change, there is strong agreement on several matters related to policy and action: (1) International collaborative action will be required, since climate change is a shared problem. No one country can take unilateral action to resolve it. The challenge, of course, is that each national government may be reluctant to take the 'first step' for fear that it might make their country less economically competitive with other nations. (2) A mix of strategies will be required, including both mitigation and adaptation.

In the discussion of the Kyoto Framework and the process to create a successor framework, we highlighted the importance of nations working collaboratively to reduce GHG emissions. We also saw that some of the most developed countries, including the United States, Canada, and Japan, have argued that they are not prepared to take a leadership role because of a concern about hindering economic growth and impairing short-term livelihoods. One consequence is that it has been very difficult to develop any meaningful international strategy to address the fundamental causes of climate change. Work will have to continue on creating an international approach. Key pieces of such an approach would be both mitigation and adaptation.

Mitigation

Mitigation involves reducing emissions of GHGs, which in turn will limit future temperature changes. Many scientists agree that a critical threshold would be an upward change of 2°C in temperature. Ensuring that global warming does not pass that threshold will require emissions of GHGs not exceeding two times current CO_2 levels, or about 560 parts per million. To achieve this goal, emissions would have to be reduced significantly below current levels, and about 75 per cent of energy production would have to be based on 'carbon-free' sources by 2010. To reach 75 per cent carbon-free energy production will require development and implementation of new technologies.

In contrast, if CO_2 emissions continue at present levels, then incremental warming of between 2°C and 6°C is likely over the next four centuries. If production of all GHGs were stopped immediately, it is most likely that because of past emissions, the Earth would still experience a warming of 1°C.

In terms of mitigation strategies, Hengeveld (2006, 30–2) explains that a mix of options exist. They include:

- **Carbon tax:** A tax would be levied on countries based on their generation of GHGs. The measure could also include creating a market for GHG emissions, with some countries not able to meet their targets purchasing emission capacity from other countries well below their limits. In February 2008, the government of British Columbia introduced the first carbon tax in Canada, as outlined in the accompanying box.

    Rivers and Sawyer (2008) have outlined the rationale for a carbon tax. They argue that a primary cause of global warming is that producers or users of goods do not have to pay any charge for the associated carbon emissions. As a result, all Canadians 'pay the price' through a degraded environment. In contrast, they note that most residents and businesses in cities accept a charge to send waste to a local landfill site. In their words, the bottom line is that Canadians need to stop using the atmosphere as a free dumping ground.

    Rivers and Sawyer (2008, 6) also argue that the necessary shift to clean, renewable energy is hindered by unfair competition from the fossil-fuel sector because it does not have to include the cost of GHG emissions in its pricing. They further state that 'a carbon price can help fix the

Carbon Tax in British Columbia

Effective July 1st [2008], we intend to put a price on carbon-emitting fuels in BC. This carbon tax will be entirely revenue-neutral, meaning every dollar raised will be returned to the people of BC in the form of lower taxes. Furthermore, that commitment will be enshrined in law. And we will provide additional support for lower-income British Columbians.

The carbon tax will start at a relatively low rate to give us all—businesses and individuals—time to adjust. As it rises, it will underline the fact that there is, indeed, a cost attached to generating greenhouse gases.

. . . The carbon tax will apply to a wide range of fossil fuels—including gasoline, diesel, coal, propane, natural gas, and home heating fuel—making it among the broadest and most comprehensive taxes of its kind in the world.

It will start at a rate of $10 per tonne of associated carbon, or carbon-equivalent, emissions. That works out to about 2.4 cents a litre at the gas pump and 2.8 cents a litre for diesel and home heating oil.

These initial rates are relatively low, and that is by design. It will give consumers, business, industry, and the economy time to adapt to this new landscape. The rate per tonne will rise by $5 a year for the next four years—reaching $30 a tonne (or 7.2 cents per litre of gasoline) by 2012.

I want to repeat: this tax will be 100 per cent revenue neutral. We expect it to raise approximately $1.8 billion over three years—so we are committed to reducing taxes by the same amount.

. . . rather than ask the people of BC to wait until they see their 2008 tax form before they see the benefits of tax reductions, today we are announcing an upfront payment of $100 to every woman, man, and child in British Columbia.

This one-tine Climate Action Dividend will be going out in June [2008], a total benefit of $440 million funded from the surplus from the year just ending.

—British Columbia 2008

imbalance between clean energy and carbon-intensive fossil fuels in two ways. First, putting a price on carbon will introduce the true cost of carbon emissions into the equation, and second, the substantial revenue generated by a carbon price can be used to pay for the massive increase in home efficiency retrofits and the deployment of green energy technology.'

- **Technologies:** One key technology involves finding alternatives to fossil-fuel combustion for heating buildings, running manufacturing and industrial equipment, and powering vehicles, aircraft, and ships. A specific example is fuel switching through which fossil-based fuels are replaced by other fuels, ranging from hydro-generated electricity to ethanol fuels. Wind-based energy is another alternative, discussed in more detail in Chapter 12. Other options include renewable biomass and developing a 'hydrogen economy'.

- **Carbon sequestration:** Carbon can be sequestered in biological sinks. Land-use practices that encourage agricultural crops and forest systems with the capacity to sequester carbon have been endorsed as a legitimate way for nations to achieve GHG targets under the Kyoto Protocol. A challenge with this option is that biological sinks are not permanent, and hence attention must be given to ways of preventing the sequestered carbon from entering the atmosphere.

Adaptation

As noted above, even if all GHG emissions were stopped tomorrow, there are enough GHGs already in the atmosphere to generate significant climate change. The change will have impacts, some positive and some negative. Given that there will be negative impacts, we need to develop adaptation strategies so that activities and regions that will be adversely affected can create capacity for resilience.

In developing countries in particular, the tendency has been to prepare adaptation strategies that emphasize recovery following a disaster, preventing starvation as a result of a loss of agriculture crops because of flooding, hail, drought, disease, or pests. Today, social scientists argue that more attention should be directed towards creating capacity to deal with disasters.

For example, climate change is likely to have an impact on agricultural production. And as Hengeveld (2006, 32) observed, some choices—such as increasing production through greater crop specialization, more water-intensive crops, more tile drainage and water competition, and less grazing—will make agriculture more vulnerable to climate extremes. In the resource-based recreational sector, warmer winters are likely to adversely affect downhill skiing operations, so an investment in snow-making equipment now would be prudent as an alternative to waiting for government support after a series of lower-than-average snowfall years. Choices are always available, and both policy-makers and individuals in a resource-based sector can opt for choices that allow greater scope for adaptation in the future when climate conditions are quite likely to be different from what they are today. However, to promote more adaptive choices, governments will have to review current systems of incentives, such as crop insurance and agriculture support programs. Any proposed changes would probably meet vocal resistance, since some people profit from such programs and can be expected to oppose any changes to them. Currently, more attention is being given to **adaptation**. Walker and Sydneysmith (2008), for example, have laid out some of the adaptive mechanisms that might be appropriate for British Columbia.

However, it is unlikely that either mitigation or adaptation strategies on their own will suffice. Both are needed, and each one creates many opportunities for those able and willing to look beyond 'business as usual' practices.

What Else?

Clearly, there is considerable uncertainty regarding the precise effects of anticipated climate change. Furthermore, concerted and co-ordinated initiatives by provinces, states, and nations will be necessary to reduce the projected negative impacts. Some of the multilateral initiatives underway have been identified in this chapter, but it is obvious that some nations, especially the United States and Canada, have been unwilling to participate in co-ordinated global activities if they could create economic disadvantage for them in the short term. But what can individuals do?

Hengeveld, Whitewood, and Fergusson (2005, 51) provide thoughtful guidance:

> How can the individual Canadian citizen influence the outcome of a global environmental issue that is already challenging the wisdom and resources of the world's governments and international agencies? The answer, simply put, is that it is the individual citizens who must create the environment of opinion which will encourage governments to act.
>
> And it is the individual citizens who can take actions themselves to reduce their personal emissions and who can support the policies that an effective response to the risks of climate change will demand.

They go on to observe that it is not the governments of countries that release the estimated 23 billion tonnes of CO_2 annually into the atmosphere through use of fossil fuels but rather the actions of the 6 billion–plus people on the Earth. In their view,

> Each time we turn on a hot water tap, or open a refrigerator door, we add to the problem. By changing our attitudes and lifestyles, by becoming more knowledgeable about the issue and rethinking our attitude to the environment, we can make a difference.

Box 7.4 presents some options. A change in one person's behaviour will not have a significant impact. However, cumulatively, many individual actions can be significant, as emphasized in Chapter 15. The challenge will be to decide whether we are prepared to make such modifications to our behaviour, since in many instances we will not be the direct beneficiaries—more likely, people one or two generations in the future, such as your grandchildren, will reap the benefits.

Box 7.4   What You Can Do

1. Much of global climate change is influenced by our society's love affair with the automobile. Find something else to love, such as a bicycle or a bus. Simply aim to reduce your consumption of fossil fuels.
2. Use energy more efficiently in your home and in all other aspects of life. Make sure your appliances, lighting, and heating are energy-efficient.
3. Encourage power utilities to invest more effort in renewable energy resources, as discussed in Chapter 12.
4. Consider getting by without air conditioning in your car and at home, or if/when you use it, do so sparingly. Plant shade trees to reduce the summer heat levels in your house.
5. Consider purchasing carbon offsets related to transportation or energy use (Box 7.5).
6. Plant trees to help absorb atmospheric carbon dioxide.
7. Let your political representatives know that you are in favour of mandatory measures to curb emissions, even if it costs more money in the short term.

Box 7.5   Carbon Offsets

Carbon offsets involve a financial transaction to compensate for the generation of greenhouse gas emissions through our activities, such as travel or use of electricity. Offsets are measured in metric tons of CO_2 equivalent, with one carbon offset equalling one ton of CO_2 or equivalent greenhouse gas.

There are two categories of carbon offsets. One involves governments and companies purchasing carbon offsets in order to comply with caps placed on their GHG emissions. The Kyoto Protocol approved offsets as one option for governments and companies to meet their emission targets. In that context, companies providing energy from wind, solar, small hydro, geothermal, and biomass all create offsets by substituting or displacing fossil fuels and their associated GHGs.

The other category is voluntary and involves individuals and companies who purchase carbon offsets to mitigate the GHG emissions generated by activities such as air travel. The next time you purchase an airline ticket, check to see whether you can purchase carbon offsets. The funds generated from such purchases are often used to plant trees that absorb GHGs.

For further information, see the David Suzuki Foundation (2007).

Summary

1. The *weather* of any place is the sum total of its atmospheric conditions (temperature, pressure, winds, moisture, and precipitation) for a *short* period of time. It is the momentary state of the atmosphere. *Climate* is a composite or generalization of the variety of day-to-day weather conditions. It is not just 'average weather', since the variations from the mean, or average, are as important as the mean itself.

2. Scientific evidence confirms that the world has been warming, with the average global temperature at the Earth's surface having increased by about 0.6°C, with an error range of plus or minus 0.2°C, since the late nineteenth century.

3. The increase in the average temperature for the northern hemisphere during the twentieth century was the largest of any century in the past 1,000 years.

4. Evidence showing increases in greenhouse gases in the atmosphere, loss of mass in glaciers, reduction in permafrost and snow cover, and rises in sea level are all consistent with global temperature increases. Local knowledge from Aboriginal people provides further confirmation.

5. There are four basic climate models, the most commonly used being the General Circulation Model. GCMs are best used for global or overall climate modelling rather than for regional representations of climate change.

6. Coarse spatial resolution, poor predictive capacity for precipitation, relatively weak simulation of oceans, lack of baseline data, and many other limitations make GCM outputs highly variable.

7. The reports from the Intergovernmental Panel on Climate Change are explicit that most warming since the mid-twentieth century is associated with human activities.

8. At a global scale, records document that mean sea level has been rising at a rate of 0.1 to 0.2 metres per century during the past 100 to 200 years. The coast of Prince Edward Island is one of the areas in Canada most vulnerable to sea-level rise.

9. There are five ways to adapt to the hazards caused by climate change: prevent the loss, accept the loss, spread the loss, change behaviour, and change the location for activity.

10. Climate change has implications for terrestrial systems, agriculture, freshwater systems, fisheries, cryosphere, ocean and coastal systems, and infectious diseases.

11. Climate change normally negatively affects the poorest people the most and can trigger out-migration, which can lead to 'environmental refugees' for whom Canada will be one destination.

12. The challenges of communicating information or understanding about global change include: (1) global change is a complex issue; (2) uncertainties exist regarding almost every aspect of the global change issue, and they increase when moving from natural to human systems; (3) the impact of global change will be disproportionately heavier on people in lesser developed countries and on future generations; (4) the basic causes of global change are embedded in current values and lifestyles.

13. To improve communication about situations such as global change, which are characterized by high uncertainty, four aspects should be addressed: (1) what do we know, with what accuracy, and how confident are we about the data? (2) what don't we know, and why are we uncertain? (3) what could we know, with more time, money, and talent? (4) what should we know in order to act in the context of uncertainty?

14. During December 1997, representatives from more than 160 countries met in Kyoto, Japan, and reached an agreement, popularly referred to as the Kyoto Protocol, with targets for 38 developed nations as well as the European Community to reduce their overall emissions of greenhouse gases by at least 5 per cent below 1990 levels by the period 2008–12.

15. Canada, which ratified the protocol in December 2002, is to reduce greenhouse emissions to 6 per cent below 1990 levels by between 2008 and 2012. In contrast, at the end of March 2001, US president George W. Bush stated that the US would not agree to the protocol and would develop its own approach.

16. Despite the commitment to the Kyoto Protocol, Canada has been unsuccessful in reducing GHG emissions.

17. Canada received the 'fossil of the day' award at various international climate change conferences because of its poor performance in reducing GHG emissions.

18. At the Bali Conference in December 2007, an effort was made to reach agreement on how to develop a new climate change agreement to replace the Kyoto Protocol. At that event, Canada argued against specific targets and wanted the timeline extended from 2020 to 2050.

19. Canada was labelled a 'climate hypocrite' at the Bali Conference.

20. Both mitigation and adaptation must be used in policies and strategies to deal with climate change.

Key Terms

adaptation

aspirational approach

atmosphere

Bali Conference, 2007

carbon offset

carbon sequestration

carbon tax

climate

climate change

climate modelling

cryosphere

El Niño

emission credits

emission trading

fossil fuels

general circulation models
  (GCMs)

global warming

greenhouse effect

greenhouse gas

Kyoto Protocol

mitigation

Montreal Protocol

ozone

ozone layer

Rio Declaration

stratosphere

uncertainty

weather

Questions for Review and Critical Thinking

1. Explain the difference between weather and climate.

2. What are some of the key natural and human causes of climate change?

3. How credible is the evidence for climate change or global warming? For which aspects is there the most uncertainty?

4. Have you noticed any indication of climate change in the area in which you live? If so, what are they, and how confident are you that they are valid and reliable indicators of climate change?

5. What would be the best sources of traditional ecological knowledge about climate change in your area?

6. Why have general circulation models (GCMs) become dominant in the modelling research focused on climate change? What are the limitations of GCMs?

7. What is viewed as the main cause of increased carbon dioxide emissions to the atmosphere in the twentieth and twenty-first centuries?

8. Which of the negative impacts of climate change should be of greatest concern to Canadians? Will the priority shift depending on which Canadian region is considered?

9. What role should Canadians have with regard to the challenge of environmental refugees?

10. The case study regarding sea-level rise and its impact on Prince Edward Island highlights possible impacts from climate change. What would other appropriate Canadian case studies be for understanding the consequences for marine and coastal resources, freshwater resources, and terrestrial resources?

11. What are the five strategies normally used to adapt to natural hazards, such as sea-level rise caused by climate change?

12. What are the basic communication challenges related to climatic change, and what, in your view, should the first steps to overcome them be?

13. What are the strengths and limitations of the Kyoto Protocol? If it did not exist and you were to start with a 'blank sheet', what type of protocol would you propose?

14. Why has the Canadian Conservative government opposed the Kyoto Protocol? What has it proposed as an alternative?

15. Why do people argue that climate change will only be addressed successfully if there is international collaborative effort? In that context, do you agree or disagree that nations such as China and India have to be signatories to any international agreement?

16. What was the significance of the Bali Conference on Climate Change in December 2007?

17. How do you think the three mitigation measures—carbon tax, technology, carbon sequestration—should be used together to reduce GHG emissions?

18. There have been different perspectives in Canada regarding the best way to reduce greenhouse gases. What do you think is the best strategy to resolve differences among federal, territorial, and provincial governments and among the provincial governments as a group?

Related Websites

Centre for International Climate and Environmental Research, Oslo:
http://www.cicero.uio.no/home/index_e.aspx

Climate Action Network Canada:
http://www.climateactionnetwork.ca

David Suzuki Foundation, Climate Change: Impacts and Solutions: http://www.davidsuzuki.org/Climate_Change

Environment Canada, climate change site:
http://www.ec.gc.ca/climate

Environment Canada, 'Climate Trends and Variations Bulletin': http://www.msc-smc.ec.gc.ca/ccrm/bulletin/national_e.cfm

Environment Canada, State of the Environment InfoBase: http://www.ec.gc.ca/soer-ree

European Union: http://ec.europa.eu/environment/climat/home_en.htm

Government of Canada, ecoACTION: http://www.ecoaction.gc.ca

Intergovernmental Panel on Climate Change: http://www.ipcc.ch

Federation of Canadian Municipalities, Partners for Climate Protection: http://www.sustainablecommunities.fcm.ca/Partners-for-Climate-Protection

Meteorological Service of Canada: http://www.msc-smc.ec.gc.ca/index_e.cfm

Natural Resources Canada: http://adaptation.nrcan.gc.ca/posters

The Climate Group: http://www.theclimategroup.org

Union of Concerned Scientists, Citizens and Scientists for Environmental Solutions: http://www.ucsusa.org/global_warming

United Nations: http://unfccc.int/2860.php

United States Environmental Protection Agency: http://www.epa.gov/climatechange

World Meteorological Organization: http://www.wmo.ch/pages/index_en.html

Yukon College, Northern Climate Exchange: http://www.taiga.net/nce/index.html

Further Readings

Note: This list comprises works relevant to the subject of the chapter but not cited in the text. All cited works are listed in the section 'References' at the end of the book.

Alberta. 2003. 'Kyoto Protocol: Our position'. http://www.gov.ab.ca/home/kyoto/Display.cfm?id'5.

Ashmore, P., and M. Church. 2001. *The Impact of Climate Change on Rivers and River Processes in Canada*. Geological Survey of Canada Bulletin 555. Ottawa: Natural Resources Canada.

Bernstein, S., and C. Gore. 2001. 'Policy implications of the Kyoto Protocol for Canada'. *Isuma* 2 (4): 26–36.

Canada. 2001. 'Canada and the Kyoto Protocol'. http://www.climatechange.gc.ca/english/whats_new.overview.html.

———. 2001. 'Viewpoint: The road ahead'. http://www.climatechange.gc.ca/english/whats_new/road_e.html.

David Suzuki Foundation. 2003. 'Climate change: Kyoto Protocol'. http:www.davidsuzuki.org/Climate_Change/Kyoto/Kyoto_Protocol.asp.

Dessler, A.E., and E.A Parson. 2006. *The Science and Politics of Global Climate Change: A Guide to the Debate*. Cambridge: Cambridge University Press.

Fleming, S.W., and G.K. Clarke. 2003. 'Glacial control of water resources and related environmental responses to climatic warming: Empirical analysis using historical streamflow data from northwestern Canada'. *Canadian Water Resources Journal* 28: 69–86.

Gore, A. 2006. *An Inconvenient Truth: The Planetary Emergency and What We Can Do about It*. Emmaus, PA: Rodale Books.

Harvey, L.D.D. 2000. *Global Warming: The Hard Science*. Harlow, UK: Prentice-Hall.

———. 2003. 'Climatic change: Addressing complexity, uncertainty, and conflict'. In B. Mitchell, Ed., *Resource and Environmental Management in Canada: Addressing Conflict and Uncertainty*, 3rd edn, 132–65. Toronto: Oxford University Press.

Jaccard, M., J. Nyboer, and B. Sadownik. 2002. *The Cost of Climate Policy*. Vancouver: University of British Columbia Press.

Lomborg, B. 2001. *The Skeptical Environmentalist: Measuring the Real State of the World*. Cambridge: Cambridge University Press.

Mote, P.W. 2003. 'Twentieth-century fluctuations and trends in temperature, precipitation, and mountain snowpack in the Georgia Basin–Puget Sound Region'. *Canadian Water Resources Journal* 28 (4): 567–85.

Parson, E.A. 2003. *Protecting the Ozone Layer: Science and Strategy*. New York: Oxford University Press.

Reidlinger, D. 1999. 'Climate change and the Inuvialuit of Banks Island, NWT: Using traditional environmental knowledge to complement Western science'. *Arctic* 52: 430–2.

Reidlinger, D., and F. Berkes. 2001. 'Contributions of traditional knowledge to understanding climate change in the Canadian Arctic'. *Polar Record* 37: 315–28.

Scott, D.N., et al. 2000. *Climate Change Communication: Proceedings of an International Conference*. Waterloo, ON: University of Waterloo; Hull, QC: Environment Canada, Adaptation and Impacts Research Group.

Simpson, J., M. Jaccard, and N. Rivers. 2007. *Hot Air: Meeting Canada's Climate Change Challenge*. Toronto: McClelland and Stewart.

Stern, N.H. 2007. *The Stern Review of the Economics of Climate Change*. Cambridge: Cambridge University Press.

Stewart, E.J., and D. Draper. 2006. 'Sustainable cruise tourism in Arctic Canada: An integrated coastal management approach'. *Tourism in Marine Environments* 3 (2): 77–88.

Van Kooten, G.C., and G. Hauer. 2001. 'Global climate change: Canadian policy and the role of terrestrial ecosystems'. *Canadian Public Policy* 27: 267–78.

Vitousek, P.M. 1994. 'Beyond global warming: Ecology and global change'. *Ecology* 75: 1,861–76.

Chapter 8

Oceans and Fisheries

Learning Objectives

- To understand the nature of oceanic ecosystems and their similarities to and differences from terrestrial ecosystems.
- To know the main challenges facing the oceans.
- To learn about some of the global management responses to these challenges.
- To realize the main characteristics of Canada's ocean regions.
- To understand the reasons behind the collapse of Canada's east coast groundfish fishery.
- To appreciate the background to Aboriginal use of marine resources and to examine some current conflicts.
- To gain an understanding of Canada's main management strategies for ocean management.
- To be aware of some of the challenges regarding aquaculture.

Introduction

At the height of the northwest Atlantic cod fishery in 1968, more than 800,000 tonnes of northern cod were landed in Canada; by 2002, the amount landed had been reduced to 5,600 tonnes. In response to this crisis, the federal government introduced a moratorium on cod fishing in 1992, but the moratorium has been totally ineffective in rebuilding stocks, which remain at historically low levels. Scientists have put forth a number of suggestions to explain the stock's recovery failure, including changes in oceanic temperatures and increased predation. But the truth is, scientists are very far from understanding the basics of the marine ecosystems that have been sustaining the people of the Atlantic provinces for centuries.

Meanwhile, on Canada's west coast, the once mighty salmon runs have dwindled over the past decade to a trickle in many areas. Some stocks have become extinct, others have been placed on the endangered species list by the Committee on the Status of Endangered Wildlife in Canada (COSEWIC), and numbers of returning salmon have so decreased in many rivers (including the world's largest salmon river—the Fraser)

that fishing has been banned or drastically reduced for many stocks. In 2007, for example, out of a predicted return of 6.4 million sockeye salmon for the Fraser River, less than 2 million showed up.

It is not only saltwater fisheries that have suffered. The freshwater fishery accounts for about 3 per cent of the total value of commercial fishing in Canada, with 88 per cent of this coming from Ontario and Manitoba. Overall, the landed volume of fish decreased by 11 per cent between 2004 and 2005, the last year for which data is available (Fisheries and Oceans Canada 2008).

When the fishing economies on both coasts collapsed, coastal economies collapsed with them. Politicians (who had consistently approved fishing quotas well in excess of those recommended by their own scientists) finally began to realize the gravity of the situation—there was no quick fix to set things right again. As a result, a way of life for countless Canadians would have to change.

Over the past decade, fisheries management in Canada has risen from relative obscurity to become a high-profile public issue. Society has finally realized that the planet on which we live is finite. For thousands of years, the world's oceans—covering 72 per cent of the Earth's surface—have been considered vast, inexhaustible resources. But this past decade's experience has soundly dismissed this idea, as fisheries around the world collapsed. The United Nations Food and Agriculture Organization (FAO) estimates that 80 per cent of the world's fisheries have reached or surpassed their limits of exploitation. Some fisheries, such as the North Atlantic cod, may never recover.

The collapse of these fisheries is not really the root of the problem; it is a symptom of the problem, just as a runny nose is a symptom of a cold. The real problem is that we do not understand our ocean ecosystems and we have yet to learn that they cannot be managed. Human use needs managing, not only the fisheries. The theme of this book, *change and challenge*, is perhaps best epitomized in our relationship with our oceans.

Oceanic Ecosystems

The oceans and their well-being are integral to sustaining life on this planet. They are key components in global cycles and energy flows (Chapter 2). Marine ecosystems are home to a vast array of organisms displaying greater diversity of taxonomic groups than their terrestrial counterparts. Marine organisms help feed us, and they are also the source of many valuable medicinal products. The value of marine ecological goods and services is estimated at $21 trillion annually, 70 per cent greater than that of terrestrial systems (Costanza et al. 1997). We use the seas to dump our waste products and to transport most of our goods around the world. The oceans enrich our cultures, and nations draw strength and inspiration from their links to the vital life-giving nature and awesome power of the seas.

One of the major difficulties working against sustainable human use of the oceans is our lack of understanding of oceanic ecosystems. In 2003, a $1-billion, 10-year expedition announced that it had described 150 new species of fish and another 1,700 plants and animals in just the first three years of its travels and anticipated that as many as 5,000 species of fish were waiting to be discovered. By 2007, the expedition had described more than 230,000 marine species and expects perhaps 10 times that number by 2010 (see Census of Marine Life, http://www.coml.org/coml.htm). Not all new discoveries are small and in distant locales. For example, one discovery was a new species of squid more than nine metres in length in the Gulf of Mexico; another was a giant jellyfish in the heavily studied waters off Monterey, CA. If we know so little about oceanic ecosystem components, it is even more difficult to understand the functional relationships among them.

Yet the general principles that govern life and energy flows and matter cycles, discussed in Chapters 2 and 4, also hold true for oceanic ecosystems (see Box 2.5). But there are important differences. On land, water is the most common limiting factor for life. In the oceans, this is obviously not the case. Here it is nutrients. While the oceans cover more than 70 per cent of the Earth's surface, they account for only 50 per cent of global primary productivity. Much of the ocean's surface is the marine equivalent to a desert, with productivity limited to the areas where nutrients are abundant. Generally, nutrient concentrations increase with depth because of the decomposition of organisms falling from the surface layer.

The most productive areas are in the coastal zones and in areas of the ocean where upwellings from the deep ocean return nutrients to the surface layers and where photosynthetic activity occurs. In terrestrial ecosystems, productivity generally increases from the poles to the tropics. In the ocean, this is not true. Although there are productive areas in the tropics (Box 8.1), some of the most highly productive marine areas in the world are situated off the coast of Canada, such as the Grand Banks and in the Arctic Ocean, where nutrient

The oceans sustain an amazing variety of life. Here are two fish with very different life strategies. The whale shark, the world's largest fish, cruises the world's oceans feeding on plankton. The stone fish rarely moves and sits on coral reefs, disguised as a stone, until it spots prey to ambush. It is highly poisonous to humans.

Box 8.1   Coral Reefs—The Rain Forests of the Sea

Coral reefs are in many ways oceanic analogues of the tropical rain forests. Found throughout tropical and subtropical seas, they are among the most diverse and productive ecosystems on Earth. Like the rain forests, they have an ancient evolutionary history, having first appeared more than 225 million years ago, with some living reefs perhaps as old as 2.5 million years. With solar radiation the primary source of energy, these habitats are found predominantly within 30° north and south of the equator at depths of less than 50 to 70 metres. Coral reefs are categorized into three main types: (1) fringing reefs that are found along the shorelines of continents and offshore islands; (2) barrier reefs that have been separated from land by lagoons; and (3) atolls that appear as their own coral islands, having developed around (now submerged) lagoons.

Coral reefs are made up of countless numbers of individual **coral polyps** and the calcium carbonate skeletons deposited by prior generations of corals and other reef-associated organisms (e.g., coralline red algae and molluscs). These limestone secretions serve as a substrate for live coral polyps to grow and flourish.

Many coral species are involved in symbiotic relationships with unicellular algae, or **zooxanthellae**, that live inside the coral's protective skeleton. These photosynthetic algae produce carbohydrates that serve as the primary food source for the corals in which they live. When water temperatures get too warm, the zooxanthellae are often expelled, leading to the eventual death of the corals. This is called **coral bleaching** and has been recorded over large areas of reef throughout the world over the past decade. In tropical shallow waters, a temperature increase of only 3°C by 2100 may result in annual or biannual bleaching events of coral reefs from 2030 to 2050. Even the most optimistic scenarios project annual bleaching in 80 to 100 per cent of the world's coral reefs by 2080. This is likely to result in severe damage and widespread death of corals around the world.

Acting as sources of food and refuge to an incredible diversity of sea life, coral reefs are vitally important habitats. These complex ecosystems provide a number of critical ecosystem services, including the regulation of environmental disturbances, the treatment of organic wastes, the production of food, and the creation of recreational opportunities—for example, for scuba diving. In monetary terms, coral reef habitats have been estimated to generate approximately US $375 billion annually (Costanza et al. 1997).

Despite their value, coral reefs are now highly threatened worldwide, having been plagued by the effects of destructive fishing practices, coastal erosion, marine pollution, and irresponsible tourism activities. Southeast Asia is the world's epicentre for coral reef diversity, and more than 80 per cent of its reefs are considered threatened, with

Islands fringed with coral reefs as well as ring-shaped atolls in the Maldives, Indian Ocean.

A coral reef that has been dynamited for aquarium fish in the Sulu Sea, Philippines.

50 per cent in the high-risk category. Over the past two decades, coral has disappeared at five times the rate of Earth's rain forests in the Indian and Pacific oceans, home to 75 per cent of the world's reefs. A recent survey spanning four decades and 2,600 reefs found that more than 3,000 square kilometres of living coral reef is lost each year. Conservationists had previously believed that accelerated declines began in the 1990s, but the researchers found reports of widespread loss dating back to the 1960s, when pollution, deforestation, and overfishing trends began. Reefs vanished at an annual rate of 1 per cent during the 1980s, with declines climbing through the 1990s to the current rate of 2 per cent—nearly five times the pace of rain forest elimination (Bruno and Selig 2007).

Globally, less than 2 per cent of coral reefs are protected from extraction, poaching, and other major threats. Local, national, and international initiatives have now been launched to counter the effects, but consumers can help too. Coral has become a fashionable accessory; avoid purchasing jewellery with coral. You can also help by not purchasing coral as a souvenir or for your aquarium. If you keep tropical fish, ensure that they have been bred in captivity and in a sustainable way. Research shows that as fish species are removed from the reefs, the whole community becomes less stable and more likely to collapse (Carr, Anderson, and Hixon 2002).

The Banggai cardinal fish, a reef species, is an example of what can happen to fish populations if they are over-exploited for the aquarium trade. The fish is found only in the Banggai Archipelago near Sulawesi in Indonesia, and its iridescent beauty and rarity has made it a favourite of aquariumists. The fish has experienced an 89 per cent drop in population since the aquarium trade began in 1995, and in 2007 it was declared an endangered species by the World Conservation Union. The fish has the highest genetic structure ever recorded in a marine fish; populations as little as five kilometres apart have as different a structure as different species. The fish has no dispersal capability, low fecundity, long gestation periods, high early mortality, and a very restricted habitat. These are conditions that make some species more vulnerable to extinction than others, as discussed in Chapter 14. Breeding programs have been undertaken, but unless unsustainable exploitation can be controlled, ultimately they will not work.

An even bigger challenge is now looming for coral reefs. Excess carbon dioxide in the atmosphere, a main source of global warming (Chapter 7), is also predicted to lead to the extinction of coral reefs. The CO_2 is absorbed by the oceans and produces carbonic acid, making the oceans more acidic. The hydrogen ions released from this acid lower the pH (like acid rain; see Chapter 4). The oceans have absorbed nearly half of the fossil-fuel CO_2 emitted into the atmosphere since pre-industrial times, causing a measurable reduction in seawater pH and carbonate saturation. If CO_2 emissions continue to rise at current rates, upper-ocean pH will decrease to levels lower than have been experienced for tens of millions of years and, critically, at a rate of change 100 times greater than at any time over this period. Scientists predict that by the end of the century, nowhere on Earth will water chemistry still permit the growth of coral reefs. Their calcium skeleton will simply be melted away by the increased acidity, and the concentration of available carbonate ions will be too low for marine calcifiers, such as coral reefs, molluscs, crustaceans, and some algae, to build their shells and skeletons. Ocean acidification will have a very visible impact on coral reefs, but its impact will be felt not only throughout the oceans but throughout planetary life support systems. It is one of the greatest challenges we now face.

Coral reefs are home to some of the most complex and colourful organisms on Earth. Here are the bangai cardinal fish from Indonesia, an endemic nudibranch from the Andaman Sea, and a clam and coral from the Maldives.

upwellings occur. These upwellings promote large phyto-plankton populations, especially in the Arctic where there is virtually unlimited light in the summer. Many species of whales and birds migrate to these waters to take advantage of this abundance. The largest whale, the blue whale, and the bird with the largest wing span in the world, the albatross, are two good examples.

Besides nutrient availability, the other major ecological influences on marine life are temperature and light. Both temperature and light decrease with depth. This means that surface waters in the euphotic zone (Box 2.4) both are warmer and have higher light levels, resulting in higher productivity. There is usually a sharp transition in temperature between the warmer surface waters and the cooler waters underneath. This is known as the **thermocline** and generally occurs at a depth of 120 to 240 metres, depending on latitude and ocean currents.

The deepest part of the ocean is more than 9,000 metres deep, but more than three-quarters is between 4,000 and 6,000 metres in depth. Most productivity is on the continental shelves at a depth of less than 200 metres and within the top 100 metres. Most fisheries are concentrated in these areas. In the 1970s, however, scientists discovered that rich biological communities, mainly made up of bacteria that derive their energy from sulphide emissions, were centred around hydrothermal vents on the sea floor. Rich communities of tube worms, clams, and mussels have now been documented at depths exceeding 2,000 metres at more than 100 sites worldwide, including off the west coast of Canada. Scientists speculate that life on Earth originated in such hydrothermal vent systems, based on chemosynthesis. Similar kinds of chemoautotrophic-based communities are found on whale skeletons at depth, nourished by sulphides produced as the carcasses decay, as discussed in Box 2.2.

Another interesting difference between terrestrial and marine ecosystems is in the shape of the **biomass pyramids** discussed in Chapter 2. In terrestrial ecosystems, you will recall, the pyramids stand upright. There is a broad base of primary producers and a reduced biomass at each subsequent trophic level. This arrangement is reversed in oceanic pyramids because the biomass of the primary producers, the phytoplankton, alive at any one time is quite small compared to their predators. The food chains are still dependent on a broad energy base, but the turnover of biomass at the first trophic level is rapid. This constantly replenished, short-lived base supports long-living predators, such as the blue whale, that store energy over a much longer period.

The **carbon balance** in oceanic ecosystems is now the subject of a lot of scientific research because of the crucial importance of the oceans in absorbing carbon dioxide, a greenhouse gas, and the role it may play in mitigating the impact of global climate change (Chapter 7). The ocean surface takes up about 2 billion metric tonnes of carbon per year by gas exchange, equivalent to one-third of annual anthropocentric emissions.

This uptake is driven by wind exchange and the imbalances between the amount of carbon in the atmosphere and the oceans. Marine primary producers get their carbon from the dissolved CO_2 in the water as bicarbonate. There is a balance between the amount of CO_2 in the atmosphere and bicarbonate in the water. If there is too much in either compartment, a gradient is created, and the carbon migrates along the gradient between the ocean's surface and the atmosphere. However, carbon is also constantly being moved out of the surface layers of the ocean into deeper water, where it is stored in dead organisms, ocean sediments, and coral reefs. The more that is stored at depth, the greater will be the gradient pulling carbon out of the atmosphere and into the surface waters to compensate for these losses. Scientists estimate that the oceans have absorbed at least half of the anthropocentric CO_2 emissions that have occurred since 1750. However, the lag time in these movements is considerable and too slow for us to hope that the process can compensate for all the extra CO_2 released into the atmosphere by human activities. The Intergovernmental Panel on Climate Change (IPCC 2007c) predicts that ocean acidity will increase by 150 per cent by 2100 (see Box 8.1).

The carbon-saturated water does not stay where it is but moves around the globe, mainly as a result of differing water densities. This is known as **thermohaline circulation** and involves warm surface water that is cooled at high latitudes sinking into deeper basins with water close to 0°C. The key to this circulation is the high salt content of the sea water that allows the water density to increase before it freezes and sinks at certain sites. The main sites for conversion are in the North Atlantic, the Arctic Ocean, and the Weddell Sea in the Antarctic. When the water sinks, it carries with it large quantities of carbon. This sinking is the main mechanism for the removal of atmospheric carbon by the oceans. The cold water is then carried along the ocean floor until it mixes with surface waters and is transported back by wind-driven currents to the conversion areas (Figure 8.2).

The system is in fact a series of interlinked and variable currents that serve to mediate the Earth's climate through the transport of heat and water around the globe (Figure 8.2). Its scale is vast, with the flow estimated at more than 100 Amazon Rivers and the heat delivered to the North Atlantic being about one-quarter of that received directly from the sun. One of the main concerns with global warming (Chapter 7) is the impact it may have on thermohaline circulation. For example, with increased temperatures, there will be increased freshwater melting from the polar ice caps. Fresh water is less dense and will not sink to the same depth as super-cooled salty water, and therefore less carbon will be sequestered for shorter periods of time.

This is a classic example of a positive feedback loop (Chapter 3): the warmer the atmosphere gets as a result of increased CO_2 emissions, the more ice will melt, resulting in less CO_2 being absorbed by the ocean, which will result in increased atmospheric warming . . . and more melting. The cycle will

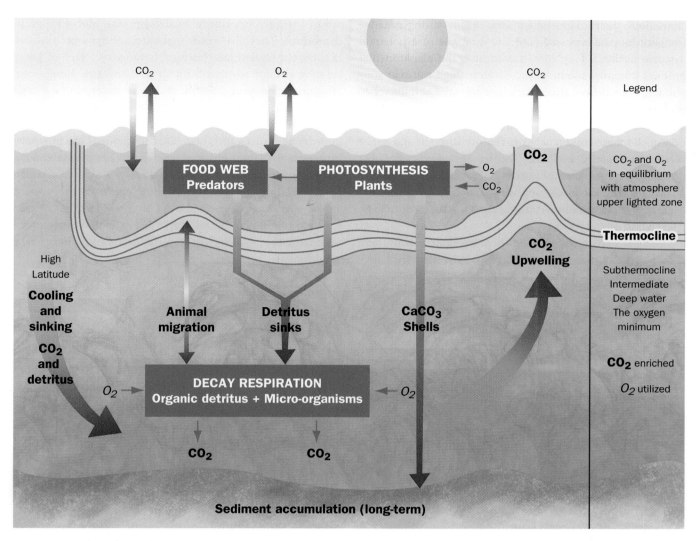

Figure 8.1 The ocean-atmosphere carbon cycle. Source: *Oceans 2020* by John G. Field et al. Copyright © 2002 Island Press. Reproduced by permission.

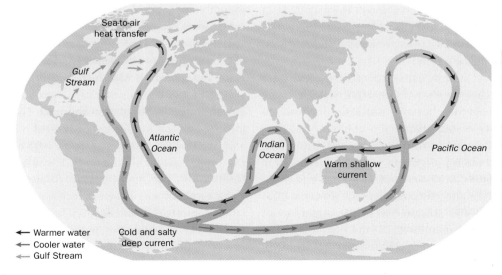

Figure 8.2 The global ocean conveyor.

continue to perpetuate itself unless equally strong negative feedback loops come into play. One example of a potential negative feedback loop in this context is the plankton that produce dimethyl sulphide gas. Given enough nutrient supply, a warmer climate should produce a greater abundance of plankton. The gas is important in enhancing the formation of atmospheric sulphate aerosol particles and cloud condensation nuclei that would tend to screen out sunlight and hence produce a negative feedback loop for the effects of global warming. Unfortunately, scientists do not feel that the effects of this negative feedback loop will be enough to counteract the effects of the numerous positive feedback loops likely to be triggered by global warming.

There is a great deal of scientific uncertainty about these global systems. Scientists know they exist and that they are of crucial importance in determining global climatic conditions. They also know that they are dynamic systems that change over time. However, because of a lack of good baseline data, it is often difficult to assess the dimensions of natural change and the mechanisms underlying it. Large spatial changes happen only rarely and slowly, requiring the collection of very long data sets of frequent measurements in order to understand them. Ocean scientists are now trying to establish these kinds of monitoring systems so that we can understand observed changes and reduce uncertainty (Box 8.2). Meanwhile, the physical evidence to support global climatic change continues to grow. In 2007, Arctic sea ice was at the lowest ever recorded extent, with scientists predicting that by the year 2010, the Arctic could be ice-free in summer.

Box 8.2 Regional Scale Nodes and the Canadian Neptune Project

The Regional Scale Nodes (RSN) program is a component of the US National Science Foundation's Ocean Observatories Initiative. Formerly known as NEPTUNE, RSN is a cabled ocean observatory that will be constructed in the northeast Pacific Ocean off the coasts of Oregon and Washington. Led by the University of Washington, the RSN program will extend continuous high-bandwidth (tens of Gigabits/second) and power (tens of kilowatts) to a network of many different types of instruments distributed across, above, and below the seafloor and ocean associated with the southern portion of the Juan de Fuca tectonic plate. NEPTUNE Canada is a complementary program building a similar cabled system on the northern portion of the plate.

Such Internet-linked, sensor-robotic networks offer novel approaches to human interaction with many remote or dangerous oceanic processes intrinsic to the habitability of our planet. Continuous, real-time information flow from the ocean via electro-optical cables will launch rapid growth in our understanding of (1) the habitats and behaviour of known and novel life forms; (2) many climate-changing processes; (3) erupting underwater volcanoes; (4) the migration of charismatic marine mammals; (5) the assessment and management of living and non-living marine resources; (6) the timing and intensity of major under-sea earthquakes; (7) the mitigation of natural disasters; and (8) a host of new discoveries continuously unfolding within the 'Inner Space' of the global ocean. Researchers, educators, and members of the public will have open web access to all imagery and information in their own laboratories, classrooms, and living rooms as the technologies and visualization software required to operate these *in situ*, submarine sensing networks become more sophisticated.

In 2009, NEPTUNE Canada (www.neptunecanada.ca) should complete installation of its regional scale observatory with an 800-kilometre cable loop from the shore station at Port Alberni, Vancouver Island, connecting five observatory nodes in coastal, continental slope, abyssal plain, and spreading ridge environments.

VENUS (www.venus.uvic.ca) is a coastal observatory in waters near Victoria and Vancouver. The first four-kilometre line was installed in early 2006 in Saanich Inlet, with the node at 100 metres depth near the oxic/anoxic transition zone within the fjord. The second 40-kilometre line with two nodes extends from the Fraser River Delta across much of the Strait of Georgia, with installation completed in early 2008. The observatories investigate ocean and biological processes and delta dynamics in waters of up to 300 metres in depth. Real-time data and imagery have been successfully relayed from Saanich

Figure 8.3 Regional cabled observatory essential elements. Source: OOI Regional Scale Nodes and Center for Environmental Visualization, University of Washington.

Inlet through the VENUS website since 2006. VENUS is used in an interactive mode for researchers to trigger experiments remotely and for educators to involve students in on-line studies.

The University of Victoria owns and operates the observatories and has established Ocean Networks Canada as a not-for-profit agency to manage NEPTUNE Canada and VENUS as national facilities.

Ocean Management Challenges

Oceanic ecosystems have been providing humans with sustenance since time immemorial. Coastal zones, including the continental shelf, occupy about 18 per cent of the Earth's surface, supply about 90 per cent of the global fish catch, and account for roughly 25 per cent of global primary productivity. Approximately 60 per cent of the world's population now lives in the coastal zone. By 2100, nearly 75 per cent of the world's population will live within 80 kilometres of the coast, mostly clustered into 'mega-cities' on just 11 per cent of the land.

The oceans are crucial to the way in which planetary ecosystems work and to the functioning of human society. Recent global research indicates that virtually no region of the ocean is untouched by humanity and more than 40 per cent is heavily affected. In 1883, Thomas Huxley, the great nineteenth-century biologist, voiced the common opinion 'that probably all the great sea fisheries are inexhaustible.' This has proven to be far from the truth. This section reviews some of the main management challenges facing ocean ecosystems as a result of fisheries and other human activities (Figure 8.4).

Fisheries

The most important fishing grounds in the world are found on and along continental shelves within less than 200 nautical miles of the shore. The distribution of these fishing grounds is patchy and very localized. More than half of the marine landings are caught within 100 kilometres of the coast in depths generally less than 200 metres covering an area of less than

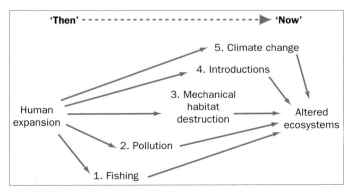

Figure 8.4 Historical sequence of human impacts on marine ecosystems. Source: Roberts 2003, 170, from Jackson et al. 2001. Reprinted with permission from AAAS. http://www.sciencemag.org/cgi/content/abstract/293/5530/629.

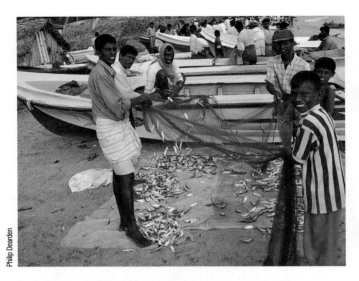

Small-scale fishers on the east coast of Sri Lanka sort the morning's catch.

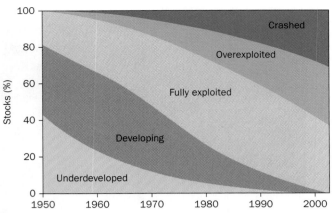

Figure 8.5 State of the world's fishery stocks. Source: Nellemann et al. 2008.

7.5 per cent of the world's oceans, and 92 per cent are caught in less than half of the total ocean area. These areas of the ocean play a major role in sustaining global populations.

Fisheries supply some 20 per cent of the world's annual animal protein supply, but in some areas of Asia and Oceania, fish provide virtually all the protein supply. More than one billion people rely on fish as their primary source of protein. Seafood supplies more than 2.6 billion people with at least 20 per cent of their average per capita animal protein intake (FAO 2006). This supply is tapped by fishers ranging from villagers using homemade canoes trying to feed their families through to multi-million-dollar offshore factory ships owned by multinational corporations. Given this great scale and variation, it is only recently that fisheries scientists have begun to understand more about what is happening in global fisheries.

There is clear evidence, obtained from bottom sediment cores going back hundreds of years, that fish populations fluctuated quite widely as a result of changing environmental conditions even before the advent of modern fishing. However, the scale and speed of current changes are unprecedented. As stated earlier, more than 80 per cent of global fisheries are now fully utilized or overexploited (Figure 8.5). Because of advances in technology and subsidies, fishing capacity is now estimated at as much as 2.5 times what is needed to harvest a sustainable yield from the world's fisheries. The graph in Figure 8.6 shows a gradual levelling off in more recent years, but many scientists argue that this levelling off masks a major unprecedented collapse of ocean fisheries. Scientists at the Fisheries Research Centre at the University of British Columbia, for example, revealed that the catch data shown in the graph may be up to 25 per cent above actual catch data. This may be the result of the routine reporting of inflated catches by China (Watson and Pauly 2001). The problem is that permitted catch levels are based largely on historical catch information. If the latter is inflated, it leads to unsustainable catch levels in the future. Some fisheries scientists have predicted the global collapse of all taxa currently fished by 2048 if current levels of exploitation continue (Worm et al. 2006), although others (e.g., Branch 2008) dispute this finding.

The fish being caught are also substantially smaller than the ones caught in the past (Figure 8.7). Research suggests that for top predators, current sizes are one-fifth to one-half what they used to be (Myers and Worm 2003). Modern fisheries management often encourages fishers to select the large

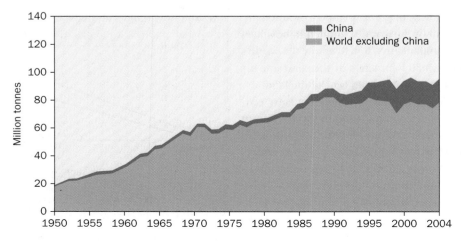

Figure 8.6 World capture fisheries production. Source: FAO 2006, http://www.fao.org/fishery/statistics/programme/3,2,1.

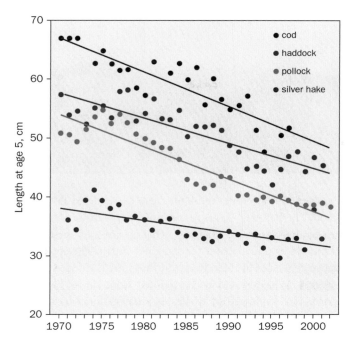

Figure 8.7 Fish size decline, 1970–2000. Source: Fisheries and Oceans Canada 2003. Reproduced with the permission of Her Majesty the Queen in Right of Canada, 2005.

individuals of targeted stocks, either by using size-selective gear or by releasing small individuals back to the water. The reasoning has been that this allows smaller, younger individuals to grow up to reproductive age, thereby sustaining the stock. Recent research, however, shows that removing the larger, older individuals of a population may actually undermine stock replenishment. Some researchers have proposed that maintaining old-growth age structure can be important for replenishing fished populations. It is termed the Big Old Fat Fecund Female Fish (BOFFFF) hypothesis. Research shows that removing BOFFFF and other large adults can result in evolutionary changes in populations (Longhurst 2002).

Changes in fish populations are not immediately obvious, since scientists tend to look only at the most recent data rather than comparing them with historical catches. This problem is known as a **shifting baseline** in which scientists have no other option than to take the current degraded state as the baseline rather than the historical ecological abundance. Removal of virtually all the large predatory fish from oceanic ecosystems will have significant implications for the structure and functioning of marine ecosystems.

In fact, some authors (e.g., Jørgensen et al. 2007) are going as far as suggesting that the impact of commercial fishing is now so great that 'evolutionary impact assessments' should be undertaken, since fishing causes changes to occur in decades rather than millennia, as would happen under normal conditions. The scientists note that fisheries-induced mortality is now reckoned to exceed natural mortality by 400 per cent and that increased mortality generally means that fish will reach sexual maturity at a younger age and increase levels of reproduction. These trends are further exacerbated when fisheries selectively target larger or more mature individuals.

Many fish species show changes in their size and age at maturation, reduced annual growth, and loss of genetic diversity. Such fisheries-induced evolution may be slow to reverse, or even irreversible, and will affect many other species through trophic and ecological interactions. Managing fisheries from an evolutionary perspective will almost certainly require a variety of strategies. For some fisheries, we may need no-take zones such as marine protected areas (MPAs), as discussed later in the chapter, where the full range of sizes and ages of a given species can thrive. For those that target migratory species, setting both maximum and minimum size restrictions might work better.

The total catch is made up of many different species, which creates several problems. In theory, fishing should be self-regulating. As the catch of the target species declines, this should result in a reduced fishing effort as it becomes unprofitable. Unfortunately, this is not what happens. Instead, fishing fleets increase their efforts towards the target species and then turn to the next most profitable species until that too is depleted. Then they pursue the next most profitable species and so on. This is a familiar foraging behaviour in ecology, known as **prey switching**. In the fishery, it leads to **serial depletion** in which one stock after another becomes progressively depleted even if the total catch remains the same. This phenomenon has been well documented in whales (Box 8.3). Unfortunately, after switching the target species, many fisheries take some of the depleted species as bycatch (Figure 8.8), making it even more difficult for the stocks to recover.

The shift in target species is not the only change in the world's fishing activities. We are progressively exploiting lower and lower trophic levels to derive our catch. Initially, we targeted large, long-lived fish high in the food chain, but as they became depleted, we sought smaller fish until they too were depleted. Fisheries in many areas have now focused on invertebrates, like the Atlantic crab. Large-sized fish no longer exist. This is known as **fishing down the food chain** and is illustrated

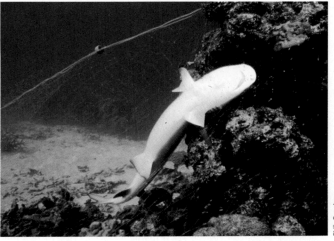

This shark was caught by a 'ghost net' snagged on the bottom but still doing the deadly function it was designed for.

Box 8.3   Leviathan

Canada has the longest coastline in the world as well as some of the richest waters. Cold oceanic waters from the north and deep upwelling currents mix with warmer waters from the tropics to create an abundance of life. Rich nutrient supplies, accompanied by shallow seas and long daylight hours in the summer, have created some of the richest waters off the east coast. Plankton flourish and provide the base for a diverse food web that supports three main groups of marine mammals—the Odontoceti or toothed whales, the Mysticeti or baleen whales, and the Pinnipedia or seals and walruses. The baleen whales all feed on plankton, small fish, and marine algae by means of plates of baleen in their mouths that serve to filter these organisms from the water. From these small food items, the baleen whales have evolved as the largest creatures on this planet. The blue whale is the largest of the baleen whales; the largest recorded was a female more than 30 metres in length and weighing 140 tonnes. A calf is typically seven metres long at birth and may weigh almost three tonnes. Blue whales, as well as other baleen whales such as the fin and sei whales, were once found in abundance off the east coast, as was the largest toothed whale, the sperm whale. These whale populations were all decimated by hunting and are only now starting to recover.

In the late fifteenth century, Europe found itself increasingly short of the oil needed to light its lanterns. Marine mammals, with a thick layer of blubber, were the solution.

Exposed to high heat, the blubber can be rendered down to oil. The Basques—a great seafaring people from northern Spain—discovered the rich whaling grounds off the east coast of Canada, and the slaughter began. The abundance of whales in these waters at that time is difficult to imagine. Whales could be harpooned from shore. Early mariners complained of whales as a navigational hazard because they were so numerous; one missionary in the Gulf of St Lawrence reported that the whales were so numerous and loud that they kept him awake all night! The limiting factor was not the number of whales but the ability to process them, and as increasing numbers of shore stations were established, another toehold of colonization began.

The rest of the story is fairly well known. As human numbers increased, whale numbers declined. First to go was the one hunted the most, the black right whale. The black right whale was targeted because it was slow, it floated when it was killed, and one whale could be rendered into more than 16,000 litres of oil. Besides the oil, the baleen was used for other indispensable purposes such as clothing supports, brush bristles, sieves, and plumes for military helmets. A Basque shipowner could pay off his ship and all his expenses and still make a good profit in one year from such whales. Although the whales gained some respite when England destroyed the Spanish Armada, other nations finished off the job.

Humpback whales were once common off both the Pacific and Atlantic coasts of Canada.

Whale harpoon in the Fisheries Museum of the Atlantic, Lunenburg, Nova Scotia.

Other whales—sperms, humpbacks, blues, fins, seis, and minkes—soon joined the right whale as commercially, if not biologically, extinct. This is a classic example of serial depletion. By the early 1970s, there were no commercially viable populations of large whales remaining. A global moratorium on whaling was announced in 1987 and still stands today, despite pressure applied by Japan and Norway, both of which continue to kill whales for 'research purposes'. The appeal of 'scientific whaling' is not only that it bypasses any internationally agreed catch limits but that it also circumvents all other rules regarding protected species, closed areas, killing of juveniles, killing methods, and so on.

Iceland resumed commercial whaling in 2006 and in 2007 issued permits to kill 100 minke whales. In late 2007, Japan's whaling fleet left port for the South Pacific with orders to kill up to 50 humpback whales—the first known large-scale hunt for the whales since a 1963 moratorium put them under international protection. Besides humpbacks, the fleet will take up to 935 Antarctic minke whales and up to 50 fin whales. The American Cetacean Society estimates the global humpback population at 30,000 to 40,000, about a third of the levels seen before modern whaling, and the World Conservation Union lists the species as 'vulnerable'. Greenpeace protested the hunt, announcing later that its peaceful interference with Japan's whaling fleet in Antarctic waters resulted in the killing of less than half of the 985 whales targeted.

The removal of such a large biomass from the top of the food chain obviously has ecological repercussions for other organisms. Unfortunately, we have interfered so much with populations of other animals that the impacts are seldom clear. One possible implication is an increase in other krill eaters that would benefit from removal of these large and efficient competitors. For example, it has been suggested that increases in the populations of krill-eating seals (e.g., fur and crab-eating seals) in the Antarctic came about as a result of whaling. More breeding sites are required to support the higher populations of seals, so seals colonize areas previously used for nesting by birds such as the albatross. Did declining whale numbers also lead to a decline in albatross? We cannot say for sure, but the example does illustrate the complexities of changes in food webs.

Populations for most species have been slow to recover because of the slow reproductive rates of these K-strategists (Chapter 3). Whaling has now been controlled throughout most of the world, and on all of Canada's coasts a resurgence is occurring, albeit very slowly for some species such as the right whales. The largest whale, the blue, is also in trouble. Although numbers still exist, finding a mate is difficult for this wide-ranging species. Scientists have now found evidence that the blues are interbreeding with the more common fin whales, the second largest whale in the world, and fear that this hybridization will result in a loss of genetic identity for the blue, which could then disappear as a species. Unfortunately, the rapacious killing by our ancestors denied all future generations the spectacle of our seas full of these mighty creatures. Are our actions today denying future generations similar opportunities?

Figure 8.8 Bycatch. Reprinted by permission of Adrian Raeside.

Figure 8.9 Fishing down the food chain.

in Figure 8.9. The gains in fish catch shown in Figure 8.6 were mainly the result of reaching further down the food chain to previously underexploited trophic levels, as well as accessing fish from greater and greater depths (Figure 8.10).

Eating lower on the food chain is not a bad idea. Few of us habitually eat wolves, tigers, and other top terrestrial predators, but in the oceans we target the top predators. In terms of protein production, the seas might not be as overexploited as we think. For example, it takes close to 60 million metric tons of potentially edible fish per year to feed the three million metric tons of the three major tropical tuna species we harvest annually because of the second law of thermodynamics, discussed in Chapter 2. If we could replace some of our tuna sandwiches with the anchovies, sardines, squid, and other species the tuna eat, we would open up a substantial supply of protein that could feed millions more.

Besides the impact on target species, fishing activities have many other ecological repercussions. Of particular concern is the impact on non-target organisms, or **bycatch**. Estimates suggest that 25 per cent of the world's catch is dumped because it is not the right species or size. Virtually all organisms dumped overboard die. The world's largest turtle, the leather-back turtle, is rated as critically endangered on the Red List of the International Union for the Conservation of Nature and

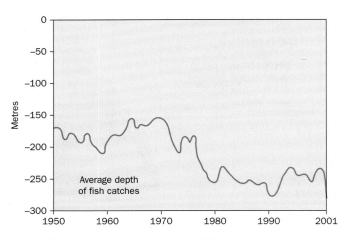

Figure 8.10 Trend in mean depth of catch since 1950. Source: Millennium Ecosystem Assessment 2005.

Rich ocean waters attract many pelagic (open ocean) bird species, such as this adult waved albatross.

encounters with sea turtles while maintaining the catch of target fish. Turtle mortality can be further reduced by changing the types of hooks used and using fish rather than squid as bait. Another winner proposed placing magnets on fishing lines to scare away sharks, which are particularly sensitive to magnetic fields.

One of the most destructive means of fishing is **bottom trawling**, in which heavy nets are dragged along the sea floor, scooping up everything in their path. This is a common method for catching shrimp, and the ratio of shrimp to other organisms caught is generally around 10 per cent (i.e., one shrimp caught for every 10 organisms caught unintentionally). Trawling has been estimated to be as damaging to the seabed as all other fishing gear combined. The damage extends to more than half of the seabed area of many fishing grounds and is worse in the inner and middle parts of the continental shelves, with particular damage to small-scale coastal fishing communities.

One noted victim of bottom trawling was the ancient sponge reefs found off the coast of BC. The sponge reefs, found at depths between 165 and 240 metres, are more than 9,000 years old. Previously, they were known only from fossils in Europe dating from 146 to 245 million years ago, and they were thought to be extinct before their discovery in 1988. Sponge reefs are widely believed to be the first multi-cellular animals on earth, appearing almost 570 million years ago. The reefs thrive in silica-rich water and are the only siliceous sponge reefs on the planet. Living at a depth of 160 to 250 metres and attaining heights of 20 metres, some reefs are as big as six-storey buildings. Some live in shallower water but are limited to individuals rather than reefs. The federal government called for *voluntary* trawl restrictions, but in 2002, after documentation of extensive damage to the most pristine reefs by trawling, mandatory closures were finally implemented to protect the northern reefs. Smaller ones in the Strait of Georgia remain unprotected.

Natural Resources (IUCN). In 1980, there were some 91,000 leatherbacks left in the Pacific; there are now fewer than 5,000. They are mainly the victims of the longline and gill-net fisheries. **Longline** fishing in all the world's deep oceans kills some 40,000 sea turtles each year, along with 300,000 seabirds and millions of sharks. Bycatch of albatross, petrels, and shearwaters in longline fisheries is one of the greatest threats to these seabirds. All 21 of the world's species of albatross are now considered at risk of extinction, along with 57 species of sharks and rays.

Some innovative ways to address bycatch are starting to be developed, in some cases helped along by cash prizes offered by NGOs. The Worldwide Fund for Nature, for example, launched the International SmartGear Competition in 2004 to encourage innovative, practical, and cost-effective gear designs that safeguard marine life while enabling fishers to better target their intended catch. One winner, knowing that most turtles are hooked at shallow depths, proposed a system for setting baited hooks deeper than 100 metres, thereby minimizing

Model of a box trawl in the Fisheries Museum of the Atlantic, Lunenburg, Nova Scotia. The large weights are dragged along the sea floor, destroying everything in their path.

One of the deep sponges found off the BC coast.

Philip Dearden

Steller's sea lions, the world's largest sea lions, are now highly endangered in the North Pacific. These are at Cape St James, off the southern tip of the Queen Charlotte Islands (Haida Gwaii), also the windiest spot in Canada.

Marine pollutants take a number of forms, originate from many different sources, and have a wide range of effects. About 80 per cent of ocean pollution comes from activities on land. The remaining 20 per cent comes from activities at sea, such as waste disposal, oil spills, vessel traffic, oil and gas exploration, and mining. Although we are all familiar with major oil spills, such as that of the *Exxon Valdez* in Alaska (discussed below and in more detail in Chapter 12), we think little of the many diffuse sources of oil, such as leaking car engines, that eventually seeps into the oceans. In fact, the total amount from non-point sources of pollution is considerably greater than from point-source pollution. During the past 20 years, many governments have made considerable progress in monitoring and regulating point-source pollution, such as effluent discharge from factories. However, addressing non-point-source pollution is a much more challenging task, since it requires a change in behaviour by millions of people.

Chemical pollutants take two main forms: toxic materials and nutrients. We live in a chemical society, with more than 100,000 chemicals used in manufacturing and released into the environment every year. Many of these chemicals are harmful to life; when they end up in the ocean, they may cause instant death if released in sufficient quantity, or they may have sub-lethal effects such as inhibition of reproduction. The chemicals are also subject to *bioconcentration*, discussed in more detail in Chapter 10. Synthetic organic chemicals and toxic metals tend to concentrate along two main interfaces in

In addition to the direct destructive effects of fishing on non-target organisms, it also has indirect effects through food chain relationships. Steller's sea lion, for example, was once very abundant in the North Pacific, with more than 300,000 animals recorded in 1960. By 1990, this number had fallen to 66,000, and the US declared the sea lion endangered. The main reason for the decline is thought to be the decline in pollock, their chief food source. As the harvests of cod diminished elsewhere, the demand for pollock increased, leading to unprecedented catches and a subsequent decline in sea lion numbers. Scientists have recently suggested that another factor may have contributed to this decline. Earlier industrial whaling activities removed the bulk of the killer whales' main prey. In response, the killer whales started 'fishing down the food chain', eating sea lions in much greater numbers. Reaching even further down the food chain, killer whales are now implicated in the decline of sea otters in the North Pacific (Chapter 3). Other scientists (e.g., Trites et al. 2007) have questioned this relationship. But sea lions and sea otters are not the only marine mammals to experience declining numbers; two-thirds of all marine mammals are now classified as endangered on the IUCN Red List (Chapter 14).

Pollution

As the recipient of all the polluted water that flows off the land as well as airborne contaminants, the oceans are the ultimate sink for many pollutants. The scale of global pollution is now astounding—even the oceans are being rendered eutrophic (Chapter 4), in some cases from farming practices thousands of kilometres away. For example, scientists suspect that iron particles picked up into the atmosphere in Africa and blown across the Atlantic are a major factor in increasing the growth of smothering algae on the coral reefs of the Caribbean.

John W. Chardine

This dovekie was killed by an oil spill. Oil penetrates through the feathers to the layer of down beneath and decreases the effectiveness of the feathers' insulating properties.

the ocean: the boundary between the seabed and water and the boundary between the water and the atmosphere.

One rapidly emerging impact relating to pollution is **endocrine disruption**. The endocrine system consists of glands and hormones that control many bodily processes such as sex, metabolism, and growth. Many chemicals in everyday use have been found to mimic these processes and may stimulate, replace, or repress the natural processes. More than 50 such endocrine disrupters have been positively identified, and scientists suspect that there are hundreds more. Many of them are in commonly used products such as soaps and detergents. The major effect of these chemicals on marine life detected so far is the feminization of various aquatic species. Hermaphroditic fish are appearing all over Europe and also in the Great Lakes. All the fish in the River Aire in England, for example, show signs of feminization. Increasing numbers of reports from the Arctic are reporting hermaphroditic polar bears. Many endocrine disrupters, such as the pesticide DDT and other organochlorines, are vulnerable to both the grasshopper effect and biomagnification (as described in Chapter 10) and would logically find their way to the top Arctic predator. There is now extensive research underway in Europe on this issue, since it is also feared that these so-called 'gender-benders' may be at least partly behind the documented fall in human male sperm counts in industrialized societies over the past 50 years.

Oxygen depletion occurs as a result of nutrient enrichment, as described in Chapter 4, and leads to large dead areas within the oceans. These areas (**hypoxic** or oxygen-deficient areas) increased from 149 in 2003 to more than 200 in 2006. Their growth could also promote the development of far more male fish than female, threatening some species with extinction. Shang, Yu, and Wu (2006) found that low levels of dissolved oxygen decreased the activity of certain genes that control the production of sex hormones and sexual differentiation in embryonic zebra fish. As a result, 75 per cent of the fish developed male characteristics versus 61 per cent of those raised under normal oxygen conditions. This gender shift decreases the likelihood that they will be able to reproduce in sufficient numbers to maintain sustainable populations. The study raises new concerns about vast areas of the world's oceans, known as 'dead zones', that lack sufficient oxygen to sustain most sea life. Fish and other creatures trapped in these zones often die. Those that escape may be more vulnerable to predators and other stresses. This study suggests that dead zones potentially pose a third threat—the inability of offspring to find mates and reproduce.

Energy

World demand for energy, particularly oil and gas, continues to rise (Chapter 12). Every time you jump into your car or go on an airplane, you are sending a financial message to industry and the government that you support further development of fossil-fuel sources. All of the world's cheap and accessible terrestrial sources have now been tapped. Many of the world's main oil-fields, such as the North Sea in Europe and Hibernia off the coast of Newfoundland, are situated in sedimentary basins under the oceans. More than 60 per cent of current global production comes from these sources. Offshore oil rigs are a source of chronic, low-level pollution caused by the disposal of drilling mud and drill cuttings, which smother the local environment and are often contaminated with oil or chemicals. Oil is pumped to the surface and loaded on tankers or piped ashore, where it is often stored close to the ocean before being refined and further distributed. There is potential for spillage at every stage. Many seabirds and marine organisms are highly vulnerable to oil pollution. Following the wreck of the *Exxon Valdez* off Alaska, for example, more than 750 sea otters were killed. Peterson et al. (2003) argue, however, that the long-term effects of the spill, including mortalities of species such as sea otters, were much greater because of ongoing impacts such as contaminated food chains. Oil is lethal to marine life by virtue of its physical effects and its chemical composition (Garshelis 1997). Cleanup attempts following oil spills can also damage many species.

Most accessible ocean oil basins have already been developed, which means that exploration and development are pushing into increasingly challenging and fragile environments, such as the Arctic Ocean and North Pacific. The most effective way to reduce the effects of oil exploration and development in these regions is by reducing demand for oil products.

Recent research has highlighted the potential for seabed-based methane hydrates to meet some energy demands. Methane hydrates are ice-like deposits found in the top few hundred metres of sediment in certain deep ocean areas of the continental margins. The methane gas is actually trapped in ice

Offshore oil rig.

iStockphoto.com/Robert Ellis

cages and can be easily extracted from it, but removing the hydrates from the seabed has proved problematic. Russian energy experts have tried using antifreeze to remove the methane from hydrates, and recent research has focused on trying to pipe warm surface water down to melt the hydrates and then piping the gas to the surface using a parallel set of pipes. However, melting the hydrates to release methane may cause the seafloor to become unstable and could have untold ecological impacts as well. In addition, if methane is lost to the atmosphere during the process, it could add to the global warming phenomenon, since methane is a potent greenhouse gas.

Coastal Development

Twenty-one of the world's 33 mega-cities are coastal. Almost half of the world's population lives within 100 kilometres of the coast, and this proportion is expected to increase to 75 per cent by mid-century. These regions also contain the highest concentration of supporting infrastructure, industrial plants, energy use, food production, and tourism in the world. Half of the world's coastal wetlands have been filled in to support these developments. Meanwhile, as in Canada, environmental decision-making and management within the coastal zone are often highly fragmented among many different agencies. These extreme pressures on fragile ecosystems with ineffective management control have often led to highly degraded coastal environments in many countries. Half of the countries in the world do not have any coastal legislation in place to address the situation.

Climate Change

Global air temperature is expected to warm by 1.4 to 5.8°C this century, affecting sea-surface temperatures and raising the global sea level by nine to 88 centimetres. The main causes of sea-level rise are thermal expansion of sea water and the melting of ice in land-based glaciers. The rate of sea-level rise has increased significantly over the past few years. Since 1961, global average sea levels have risen at an average rate of 1.8 (1.3 to 2.3) millimetres annually and since 1993 at 3.1 (2.4 to 3.8) millimetres annually (IPCC 2007c). Higher sea levels increase the impact of storm surges, accelerate habitat degradation, alter tidal ranges, exacerbate flooding, and change sediment and nutrient circulation patterns. A case study of coastal areas in Prince Edward Island in Chapter 7 highlights some of these challenges. Rising levels will displace approximately 1 billion people, many of them among the poorest in the world.

Estimates suggest that an increase in mean sea-surface temperature of only 1°C could cause the global destruction of coral reef ecosystems. The waters off British Columbia have already experienced a rise in sea-surface temperature of between 0.3 and 0.9°C over the past 50 years. Increases in temperature may slow or shut down the thermohaline circulation (Figure 8.2), causing widespread climatic change, changes in the geographic distributions of fisheries, and increased risk of hypoxia in the deep ocean. Scientists have already detected changes in the distribution patterns of many marine species as a result of changes in ocean temperatures attributed to global warming. Early break-up of sea ice is thought to be one of the main contributing factors in the increased stress experienced by polar bear populations (Regehr et al. 2007).

The Arctic Ocean is one of the most sensitive indicators of global climate change. Significant changes are taking place very rapidly. In addition to the well-documented loss of perennial ice cover as a whole, the amount of oldest and thickest ice within the remaining multi-year icepack has declined significantly. The oldest ice types have essentially disappeared, and 58 per cent of the multi-year ice now consists of relatively young two- and three-year-old ice, compared to 35 per cent in the mid-1980s. Ice coverage in the summer of 2007 was at a record minimum, with ice extent declining by 42 per cent compared to conditions in the 1980s.

Global Responses

International Agreements

Strong international action can help to address these problems, and there are many examples of international treaties concerning the oceans. The overall international legal framework is provided by the United Nations Convention on the Law of the Sea (UNCLOS), which entered into force in 1994. Activities under its jurisdiction include protection and preservation, navigation, pollution, access to marine resources, exploitation of marine resources, conservation, monitoring, and research. One of the most important provisions was the agreement for coastal nations to establish **exclusive economic zones** (**EEZS**). The convention also established 45 per cent of the seabed as common property. Canada ratified the convention in November 2003.

More specific agreements include the 1972 London Dumping Convention that has led to a gradual decline in the amount of sewage sludge and industrial waste dumped into the oceans. In terms of fisheries, the UN Moratorium on High Seas Driftnets was quite successful. The same kind of approach now needs to be applied to reduce the impact of other highly damaging types of fishing gear, especially longlines and industrial trawlers with high rates of bycatch. The United States has already taken some steps to protect embattled marine species by closing the west coast to longlining altogether and by prohibiting the Hawaii longlining fleet from fishing for swordfish. There are also temporal and spatial closures for gill-netting and moves to ban bottom trawling, which may be the most destructive fishing method of all.

Some strong ecologically based marine agreements have emerged over the past decade, including the UN Convention on Straddling and Highly Migratory Fish Stocks and the Global Program of Action for the Protection of the Marine Environment from Land-Based Activities. However, some of these new agreements have not yet been fully implemented.

For example, 10 of the world's top 15 fishing nations have not yet ratified the Straddling Stocks agreement, although Canada has done so.

In addition to these international agreements, international programs are aimed at creating a more comprehensive, global approach to oceans management. At the World Summit on Sustainable Development (WSSD) held in Johannesburg in 2002, governments committed to restore most of the major global fisheries to commercial viability by 2015.

However, Daniel Pauly, director of the Fisheries Centre at the University of British Columbia, suggests that to achieve real sustainability, we need to rethink the way that fishing is undertaken. Policy-makers have concentrated on promoting industrial fisheries under the mistaken belief that they catch the vast majority of fish. Pauly (2006) argues, however, that not only are small-scale inshore fisheries (**artisanal** fisheries) as productive as industrial fisheries but they are much more efficient and less damaging and provide more support for local communities (Table 8.1). A wealth of knowledge has built up around the world that reflects the intimate relationship that artisanal fishers have with the resource and the more geographically based management schemes that have evolved to reflect this relationship.

Pauly and others say that the quickest way to effect a change in fishing methods is simply to end the subsidies that support commercial fisheries. In an analysis, they found that subsidies make up about 25 per cent of the total landed values from the catch of these fleets, whereas the reported profit per landed value is no more than 10 per cent. Were the subsidies removed, the fisheries would die as uneconomical (Sumaila and Pauly 2007). Since more than half the subsidies are for fuel, eliminating fuel subsidies would be simple and effective while contributing to a reduction in fossil-fuel consumption.

Marine Protected Areas

Compared to the terrestrial environment, where ecological communities are often associated with areas defined by geographical features (such as mountains and rivers), precise boundaries of distinct communities or processes in oceans are rare. Geographic scales are large, and biological processes are not self-contained within a given area. The water overlying the seabed is a mobile third dimension that provides nourishment for much of ocean life. These characteristics of the ocean environment mean that surveys take time and are costly, and because of the high variability in sizes of marine populations, surveys are typically uncertain. Until recently, these difficulties prevented scientists from being able to assess the effectiveness of **marine protected areas** (MPAs). However, in 2001, following two and a half years of rigorous scientific review, 161 of the world's top marine scientists signed a consensus statement (http://www.nceas.ucsb.edu/Consensus) that marine reserves:

- conserve both fisheries and biodiversity;
- are the best way to protect resident species;
- provide a critical benchmark for the evaluation of threats to ocean communities;
- are required in networks for long-term fishery and conservation benefits;
- are a central management tool supported by existing scientific information.

Table 8.1 Comparative Benefits of Large- and Small-Scale Fisheries

Benefits	Large-Scale Fishery	Small-Scale Fishery
Number of fishers employed	About ½ million	More than 12 million
Annual catch of marine fish for human consumption	About 29 million tonnes	About 24 million tonnes
Capital cost of each job on fishing vessels	US $30,000–$300,000	US $250–$2,500
Annual catch of marine fish for industrial reduction to meal, oil, etc.	About 22 million tonnes	Almost none
Annual fuel oil consumption	14–19 million tonnes	1–3 million tonnes
Fish caught per tonne of fuel consumed	2–5 tonnes	10–20 tonnes
Fishers employed for each US $1 million invested in fishing vessels	5–30	500–4,000
Fish and invertebrates discarded at sea	10–20 million tonnes	Little

Source: Pauly 2006.

The establishment of MPAs has lagged substantially behind their terrestrial counterparts—currently less than 0.65 per cent of the ocean has been designated as protected. However, with increasing political awareness of the importance of the oceans and their highly degraded state, MPAs are starting to receive some attention. The 2002 World Summit on Sustainable Development set 2012 as the target date for completion of an effectively managed, ecologically representative network of marine and coastal protected areas within and beyond areas of national jurisdiction. The latter occur in the 64 per cent of the oceans beyond the 200-nautical-mile limit of the EEZS of coastal states. This area is known as the High Seas, and because of a lack of an established legal framework for their allocation and management other than UNCLOS, it will be necessary to take concerted international action to achieve their protection. Researchers have recently calculated the impact on fisheries of establishing such areas outside of territorially defined waters and conclude that closing 20 per cent of waters on the High Seas to fisheries might lead to a loss of only 1.8 per cent of the current global reported marine fisheries catch (Sumaila et al. 2007).

Canada's Oceans

Canada has the longest coastline of any country and the second largest continental shelf, equal to 30 per cent of Canada's land mass (Figure 8.11). There are some 1,200 species of fish and many globally important populations of marine mammals, including grey, bowhead, beluga, right, minke, humpback, and killer whales. Unfortunately, several of these species are also on Canada's list of species at risk (see Chapter 14), including the beluga, bowhead, northern right, and Georgia Strait killer whales. More than 7 million Canadians live in coastal communities.

The federal government largely holds jurisdiction for the marine environment below the high-water mark. Twenty-seven different federal agencies and departments are involved. The lead agency is the Department of Fisheries and Oceans (DFO), which has traditionally focused its efforts on commercial fishery management but is also responsible for all marine species (except seabirds, which come under Environment Canada). DFO is also the lead agency for Canada's Oceans

Figure 8.11 Canada's coastline and continental shelf.

The population of the southern pods of killer whales has fallen 20 per cent over the last 10 years and is now considered endangered.

The Pacific coastal waters also support large populations of sea lions and seals and a variety of whale species. However, many of these populations have yet to recover from hunting activities. The Pacific (or California) grey whale is the most abundant whale and can be readily seen on its yearly migrations from the calving grounds of Baja California to the feeding areas in the Arctic Ocean as it passes through and sometimes lingers and feeds in BC coastal waters. The Pacific grey has recovered twice from excessive hunting pressure to become one of the most populous whales, with a current population of around 25,000 individuals (Box 8.4). Humpback whales are also sometimes seen, but the best-known sighting is of the killer whales that ply the waters around Vancouver Island, taking advantage of the salmon coming back to spawn. The sea otter was once common on this coast, but was *extirpated* through hunting in the previous century and is only now beginning to recover in numbers as a result of reintroductions (Chapter 3).

However, extirpations are not the only problem. Scientists are finding many unwanted introductions spreading through marine ecosystems. One example in this zone is the green crab from Europe, which has spread rapidly up the coast from California and Oregon into BC. The small crabs are voracious feeders in the inter-tidal zone and can play havoc with shellfish populations, as they have already demonstrated in areas as far apart as Tasmania and the eastern US. Now it seems Canada's turn.

Strategy, discussed later. Provincial governments have responsibility for shorelines, some areas of seabed, and some specific activities, such as aquaculture. Municipal governments also influence the coastal zone, since they have responsibility for many land-based activities affecting the oceans. There are quite a few areas of overlap between these different levels of jurisdiction, and inevitably, conflicts have arisen. As a result, coastal and marine resource management is typically fragmented and often ineffective. In addition, the very size of Canada's coastal zone, ranging from the Mediterranean climate of BC's Gulf Islands through to the High Arctic, means that a wide range of conditions and activities exists. The surface area of Canada's ocean estate is 7.1 million square kilometres. The next section provides an overview of the main marine ecozones in Canada.

Marine Ecozones

Pacific Marine

From the continental shelf, with depths of less than 200 metres, this zone descends to the Abyssal Plain offshore, with depths in excess of 2,500 metres. Sea temperatures and salinities are higher than those in the Arctic Ocean, and this zone acts as a transition area between cold polar waters and the temperate waters of mid-latitudes. The zone is of particular importance for the feeding, staging, and wintering of waterfowl, seabirds, and shorebirds. Pollution is a constant concern. Ten pulp and paper mills discharge effluent into adjacent marine coastal waters, and large areas have been closed to shellfish harvesting as a result. One small oil spill off the Washington coast in 1988 resulted in the death of at least 56,000 birds on the BC coast.

A cormorant killed by an oil spill.

Box 8.4   The Makah and the Whale Hunt

In May 1999, members of the Makah Indian tribe in Neah Bay in the state of Washington returned to the whale-hunting practice of their ancestors, with the help of powerboats and a high-calibre rifle. On that day, a crew of tribal members set out in a traditional dugout canoe and proceeded to fire three harpoons into a migratory grey whale off the Olympic coast of Washington and across the international boundary from Victoria, the site of a thriving whale-watching industry. The wounded mammal then dragged the canoe farther out to sea before a motorized support crew closed in and fired two high-calibre rifle shots to complete the kill. This day marked the first and only legal whale hunt by the Makah in more than 70 years.

The hunting of grey whales in the region had been officially banned since the 1920s when their numbers had plummeted, landing them on the endangered species list. After a strong recovery, the grey whale was removed from the endangered species list in 1994 but only to be met with increasing pressure from the Makah to reinstate the whale hunt, with passionate claims that the hunt is vital to the survival of the Makah culture. Eventually, the US government agreed to the Makah requests and acknowledged their 'right' to hunt, which had been guaranteed in an 1855 treaty. In accordance with the International Whaling Commission, the US government authorized the Makah to resume whaling for 20 days each year and to harvest a maximum of four grey whales.

The decision to resume the hunt was greeted by opposition from environmentalists and animal rights activists. Many believe that the Makah no longer have any real subsistence need and that allowing a renewed hunt will be the first step towards a disintegration of a world ban on whale hunting. Many activists protested the hunt by demonstrating at the reservation boundary and others by actively interfering on the water, scaring off the whales or irritating hunters.

In 2004, a court ruled that the Makah must comply with the process prescribed in the Marine Mammal Protection Act (MMPA) for the take of marine mammals otherwise prohibited by a moratorium. In 2005, the Makah requested a waiver of the MMPA's take moratorium, including the issuance of regulations and any necessary permits, and that request is now under evaluation. Dissatisfied by the slow progress, several Makah tribal members shot and killed a grey whale in September 2007 without any authorization or permit from the National Oceanic and Atmospheric Administration Fisheries Service or any formal tribal authorization. After struggling for some 10 hours, the whale died and sank to the bottom of the Strait of Juan de Fuca.

No reasonable person would deny that the Makah people, or the whale-hunting tribes of Vancouver Island, have endured severe cultural losses or that the grey whale had a significant role in their way of life. The question is whether the act of killing whales is necessary for revitalizing Makah culture and whether this objective outweighs the numerous moral and political ramifications of such action. Will resumption of the hunt spur the efforts of some coastal indigenous people in British Columbia to resume hunting? Will the Makah hunt be used as a wedge to break down international protection against whaling?

One reason that the conflict has had such a high profile is the multi-million-dollar whale-watching industry that has developed on the west coast. The industry helps to introduce thousands of people every year to the oceans, although there are some very real questions as to how much educational information people receive and retain and whether or not the whale-watching experience (and other nature-based tourism experiences) can alter consumer behaviour (e.g., using fewer chemicals at home) enough to assist in ocean conservation (Malcolm 2003). Nonetheless, many people have been whale-watching and do not take kindly to the graphic images of whales being shot, dragged up on a beach, and butchered, as shown on television.

In contrast, Canada has been licensing the Inuit to kill the much more endangered northern bowhead for many years, with relatively little public response. Every year, the International Whaling Commission, of which Canada is no longer a member because of this practice, sends a note of protest to the Canadian government.

These hunting activities also have to be considered against the backdrop of global change and the potential impact it might have on whale populations. The current estimate of the population of grey whales, for example, is about 22,000. However, recent research (Alter, Rynes, and Palumbi 2007) indicates that historically, populations may have been as high as 118,000 before whaling reduced their numbers. Researchers speculate that large-scale ocean changes may be responsible for limiting current populations and that these changes will become increasingly significant as the Arctic oceans warm up.

Whale-watching has become a thriving business on both the Pacific and Atlantic coasts.

The mammoth bones of the California Gray lie bleaching on the shores of those silvery waters, and are scattered along the broken coasts, from Siberia to the Gulf of California; and ere long it may be questioned whether this mammal will not be numbered among the extinct species of the Pacific.

—Captain Charles Scammon, 1874

Arctic Basin

This zone covers the Arctic Ocean basin between Canada and Russia, Greenland, and Norway. Depths are mostly greater than 2,000 metres, and the basin is largely covered in permanent pack ice. Biological productivity and diversity are low. Three species of whale, the bowhead, beluga, and narwhal, are found up to the edge of the pack ice, and the ringed seal, the smallest of all the pinnipeds or fin-feet (seals, sea lions, walruses), is the most abundant marine mammal. The seals have claws on their front flippers to help scrape breathing holes through the ice, and in the winter their bodies are composed of up to 40 per cent blubber to insulate against the cold. Seals are mainly krill-eaters, but they also eat small fish. So well adapted are they to their ice environment that even mating occurs under the ice. In spring, the female finds a crevice through the pack ice and

excavates a den for pupping in the snow above. Polar bears are the seals' main predator.

Perhaps of greatest concern for the future state of the two polar marine systems must be the potential for cumulative impacts from multiple sources, combining global pressures in the form of climatic and ozone changes with local pressures in the form of pollution and unsustainable exploitation of living resources. While species may be able to withstand and adapt to one or even several of these pressures, the cumulative impact of a number of impacts operating synergistically may be too much to bear. Moreover, should severe impacts occur at a critical point in the food chain, dependent and associated species are likely to follow suit. In this sense, the polar marine ecosystems are particularly vulnerable and exposed to risk.

—Clarke and Harris 2003, 21

Arctic Archipelago

This maze of channels and islands extends from Baffin Island and Hudson Bay in the east to the Beaufort Sea in the west. Ice cover is not permanent, and open water exists in most areas for two to three months every year. There are also areas of permanent open water, known as **polynyas**, that are biologically productive, allowing regionally significant concentrations of marine life. Resident birds such as black guillemots, thick-billed murres, ivory gulls, and dovekies congregate in large numbers at these sites in winter to obtain food in the open leads. The polynyas vary greatly in size from 60 to 90 metres in diameter to as large as the North Water polynya between Ellesmere Island and Greenland that may cover as much as 130,000 square kilometres. These areas remain ice-free as a result of various combinations of tides, currents, ocean-bottom upwellings, and winds, all of which promote constant water movement. This long-term experience has given the Inuit some useful insights, for example, into bowhead whale numbers and distributions (Box 8.5).

Perhaps contrary to expectation, Arctic marine mammals live in a less extreme environment than their terrestrial counterparts. Polar waters seldom fall below the freezing point. The temperature gradient between external and internal temperatures that animals living in this environment must cope with is therefore only about half that of land animals. Animals such as whales, seals, and walruses have virtually no hair but instead rely on a thick layer of blubber to protect them from the cold. Their skin is directly exposed to the water, and they therefore need a very finely tuned thermal system to maintain a skin temperature warm enough to resist freezing but cool enough to avoid excessive waste of energy.

Inuit people engage in hunting and fishing activities throughout the zone. Potential also exists for increased exploitation of fossil fuels, which could cause serious resource conflicts between traditional Inuit ways of life and the populations of wildlife upon which they depend and the desire for economic development through exploitation of oil and gas reserves.

Northwest Atlantic

Extending from Hudson Strait and Ungava Bay south to the St Lawrence, this zone mostly comprises continental shelf where

Box 8.5   So How Many Whales Was That? Traditional Ecological Knowledge (TEK) and Science in Canada's Arctic

DFO scientists believed that bowhead whales numbered only in the hundreds and were divided into two separate populations. Since 1996, their figure of 345 bowhead whales was used to determine an Inuit bowhead whale quota in Nunavut of about one every two years. But the scientists' new, much higher bowhead whale estimate—showing a population that could run as high as 43,105—supports an annual hunt of between 18 and 90.

Inuit have said for years that the eastern Arctic's stock of bowhead whales belong to one large and healthy population. The first sign that bowhead whales were more numerous than scientists first thought came from a study of Inuit bowhead knowledge completed in 2000. That study was based on interviews with 252 Inuit hunters and elders in 18 communities. In those interviews, most Inuit informants said they see far more bowhead whales now than in the 1950s. It took more than seven years for the DFO's science to catch up. DFO estimates of the bowhead population jumped from 345 in 2000 to about 3,000 in 2003, then to 7,309 in 2007, and to 14,400 in 2008. The DFO's most recent stock assessment says this latest number is only a 'partial estimate' and that there could actually be as many as 43,105 bowhead whales in the eastern Arctic.

The DFO also concedes that rather than two populations of bowhead whales in the eastern Arctic, there is one. Based on tagging and genetic studies, scientists now say that bowhead whales off Canada and west Greenland share the same summering grounds along east Baffin Island and the Canadian High Arctic and the same wintering grounds in Hudson Strait. The DFO stock assessment says the bowhead whale population, downlisted in 2005 from 'endangered' to 'threatened' under the federal Species at Risk Act, may have completely regained its health. This assessment suggests that based on the most recent numbers, an annual hunt of 18 bowhead whales is realistic and 'conservative'.

However, the message to take from the reassessment is not that one side was right and the other wrong. It is that population estimates in remote locations are very difficult to make and all sources of information should be considered fairly. It is also appropriate to take a precautionary approach, especially when endangered species are being considered.

depths are less than 200 metres, although depths in the Labrador Sea may reach 1,000 metres. In winter, ice formations move along the coast as far as the south coast of Newfoundland. Sea temperatures in this zone may be 20°C higher than in the Arctic because of warm ocean currents. Marine life is generally abundant. A deep trench channels cold oceanic water up the St Lawrence as far as the entrance to the Saguenay Fjord, creating a productive feeding zone for beluga and various baleen whales in the upwelling currents. Some 22 species of whales exist in the zone, as well as six species of seals.

This zone was the site of some of the most controversial international exposure for Canada because of the seal hunt that used to occur annually off these shores. Since the mid-eighteenth century, the harp seal had been the target of hunting, mainly for pelts. Between 1820 and 1860, for example, about half a million harp seals were killed every year. In 1831, more than 300 ships and 10,000 sealers pursued the hunt; 687,000 pelts were taken. Publicity over the hunt in the early 1980s led to bans on the importation of sealskins into Europe. Subsequently, the hunt all but collapsed (Box 8.6).

Atlantic

This zone includes the highly productive waters of the Grand Banks, Scotian Shelf, St Lawrence Trough, and Bay of Fundy, as well as the Northwest Atlantic Basin. Generally ice-free, this

Icebergs, such as this one in Trinity Bay, Newfoundland and Labrador, are characteristic of the Northwest Atlantic.

zone is a mixing ground for warmer waters moving north from the tropics and the cold water of the Labrador Current. Waters are generally shallow, with large areas less than 150 metres in depth. There is an abundant and diverse population of marine mammals, such as grey, hooded, and harp seals and Atlantic pilot, killer, and northern bottlenose whales. The fisheries off these coasts were at one time among the most abundant on Earth, and a large population of more than 300

Box 8.6   The Seal Hunt

The annual hunt for seals off the east coast has probably been Canada's biggest foray on the international stage regarding environmental matters. Celebrities such as French actress Brigitte Bardot appeared on television across the world as they tried to protect seemingly helpless white-coated seal pups from being clubbed to death. Eventually, following bans on the import of sealskins by the US and the European Union, the Canadian government banned the hunt in 1987, and the number of seals is estimated to have tripled. Now, however, a new hunt has begun, ostensibly to help in the recovery of the endangered cod stocks. International environmental groups have suggested that the real motive for reinstituting the seal hunt is not related to seal predation on cod but rather to the need for a political scapegoat in economically depressed areas and the demand for seal penises on the Asian market.

Seals eat fish. On that everyone agrees. What is contentious, however, is the numbers of seals and how many fish of various species they are eating. Federal scientists estimate a harp seal population of about 5.5 million consumes more than 140,000 tonnes of cod per year, enough to put a considerable dent in the recovery of the fish stock. Widespread disagreement exists concerning all of these figures, however. Some estimates of seal numbers are as low as two million. And federal researchers have indicated that virtually all the cod eaten by harp seals are the smaller Arctic cod, which are not commercially fished.

In 2006, Fisheries and Oceans Canada (DFO) introduced a five-year management plan (2006–10). The total allowable catch of seals is set on an annual basis to allow for adjustments to changing environmental conditions and changes in harvest levels in Arctic Canada and

Greenland. A one-year total allowable catch (TAC) of 270,000 harp seals was set for 2007. This included 2,000 seals for the personal-use hunt, 4,860 for Aboriginal initiatives, and a carry-forward of 19,000 for Newfoundland and Labrador Front fleets. An increase of 5,000 seals was allowed for 2008.

While markets for seal pelts are subject to significant variation from one year to the next, the 2006 seal hunt was one of the most profitable in memory. The landed value of the harp seal hunt was $33 million. The average price per pelt received by sealers was $97, an increase of 77 per cent over the 2005 average value of $55. Income from sealing may account for 25 to 30 per cent of total annual income for some families, and many of their communities have unemployment rates more than 30 per cent higher than the national average.

A beater, like this harp seal pup, is one that has moulted its white fur.

different seabird species was also supported by the marine bounty. These species include the northern gannet, which almost followed the great auk into extinction (see Chapter 14) as a result of hunting before protective laws came into effect in the early part of the twentieth century. Other species include the Atlantic puffin, murres, great and double-crested cormorants, guillemots, great black-backed gulls, and many others. The region also contains coral reefs. Scientists on board a DFO research ship in 2003 discovered the reef between Cape Breton Island and Newfoundland. Unfortunately, the reef, made up mainly of *Lophelia pertusa*, had already been badly damaged by fishing activities.

Cold-water corals are found in waters ranging from 3 to 14°C and have now been recorded in 41 countries worldwide (Freiwald et al. 2004). They occur where cold, clear, nutrient-rich waters are present. Living mostly in perpetual darkness, cold-water corals do not possess symbiotic, single-celled algae like warm-water corals (Box 8.1) and rely solely on zooplankton and detritus for sustenance. Some species, such as *Lophelia*, can form large, complex, three-dimensional reef structures several metres high. Other soft corals living in colder waters, such as *Gorgonia*, do not form reefs but large 'gardens' covering vast areas—for example, around the Aleutian island chain in the North Pacific. The ecological functions of such reefs and gardens in the deeper waters are very similar to those of tropical reefs: they are biodiversity hotspots and home, feeding, and nursery grounds for a vast number of other organisms, including commercial fish and shellfish species. It is predicted that the cold-water corals will be the first to be destroyed by increased oceanic acidification, as discussed in Box 8.1.

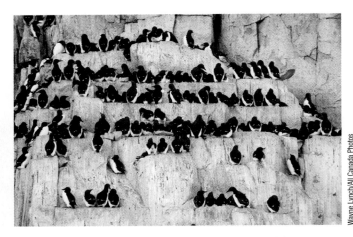

Thick-billed murres nest along sea cliffs.

Northern gannets range over the North Atlantic and are often seen well offshore.

The tropical connection was confirmed by the discovery that the region was a critical habitat for endangered leatherback turtles. Although we tend to think more of tropical beaches when thinking of turtles, the leatherback, the world's largest turtle, forages for much of its life in food-rich temperate waters. However, what little is known of their distribution in these waters has usually come from their entanglement in fishing gear rather than from surveys of distributions. More comprehensive surveys involving volunteer reporting, telemetry, and aerial surveys suggest high numbers of leatherbacks in eastern Canadian waters and particularly on the Scotian Shelf. Food resources are plentiful, and the orientation of the coast seems to funnel the species onto the shelf.

Major gas and oil discoveries have been made throughout the zone. The Hibernia field on the eastern edge of the Grand Banks, some 300 kilometres east-southeast of St John's, was the first field put into production. In 2002, it was joined by the Terra Nova field, with a third, White Rose, coming into production in 2005.

The White Rose oilfield provided an additional 30 million barrels per year and increased total production to more than 130 million barrels from 112 million in 2005. Chapter 12 provides more detail about these oil and gas operations.

Fisheries

Canada has been blessed with some of the most productive marine environments in the world, and these environments were critical in sustaining populations of Aboriginal peoples and attracting European attention on both coasts. Unfortunately, the squandering of this rich biological heritage stands as one of the sharpest reminders of our inability to manage ourselves in a way that sustains resources over a long period of time. Since 1992—the year of the cod fishery closure—commercial landings have declined by 19 per cent overall, including cod by 82 per cent, haddock by 32 per cent, turbot by 35 per cent, redfish and rockfish by 69 per cent, and hake by 43 per cent. Bear in mind, too, that these percentages reflect a shifting baseline, as discussed earlier. The number of vessels in Canada's commercial marine fishing fleets has decreased by 31 per cent, but the average income earned within fleet sectors has doubled. In Atlantic Canada, for instance, the average vessel earned $18,400 in 2002, more than double the average income of $8,700 in 1992. Overall, the catch has doubled in value over the past decade. This is mainly due to the growth in shellfish catches, up by 77 per cent, which have been spurred by the closure of the Atlantic cod fishery. Similarly, in freshwater fisheries, although catch levels continue to decline, the value of the catch has increased because of higher prices.

The Fisheries Act is federal legislation dating back to Confederation and was established to manage and protect Canada's fisheries resources. It applies to all fishing zones, territorial seas, and inland waters of Canada and is binding on federal, provincial, and territorial governments. As federal legislation, the Fisheries Act overrides provincial legislation when the two

conflict. Consequently, approval under provincial legislation may not necessarily mean approval under the Fisheries Act. However, the Act is badly outdated, and a new Act has been formulated, although it has yet to be approved by Parliament. The new Act strengthens measures for conservation, provides more opportunity for input by fishers but also more responsibility, affords greater long-term security for fishers, and creates an administrative sanctions tribunal, thereby reducing reliance on the criminal courts. Importantly, it also reduces the discretion of the minister in several important matters that led to many of the problems described below in the collapse of the east coast fishery.

Case Study: East Coast Fisheries

The marine fishery has been an essential component of the economy and culture of Atlantic Canada for centuries. After 1977, when Canada declared a 200-nautical-mile (370-kilometre) fishing limit off its coasts, cod were the mainstay for more than 50,000 fishers and 60,000 fish-plant workers in Atlantic Canada. In Newfoundland and Labrador alone, about 700 communities depended entirely on the cod fishery, which had a 1991 value to fishers of more than $226 million.

However, between July 1992 and December 1993, decisions to cut back on the rate of harvesting resulted in employment losses for 40,000 to 50,000 people in Newfoundland, the Maritime provinces, and Quebec. How could this dramatic collapse of a renewable resource occur in such a relatively short period of time? Why did fisheries scientists fail to anticipate the collapse? Why was action not taken earlier to avoid degradation of the fishery and economic disruption? Is it possible for the fishery to rebound and become a mainstay of the regional economy once again?

The Nature of the Collapse

There are four main areas for this fishery in Atlantic Canada: the Scotian Shelf, the Gulf of St Lawrence, the Grand Banks, and the Labrador coast (Figure 8.12). The two areas most affected by the harvesting cutbacks have been the Scotian Shelf, which extends from the mouth of the Bay of Fundy to the northern tip of Cape Breton Island, and the Labrador coast.

These areas supported two different kinds of fishery. The fishery on the Scotian Shelf is readily accessible to the inshore fishers along the coast of Nova Scotia and New Brunswick and included a wide mix of species, including cod, haddock, flounder, pollock, hake, herring, redfish, crab, scallop, and lobster. In contrast, the Labrador coast fishery was dominated by the northern cod stock, extending east of the Labrador coast and north and east of Newfoundland. The northern cod traditionally yielded about half of Atlantic Canada's cod catch and one-quarter of all groundfish landings in the region. The northern cod has formed the backbone of the Atlantic fishery. This explains why stock depletion has been such a blow to regional economies, where fishing has provided a significant percentage of the jobs—and almost all jobs in small outport communities.

The northern cod were caught by larger inshore vessels and especially by offshore draggers or factory trawlers,

Figure 8.12 Major fishing areas in Atlantic Canada. Source: Cameron 1990, 30.

Small outport communities in Newfoundland have always relied heavily on harvesting marine products, from seals to fish.

multi-million-dollar boats that drag huge nets across the bottom of the ocean and stay on the fishing grounds for extended periods without landing. However, the northern cod migrate to the shores of Newfoundland in the summer and thus also supported an inshore fishery that relied on much smaller boats using traps, hooks, and nets. The inshore fishery has been an important one. Until the late 1950s, the inshore catch was typically higher than 150,000 tonnes. By 1974, as a result of overfishing by boats from more than 20 nations, the inshore catch had fallen to 35,000 tonnes. After Canada declared its exclusive fishing zone in 1977 and banned fishing by foreign draggers in that area, the inshore catch increased. It peaked at 115,000 tonnes in 1982, but by

1986 the catch had fallen to 68,000 tonnes, and inshore fishers were complaining that the fish caught were very small. Local fishers had identified the first signs of serious problems. Unfortunately, the models used by the fishery scientists indicated that stocks were still abundant, so these early warnings were not heeded. Total catches of northern cod increased, reaching 252,000 tonnes in 1986, almost twice what they were in 1978. However, by 1989, on the basis of new scientific advice, the minister reduced the **total allowable catch** (TAC) for northern cod to 235,000 tonnes. By 1991, it was clear that the stock was in trouble, the TAC was reduced to 120,000 tonnes, and in July 1992 a moratorium on northern cod until May 1994 was announced. Ottawa agreed to provide $500 million (later rising to $912 million) to compensate the 20,000 fishers and plant workers expected to lose their jobs (Figure 8.13).

In 1993, the government banned cod fishing in five more areas and sharply reduced quotas for other valuable species. The result was a total loss of about 35,000 to 40,000 fisher and fish-plant jobs in Atlantic Canada since the closures began in July 1992. In 1994, the federal government announced that Atlantic cod fishers would receive $1.7 billion over five years. In Atlantic Canada, there were 17,200 groundfish licence holders in 1992 and 10,783 in 2000, representing a decline of roughly 35 per cent. In total, the government spent $3.9 billion for income support, industry adjustment measures, and economic development assistance programs for the Atlantic fishing industry between 1992 and 2001.

In 2003, almost 11 years after the moratorium on cod fishing, Fisheries Minister Robert Thibault announced the closure of what remained of the cod fishery in Newfoundland, the Maritime provinces, and Quebec. Atlantic cod was officially listed as endangered as COSEWIC (see Chapter 14) estimated a 97 per cent decline in cod off the northeast coast of Newfoundland and Labrador over the previous 30 years. In 2007, the Atlantic Council of Fisheries and Aquaculture Ministers (ACFAM)

established a federal-provincial-territorial working group to explore opportunities for strategic marketing of seafood products, analyze international market trends, improve coordination between marketing efforts and harvesters, and gain access to additional funding for marketing from other sources.

Some Reasons for the Collapse

At the time of the collapse, a single cause for the depletion of the groundfish stocks seemed unlikely. It was suggested that changing environmental conditions, creating colder, less hospitable water temperatures, had driven the cod away, while growing seal populations had devoured entire stocks of both cod and capelin, the favourite food source for cod. However, ongoing research shows that environmental factors played only minor roles in the disappearance of the fish. With study after study, it has become increasingly likely that 'the northern cod stock had been overfished to commercial extinction' (Steele, Anderson, and Green 1992, 37). The politicians and bureaucrats running Canada's Atlantic fisheries created opportunities for overfishing through the provision of inappropriate incentives for processing plants and lucrative subsidies (unemployment insurance) to all fishers and plant workers involved in the fishery.

Foreign Overfishing

Once Canada established the 200-mile fishing limit in 1977, foreign fleets were required to fish outside that boundary or to fish inside the boundary for that portion of the domestic quota not taken by Canadian vessels. Foreign fishing fleets were monitored by the Northwest Atlantic Fisheries Organization (NAFO), and during the 1970s and early 1980s, foreign fleets for the most part followed the rules for harvesting. However, in 1986, Spain and Portugal entered the European Community (EC), and that year the EC unilaterally established quotas considerably higher than those set by NAFO. Furthermore,

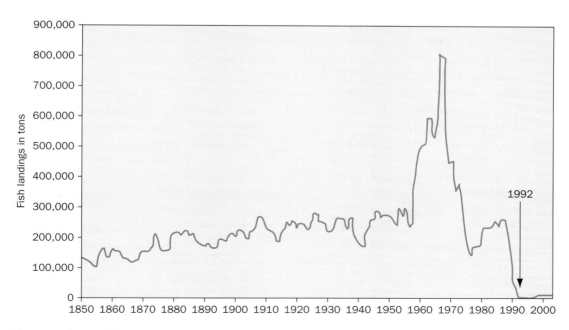

Figure 8.13 Collapse of Atlantic cod stocks off the east coast of Newfoundland, 1992. Source: Millennium Ecosystem Assessment 2005.

the EC boats harvested fish well beyond the EC limits. The EC then raised the quota the following year, and again the NAFO quota was exceeded by the actual catch. In 1988, just half of the EC target was achieved, even though it was 4.5 times higher than the target recommended by NAFO. The EC, now known as the European Union (EU), later rejected a NAFO northern cod moratorium. In 1993, however, the EU finally accepted all NAFO quotas, after having set its own quotas at a much higher level since the mid-1980s.

Thus, strong evidence exists that foreign vessels, especially those from Spain and Portugal, were overfishing at least during the mid- and late 1980s. Since cod migrate towards the coast in summer and then move offshore in winter to spawn in deeper waters, the fish are more vulnerable to foreign fishing.

Domestic Overfishing

Despite the pressure placed on the stocks by foreign fishing vessels, most of the principal fishing grounds have been under Canadian control since the 200-mile limit was set in 1977. Two fisheries—both inshore and offshore—must be managed, which has been and continues to be a challenge.

For hundreds of years, the inshore fishery consisted of many fishers (particularly from Newfoundland) relying on small wooden boats, lines, traps, and nets to catch cod during the spring and summer months when the cod move close to shore. Until the mid-1950s, the inshore fishery, combined with limited offshore fishing by Canadian boats, resulted in annual landings of 200,000 tonnes or more. Foreign fishers were harvesting another 30,000 to 50,000 tonnes each year. Such harvesting did not appear to adversely affect the then estimated breeding stock of 1.6 million tonnes in the North Atlantic.

In the mid-1950s, the introduction of large offshore trawlers that operated year-round in the North Atlantic was a significant change to this pattern. Initially, catches were very high, but the spawning stocks were placed under great pressure because cod require seven years to reach maturity. For example, in the 1970s, yields reached a high of 800,000 tonnes per year before they started to drop. Until 1977, foreign trawlers did most of the offshore fishing. Following the establishment of the 200-mile limit, the Canadian offshore fleet expanded, and Canadian-based offshore trawlers became the main harvesters of northern cod. By the time the moratorium was placed on the fishery in the summer of 1992, Newfoundland was the base for some 55 large and 30 medium-sized offshore trawlers. Thus, Canadian offshore draggers, operating on a year-round basis, placed considerable pressure on groundfish stocks.

Critical in this regard are the ecology and behaviour of the northern cod. Initially, the harvesters caught a mix of ages and sizes of fish. However, market demand and net mesh sizes led to a focus on larger, older fish. Cod swim in groups or schools of similar ages primarily because the larger cod will eat the smaller and younger cod. The emphasis on larger fish had two consequences. First, the northern cod normally do not begin to spawn until they reach seven years of age. Second, older

fish produce more eggs. As the larger fish became scarce, the fishery then concentrated on fish in the five-to-seven-year age range. The result was that by the early 1990s, most of the older fish had been overharvested and attention had shifted to pre-adolescent fish, which were being caught before they were old enough to spawn. The consequence was a dramatic decline in the fish stock.

A further complication was that many domestic fishers overharvested. Some estimates suggest that up to 50 per cent more fish were being landed than were being reported (Cameron 1990). Often, this cheating was driven by a need to catch enough fish to pay for the costs of operating increasingly sophisticated vessels.

Imperfect Science and Management

Fishery scientists did not anticipate the collapse of the Atlantic fishery, especially the northern cod stocks. One reason was that sampling procedures do not provide sufficient ecological information about fish stocks. Until 1991, the total allowable catch was based on the assumption of a biomass of 1.1 million tonnes of cod. However, in 1991, the sampling from DFO research ships indicated only 600,000 tonnes. Sampling in distant areas did not reveal that the cod had migrated to other areas. Significant numbers of diseased or dead fish had not been found. The scientists simply did not know what had happened. They had been receiving warnings for a number of years from the inshore fishers that the fish being caught were fewer in number, smaller in size, and lower in weight. However, the scientists, who put much greater credence in the quantitative data gathered by DFO research vessels and from what were likely under-reported offshore landings, considered the observations of inshore fishers as anecdotal and less credible.

Inappropriate Incentives for Processing Plants and Fish Workers

By the early 1990s, Newfoundland had about 100 large and small fish-processing plants, some two-thirds of which processed northern cod. In Atlantic Canada as a whole, the number of plants increased from about 500 in 1977 to nearly 900 in 1988, and employment grew from approximately 25,000 full-time jobs to about 33,000. The provincial governments, which license onshore fish-processing plants, provided incentives for the development of new processing plants as a way of creating new jobs in small communities. This placed political pressure on DFO to keep increasing the total allowable catch.

Another incentive for people to enter or stay in the fishing industry was the federal unemployment insurance program. After working for 10 weeks, fish-plant workers were permitted to collect unemployment insurance for the other 42 weeks of the year. This arrangement resulted in several individuals in a community sharing one job but all qualifying for separate benefits. For fishers, the unemployment benefits were based on the sale value of fish caught during the May–November season (26 weeks), which created the potential for receiving unemployment benefits for the other 26 weeks of the year.

The state of our ignorance is appalling. We know almost nothing of value with respect to behaviour of fish. We don't even know if there's one northern cod stock, or many, or how they might be distinguished. We don't know anything about migration patterns or their causes, or feeding habits, or relationships in the food chain. I could go on listing what we don't know. . . . Our technology has outstripped our science. We have underestimated our own capacity to find, to pursue, and to kill.
—Leslie Harris, chairman of the Northern Cod Review Panel, 1990, quoted in Cameron 1990, 35, 29

This program, which was intended to provide a social safety net, encouraged more people to become involved in the Atlantic fishery than could be justified economically. There was little incentive to consider other types of work, and the program also helped to reinforce an outlook in which little value was placed on education. By the time the fishery was closed, 50 per cent of Newfoundland's 19-year-olds were already on unemployment insurance, and 80 per cent of the fishery workers did not have a high school diploma. Thus, the fishery involved more people than realistically could be supported over

the long term, yet unemployment insurance programs provided little incentive for individuals to consider alternatives. This made the trauma of the 1992 moratorium even greater than it might otherwise have been. In 2008, the World Trade Organization started to examine whether unemployment insurance constituted an unfair subsidy to fishers.

Changing Environmental Conditions

One theory for the depletion of groundfish in the North Atlantic fishery is based on the idea of environmental change. Records show that in 1991, the ocean temperatures off Newfoundland were the coldest ever measured. It has furthermore been suggested that there had been several years of very cold water temperatures before that. The water warmed slightly in 1992, then cooled again in 1993. It is possible that colder waters, combined with overfishing in the previous two decades, have been preventing or inhibiting the shrunken stocks from regenerating. However, since relatively little is known about the migratory patterns of the northern cod, it is difficult to determine what the specific implications of changing water temperatures might have been. Furthermore, water temperatures began to heat up and stabilize around 1998, restoring a theoretically favourable habitat for groundfish. But despite this warming trend, no significant recovery has been observed within the affected groundfish stocks.

Predators

At the time of the fisheries collapse, it was also popular to blame predation. Seals in particular were identified because of their 'voracious appetites' and their growing numbers because of the closure of the seal hunt in the early 1980s (Box 8.6). There is no scientific evidence to support this view.

However, there is some evidence to suggest that seal predation may be a factor in the slow recovery of the east coast groundfish stocks. The seal population has more than doubled in the past three decades, and although cod represents only a small percentage of seals' diets, they are consuming more northern cod than fishers are catching. In fact, the total allowable catch for northern cod in 1999 was only 9,000 tonnes, or less than 20 per cent of corresponding predation by harp seals in that year (Doubleday 2000)

The science of seals and cod is inconclusive and will remain inconclusive for the foreseeable future. Seals are a significant source of mortality for northern cod in the northwest Atlantic and the Gulf of St Lawrence. Nevertheless, reducing the abundance of harp seals may or may not lead to more rapid recovery of depleted cod stocks. Even if a reduced seal population resulted in an increased number of fish in the ocean, there are other predators in marine ecosystems. Any increase in the size

It is difficult for urban dwellers to imagine the close relationship that built up over the centuries between the people in the outports of Newfoundland and the sea. Virtually every family would be involved in some way with fishing. When the fish were exposed to such fishing pressure that they could no longer be caught in any numbers, it was not just the economy that suffered but a whole way of life.

We reject hypotheses that attribute the collapse of the northern cod to environmental change. . . . We conclude that the collapse of the northern cod can be attributed solely to overexploitation [by humans]. . . .

— Hutchings and Myers 1994

of a commercially important fish stock could well be eaten by these other predators before being caught by fishers. Furthermore, seals eat predators of commercially important fish, and so fewer seals could actually mean fewer fish for fishers.

Fishing down the Web

Fishing down the food chain has been discussed earlier and is exemplified in Atlantic Canada with the replacement of cod fisheries by shrimp and crab fisheries. Recent observations of the poor physiological condition of many predator fish in the area also suggest an overall lack of prey, forcing predator species such as the Atlantic cod to the same alternative as human fishers, 'fishing down the web'. After the cod were fished down, fishing pressure increased on shrimp. 'Cod feed on shrimp. If you remove the shrimp, how will the cod ever recover?' asks Daniel Pauly of the British Columbia Fisheries Centre.

Lessons

The collapse of the Atlantic groundfish fishery highlights how some contemporary resource management practices may

The disaster in the cod fishery is now worse than anyone expected. . . . It may be a generation before we see a recovery of the cod. That a five-hundred-year-old industry could be destroyed in fifteen years by a bureaucracy is a tragedy of epic proportions.

— Ransom Myers, quoted in Harris 1998, 332–3

encourage resource liquidation. In theory, fisheries managers around the world and in Canada seek to manage the fishery according to three priority areas: ecological sustainability, economic goals, and social outcomes. In reality, these goals are often in conflict, at least in the short term, and the second goal often becomes paramount because of political interference. The case study illustrates that fisheries management requires scientific understanding of the biophysical resource system, a greater appreciation of traditional or local ecological knowledge, and parallel understandings of the history, culture, economy, and politics of the region, as well as federal and provincial fisheries and regional development policies. The Atlantic fishery also provides an excellent example of how inexact science often is and the extent to which complexity and uncertainty are paramount. It demonstrates conflict among different values and interests (Chapter 5) and the manner in which conditions can change dramatically over a relatively short time period.

The situation is readily comparable with the framework introduced in Chapter 1. Not only were our simplified models of the complex biophysical system inadequate for a proper understanding of the east coast fisheries, but our resulting attempts to assess the status of the system were inadequate and there was a lack of clarity regarding societal expectations and management directions. Davis and Wagner (2006) have provided a useful analysis of some of the social dimensions of fisheries regulation in the Maritimes. Some of the approaches suggested in Chapter 6, relating to identification of stakeholders, resolving conflict, and taking more ecosystem-based and adaptive approaches, could usefully be applied to the resolution of these problems.

However, one of the chief proponents of adaptive management, Dr Carl Walters of Simon Fraser University, warns in a recent paper (2007) that adaptive management (Chapter 6) that was designed to address some of these problems has not been applied successfully in the vast majority of cases. He traces the failures to three institutional problems: first, a lack of management resources for the expanded monitoring needed to carry out large-scale experiments; second, the unwillingness of decision-makers to admit and embrace uncertainty in making policy choices; and third, lack of leadership in the form of individuals willing to do all the hard work needed to plan and implement new and complex management programs.

Aboriginal Use of Marine Resources

One of the most challenging aspects of fisheries management in Canada is allocation of catch, especially when catches are declining. This is particularly difficult when allocation involves Aboriginal communities. There is a patchwork of treaties with First Nations in different regions of Canada, and the rights to sustenance from fishing were often written into these treaties. However, it has never been clear to which regulations

First Nations should be subject and how broad a range sustenance might cover. Over the past decade, many important court cases have helped to clarify some of these issues. Nonetheless, high-profile conflicts still occur on both coasts, as detailed in the two examples below.

In the fall of 1999, a violent and complex dispute erupted between Native and non-Native fishers in Miramichi Bay in northeastern New Brunswick. Lobster traps were cut and damaged, threats were exchanged, boats were rammed, and multiple shots were fired.

The crux of the dispute lay in the Supreme Court's 1999 decision in the case of a Nova Scotia Mi'kmaq, Donald Marshall Jr, who had caught and sold eels out of season and claimed protection under a 1760 treaty. The Court's ruling upheld the 1760 treaty, which effectively gave Mi'kmaq, Maliseet, and Passamaquoddy bands the right to earn a 'moderate livelihood' from year-round fishing, hunting, and gathering. As news of this decision spread, it spurred an immediate reaction among the Natives of the Burnt Church band of Miramichi Bay to resume catching and selling species like lobster, even though the season was formally over. Within days, the local Native bands had more than 4,000 traps in the water. This infuriated the non-Native fishers, who were confined to the short summer season and who believed that this continued harvest would lead to the destruction of the lobster fishery.

Despite the rising tension and conflict between the two groups, the Department of Fisheries and Oceans was hesitant to intervene, stating that Natives now had a right to fish that had been denied for more than two centuries. However, DFO did promise a peaceful solution to resolve the ambiguity of the *Marshall* decision. The *Marshall* case specifically indicated that the federal government still retained the right to regulate the fishery but was equally firm that Ottawa's authority to regulate treaty rights was limited to those actions that could be 'justified'. This led to an examination of the term 'justification' and, more specifically, of whether or not the DFO's limit on the number of lobster traps and length of fishing season was reasonable and 'justifiable' according to the rights of the Mi'kmaq in their 1760 treaty.

After several years of research into the affected lobster stocks, commissioned in order to 'justify' the decision scientifically, the minister of fisheries and oceans announced agreement with the Burnt Church band. The $20-million agreement included enhanced commercial fishery access for Native fishers, including additional lobster licences and extra boats and gear. However, a quota was set for the Native fall fishery of 25,000 pounds of lobster for food and 5,000 pounds of lobster for ceremonial use. Furthermore, the fishery would be limited to six weeks or until the quota was filled, and the sale of lobster would be strictly prohibited at all times.

Since the federal decision, there has been a significant decrease in violent conflicts on the water, and an improved relationship has emerged among all parties. The non-Native lobster fishers and many fisheries scientists still maintain that any kind of fall fishery needs to end because, they believe, it exploits the lobster at their most vulnerable stage. Yet they recognize that while the negotiated agreement does not represent a definitive solution to the conflict, it is a vital step towards the government providing a more appropriate degree of oversight and accountability in the management of the lobster fishery.

The Atlantic Integrated Commercial Fisheries Initiative (AICFI) was announced in 2007. This program aims to ensure that Mi'kmaq and Maliseet First Nations in the Maritimes and Gaspé region of Quebec (MMFNs) have the capability to manage and maximize access to the integrated commercial fishery that has been provided through the Marshall Response Initiative. The new $20-million, two-year program assists MMFNs to continue building their capacity, to plan and manage successful commercial fishing enterprises, and to participate in fisheries co-management.

On the west coast, Native communities have unemployment rates of between 65 and 85 per cent, and the fishery plays a critical role. There is an Aboriginal right, established by the courts, to fish 24 hours a day, seven days a week, wherever Aboriginal people wish to, for food and ceremonial purposes. In the past, this right was sometimes abused, with fish being caught for commercial use. This prompted DFO to establish a special Natives-only commercial salmon fishery for some areas. In 2003, the BC Appeal Court, following complaints by non-Native fishers, struck down the Natives-only commercial salmon fishery. The judge said that it amounted to 'legislated racial discrimination' and was against the Charter of Rights. DFO then cancelled the program, but Native fishers vowed to continue catching and selling the salmon as they always have. This decision was the second court setback for the Natives-only fishery. A few months earlier, another judge had given absolute

Native fishers use dip nets to intercept salmon heading up the Fraser River.

Philip Dearden

Box 8.7   Salmon: The Stories They Tell

From the shores of Japan to almost 2,500 kilometres up the Yukon River, a tangible thread exists—the Pacific salmon. Every year, millions of salmon make their way back from the other side of the Pacific Ocean to the streams of their birth. The five species of Pacific salmon—chum, coho, chinook, pink, and sockeye—are **anadromous**—that is, they spend part of their lives in fresh water and part in salt water. They depend on a wide range of conditions that link the mountains to the seas: the amount of snowpack to feed the streams, the lack of floods to wash away spawning gravel, unpolluted rivers and estuaries, the right temperature for entry into the marine environment, avoidance of predators, and avoidance of fishing nets. If any of these myriad factors go awry, then higher mortality rates can drastically reduce the numbers of fish returning to spawn in subsequent years. These factors are the links in a chain reflecting the limiting factor discussed in Chapter 2. It also means that salmon are good indicators of the overall health of our environment and our resource management practices.

What have these indicator species been telling us? The story is not a good one. Salmon in their millions sustained populations of coastal Aboriginal peoples. Early descriptions of the Fraser River by explorers talk about a river that could be crossed on the backs of the salmon. But early logging and mining practices, along with wasteful fishing practices, soon made a considerable dent in these numbers by negatively affecting spawning grounds. Habitat destruction and overfishing led to the virtual closing of the fishery in many areas by the 1990s. Scientists estimate that the salmon biomass has been diminished by half from pre-commercial fishing levels. Some stocks have been declared extinct, while others are now on the official endangered species list in both Canada and the US. The weight of salmon landings fell by more than 50 per cent between 1997 and 2006. In 2008, the Pacific Salmon Commission proposed cutting the chinook harvest by a third until 2018 in an effort to rebuild stocks.

The salmon have also been trying to tell us something that scientists are only now starting to realize. Salmon spend anywhere from two to seven years in the ocean environment before returning to spawn and die. When they die, the nutrients they have collected over this sojourn do not disappear (law of conservation of matter) but are released into the surrounding environment. The salmon provide food and nourishment not only for the plankton and insects that feed the next generation of fish and propel their journey to the sea but also for the terrestrial riverine environment. When the fish die, they provide a feast for many other species, including eagles, raccoons, and bears. As the fish are digested, their nutrients are distributed throughout the forest as feces, providing the rich fertilizer on which some of the tallest trees in the world depend. Tom Reimchen of the University of Victoria estimates that BC's bears could be transferring 60 million kilograms of salmon tissue into coastal forests each year. When we take away the fish, we take away this fertilizer, and forest growth suffers. Scientists have found that up to 40 per cent of the nitrates in the old-growth forests in coastal BC originate from marine environments. Rivers with barriers to salmon, such as waterfalls, have noticeably poorer forest growth.

Scientists have now discovered another aspect of this remarkable linkage. Not only do salmon collect nutrients from the ocean, they also collect pollutants and

A grizzly bear track. Research now shows that the nutrients bears carry back from the ocean and the rivers are key in promoting rich forest growth along many coastal streams.

concentrate them within their bodies (Chapter 10). Scientists in Alaska have found that when the fish die in their millions after spawning, there is a sevenfold increase in the concentration of PCBs in these remote, pristine, freshwater lakes. Lakes with the highest numbers of spawning salmon also have the highest concentrations of PCBs.

However the biggest challenge may be the one created by global warming and the changes it will bring to all aspects of the salmon habitat. From the amount of snowpack in the mountains that controls the water in the rivers, through river temperatures, to changes in oceanic currents and predator-prey relationships, the salmon will be very vulnerable. Already, some species are at the edge of their temperature tolerance range (Chapter 2) for fresh water, and the most productive salmon river in the world, the Fraser, may soon be too warm for them. Over the past 50 years, sea-surface temperatures have risen between 0.3°C and 0.9°C on the BC coast. In response, the annual cycles of phytoplankton bloom that nourish the entire food chain are occurring earlier, raising the possibility of a growing mismatch between food supplies and emergence of salmon smolts into the ocean.

The salmon have an eloquent and tragic story to tell about how we are treating their environment, and they will be one of the best indicators of the impact of global change on coastal and marine environments.

discharges to 40 non-Native fishers who had cast their nets during a Natives-only fishery on another part of Vancouver Island. However, many argue with this stance, pointing out that First Nations are being allocated fishing rights not because of their race but because their fisheries were wrongfully appropriated in the first place. In July 2008, a Supreme Court decision ruled that such retribution was acceptable. Others have argued that non-Native fishers should not bear the costs for wrongs perpetrated in the past by society as a whole. What do you think?

In many ways, the story of the BC coastal Aboriginal cultures is a story of the sea in general and of salmon in particular (Box 8.7). Nowhere else on Earth did hunter-gatherer societies develop such complex social structures, rigid hierarchies, and dense populations in permanent winter villages. From these villages, the people developed complex and effective hunting practices for whales, sea lions, seals, sharks, tuna, wolf eels, sole, oolichan, greenlings, herring, halibut, crabs, clams, mussels, skate, sturgeon, and above all, salmon. The salmon fishery was managed effectively; no stocks crashed. And the salmon was venerated through myth and legend among the coastal peoples.

Examples such as the two discussed above will continue to arise as Canada strives to achieve equitable solutions to fish resource allocation problems involving First Nations peoples. The clock cannot be rolled back to pre-treaty times, yet there must be some recognition of the central life that fish and fishing have played in the societies of many First Nations people in Canada. Roughly three out of four British Columbians told a federal government–commissioned pollster that they oppose special commercial fishing rights for Aboriginal Canadians, and Prime Minister Stephen Harper stunned Aboriginal groups, federal bureaucrats, and the BC government by declaring that his government is opposed to 'racially divided' fisheries. Yet the federal government has since gone ahead with west coast treaties that include exclusive rights to the commercial fishery and have developed a BC First Nations Fisheries Action Plan that will allow greater involvement of First Nations in commercial fisheries.

Pollution

In 1999, the Canadian Environmental Protection Act (CEPA) set a goal for the government of Canada to categorize all 23,000 chemical substances approved for use in Canada. This task was completed by September 2006, and substances were categorized to identify those that were:

- inherently toxic to humans or to the environment and that might be
  - persistent (take a very long time to break down) and/or
  - bioaccumulative (collect in living organisms and end up in the food chain);
- substances that had the greatest potential for exposure to humans.

Approximately 4,000 chemical substances were identified as needing further attention.

The main sources of marine toxic pollution in Canada originate with the deposition of airborne pollutants from fossil-fuel combustion, agricultural runoff, inadequately treated sewage, and by-products or waste materials from refining processes (e.g., effluent from pulp and paper mills). Some chemicals, known as POPs (persistent organic pollutants), including PCBs (polychlorinated biphenyls) and DDE (the breakdown product of the now-banned pesticide DDT, discussed in Chapter 10), can take decades or even centuries to degrade and tend to bioaccumulate in the fatty tissues of organisms over time. The concentrated contaminants are then passed along through the food chain (biomagnification—see Chapter 10) and can reach very high concentrations in the tissues of animals in the top trophic levels (such as polar bears, whales, and humans).

A study by the Department of Fisheries and Oceans concluded that the killer whales of the Strait of Georgia are among the most contaminated mammals on the planet (Ross et al. 2000). Although PCBs have been the main concern, levels of a toxic flame retardant (PBDE) are growing so quickly that they are expected to surpass PCBs as the leading contaminant in endangered southern resident killer whales. At the current rate of increase, PBDE levels are predicted to exceed those of PCBs in killer whales by 2020. Unlike PCBs, which were largely used as coolants in industrial transformers before being banned 30 years ago, PBDEs are widely used as flame retardants in polymer resins and plastics and found in consumer products such as furniture, TVs, stereos, computers, carpets, and curtains. PBDEs find their way into the marine environment through the air or through runoff and effluent and pose a risk to the endocrine system, reproductive health, and the immune system. PBDE levels in harbour seals in Puget Sound in the state of Washington have increased steadily: 14 parts per billion in 1984, 281 in 1990, 328 in 1993, 644 in 1996, and 1,057 in 2003. Killer whales carry 10 times the contaminants of harbour seals, which means an increase in PBDEs in seals is an immediate cause for concern about the whales.

The dangers of pollutants are further exacerbated by the long-range polar transport of toxins in the atmosphere, and many Aboriginal people in Canada's North have bioaccumulated extraordinarily high levels of toxins in their bodies because of their dietary reliance on marine mammals. Some of these persistent toxins are also endocrine disrupters, which have been linked to severe growth, development, and reproductive problems in wildlife populations, as discussed earlier in this chapter. However, even toxic substances that are not persistent or bioaccumulative (such as benzene) can have significant harmful effects on the health of the marine environment.

In recent years, Canada has made progress in reducing emissions from a number of marine toxic pollution sources. Agricultural industries have been forced to develop and use more environment-friendly pesticides and fertilizers and to increase conservation tillage to reduce runoff pollution. There has also been a significant decrease in the amount of toxic pollutants coming from other industries such as pulp and paper, petroleum refining, and aluminum. For example, discharges of dioxins and furans from Canada's forest products industry have decreased by 99 per cent since 1988 (Environment Canada 2003d) because of new regulations under federal and provincial legislation.

Already, some of these reductions are apparent in ocean life. Temporal trends of persistent, bioaccumulative, and toxic (PBT) chemicals were examined in beluga whales from the St Lawrence estuary. Blubber samples of 86 stranded adult belugas were collected between 1987 and 2002 and analyzed for several regulated PBTs, including PCBs, DDT and its metabolites, chlordane (CHL) and related compounds, HCH, HCB, and Mirex. Concentrations of most of the PBTs examined had exponentially decreased by at least a factor of two in belugas between 1987 and 2002, while no increasing trends were observed for any of the PBTs measured (Lebeuf et al. 2007).

Although the concentration of toxins in the Canadian environment has declined, they have not disappeared. Existing toxic residues will be recycled and dispersed throughout ecosystems for some time. In addition, toxic substances from sources outside Canada continue to enter our ecosystems through oceanic and atmospheric transport. One study of contaminant concentrations in the eggs of double-crested cormorants shows a significant decrease (Figure 8.14) in the past 30 years (Canadian Wildlife Service 2003). However, the lack of further declines, despite the banning of these chemicals in Canada, leads scientists to believe that it may be the result of long-range transport of POPs used outside of Canada, as well as the slow release of contaminant residues from bottom sediments and dump facilities.

In Canada, federal law and policy have declared the management and reduction of toxic substances in the environment 'a matter of national priority'. Under the Canadian Environmental Protection Act, the minister of environment is mandated to virtually eliminate the production of POPs and manage the discharge of other pollutants and wastes into the environment. Canada has also been active on an international scale and was the first nation to ratify an international treaty, known as the Stockholm Convention on Persistent Organic Pollutants, that aims to identify problematic substances for which comprehensive global action is required.

Organic pollution is also of concern in some areas (the overall process of organic decay was outlined in Chapter 4). Given the immense volume of water in the ocean, it might be thought that by and large, an infinite adsorption capacity exists for receiving and breaking down organic wastes. However, where there are dense populations and waste is deposited in a site with low adsorptive capacity, even the ocean can become polluted. On the east coast, for example, 52 per cent of all towns and cities lack any sewage treatment. The problem became quite obvious in Halifax, where sewage has been deposited directly into the harbour since 1749 and toilet paper, tampon applicators, and condoms have become a familiar sight. In late 2003, Halifax made a historic move after 30 years of delay, announcing that it would install a network of treatment plants for the 181 million litres of raw sewage pumped out every day. The total cost is estimated at $330 million, and the first of three treatment facilities opened in the spring of 2008, with the others expected to open before the end of the year.

At the other end of the country, Victoria has taken a different approach. The city pumps out 100 million litres of raw

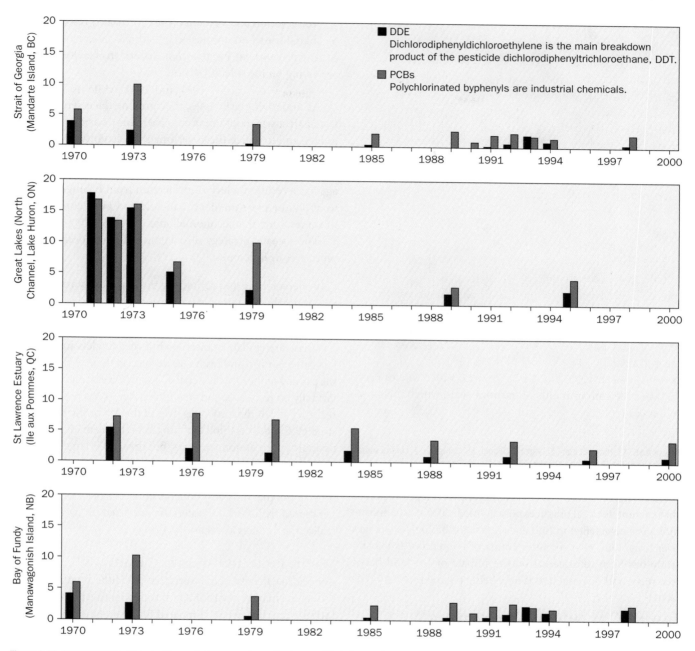

Figure 8.14 Contaminant levels in double-crested cormorant eggs (parts per million). Source: Environment Canada 2003b. Reproduced with the permission of the Minister of Public Works and Government Services, 2005.

sewage into Juan de Fuca Strait every day through two deep-sea pipes that extend more than a kilometre offshore and are more than 60 metres deep. Victoria's situation is rather different from that of Halifax in that the large volume of fast-moving water in the strait breaks down and disperses the sewage very quickly. The plume from the discharge never reaches the surface in summer and only 1 per cent of the time in winter. Marine biologists have monitored the situation for years and not been able to detect virtually any negative impacts. Most biologists are satisfied that the dilution of organic waste is acceptable. However, there is much greater concern over the non-organic wastes that are disposed of illegally through the sewage system. Over a two-year period, estimates

suggest that these wastes include 2,920 kilograms of oil and grease, 17,400 of zinc, 9,000 of copper, 2,560 of cyanide, and 1,360 of lead. The regional government has introduced educational programs on waste disposal, since the most effective way to deal with these substances is to halt their entry into the sewage system rather than trying to treat them once they are there (Figure 8.15).

In 2005, Victoria commissioned an independent study to review liquid waste practices, which suggested that an increasing population would soon require increased treatment. A plan was devised and is now being implemented at a cost of more than $1 billion. Many scientists and health professionals question whether this is the wisest investment for that amount of

Figure 8.15 Ad from CRD (Capital Regional District, BC) campaign to control source pollution (courtesy the Capital District, Regional Source Control Program).

money. There is a strong consensus among health professionals that current disposal practices pose no medical risk, and marine biologists have failed to find any significant biological changes over time. One of their main points of contention is that no study has been undertaken on the impact of the land-based treatment and disposal that is now being planned.

Some Canadian Responses

Canada's Oceans Strategy

The Oceans Act was passed in 1998 to provide a comprehensive and co-ordinated approach to marine resource management in Canada. One of the main requirements of the Act was for the minister of fisheries and oceans to develop a national Oceans Strategy. This strategy established three principles to guide *all* ocean management decision-making:

1. *Sustainable development* 'recognizes the need for integration of social, economic, and environmental aspects of decision-making and that any current and future ocean resource development must be carefully undertaken without compromising the ability of future generations of Canadians to meet their needs.'

2. *Integrated management* 'is a commitment to planning and managing human activities in a comprehensive manner while considering all factors necessary for the conservation and sustainable use of marine resources and shared use of ocean spaces.'

3. The *precautionary approach* is defined in the Oceans Act as 'erring on the side of caution'.

As underlying principles, they provide the essential litmus test against which all ocean management decisions should be judged and to which the government is accountable. Unfortunately, this has not proved to be the case. Globally important glass-sponge reefs off the coast of BC have been heavily damaged as a result of a lack of protection from fishing—a failure to apply the precautionary principle and poor consideration of values other than economic ones.

The Oceans Strategy also identified three desired policy objectives or outcomes:

- understanding and protecting the marine environment;
- supporting sustainable economic opportunities;
- providing international leadership.

These are laudable objectives but meaningless unless progress can be shown and resources devoted to them. A decade after the passing of the Oceans Act, the Oceans Directorate of DFO has virtually no resources, and very little progress has been made on realizing any of the lofty objectives of the Act or the subsequent strategy. Canadians think that progress is being made because of periodic policy announcements, but in fact it is largely business as usual. A public poll undertaken in Atlantic Canada and New England on behalf of a consortium of conservation organizations found that most respondents think that between 20 and 23 per cent of their offshore waters are protected (Seaweb 2003). In reality, the figure is less than 1 per cent.

Marine Protected Areas in Canada

Canada, like the rest of the world, has paid little attention to protecting the marine environment through marine protected areas (MPAs), especially when compared to the attention and protection given the terrestrial environment (see Chapter 14). Depending on the definition applied, estimates suggest that Canada has more than 560 protected areas, with a marine component covering approximately 3 million hectares, or 0.4 per cent of Canada's ocean area (Environment Canada 2006b). However, almost all of this area is in terrestrial protected areas that happen to be coastal. This figure compares with more than 7 per cent of the terrestrial environment under protection.

In response to this situation, three programs have been created at the federal level to establish MPAs (Table 8.2). The first is an MPA program established under the Oceans Act within the Department of Fisheries and Oceans. The purpose of these MPAs is to conserve commercial and non-commercial fisheries, protect species at risk, and conserve unique habitats—i.e., areas of high biodiversity or biological productivity. Five Atlantic MPAs have been designated: The Gully near Sable Island designated in 2004; Basin Head in the Gulf of St Lawrence, Eastport in Bonavista Bay, and Gilbert Bay in the Labrador Sea, all

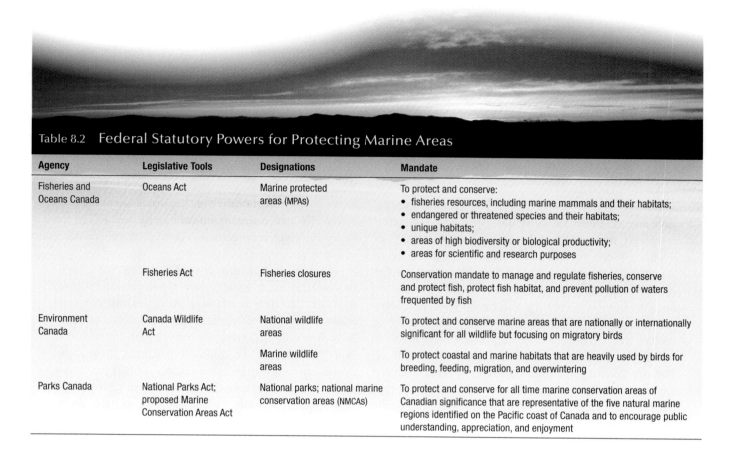

Table 8.2　Federal Statutory Powers for Protecting Marine Areas

Agency	Legislative Tools	Designations	Mandate
Fisheries and Oceans Canada	Oceans Act	Marine protected areas (MPAs)	To protect and conserve: • fisheries resources, including marine mammals and their habitats; • endangered or threatened species and their habitats; • unique habitats; • areas of high biodiversity or biological productivity; • areas for scientific and research purposes
	Fisheries Act	Fisheries closures	Conservation mandate to manage and regulate fisheries, conserve and protect fish, protect fish habitat, and prevent pollution of waters frequented by fish
Environment Canada	Canada Wildlife Act	National wildlife areas	To protect and conserve marine areas that are nationally or internationally significant for all wildlife but focusing on migratory birds
		Marine wildlife areas	To protect coastal and marine habitats that are heavily used by birds for breeding, feeding, migration, and overwintering
Parks Canada	National Parks Act; proposed Marine Conservation Areas Act	National parks; national marine conservation areas (NMCAs)	To protect and conserve for all time marine conservation areas of Canadian significance that are representative of the five natural marine regions identified on the Pacific coast of Canada and to encourage public understanding, appreciation, and enjoyment

designated in 2005; and Musquash Estuary in the Bay of Fundy, designated in 2006. In addition, the Endeavour Hydrothermal Vents MPA has been created in the Pacific. DFO committed to establishing nine MPAs between 2005 and 2010. To date, four have been designated within this time period, and there are currently five proposals under consideration: Manicouagan (QC), St Lawrence Estuary (QC), Beaufort Sea (NWT), Bowie Seamount (BC), and Race Rocks (BC). DFO is projecting that a total of 16 to 17 sites will have been established by 2012.

Second, through the Canadian Wildlife Service (CWS), Environment Canada has several programs that may include designation of marine sanctuaries such as national wildlife areas and migratory bird sanctuaries. A new designation, called protected marine areas, has also been introduced. At present, the Scott Islands at the north end of Vancouver Island is the only site being considered for protection under this designation.

The third program involves national marine conservation areas (NMCAs), developed by Parks Canada. These areas differ from terrestrial national parks and other Canadian marine initiatives in that they are managed for sustainable use. NMCAs will generally be larger than the MPAs established by DFO, will be selected to represent Canada's ocean heritage, and will also contain an explicit mandate for recreation and education. Although some of the CWS sanctuaries are large, especially in the Arctic, they are designed primarily to protect specific species (particularly seabirds) rather than eco-systems, and they have no minimum standards to control extractive activities.

NMCAs will contain zones where special protection measures, such as no-take fishing zones, will be implemented. In many jurisdictions, such areas are often called marine reserves or sanctuaries. The location and size of these zones will be decided through consultation among fishers, scientists, conservationists, government agencies, and other stakeholders. These decisions will be crucial. Setting aside small fragments in unproductive areas will produce few benefits.

Apart from no-take zones, commercial and recreational fishing will continue in NMCAs, although additional conservation measures may be stipulated. Some activities, such as exploration or exploitation of hydrocarbons, minerals, aggregates, or any other inorganic material, will be prohibited. Dumping will not be allowed. Conservation interests sought to have bottom trawling, dragging, and fin-fish aquaculture prohibited as well because of their destructive impact on ocean ecosystems, but such prohibitions were not included in the National Marine Conservation Areas Act.

The goals of NMCAs are to conserve areas representative of the ocean environment and the Great Lakes and to foster public awareness, appreciation, and understanding of our marine heritage. The interpretation aspect of these areas may be their greatest contribution and a unique Parks Canada mission. Canadians are poorly informed about the marine

Box 8.8   The Endeavour Hydrothermal Vents Marine Protected Area

The Endeavour Hydrothermal Vents Marine Protected Area lies in water 2,250 metres deep, 250 kilometres southwest of Vancouver Island. As part of the Juan de Fuca Ridge system, the Endeavour segment is an active sea floor–spreading zone where tectonic plates diverge and new oceanic crust is extruded onto the sea floor. In this zone, cold sea water percolates downward through the crust, where it is heated by the underlying molten lava, eventually emerging through the sea floor as buoyant plumes of particle-rich, superheated fluid. The five known vent fields on the Endeavour Segment are separated from one another by about two kilometres along the ridge. Their associated plumes rise rapidly about 300 metres into the overlying water column.

Hydrothermal vents in the Endeavour area consist of large, hot, black smokers (chimney-like structures) and surrounding lower-temperature sites. The fields span a wide range of hydrothermal venting conditions characterized by different water temperatures and salt content, sulphide structure morphologies, and animal abundance. Temperatures associated with black smokers are typically in excess of 300°C. Formation of the large polymetallic sulphide chimneys takes place when dissolved minerals and metallic ions carried upward by the smokers precipitate on contact with the cold sea water. Cooler waters below 115°C on the sea floor and along the flanks of the chimneys support an abundance of flora and fauna. This rich ecosystem is supported by microbes whose life processes are fuelled by the chemical energy from the emerging fluids in the hydrothermal vents in the process of chemosynthesis (Chapter 2).

Hydrothermal venting systems host one of the highest levels of microbial diversity and animal abundance on Earth. The deep ocean near the Endeavour area normally only supports sparse animal abundance of about 20 worms and brittlestars per square metre. In the diffuse vent flows around the sulphide structures, these abundances can range up to half a million animals per square metre. An amazing abundance of life is in concentrated areas around the vents, surrounded by a veritable desert in the deep oceans.

Globally, hydrothermal venting systems foster numerous unique species of animals. Some 60 distinct species are native to the Juan de Fuca Ridge. Many of these species were first identified in this area. Hydrothermal vents at Endeavour are home to 12 species that are not known to exist anywhere else in the world.

The Endeavour Hydrothermal Vents Marine Protected Area was designated to ensure the protection of the vents and the unique ecosystems associated with them. Damage to or removal, disturbance, or destruction of the venting structures or the marine organisms associated with them is prohibited. The regulation does permit scientific research that will contribute to the understanding of the hydrothermal vents ecosystem.

environment. Creating ocean literacy among Canadians, who must support public policies for the sustainable use and protection of Canada's marine environment, will pose unique challenges to interpreters.

One of the main challenges with all these programs is to actually designate an area. It is essential that local communities support these conservation measures, and gaining support can be time-consuming. Prime Minister Jean Chrétien made a commitment at the World Summit on Sustainable Development in Johannesburg in 2002 to create five NMCAs by 2007. Unfortunately, none have yet been designated, although some progress has been made at sites in Lake Superior, Haida Gwaii (the Queen Charlotte Islands), the southern Strait of Georgia in BC, Lancaster Sound in the Arctic, and the Magdalen Islands in the Gulf of St Lawrence. Parks Canada has a marine system plan analogous to its terrestrial system plan (Chapter 14), with 29 marine regions. The goal is to have representation within each of these regions. Prime Minister Chrétien also committed the government to completing the system within 10 years, and Canada has international treaty commitments under the Convention on Biodiversity to complete the system by 2012. It seems that at best, we will achieve no more than a third of this goal by that date.

Guest Statement

Public Awareness and Ocean Management
Sabine Jessen

In May 2003, scientists at Dalhousie University reported that only 10 per cent of the large ocean fish still remain in the world's oceans. In their 10-year study of global fisheries, they found that 90 per cent of tuna, sharks, swordfish, and cod have been fished out. Other scientists point out that this is having huge impacts on marine ecosystems, which are often regulated from the top by these large predators. This study followed on previous ones that showed how we are fishing down marine food webs on a systematic basis, moving from these large predators to the smaller fish further down on the food chain.

Scientists point out that there is a simple solution to these problems—reduce fish mortality by reducing quotas, reducing overall effort, cutting subsidies, reducing bycatch, and creating networks of marine reserves. They acknowledge that it is not an easy task. But why has it been so difficult for most countries to make these needed changes in the face of the overwhelming evidence of the impact of overfishing?

Part of the answer to this question lies in the serious disconnect between what the scientists are telling us and what the public believes is happening in the oceans. When the public is asked what the greatest threat to ocean ecosystems is, the answer is usually pollution. But these recent scientific studies clearly show that the greatest threat is overfishing—the most serious problem is in what we are taking out of the oceans, not what we are putting in. Couple this with the results of another recent poll of Americans, asking them what the most important decision was that they make every day. You might be shocked to hear that deciding what to wear was considered the day's most difficult decision for one in every 10 people.

Why have I highlighted these points in talking about ocean management? Because without public understanding of the issue, and public demand for change, it is difficult to persuade politicians to make the difficult decisions that will lead to the much-needed fundamental change in how our oceans are managed. Political support determines the priorities and the resources allocated to address these issues. While massive amounts of money are spent on fishery allocation decisions, broader ocean management programs and marine protected areas are receiving very few resources.

Over the past few years, the plight of our oceans has been receiving increasing attention. In 2002, Canada released an Oceans Strategy, and more recently the government has committed to the development of an oceans action plan. In the US, two national-level commissions have examined oceans policy and highlighted the need for concerted attention and reform of ocean management. These are reasons for hope. But until concrete steps are taken, we will continue to witness the ongoing destruction of the blue frontier. And by the time the public really understands and pushes for change, it could be too late.

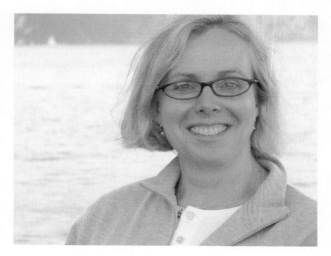

Sabine Jessen

Sabine Jessen is the conservation director of the Canadian Parks and Wilderness Society, British Columbia chapter, and one of the foremost activists pushing for increased conservation of Canada's ocean environment.

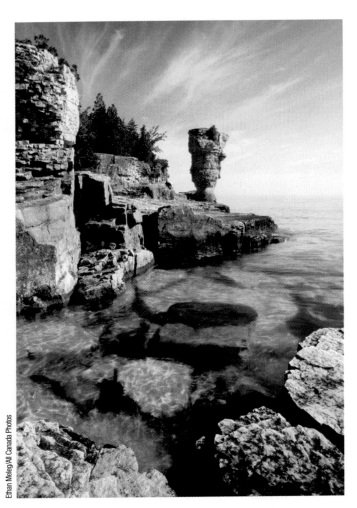

Ethan Meleg/All Canada Photos

Canada has not done very well in protecting the marine environment, with Fathom Five, at the tip of the Bruce Peninsula in Lake Huron, being only one of two Parks Canada national marine conservation areas. However, the federal government has committed to the establishment of five new areas by 2010.

Aquaculture

One response to the declining catch in wild fisheries is to produce more seafood through farming or **aquaculture**. Aquaculture is the fastest-growing food production sector in the world and is expected to surpass beef production by 2010. Aquaculture accounts for almost 30 per cent of the volume and 39 per cent of the value of global fish landings. It is expected that by 2030, it will be the dominant source of fish and seafood. Aquaculture could play a critical role in reducing world hunger.

Canada has been part of this growth and currently ranks 22nd in the world in aquaculture production, although DFO predicts that Canada has the potential to be among the top three. In 2005, wild harvest and aquaculture production totalled more than $2.9 billion. Farmed fish and seafood production was valued at $715 million in 2005, comprising approximately 25 per cent of all Canadian fisheries and

aquaculture. Salmon is the predominant farmed species in Canada, producing 98,441 tonnes in 2005 and generating more than $543 million. Ninety-five per cent of Canada's farmed fish and seafood is exported to the United States. Some 14,000 people are employed in aquaculture, and it is predicted that employment levels will quadruple over the next 15 years.

BC has Canada's largest output, worth $429 million in 2006 (Figure 8.16), followed by New Brunswick, PEI, and Nova Scotia. Most of BC's production comes from salmon. Salmon production contributes 88 per cent of all cultured seafood in BC. The 2006 cultured salmon harvest of 78,000 tonnes was up 11 per cent from the previous year, while the farm-gate value of $407.4 million represented an increase of 28 per cent. The increasing importance of cultured salmon to BC fisheries production can be seen in Figure 8.17. British Columbia is the fourth largest producer of farmed salmon in the world after Norway, Chile, and the United Kingdom.

A typical salmon farm consists of 10 to 30 cages, each 12 or 15 metres square, and contains on average 20,000 fish. The cages are made of open nets that allow water to flow through and antibiotics, uneaten food, feces, and chemicals used to prevent excessive marine growth on the cages to flow out.

There are currently more than 140 fish farms on the BC coast, mainly concentrated in three small areas. In addition, many applications await approval to expand operations to other areas along the coast. BC's salmon farming industry employs approximately 3,000 people in full-time, year-round jobs either directly (on farms) or indirectly (in processing). More than 92 per cent of the direct jobs are located in coastal communities outside of Greater Victoria and Vancouver. These economic opportunities can be life-savers for some remote communities, especially First Nations communities.

Environmentalists have raised a number of concerns about aquaculture:

Escapement. Salmon farms in BC mainly raise Atlantic salmon, primarily because the Norwegian-dominated industry had more experience with and well-developed markets for Atlantic salmon. The Atlantic salmon are also more efficient in converting feed into flesh, are less aggressive, and tolerate crowded conditions. The farming of Atlantic salmon is an environmental concern because escapement from farm fish cages is a regular occurrence and escapes often occur in high numbers. It has been estimated that up to now, more than 1 million Atlantic salmon have escaped. In July 2008, more than 30,000 ready-for-harvest Atlantics escaped after a net anchor was apparently displaced by high tides. There is irrefutable proof that Atlantic salmon are now spawning wild in Pacific rivers. DFO maintained for many years that this was impossible, until scientists proved otherwise (Volpe et al. 2000). There are concerns that these escapers, an invasive species (Chapter 3), may displace the native salmon. Atlantic salmon have been found in more than 80 rivers on the BC coast, and spawning has occurred.

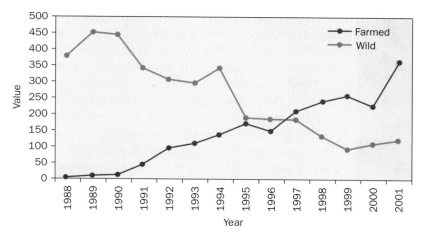

Figure 8.16 Value of BC salmon exports in millions of dollars.

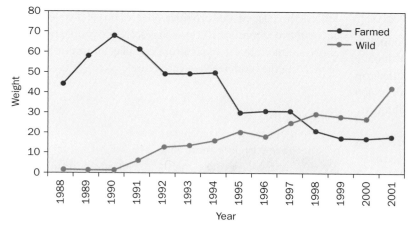

Figure 8.17 Weight of BC salmon exports in millions of kilograms. Source: BC Salmon Market Database, http:www.bcsalmon.ca/database/export/summary/sumwtpd.htm.

considerable scientific uncertainty entailed in all these issues. And although DFO's stated policy is to 'err on the side of caution', the department has failed to do so with respect to the dangers associated with escapement.

Disease. The high stocking levels of fish in netted areas promote rapid spread of infectious diseases and parasites. Since fish farms are along migration routes for wild salmon, diseases and/or parasites can be passed along easily, with a detrimental impact on wild populations. In Clayoquot Sound on the west coast of Vancouver Island, a viral disease, infectious hematopoietic necrosis (IHN), swept through fish farms in the fall of 2002. As a result, the main operator in the region—Pacific National Aquaculture—lost some US $5.7 million in 2002 and closed down its processing plant in Tofino.

Lice. There is strong scientific evidence that pink salmon smolts in certain areas of the coast are being weakened by excessive sea lice coming from farms near their migration routes. A recent study (Krkoek et al. 2007) found 90 per cent mortality among populations of juvenile pink salmon and predicted that stocks in some rivers will be extinct within 10 years if no mitigating actions are taken.

Pollution. To combat the diseases mentioned above, farmed salmon are treated with antibiotics. More antibiotic per weight of livestock is used by salmon aquaculture than by any other form of farming. Antibiotics can harm other marine organisms, since they are released directly into the ocean. As well, excess food and feces create a large amount of organic pollution. The substances build up on the ocean floor, depleting oxygen levels, releasing noxious gases (as a by-product of decomposition), and smothering benthic organisms. On a daily basis, the aquaculture industry in BC dumps the same amount of sewage in the ocean as a city of half a million people.

If salmon farming were confined to native Pacific species alone, it might not help. At the moment, there is no evidence that Atlantic and Pacific salmon interbreed. However, if Pacific salmon were to be found both wild and in farms, there would undoubtedly be interbreeding. This genetic introgression could have a devastating effect on wild stocks, as determined by McGinnity et al. (2003) on the east coast. There is

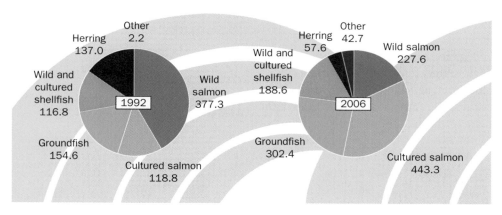

Figure 8.18 BC seafood wholesale value species shares, 1992 and 2006 ($ millions).

A small fish farm on the BC coast.

Predator control. Predators such as seals and sea lions are one of the main problems for the farmers, since they literally eat profits. Farmers are permitted to shoot animals that rip nets open. In 2001, farmers reported killing more than 400 seals and sea lions. Many observers think that these killings are grossly under-reported.

Energetics. Unlike the herbivorous fish produced in the vast majority of fish farms around the world, salmon are carnivorous. As a result, farmed salmon are mainly fed other fish, such as anchovies and mackerel that are caught as far away as South America. As dictated by the second law of thermodynamics (Chapter 2), only one kilogram of farmed salmon is produced for every three to four kilograms of feed fish. This is a poor use of fish protein and leads to the reduction of fish stocks elsewhere.

Social dimensions. Most profits from production go to five multinational companies that control 80 per cent of the industry in BC. As a result, a high percentage of the economic benefits attached to salmon aquaculture are exported out of the province. Increased mechanization is leading to lower employment figures, further limiting the economic benefits accruing to local communities. It is also feared that further growth in the industry will be detrimental to the wild fishery and reduce the health of communities dependent on wild fish. Increased supply of farmed salmon may continue to depress the price of BC's wild salmon.

Human health. To turn the white flesh pink, farmed salmon are fed artificial colouring. The most commonly used colourants are synthetic astaxanthin and canthaxanthin. In 2003, the European Union reduced the amount of canthaxanthin that can be fed to salmon by two-thirds because of concerns over retinal damage caused by ingesting too much of the chemical. A study published in *Science* (Hites et al. 2004) found that farmed salmon contained 11 times the amount of toxic contaminants found in wild salmon.

Most of these problems are not insurmountable. Salmon can be produced in closed, land-based systems that all but eliminate some of the problems. One large grocery chain on Vancouver Island, Thriftys, now buys salmon from closed-pen systems for approximately 20 to 30 per cent more than for salmon from net-pen fish but sells them at roughly the same price in order to promote more sustainable practices. The main factor in why more environmentally and socially sound farming techniques are not being more widely adopted is the consumer. If people were willing to pay more for salmon produced using techniques that avoided the problems outlined above, then producers would not be so resistant to adopting these more sustainable systems (Box 8.9). At the moment, however, the environment pays those extra costs.

In response to these concerns and vocal opposition from many communities, the BC government announced in 2008 that salmon aquaculture would not be allowed to expand to the north coast of the province. The main challenge to the aquaculture community is how to resolve the problems raised above so that salmon aquaculture can play an important role in meeting the food needs of the future.

The combined impact of hybridization and competition means that when a large number of farm salmon spawn in a river, the number of adult salmon returning to the river and the potential offspring production in the next generation are reduced. . . . As repeated escapes are now a common occurrence in some areas, a cumulative effect is produced generation upon generation, which could lead to extinction of endangered wild populations as a result of this 'extinction vortex'.

—McGinnity et al. 2003

Implications

Experience over the past five years has begun to provide proof of the extent of the extreme degradation of the oceans of this planet, something that has been suspected for many years. Scientific efforts have intensified, and understanding has increased, yet much remains to be done. The very visible collapse of fishing stocks around the world and on the east and west coasts of Canada has helped to direct a little more

Box 8.9   What You Can Do

1. Fish are an important dietary component for many people and a healthy one. However, it is important that the fish you eat are not endangered or caught with a method that involves killing other species as bycatch. Use Canada's Seafood Guide produced by Sustainable Seafood Canada to inform your consumption (www.seachoice.org).

2. Buy only certified brands where they are available, such as dolphin-free tuna.

3. If you buy farmed seafood, consider paying a little more for products that have been produced using low-impact methods. For example, Thrifty Foods in BC sells farmed salmon produced using land-based, closed-system methods.

4. Ensure that you dispose of any toxic materials in the correct manner, not down the drain.

5. Use natural cleaners, such as vinegar and water, rather than commercial cleaners.

6. Using less water for your own needs leaves more water in rivers for fish such as salmon.

7. Support NGOs, such as Oceans Blue and the Canadian Parks and Wilderness Society, that are working for ocean conservation and the development of marine protected areas.

8. Whenever you do something that involves carbon emissions, whether travelling or buying a product, your actions are leading to ocean acidification. The car you drive and the coral reef in the South Pacific are intimately connected.

political attention to oceans in general and fisheries in particular. Commitments have been made at both international and national levels to adopt more sustainable ocean practices, including encouraging and enabling sustainable fisheries, limiting pollution, and establishing systems of marine protected areas. At the moment, most of these measures are in the embryonic stage. Some plans, such as Canada's Oceans Strategy, have shown little progress. Only time will tell whether international and national commitments will be successful in turning around the trends described in this chapter.

Summary

1. Throughout history, the resources of the oceans have been thought of as vast and undiminished. The past decade has furnished conclusive proof that this view is far from correct. More than 70 per cent of global fisheries are now at or over their maximum exploitation levels.

2. Oceanic ecosystems are controlled by the same general principles that influence terrestrial ecosystems, but their manifestations may be different. There may be up to 5,000 species of fish still awaiting discovery.

3. The carbon balance of the oceans is of great interest because of its relationship with global climatic change.

4. Ocean fisheries supply about 20 per cent of the world's annual animal protein. Catch statistics showed very large increased catches over the past 50 years, but they have now levelled off considerably.

5. The oceans are the ultimate sink for many pollutants, and about 80 per cent of ocean pollution comes from activities on land.

6. More than 60 per cent of global oil production originates under the oceans. Exploration, drilling, transporting, and processing this oil is a major source of contamination.

7. Almost half the world's population lives within 100 kilometres of the coast, a proportion that is expected to increase to 75 per cent by mid-century.

8. Global climatic change will lead to increases in sea level of between nine and 88 centimetres this century. This will create severe challenges for many coastal communities.

9. There are many international agreements and programs on ocean management. Most have yet to fulfill their true potential in improving oceanic conditions.

10. Marine protected areas have been endorsed by the scientific community as necessary to improving ocean conservation.

11. Canada has the longest coastline of any country and the second largest continental shelf, equal to 30 per cent of Canada's land mass.

12. Since 1992, commercial fishery landings in Canada have declined by 19 per cent and the number of vessels by 31 per cent. However, the value of the catch has doubled.

13. The east coast fisheries have experienced profound changes over the past decade with the total collapse of the northern cod stocks.

14. Management of Aboriginal use of marine resources has emerged as an important concern on both coasts.

15. Exploitation of offshore hydrocarbons in Canada has taken place over the past decade, mainly off the east coast. Increased attention is now being given to the Pacific and Arctic coasts.

16. Pollution levels of many substances have declined over recent years in response to government controls. However, a recent study concluded that the killer whales of Georgia Strait in BC are among the most polluted animals on the planet.

17. Canada passed a comprehensive Oceans Act in 1998, but to date it has been ineffectual because of a lack of political support and funding.

18. There are three federal programs for establishing marine protected areas (MPAs) in Canada. However, less than half of 1 per cent of the area of Canada's marine environment is currently protected.

19. Aquaculture accounts for almost 30 per cent of the volume and 39 per cent of the value of global fish landings. Aquaculture is the fastest-growing food production sector in the world and is expected to surpass beef production by 2010.

20. BC has Canada's largest share of the total value of aquaculture production, focusing mainly on salmon. Although economically important to some communities, salmon farming also raises concerns over escapement, disease, proliferation of sea lice, killing of predators, energetics, and pollution.

Key Terms

aquaculture
biomass pyramid
bottom trawling
carbon balance
coral bleaching

coral polyps
endocrine disruption
exclusive economic
    zones (EEZs)
fishing down the food chain

longline
marine protected areas (MPAs)
polynyas
prey switching
serial depletion

shifting baseline
thermocline
thermohaline circulation
total allowable catch (TAC)
zooxanthellae

Questions for Review and Critical Thinking

1. In what ways are oceanic and terrestrial ecosystems the same, and in what ways do they differ?

2. What are the most biologically productive areas of the ocean?

3. What is thermohaline circulation, and why is it important?

4. Give an example of a positive feedback loop related to global climate change and the oceans. Are there any negative feedback loops?

5. Explain the concepts of shifting baselines, serial depletion, and fishing down the food chain.

6. Give an example of the destructive effects of bottom trawling.

7. What are the two main forms of chemical pollutants in the oceans, and what are their main effects?

8. What are some of the main international conventions concerning ocean management?

9. What are the jurisdictional arrangements for ocean management in Canada?

10. Characterize two of Canada's marine ecozones.

11. Discuss the principal reasons behind the collapse of the Atlantic groundfish stocks and some of the lessons to be learned from this experience.

12. Outline some of the challenges involving the Aboriginal use of marine resources in Canada.

13. Discuss the differing approaches of Halifax and Victoria to ocean pollution resulting from sewage.
14. What are the main principles underlying Canada's Oceans Strategy?
15. Outline the three federal programs for creating marine protected areas in Canada and their similarities and differences.
16. Discuss the positive and negative aspects of aquaculture production.

Related Websites

CPAWS: http://www.cpaws.org

David Suzuki Foundation, oceans, fishing, aquaculture, MPAs: http://www.davidsuzuki.org/Oceans

Endeavour Hot Vents Marine Protected Area: http://www.pac.dfo-mpo.gc.ca/oceans/mpa/Endeavour_e.htm

Fisheries and aquaculture: http://www.agf.gov.bc/fisheries/studies_rpts.htm

Fisheries crisis: http://www.fisherycrisis.com

Marine protected areas, Canada: http://www.pac.dfo-mpo.gc.ca/oceans/mpa/default_e.htm; http://parkscanada.pch.gc.ca/progs/amnc-nmca/index_E.asp

Save Our Seas: www.saveourseas.org

Seafood consumption choices: http://www.seafoodchoices.com; http://www.oceantrust.org; http://environet.policy.net/

marine/csb; http://www.legalseafoods.com; http://www.montereybayaquarium.org

Sierra Club, oceans, fishing, aquaculture, MPAs: http://www.sierraclub.ca/bc/programs/marine/publications.shtml

United Nations Food and Agriculture Organization, Fisheries: http://www.fao.org/fi/default_all.asp

University of British Columbia Fisheries Centre: http://wwwl.fisheries.ubc.ca

Value of marine protected areas: http://www.nceas.ucsb.edu/Consensus

Watershed Watch Salmon Society, publications: http://www.watershed-watch.org/ww/publications.html

Further Readings

Note: This list comprises works relevant to the subject of the chapter but not cited in the text. All cited works are listed in the section 'References' at the end of the book.

Canadian Environmental Protection Agency (CEPA). 2008. 'Environmental Registry, Toxic Substances List'. http://www.ec.gc.ca/CEPARegistry/subs_list/Toxicupdate.cfm.

Dearden, P., and R. Canessa. 2008. 'Marine parks'. In P. Dearden and R. Rollins, Eds, *Parks and Protected Areas in Canada*, 3rd edn, 403–31. Toronto: Oxford University Press.

Fisheries and Oceans Canada. 2003. 'Canadian Science Advisory Secretariat status report. State of the Eastern Scotian Shelf ecosystem'. http://www.dfo-mpo.gc.ca/csas/csas/status/ESR2003_004_E.pdf.

Glavin, T. 2000. *The Last Great Sea: A Voyage through the Human and Natural History of the North Pacific Ocean*. Vancouver: Greystone Books.

Naylor, R.L., J. Eagle, and W.L. Smith. 2003. 'Salmon aquaculture in the Pacific Northwest: A global fishery with local impacts'. *Environment* 45: 18–39.

Pauly, D., et al. 1998. 'Fishing down marine food webs'. *Science* 279: 860–3.

———. 2001. 'Fishing down Canadian aquatic food webs'. *Canadian Journal of Fisheries and Aquatic Sciences* 58: 51–62.

———. 2003. 'The future for fisheries'. *Science* 302: 1,359–61.

Rose, G.A., and R.L. O'Driscoll. 2002. 'Capelin are good for cod: Can the northern stock rebuild without them?' *ICES Journal of Marine Science* 59: 1,018–26.

Safina, C. 1997. *Song for the Blue Ocean*. New York: Henry Holt.

———. 2002. *Eye of the Albatross: Visions of Hope and Survival*. New York: Henry Holt.

Smedbol, R.K., and J.S. Wroblewski. 2002. 'Metapopulation theory and northern cod population structure: Interdependency of subpopulations in recovery of a groundfish population'. *Fisheries Research* 55: 161–74.

Task Force on Incomes and Adjustments in the Atlantic Fishery. 1993. *Charting a New Course: Towards the Fishery of the Future*. Ottawa: Minister of Supply and Services.

Watling, L., and E.A. Norse. 1998. 'Disturbances of the sea bed by mobile fishing gear: A comparison to forest clearcutting'. *Conservation Biology* 12: 1,180–97.

Chapter 9

Forests

Learning Objectives

- To understand what the boreal forest is, its significance to Canada, and the main threats it is facing.
- To describe the eight main forest ecozones of Canada.
- To discuss the economic and non-economic values of Canada's forests.
- To appreciate the management arrangements and different approaches for harvesting Canada's forests.
- To understand some of the environmental and social aspects of forest management practices.
- To discuss the theory and practice of 'new forestry'.
- To describe current directions for forest use in Canada.

The Boreal Rendezvous

When the combined effects of climate warming, acid deposition, stratospheric ozone depletion, and other human activities are considered, the boreal landscape may be one of the global ecoregions that changes the most in the next few decades. Certainly, our descendants will know a much different boreal landscape than we have today.

— *D.W. Schindler 1998*

In July 2003, Justin Trudeau, the son of former prime minister Pierre Trudeau, broadcast live on CBC from Virginia Falls on the Nahanni River, one of Canada's, and the world's, most spectacular sites. Justin was part of a team canoeing down the Nahanni, just as his father had done before him. The prime minister's journey resulted in a directive to establish Nahanni National Park in 1976, later recognized as one of the first World Heritage Sites. Justin was on a similar mission, but this time as part of a Canadian Parks and Wilderness Society (CPAWS) campaign to draw greater attention to the great boreal forests that stretch across the roof of Canada.

Justin was not the only celebrity taking part. Others included Tom Cochrane, Silken Laumann, Rick Mercer, Ken Dryden, and David Suzuki, all part of expeditions down 10 of Canada's most spectacular boreal rivers. Participants were united in their recognition of the ecological, recreational, and spiritual values of the boreal forest and the need to create a sustainable development path that recognizes these values, in addition to the economic values that have dominated land-use allocations in the past. All would later meet at the 'Boreal Rendezvous' in Ottawa as part of a national campaign to give voice to these concerns.

Why are these and many other well-known Canadians so concerned about the boreal forests? Stretching 3,800 kilometres from the eastern tip of Newfoundland to western BC, the Boreal Shield, which is coterminous with the geological formation known as the Canadian or Precambrian Shield, houses roughly one-quarter of the world's remaining original forests. The Boreal is Canada's largest ecozone, covering almost 58 per cent of the country's land mass and stretching through all provinces except PEI, Nova Scotia, and New Brunswick. The forests are home to a wide diversity of terrestrial and aquatic wildlife, and 30 per cent of North America's bird population relies on the Boreal for breeding. Many First Nations people depend on the resources of the forests for subsistence, and more than 600 First Nations communities retain their roots in the forest.

Boreal *is a term that literally means 'of the North'. It comes originally from the Greek god of the north wind, Boreas. The term is now applied to many northern phenomena, perhaps the most famous being the aurora borealis, or northern lights. Many animal and plant species that live in the North have 'borealis' as part of their Latin name, such as the delicate twinflower* Linnaea borealis, *which is found all across the country. It is also the name used to characterize the great northern forests that stretch not only across Canada but all across the northern hemisphere.*

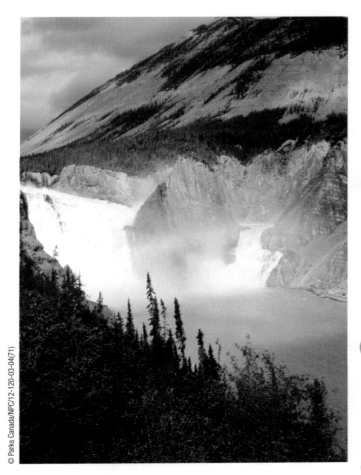

Virginia Falls, Nahanni River, NWT.

The boreal forests also support commercial activities such as logging, wood fibre and sawlog production, pulp and paper mills, and fibreboard production. Its wealth of minerals supports prospecting, mining, and smelting activities. There are large-scale hydroelectric developments, and the abundant fish and wildlife resources support subsistence, sport, and commercial harvesting activities, as well as a growing tourism industry. Recreation-related activities, such as canoeing, hiking, and birding in the Boreal, contribute more than $4 billion to the economy every year. In addition, more than 186 billion tonnes of carbon are stored in the Boreal's trees, soils, water, and peat—equivalent to 913 years' worth of greenhouse gas emissions in Canada. The global Boreal is the largest terrestrial carbon 'bank account' on the planet.

Almost 50 per cent of the boreal forest is currently allocated to industry, and multinational logging companies are currently eyeing the remaining intact forests as the next cutting frontier. Although the Boreal Shield is the largest of Canada's 15 terrestrial ecozones, it has one of the lowest proportions of land (6 per cent) dedicated to protected areas in which all forms of industrial activity are prohibited. Approximately 4 per cent of the ecozone includes additional protected areas where forestry, mining, and other activities may be permitted. The Canadian Boreal Forest Conservation Framework is an initiative supported by more than 1,500 scientists and thousands of others to gain permanent protection of at least 50 per cent of the Canadian Boreal as parks and wildlife reserves and application of strict protective management standards in any other areas where development will occur. The framework has received a great deal of support, including support from more than 25 First Nations and some major forest companies such as Domtar and Tembec. It is hoped that it will lead to some mitigation of the vast range of challenges now facing the Boreal (Table 9.1).

Large areas of the Boreal are now experiencing a number of serious environmental stresses. In many ways, these stresses are no different from those experienced elsewhere in Canada, and they epitomize the challenges of developing sustainable strategies for the management of Canada's forest ecosystems into the future. This chapter outlines the main challenges and some of the strategies developed to address them.

An Overview of Canada's Forests

Canada is a forest nation. The symbol on our national flag is a maple leaf. Along with our northern latitude, the forests have provided the historical context for our national identity. Canada has one-tenth of the world's forests, and these forests cover almost half of the nation's land area (Figure 9.1) and a much higher proportion of southern Canada, where most Canadians live. These forests include one-quarter of the world's temperate rain forests and more than one-third of the world's boreal forests. Furthermore, estimates suggest

Table 9.1 Environmental Stresses in the Boreal Shield Ecozone

Logging

- The 1990s witnessed a dramatic northward expansion of forestry-related activities. Approximately 400,000 hectares of forest are harvested annually. About 90 per cent of all harvesting is carried out through clear-cutting.
- Technological advances have encouraged the industry to harvest smaller trees and additional trees (poplar, birch) that were formally considered uneconomical. Many areas have been cut over by small northern operators who lack the capacity or the responsibility for replanting after harvest.
- The long-term and cumulative impacts of intensified forestry on the fragile areas of the northern Boreal Shield are entirely unknown—the science behind current forest management practices comes from research in the southern areas of the ecozone, where ecosystems, growth rates, and nutrient dynamics may vary considerably from those in northern portions.
- Logging continues to threaten boreal wildlife, particularly large mammals and migratory birds. A recent report suggested that as many as 85,000 migratory bird nests were destroyed in 2001 by logging in Ontario alone (CPAWS 2003). A number of socio-economic impacts have been reported by indigenous communities, including decreased trapping returns, destruction of traditional trails, significant noise pollution, and erosion of cultural values.
- Numerous pulp and paper mills (27) have been major point sources of aquatic pollution by mercury, dioxins and furans, and other organochlorines. The presence of organochlorines such as polychlorinated biphenyls (PCBs) has been linked with liver enzyme dysfunction in fish. Highly toxic leachate from both hardwoods and softwoods stored near the water's edge has been identified. The long-term cumulative effects on lakes and rivers are not well known.

Mining

- Exploration companies hold more than 1,200 mineral leases or claims covering more than 2.2 million hectares in the boreal forest. The Boreal Shield produces 75 per cent of the total Canadian production of iron, copper, nickel, gold, and silver, generating more than $6 billion annually and supporting some 80 communities. Mercury and cadmium are released into the atmosphere through smelting and other industrial processes, then deposited across the landscape. Mine effluents discharged into water systems contain high concentrations of heavy metals as well as cyanide, arsenic, or radioactive compounds.
- Large tracts of land are disturbed by ongoing mineral exploration, which involves road construction, trenching, blasting, and exploratory drilling.
- Metal smelters are point sources of acid and heavy metal pollution that cause widespread aquatic contamination via atmospheric transport. Mercury and cadmium enter the food chain and accumulate in tissues of organisms, reaching toxic levels (Chapter 12). Improper disposal of mine tailings and associated acid mine drainage has caused acute toxicity problems in localized areas and more widespread chronic effects downstream. Such impacts can continue for decades after a mine is abandoned. There are 6,000 abandoned mine sites in Ontario alone.

Hydroelectric Development

- Eighty-five per cent of the drainage basins contained whole or in part in the Boreal Shield have been altered by hydro development in one way or another; 77 per cent of the drainage areas contain major dams, 25 per cent have major reservoirs, and 33 per cent have rivers whose flow has been either augmented or diminished by water transfers.
- Impacts associated with dam construction and reservoir impoundment include soil erosion, scouring of river channels, and loss of terrestrial habitats. Transmission lines reduce the wilderness character of remote regions, especially when they create long, wide swaths through undisturbed terrain. Increased hunting access provided by transmission corridors may be linked to local depletions of wildlife. For example, abrupt declines in moose and other large game animals have been linked to increased access associated with hydroelectric projects.
- Aquatic ecosystems are affected in several ways: increases in water volume caused by river flooding and reservoirs disperse existing species, thus altering competitive and predatory regimes; lake and brook trout spawning beds are adversely affected by reservoir impoundments, causing abandonment; river diversion can result in high levels of turbidity owing to increased flow rates and bank erosion; and flooding of organic matter by reservoirs results in the release of naturally occurring mercury, with toxic effects on aquatic and terrestrial food chains (Chapter 11). Extensive flooding of land also causes the release of significant quantities of carbon dioxide and methane, two contributors to climate warming (Chapter 7).

Climate Change

- Atmospheric models predict that forests and wetlands within the Boreal Shield are likely to be highly vulnerable to the effects of global warming. Many distinctive characteristics of the Boreal Shield are closely tied to climate, including plant productivity, carbon storage, insect populations, hydrology, and fish habitat. Changes in climate include a shift towards shorter periods of ice cover, increases in air temperature, decreases in precipitation, increases in evapotranspiration rates, increases in forest fire frequencies, and increases in wind velocity.
- Possible impacts include decreased stream flow resulting in decreased transportation of chemicals such as dissolved organic carbon (DOC) and phosphorous (important elements in food webs; see Chapters 2–4), decreased phytoplankton abundance, increased transparency (caused by reduced DOC) resulting in deeper penetration of UV-b radiation, and deeper thermoclines, reducing habitat for cold-adapted invertebrates and fish. Climate change is also affecting the life cycle, behaviour, and range of many boreal insect species—there are numerous examples of insect outbreaks associated with warmer temperatures resulting in changes to tree species composition. The range of most boreal insects is controlled by low threshold temperatures, and current climate trends may remove this climatic control in some areas, resulting in more widespread outbreaks of some insects.

Acid Precipitation

- Acid precipitation, the result of human emissions of sulphur and nitrogen oxides to the atmosphere, has caused considerable stress to ecosystems throughout eastern North America (Chapter 4). The poorly buffered lakes of the eastern half of the Boreal Shield are particularly sensitive to acidification.
- Thousands of fish populations and perhaps millions of invertebrate populations from Canada's boreal region have been lost as a result of acid precipitation. Widespread forest decline, growth reductions, and degraded soils have already been documented in southern regions of the Boreal Shield. Increased acidity may contribute to increased mercury contamination in some boreal fish species, limiting human consumption of fish and impairing reproduction in fish-eating birds, particularly the common loon. Low pH in lakes is also linked to high concentrations of cadmium and lead in fish, as well as to loss of DOC in surface waters.
- The impacts of acid precipitation reduce the overall ability of forests to cope with climate warming, drought, insect outbreaks, disease, increased UV radiation, and other stressors.

Sources: Environment Canada 2000; CPAWS 2003.

The use of wood products is an integral part of the livelihood of many Canadians, as illustrated by this local boat-building in Newfoundland.

that more than half of Canada's forest area consists of tracts that are more than 50,000 hectares in size, as yet undisturbed. More than one-third of Canada, however, is naturally treeless, and most of it occurs in the North. Together, Quebec, the NWT, Ontario, and British Columbia account for almost two-thirds of the country's boreal forest. Canada clearly has a major international role to play in forest conservation and management.

Ecosystem and Species Diversity

Although all the provinces are dominated by forest land, forest types differ significantly. Canadian forest ecosystems are a mixture of forest, woodlands, wetlands, lakes, glaciers, and rock, providing habitat for a great variety of plants, animals, insects, fungi, and other organisms. Of the estimated 140,000 species in Canada, approximately 66 per cent occur in forests, although many species currently occupy only a portion of their former range owing to habitat modifications (Chapter 14). There are 15 terrestrial ecozones in Canada (Figure 9.2), although the majority of Canada's forests lie within the eight ecozones discussed below.

The **Boreal Cordillera** ecozone, to the south and west of the Tundra Cordillera, is found in northern BC and southern Yukon. Physiographically, this zone is made up of the mountains in the west and east, separated by intermontane plains. The wetter climate gives rise to more tree growth, hence the boreal designation. Much of the terrain along the western side of the ecozone is covered by permanent ice and snow. In parts of the ecozone located in BC, south-facing slopes are covered with meadows, and north-facing slopes are

Arctic tundra and icefields (27.4%)

Open lands: Alpine tundra, wetlands, etc. (14%)

Forests (43.7%)

Agricultural land (6.6%)

Urban areas (0.3%)

Transportation facilities (0.4%)

Fresh water (7.6%)

Figure 9.1 Canada's physical makeup.

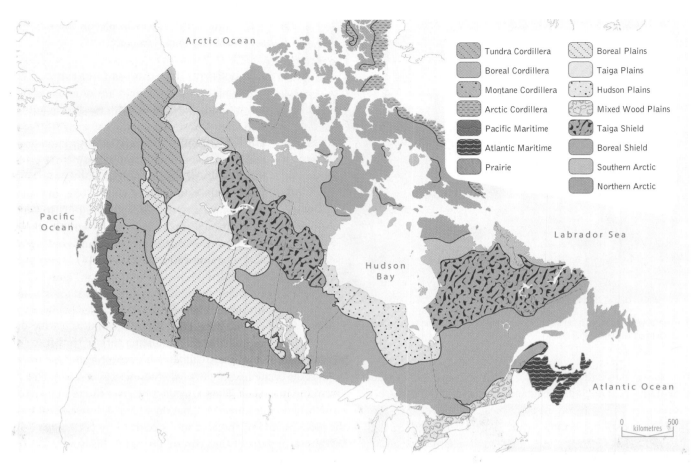

Figure 9.2 Terrestrial ecozones of Canada. Source: Wiken 1986.

covered with boreal forest. Vegetative cover ranges from dense to sparse on a large part of the plateaus and valleys. Tree species include trembling aspen, balsam poplar, white birch, black and white spruce, alpine fir, and lodgepole pine. At higher elevations, there are large areas of rolling alpine tundra, with sedge meadows and stone fields colonized by lichens. Characteristic mammals include woodland caribou, moose, Dall's sheep, mountain goats, black and grizzly bears, martens, lynx, hoary marmots, and Arctic ground squirrels. Representative bird species include willow, rock, and white-tailed ptarmigan, spruce grouse, and a range of migratory songbirds and waterfowl. The rich resources contained within this ecozone have fostered mining, forestry, and tourism industries, in addition to hydroelectric development and localized agriculture.

Stretching along the entire BC coast, the **Pacific Maritime** ecozone is characterized by the influence of the Pacific Ocean. Westerly winds sweeping across the ocean pick up considerable amounts of water. Most of this is precipitated on the windward side of the coastal mountains, producing the highest rainfall figures in Canada (up to 3,000 millimetres). The maritime influence also results in the warmest average temperatures in the country, with mild winters and relatively cool summers. These climatic characteristics provide ideal growth conditions and give rise to the most productive growth conditions in the country in the temperate rain forests at lower elevations. Some of the tree species in the temperate rain forests are long-lived

and grow to impressive heights—yellow cedars can live more than 1,000 years, Douglas firs 750 years, and Sitka spruces 500 years, and Douglas fir and Sitka spruce can grow to more than 70 metres in height. The most common tree species in this ecozone are western red cedar, western hemlock, amabilis fir, Sitka spruce, and yellow cedar. The Pacific Maritime ecozone is also home to a variety of mammals and birds, including black-tailed deer, elk, otters, black and grizzly bears, cougars, fishers, wolves, American black oystercatchers, mountain and California quails, blue grouse, tufted puffins, pygmy owls, Steller's jays, and northwestern crows.

Forestry is the major industry in most of the ecozone. Douglas fir and western red cedar form the backbone of the BC lumber industry, but they are no longer in the infinite supply that they were once assumed to be. Area estimates show that approximately 25 per cent of the coastal temperate rain forest within BC has been logged and reforested and an additional 3 per cent of the total area has been logged and urbanized. Approximately 50,000 hectares of coastal temperate rain forest are logged annually.

As in the Boreal Cordillera ecozone, the great altitude variation in the **Montane Cordillera** ecozone leads to considerable contrast between the summits of the snow-bound peaks, the high montane valleys, rolling plateaus, and deeply entrenched desert-like conditions in the BC Interior. The climate is similarly varied but is generally characterized by long, cold winters and short warm summers. Precipitation ranges from highs of more than 1,200 millimetres along the mountain summits to as little as 205 millimetres in the valleys in the rain shadow. Vegetation varies according to these conditions and can be thought of as a series of vertical zones increasing in altitude. At the summits, the vegetation is alpine, characterized by lichens, herbs, and

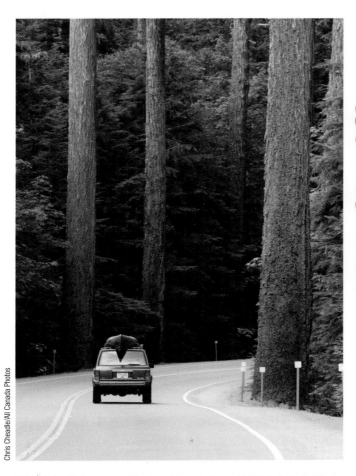

British Columbia has some of the largest trees in the world. This is Cathedral Grove, a stand of large Douglas fir and western red cedar on Highway 4 en route to Pacific Rim National Park Reserve.

small shrubs. In the lower subalpine environment, trees such as alpine fir, Englemann spruce, and lodgepole pine become more common. Below this zone, there is considerable variation, depending on local conditions. Ponderosa and lodgepole pines, Douglas fir, and trembling aspen are found towards the north, and in moister conditions in the southeast, western hemlock, red cedar, and Douglas fir may be found. Characteristic mammals include black and grizzly bears, woodland caribou, mule deer, moose, mountain goats, California bighorn sheep, wolverines, and fishers. Common bird species include the pileated woodpecker, Clark's nutcracker, and red crossbill.

Commercial forest operations have been established in many parts of the Montane Cordillera ecozone, particularly in the northern interior. Other significant industries include mining, oil and gas production, tourism, and agriculture. The southern valleys are well-known for their orchards and vineyards.

The **Boreal Plains** ecozone extends from the southern part of the Yukon in a wide, sweeping band down into southeastern Manitoba. The underlying glacial moraine and lacustrine deposits give a generally flat to undulating surface similar to the Prairie zone to the south. Climatic differences, making it wetter and cooler than the Prairie zone, produce vegetation dominated by trees rather than grasses. Coniferous trees include tamarack, Jack pine, and black and white spruce, although deciduous trees such as trembling aspen, white birch, and balsam poplar are common, especially at the transition zone into the true Prairie. Mammals found in the Boreal Plains include woodland caribou, white-tailed deer, bison, wolves, black bears, mule deer, and elk. Characteristic bird species include boreal and great horned owls, blue jays, a number of warblers, grouse, red-tailed hawks, cormorants, gulls, herons, and terns. One of Canada's most famous endangered species, the whooping crane, nests in wetlands in Wood Buffalo National Park at the extreme north of the ecozone. A number of species considered at risk of extinction can be found in this ecozone, including woodland caribou, wolverines, grizzly bears, and wood bison. The southern and northwestern areas of the ecozone have been transformed by agricultural development and timber, mining, and oil production.

Many Canadians are familiar with the spectacular mountain landscapes of the Montane Cordillera ecozone, which includes Banff, Canada's first and most visited national park.

Trembling aspen is characteristic of the Boreal Plains.

Taiga, a Russian word, is used to describe coniferous forests in that country. In North America, it describes that portion of the boreal forest lying between the southern boundary of the tundra and the closed-crown coniferous forest to the south. Topography in the **Taiga Plains** is gently rolling, with a high proportion of surface water storage, wetlands, and organic soils. The climate is cold and relatively dry, with as little as 200 millimetres of precipitation in the northern sections. Canada's largest watercourse, the Mackenzie River and its tributaries, is located in the Taiga Plains. Winters are long and cold, with mean daily January temperatures ranging from −22.5 to −30°C. These conditions, plus the topography, give rise to large areas of wetlands dominated by species such as Labrador tea, willow, dwarf birch, mosses, and sedges. On better-drained localities and uplands are found mixed coniferous-deciduous forests containing white birch, trembling aspen, balsam poplar, lodgepole pine, tamarack, and black and white spruce. A number of large mammals are found in this ecozone, including moose, caribou, black bear, wolf, and wood buffalo. Common bird species include the bald eagle, the peregrine falcon, and the osprey. The Mackenzie Valley forms one of North America's most travelled migratory corridors for waterfowl breeding along the Arctic coast. Economic activity in the area is based on subsistence hunting, trapping, and fishing, in addition to a few industrial activities, such as mining and oil extraction.

The **Boreal Shield** is the largest ecozone in Canada, stretching along the Canadian Shield from Saskatchewan to Newfoundland. It is also part of one of the largest forest belts in the world, the **boreal forest**, extending all across North America and Eurasia, encompassing roughly a third of the Earth's forest land and 14 per cent of the world forest biomass. It is the belt that generally separates the treeless tundra regions to the north from the temperate deciduous forests or grasslands to the south. Winters are cold and summers warm to hot, with moderate precipitation. In Canada, the zone is influenced by cold Hudson Bay air masses that yield relatively high precipitation, ranging from 400 millimetres annually in the east to more than 1,000 millimetres in the west. Average January temperatures range from −10 to −20°C, with a July range of 15 to 18°C. The terrain is characteristically rolling, with bedrock outcrops, glacial moraine, and many lakes dominated by coniferous forests of black and white spruce, balsam fir, tamarack, and Jack pine but with significant cover by aspens and white birch. This forest cover is interrupted by large areas of wetlands, particularly moss-dominated bogs. Overall, the boreal forest is characterized by its lack of diversity of tree species; large areas are covered in just one or two species, particularly the spruces. Balsam fir becomes more dominant in areas of heavier precipitation towards the east. Further south, hardwoods such as white birch, yellow birch, and trembling aspen are more common, along with softwoods such as red pine, Jack pine, and eastern white pine.

Characteristic mammals include the woodland caribou, moose, white-tailed deer, black bear, wolf, marten, snowshoe hare, striped skunk, Canadian lynx, and bobcat. Characteristic bird species include the common loon, boreal owl, great

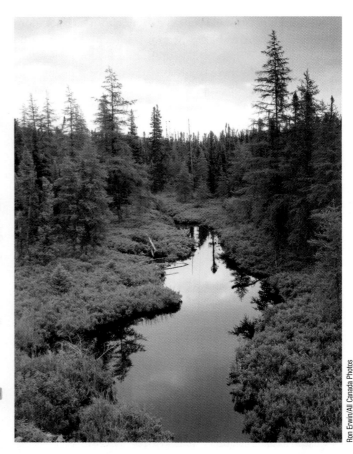

The boreal forest in Newfoundland has bedrock outcrops, lakes, and muskeg.

Ron Erwin/All Canada Photos

horned owl, and evening grosbeak. The rich natural resource base of the Boreal Shield ecozone supports various industries, including mining, forestry, energy, and tourism, in addition to commercial and subsistence hunting, trapping, and fishing.

The **Mixed Wood Plains** ecozone is the most urbanized in Canada, spreading from the lower Great Lakes north through the St Lawrence Valley. The topography is gentle, mainly resulting from the lacustrine, marine, and morainic deposits. The climate is continental, with warm, humid summers and cool winters. Precipitation ranges from 720 to 1,000 millimetres per year. These conditions have produced the most diverse tree coverage in Canada, with more than 64 species, according to Schueler and McAllister (1991). However, few intact areas of natural vegetation remain. In the northern part of the ecozone, the mixed coniferous-deciduous forest is dominated by red and white pine, oak, maple, birch, and eastern hemlock. Further south, the warmer zones contain deciduous species such as sugar maple, beech, basswood, and red and white oak. The white elm is also native to this zone. Unfortunately, many of them have been devastated by an imported fungus that causes Dutch elm disease. First noticed in 1944, the fungus is spread from tree to tree by beetles. It is a good example of the problems created by alien organisms, discussed in more detail in Chapter 3.

White-tailed deer and black bears were once common in this ecozone, but small mammals such as the raccoon, striped skunk, black squirrel, groundhog, and eastern cottontail

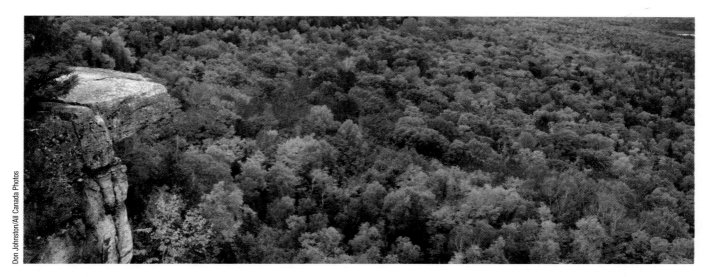

The Mixed Wood Plains ecozone has the highest tree diversity in Canada.

rabbit now dominate. Common bird species include the whippoorwill, blue jay, red-headed woodpecker, great blue heron, cardinal, and northern oriole.

The Mixed Wood Plains is the most densely populated ecozone in the country. Most of the deciduous forest has been cleared for agriculture and urban development, with small patches of forest cover found scattered throughout the zone. Two tree species, eastern white pine and eastern hemlock, were overharvested and are now under-represented. Service industries and the manufacturing sector are the largest employment sectors.

The **Atlantic Maritime** ecozone, stretching from the mouth of the St Lawrence River across New Brunswick, Nova Scotia, and Prince Edward Island, is heavily influenced by the Atlantic Ocean, which creates a cool, moist maritime climate. However, conditions vary considerably between the upland masses of hard crystalline rocks, such as the Cape Breton and New Brunswick highlands, through the coastal lowlands that support most of the population. Mean annual precipitation is as high as 1,425 millimetres on the coast but falls to less than 1,000 millimetres further inland. Temperatures are also moderated by the ocean, with mean daily January temperatures of −2.5 to −10°C and a mean July daily temperature of 18°C. Forests are generally mixed stands of deciduous and coniferous species, such as balsam fir, red spruce, yellow birch, sugar maple, eastern hemlock, and red and white pine, mixed in with boreal species such as black and white spruce, white birch, Jack pine, and balsam poplar. Characteristic animal species include the white-tailed deer, moose, black bear, bobcat, snowshoe hare, wolf, eastern chipmunk, mink, whippoorwill, blue jay, eastern bluebird, and rose-breasted grosbeak.

Major land-oriented activities include forestry, agriculture, and mining. Forest landscapes in the region are undergoing significant change because of overharvesting on private woodlots

and spruce budworm outbreaks, which have increased dramatically in frequency, extent, and severity.

Forest Ecosystem Services

Canada's forest ecosystems provide an array of beneficial services arising from ecological functions such as nutrient and water cycling, carbon sequestration, and waste decomposition. For example, plant communities are important in moderating local, regional, and national climate conditions. Biological communities are also of vital importance in protecting watersheds, buffering ecosystems against extremes of flood and drought, and maintaining water quality. Indeed, the contributions of forest lands to the maintenance of ecological processes (Chapters 2 and 4) within Canada are substantial. However, the sheer scale of Canada's forests means that they are significant contributors on a global scale. It is estimated that 20 per cent of the world's water originates in Canada's forests. The forests are also major carbon sinks, with an estimated 224 billion tonnes stored and a

Cape Breton Highlands.

Many people find forest lands a source of recreational and spiritual fulfillment.

yearly accumulation of some 72 million tonnes. It is likely that this accumulation rate will grow in the future as a result of re-forestation, improved fire suppression, reduced slash burning, and increased recycling of forest products.

Forests are also places of exceptional scenic beauty, and millions of Canadians travel each year to participate in nature-related recreational activities such as wildlife viewing in parks and protected areas, nature walks, and birdwatching. The monetary value of these activities can be significant. For example, more than 20 million Canadians participate in nature-related recreational activities, creating approximately 245,000 jobs and contributing more than $12 billion to Canada's GDP (NRTEE 2003b).

Forest Ecosystem Products

Non-timber forest products. In addition to the important 'services' that forest ecosystems in Canada provide, forests are also a valuable source of commodities. Wild rice, mushrooms and berries, maple syrup, edible nuts, furs and hides, medicines, ornamental cuttings, and seeds—collectively known as **non-timber forest products (NTFPs)** are typical examples. Their total value is unknown, but food products alone were estimated at more than $725 million (Table 9.2) by Natural Resources Canada in 2005. Some NTFPs are harvested commercially and are allocated by licence, while others are freely available and contribute significantly to recreational values, including tourism. These commodities are also important in sustaining First

Nations communities. With careful management, NTFPs are renewable. A useful reference website on NTFPs is maintained by the Centre for Non-Timber Resources at Royal Roads University in Victoria, BC (http://cntr.royalroads.ca)

Table 9.2   Estimated Current Output of Selected Non-Timber Forest Products in the Canadian Economy

NTFP	Output in tonnes or litres (thousands)	Current economic value (thousand $)
Honey	37,072	$160,805
Tree saps	34,761	$163,968
Berries	149,373	$278,654
Mushrooms	1.14	$43,000
Understorey plants	2.30	$75,321
Wild rice	1,013	$3,492
Total		**$725,240**

Source: http://foretscanada.rncan.gc.ca/articletopic/20.

Timber forest products. Despite recognition from federal and provincial governments that Canadian forests provide a broad range of values (wilderness, recreation, wildlife habitat, etc.), forest management paradigms over the past century have focused on the management of Canadian forests to supply wood. The economic benefits arising from timber products are substantial. For some 324 communities, the forest sector makes up at least 50 per cent of the economic base. Direct employment in 2005 was 339,900, or 2.1 per cent of total employment in Canada. Wood industries accounted for 169,200 jobs, paper and allied industries 101,200, logging 45,500, and forestry services 24,000. Employment is spread across Canada but is primarily in Quebec (113,000), British Columbia (79,700), and Ontario (84,500). However, between January 2003 and March 2007, more than 22,000 jobs were lost in Canadian mills (http://www.nrcan.gc.ca/statistics/forestry/default.html).

Canada is the world's leading forest-product exporter, accounting for 15.9 per cent of global trade. In 2006, Canada's forests contributed $36.3 billion to national GDP (Table 9.3). The forestry industry is the largest single contributor to Canada's balance of trade, with exports totalling almost $38 billion in 2007. The value of Canadian forest-product exports decreased by 6.1 per cent to $41.9 billion from 2004 to 2005. In 2005, British Columbia accounted for $13.7 billion (32.7 per cent), Quebec $11.6 billion (27.7 per cent), Ontario $8.4 billion (20.1 per cent), and other provinces $8.2 billion (19.5 per cent). The United States is by far the largest buyer of Canadian forest products. As the Canadian dollar strengthened against the American dollar in 2007 and 2008 and with severe pressure on US housing markets, this reliance on the US market led to a collapse in the forest industry in much of Canada. For

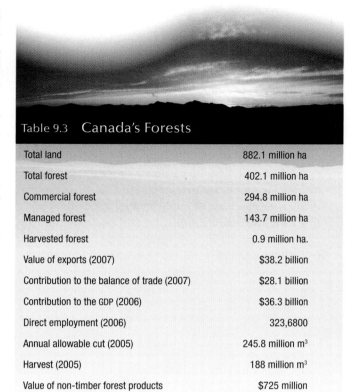

Table 9.3   Canada's Forests

Total land	882.1 million ha
Total forest	402.1 million ha
Commercial forest	294.8 million ha
Managed forest	143.7 million ha
Harvested forest	0.9 million ha.
Value of exports (2007)	$38.2 billion
Contribution to the balance of trade (2007)	$28.1 billion
Contribution to the GDP (2006)	$36.3 billion
Direct employment (2006)	323,6800
Annual allowable cut (2005)	245.8 million m³
Harvest (2005)	188 million m³
Value of non-timber forest products	$725 million

Source: NRCAN 2007b.

example, between January 2007 and March 2008, 34 mills in BC closed, with another 23 either introducing job-sharing or cutting shifts, putting more than 10,000 workers out of jobs.

The volume of wood produced per unit area differs across the country, rising to highs in excess of 800 cubic metres per hectare on the most productive sites in coastal British

Box 9.1   The Competitiveness of Canada's Forest Industry

Canada is the world's largest exporter of forest products. In the international marketplace, Canada has a number of assets, including:

- abundant forest resources;
- high-quality forest resources;
- access to low-cost energy to transform industrial roundwood into secondary forest products;
- proximity to major markets;
- access to well-established export markets;

- rapid expansion of the value-added products sector;
- world-class forestry product research.

Despite these many advantages, the forest industry is vulnerable. Forest products are highly sensitive to economic fluctuations and can suffer severe losses during periods of recession. In addition, the imposition of non-tariff barriers, including codes, standards, eco-labelling requirements, and policies regarding recycled content, can retard growth of the forestry industry, even though they are environmentally sound things to do. The cost of production commodities such as energy, labour, and transportation also plays a key role in competitiveness. As commodity prices continue to rise, the industry must increase efficiency to maintain market share. However, improved efficiency often entails an investment in technology, leading in turn to increased skill requirements for workers and/or a loss of jobs. Furthermore, rising environmental standards, including international certification schemes, are putting significant pressure on the Canadian government to develop ways of implementing standards at minimal cost. To remain competitive and retain a dominant position in the world marketplace, Canada's forest industry will have to tailor products to consumers' increasing expectations, develop new value-added products, develop new markets, and cultivate an image as an environmentally responsible industry.

In 2006 and 2007, three major reports were compiled on the competitiveness of Canada's forest industry by the Canadian Council of Forest Ministers (CCFM 2006), the Forest Products Industry Competitiveness Task Force (2007), and the Conference Board of Canada (2007). All had similar messages, ones that had been known but not given attention for many years. These messages included paying greater attention to research and development, developing new products and opportunities, encouraging new investment, developing human resources, and cutting input costs. The challenge now is that given the forest industry's poor condition in most parts of the country, there is little time or resources available to make these changes. The underlying message here is to continually develop better practices for the future, even when times seem good.

Source: NRCAN 2003; 2007b.

Columbia, where mild temperatures, deep soils, and abundant rainfall create some of the most productive growing sites in the world. Volumes harvested also vary by province. For example, in 2005 British Columbia harvested close to double the volume of any other province while less area was involved in BC than in either Quebec or Ontario (Table 9.4).

The Canadian forestry industry is also a frequent flashpoint of conflict. Names such as Carmanah, Temagami, and Clayoquot became well known across the country in the 1990s as they appeared in newspaper headlines and on national news broadcasts. All these conflicts revolved around questions of whether particular areas should be logged or preserved. These conflicts reflected the increasing appreciation of the many values provided to society by forests besides the usual economic benefits. Few of these non-economic values are easy to calculate in monetary terms and compare against the financial returns of the forest industry. However, these values are gaining increasing recognition on the part of the public and decision-makers as the process of converting old-growth forests across Canada into managed forests continues.

The sheer scale of the industry has also attracted international attention. Concern over the destruction of the tropical forests (Box 9.2) and resulting impacts on the biosphere and on biodiversity has spread from countries such as Malaysia, Indonesia, and Brazil to temperate forest nations and Canada in particular, which

has been dubbed, rightly or wrongly, 'the Brazil of the North'. Governments in Canada have launched large public relations missions around the world to combat this perception. They have also started to look more closely at the basis for these accusations, and substantial changes are underway in many aspects of the forest industry in Canada.

When considering the allocation of any resource to different uses, it only makes good sense to evaluate the relative values that will result from the various allocation decisions. However, resources are commonly allocated in society with little

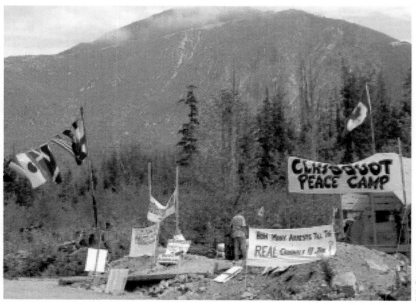

The Clayoquot Peace Camp became the centre for the Clayoquot protests.

Philip Dearden

Table 9.4   Selected Forest Resource Statistics 2005 (*2006)

Province/Territory	Land Area (million ha)	Forest Land (million ha)	AAC (million m³)	Harvest Volume (million m³), Industrial Roundwood	Harvest Area (ha), Industrial Roundwood	Area (ha) Defoliated by Insects	Area (ha) Burned*
Nunavut	201	0.9	n/a	n/a	n/a	n/a	n/a
Northwest Territories	128	33	n/a	24,000	133	35,040	53,397
Quebec	152	85	48	38.5	356,920	129,976*	124,175
British Columbia	95	64	93	87	197,599	11.6 million*	135,634
Ontario	107	68	34	23.4	185,842*	2 million*	149,517
Alberta	65	36	29.2	27.5	79, 797 (2004)	2.7 million	118,782
Saskatchewan	65	24	9.7	5.3	29,183*	140, 088*	1.2 million
Manitoba	64	36	9.6	2.1(2004)	13, 648	47, 168	166,050
Yukon	48	23	238,500	240, 000	1,200*	82, 620	95, 033
Newfoundland and Labrador	40	20	2.5	2.4	22, 890	57, 845	3, 436
New Brunswick	7.3	6.2	11.3	11(2004)	111,348 (2004)	1, 504 (2003)	506
Nova Scotia	5.5	4.3	7.6	6.2	54, 326	17, 561 (2003)	1,576
Prince Edward Island	0.58	0.27	0.46	0.6*	2, 195*	1,110	51
Canada	979	402	245.8	191	1.04*	15.9	2.1 million

Sources: http://foretscanada.rncan.gc.ca/statsprofile; http://foretscanada.rncan.gc.ca/quickfacts/landclassification#ca.

Box 9.2   Forests: A Global Perspective

- Some 40 per cent of the land surface of the Earth supports trees or shrubs. Approximately 27 per cent is covered in trees, with the rest in open scrub.
- The largest forested areas are in Europe (1,040 million hectares), South America (874 million hectares), Africa (650 million hectares), Asia (548 million hectares), and North America (471 million hectares). Four countries—Canada, Russia, the US, and Brazil—contain more than 50 per cent of the world's forests.

- Each year, about 13 million hectares of the world's forests are lost to deforestation, but the rate of net forest loss is slowing down, thanks to new planting and natural expansion of existing forests.
- From 2000 to 2005, the net forest loss was 7.3 million hectares per year—an area the size of Sierra Leone or Panama and equivalent to 200 square kilometres per day.
- Deforestation accounts for up to 20 per cent of the global greenhouse gas emissions that contribute to global warming.
- Primary forests are lost or modified at a rate of 6 million hectares per year through deforestation or selective logging.
- Plantation forests are established at a rate of 2.8 million hectares per year.
- Plantations cover more than 187 million hectares (less than 5 per cent of total forested area) and account for 20 per cent of wood production.
- The 10 countries with the largest net forest loss per year between 2000 and 2005 (Brazil, Indonesia, Sudan, Myanmar, Zambia, United Republic of Tanzania, Nigeria, Democratic Republic of the Congo, Zimbabwe,

Bolivarian Republic of Venezuela) had a combined net forest loss of 8.2 million hectares per year.
- The 10 countries with the largest net forest gain per year between 2000 and 2005 (China, Spain, Vietnam, United States, Italy, Chile, Cuba, Bulgaria, France, and Portugal) had a combined net forest gain of 5.1 million hectares per year as a result of afforestation and natural expansion of forests.
- Estimates suggest that more than 80 per cent of the world's terrestrial species are found in forests. The tropical forests are our richest terrestrial biome. Tropical forests occupy only 7 per cent of the world's land area, but they contain more than half of the world's species.
- Developing countries consume more than 80 per cent of their wood as fuel; in developed countries, only 16 per cent goes to fuel, with the rest being processed as wood products. Approximately 1.5 billion tonnes of wood is harvested for fuel annually worldwide.
- Most wood products (85 per cent) are used domestically.
- Global trade in primary wood products totalled US $186 billion in 2005. Developing countries are responsible for only about 13 per cent of world trade in wood products.
- Global forests provide wage employment and subsistence equivalent to 60 million work-years worldwide, 80 per cent of which is in developing countries. More than 1.6 billion people depend to varying degrees on forests for their livelihoods (e.g. fuelwood, medicinal plants, and forest foods).
- People in developing countries consume much less wood products (30 cubic metres per 1,000 people) and paper (12 tonnes per 1,000 people) than people in developed countries (300 cubic metres of wood products per 1,000 people and 150 tonnes of paper per 1,000 people).
- It is estimated that an additional 50 to 100 million hectares of forest will be needed to meet the projected wood demands of developing countries by 2010.
- Worldwide, an estimated 290 million hectares of forested land are under protection from logging. However, of 200 areas of high biological diversity, 65 per cent are threatened by illegal logging. Illegal logging is estimated to cost governments approximately $15 billion annually.

Can tropical forests, such as these in Sri Lanka, sustain the needs of both the people and animals that depend on them?

Sources: http://www.fao.org/forestry/site/sofo/en; http://www.fao.org/forestry/site/fra/en.

appreciation of their true value. Anielski and Wilson (2005) undertook a study to assess the total economic values of Canada's boreal forest. They found that the value of the non-timber forest products outweighed that of the timber products by a ratio of 2.5 to 1 (Table 9.5), yet these values are rarely if ever taken into account in decisions about how to allocate the boreal forests among competing uses.

Table 9.5 Boreal Ecosystem Wealth: Monetary Economic Values and Regrettable Costs[a] for Canada

Natural Capital Accounts (2002$ per annum)[b]

Forest market values
- $14.9 billion in estimated market value of forestry-related GDP in the boreal region (est. 2002)

Costs
- $150 million in estimated cost of carbon emissions from forest industry activity in the boreal region (deduction against forestry-related GDP)

Non-market values
- $5.4 billion in value of pest control services by birds
- $4.5 billion for nature-related activities
- $1.85 billion for annual net carbon sequestration (excludes peatlands)
- $575 million in subsistence value for Aboriginal peoples
- $79 million in non-timber forest products
- $18 million for watershed service (i.e., municipal water use)
- $12 million for passive conservation value

Wetlands and peatlands non-market values
- $77.0 billion for flood control and water filtering by peatlands only
- $3.4 billion for flood control, water filtering, and biodiversity value by non-peatland wetlands
- $383 million for estimated annual replacement cost value of peatlands sequestering carbon

Minerals and subsoil assets market values
- $14.5 billion in GDP from mining and oil and gas industrial activities in the boreal region (est. 2002)

Costs
- $541 million in federal government expenditures as estimated subsidies to oil and gas sector in the boreal region
- $474 million in government expenditures as estimated subsidies to mining sector in the boreal region

Water resources market values
- $19.5 billion in GDP for hydroelectric generation from dams and reservoirs in the Boreal Shield ecozone (est. 2002)

Waste production costs (emissions to air, land, and water)
- $9.9 billion in estimated air pollution costs to human health

TOTAL market values (forestry, mining, and oil and gas industrial activity and hydroelectric generation)
- $48.9 billion

Less cost of pollution and subsidies
- $9.9 billion in air pollution costs
- $474 million in government subsidies to mining sector
- $541 million in federal government subsidies to oil and gas sector
- $150 million in forest sector carbon emissions

NET market value of boreal natural capital extraction
- $37.8 billion

TOTAL non-market value of boreal ecosystem services
- $93.2 billion

RATIO of non-market to market values
- 2.5

Notes:
[a] These are either environmental or societal costs associated with market-based activities (e.g., forest industry operations).
[b] A GDP-chained, implicit price index was used throughout the study to standardize to 2002 dollars.
[c] Based on European Union air pollution cost estimates for SO_2, NO_x, $PM_{2.5}$, and VOC for 2002.

Source: Anielski and Wilson 2005. Reprinted by permission of the publisher.

Forest Management Practices

The provincial governments are responsible for 71 per cent of the nation's forests, with the federal and territorial governments responsible for 23 per cent. The remaining 6 per cent is managed by 425,000 private landowners. These figures are likely to change over the next decade as an increased number of land claims by Aboriginal peoples are settled and more land comes under their control. Areas under current land claims account for about one-quarter of large, intact forest landscapes in Canada. Currently, forest management is primarily a provincial responsibility, with governments managing forest resources on behalf of the public through agreements with private logging companies. Different forms of tenure exist, but generally all involve the submission of plans by the logging company that outline where it intends to cut, the details of the harvesting process (including the location of roads), and

reclamation plans. The governments provide regulations and guidelines for these practices and have the authority to ensure that they are followed.

Of the total of 402 million hectares of forest land in Canada, some 143 million hectares is currently managed for timber production (Table 9.3). On these lands, forest ecosystems are being transformed from relatively natural systems to control systems, as described in Chapter 1, in which humans, not nature, influence the species that will grow there and the age that they will grow to. Over the past decade, increasing awareness of and concern about the environmental impact of forest harvesting has prompted questions about the environmental sustainability of forestry and about the different kinds of management approaches that might lead to sustainability. Key questions relate to the amount of forest protected from logging, the amount of fibre harvested over a specific period, the way by which it is logged, and what happens to the land after harvesting.

Guest Statement

The Forest Tenure Reform Imperative
Kevin Hanna

Forest tenure refers to the conditions that govern forest ownership and use. It is a basic element in shaping forest policy and management. Tenure reform must play a central role in creating solutions to significant challenges in Canadian forestry—such as investment in reforestation and innovative management and product development. Because it is a difficult subject, governments, industry, labour, and environmental groups in Canada have tended to ignore tenure. Sadly, there is little evidence that this is about to change.

About 94 per cent of Canada's forest land is owned by governments, largely the provinces. About 80 per cent of Canada's private forestland is located east of Manitoba, most of it in the Maritimes. British Columbia has the highest level of provincial forest ownership—96 per cent. Compared to other major forest nations, Canada's tenure outline is unique. Our two closest export competitors, Finland and Sweden, have very different tenure profiles. Both are also markedly smaller in area than Canada. But

they have developed forest industries that are globally competitive and innovative in their product development and that have become leaders in international investment. In contrast, Canada's forest industry lags.

Finnish forests are mainly privately owned. Individuals, mostly pensioners and farmers, own about 62 per cent of forest land, timber companies 6 per cent, and the Finnish government 31 per cent. There are some 280,000 private forests, with an average size of 37 hectares. But these small holdings are particularly productive and characterized by high levels of forest management investment. They supply the bulk of Finland's industrial wood needs and provide about 80 per cent of the stumpage income and 80 per cent of annual growth and cut. In Sweden, small-scale landowners own about 50 per cent of the forest land, while the state and forest companies each have about 25 per cent. There are about 240,000 private forests in Sweden, and about 30 per cent of them are less than 50 hectares. They also supply a major part of Sweden's timber needs. In Canada, while wood production on private lands has grown, the vast bulk of timber still comes from provincial forests.

In Sweden and Finland, forestry investment levels (regeneration, tending and harvesting techniques, and worker training) tend to be relatively high. Forests also hold a special place in the psyche of each nation. But there are problems and controversies in Scandinavian forestry—simmering conflict over preservation, the ecological weaknesses of the national 'plantation forests', and the impact of mechanization on employment. But overall, a strong culture of forest stewardship permeates the policies of these two Nordic countries.

Many who work in Canada's forest industry would say that a culture of stewardship also exists here. Workers care about sustaining the resource on which their livelihoods depend. They might also suggest that it is the short-term vision of companies and governments that limits the potential for such a culture to really flourish. Canada's forest industry seems to be among the least innovative sectors in our business community. But industry will counter that tenures are unstable, the times too short, or conditions too uncertain to increase investment levels, develop non-timber forest products, or acknowledge the services that forests provide beyond timber. So we might ask: is tenure reform necessary to make our forest industries winners and world leaders in forest management, product innovation, and investment in sustainability? Yes, tenure reform is an essential start.

How could tenure be changed in Canada? A solution offered at various times has been to emulate parts of the Scandinavian model. Large companies might be more willing to invest in innovative forestry practices if they owned the land. But there might not be much public support for selling provincial forests, certainly not to large companies, nor would many large firms necessarily want to buy forests. After all, the present system has served them well, and some firms have been innovative under existing tenure forms. Another option is to increase the management role of communities through community-held tenures. Tenure opportunities for small firms and individuals could be provided or expanded, perhaps creating many small

Kevin Hanna

private forests. But would small holdings, community forestry, or corporate ownership result in better forestry practices?

Experience in the US and Canada shows that some companies with private forests are not always good forest managers. Some have logged their lands quickly for short-term profit, made few investments in regeneration, and moved to sell denuded forest land as 'development land'. But others treat their forests as the foundation of their long-term survival. Community tenures can also result in a more stable, long-term vision of forest management. Alternatively, some communities might support a short-term timber emphasis as they seek immediate employment and prosperity. Small private holdings would require capacity-building and stable investment sources. And just as some large firms have, individuals may also log their lands quickly for the sake of short-term profit. In Scandinavia, the governments act to dull some of these negative effects by regulating private forestry practices, requiring forest plans, guiding cutting rates, and financing forestry—all while still supporting private land ownership.

In Canada, tenure reform will require a careful consideration of the lessons seen in other places. While we can look to other jurisdictions for information and experience, Canada's forests, geography, and cultures are distinct. As part of addressing Canada's forest management problems, tenure reform will entail innovation. Outlining the various solutions and evaluating their potential is more than can be done here; each tenure alternative has its promises and pitfalls.

The answer certainly lies in creating a tenure context that provides new opportunities for individuals, community-based tenures, building conditions that encourage stable long-term business investments, and developing a new and complex vision of the diverse benefits our forests provide. In other words, we need a balance between regulation, state ownership, market mechanisms, and more private forest land control. Tenure reform is a necessary part of realizing sustainable forests and stable forest communities across Canada.

Kevin Hanna teaches in the Geography and Environmental Studies Department at Wilfrid Laurier University in Waterloo, ON, where his main research interests are in community-based resource management.

Rate of Conversion

The rate of conversion of natural to managed forests is one of the most controversial issues in Canadian forestry. Each province establishes an **annual allowable cut (AAC)**, which is based on the theoretical annual increment of merchantable timber, after taking into account factors such as quantity and quality of species, accessibility and growth rates, and amounts of land protected from harvesting because of other use values, such as parks and wildlife habitat (Table 9.6). The AAC should reflect the **long-run sustained yield** (LRSY) of a given unit of land, or what that land should yield in perpetuity. This is ultimately limited by the growth conditions, the biological potential of the site, and how that potential can be augmented by silvicultural practices. It is not sustainable to have an AAC that consistently exceeds this biological potential. Economists, however, often argue for the need to maximize the monetary return of the first cut in order to invest in other wealth-producing programs and to provide social services. The dominance of this line of thought has led to rates of conversion significantly higher than can be supported biologically.

To calculate AACs, it is also necessary to know the rotation period for each forest type. This is the age of economic maturity of the tree crop and varies widely but usually falls within the 60- to 120-year range in Canada. Foresters call this the **culmination age**.

The AAC will also vary substantially depending on the proportion of old-growth to second-growth timber included in the proposed cutting unit. **Old-growth forests** have very high timber volumes—for example, up to 800 cubic metres per hectare in BC's coastal forests. However, at the culmination age for **second growth** on these sites, volumes will be much lower, in the region of 500 cubic metres. This is known as the **falldown effect** and results in AACs up to 30 per cent lower as old-growth forests are eliminated.

For Canada, the total AAC is calculated by adding together all the provincial and territorial AACs where these figures are available. In 2005, the AAC stood at 245.8 million cubic metres; 191 million cubic metres of timber were harvested, or 20 per cent less than the estimated 2005 wood supply. Canada's wood supply is estimated by combining provincial AACs with the wood supply estimated for private, federal, and territorial lands. Harvest levels on private, federal, and territorial lands are not regulated by legislation, although the managers of these lands often set harvest targets. This means that wood supply from these lands can only be estimated based on the sum of these targets and, for lands where targets have not been set, the average of past harvest levels. Thirty-eight per cent of Canada's wood supply is in BC, followed by Quebec (20 per cent), the Prairies (20 per cent), Ontario (14 per cent), and the Maritimes (9 per cent).

Silvicultural Systems

Silviculture is the practice of directing the establishment, composition, growth, and quality of forest stands through a variety of activities, including harvesting, reforestation, and site preparation.

Harvesting Methods

Perhaps no aspect of resource or environmental management has created as much conflict in Canada as the dominant forest harvesting practice of **clear-cutting**, and this method is consistently used on much of the forest lands harvested in Canada (Figure 9.3). The size of the clear-cuts varies widely, from approximately 15 hectares to more than 250 hectares, and in some cases clear-cuts extend for many thousands of hectares. Although openings of this size have few ecological advantages, it should not be assumed that more but smaller clear-cuts are necessarily superior to fewer, larger clear-cuts. More clear-cuts create more fragmentation and less undisturbed forest area, to the detriment of 'interior' forest species.

Not only are clear-cuts aesthetically unappealing to many Canadians, but their environmental impact, especially cumulatively as they spread across the landscape, can be substantial. For example, Nova Scotia has implemented guidelines for the establishment of wildlife corridors to reconnect habitat. If clear-cuts are in excess of 50 hectares, the guidelines

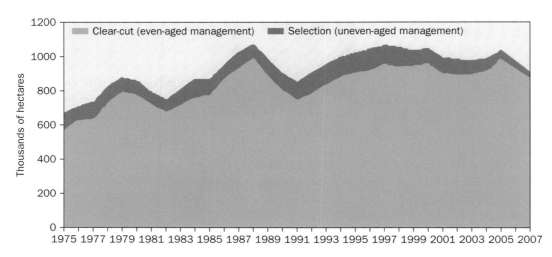

Figure 9.3 Silviculture: Area harvested, 1975–2007. Source: CCFM 2007.

recommend that at least one corridor be created, with irregular borders and a minimum width of 50 metres. While this is a step in the right direction, it is the implementation of such guidelines, not just their specification, that is important. British Columbia tried to take a tougher approach by legislating forest practices in detail through the Forest Practices Act. Yet current government policy has seen considerable weakening in the implementation of the legislation as economic conditions have become less favourable to the forest industry.

The rate of conversion from natural to managed forests has alarmed many environmentalists, who claim that forest companies have been allowed to extract too much wood too quickly. This will result not only in environmental problems but also in a lack of adequate fibre for industrial use in the future.

A clear-cut area.

Clear-cutting is the preferred means of harvest on vast areas. It is the most economical way for the fibre to be extracted and also allows for easier replanting and tending of the regenerating forest. In certain types of forests, it may mimic natural processes more closely than selective or partial cutting systems. This is especially true where natural fires have created even-aged stands of species such as lodgepole and Jack pine, black spruce, aspens, and poplars. Researchers working in the eastern boreal forest, however, are starting to question this assumption (Bergeron et al. 2001; Lesieur, Gauthier, and Bergeron 2002). They have found a dramatic decrease in fire frequency in this area over the past 150 years as a result of climatic change, and they suggest that this trend will continue into the future, leading to a higher proportion of old-growth forest in the landscape. In turn, this will lead to natural species replacement in these forests, with deciduous and mixed stands replaced by balsam fir and Jack pine by black spruce. Clear-cutting would counteract these changes rather than mimic natural processes and would lead to a dramatic decrease in stand diversity at the landscape level. These authors suggest that in the future, forest management will have to employ mixed harvesting systems (Table 9.6) if forest management is to more closely emulate natural systems.

Reforestation

Until 1985, Canada's forests were considered to be so extensive that little effort was given to reforestation. Sites, once logged, might be burned to facilitate rapid nutrient return to the soil but then were abandoned in the hope that they would be satisfactorily recolonized by seeds from the surrounding area. Sometimes this was successful, but often it was not. Thus, with increased harvesting levels, the amount of land that no longer supported trees that could be harvested in the future gradually grew. In 2005, 0.9 million hectares of land were harvested, a further 2.1 million hectares were burned by fire, and 15.9 million hectares were defoliated by insects, yet less than 450,000 hectares were planted and reseeded (Figure 9.4), and the success of these plantings cannot be guaranteed. Given this annual deficit between what is cut and what is replanted, many conservationists have difficulty understanding further allocation of old-growth timber to the forest industry.

Site Preparation—Biocide Use

Biocides are used on forest lands in Canada to reduce competition for seedlings on replanted sites and to protect seedlings from insect damage. Sites regenerating from forest harvesting return to an earlier successional phase (Chapter 3), and under natural conditions, a vigorous and diverse secondary succession takes place. However, for many years the community will not be dominated by the commercial species desired by foresters. Chemicals are used to suppress early successional species, to compress the successional time span, and to maximize the growth potential of the more commercially desirable species, usually conifers. Chemical use is generally quicker, easier, and more effective than using mechanical alternatives for weed

Table 9.6   Main Characteristics of Common Silvicultural Systems Practised in Canada

Clear-cutting

The most commonly applied silvicultural system in Canada, clear-cutting involves the removal of all trees in a cutblock, in one operation, regardless of species and size. Some trees are left along riparian zones to protect streams. The objective is to create a new, even-aged stand, which will be regenerated naturally or through replanting.

Advantages	Disadvantages
• Cost-effective. Clear-cut areas are easily accessed for site preparation and tree planting.	• Nutrients stored in the bodies of trees are removed from the ecosystem.
• Stands of even-aged trees are created, producing wood products with more uniform qualities.	• Loss of habitat for some wildlife species. Loss of biodiversity. New vegetation does not maintain the complexity and stability of mature forests.
• Newly planted seedlings quickly take root and grow in the sunlight reaching the ground. This can benefit certain animal species.	• Clear-cutting in sensitive ecosystems can cause soil erosion, landslides, and silting, which can damage watersheds, lead to flooding, and inhibit successful fish reproduction.
• In some respects, clear-cutting simulates natural disturbances such as wildfire and insect disease outbreaks.	• Large gaps are opened up, fragmenting the forest and exposing more area to the edge effect.
• Safest harvesting method with least risk of worker injury.	• Aesthetically unattractive. Clear-cutting can conflict with other forest values.
	• No timber products for a long period of time (e.g., 50 to 70 years).

Seed tree

Method of clear-cutting in which all trees are removed from an area in a single cut, except for a small number of seed-bearing trees, which are intended to be the main source of seed for natural regeneration after harvest. Tree species that have been managed under this system in Canada include western larch, Jack pine, eastern white pine, and yellow birch.

Advantages	Disadvantages
In addition to the above:	In addition to the above:
• Next to clear-cutting, this system is the least expensive to implement.	• Regeneration can be delayed if seed production and/or distribution are inadequate.
• The system can result in improved distribution of seedlings and a more desirable species mix, since the seed source for natural regeneration is not limited to adjacent stands.	

Shelterwood

Mature trees are removed in a series of two or more partial cuts. Residual trees are left to supply seed for natural regeneration and to supply shelter for the establishment of new or advanced regeneration. The remaining mature cover is removed once the desired regeneration has been established. Thirty to 50 per cent canopy removal on the first cut is common. In Canada, this system has mainly been applied to conifers (e.g., red spruce in the East, white pine in Ontario, interior Douglas fir in the West).

Advantages	Disadvantages
• Trees left after the first cut grow faster and increase in value.	• Complex and costly to plan and implement.
• For eastern white pine, this system can be an effective management tool against white pine weevils, which are more attracted to pine shoots under full light exposure than under shade.	• Young trees can be damaged during removal of mature trees.
• Visually more appealing than clear-cutting.	• Windthrow is a serious concern. Uprooted stems can displace significant amounts of soil.

Selection

Involves the periodic harvest of selected trees of various ages in a stand. Trees are harvested singly or in groups as they reach maturity. Valuable, mature trees, along with poorly shaped, unhealthy, crooked, and leaning trees and broken or damaged trees, are selected for removal. The objective of this method is to create and maintain an uneven-aged stand. Small gaps created by harvesting leave room for natural seeding.

Advantages	Disadvantages
• This method is often favoured in areas where recreation or scenic values are important, since the harvested area is less visually offensive.	• This system can only be successfully applied to stands containing shade-tolerant tree species (e.g., sugar maple, western red cedar, red spruce, balsam fir, eastern/western hemlock).
• The method results in a continuous, regular supply of mature trees over time. Overall stand quality should improve after each harvest cycle.	• Requires skilled workers to implement successfully.
• Biodiversity loss is minimized.	• Can require the maintenance of more roads and skid trails per unit area.
	• Complex and costly system to plan and implement.
	• In some instances, landowners take the best trees in a forest, leaving poorly shaped and unhealthy trees to provide seed for the next generation, resulting in forest degradation over the long-term.

Source: NRCAN 1995; Alberta Food and Rural Development 2001; Canadian Institute of Forestry 2003.

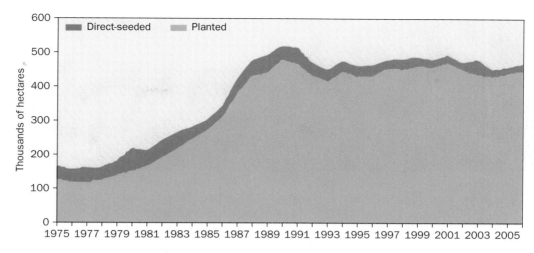

Figure 9.4 Silviculture: Area planted or direct-seeded, 1975–2006. Source: CCFM 2006.

suppression. Three herbicides (2,4-D, glyphosate, hexazinone) are registered for forest management in Canada. Glyphosate (or 'Roundup'), the most widely used, affects a broad spectrum of plants but degrades quickly and is relatively non-toxic to terrestrial animals.

Research shows that early colonizers often compete more effectively for soil nitrogen than conifers (Hangs, Knight, and Van Rees 2003). Where nitrogen is the most limiting factor, as it is in the boreal forest, this can inhibit conifer growth over the short term. However, the law of conservation of matter (Chapter 4) tells us that these nutrients have not disappeared—they are simply being held by different species. As these species die and decay, the nutrients will once more be returned to the soil and become available for uptake. Furthermore, by holding nutrients in this way, early colonizers often slow down the loss of nutrients from the site that might otherwise occur as a result of leaching. They act as a biological sponge over the short term. Herbicide application may also eliminate species that are ecologically advantageous, such as nitrogen fixers like the red alder, exacerbating nutrient loss from logged sites.

The balance between these effects needs to be evaluated over long periods and probably differs from site to site. Lautenschlager and Sullivan (2002) reviewed the literature on the effects of forest herbicide applications on major biotic components of regenerating northern forests. They conclude that there is currently very little evidence to support any long-term negative consequences. Indeed, they suggest that at the landscape scale, application of herbicides might help in the return to a more natural species composition in northern forests. This is because since colonization, hardwoods have expanded in these areas as coniferous species have been harvested. Use of herbicides may help to reverse this process. Nonetheless, Lautenschlager and Sullivan caution that there are still large gaps in knowledge before we can truly understand all the biotic implications of herbicide application.

Insecticides are used to attack pests such as the spruce budworm, Jack pine budworm, hemlock looper, mountain pine beetle, gypsy moth, and forest tent caterpillar. The amount sprayed varies, depending on the population dynamics of these insects, which changes in response to environmental factors. Most spraying has occurred in eastern Canada; the spruce budworm spraying program in the Maritime provinces is the best-known incidence of such spraying and has caused considerable controversy (Box 9.3). However the total area treated with forest chemicals in Canada has steadily declined since 1990 (Figures 9.5 and 9.6).

The health and ecological concerns arising from widespread application of chemicals (discussed further in Chapter 10) have led to several high-profile confrontations. Attention is being increasingly directed towards the replacement of synthetic insecticides with biological control agents such as *Bacillus thuringiensis* (Bt). Bt now accounts for almost all the insecticide used in Canadian forests and is non-toxic to humans and most wildlife, although it does affect moth and butterfly larvae of some non-target species. Plants also manufacture many chemicals themselves as protection against insects. Several of these chemicals appear to be good prospects for the development of insecticides for forestry use in the future.

Pheromones are volatile compounds, or 'scents', used by insects of a given species to communicate with each other. There are various types of pheromones serving different purposes—e.g., alarm, aggregation, territorial marking, tracking or recognition, and sex pheromones. Knowledge regarding the identification of pheromone components and determining how they are used by insects is now being applied to developing synthetic pheromones that can be used for tracking and monitoring insect pests or for controlling such pests through mass captures or mating disruption.

Installing a few pheromone traps as opposed to systematically sampling trees substantially reduces the time spent on

Box 9.3   The Spruce Budworm Controversy

Since 1952, aircraft have been dousing the forests of eastern Canada, particularly New Brunswick and Nova Scotia, with an array of chemicals in the competition to see who would harvest the area's lumber—humans or the eastern spruce budworm. A native in Canada, the budworm feeds primarily on balsam fir but will also eat white spruce, red spruce, and, to a lesser extent, black spruce. Its range extends wherever balsam fir and spruce are found, from Atlantic Canada to the Yukon. Damage varies considerably from one stand to another. Mortality rates are related to stand composition and age, as well as site quality (soil, water, climate). Ordinarily, the influence of the insect goes unnoticed in a forest, but its impact occasionally reaches epidemic levels, with massive damage to the host trees. Outbreaks last between six and 10 years and have been documented for the past two centuries as the product of a long-term budworm-fir ecological cyclic succession. However, spruce budworm epidemics are occurring with increasing frequency as a result of human intervention related to forest harvesting. Human intervention has reduced the natural diversity of the forest through removal of preferred species such as white pine, creating a less diverse forest composed of large areas of mature balsam fir, the budworm's preferred food. Extensive mortality occurs in stands that have suffered defoliation for several years. In 1975, at the height of infestation, 54 million hectares of forest were defoliated, resulting in serious economic losses for the forestry industry.

Population levels of the spruce budworm have been declining, following a downward trend that started in 2003 and a precipitous drop in 2004. The area affected also declined, from 755,325 hectares in 2004 to 696,483 in 2005. The spruce budworm is still present in central Canada, although at reduced levels. However, the population is expected to grow over the next five to 10 years as habitat and environmental conditions become favourable.

The long-term ecological stability of the budworm-fir system does not match the shorter-term demands of the economic system dependent on the forests for products. As a result, the forests have been extensively sprayed to limit defoliation and mortality. When spray programs started in 1952, DDT was the chemical of choice, and some 5.75 million kilograms was sprayed in New Brunswick alone between 1952 and 1968, when use was suspended.

Other chemicals replaced DDT, such as phosphamidon, aminocarb, and fenitrothion, until questions were raised about their ecological and health impacts. Phosphamidon, for example, is very toxic to birds. Up to 1985, 118.5 million hectares (mostly in New Brunswick) was sprayed, some areas on an annual basis. Continual spraying appears necessary once natural controls are disrupted.

The organophosphate fenitrothion then became the most popular chemical. It also became the source of heated controversy as to its health and ecological impacts. In particular, there was concern over the link between the chemical and Reye's Syndrome, a rare and fatal children's disease. As a result, and because of unfavourable reviews of the ecological impact of the chemical, particularly on songbirds, the use of fenitrothion in aerial applications was cancelled by 1998. The biological control *Bacillus thuringiensis* (Bt) is now being used more extensively. In some areas, it is supplemented by an insecticide called Mimic that kills the budworm through interruption of the moulting process, starving the larvae to death. Although Mimic is not a broad-spectrum biocide, it has the same lethal effect on the larvae of butterflies and moths. Nonetheless, it has been used in some areas, such as Manitoba, since 1998.

Attention is also being devoted to other less toxic approaches, such as species and landscape diversification, so that large stands of mature balsam fir do not dominate the landscape. Biological control is also being investigated with a parasitic wasp that attacks the larvae of the budworm.

Spruce budworm.

Western provinces also use chemicals to try to limit damage from the spruce budworm. Manitoba, for example, uses dimethoate and malathion, and Bt is commonly used in Saskatchewan.

Management of spruce budworm outbreaks provides a graphic example of the challenges presented by the conflict between longer-term ecological cycles and shorter-term economic dependencies.

tracking and monitoring. This technique also makes it possible to increase the area covered at a cost lower than or equivalent to that of traditional sampling. Tracking is used to determine whether an insect population has exceeded a critical threshold, whereas monitoring is designed to estimate the extent of the damage that the next generation of larvae could cause. Armed with this information, researchers can determine whether phytosanitary measures are required and can select the best time for implementing such measures. The most common types of traps are sticky traps, the capacity of which is limited to the size of the sticky surface, and large-volume traps. Sticky traps are inexpensive and easy to use. Large-volume traps have a receptacle containing an insecticide that kills the insects that gather at the base of the trap.

Intensive Forest Management

After a new crop of trees has been established and reaches a free-growing condition, future timber resource values can

be further enhanced by intensive silvicultural practices. Activities undertaken by foresters to improve stand growth include:

*   pre-commercial thinning (homogenizes the stand, increases mean tree size, and lowers the age at which the stand can be harvested);
*   commercial thinning (attempts to recover lost volume production in a stand as a result of competition-induced mortality);
*   scarification (physical disturbance of the forest floor to create improved seedbeds for natural regeneration);
*   prescribed burning (removes slash and woody debris, sets back competing vegetation, provides ash as fertilizer, and increases nutrient mobilization and availability through increased soil temperatures);
*   pruning and shearing (increases the value of individual trees by prematurely removing the lower branches so

*Other includes all provinces and territories except BC, Ontario, Quebec, and NB.

Figure 9.5 Forest insecticide use by product and province, 1990–2006. Source: National Forestry Database.

*Other includes all provinces and territories except BC, Ontario, Quebec, and NB.

Figure 9.6 Forest herbicide use by product and province, 1990–2006. Source: National Forestry Database.

that clear wood, free of knots, is laid down around an unpruned knotty core);
- timber stand improvement (cutting down or poisoning all deformed and unwanted trees within older stands).

The long-term impacts of these activities are generally not well known. Thompson, Baker, and Ter-Mikaelian (2003) reviewed the published information on some of the likely impacts on vertebrate wildlife in boreal forests in Ontario. One important impact identified was the structural simplification that occurs with increased tending. However, overall they conclude that intensive forest management will benefit some species and have negative effects on others and that there is currently little evidence to suggest that these techniques will have any greater impact than more traditional silvicultural activities.

Fire Suppression

In certain areas, fire is a frequent occurrence and necessary to the reproduction of forest tree species. Fire is part of the long-term dynamics of these ecosystems. Fire initiates secondary successions, renewing vegetation through regeneration involving a complete change of species, regeneration of the same species, or diversification of the species. Jack pine, birch, and trembling aspen are common in areas where fires have recently occurred. Fire suppression, viewed as essential to protect lives, property, and commercially valuable timber, has resulted in ecological changes not characteristic of fire-dominated ecosystems. For example, it has contributed to the very dense regeneration of almost pure Douglas fir in old-growth ponderosa

pine stands in the interior of BC. Old-growth ponderosa pine stands are maintained by low-intensity, naturally occurring surface fires that burn brush and prevent maturation of the more shade-tolerant Douglas fir.

In the absence of recurring fires, ground fuel may accumulate over time, increasing the risk of a major wildfire event (Box 9.4). The area affected by wildfires in Canada each year is immense: during the 1990s, an average of 8,248 fires burned 3.2 million hectares annually, including more than 700,000 hectares of commercial forest land, or 74 per cent of the annual area harvested (Neave et al. 2002). From 2001 to 2006, there were 48,102 fires, burning 12.3 million hectares,

Alder is a common pioneer of secondary succession on logged-over sites. The smaller species in the foreground is broom, a non-native plant from Europe. Both these species biologically fix nitrogen into the soil and help fertility levels to recover.

an average of 2 million hectares and 8,017 fires per year (http://canadaforests.nrcan.gc.ca/statsprofile/ca).

Environmental and Social Impacts of Forest Management Practices

Change in species and age distributions arising from forest management practices has a major impact on ecological processes such as energy flows, biogeochemical cycles and the hydrological cycle, and the habitat for other species. We have only a rudimentary knowledge of how forest ecosystems function. It is therefore difficult to be precise about the possible impacts of wholesale conversion from complex natural to more simple human-controlled systems. In addition, there are important differences among

New forestry would devote much more attention to stand tending on commercial sites, such as this one, where stand limbing and thinning enhances growth.

Philip Dearden

Box 9.4   British Columbia's Burning

Imagine a landscape in only black and white . . . a place between a beginning and an end where the only sound is the whisper of flames and the only colour is fire-red.
—Firefighter, McLure, BC, 2003

This was the sinister scene as catastrophic flames surged across the BC Interior, consuming more than 175,000 hectares of forest, demolishing more than 350 homes and businesses, and forcing the evacuation of more than 50,000 people. By midsummer of 2003, a state of emergency had been declared. Highways and parks were closed as thousands of firefighters, Canadian military personnel, emergency response workers, and community volunteers battled some 800 forest fires burning throughout the province. The areas that suffered the greatest hardship and adversity from the disaster were the smaller communities, such as Barriere, McLure, Louis Creek, and Kelowna. The total economic impact on the province was estimated at more than $500 million.

Although many factors contributed to the violent summer fires, including several years of drought, global warming, and carelessly discarded cigarettes, perhaps the greatest culprit was decades of poor forest management practices that focused on timber production and fire

Firefighting in Kitimat, British Columbia.

Andrew Farquhar/Valan Photos

suppression. For example, the 2003 fires may well have been aggravated by the fact that naturally occurring, low-intensity ground fires, which normally release nutrients into the soil, kill insect infestations, and clear the forest floor of debris and small trees, have been suppressed for generations. When a fire finally started, instead of being a restorative force, the excessive debris caused an environmental disaster, burning the entire forest area to the ground.

Catastrophic fires like those that swept through the BC Interior do not just move along the ground—they race through the treetops and the root networks, killing even tree species, like ponderosa pine, western larch, and Douglas fir, that have evolved thick bark to survive fire. Furthermore, such huge fires do not release nutrients into the soil but vaporize them into the air, generating enough heat to create their own weather system with conditions that tend to spark even more fires.

In future, we need to recognize that fire is a natural and essential part of the forest ecosystem. Rather than trying to stop all fires, or letting every fire burn, we need to use selection, cutting, and controlled burns as a way to thin out fuel and restore vigour and diversity to our over-managed forests. In short, we need to start managing forests for overall ecosystem health rather than for timber production alone.

forest ecosystems. Some, such as the boreal forest, have naturally evolved with periodic disturbances such as fire or insect attack that stimulate forest renewal. Others, such as the rain forests of the west coast, have little history of disturbance. The difference between disturbance through forestry and natural processes (Box 9.5) must be considered against this background. One essential difference is that natural disturbances such as fire or insect attack do not result in the physical removal of the biomass from the site; it is merely converted from one form to another at that site, consistent with the law of conservation of matter (Chapter 2). In contrast, logging results in the physical translocation of nutrients from the site. The closer that forest

Box 9.5   Forest Disturbance: Natural versus Clear-cut

Some forests are more susceptible to disturbance than others, and different disturbances have differing impacts. There are similarities between some disturbances and clear-cutting, but there are also important differences. Some of the differences between the effects of fire and clear-cutting are:

- Openings created by fire are generally irregular in shape, with high perimeter-to-edge ratios that facilitate natural reseeding. Boundaries tend to be gradual rather than the abrupt edge of a clear-cut.
- Fires leave standing vegetation in wet areas that continues to provide habitat for wildlife and acts as a natural seed source. Clear-cuts remove all trees.
- Fires tend to kill pathogens; clear-cutting allows many pathogens to survive.

- Fire releases nutrients into the soil; clear-cutting removes nutrients in the bodies of the trees.
- Fire helps to break up rock that aids in soil formation; clear-cutting tends to physically disturb the site, leading to compaction and erosion.
- Fire stimulates growth of nitrogen-fixing plants that help to maintain soil fertility; clear-cutting does not.
- Fire encourages the continued growth of coniferous species in many areas through stimulating cone opening; clear-cutting often leads to dominance by shade-intolerant hardwoods. Ultimately, this changes species composition, as was found in studies in Ontario where in 1,000 sampled boreal sites, regenerating poplar and birch had increased by 216 per cent and spruce had fallen by 77 per cent (Hearnden et al. 1992).

harvesting approximates the conditions of natural perturbations, the less disturbing it will be to ecosystem processes.

Chapter 2 described how energy flows through ecosystems. It is also stored in various compartments. Trees, for example, represent energy stored in the autotrophic component of the ecosystem; deer represent storage in the herbivorous component. Thus, a forester may wish to maximize the energy storage in trees and minimize energy losses to herbivores, whereas a wildlife or recreation manager may wish to move the energy storage further up the food chain to support more wildlife. Conflicts therefore arise about where energy should be stored in ecosystems to optimize societal values. This section describes some of the environmental and social implications of forestry activities.

Forestry and Biodiversity

Most natural forest land is dominated by forests with old-growth characteristics, although the age of the trees and degree of structural and compositional attributes described for old growth vary greatly across Canada. Old-growth forests have ecological attributes that tend to be absent from forests that have been harvested.

Various definitions of old-growth forests have emerged. This one was suggested by the Forest Land Use Liaison Committee of British Columbia: 'Old-growth forests include climax forests but do not exclude sub-climax or even mid-seral forests.' The age structure of old growth varies significantly by forest type and from one biogeoclimatic zone to another. The age at which old-growth forests develop their characteristic structural attributes varies according to forest type, climate, site characteristics, and disturbance regime. However, old growth is typically distinguished from younger stands by several of the following attributes:

- large trees for species and site;
- wide variation in tree sizes and spacing;
- accumulations of large dead, fallen, and standing trees;
- multiple canopy layers;
- canopy gaps and understorey patchiness;
- decadence in the form of broken tops or boles and root decay.

Old-growth forests typically contain trees that span several centuries. Based on age, approximately 18 per cent of Canada's forests can be classified as old-growth. In Canada, the longest-lived tree species are yellow cedar on the west coast and eastern white cedar in central Canada. Trees of both these species may live more than 1,000 years.

Old-growth forests supply high-value timber, contain large amounts of carbon, contain a large reservoir of genetic diversity, provide habitat for many species, regulate hydrologic regimes, protect soils and conserve nutrients, and have substantial recreational and aesthetic values.

Logged forests undergo a number of changes, including modifications of the physical structure of the ecosystem and changes in biomass, plant species mixtures, and productivity. These changes affect biodiversity directly and indirectly by changing the nature of the habitat. Since more than 90,000 species depend on forest habitats in Canada, this is obviously of concern. About 65 per cent of species considered at risk of endangerment in Canada are forest-related species (NRCAN 2005b). The largest concentrations of these species are in the Carolinian forests and the coastal forests of BC.

Direct changes arising from forestry practices include the effects on the genetic and species richness of a biotic community. Most species contain a wide range of genetic variability that helps them adapt to changes in the environment. As natural forests are replaced by plantations, this natural variability is reduced, since most plantation-grown trees are specially selected from the same genetic base to have desirable characteristics. This makes them more susceptible to pest infestations and disease and less able to adapt to future environmental changes (Box 9.6). For example, it has been suggested that the emergence of the spruce forest moth in central and eastern Canada since 1980 is at least partly due to the establishment of white spruce plantations. More than 85 per cent of the forest harvesting in Canada is done by clear-cutting in which large blocks of forest are removed at the same time. This does not destroy the ecosystem per se in that an ecosystem still exists on that unit of land, but it does dramatically alter the attributes of that ecosystem. For vegetation, changes include removal of the previously dominant trees and their ecological influence, followed by the vigorous growth of other assemblages of plants (where herbicides have not been applied) as the successional process starts again.

Early successional species dominate in the immediate post-harvesting phase but are much reduced over time as the canopy closes. Tree species such as alder, birch, cherry, Jack pine, poplar, and aspen often fit into this category, along with semi-woody shrubs such as elderberry and blackberry, annual and short-lived perennial herb species such as members of the aster family, and various grasses and sedges. Other species that existed in low abundance in the original forest may survive the harvest and, freed from the competitive suppression of the harvested trees, may dominate the community for some time. Trees such as red maple, yellow birch, and white pine often fall into this category in central and eastern Canada. As the canopy closes, most of these species will eventually be out-competed. Some species that survive the harvest or invade from surrounding areas may be found through all stages of succession. They have low light compensation thresholds, allowing survival under heavy shading. Balsam fir, hemlock, sugar maple, and beech fall into this category.

Once clear-cuts have had the opportunity to start regenerating, species diversity increases rapidly and usually results in a plant community that is more species-rich and diverse than the harvested community. The exception is where replanting takes place, which involves few species and where steps are taken to reduce competition for plantation trees. Herbicides such as Roundup are commonly applied to achieve this and

Box 9.6   The War against the Mountain Pine Beetle—Who's Winning?

The outbreak of mountain pine beetles in the interior forests of BC is one of the most dramatic changes ever to happen to BC's forest landscape. The mountain pine beetle epidemic is spreading throughout BC's range of lodgepole pine forests as a result of a combination of natural beetle population cycles, continuous mild winters, and an abundance of uniformly mature pine stands. Since 1997, mountain pine beetles have infested more than 9.2 million hectares of pine forests in BC, totalling some 582 million cubic metres of wood by 2006. Defying predictions, the beetles have spread as far east as Great Slave Lake in Alberta, and by 2007 an estimated minimum of 3 million trees in Alberta had been affected, up from 20,000 in 2005. The extent in BC is more than double that predicted by the models in 2001, and there is no sign of a slowing down of the invasion. By 2013, more than 80 per cent of the mature pine will have been consumed, and it is likely that the epidemic will only be over once it has infested most of the mature pine in BC. There is no doubt who is winning the war!

Bark beetles are small, cylindrical insects that attack and kill mature trees by boring through the bark and mining the phloem (the layer between the bark and wood of a tree). The mountain pine beetle and other bark beetles are native species and natural and important agents of renewal and succession in BC Interior forests. However, when the beetles reach epidemic levels, natural predators like woodpeckers cannot reproduce quickly enough to maintain the insect population at manageable levels.

The beetles have damaged and/or killed more than $9 billion worth of lodgepole pine, enough to build 3.3 million homes. The dead and dying trees, visible from the air as huge swaths of red and grey forest throughout central BC, are the source of a hotbed of controversy as government agencies try to combat the rate of infestation through increases in the annual allowable cuts for some areas, reductions in environmental regulations and planning requirements on treatment units, and reduced stumpage (fees paid for logging on public lands). Beetle-infested trees retain economic value and can be salvage-harvested, but since wood quality and value

decline with time since attack, there is significant pressure to cut beetle-infested timber promptly to maximize the economic value of salvaged trees. The recommended size of clear-cut areas is set at 60 hectares in the BC Interior, but exemptions for salvage loggers have resulted in cutblocks as large as 1,300 hectares, more than 20 times the recommended maximum. The salvage clear-cuts can be placed adjacent to previously logged areas that have not yet regenerated, creating potentially very large contiguous openings. Furthermore, little attention has been devoted to the hydrological implications of suddenly deforesting large areas in the Interior, and now greater flood damage is predicted as a result.

Accelerated cutting of large attacked areas can prevent the decline of timber value in the short term, but a number of social and environmental costs are associated with this approach. Increases in AACs in response to insect infestations can disrupt forest plans, oversupply markets and cause short-term decreases in timber value, and affect employment levels when workers are no longer needed to support the temporary increase in harvest levels. Research in forest ecology has increasingly recognized the essential role that natural disturbances play in shaping forest structure and maintaining ecosystem processes. Small infested areas create patchiness at the stand level, increasing local diversity, and generate habitat and food for a large number of species. In addition, outbreaks accelerate forest succession (Chapter 3) to other tree species in some areas, as well as creating the conditions for stand-replacing fires that maintain the dominance of lodgepole pine.

The approach of the Ministry of Forests has received considerable attention from environmental groups, which view the response of government agencies as drastic, particularly in light of a government announcement that BC parks and protected areas would not be exempt from the war against pine beetles. Environmentalists generally advocate natural control methods. Cold winters or fires will kill mountain beetle larvae, reducing the size of beetle populations. However, winters are now warmer than ever before, and many scientists point to the beetle outbreak as an indicator of the kinds of disturbance that will happen more commonly in the future. Some environmental groups, such as the David Suzuki Foundation, are calling for an **ecosystem-based management** (EBM) approach for outbreaks of the pine beetle. An EBM approach (Chapter 6) would seek to decrease the size of the affected area as well as the duration of outbreaks while not compromising forest biodiversity and ecosystem integrity.

Despite nearly a century of actively managing the mountain pine beetle, efforts to suppress the outbreaks across BC and other parts of North America have been largely unsuccessful. Why do you think this is? What approach would you recommend, given the value of lodgepole timber stands? Do you think provincial parks should be open to clear-cutting as a response to the beetle infestation?

The mountain pine beetle.

A view of mountain pine beetle–damaged landscape.

may be applied consistently until the planted trees become established. Under these circumstances, an artificial lack of diversity is created, just as a farmer creates a similar system to maximize the amount of energy stored in the particular component that he or she wishes to harvest (Chapter 10).

Naturally regenerating clear-cuts may also produce a higher biomass of herbivorous species such as white-tailed and mule deer that require brushy habitats for at least part of the year. Both the quantity and quality of browse is greater in regenerating clear-cuts up to the time that the canopy starts to close and is usually optimal during the first eight to 13 years. In many parts of their range, white-tailed deer are more abundant than they were before European colonization when the landscape was covered mostly in mature forest. However, even these species benefit most from a pattern of small clear-cuts, since they prefer edge habitats where the protective cover of the forest is not too far distant. Optimal clear-cut size for deer in southern Ontario, for example, is around two hectares.

Unharvested forests in most areas across Canada comprise a patchwork of forest stands of different ages and different diversities regenerating from the effects of various natural disturbances such as fire and insect attack. Clear-cutting at a certain rate and scale may not be inappropriate in some of these ecosystems. However, some forests, such as the coastal forests of BC and the mixed deciduous forests of southern Ontario and Quebec, are heavily influenced by the pattern of death of individual canopy trees. For example, more than half of the coastal rain forest is more than 250 years in age, and much smaller interventions are required to mimic natural processes.

Some animal and bird species require forests with old-growth characteristics—for example, ample lichen growth for woodland caribou or the presence of dead trees to provide nesting cavities for birds such as woodpeckers (Table 9.7). If forest harvesting takes place at a rate and scale that eliminates stands with these characteristics, these species will decline in numbers and may become extirpated. The case study presented in Box 9.7, the spotted owl, is a good example of the difficulties associated with maintaining populations of species dependent on old-growth habitat. Other species that have suffered in this regard include the woodland caribou, American pine marten, and marbled murrelet (Neave et al. 2002).

The pressure on high-profile species such as the grizzly bear (Box 9.8) is not the only concern in considering the destruction of old-growth forests; less well-known species, even unknown ones, and the kinds of ecological functions undertaken by these species are also at risk as the old-growth forests are cut away. Some 85 per cent of the species in some forests are arthropods such as insects and spiders. Only recently is the richness of this fauna being realized. There may be more than 1,000 species of invertebrates within a single forest stand. Many of these species are new to science, and we have little idea of their ecological role in maintaining healthy ecosystems. However, research has indicated a clear pattern of different spider species associated with different stages of the successional process, with at least 30 years required in even the fastest-growing forests for the spider community to recover from clear-cutting (McIver, Moldenke, and Parsons 1990). Studies of the upper branches of coastal forests in BC are revealing very complex predator-prey relationships among many different species. These relationships are not replicated in managed forests, suggesting that continued elimination of old-growth habitat will lead to species extinctions, a decrease in genetic diversity in these communities, and removal of natural controls on forest pests.

One kind of impact on biodiversity that is often overlooked results from human intrusion to undertake silvicultural and other activities. For example, after looking at the impact of forestry on grizzly bears in the Selkirk Mountains of

Table 9.7   Examples of Forest-Dwelling Species at Risk

Mammals	Birds	Plants	Reptiles
Endangered			
American marten	Acadian flycatcher	American ginseng	Blue racer (snake)
Vancouver Island marmot	Kirkland's warbler	Bashful bulrush	Night snake
Wolverine (eastern population)	Northern spotted owl	Blunt-lobed woodsia	Rocky Mountain tailed frog
Woodland caribou	Prothonotary warbler	Cucumber tree	
	White-headed woodpecker	Deltoid balsamroot	
	Western yellow-breasted chat	Drooping trillium	
		Heart-leaved plantain	
		Large whorled pogonia	
		Nodding pogonia	
		Prairie lupine	
		Purple twayblade	
		Red mulberry	
		Seaside centipede (lichen)	
		Small whorled pogonia	
		Spotted wintergreen	
		Tall bugbane	
		Wood-poppy	
Threatened			
Ermine haidarum subspecies	Hooded warbler	American chestnut	Black rat snake
Pallid bat	Marbled murrelet	Deerberry	Blanding's turtle
Wood bison	Queen Charlotte goshawk	Goldenseal	Eastern Massasauga rattlesnake
Woodland caribou		Kentucky coffee tree	Jefferson salamander
		Lyall's mariposa lily	Pacific giant salamander
		Phantom orchid	
		Purple sanicle	
		Round-leaved greenbrier	
		Scouler's corydalis	
		White wood aster	
		White-top aster	
		Yellow montane violet	

Sources: Natural Resources Canada 1995; Alberta Food and Rural Development 2001; *Canadian Institute of Forestry 2003.*

Box 9.7   Case Study: The Northern Spotted Owl

In the latter part of the 1980s, the northern spotted owl in the western United States became the focal point of high-profile conflicts between conservationists and logging interests. The owl was accorded threatened status throughout its entire range in the US under the US Endangered Species Act. A 'threatened status' designation means that the owl is likely to become an endangered species within the foreseeable future throughout all or a significant part of its range. This designation requires that critical habitat be identified and a recovery plan implemented. Because the owl is dependent on old-growth forests, this decision led to severe conflicts with forest harvesting activities. In 1994, the Clinton administration established reserves on more than 4 million hectares where harvesting would be severely restricted.

The spotted owl also occurs in Canada in the old-growth forests of southwestern British Columbia. This northern extension of the range is important, since individuals at the extremes of a species' range are often the most important to protect because they may have the genetic diversity best suited for future adaptability. Furthermore, just as in the US, the old-growth forest on which the owl depends was allocated for harvest. The spotted owl therefore provides a good case study of the impact of forest harvesting on biodiversity.

Biology and Range

The northern spotted owl is found from northern California to southwestern BC. Scientists estimate that historically, 500 pairs of spotted owls once inhabited the old-growth forests of southwestern BC. Between 1985 and 1993, some 39 active spotted owl sites were recorded, totalling a minimum of 71 adult owls. By 2003, research reported only 25 breeding pairs left in the province; by 2007, this number was down to 11. The historic range of the species in the province is probably not that much different from the current range, although the distribution of the owl within that range has changed significantly, primarily because of destruction of its prime habitat, old-growth forests.

Superior habitat for the owls has old-growth characteristics, such as 'an uneven-aged, multi-layered multi-

species canopy with numerous large trees with broken tops, deformed limbs, and large cavities; numerous large snags, large accumulations of logs, and downed woody debris; and canopies that are open enough to allow owls to fly within and beneath them' (Dunbar and Blackburn 1994, 19). Main prey items for northern spotted owls are small mammals such as the northern flying squirrel and dusky-footed and bushy-tailed wood rats. Both prey species are abundant in old-growth forests.

Threats

The single greatest threat to the spotted owl is the continued logging of old-growth forests, leading to loss of suitable habitat. Estimates suggest that probably less than 50 per cent of the old-growth habitat that once covered the Lower Mainland of BC is suitable habitat, and much of it is highly fragmented. The provincial government has approved logging in at least three and as many as six of the 10 areas in which the owl was detected in 2003. Meanwhile, even British Columbia's largest logging companies have agreed to a voluntary halt to logging in spotted owl habitat.

Provincial and regional parks provide some protection in two main blocks, covering some 110,000 hectares in total. Unfortunately, these blocks are about 85 kilometres apart, with little interconnecting habitat. It is extremely unlikely that the two populations will be able to interbreed. Provincial forests managed for timber production contain much larger amounts of suitable habitat, but much of this land is scheduled to be harvested over the next 100 years. In 2001, the provincial government gave the logging company Cattermole Timber preliminary permission to log 88 hectares of old-growth forest near Anderson River, north of Hope, a decision that was challenged and later upheld in the BC Supreme Court of Appeal. The northern spotted owl is listed as endangered by the Committee on the Status of Endangered Wildlife in Canada (COSEWIC), but nothing in BC law 'precludes approval of a forest development plan if there is any element of risk to a forest resource, even where that forest resource is an endangered species' (*Vancouver Sun* 2003).

Management Options

Concern over the future status of the spotted owl resulted in the formation of a Spotted Owl Recovery Team in 1995 to develop a recovery plan for the species. The primary goal was to outline a course of action for stabilizing the current population and leading to an improvement of the status of the species so that it could be removed from the endangered species category. Sixteen management options, which correspond with the different categories of abundance status outlined by COSEWIC, are discussed in detail in Chapter 14. The options range from banning all timber harvesting that could degrade suitable owl habitat within the entire range of the owl to no owl management beyond existing parks and protected areas. The costs associated with the former option in terms of withdrawing this much land from the operable forest have yet to be estimated, but they are likely to be significant. This, however, is the price that society must pay in terms of forgone timber revenues if an old growth–dependent species such as the spotted owl is to survive. Unfortunately, the spotted owl is not alone in its predicament. Of the forest-related species reassessed by COSEWIC (see chapter 14) between 1999 and 2007, 17 per cent were placed in a higher category of endangerment, and only 4 per cent were assigned to a lower category or removed from the list. Although these reassessments may in part result from new information coming to light, they are the exact opposite of what is desired.

The northern spotted owl.

Jared Hobbs/All Canada Photos

BC, Wielgus and Vernier (2003) reported that the main impact seemed to be disturbance from roads rather than avoidance of clear-cuts or young forests.

From this brief review of the impact of forest harvesting activities on biodiversity, we can conclude that substantial changes can occur. These changes are beneficial to some species and detrimental to others. It is critical, therefore, to understand the complex ecological relationships involving forests if these changes are to be fully evaluated and taken into account in designing harvesting activities. And it is essential to maintain some areas in their original state to maintain landscape biodiversity. Currently, about 8 per cent of Canada's forest area is protected by legislation.

Cameron (2006) has reviewed some of the interactions between protected areas and the so-called 'working forest' in Canada and suggests:

- retaining or restoring natural climax forest species composition;

- reducing edge contrast between working forest and protected areas;
- maximizing protection of watercourses draining into protected areas;
- planning road networks to minimize undesirable effects on nearby protected areas.

Many conservationists do not like using the term 'working forest' to differentiate forests that are logged from forests that are not. They maintain that it is the unlogged forests that are in fact working the hardest for society and to imply that they are not working does not reflect modern scientific understanding of the ecosystem value of unlogged forests.

When enlightened forestry companies realize how whole forests work, they come up with innovative solutions. For example the Canadian Parks and Wilderness Society (CPAWS) and the forestry corporation Tembec have negotiated a minimum 50-year halt on logging in an area used extensively by woodland caribou on the east side of Lake Winnipeg. Habitat

Box 9.8   Forestry and Grizzlies

The grizzly bear once extended across most of North America. Unlike its smaller cousin, the black bear, however, it is not able to tolerate human disturbance, and as human populations grew, the grizzly populations shrank. Some of the densest remaining populations are in the lowlands of the coastal valleys in BC, where the bears are attracted not only to the annual salmon runs but also to the abundant berry-producing shrubs and other nutritious vegetation found on the flood plains. These nutrient-rich sites also constitute some of the most productive forestry sites.

Studies over the past 10 years have documented declines in grizzly populations following logging activities. One of the main problems was the changes in vegetation that occurred as a result of logging. The favourite forage foods of the grizzly are those that compete with the re-establishment of trees as secondary succession takes place (Chapter 3), and because the re-establishment of trees has meant profit for logging companies and governments down the road, the mixed vegetation has often been suppressed by the use of chemicals. In turn, this helps to create dense stands of conifers, which have the long-term effect of shading out preferred forage for the grizzly. The result is a critical shortfall in grizzly food over an extended period.

Concern about these changes has led to new approaches. For example, lower replanting densities for conifers and more open space between tree clusters are being implemented in an effort to mimic the natural environment. The use of chemicals is restricted to the tree stands. A new adaptive management approach has been formulated for silviculturists dealing with grizzly bear habitats. It involves not only the measures outlined above for new logging activities but also revisiting past sites to undertake remedial action. The approach not only is adaptive but also exemplifies an ecosystem approach, as discussed in Chapter 5, which takes a more holistic view of resource management by accounting for the limits of tolerance (Chapter 4) of one of the most spectacular animal species in Canada. Although it is too early to assess the success of these efforts, in combination with the absolute protection of some of these sites, such as a grizzly reserve established in the Khutzeymateen Valley, they should help to ensure that grizzly bear populations survive in the coastal valleys of BC.

The grizzly bear is not very tolerant of human disturbance and has suffered large declines in number and distribution since European colonization.

protection is key to maintaining populations of this threatened species, since they are extremely sensitive to human developments. Estimates suggest that the Manitoba woodland caribou population has decreased by 50 per cent since 1950, and in 2006 the species was listed as threatened under the Manitoba Endangered Species Act. The 26,000-hectare area where harvesting is deferred is the 'winter core zone' of the Owl Lake woodland caribou herd—in other words, the lands the herd uses most during Manitoba's cold months, the most critical time of year.

Forestry and Site Fertility

Forest harvesting removes nutrients from the harvested site (Figure 9.7). The amount of nutrients removed depends on the kind and extent of harvesting. Selective tree whole-stem harvesting removes relatively few nutrients compared to large clear-cuts of complete-tree (above- and below-ground biomass) harvesting. The latter maximizes the short-term yield of biomass from the forest but may compromise the potential of that site over the long term to produce further harvests. More than half of the harvesting in Canada is done with complete-tree systems.

To judge the potential effects of forest harvesting on site fertility, it is necessary to consider the size of the soil nutrient pool, the amount of nutrients being removed, the net accretions and depletions of nutrients in the forests, and the ways in which these variables interact. The process of site impoverishment over time as a result of harvesting is shown in Figure 9.8. On some sites, the proportion of nutrient capital removed in the biomass will be relatively minor, while on others it may be substantial. This depends to a large degree on the existing nutrient capital of the site. Areas with high soil fertility will be less affected. Some sites will recover quickly from harvesting and can sustain relatively short rotations. Other sites will not recover adequately between rotations, and site nutrient capital will fall, making tree regeneration difficult and in some cases impossible. Thus, nutrient-deficient sites should have long rotations with just stem harvesting in order to maintain productivity.

The amount of nutrients removed by harvesting is influenced by tree species, age, harvesting method, season of harvesting, and other factors. Older trees contain larger amounts of nutrients—such as nitrogen, phosphorus, potassium, calcium, and manganese—than younger trees. There is also considerable variation among species in the amount of nutrients organically bound and the nutrients preferentially held by different species. The differences between whole-tree and stem-only harvests can also be significant. In a study of black spruce in Nova Scotia, Freedman et al. (1981) found an almost 35 per cent increase in biomass take for whole-tree harvesting. The loss amounted to 99 per cent for nitrogen, 93 per cent for phosphorus, 74 per cent for potassium, 54 per cent for calcium, and 81 per cent for magnesium, derived mainly from the nutrient-rich foliage and small branches. When deciduous trees are harvested, the loss of nutrients can be reduced significantly by cutting during the dormant period when leaves are not present.

Forest harvesting can also lead to dramatically increased rates of nutrient loss through leaching (the downward movement of dissolved nutrients) to the hydrological system. The amount of loss varies according to the intensity and scale of the harvest and the particular ecosystem. Loss of nitrate is of most concern, since it is not only often a dominant limiting factor (see Chapter 2) but also the nutrient that is lost most often in large quantities. One reason for this is disturbance in the nitrogen cycle (Chapter 4) by logging, which results in an increase in the bacterial process of nitrification turning ammonium to nitrate. Nitrate is highly soluble, resulting in significant losses of nitrogen site capital in some ecosystems, particularly if the soil is not too acidic. Other factors, such as warmer soil temperatures, decreased uptake by vegetation, and abundant decaying organic matter on the forest floor, also contribute to increased losses of soluble nutrients following logging. In addition, younger stands (less than 145 years old) support a smaller biomass of lichens with nitrogen-fixing abilities compared to old-growth trees on the west coast of BC.

It is also important to consider nutrient inputs. Precipitation adds substantial amounts of nutrients over time. For a maple-birch stand in Nova Scotia, for example, Freedman, Duinker, and Morash (1986) calculated that it would take 96 years of precipitation to replace the nitrogen lost through whole-tree removal, 83 years for potassium, 166 years for calcium, and 41 years for magnesium. Other nutrient inputs occur through dry deposition of gases and particulate matter, the weathering of minerals, and the fixation of atmospheric dinitrogen. Soil mycorrhizae are particularly important for the fixation of atmospheric dinitrogen (Chapter 4), but land treatments after clear-cutting, such as slash burning and pesticide use, may adversely affect fixation rates.

Forestry and Soil Erosion

Besides its direct influence on nutrients described above, forest harvesting can have a substantial impact on soil through erosion, especially on steep slopes in areas of heavy precipitation. Such losses also contribute to loss of site fertility, remove substrate for further regrowth, and contribute to flooding and the destruction of fish habitat. Poor road design and maintenance are often key factors behind accelerated soil erosion. Cutting roads across steep terrain exposes large banks of unprotected topsoil. Compacted road surfaces encourage

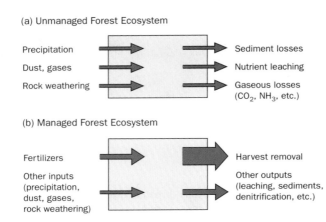

Figure 9.7 Nutrient inputs and outputs from managed and unmanaged forest ecosytems.

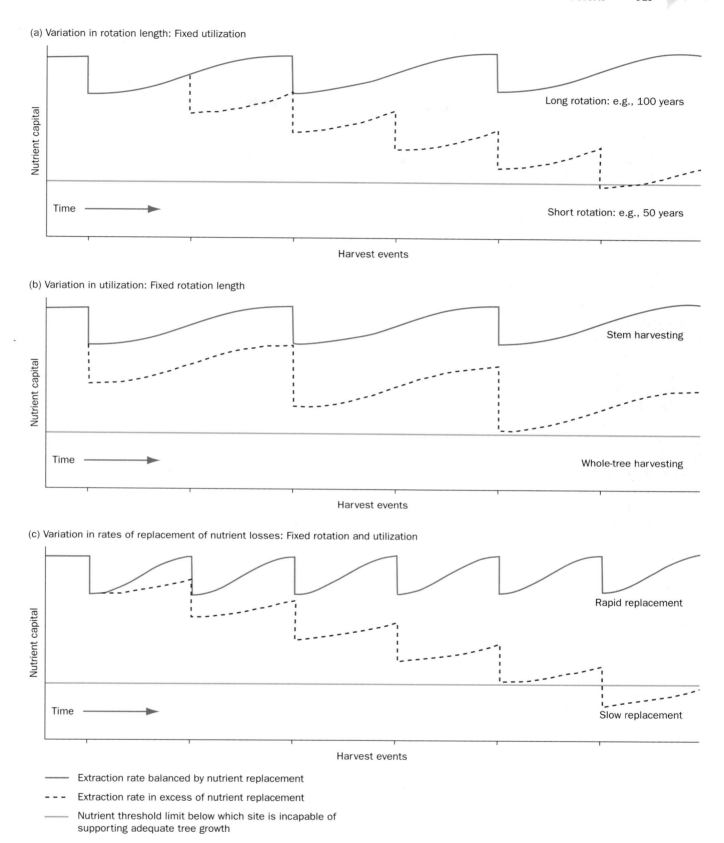

(a) Variation in rotation length: Fixed utilization

Nutrient capital

Time ⟶

Long rotation: e.g., 100 years

Short rotation: e.g., 50 years

Harvest events

(b) Variation in utilization: Fixed rotation length

Nutrient capital

Time ⟶

Stem harvesting

Whole-tree harvesting

Harvest events

(c) Variation in rates of replacement of nutrient losses: Fixed rotation and utilization

Nutrient capital

Time ⟶

Rapid replacement

Slow replacement

Harvest events

—— Extraction rate balanced by nutrient replacement

- - - Extraction rate in excess of nutrient replacement

—— Nutrient threshold limit below which site is incapable of supporting adequate tree growth

Figure 9.8 Site impoverishment as a result of forest harvesting. Source: Adapted from Kimmins 1977.

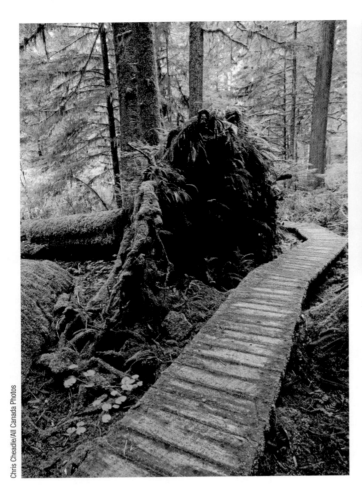

Chris Cheadle/All Canada Photos

When an old-growth tree such as this giant Sitka spruce on the west coast of Vancouver Island dies, the nutrients it contains recycle to fuel new tree growth. When logging removes the whole tree, these nutrients are lost to the ecosystem.

Philip Dearden

Logging has resulted in severe changes in the morphology of many coastal streams in BC, with large amounts of logging debris accumulating following flooding.

overland flow with high erosive power. Many jurisdictions are implementing much stricter regulations on road construction. Other regulations to minimize soil erosion losses include leaving buffer strips along watercourses, using minimal impact logging techniques, and pursuing selective harvesting rather than clear-cutting.

Forestry and Hydrological Change

Large-scale forest harvesting can have a significant impact on hydrology. Under natural conditions, large amounts of water are returned into the atmosphere by the trees through the process of **transpiration**. **Evapotranspiration** includes transpiration plus evaporation from non-living surfaces. These processes are normally greatest during the summer. Removal of the trees significantly reduces this mechanism and other storage capacities, releasing large amounts of water into stream flow. Flooding in high-discharge periods can often be the result. Furthermore, without the delaying mechanism of the trees, and with soils that have suffered compaction through harvesting, the speed of flow is often increased, again raising the potential for flooding as well as for erosion. In turn, sediment from erosion can damage fish spawning beds.

Forestry and Climate Change

Forests are a carbon sink. They take in carbon dioxide and convert it to wood, leaves, and roots. They are also a carbon source. They release stored carbon into the atmosphere when they decompose or burn. Because of this ability to both absorb and release huge amounts of carbon dioxide (a major greenhouse gas), forests play a major role in the global carbon cycle (Chapter 4), the exchange of carbon between the atmosphere and the biosphere. More carbon is stored in forest biomass (trees and other living plants), dead organic matter, and soil than is contained in the atmosphere. That is why forests are a key part of the global carbon cycle.

Large changes in forest carbon sinks and sources, whether due to human or natural causes, can affect the climate by altering the amount of carbon dioxide in the atmosphere. As the climate changes, forest carbon storage will be affected. A warmer climate can speed up vegetation growth, which means more carbon storage. However, it can also accelerate decomposition, resulting in more carbon emissions, and boost the risk of drought, pest outbreaks, and fire, all of which can significantly reduce carbon storage. The extent of these effects is also influenced by the amount and/or timing of precipitation changes.

A rapidly changing climate has important implications for the forest sector and the more than 300 communities whose livelihood is closely associated with forests. One example is the effects on timber supply. Growth and yield databases used in timber supply forecasting will need to be re-evaluated because of changing tree growth and productivity. Long-term timber supply planning may also need to take into account changes in species composition over time. More frequent large-scale disturbances will cause timber supply fluctuations and result in more salvage-harvesting of trees killed by disturbances, which affect fibre quality.

Because wood continues to store carbon even after it is made into products (such as lumber and paper), only a fraction of the carbon removed from the forest is actually emitted into the atmosphere. As well, some of the wood-waste from product manufacture is burned to produce energy, offsetting fossil-fuel use. After harvest, 40 to 60 per cent of the carbon remains in the forest in the roots, branches, and soil and decomposes slowly, providing nutrients for the newly regenerating forest. Natural disturbances such as forest fires and insect infestations release large amounts of carbon dioxide into the atmosphere, although the areas affected and emission levels can vary considerably from year to year. The area of forest burned each year is on average 2.5 times the area harvested and is projected to increase under warmer, drier climate conditions. A significant difference between fires and harvesting is that with harvesting, much of the carbon ends up being stored in long-lived products, while with fires, the carbon goes into the atmosphere.

Analysis using the Canadian Forest Service's Carbon Budget Model indicates that Canada's managed forests were a net carbon sink in most years between 1990 and 2005. But in some years they were a source, mostly because of wildfires (Figure 9.9). From 1990 to 2005, Canada's managed forests sequestered an average of 50 million tonnes of CO_2 per year. Annual amounts ranged from a large sink of 174 million tonnes of carbon dioxide equivalents (CO_2e) in 1992 to a large source of 137 million tonnes of CO_2e in 1995, mostly because of wild-fires. There has also been a slight increase in carbon losses resulting from forest harvesting over the years. International accounting rules dictate that they be included with other carbon losses, although in fact the carbon will remain stored in timber products for many years. The impact of the mountain pine beetle infestation in western Canada increased towards the end of the 1990–2005 period. In 2006, BC Ministry of Forests and Range surveys detected 9.2 million hectares under attack by the mountain pine beetle. This infestation will affect the sink/source balance and trends for many years to come.

Deforestation, the permanent conversion of land from forest use to other uses such as agriculture and urban and industrial use, has slowed over the years (Figure 9.10), although it remains a subject of major international concern and negotiations (Chapter 7). Worldwide, deforestation creates about 20 per cent of human-generated greenhouse gas emissions, more than is produced by the transportation sector. In Canada, deforestation accounts for less than 3 per cent of national emissions, a figure that is declining. In 2005, an estimated 56,000 hectares of forest was converted to other land uses, equivalent to 21 million tonnes of CO_2e, down from 70,000 hectares (29 million tonnes) in 1990. However, this is still considerably in excess of the sink amount of roughly 1 million tonnes of CO_2 per year created by afforestation (planting forests on land previously used for other purposes, usually agriculture).

Considering carbon sequestration in timber management is different from any concept that has previously been applied, but it might become a major factor in how we manage our forests. Neilson et al. (2007), for example, point out that hardwood stands in New Brunswick typically contain 10 to 20 per cent more carbon per hectare than a similar volume of softwoods of similar age. Forest managers might well change their management practices in the future as global climate change becomes more severe and grow multi-aged hardwood stands that promote carbon sequestration instead of the single-aged softwoods that now dominate their thinking.

Forestry and Non-Timber Forest Products and Services

Historically, our forests' non-timber products and services have received little attention in Canada. However, as timber harvest levels have increased and the public has become more aware of and vocal about declines in these other forest values, forestry companies are being required to take these values into

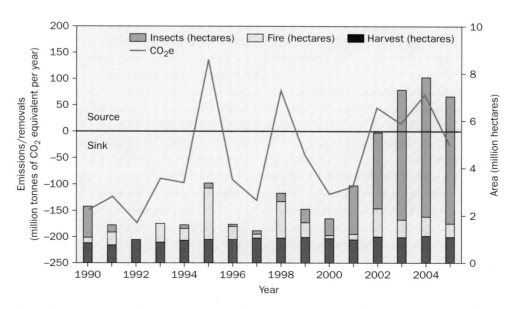

Figure 9.9 Greenhouse gas emissions/removals in Canada's managed forests and area disturbed by harvesting, forest fires, and insect pests. Source: Environment Canada 2006c.

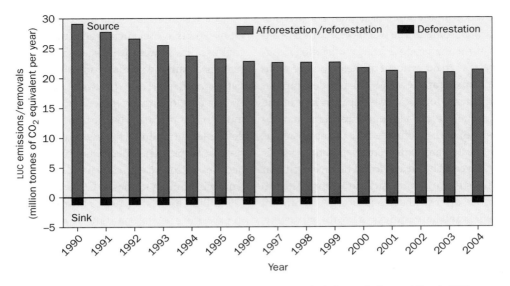

Figure 9.10 Greenhouse gas emissions/removals due to land-use change (LUC). Source: Environment Canada 2006c.

account in their cutting plans. In other words, they are having to take a more ecosystem-based approach, as described in Chapter 6. Non-timber forest products (NTFPs) are also seen as bringing diversification to rural economies and can yield valuable economic returns (see Table 9.2). Canada has a long way to go to catch up to most other countries in introducing this kind of product diversification to the forest land base.

NTFPs may be wild or managed and may come from both natural and managed forests. It is important to understand these differences if NTFPs are to play a fuller role in forest valuations and decision-making. For example, the harvesting of some wild stocks, such as mushrooms, from managed forests may conflict with timber production activities. However, harvesting some managed NTFPs can be encouraged alongside timber production and raise the overall level of return from the land. Such an approach has been adopted for blueberries in some areas of Quebec. It is an example of symbiotic use between the different resources in which both kinds of resource use can benefit. We often see the approach used in agroforestry ecosystems in the tropics but rarely in Canada.

Other kinds of relationships include complementary, competitive, and independent resource use. Complementary relationships occur when NTFPs and timber are extracted from the same land base in non-conflicting ways. Craftspeople, for example, may get improved access to their raw materials (e.g., tree bark, boughs) because of the development of logging roads. In contrast, competitive relationships often involve mutually exclusive uses. Logging old-growth forests in western Canada, for example, would devastate the lucrative pine mushroom industry. Finally, independent systems develop when the two uses operate on different units of land—for example, in commercial and non-commercial forests.

Besides tangible non-timber forest products, forests fulfill a host of less tangible values related to cultural and spiritual fulfillment and knowledge and understanding. Such values

are difficult to assess, let alone manage. The aesthetic value of forests in particular has become prominent over the past decade. Our forested areas include some of our most scenic landscapes. Not just an attraction for tourists, they also offer recreational and spiritual satisfaction. Most provinces have introduced procedures for including assessments of aesthetic quality into harvesting plans. Unfortunately, many of these procedures still leave it up to the people in charge of timber extraction to decide what interests should be considered in assessing scenic value and what harvesting regime might follow. As a result, modifications to cutting plans to take aesthetics into consideration often tend to be minimal.

New Forestry

Concerns over the impacts and sustainability of forest practices have given rise to calls for what has been termed **new forestry**, which involves new ways of looking at the management of forest ecosystems. Current approaches usually emphasize economic maximization over the short term through intensive forest management subsidized by auxiliary energy flows, leading to a simplification of forest biology. This entails genetic simplification through the exclusion of non-commercial species from regrowth areas and genetic manipulation to homogenize the species grown. Intensive forestry emphasizes production of a young, closed-canopy, single-species forest, usually the least diverse of all successional stages. Moreover, the strength and reliability of the wood produced from such forests have been questioned, and these plantations are susceptible to **windthrow**, insect infestations, and gradual nutrient depletion. Structural simplification also takes place as the range of tree sizes and growth forms is reduced, snags and fallen trees are removed, and trees are regularly spaced to optimize growth. At the landscape scale, simplification occurs as old growth is removed and the irregularity of wind- and fire-created openings is replaced by the regularity of planned clear-cuts. Successional simplification also takes place, since intensive management aims at eliminating early and late successional stages from the landscape.

New forestry embraces an approach that mimics natural processes more closely, emphasizing long-term site productivity by maintaining ecological diversity. This includes rotation periods sometimes longer than the minimum economic periods, reinvesting organic matter and nutrients in the site through snag retention and stem-only harvesting, minimizing chemical inputs, and diversifying the range of tree species and other forest products. Growth of traditionally non-commercial species such as alder and other early successional species, particularly nitrogen fixers, is permitted, and all stages of the successional process are accommodated. Old-growth big-leaf maple, for example, provides excellent growth sites for many epiphytes (plants that use other plants for physical support but not nourishment), which supply valuable nutrient accumulation and water retention. Riparian or riverbank habitats receive special attention; litter from streamside vegetation provides the primary energy base for the aquatic community, and management of coarse woody debris is particularly important for the structure of smaller streams. Needless to say, large woody debris cannot be produced by a forest that no longer contains large trees.

New forestry also emphasizes the maintenance of non-timber parts of the forest community. Special attention is given to the impact of the size, shape, and location of forest patches on wildlife and how these patches can be connected to sustain populations. The ecological complexities of forests are only just starting to be revealed. Recent research, for example, suggests that the younger the forest, the less conifer seed production. Species that rely on these seeds, such as crossbills, also experience a decline. In western Canada, five species of crossbill have evolved, each of which specializes on a different species of conifer and even different varieties of the same species. Protection of this diversity of crossbills will require protection of old-growth stands and an increase in rotation ages throughout the range of each conifer. Similar consideration must be given to the entire range of forest biodiversity if it is to be maintained into the future.

The kinds of changes suggested by new forestry make it unlikely that the dominant practices of today, such as extensive clear-cuts, can continue and indicate that other harvesting systems (Table 9.6) will play a larger role as an ecosystem-based perspective becomes more widespread. New forestry may not suit the need for short-term economic return on the part of the large corporations that now dominate the industry, and it is likely that more, smaller, and community-based companies will emerge (Box 9.9). Monetary returns over the short term will probably fall as the amount of wood fibre extracted from

Box 9.9   New Forestry in Action

The ideas contained in new forestry must be put in motion if change is to occur. Several examples of alternatives to the dominant way of managing our forests are already in operation, both in regard to individual woodlots and for management of more extensive areas by communities.

At the individual scale, one well-known example is Merv Wilkinson and his 55-hectare woodlot on southern Vancouver Island. Since 1936, Wilkinson has been practising **sustained yield** forestry, and despite the removal of more than 4,000 cubic metres of timber, his woodlot still contains as much wood as it did when it was first assessed in 1945. He removes forest products by cutting in five-year rotations. The straightest, most vigorous trees with good foliage and abundant cone production are left as seed trees, including some estimated to be as old as 1,800 years. There is no clear-cutting, slash burning, or use of chemicals. The canopy is left intact

to shield seedlings but thinned a little to promote good growth. Sheep are used for brush control. Wilkinson's model may not apply everywhere, but it has worked for him and offers a good example of how a forest can be maintained while still retaining its essential ecological characteristics.

At a regional scale, attention has recently focused on 'community forests'. Decisions on forest use are often made in boardrooms at the dictates of international capital flow. Such decisions may not be to the benefit of the local communities dependent on the forests for their livelihoods. Concern over this situation has prompted interest in how to manage forests to maximize the benefits to local communities. In an overview of the situation in Canada, Duinker et al. (1994, 712) defined a community forest as 'a tree-dominated ecosystem managed for multiple community values and benefits by the community'.

There are many types of community forests in Canada, with different forms of land tenure and administrative arrangements. However, they are all aimed at achieving benefits for the community and encouraging local involvement in decision-making. In BC, for example, the provincial government amended the Forest Act to create Community Forest Agreements, a new form of tenure designed to enable more communities and First Nations to participate directly in the management of local forests. In 1998, the Ministry of Forests launched a pilot project involving a special form of the tenure, called a community forest pilot agreement (CFPA). To date, 12 agreements have been issued to a range of communities and First Nations. The pilot agreements are located throughout the province and range in size from about 400 hectares to more than 60,000 hectares. The BC government has a number of objectives for the program:

- provide long-term opportunities for achieving a range of community objectives, including employment, forest-related education and skills training, and other social, environmental, and economic benefits;
- balance the use of forest resources;
- meet the objectives of government with respect to environmental stewardship, including the management of timber, water, fisheries, wildlife, and cultural heritage resources;
- enhance the use of and benefits derived from the community forest agreement area;
- encourage cooperation among stakeholders;
- provide social and economic benefits to British Columbia.

Community forests are typically managed to reflect community goals:

- supporting the local economy by hiring and buying supplies locally and selling timber to local timber-processing facilities;
- diversifying the local economy by making small volumes of wood fibre available to new and existing niche markets such as small-scale, value-added manufacturers and artisans;
- maintaining and enhancing local recreational opportunities;
- protecting drinking water, viewscapes, wildlife, and other environmental attributes;
- providing a source of income to support local community initiatives.

Under the Forestry Revitalization Plan announced in 2003, the BC government committed to significantly increasing the volume of timber allocated to community-based forest tenures. New community forest pilot agreements are no longer issued. Instead, 'probationary' agreements, carrying a five-year initial term, allow both communities and the ministry an assessment period. If successful, the agreement-holder is offered a long-term community forest agreement, which has a 25- to 99-year term and is replaceable every 10 years. Existing pilot agreements can also be rolled over into long-term agreements, if successful. Burns Lake Community Forest was offered the first long-term (25-year) licence in 2004. As of January 2007, approximately 350,000 hectares in BC were being managed as community forests.

More partnerships are now being made with First Nations communities in BC. The government currently has agreements with 127 First Nations, covering 8 per cent of the AAC and producing more than $166 million in revenue.

Clearly, alternative approaches to forestry do exist. It is essential, however, to specify the goals of forestry activities before the most appropriate approach can be chosen, as emphasized in the framework in Chapter 1. Current models have evolved to maximize economic returns over the short term; the alternatives described above have different goals, which are more consistent with the demands of today. However, as Bullock and Hanna (2008) point out, community forestry is not a panacea for resolving conflict in forest management.

the forest declines. Proponents of new forestry argue, however, that these changes will have to occur anyway. Continuing the old approaches will simply lead to an abrupt decline in the amount of timber available and consequently will diminish future prospects. This way of thinking is gaining wider acceptance. The development of Canada's National Forest Strategies, described in the next section, is one result.

Canada's National Forest Strategies

During the 1980s, it became increasingly clear that forestry in Canada could not continue as it had in the past and that new ways had to be found to develop more sustainable management practices. This realization led the Canadian Council of

The belief that community forests will enable local control, sustainable community development, and multiple-use management plans that reflect diverse values and interests has heightened expectations from community forests. As Duinker et al. (1991) warn, we cannot assume that all communities are suited to or have the capacity for this approach to forest tenure and management. The community forest is not a panacea for all the difficulties associated with industrial forestry or the failures of existing management systems, but it does hold promise.

—Bullock and Hanna 2008, 84

Forest Ministers (CCFM), a body composed of all the provincial and territorial forest ministers and the Canadian minister of natural resources, to sponsor a series of public forums across the country. These discussions resulted in the formation of a National Forest Strategy (NFS), Sustainable Forests: A Canadian Commitment (1992–7). Canada was the first country in the world to develop such a national strategy. It was revised and extended in a second strategy covering 1998 to 2003. The National Forest Strategy Coalition (NFSC), composed of 52 governmental and non-governmental agencies, was formed to oversee implementation of the strategy. The strategy made 121 commitments under nine strategic directions to move Canada along the road to a more sustainable use of forest ecosystems.

Such an approach is only useful, however, if the commitments are meaningful and some progress is made to achieve them. An independent evaluation concluded that there had been substantial progress on 37 commitments, some progress on 76 commitments, little progress on six commitments, and no progress on two commitments. The evaluation commended Canada for showing international leadership but suggested a simplification of any future strategy, as well as the inclusion of clear targets and timetables.

In 2003, a further five-year strategy was adopted, along with a commitment on the part of the NFSC partners to work towards its completion. However, Alberta, Quebec, and the Forest Products Association of Canada did not sign the strategy. Eight themes were outlined, each of which specifies an objective and action items to be undertaken. Two noteworthy themes are the urban forest, which emphasizes the need to actively engage more of Canadian society in forest questions, and recognition of the importance of private woodlots

to sustainability. The latter is particularly important in the Maritime provinces, where more than half of the forests are in private hands and not subject to provincial forestry regulations. Over-cutting and neglect of these lands have been problems in the past, and the action items in the strategy include providing more incentives and support for landowners to manage their woodlots on a sustainable basis.

The objectives of the 2003–8 NFS are highly laudable and seem to herald a new and progressive approach to forest management in Canada. A self-evaluation in 2005 concluded that 'National Forest Strategy Coalition members and other partners across Canada's forest community are leading change, as highlighted in their two-year accomplishments report. Collectively, the achievements show a strong commitment to work together to face the challenges along the path towards a sustainable forest, nationally and globally' (Jean Cinq-Mars, chair, NFSC [NFSC 2005, iii]). However, others were less convinced. The Sierra Club withdrew in 2006, concerned about the lack of progress, lack of transparency, and efforts to select inappropriate indicators for assessing progress. The Sierra

Different Perspectives on the National Forest Strategy

The current NFS [2003–8] fulfills an important role of informing interested and concerned Canadians and the international community about the state of Canada's forests and forest practices.

—NFSC 2005

After three years of working to implement the National Forest Strategy from within the National Forest Strategy Coalition, the Sierra Club of Canada has decided to disengage.

—Sierra Club of Canada 2006

What cannot be determined in any traceable way is the process of taking the civil society inputs and translating them into final texts in the NFS. Thus, the processes of generating NFSs have not been entirely transparent, and in the case of the current NFS the outcomes were not entirely satisfactory.

—KBM Forestry Consultants 2007

When we look at the international milestones in sustainable forest management, Canada's leadership stands out.

—NRCAN 2007b

Table 9.8   A Vision for Canada's Forests: 2008 and Beyond

Published by the Canadian Council of Forest Ministers, *A Vision for Canada's Forests: 2008 and Beyond* takes a broader and more aspirational approach than previous national forest strategies.

Overall Vision Statement

Be the best in the world in sustainable forest management and a global leader in forest sector innovation.

- Take a non-prescriptive approach and increase awareness of forest issues at home and abroad.
- Encourage domestic and international engagement; promote partnerships among traditional and non-traditional forest interests.
- Reflect the collective ambitions of Canadians for their forests and communities; open opportunities for all to draw on each other's strengths.
- Mobilize the talents of all Canadians in the search for creative solutions.

Commitment to Sustainable Forest Management

Demonstrate commitment to the sustainable management of forests by continuing to implement the principles of sustainability (including ecosystem-based and integrated landscape management) and embrace the principles of stewardship, innovation, partnership, transparency, and accessibility.

- Recognize the evolving nature of sustainability; for example, give increasing consideration to the effects of climate change, the provision of environmental good and services, and the importance of urban forestry to inhabitants of towns and cities.
- Strive to improve the forestry sector's environmental record while considering the social implications of sustainable forest management practices and expanding the contribution of forest resources to wealth and prosperity.
- Maintain the variety, quality, and extent of forest types; conserve biological diversity and soil and water resources; enhance the resilience of forests by managing carbon balances and adopting innovative forest-protection strategies.
- Innovate to facilitate the establishment of partnerships among the various forest sector members to maintain and expand the contribution of renewable forest resources, bioeconomy, and new market mechanisms for environmental good and services to Canada's economy.
- Include communities who depend on forest resources for their well-being and livelihood in helping decide how forest resources are managed.
- Diversify Aboriginals' role in development and implementation of sustainable forest management to include development of non-timber forest products and tourism initiatives.
- Consider urban populations' influence on the management of Canada's forests, including the need for urban forests that reduce the negative effects of air pollution, conserve energy, reduce soil erosion, provide wildlife habitat, and offer a place for recreation and spiritual renewal.

Priorities of National Importance

1. *Forest Sector Transformation: Ensure a prosperous and sustainable future for Canada's entire forest sector.*

- Develop new ideas, technologies, processes, and markets through a systematic engagement of science and technology organizations in collaborative research and public–private partnerships.
- Maximize economic value from forest resources through the diversification of uses.
- Make products from Canada's forests recognizable as an environmentally and socially responsible choice for consumers around the world.
- Take the provision of environmental goods and services into account in the sustainable management of Canada's forest resources, including through the development of new markets.
- Ensure meaningful Aboriginal participation in an innovative forest sector, including use of their insights and expertise.
- Have highly skilled workers contribute to the expansion of knowledge-based forest industries through education and training.
- Put into place creative public policies that facilitate forest sector transformation.

2. *Climate Change: Become a world leader in innovative policies and actions to mitigate and adapt to the effects of climate change on our forests and forest communities.*

- Include climate-change considerations in all aspects of the sustainable management of Canada's forests.
- Recognize, harness, and manage the economic value of carbon in trees, forests, and wood products.
- Identify and address the knowledge gap in the impacts of climate change on forests, industries, and communities.
- Ensure policies and institutions provide means for forests, industries, rural and urban communities, and private woodlot owners to adapt to changing conditions and mitigate the effects of climate change.

- Develop, share, and implement innovative adaptation and mitigation practices, including those that integrate Aboriginal knowledge.
- Involve many forest-resilient communities, including Aboriginal ones, in the development, sharing, and implementation of forest mitigation and adaptation strategies.
- Enable innovators and entrepreneurs, through institutions and creative policies, to take advantage of transformative and sustainable bioenergy opportunities that contribute to broad climate-change objectives.

Realizing the Vision through Partnerships

To be effective, the Vision needs to be embraced by Canada's entire forest sector, including future players.

1. *Canadian Council of Forest Ministers (CCFM)*

- Champion the Vision to generate greater public awareness of forest issues and communicate the Vision's goals and outcomes.
- Convey progress on the Vision, including advancement of sustainable forest management in Canada and progress on priority issues.
- Facilitate the consolidation of progress reports from forest sector partners and ensure coordination with national initiatives and activities of other ministerial councils, and with its own initiatives and strategies.
- Support selected events, including conferences and workshops that advance the Vision.
- Host a national workshop every three years to showcase the initiatives that advance the Vision and to help inform the review process.

2. *Federal, Provincial, and Territorial Governments*

- Act individually to promote sustainable forest management and bring focus to priority issues.
- Use a variety of arrangements to consult and involve forest sector members and the general public.

3. *Private Owners and Forest Companies*

- Continue to manage owned forests through stewardship, working within a framework of norms, regulations, and incentives defined by governments.
- Continue to manage public lands with oversight from public agencies and innovate within new approaches that give regional authorities or communities more responsibilities for managing public forest.

4. *Aboriginals*

- Continue to maintain a relationship with the forests that represent a source of food, medicine, and economic activity as well as provide opportunities for recreation, social interaction, and spiritual growth.
- Play a role in forest economy, including involvement in the development of sustainable forest management practices, notably through the application of their knowledge and practices.

5. *Other Forest Sector Members*

- Have consultants, land-use planning experts, environmental and conservation groups, communities, academics, professional associations, and other organizations play a role in helping to develop and deliver effective sustainable forest management policies and practices.

6. *International Partners*

- Collaborate with those at home and abroad on science and technology issues and share expertise and resources to advance global forest development.
- Remain engaged in multilateral and bilateral international forestry initiatives and conventions.
- Remain engaged in international dialogues and dedicated to international cooperation to further the sustainable management and development of forest resources.

Assessing and Communicating Progress

Communicating progress on the Vision requires coordinating reporting activities and sharing information.

- Use CCFM's Criteria and Indicators Framework as an evaluation tool to gauge improvements in the sustainable management of Canada's forests.
- Use federal and provincial State of the Forests reports in addition to collaborative tools available on the Internet.

Source: CCFM 2008. A Vision for Canada's Forests: 2008 and Beyond. http://www.ccmf.org/pdf/Vision_EN.pdf.

Club published its own set of indicators and an evaluation report (Sierra Club of Canada 2006). The Ontario Association of Anglers and Hunters also withdrew, feeling that inadequate attention was being devoted to non-timber values in the forests and that the situation was really 'business as usual'.

A final evaluation was undertaken in 2007 by an independent team of evaluators (KBM Forest Consultants 2007). The team undertook an evaluation of process-based indicators and also outcome-based indicators for objective 1 on ecosystem-based management (Figure 9.11). The table shows

Final Report on the National Forest Strategy (2003–2008) Evaluation

Criteria	Indicator	Description	Score
Biological Diversity	1.1.1	Area of forest and wetlands in each ecozone	=
	1.1.2	Area of forest and wetlands in protected areas by ecozone	−
	1.2.1	The status of forest-associated species at risk	−
	1.2.2	Population levels of selected forest-associated species	=
	1.2.3	Distribution of selected forest-associated species	=
	1.2.4	Number of invasive, exotic forest-associated species	−
	1.3.1	Genetic diversity of reforestation seed-lots	=
	1.3.2	Status of conservation efforts for native tree species	=
Ecosystem Condition and Productivity	2.1.1	Total growing stock of tree species on forest land	=
	2.2.1	Additions and deletions of forest area by cause	=
	2.3.1	Area disturbed by fire, insects, disease, and timber harvest	−
	2.4.1	Area with impaired function due to ozone and acid rain	=
	2.5.1	Proportion of timber harvest area successfully regenerated	=
Soil and Water	3.1.1	Rate of compliance with soil disturbance standards	=
	3.2.1	Compliance with road construction, stream crossing standards	=
	3.3.1	Watersheds with stand-replacing disturbance in the last 20 years	−
Role in Global Ecological Cycles	4.1.1	Net change in forest ecosystem carbon	=
	4.1.2	Forest ecosystem carbon storage by forest type and age class	IE
	4.1.3	Net change in forest products carbon	IE
	4.1.4	Forest sector carbon emissions	+
Economic and Social Benefits	5.1.1	Contribution of timber products to the gross domestic product	−
	5.1.2	Value of secondary manufacturing of timber products	IE
	5.1.3	Production, consumption, imports, and exports of timber products	−
	5.1.4	Contribution of NTFP and forest-based services to the GDP	+
	5.1.5	Value of NTFP and forest-based services	+
	5.2.1	Forest area by timber tenure	=
	5.2.2	Distribution of benefits from the timber products industry	=
	5.3.1	Annual harvest of timber relative to the sustainable level	=
	5.3.2	Annual harvest of NTFP relative to the sustainable level	IE
	5.3.3	Return on capital employed	−
	5.3.4	Productivity index	IE
	5.3.5	Direct, indirect, and induced employment	−
	5.3.6	Average income in major employment categories	=
Society's Responsibility	6.1.1	FN consultation in FMP and forest policy development	+
	6.1.2	Area of forest land owned by Aboriginal peoples	+
	6.2.1	Area of forested Crown land with traditional land-use studies	IE
	6.3.1	Economic diversity index of forest-based communities	NA
	6.3.2	Education attainment levels in forest-based communities	+
	6.3.3	Employment rate in forest-based communities	=
	6.3.4	Incidence of low income in forest-based communities	=
	6.4.1	% of population satisfied with public involvement in FMP	=
	6.4.2	Rate of compliance with forest laws and regulations	=
	6.5.1	Completeness and statistical reliability of forest inventories	+
	6.5.2	Availability of forest inventory information to the public	+
	6.5.3	Investment in research, development, and education	+
	6.5.4	Status of new or updated FMP guidelines and standards	+

Figure 9.11 Performance and evaluation summary for outcome-based indicators. Key: + positive trend; − negative trend; = stable; IE Inadequate evidence.
Source: KBM Forestry Consultants 2007.

a preponderance of stable trends. In many cases, the evaluating team took this to be a positive finding in that things were no longer deteriorating. However, they were generally critical of the NFS. They found an overall lack of engagement and financial support. For example, only 15 out of 66 coalition members attended the 2007 annual meeting. They were also highly critical of the evaluation system designed by the NFS:

- 'The action items were formulated either to represent activities already underway in Canada's forest sector . . . or written so vaguely and ambiguously that it is virtually impossible to determine what the action item actually means.'
- 'Many of the indicators seem (a) disconnected from the action items, and (b) difficult to measure with confidence.'
- 'No one has set any desirable levels or targets for the indicators. Without targets, how can one determine adequacy of performance?' (KBM Forestry Consultants 2007, 48)

The report notes that there are some positive trends in Canadian forestry but 'these trends would no doubt continue with or without an NFS' (KBM Forestry Consultants 2007, 50). The main conclusion is that 'the NFS is successful at communicating a consistent pattern of behaviour of Canada's forest jurisdictions for the benefit of its citizens and its trading partners around the world' (50). Unfortunately, what this seems to indicate is that rather than the NFS creating real progress in forest management in Canada, it is mainly a communication vehicle. Understandably, NGOs interested in generating real change may feel a little co-opted by such a process. The Sierra Club (2006, 7) advises other NGOs: 'Our experience has shown us that, overall, it's not worth our while to work within the NFS. While the NFS umbrella itself is good, efforts spent working outside of the NFS proved more fruitful than the inside approach.'

Both the Sierra Club and the independent evaluation indicate that the future trend of the government seems to be to limit inclusivity and return more to a top-down process of input to forest land decision-making. The newest draft of the 2008–18 strategy seems to confirm this approach, with the Canadian Council of Forest Ministers reassuring Canadians that they will lead the process 'on behalf of all Canadians' (CCFM 2008, ii) and therefore will not need the participatory approach adopted previously. The new strategy has two themes: transforming the forest sector and mitigating and adapting to climate change. There seems to be little appetite for re-engaging in the more detailed aspects of forest management change that emanated from the previous strategy.

The Model Forest Program

One commitment from Canada's 1992–7 National Forest Strategy that has borne fruit is to develop a system of model forests in the major forest regions. The objectives of the program are:

- to increase the development and adoption of sustainable forest management systems and tools within and beyond model forest boundaries;
- to disseminate the results of and knowledge gained through Canada's Model Forest Program at local, regional, and national levels;
- to strengthen model forest network activities in support of Canada's sustainable forest management priorities;
- to increase opportunities for local-level participation in sustainable forest management.

Proposals were solicited for areas between 100,000 and 250,000 hectares where partners would develop a management structure to facilitate cooperation and include a vision and objectives to balance a variety of values, as well as actions to demonstrate sustainable forest management. Key attributes of model forests include:

- a partnership that includes principal land-users and other stakeholders from the area;
- a commitment to sustainable forest management, based on an ecosystem-based approach;
- operations at the landscape or watershed level;
- activities that reflect stakeholder needs and values;
- a transparent and accountable governance structure;
- commitment to networking and capacity-building.

Eleven model forest agreements in six major forest regions across the country were initiated, each with a unique management structure designed to address the particular situation. Each model forest is an independent, not-for-profit organization, and in 2006 the Canadian Model Forest Network became an independent, not-for-profit organization, although Natural Resources Canada remains a key partner and supporter. Core issues relate to ecosystem management, Aboriginal participation, public participation, science and innovation, and the integration of non-market values into decision-making.

Model forests have a deliberate strategy of intra-site and inter-site demonstration and networking. This strategy has expanded to the international context, with sites in Mexico, Chile, Argentina, China, the US, Japan, Indonesia, Thailand, Myanmar, the Philippines, and Russia as part of an International Model Forest Program. These initiatives have been supported by Canadian aid programs totalling more than $11 million, with additional support from other donors of more than $7.5 million. This support reflects recognition that the programs constitute a tangible demonstration of the value of cooperatively working together towards sustainability. They illustrate many of the approaches outlined earlier in Part C.

The Model Forest Program terminated in 2007, but the Canadian government has announced the Forest Communities Program, patterned on the Model Forest Program. The new program is intended to help communities make the most of new resource-based economic opportunities.

Global Forest Strategies

Most of this chapter has concentrated on the Canadian situation, but as outlined in Box 9.2, challenges also exist at the global level. One of the major disappointments of the United Nations Conference on Environment and Development (UNCED) in 1992 was the failure to establish an international convention on forests. However, a set of non-binding forest principles was adopted and fed into the development of Canada's national forest strategies. Some legally binding outcomes related to forestry did emerge from UNCED, in particular the Convention on Biological Diversity, which committed signatories to prepare and adhere to a national biodiversity strategy, including the designation of representative samples of their forest lands as protected areas and ensuring that forest management does not impair biodiversity. Chapter 14 delves into these commitments.

Forests were also included in Agenda 21, the non-binding principles that emerged from UNCED and set an agenda for development in the twenty-first century. In 2000, the United Nations Forum on Forests (UNFF) was created with the objective of promoting the forest principles focusing on sustainable management contained in Agenda 21. In 2006, at its sixth session, the forum finally agreed on four shared Global Objectives on Forests:

- reverse the loss of forest cover worldwide through sustainable forest management (SFM), including protection, restoration, afforestation, and reforestation, and increase efforts to prevent forest degradation;
- enhance forest-based economic, social, and environmental benefits, including by improving the livelihoods of forest-dependent people;
- increase significantly the area of sustainably managed forests, including protected forests, and increase the proportion of forest products derived from sustainably managed forests;
- reverse the decline in official development assistance for sustainable forest management and mobilize significantly increased new and additional financial resources from all sources for the implementation of SFM.

In 2007, another breakthrough was achieved with agreement on the Non-Legally Binding Instrument on All Types of Forests. This was the first time that member states had agreed to an international instrument for sustainable forest management, and it is expected to have a major impact on international cooperation and national action to reduce deforestation, prevent forest degradation, promote sustainable livelihoods, and reduce poverty for all forest-dependent people.

Other global initiatives are expanding. Approximately 7 per cent of the world's forest area is certified (see Box 9.10

Box 9.10   The Seal of Approval

Increasingly, customers for wood products around the globe are asking for guarantees that the products they buy (kitchen chairs, 2 x 4s, paper products, etc.) come from forests that are managed and logged according to ecologically responsible standards. The trend towards responsible consumerism is supported by certification and labelling, a process in which an independent audit of a forestry company is conducted to assess whether it meets internationally and/or nationally recognized guidelines for responsible forest management. Certification is designed to enable consumers and participants to measure forest management practices against approved standards and also provides forest owners with an incentive to maintain and improve forest management practices.

In Canada, there are three main certification systems:

- Forest Stewardship Council (FSC)
- Canadian Standards Association (CSA)
- Sustainable Forestry Initiative (SFI)

There are important differences among these certification systems related to standards, policies, procedures, and on-the-ground results. The FSC is an international system with strict standards; the others are largely generated and controlled by the forest industry. Canada boasts that it has the world's largest area of certified forest (NRCAN 2007b), but only a small proportion of this certified forest meets international standards specified by the FSC (Figure 9.11). A recent independent report (ÉEM inc. 2007) found that FSC was the only system prohibiting the use of genetically modified trees, preventing the conversion of natural forests to plantations, and requiring a precautionary approach to the management of areas with high conservation value. Although it found that certification systems often mentioned similar requirements, the study raised concerns that under some systems, it was left to the individual forest manager to decide what to do on the ground. The study found that FSC was generally more rigorous in its performance requirements.

Forest certification systems are designed to link environmentally and socially conscious consumers with like-minded producers, retailers, and distributors and typically involve:

- independent third-party auditing;
- chain-of-custody procedures (verification of compliance from the forest through to the final product);
- on-the-ground inspections of forested areas to determine whether they are managed according to established sets of environmental and social standards;
- certified product labelling;
- multi-stakeholder involvement.

The range of issues considered in defining responsible forest management include wildlife habitat protection, endangered forests identification and maintenance, riparian and water utility protection, indigenous peoples' rights, and the equitable sharing of benefits with forestry-dependent workers and communities. For example, the FSC advocates that all functions of a forest ecosystem remain intact after an area is logged. This requires that a mix of tree species of different ages still remains standing after the forest is logged and that the functions of trees and other plant species also remain intact.

Forest products given the seal of approval should give consumers confidence that the products they purchase are derived from responsibly managed forests. As of 2006, 123.7 million hectares in Canada had been certified under one or more of the main certification systems listed above (Figure 9.12).

More and more, Canadian companies are trying to meet international standards. One company operating mainly in the boreal forest of northeastern Ontario has earned the FSC logo for voluntarily meeting its high standards for forest management. Clear-cutting once dominated the 2-million-hectare forest managed by Tembec Inc., but today considerable patches of trees are left standing, large tracts of old-growth forest are being protected, and selections of all forest types are being set aside to serve as wildlife habitat. Tembec is a leader in environmental forestry and now has almost 10 million hectares of FSC-certified forest in Canada.

Tembec has demonstrated that it is possible to dramatically improve forest management practices to attain the coveted FSC logo. In a climate in which consumers are more concerned about environmental and social issues surrounding primary resource extraction, securing a seal of approval from a highly rated certification program will become increasingly important.

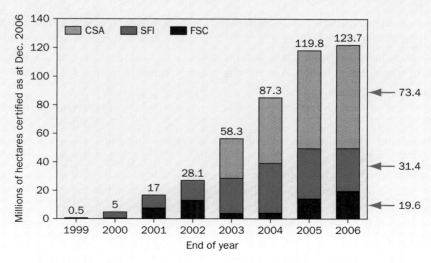

Figure 9.12 Sustainable forestry certification in Canada, 1999–2006. Source: Canadian Sustainable Forestry Certification Coalition. December 2006. www.certificationcanada.org.

for the implications of certification), a nearly fivefold increase since 2000. The two largest **certification** organizations worldwide, with 28 and 26 per cent of global certified forest area respectively, are the Forest Stewardship Council (FSC), a membership organization dedicated to sustainable development principles, and the Pan-European Forest Certification Framework (PEFC), a voluntary initiative led by the forest industry to promote an internationally credible certification framework. Both organizations develop principles and criteria for SFM using stakeholder participation and accredit third-party auditors to verify compliance through annual audits. Certifiers may issue a Forest Management Certificate for forest stewards or a Chain-of-Custody (explained in Box 9.10) Certificate for forest product manufacturers and distributors. Consumers can then identify certified wood products through a certification logo.

Although the early proponents of forest certification hoped to target tropical deforestation, the temperate and boreal forests of industrialized countries account for 87 per cent

of all certifications (Box 9.10). Some argue that this is because certification is not conducive to forest management schemes involving communities or small enterprises, which are typical in developing countries. However, tropical forest certifications in the developing world are continuing to grow.

Perhaps the biggest factor that might affect the way we look at global forests is their role in mitigating the impact of global climate change. In box 9.11, someone actively involved in conservation through carbon trading discusses the issue.

Implications

Forestry is at a watershed in Canada in terms of how forests, their values, and their management are viewed. The next decade will be crucial in determining whether we as Canadians will still consider ourselves a forest nation in another 50 years. Although society in general and government and industry in particular now have a much greater appreciation of the changes

Box 9.11   Can the Private Sector Help Slow Deforestation?

Gijsbert Nollen

The private sector is usually looked at as being a main cause of deforestation, and in the past it has been. The price mechanism dictates that if a forest is worth more money cut down than standing, then the chances of it being cut down are very high. However the way in which the value of forests is calculated may now be changing so that higher values are achieved through leaving the trees standing. Of particular importance is the attention now being given to carbon sequestration to help mitigate the effects of global warming, as discussed in Chapter 7.

More than 20 per cent of total annual carbon emissions are emitted by land-use change, most of which is tropical deforestation, and approximately 14 million hectares of forest are lost annually (Kanninen et al. 2007). Inclusion of Reduced Emissions from Deforestation and Forest Degradation (**REDD**) in any future global climate change agreements is a must for mitigating the impacts of global climate change. Although for policy-makers, 2012 seems to be in the near future, for our forests every minute counts. By the time you have read this article, approximately 250 ha of forest has been lost. Assuming

an equilibrium state of carbon sequestration of 75 tC per hectare, this results in a net emission of 18,750 tC (70,000 carbon credits; one tC is 3.6 CO_2).

Current lack of proper platforms for payments for ecosystem services, such as carbon and water, and global failure to account for all costs, specifically environmental costs, in supply chain and production processes and unwillingness of policy-makers to address both issues set the protection of ecosystems including forests at a huge disadvantage in marketplace economics and makes their liquidation more attractive.

The ideal approach would be to work on an ecosystem approach which would develop a framework addressing the deforestation drivers, integrating ecosystem services into a profitable deal structure with a conservation goal, going hand in hand with sustainable development. This approach incorporates a variety of methods, including ecosystem service dependency and impact assessment, valuations of ecosystem services, land-use scenarios, development requirements, and policies to support the transaction. This approach could also act as a catalyst to change contemporary economic land use and revenue generation from resource extraction to sustainable land use and sustainable development and provision of ecosystem services (based on resource conservation).

Therefore, if we want to address carbon emissions stemming from land-use change effectively and curb unsustainable land-use change (such as deforestation), returns from REDD should be more attractive than the alternative. Inclusion of payments for other ecosystem services, such as water, siltation, and biodiversity, could add to the viability of such a project. However, given the complexities of structuring these transactions, a deal should be viable and attractive on carbon trade alone.

Estimates of the potential global value of REDD payments vary depending on the underlying assumptions. Assuming a conservative carbon value of $10 per ton of CO_2e, estimates include a net present value of $150 billion (Chomitz et al. 2007) and annual revenue of $2.3 billion to $12 billion (Ebeling 2006; El Lakany, Jenkins, and Richards 2007). These large values make an interesting market for financial institutions, emerging forested countries, investors, and transaction developers. As such, the private sector has already moved into the market, and deals have been struck in Indonesia (Aceh by Merrill Lynch), Guyana (Iwokrama by Canopy Capital), Brazil (Amazonas by Marriott Hotel Chains), and Bolivia (Noel Kempff, traded on the Chicago Climate Exchange).

A lot of research is still required to develop rigorous scientific assessments and trade forest carbon effectively. However, time is not on our side; the drivers for deforestation are winning, and conservation is losing. If deforestation continues at current rates, approximately another 52 million hectares will be lost by 2012. Therefore, it is imperative that we do not wait until the policy-makers finally draw up a framework. Under the current regulated carbon forestry market under the Kyoto Protocol, the Clean Development Mechanism (CDM) and Joint Implementation (JI), REDD is not an approved methodology, and only one reforestation/afforestation project was registered under the CDM by the end of 2007. Even though REDD is already being traded on the Voluntary Markets (approximately 5 per cent of total voluntary market value), the future of this market is dependent on what happens post-2012. Therefore, if REDD were to be traded on a post-Kyoto United Nations Framework Convention on Climate Change lead platform, the regulator has to prove itself capable of dealing with these transactions effectively and efficiently. So far, the CDM has been too expensive and bureaucratic to be a trading platform addressing land use, land-use change, and forestry project proposals effectively and leading to successful carbon return–based land management transactions.

Policy-makers should lower the transaction costs, simplify the verification and monitoring of the transactions, increase benefits to developing nations trading carbon return–based land management transactions, and address the demand for unsustainable forest products and biofuels at home. An effective framework should be simple for developing countries willing to look at carbon trading from land management transactions, developers of the transactions, investors, communities in the transaction area, and purchasers of the carbon.

The private sector is taking a lead in this, and several projects are already being developed and traded on a voluntary basis. It will take visionary entrepreneurs, nations, and policy-makers to curb current deforestation rates and combat climate change in the land-use change arena. Let us hope that policy-makers follow the trend by drawing up an effective framework and trading platform, allowing REDD transactions to enter the mainstream of economic transactions in the future. The market has been a key force in creating global deforestation; it can also play a key role in preventing future deforestation.

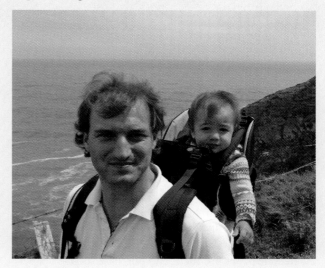

Gijsbert Nollen

Gijsbert Nollen is a principal consultant for ICE BV. He holds a master's in law from Leiden University and now focuses on project management and development as his professional specialty. Gijsbert has 10 years of work experience in Asia, seven of which were with ICE BV in Sri Lanka, where he headed and directed the local office and projects. He is inspired by a wide range of varying interests and opinion that drive his passion to advance sustainable development causes. He is now based in Bangkok and can be reached at gijsbert@icebv.net.

that need to be made in the industry to move towards more sustainable practices, actually making these changes will take some time. However, the world is watching, and the power of consumers to effect change is well in evidence. Box 9.12 offers some suggestions on things that you can do in this regard. Large companies such as Ikea and Home Depot have reacted to consumer pressure by agreeing not to sell products from forestry operations not certified as sustainable. One can only hope that similar changes in other areas of forest product use will occur (Box 9.13).

Forestry will no longer be the main economic backbone of so many communities that it was in the past as mills continue to close. In 2007, the industry lost $750 million, and the Conference Board of Canada expected a similar loss in 2008. Some places will be hit harder than others. In BC, which accounts for 20 per cent of North America's lumber production, the annual timber harvest is expected to drop by one-third over the next decade. This decline is partly due to the mountain pine beetle, but much is also due to the failure of forest management to heed warnings about overcutting during previous decades. It is to be hoped that changes will arise as a result of this situation and help to generate a more sustainable industry in the future.

Some changes are already evident. The United Steelworkers Union, representing thousands of forestry workers, is now joining hands with environmental groups such as the Western Canadian Wilderness Committee—new allies that were bitter foes 10 years ago—in demonstrations to draw attention to the situation. Communities like Ucluelet, Port Alberni, and Port Hardy on Vancouver Island that in the past strongly resisted attempts to preserve forest lands for other values, such as recreation, are now looking at forests with a much greater appreciation of the multiplicity of values they contain. Similar changes are occurring across the country. Forestry will continue to play a role in many economies, but the days of the one-industry town are gone.

Box 9.12   What You Can Do

A number of environmental stresses threaten the health of Canada's forests, but there are ways in which individuals can help:

1. Reduce your consumption of paper products. For example, use all available space on your paper (write in the margins and on both sides); carry fabric with you and use it instead of paper tissues, towels, and napkins; resist the temptation to print materials from the Internet—read on-line instead; insist that no unsolicited flyers be delivered to your home.
2. Recycle paper products, and purchase recycled paper products.
3. Choose unbleached paper products whenever possible. Unbleached paper is less harmful to the environment, since it does not require the toxic chemicals used to whiten paper.
4. Purchase certified wood products. Make an effort to purchase wood that has been certified by at least one certification agency. This will help to ensure that the wood you purchase comes from a logging company that has introduced measures to promote long-term ecological, social, and economic sustainability.
5. Reduce the risk of human-caused fires. Many forest fires are started by human carelessness. Obey fire restrictions when visiting parks and protected areas, and do not discard fire accelerants (e.g., cigarettes) along highways.
6. Join one of the NGOs that support sustainable use of forest resources.
7. Write or phone your MP or provincial or territorial representative, and/or e-mail a letter to the editor of your local newspaper. Canadians own Canadian forests. If you are unhappy about how forests in your province or territory are being managed, voice or otherwise publicize your discontent.

Box 9.13   Using Waste to Save Trees

Seeing the Light

If you can tear your eyes from these words for a moment, have a look at the paper on which they are printed. All paper stock made in Canada is manufactured from wood fibre, most of which is harvested from the boreal forest. And about 70 per cent of the paper we normally use in this magazine is made of virgin wood pulp harvested for that purpose. The remainder is from recycled paper, which was also originally made from virgin pulp.

But locked into the paper you are now holding is something revolutionary, at least in Canada: wheat straw. We call this our 'wheat sheet', and it is the fruit of a four-year project we have been working on with Markets Initiative, a Canadian environmental group devoted to the protection of the boreal forest, the Alberta Research Council (ARC), and our printer, Dollco Printing, which is based in Ottawa. The goals of the project are to alleviate the stress on our species- and carbon-rich forests and to demonstrate to the magazine industry and the pulp and paper producers in Canada that high-quality magazine-grade paper can be made from a mix of fibres, including agricultural waste such as wheat straw.

Now, the ironic part of this story is that the wheat-straw pulp we used was imported from China, where papermakers have been using wheat and rice straw for many years. There is plenty of excess wheat straw in Canada, but nobody pulps it here. Canada's pulp and paper producers, with their historic and virtually unlimited access to the boreal forest, don't use it or any other non-wood fibre in their papers.

Markets Initiative is suggesting it's time for a change, time for Canadians to begin harvesting the boreal forest in a more sustainable manner.

Why wheat straw? Unlike oats and barley, the two other major cereal grains grown on the Prairies, wheat straw is not used for animal feed. It becomes bedding for livestock, and some of it is ploughed back into the soil. But roughly 30 per cent of it is simply the waste by-product of a crop that is valued for its grain alone. Canada remains one of the world's great wheat growers, which means there is a vast amount of unused wheat straw—an estimated 21 million tonnes, which could be turned into eight million tonnes of pulp and enough paper for 20 million magazines. That straw could be a new source of revenue for farmers willing to bale and sell it to pulp and paper companies.

Yet the expert in Canada in producing pulp from wheat straw is not a pulp and paper company but the ARC, a not-for-profit applied research and development corporation owned by the Alberta government. ARC scientists have been experimenting with wheat straw since 1999 and believe they have solved the major obstacle to its widespread use as a source of pulp. Wood fibre has a silica content of less than 1 per cent. Wheat straw, by contrast, has a silica content of between 3 and 7 per cent. When wheat straw is cooked for pulp, the silica turns into glass, which is a significant production problem. But the ARC scientists have developed a clean process for removing most of the silica before the pulp is cooked. They've taken their process from bench top to a 10-tonne-a-day pilot project. Encouraged by the results, ARC would like to move to the next stage and work with a mill to turn the cleaned wheat pulp into paper. It is currently seeking a partner, and that will likely be a mill owner in China. Canadian papermakers aren't exactly leaping into the breach despite the fact that adding wheat pulp is neither complicated nor ruinously expensive. As Robert Wellwood, ARC's manager of forest products, puts it, the biggest investment needed 'is in changing the mindset of pulp-mill owners in Canada'.

That's what we hope this copy of *Canadian Geographic* will do. About 20 per cent of the pulp used for this issue was made from wheat straw. We'd like to convince the magazine industry and, ultimately, the pulp and paper industry that adding agricultural waste to the pulp mix is a step forward, that it will ease some of the demand for virgin pulp from the boreal forest and offer grain farmers a new source of revenue for what is now largely a waste by-product. The next time we print on a wheat sheet, we want the straw to be purchased from Canadian farmers and pulped in a Canadian mill. We'll provide regular updates on how the mills respond to our challenge.

—*Rick Boychuk*

Source: *Canadian Geographic.* Editorial, June 2008.

Summary

1. Some 40 per cent of the land surface of the Earth supports trees or shrubs. Four countries—Canada, Russia, the US, and Brazil—contain more than 50 per cent of the world's forests. Each year, about 13 million hectares of the world's forests are lost to deforestation, but the rate of net forest loss is slowing down, thanks to new planting and natural expansion of existing forests. From 2000 to 2005, the net forest loss was 7.3 million hectares per year—an area the size of Sierra Leone or Panama and equivalent to 200 square kilometres per day.

2. Canada is a forest nation. Not only do we have 10 per cent of the world's forests, but we are also the largest exporter of forest products in the world, accounting for 15.9 per cent of global trade. In 2006, Canada's forests contributed $36.3 billion to national GDP. The forestry industry is the largest single contributor to Canada's balance of trade. The forests, along with the North, are dominant elements in the history and culture of the nation.

3. In 2007, the industry lost $750 million, with a similar expectation for 2008. Many communities have lost their main source of income and are beginning to look at a broader array of values emanating from the forests than timber products.

4. There are 15 terrestrial ecozones in Canada. Most of Canada is within eight of these ecozones: Boreal Cordillera, Pacific Maritime, Montane Cordillera, Boreal Plains, Taiga Plains, Boreal Shield, Mixed Wood Plains, and Atlantic Maritime. They differ with respect to climate, geology, genetic diversity, and level of human development. The Boreal Shield, Canada's largest ecozone, is threatened by a number of environmental stresses, including logging, stand replacement, mining, hydroelectric development, acid precipitation, and climate change.

5. Ecosystems provide an array of beneficial services arising from ecological functions such as nutrient and water cycling, carbon sequestration, and waste decomposition. Forests are also places of exceptional scenic beauty, and millions of Canadians participate in nature-related recreational activities each year.

6. Forests are also a valuable source of commodities. Non-timber forest products contribute millions of dollars to the Canadian economy and are also an important aspect of First Nations' subsistence economies. Despite government recognition that Canadian forests provide a broad range of values, forest management paradigms have traditionally focused on the management of forests to supply wood.

7. The provincial and territorial governments are responsible for 71 per cent of the nation's forests and the federal government for 23 per cent on behalf of the owners, the people of Canada. The remaining 6 per cent is owned privately. Governments enter into contract arrangements with private companies in which they can specify the forest management practices to be followed.

8. Approximately 143 million hectares is currently managed for timber production. On these lands, forest ecosystems are being transformed from relatively natural systems to control systems in which humans, not nature, influence the species that will grow there and the age to which they will grow.

9. The rate of conversion from natural to managed forest is controlled by provincially established annual allowable cuts (AACs). In theory, the AAC should approximate what the land should yield in perpetuity. It is not sustainable to have an AAC that consistently exceeds this biological potential. In most provinces, the AAC is greater than the harvest, but in some regions, the AAC is approaching or exceeds the harvest for softwoods.

10. In 2005, Canada harvested 191 million cubic metres of timber, which was 49 million cubic metres, or 20 per cent, less than the estimated 2005 wood supply.

11. Silviculture is the practice of directing the establishment, composition, growth, and quality of forest stands through harvesting, reforestation, and site preparation.

12. Clear-cutting is the dominant harvesting system in Canada. It is the most economical way for extracting fibre and also allows for easier replanting and tending of the regenerating forest. In certain types of forests, it may mimic natural processes more closely than selective or partial cutting systems. However, clear-cutting may not be the most appropriate way to harvest timber in some areas. Clear-cuts are aesthetically unappealing to many Canadians, and their environmental impact can be substantial.

13. Biocides are used in forestry to control populations of vegetation and insect species that compete with or eat commercial species. Several high-profile conflicts have arisen over application of chemicals. Concern over spraying to control the spruce budworm in the Maritime provinces is one of the most significant. The biological control agent *Bacillus thuringiensis* is increasingly being used against insect attacks in Canada.

14. Intensive forest management techniques are used to further enhance future timber resource values. Intensive

silvicultural practices include pre-commercial thinning, commercial thinning, scarification, prescribed burning, pruning and shearing, and timber stand improvement. The long-term impact of these activities are not well understood.

15. Fire suppression has resulted in ecological changes not characteristic of fire-dominated ecosystems, as well as a gradual increase in the area burned over the past 30 years.

16. Various environmental impacts are associated with current forestry management systems, including changes to ecosystem, species, and genetic diversity; changes to biogeochemical and hydrological cycles; and soil erosion.

17. Timber harvesting can significantly alter species composition and abundance as the proportion of forest with old-growth characteristics is reduced. Species such as the woodland caribou and marten that depend on old-growth characteristics decline in abundance. Other species, such as deer, may increase as regenerating cut areas produce more forage for them.

18. The spotted owl is perhaps the best known example of the potential impact of logging on biodiversity. The spotted owl requires old-growth forests to maintain populations, but logging in BC's old-growth forests continues to threaten this endangered species. Management options to maintain populations are currently being considered.

19. Forest harvesting removes nutrients from the site. The significance of this for future growth varies, depending on the nutrient capital of the site and type of harvesting system used. Sites with abundant capital and/or selective harvesting systems that leave branches behind will suffer less growth impairment of future generations than nutrient-poor sites or sites that are clear-cut with complete-tree removal.

20. Forest harvesting may also contribute to increased soil erosion and water flows.

21. Forests are a carbon sink. They take in carbon dioxide and convert it to wood, leaves, and roots. They are also a carbon source. They release stored carbon into the atmosphere when they decompose or burn. Because of this ability to both absorb and release huge amounts of carbon dioxide (a major greenhouse gas), forests play a major role in the global carbon cycle.

22. Canada's managed forests were a net carbon sink in most years between 1990 and 2005. But in some years, they were a source, mostly because of wildfires. From 1990 to 2005, Canada's managed forests sequestered an average of 50 million tonnes of CO_2 per year.

23. Carbon sequestration was not considered in timber management in the past, but it might become a major influence on how we manage our forests. Hardwood stands in New Brunswick, for example, typically contain 10 to 20 per cent more carbon per hectare than a similar volume of softwoods of similar age. Forest managers might well change management practices in the future as global climate change becomes more severe and grow multi-aged hardwood stands that promote carbon sequestration instead of the single-aged softwoods that now dominate thinking.

24. Forests produce many values for Canadians. In the past, attention focused almost exclusively on the monetary returns from forest harvesting. However, as the amount of forest brought under management has increased and as the public becomes increasingly aware of the changes occurring in Canadian forests, more attention is being devoted to the assessment and management of other values besides timber production. An ecosystem perspective is being adopted.

25. Concern over the impact and sustainability of forest practices has given rise to calls for what has been termed 'new forestry'. Such an approach embraces an ecosystem and adaptive management perspective that seeks to mimic natural processes more closely and give greater attention to the full range of values from the forests.

26. Management of Canada's forests is directed by a National Forest Strategy, developed by provincial and territorial forest ministers and the Canadian minister of natural resources. The aim is to develop and implement more sustainable management practices. Independent reviews of the NFS have found many weaknesses. Despite this, Canada continues to boast of its international leadership in sustainable forest management.

27. Canada's Model Forest Program is one commitment arising from Canada's 1992–7 National Forest Strategy. Eleven model forest agreements were developed in six forest regions across the country, but the program was terminated in 2007. Core issues relate to ecosystem management, Aboriginal participation, public participation, science and innovation, and the integration of non-market values into decision-making.

28. Internationally, the Convention on Biological Diversity commits signatories to prepare and adhere to a national biodiversity strategy, including the designation of representative samples of their forest lands as protected areas and ensuring that forest management does not impair biodiversity.

29. The United Nations Forum on Forests (UNFF) has agreed on four shared Global Objectives on Forests relating to reversing the loss of forest cover worldwide through

sustainable forest management, enhancing forest-based economic, social, and environmental benefits, increasing the area of sustainably managed forests, and reversing the decline in official development assistance for sustainable forest management.

30. Approximately 7 per cent of the world's forest area is certified—a nearly fivefold increase since 2000.

Certification is designed to enable consumers and participants to measure forest management practices against approved standards and also provides forest owners with an incentive to maintain and improve forest management practices.

Key Terms

annual allowable cut (AAC)
Atlantic Maritime
biocides
Boreal Cordillera
boreal forest
Boreal Plains
Boreal Shield
certification
clear-cutting

culmination age
DDT (dichlorodiphenyl-
    trichloroethane)
ecosystem-based manage-
    ment (EBM)
evapotranspiration
falldown effect
forest tenure
long-run sustained yield

Mixed Wood Plains
Montane Cordillera
new forestry
non-timber forest products
    (NTFPs)
old-growth forests
Pacific Maritime
reclamation
REDD

second growth
silviculture
sustained yield
taiga
Taiga Plains
transpiration
windthrow

Questions for Review and Critical Thinking

1. Outline some of the ways in which forests are important to Canada.
2. Compare and contrast the physical and biological attributes of the Boreal Shield and Pacific Maritime ecozones. What major economic activities affect the health of these ecozones?
3. How is forestry an ecological process?
4. What is an AAC?
5. Outline some of the advantages and disadvantages of clear-cutting.
6. Is Canada reforesting all lands that are harvested? What are some of the issues associated with current replanting schemes?
7. Outline some of the pros and cons of using chemical sprays to control insect infestations in Canada's forests.
8. List all the different values that society realizes from forests. What do you think the priorities should be among these different and sometimes conflicting uses?
9. Name some species that might increase in abundance as a result of forest harvesting and others that might

decline. What are the characteristics of these species that would encourage this response?
10. What are the impacts of forest harvesting on site fertility, and how do they differ between sites?
11. What attributes of old-growth forests appear to explain their use by spotted owls?
12. What are the implications of global climate change for forest management?
13. How is forest management administered in Canada? What are the main strengths and weaknesses of this approach? What alternatives might you suggest? Do examples of such alternatives exist in your region?
14. What is 'new forestry'?
15. What tools are used to evaluate the sustainability of Canadian forests? Are these tools adequate to the task?
16. Natural Resources Canada (NRCAN 2007b) states that 'When we look at the international milestones in sustainable forest management, Canada's leadership stands out.' Do you agree?

Related Websites

BC Ministry of Forests: http://www.for.gov.bc.ca/hfp/forsite/Forest_Health.htm

Canadian Boreal Initiative: www.borealcanada.ca; www.borealbirds.org

Canadian Forest Service: http://www.pfc.forestry.ca

Canadian Forest Service, Criteria and Indicators Report: http://nrcan.gc.ca/cfs/proj/ppiab/ci/indica_e.html

Canadian Institute of Forestry: http://www.cif-ifc.org

Canadian Model Forest Network: http://www.modelforest.net

Canadian Parks and Wilderness Society (CPAWS): http://www.cpaws.org

Certification Canada: http://www.certificationcanada.org

David Suzuki Foundation, forests and lands: http://www.davidsuzuki.org/Forests

Environment Canada: What you can do: http://www.ec.gc.ca/eco/main_e.htm

Food and Agriculture Organization of the United Nations: http://www.fao.org/forestry/site/sofo/en

Foothills Model Forest: http://www.fmf.ca/pa_GB.html

Forest Stewardship Council: www.fsccanada.org

Government of Manitoba: http://www.gov.mb.ca/conservation/forestry/forest-health/fh-budworm-03.html

Government of Saskatchewan: http://www.gov.sk.ca/news?newsId=f31337b1-85f1-4100-b2ad-fc1b07990e9e

Industry Canada: http://www.ic.gc.ca/epic/site/fi-if.nsf/en/oc01496e.html

Model Forest Program: http://www.modelforest.net/cmfn/en/forests

National Forest Strategy Coalition: http://nfsc.forest.ca

National Forestry Database Program: http://nfdp.ccfm.org

Natural Resources Canada, silvicultural terms in Canada: http://nfdp.ccfm.org/silviterm

Natural Resources Canada, ecozone reports: http://www.health.cfs.nrcan.gc.ca

Sierra Club: http://www.sierraclub.ca/national/programs/biodiversity/forests

Species at Risk: http://www.speciesatrisk.gc.ca

United Nations Forum on Forests: http://www.un.org/esa/forests/index.html

Wilderness Committee: http://www.wildernesscommittee.org

Further Readings

Note: This list comprises works relevant to the subject of the chapter but not cited in the text. All cited works are listed in the section 'References' at the end of the book.

Benkman, C.W. 1993. 'Logging, conifers, and the conservation of crossbills'. *Conservation Biology* 7: 473–9.

British Columbia Ministry of Forests. 2003. 'Community forest pilot project'. www.for.gov.bc.ca/hth/community.

Buchert, G.P., et al. 1997. 'Effects of harvesting on genetic diversity in old-growth eastern white pine in Ontario, Canada'. *Conservation Biology* 11 (3): 747–58.

Canadian Boreal Initiative. 2005. *The Boreal in the Balance: Securing the Future of Canada's Boreal Region*. Toronto: Canadian Boreal Initiative.

Côté, M.A., and L. Bouthillier. 2002. 'Assessing the effect of public involvement processes in forest management in Quebec'. *Forest Policy and Economics* 4: 213–25.

Dufour, J. 2004. 'Towards sustainable development of Canada's forests'. In B. Mitchell, Ed., *Resource and Environmental Management in Canada: Addressing Conflict and Uncertainty*, 3rd edn, 265–86. Toronto: Oxford University Press.

Lee, P. 2004. *Boreal Canada: State of the Ecosystem, State of Industry, Emerging Issues and Projections*. Report to the National Round Table on the Environment and the Economy. Ottawa: NRTEE.

McCarty, J. 2005. 'Neoliberalism and the politics of alternatives: Community forestry in British Columbia and the United States'. *Annals of the Association of American Geographers* 96: 84–104.

McLachlan, S.M., and D.R. Bazley. 2003. 'Outcomes of long-term deciduous forest restoration in southwestern Ontario, Canada'. *Biological Conservation* 113: 159–69.

McRae, D.J., et al. 2001. 'Comparisons between wildfire and forest harvesting and their implications in forest management'. *Environmental Reviews* 9: 223–60.

NFSC (National Forest Strategy Coalition). 1992. *Sustainable Forests: A Canadian Commitment*. Ottawa: Canadian Council of Forest Ministers.

———. 2003. *National Forest Strategy (2003–2008), A Sustainable Forest: A Canadian Commitment*. Ottawa: Canadian Council of Forest Ministers.

———. 2005. *Boreal Futures: Governance, Conservation and Development in Canada's Boreal*. Ottawa: National Round Table on the Environment and the Economy.

Rayner, J., and M. Howlett. 2007. 'The National Forest Strategy in comparative perspective'. *Forestry Chronicle* 83: 651–7.

Timoney, K.P. 2003. 'The changing disturbance regime of the boreal forest of the Canadian Prairie provinces'. *Forestry Chronicle* 79: 502–616.

Chapter 10

Agriculture

Learning Objectives

- To understand the environmental and social impacts associated with the growth of agriculture.
- To appreciate the global food situation and some of the factors that influence it.
- To understand the nature and importance of biofuels and some of their advantages and disadvantages.
- To gain an understanding of the role of energy inputs in agriculture and the Green Revolution.
- To realize the main trends in Canadian agriculture and Canada's contribution to the global food supply.
- To know some of the main environmental implications of agriculture in Canada.
- To appreciate the contributions of agriculture to global climate change.
- To understand some of the main problems arising from the use of agricultural chemicals.
- To analyze the implications of a diet with a high level of meat consumption.
- To discover some of the changes that have to be made to move towards more sustainable modes of agricultural production.

Introduction

The origins of agriculture have been traced back at least 9,000 and perhaps 11,000 years to a few regions in which societies domesticated both plant and animal species. Through domestication, such desired traits as increased seed/fruit concentration and fleshiness, reduced or increased seed size, controlled seed dispersal, and improved taste could be achieved. Various agricultural practices such as seedbeds, improved animal nutrition, and water management were also devised. In turn, the increased availability of food, feed, and fibre provided the impetus for societies to prosper and support a larger non-farming population. Societies around the globe flourished by improving their capacity to expand agricultural production (Box 10.1).

The domestication of plants and animals continues today but under a much different set of social, economic, and environmental conditions than existed even a century ago. Agriculture is now a dominant influence on the global landscape outside the major urban centres, if not *the* dominant influence.

Over much of history, agricultural output has been increased by bringing more land into production. For example, the global extent of cropland increased from around 265 million hectares in 1700 to around 1.2 billion in 1950, predominently at the expense of forest habitats, and now stands at more than 1.5 billion hectares. However, the opportunity for further geographic expansion of cropland is small because of the comparatively limited amount of land well-suited for crop production, the increasingly concentrated patterns of human settlement, and growing competition from other land uses (Box 10.2).

Intensification of production—obtaining more output from a given area of agricultural land—has become a

Box 10.1   Social Implications of the Development of Agriculture

Agriculture has had a profound influence on society, which in turn has further implications for ecosystems.

- More reliable food supplies permitted growth in populations.
- A sedentary life became more possible as a result of these food supplies and the ability to store food; this allowed the establishment of larger, permanent settlements.
- Permanent settlements allowed greater accumulation of material goods than was possible under a nomadic lifestyle.
- Agriculture allowed food surpluses to be generated so that not all individuals or families had to be involved in the food-generating process and specialization of tasks became more clearly defined. One end result is that only some 4 per cent of Canada's population is directly involved in food production today, permitting the rest of the population to direct their energies to other tasks, usually the processing of raw materials

into manufactured goods and thereby increasing the speed of flow-through of matter and energy in society. As indicated in Chapter 4, this high rate of throughput is at the core of many current environmental problems.

- The creation of food surpluses and more material goods promoted increasing trade between the now sedentary settlements. This led to the development of road and later rail connections to facilitate the rapid transport of materials, involving the consumption of large amounts of energy.
- Land and water resources became more important, leading to increased conflict between societies regarding control over agricultural lands.
- Aggregation of large numbers of people together in sedentary settlements also served to concentrate waste products in quantities over and above those that could be readily assimilated by the natural environment. Today we call this pollution.

iStockphoto.com/Ashley Whitworth

Huge monuments, such as Angkor Wat in Cambodia, could not have been constructed, nor the large cities that surrounded them, without the development of agriculture.

Box 10.2   Urbanization of Agricultural Land

Despite the fact that Canada is the second largest country in the world and one of the biggest exporters of foodstuffs worldwide, only 676,000 square kilometres, or about 7 per cent of Canada's overall land mass, is used for agricultural production. The amount of arable land free from severe constraints on crop production is even smaller, totalling less than 5 per cent of the land base. Limitations such as climate and soil quality reduce the amount of land that can be relied on for agricultural use; consequently, about 40 per cent of agricultural activities occur on marginal or poorer-quality land, which may not be dependable for long-term agricultural activity (Statistics Canada 2001).

Of the total of dependable agricultural land (Classes 1, 2, and 3) in Canada, Saskatchewan has the largest amount (40 per cent), followed by Alberta (24 per cent), Ontario (16 per cent), and Manitoba (11 per cent). Most of the prime agricultural land is in southern Canada, where 90 per cent of Canadians live. In Ontario, for example, more than 18 per cent of Class 1 farmland is now being used for urban purposes. This juxtaposition of prime agricultural land and the main urban centres has meant that suburban expansion invariably leads to losses in agricultural land. By 2001, about half of Canada's urbanized land was located on dependable agricultural land, and 7.5 per cent was on our very best agricultural land (Hofmann, Filoso, and Schofield 2005).

Concern over the loss of agricultural land to urbanization led to a monitoring program covering 70 Canadian cities with populations of more than 25,000 between 1966 and 1986. During that period, 301,440 hectares of rural land were converted to urban use. Furthermore, some 58 per cent of the converted land was of prime agricultural capability. To replace the productivity of this land would require bringing twice as much land under cultivation on the agricultural margins. A further study covering 1971 to 2001 found that urbanization had consumed 15,200 square kilometres of surrounding lands, an increase of 96 per cent over the period. In Ontario alone, urbanization increased by 4,300 square kilometres, a growth of almost 80 per cent (Hofmann, Filoso, and Schofield 2005).

Whether these losses should be of concern is open to debate. However, in some regions, urbanization of agricultural land affects specialty crops that have a limited ability to flourish in Canada. These crops often represent an important resource to the local economy (e.g., the fruit belts in the Niagara and Okanagan regions). Cities also affect the use of surrounding lands in indirect ways—golf courses, gravel pits, and recreational areas are often located on agricultural land in areas adjacent to urban areas, and as a result, the effects of urban areas extend beyond their physical boundaries.

The Canada Land Inventory (CLI) indicates that only 18 per cent of the renewable resource base of the country has prime capability for crops and less than 2 per cent is categorized as having the highest capability (Class 1). In an effort to slow the rates of conversion, several provinces have enacted legislation regarding the protection of agricultural lands. In 1972, for example, British Columbia enacted the Agricultural Land Reserve (ALR). At the time of its inception, some 6,000 hectares of prime agricultural land were being lost to urbanization each year.

Most Canadian cities of any size are surrounded by good agricultural lands. As the cities, such as Montreal, increase in size, they invariably encroach on the surrounding lands.

The annual loss has now fallen to less than one-seventh of that amount. The ALR initially covered 4.7 million hectares (5 per cent of BC); despite boundary changes over the decades, its current area remains approximately the same. A similar program exists in Quebec. Quebec's territory is vast, but less than 2 per cent of its land is suited to agriculture, and less than 5 per cent of Canada's dependable agricultural land is in the province. Farmland was constantly disappearing in the face of expansion and rampant urbanization. In 1978, the Quebec government introduced the Agricultural Land Preservation Act, which now protects more than 63,000 square kilometres of prime agricultural land.

However, these programs do have their problems. Often, farmers whose lands fall within the program and cannot be sold for non-agricultural uses cannot compete with cheap agricultural imports from elsewhere. Hence, they cannot make a decent living in agriculture and yet cannot sell their lands for non-agricultural purposes. And as a nation, should we be concerned about becoming increasingly dependent on other countries for our food? What do you think?

key development strategy in most parts of the world. World grain production has almost tripled since 1961 (Figure 10.1), mainly because farmers are harvesting more grain from each hectare—only slightly more land today is planted in grain: 671 million hectares in 2002 compared with 648 million in 1967 (Worldwatch Institute 2003b). Worldwide, intensification of production, dubbed the 'Green Revolution', has doubled the average harvest of grain from a given hectare from 1.24 tons in 1961 to 2.82 tons in 2002 (Worldwatch Institute 2003b). The amount of grain produced per person grew from 285 kilograms in 1961 to a peak of 376 kilograms in 1986. In recent decades, however, per capita production has fallen (Figure 10.2). On average, humans get about 48 per cent of their calories from grains, a share that has declined just slightly, from 50 per cent, over the past four decades. Grains, particularly corn, in conjunction with soybeans, also form the primary feedstock for industrial livestock production.

Following several years of declining harvests, the world's farmers reaped a record 2.316 billion tons of grain in 2007, mainly because of good weather in growing areas and an estimated 5 per cent increase in fertilizer use. Despite this jump of 95 million tons, or about 4 per cent, over the previous year, commodity analysts estimate that global demand will consume all of this increase and prevent governments from replenishing cereal stocks that are at their lowest level in 30 years (Figure 10.3).

However, future capacity to deliver agricultural outputs also depends on the continuing ecological viability of agro-ecosystems, and the Green Revolution carries with it a number of significant long-term negative environmental impacts that will ultimately lead to long-term productivity losses. Despite past successes, affordably feeding the current world population—and the more than 70 million people per annum by which that population will continue to grow over the next 20 years—remains a formidable challenge (Box 10.3). The International Food Policy Research Institute (IFPRI) suggests that between 1995 and 2020, global demand for cereals will increase by 40 per cent, while meat demand is projected to grow by 63 per cent and the demand for roots and tubers by

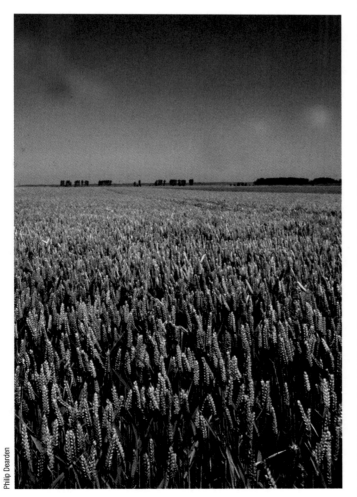

Intensification has resulted in a doubling of grain production over the last 40 years.

Philip Dearden

Figure 10.1 World grain production, 1950–2006. Source: Lester R. Brown, 'Supermarkets and Service Stations Now Competing for Grain', *Eco-Economy Update* (Washington, DC: Earth Policy Institute, 13 July 2006).

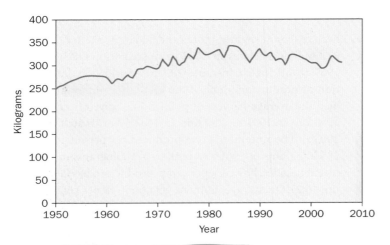

Figure 10.2 World grain production per person, 1950–2006. Source: Lester R. Brown, 'Supermarkets and Service Stations Now Competing for Grain', *Eco-Economy Update* (Washington, DC: Earth Policy Institute, 13 July 2006).

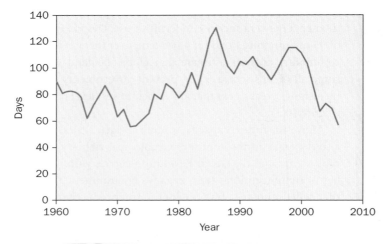

Figure 10.3 World grain stocks as days of consumption, 1960–2007. Source: US Department of Agriculture.

40 per cent. Between 2000 and 2006, world demand for cereals rose by 8 per cent while the price increased by 50 per cent, and prices more than doubled by early 2008 compared to 2000.

Slow-growing supply, low stocks, and supply shocks at a time of surging demand for feed, food, and fuel have led to drastic price increases, and these high prices will not fall soon, according to the IFPRI (von Braun 2007). In 2007, the US corn export price was up about 70 per cent from the previous year, while the American hard wheat price averaged 65 per cent more than a year earlier. Wheat prices in Argentina, another major exporter, have doubled since 2006. Important wheat exporters like Ukraine and Russia imposed export restrictions to ensure a sufficient domestic supply, as did rice exporters. The price of rice in Thailand, the world's biggest rice exporter, doubled in the first four months of 2008. Major importers reacted to high prices by purchasing grain early, which further tightened supplies and boosted prices. Developing countries are estimated to have spent a record $52 billion on imports of cereals in 2007, up 10 per cent from 2006. This follows a 36 per cent hike in the previous season, and the costs will no doubt increase further in the future.

International food aid agencies also purchase their supplies on the world market and have been forced to scale back. The volume of aid provided through the largest assistance program in the United States, Food for Peace, has dropped by nearly half since 2005 to 2.4 million tons in response to a 35 per cent increase in the cost of agricultural commodities as well as the rising costs of fuel for shipping. The combination of rising food costs and declining aid can be fatal for the millions of people worldwide who experience hunger on a regular basis.

All regions will face difficulties in meeting the growing demand for agricultural products while also preserving the productive capacity of their agro-ecosystems. Global climate change will be an additional stress. In more than 40 developing countries, mainly in sub-Saharan Africa, cereal yields are expected to decline, with mean losses of 15 per cent by 2080 (Fischer et al. 2005). Grain yields are likely to decline by 10 per cent for every 1°C increase over 30 years. Increased carbon dioxide and higher temperatures may promote lush growth, but they are deadly at the pollination stage, reducing some yields by 30 per cent. The International Rice Research Institute has found that the fertility of rice flowers falls from 100 per cent at 34°C to virtually zero at 40°C (Sanchez 2001). In addition, warm, wet weather promotes diseases like blight, and pests survive warmer winters, which means that longer growing seasons lead to an increased incidence of pest attacks.

World agricultural GDP is expected to fall by an average of 16 per cent because of global warming. However, the impacts are not evenly distributed, with

Box 10.3   Hunger

Agriculture provides approximately 94 per cent of the protein and 99 per cent of the calories consumed by humans. On average, people need about 2,500 calories per day, although this amount varies from person to person depending on weight, age, level of activity, and other factors. These calories come mainly from carbohydrates, such as potatoes and rice, but to remain healthy, humans also need protein, vitamins, fatty acids, and minerals. Fatty acids are essential for hormonal control, and minerals are necessary for bone formation and growth.

People are considered undernourished if their caloric intake is less than 90 per cent of the recommended level for their size and occupation. When people meet only 80 per cent of their recommended caloric intake requirement, they are considered severely undernourished. Malnourished people have adequate caloric intake but are deficient in other requirements. Protein deficiencies in children are obvious in their pot-bellied, skinny-limbed, and undersized appearance. TV news coverage of famines, mainly in Africa, have made such images very familiar to us. Vitamin deficiencies lead to many diseases, depending on the vitamin that is lacking. The deficiency can be fatal, such as the vitamin B deficiency that causes beriberi disease. Other effects include blindness, loss of hair and teeth, and bow-leggedness. Malnourishment also increases susceptibility to other diseases. Vitamin A deficiency, for example, impairs the immune system and is a significant contributor to the 1.5 million deaths from diarrhea each year.

The effects of malnutrition cross generations as well. Infants of malnourished, underweight women are likely to be small at birth and more susceptible to death and disease. Overall, 60 per cent of women of childbearing age in South Asia, where half of all children are underweight, are themselves underweight. In Southeast Asia, the proportion of underweight women is 45 per cent, in sub-Saharan Africa 20 per cent. Globally, the proportion of underweight children under five declined by one-fifth over the period 1990–2005. However, if current trends continue, the Millennium Development Goals target of halving the proportion of underweight children will be missed by 30 million children, largely because of slow progress in South Asia and sub-Saharan Africa.

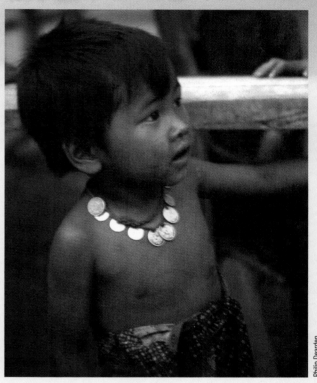

Philip Dearden

Hunger is a daily reality for many of the world's children.

Most Canadians cannot appreciate what it is like to go hungry, but for many of the world's population, hunger seldom goes away. Although global food supply could provide adequate nutrition for the entire global population, more than 854 million people (12.6 per cent of the global population) are currently malnourished, of whom 820 million live in developing countries. More than half of the undernourished people (61 per cent) live in Asia, while sub-Saharan Africa accounts for almost a quarter (24 per cent). Hunger is most pronounced where there are natural disasters and, more commonly, where warfare is ongoing. Every year, more than 3 million children die as a result of starvation. In stark contrast, the World Health Organization (WHO) reports that more than 300 million people are clinically classified as obese, and about half a million people die from obesity-related diseases every year. Obesity rates have tripled since 1980.

The 1974 World Food Summit promised to eradicate hunger within the next decade. The 1996 summit was a little more realistic and promised to cut the number of hungry people in half by the year 2015, and this was the target adopted by the Millennium Development Goals. Meeting the goal would require a reduction in the number of hungry people by 22 million each year; the actual number so far has been less than 6 million. The time frame to meet the goal has now been extended to 2060.

World population is expected to increase by 2.6 billion over the next 45 years, from 6.5 billion today to 9.1 billion in 2050. Almost all growth will take place in the less developed regions, where today's 5.3-billion population is expected to swell to 7.8 billion by 2050. This will pose significant challenges for global food production systems in the future. By early 2008, riots and deaths were occurring in countries such as Egypt, Cameroon, Haiti, and Burkino Faso in response to sharply rising food prices. In these countries, more than half of family income goes to food, and with a doubling of food prices, people will have few options.

No single solution can address the challenge of increasing the quantity and quality of affordable foods. The solution will involve a variety of approaches, including: exploration of new marine and terrestrial food sources, continued research to increase yields of existing crops, improvements in the efficiency of natural resource use, family planning programs aimed at reducing population growth rates, elimination of global agricultural tariffs, more efficient food distribution systems to address chronic hunger, and a moderation in demand on the part of the already overfed countries.

Sources: FAO 2007b; von Braun 2007.

a 20 per cent decline in less developed countries compared to a 6 per cent decline in industrial countries (Cline 2007). Agricultural prices will also be affected, with price increases of up to 40 per cent expected if temperature increases more than 3°C (Easterling et al. 2007). Distribution will also remain a major problem. For example, output per person for grain varies dramatically by region, standing at roughly 1,230 kilograms per year in the United States, most of which is fed to livestock, compared to 325 kilograms in China and just 90 kilograms in Zimbabwe.

The very large subsidies involved in global agriculture are a major confounding factor in examining food production capabilities. Industrial nations collectively pay their farmers more than $300 billion each year in subsidies, and this amount goes to the largest farmers in the richest countries, promotes chemical dependencies, inhibits change, and discriminates against producers in less developed countries. However, in 2008 World Trade Organization members agreed to significant commitments. One obliges the European Union to eliminate agricultural subsidies, but a time frame was not set, and France's agriculture minister estimated that European export subsidies might not be finally eliminated until 2015 or 2017.

As the world becomes more crowded and as pressures on biological systems and global biogeochemical cycles mount (Chapter 3), it is no longer sufficient to ask whether we can feed the planet. We need to ask ourselves more difficult questions. Can the world's agro-ecosystems feed today's planet and remain sufficiently resilient to feed tomorrow's hungrier planet? Will intensive production systems cause some agro-ecosystems to break down irreversibly? Presuming we can maintain current food production, are we paying too high a price in terms of the broader environmental effects of agriculture?

Agriculture is fundamentally an ecological process as solar radiation is converted through one or more transformations into human food supplies. Rapid growth in human population has entailed increasing disruption of natural systems in order to feed burgeoning populations and, particularly in developed countries, growing appetites. This chapter provides some context for this ecological process before considering some of the environmental challenges facing Canadian agriculture.

Agriculture As an Ecological Process

Agriculture can be considered a food chain, with humans as the ultimate consumers. Energy flows through this food chain in a manner similar to the way it does in natural food chains. Thus, the second law of thermodynamics is also important to agricultural food chains—the longer the food chain, the greater the energy loss (Chapter 2). This fact is one of the arguments for a vegetarian diet. By eating at the lowest level on the food chain as herbivores, humans will maximize the amount of usable energy in the food system. There are, however, other aspects of food production at higher levels of the food chain that should also be considered.

Food meets more than energy requirements alone. Important protein and mineral demands must also be satisfied. Animal products, by and large, are the main suppliers of these proteins and minerals, including such elements as calcium and phosphorus. Areas currently used as rangeland to support animal production often cannot be used as cropland. They may be too dry or otherwise ecologically marginal, and severe problems, such as excessive soil erosion, may have arisen in the past when humans tried to bring such lands under cultivation. Thus, it is not always correct to presume that rangelands can produce more food under tillage. Furthermore, grazing animals in many parts of the world provide not only food supplies but also other products and services, such as their energy as draft animals and their hides and other animal parts for clothing.

However, there are many valid arguments from an ecological, health, moral, and spiritual point of view for reducing meat consumption in favour of a diet with a higher vegetable content, especially in industrialized nations. While 43 kilograms of meat is produced annually per person, meat consumption still varies widely by region and socio-economic status. In the developing world, people eat about 32 kilograms of meat per year, compared to almost 85 kilograms per person in the industrial world.

The domestication of plants and animals thousands of years ago led to profound changes to the global land base. Complex natural systems that once dominated the landscape have been replaced by relatively simple control systems in which humans are in command of the species and numbers that exist in a given area. However, unlike the process that took place during the Industrial Revolution, these changes occurred over an extended time period, allowing greater potential for adaptation. Furthermore, until the past 150 years or so, energy inputs were largely limited to photosynthetic energy from the sun, the energy of domesticated draft animals such as oxen, camels, and horses, and human energy input. It was not until the Industrial Revolution unlocked past deposits of photosynthetic energy in the form of coal and later oil that industrial agriculture began and energy inputs and environmental impacts increased dramatically. Indeed, only during the latter half of the twentieth century did some of the most damaging impacts of agriculture come into play.

One of the most significant impacts is in the concentration of greenhouse gases (GHGs) in the atmosphere. Research on signals in atmospheric trace gas concentrations has revealed that even early agriculture had a significant impact (Ruddiman 2003). An increase in atmospheric carbon dioxide (CO_2) concentration of 20 to 25 parts per million between 8,000 and 2,000 years ago appears to be related to increased forest clearing for agriculture. About 5,000 years ago, there was an increase of 250 parts per billion in atmospheric methane (CH_4) concentration, coinciding with the adoption of 'wet rice' farming in Asia. These changes demonstrate that even with low populations, small changes over long periods can

Horse power is one form of auxiliary energy used in agricultural production, but the environmental impacts of this system of ploughing with horses in British Columbia are much less than those resulting from the fossil-fuelled, mechanical processes favoured by most Canadian farmers.

be as important as large changes caused by high populations over a short time.

In Canada since cropland cultivation began, an estimated 1 billion tonnes of soil organic carbon has been lost. Early cultivation would have removed most of it, and further losses continued as a result of intensive tillage, biomass burning, and removal of residues. Canada also has extensive grasslands, much of which is used for seasonal grazing. If pastures are overgrazed, they store less carbon. Paradoxically, undergrazing can produce the same effect through shortening vegetative growth during the growing season, because grazing stimulates growth up to a point. Since cultivation began, up to 30 per cent of the carbon originally present in the surface soil layer has been lost. However, the rate of loss appears to be falling because of smaller amounts of land being converted into cropland, decreases in summer fallow, increases in no-till farming, and increased fertilizer use in the Prairie provinces (Hengeveld et al. 2008).

Of course, CO_2 is not the only GHG affected by land-use change (Chapter 7). Fertilizers have a significant impact, as discussed in Chapter 4, as do the decomposition of crop residues and soil organic matter and the transport of nitrogen off-farm during leaching, runoff, and volatilization. Livestock is also a significant source, particularly of CH_4 and nitrous oxide (N_2O). Up to 10 per cent of feed energy is lost in ruminant animals through belching and other gaseous contributions, and additional amounts come from manure. The contribution of these gases has been increasing, and by 2005 the agriculture sector accounted for 66 per cent of Canada's N_2O emissions and 25 per cent of CH_4 emissions (Desjardins et al. 2008). These changes are driven mainly by increases in livestock numbers, discussed later in the chapter.

Opportunities to reduce the contributions of agriculture to global warming are many, but an obvious one is to reduce the numbers of livestock by eating less meat. Carbon sequestration in soils can be improved through such measures as no-tillage cultivation and reducing summer fallow, as discussed later in the chapter. Planting perennial crops such as trees also helps. On grasslands, grazing intensity can be reduced and productivity increased by adding nutrients and water. However, fertilizer has to be applied more precisely than in the past to ensure that the amounts added are taken up by the crops and will not denitrify into N_2O.

The complexity of finding the optimum solution for reducing GHG emissions from agriculture is illustrated by the fact that where soils are fine and it is relatively humid, such as in eastern Canada, no-tillage agriculture may result in increased N_2O emissions as a result of higher surface temperatures (Rochette et al. 2008). However, reducing no-tillage agriculture conflicts with the desire to maintain soil carbon levels through no-tillage systems. In the Prairies, for example, because of the drier environment and heavier soils, no-tillage is a very appropriate way to reduce overall GHG emissions from agriculture. Solutions obviously have to take such geographical differences into account.

Modern Farming Systems in the Industrialized World

The Green Revolution

Dramatic changes in food production systems have occurred through a variety of technological advances that were in turn influenced by changes in demographics (e.g., increases in population densities), social structure (e.g., urbanization, social stratification), and economic conditions (e.g., global trade). Early food production systems were characteristically small in scale, since they were largely dictated by fixed environmental conditions—the quantity and quality of food produced depended heavily on existing local climatic conditions (e.g., annual rainfall), native vegetation (i.e., plants indigenous to the region), and availability of human and/or animal labour. Soil fertility was maintained or enhanced by using locally available, natural elements such as manure, bones, and ashes.

Today, local conditions are manipulated to improve both the quantity and quality of outputs. The amount of food produced per unit of land has increased dramatically as a result of a variety of technological advances, including genetic engineering, greater mechanization (e.g., tractors for ploughing and seed sowing, mechanized food processing), and the creation of auxiliary energy flows, as discussed in Chapter 2. The package of inputs or agricultural techniques, which together are referred to as the **Green Revolution**, includes the introduction of higher-yield seeds (e.g., shorter maturation, drought resistance) and a reliance on auxiliary energy flows.

The development and commercialization of higher-yielding seeds through **hybridization** led to significant gains in grain yields throughout the world. In the 1940s, scientists developed a 'miracle wheat seed' that matured faster, producing wheat that was shorter and stiffer than traditional breeds and less sensitive to variation of daylight. India more than doubled its wheat production in five years with the new technology, and other Asian and Latin American countries recorded similar productivity increases. Miracle rice seeds and high-yield maize (corn) were developed by scientists soon afterwards, and their use has diffused rapidly around the world.

Were it not for these developments, there would no doubt be many more people in the world suffering chronic food shortages. However, they also pose challenges. Most of the new hybrid seeds grew better than their native counterparts only if fertilizers and biocides were applied with sufficient frequency and in sufficient quantity. As these chemicals became more expensive, many poorer farmers were unable to take advantage of the high-yield seeds. The Green Revolution also encouraged a narrowing of the genetic base of the crop, and with each farmer growing exactly the same strain, any disease or pest that managed to adapt to the strain had an almost unlimited food supply.

Today, critics of the Green Revolution point to stagnation in scientific progress—the gains that can be made through technology have already been made, and the damage done to agro-ecosystems by modern production systems will limit the ability of farmers to increase yields significantly. For example, macronutrients in the soil, such as nitrogen, phosphorus, and potassium, can be replenished by fertilizers, but many of the micronutrients required in trace amounts cannot be replaced. As we take out more crops from the soil, these nutrients may become exhausted, leading to greatly diminished returns in the future. Signs of such declines are already appearing in some areas. The increasing cost of fossil fuels also leads to increasing fertilizer costs (Box 10.4).

However, some advances are still being made. In 2001, for example, a new rice strain, called NERICA, was produced from hybridization of African and Asian rice varieties. According to the United Nations Development Programme (UNDP), the project's main supporter, NERICA produces 50 per cent more yield, uses less fertilizer, is richer in protein, and is more resistant to drought, disease, and pests. It was also developed in full consultation with farmers and consumers and is now being successfully grown in a wide range of African countries. In Guinea, thanks to the success of NERICA varieties, farmers are now able to gross US $65 per hectare with minimal inputs and $145 per hectare with a moderate level of inputs. The country saved more than US $13 million on rice imports in 2003. NERICA is now being used as a 'success story' for the Millennium Development Goals (http://www.mdgmonitor.org/goal1.cfm#).

The development of these 'miracle seeds' relied only on genetic combinations found in nature. The development of **genetically modified organisms (GMOs)**, on the other hand, involves combining genes from different and often totally

Box 10.4   Soil Fertility in Africa

Soil fertility depletion is the fundamental cause of low per capita food production in Africa. Centuries of nutrient removals in crops and through land degradation means that few nutrients remain to support crop growth, let alone the high-demanding Green Revolution species. Estimates suggest that over the past 30 years, extraction has amounted to 22 kilograms of nitrogen, 2.5 of phosphorus, and 15 of potassium per hectare, equivalent to a loss of \$4 billion in fertilizer. Ironically, fertilizers cost two to eight times more at the farm gate in Africa than they do in Europe. A metric ton of urea costs about \$100 in Europe but more than \$800 in Malawi.

The problem is being dealt with in three main ways:

- Improved fallows: Leguminous trees are planted with maize to boost soil nitrogen through biological nitrogen fixation (BNF) (Chapter 4). Following the dry season, the wood is harvested and the leaves and pods ploughed back into the land. Nitrogen increases are roughly equal to those from fertilizer application, and maize yields increase by a factor of two to four as nitrogen depletion is eased.

- Indigenous rock phosphates: Phosphorus is a main limiting factor in East Africa and the Sahel. Farmers are now crushing and applying indigenous rock phosphates as an alternative to importing superphosphate fertilizers, and they have been found to be 90 per cent as efficient as the vastly more expensive imports in boosting maize yields.

- Biomass transfers: The shrub *Tithonia diversifolia* accumulates nutrients within its biomass, which is then spread on farm fields and may double yields without the application of further fertilizers. This method also has the advantage of adding not only to the macronutrients but also to the micronutrients in the biomass.

Source: Sanchez 2001.

unrelated species. Such genetic manipulation is now a multi-billion-dollar industry, and Canada is one of the world's leading participants (Box 10.5).

The reliance on large auxiliary energy flows in modern industrialized agricultural systems is one of the main differences between natural and agro-ecosystems. Auxiliary energy flows include fertilizers, biocides (insecticides, herbicides, fungicides), fossil fuels, and irrigation systems. These energy subsidies have significantly increased crop yields over the past century, particularly over the past two decades. **Subsistence farming**, in which the production of food is intended to satisfy the needs of the farm household, relies on natural energy supplies and may produce 10 food units for every unit of energy invested. However, in the most energy-intensive food systems, such as those in Canada and the US, 10 times as much energy on average needs to be invested through auxiliary energy flows for every unit of food produced.

Improved water management, a key component in Green Revolution technologies, helped to boost worldwide productivity or output of 'crops per drop' by an estimated 100 per cent since 1960 (FAO 2003b). Agriculture accounts for 70 per cent of fresh water withdrawn from natural sources for human use and as much as 90 per cent in many developing countries. The water needs of humans and animals are relatively small—the average human drinks about four litres a day. But producing the same person's daily food can take up to

Rice yields, where sufficient water was available, tripled in some locations.

Philip Dearden

Box 10.5   Genetically Modified Organisms

Yields of many crops in the developed world will not increase significantly with conventional agricultural techniques, even those related to the Green Revolution. Instead, farmers are turning to biotechnology, or the genetic modification of crops, to increase production. The technology already dominates the production of a few crops in several countries, yet other countries ban its use. Some nations require labelling of foods that have been genetically modified, while others such as Canada have rejected such openness. Why are there such differences in approach towards this new 'genetic revolution'? Before we answer that question, it is necessary to explain what GMOs entail.

Advances in understanding of DNA have shown that the basic building blocks of life are all very similar. Traits controlled by single genes can be transplanted from one species to another. Transgenic crops are produced when a single species contains pieces of DNA from at least one other species. One of the most common examples is the implantation of genes containing the toxin from *Bacillus thuringiensis* (Bt). For years, farmers have sprayed this naturally occurring toxin on their fields as an insecticide. Geneticists were able to isolate the toxic gene in the bacteria

and are now inserting it into crops such as corn and soybeans. The crops then produce their own toxin. The other common genetic modification is to produce crops with genes that make them resistant to a particular herbicide. When a farmer sprays the crop, all competing plants are killed.

The growth in use of such transgenic crops has been very rapid. At the time of the Earth Summit in Rio de Janeiro in 1992, production of transgenic crops had not yet begun, but it now stands at more than 100 million hectares. About 75 per cent of this area is in industrialized countries. In 2006, the US, followed by Argentina, Brazil, Canada, India, and China, continued to be the principal adopters of biotech crops globally, with 54.6 million hectares planted in the US (53 per cent of global biotech area), of which approximately 28 per cent produced stacked crops containing two or three traits.

Biotech soybean remained the principal biotech crop in 2006, occupying 58.6 million hectares (57 per cent of global biotech area), followed by maize (25.2 million hectares at 25 per cent), cotton (13.4 million hectares at 13 per cent), and canola (4.8 million hectares at 5 per cent). About 16 per cent of the total area planted

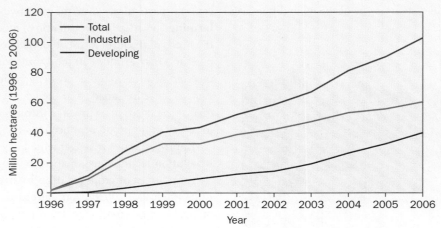

Increase of 13%, 12 million hectares (30 million acres) between 2005 and 2006.

Figure 10.4 Global area of biotech crops. Source: James 2006.

with these crops is now devoted to genetically modified varieties, and two traits—insect resistance and herbicide tolerance—dominate. Throughout the world, several thousand GMO field tests have been conducted or are underway, and many more crop-trait combinations are being investigated, with greater focus on virus resistance, quality, and in some cases tolerance to abiotic stresses.

In Canada, scientists at the University of Victoria have developed a potato that resists bacteria and fungi, allowing it to be stored for 10 times longer than 'normal' potatoes. In Guelph, ON, scientists have engineered a pig that produces manure 20 to 50 times lower in phosphorus content than that of 'regular' pigs. Since phosphorus is a main cause of eutrophication (Chapter 4), the number of pigs that can be kept in a given area is often limited. The new pig can be stocked in higher densities.

The accumulated global value of biotech crops for the 11-year period since they were first commercialized in 1996 is estimated at $35.5 billion. The global value of the biotech crop market was projected at more than $6.8 billion for 2007. There is no doubt that GMOs hold great potential for the future. It is estimated that they will generate an industry worth US $350 billion covering more than 200 million hectares by 2015. However, development has been very rapid, and several areas of uncertainty regarding their effects remain:

- *Pleiotropic effects.* These are unexpected side effects that might be suffered by the target organism as a result of incorporation of the new gene. For example, there may be a change in the toxins produced or the nutrient content.

- *Environmental effects.* There will be impacts on natural processes, such as pollination and biogeochemical cycles, as a result of creating new crops. It is also feared that there may be unanticipated gene flow to other organisms and perhaps interbreeding with wild relatives. This has the potential to produce 'super-weeds', as has happened in Canada with canola, where the biocide-resistant superweeds are now growing in wheat fields. The only way to get rid of them is to resort to broad-spectrum herbicides, those that will kill everything.

- *Unintentional spread.* Pollen and seeds from transgenic crops may spread onto lands where they are not intended to grow. This is already a problem for organic canola and honey producers in the Prairies, who can no longer guarantee that their products are transgenic-free.

There is little consensus among governments, consumers, farmers, and scientists concerning the benefits and risks associated with biotechnology or genetic engineering. A lack of perceived benefits for consumers and uncertainty about their safety have limited their adoption in some countries, while others have developed legislation requiring mandatory labelling on genetically modified foods. Canada has chosen not to do so. The government voted down a bill that would have required mandatory labelling for all foods containing more than 1 per cent genetically engineered ingredients, even though polls suggested that a large majority of Canadians would prefer to have a choice. What is your view about whether GMO food products should be explicitly labelled?

The agricultural system of these hill tribe people in northern Thailand, based on dry rice cultivation with livestock such as pigs and chickens, may seem primitive but yields a much higher return than modern farming systems in terms of the energy budget.

Philip Dearden

5,000 litres of water. For example, it takes 1,000 tonnes of water to produce one tonne of grain.

Although agriculture is not Canada's largest user of water, it is its largest consumer. Agriculture removes significant quantities of water from the landscape, tying it up in agricultural products or evaporating it back into the air rather than returning it directly to streams or groundwater. Agriculture relies on a reliable supply of good-quality water for growing crops, raising livestock, and cleaning farm buildings. Approximately 75 per cent of all agricultural water withdrawals in Canada occur in the semi-arid prairie region. More than 780,000 hectares of cropland in Canada are under irrigation, and the amount is increasing. Alberta accounts for 63 per cent of the national total. Alberta has the largest irrigated field crop and irrigated hay and pasture areas. Most of the vegetable irrigation is in Ontario, with 45 per cent of the

Irrigation water is critical for producing crops on much of Canada's landscape. Being irrigated here is a potato crop.

total irrigated vegetable area. British Columbia applies most of the irrigation to keep fruit areas, such as the orchards in the Okanagan Valley, thriving. As a whole, the province accounts for 52.8 per cent of total irrigated fruit area in the country.

Agricultural practices consume large quantities of water, but they also have a negative impact on water quality. Box 10.6 details some of agriculture's impact on water resources and aquatic ecosystems.

Fertilizers and biocides are also important inputs to modern farming systems. New strains of crops, such as wheat, will only produce superior yields if fertilized adequately. The Food and Agriculture Organization (FAO) estimates that world fertilizer supply (nitrogen, phosphate, and potash nutrient) will increase at an annual rate of

Box 10.6   Agricultural Impacts on Canada's Water Resources and Aquatic Ecosystems

Technological advances over the past 50 years have allowed Canadian farms to become more productive, but some farming practices have contributed to environmental degradation, including the decline of water quality. The main pollutants of water coming from farmland are nutrients (particularly nitrogen and phosphorus), pesticides, sediment, and bacteria. Agriculture has also changed the physical presence of water across the Canadian landscape through the construction of dams and reservoirs, distribution of irrigation water, drainage of wetlands, and sedimentation of streams and lakes.

Nutrient losses. In certain parts of Canada, the use of additional nutrients in the form of mineral fertilizer, manure, compost, and sewer sludge to increase crop productivity has led to a nutrient surplus in the soils, with the potential loss of nutrients to surface and groundwater. Nutrient addition to aquatic ecosystems promotes eutrophication (Chapter 4). Other indirect consequences include changes in the abundance and diversity of higher trophic levels (e.g., benthic invertebrates and fish), increased abundance of toxic algae, and fish kills caused by loss of oxygen from the water. One factor contributing to the decline in population numbers for 17 of

Canada's 45 frog, toad, and salamander species is nitrate concentrations in agricultural streams and runoff water that exceed thresholds for chronic and acute toxicity of amphibians.

Pesticide losses. The use of pesticides for disease, weed, and insect control has resulted in pesticide losses to the atmosphere and subsequent deposition from the atmosphere to surface water and non-agricultural lands, as well as runoff and leaching to surface and groundwater. Depending on the compound and the concentrations involved, pesticides introduced into surface waters can kill fish and other aquatic organisms; cause sub-lethal effects on reproduction, respiration, growth, and development; cause cancer, mutations, and fetal deformities in aquatic organisms; inhibit photosynthesis of aquatic plants; and bioaccumulate in an organism's tissues and be biomagnified through the food chain. Pesticide concentrations in surface waters sometimes exceed Canadian water quality guidelines for irrigation water or protection of aquatic life.

Sedimentation. Agricultural practices such as tillage and allowing livestock access to streams increase erosion and the movement of soil from farmland to adjacent waters. Alterations to soil conditions caused by tillage and cropping patterns may cause soil degradation, which can lead to less infiltration of water into soil and increased runoff and movement of nutrients and pesticides to surface waters. The introduction of soil into aquatic eco-systems can increase turbidity and thereby reduce plant photosynthesis, interfere with animal behaviours depen-dent on sight, impede respiration (by gill abrasion) and feeding, degrade spawning habitat, and suffocate eggs. Over the past 40 years, tillage practices have moved to-wards greater use of conservation practices, including no-till. These practices tend to encourage infiltration rather than runoff.

Pathogens. Pathogens from farm livestock can migrate into groundwater or be carried in runoff into surface wa-ters. Pathogen contamination of irrigation waters, drink-ing waters, or shellfish areas can create significant risks to human food supplies and can also pose a threat to aquatic ecosystems and biodiversity. Regions of the country with high livestock densities are reporting fecal coliform counts that exceed Canadian water quality guidelines for both drinking and irrigation water. For example, in On-tario, the number of wells with E. coli counts exceeding the guideline almost doubled over a 45-year period, from 15 per cent in 1950–4 to 25 per cent in 1991–2. The inter-action between agriculture and water quality came to the attention of Canadians in 2000 through the tragic deaths at Walkerton, ON, discussed in Chapter 11.

Wetland drainage. Drainage of wetlands has increased the area of agriculturally productive soils, but it has also modified local ecosystems and changed the pattern of water partitioning between evaporation, stream flow, and infiltration. Wetland drainage destroys the habitat of a number of species of arthropods, reptiles, amphibians, birds, mammals, and fish. More than 85 per cent of the decline in Canada's original wetland area has been attrib-uted to drainage for agriculture.

Source: Chambers et al. 2001.

3 per cent between 2007–8 and 2011–12, sufficient to cover demand growth of 1.9 per cent annually. Total production is expected to grow from 206.5 million tonnes in 2007–8 to 241 million tonnes in 2011–12. Fertilizer demand will increase from 197 million tonnes today to 216 million tonnes in 2011–12.

Large increases in fertilizer inputs, particularly nitrogen, have occurred over the past few decades in Canada (Fig-ure 10.5). Western Canada, for example, saw a fivefold increase in fertilizer applications between 1970 and 2000, and the three Prairie provinces now account for 69 per cent of Canada's total fertilizer sales, up from 29 per cent in 1970.

In 1970, Canadian farmers spent $1.37 million on fertilizers, compared to $1.98 billion in 2000. Canadian farmers are spending more on fertilizers, but the trend since the mid-1990s has been towards an overall decline in the total area treated with commercial fertilizers, with a decrease of 3.7 per cent between 1995 and 2000. Prince Edward Island, Ontario, and Saskatchewan reported the highest declines in crop area treated. However, these statistics mask an important trend. The *amount* of fertilizers applied *per hectare* (kg/ha) has increased considerably since the early 1970s, although the rate of increase has slowed since the mid-1980s. In 1970, 1980, 1990, and 2000, fertilizers were applied at a rate of 18.4, 42.4, 45.1, and 54.2 kg/ha, respectively. These rates are not high by international standards, with the United States, Australia, and Japan applying 103.4, 151.7, and 301.0 kg/ha, respectively (Worldwatch Institute 2003a). Nonetheless, fertil-izer application is of environmental concern because fertiliz-ers are a main contributor to the speed of the eutrophication process (Chapter 4) as well as to groundwater pollution in some areas.

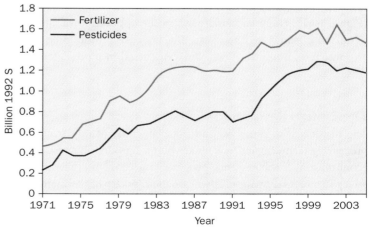

Figure 10.5 Chemical input usage in farming in Canada, 1971–2005. Source: Statistics Canada and AAFC calculators.

Not all crops require the same level of chemical inputs. Tomatoes are particularly demanding compared with, for example, cereal crops.

A central strategy in improving agricultural output is to limit losses from the effects of pests and diseases and from weed competition. Since the mid-1990s, the approach to crop protection has relied increasingly on the use of biocides (insecticides, nematocides, fungicides, and herbicides). Pesticides should be referred to as biocides, since their application often affects more than just the target species. Biocides are applied to boost yields. Yields are improved by reducing the amount of energy flowing to the next trophic layer of the food chain through the respiration of heterotrophs (often insects) and by eliminating non-food plants that compete with the crop plants for available growth resources.

Biocide use continues to increase dramatically, indicating that farmers find biocides cost-effective from a production perspective, particularly where alternative forms of crop protection are labour-intensive and labour costs are high, as they are in Canada. Canada and the United States lead the world in the consumption and use of biocides, accounting for 36 per cent of world biocide use (UNEP 2002). The biocide market has been growing at approximately 6 per cent per year since 1990 (UNEP 2002), with significant increases in Canada (Figure 10.5). These large increases in the application of agricultural biocides over the past 20 years have profound environmental implications, as will be discussed in more detail later.

The Biofuel Revolution

The rising cost of crude oil and the availability of improved processing technologies have made biofuel production more economically feasible. Some people characterize biofuels as an environmental saviour; others see them as one of the main challenges to environmental sustainability. What are biofuels, and why do views on them diverge so radically?

Biofuels are derived from plants and other organic material and seem to have great potential to help curb global greenhouse gas emissions. About one-quarter of GHG emissions are produced by the transport sector, and biofuels, including ethanol and biodiesel, are the only existing renewable fuels compatible with the current transportation infrastructure. Biofuels generate a fraction of the pollutants of traditional petroleum-based fuels, and the plants that produce them remove carbon, a climate-altering GHG, from the atmosphere. Biofuels also have the potential to reduce foreign oil dependency, lower fuel prices, increase income for farmers, and provide a host of new jobs. Of the 47 poorest countries, 38 are net importers of oil, and 25 import all of their oil; for these nations, the tripling of oil prices has been an economic disaster. Nations that develop domestic biofuel industries will be able to purchase fuel from their own farmers rather than spending scarce foreign exchange on imported oil.

These advantages led experts at an African Union Conference to conclude that 'Promotion of biofuels industry in developing countries has the capacity to propel such countries to achieve the MDGs [Millennium Development Goals] through poverty reduction (especially job creation and economic enhancement), health impact, and climate change' (http://www.africa-union.org/root/au/index/index.htm).

In response, biofuel production is growing at a rate of roughly 15 per cent per year, more than 10 times that of oil, although oil still accounts for 95 per cent of the global transportation fuel market. Ethanol production more than doubled between 2000 and 2005 to 1.2 per cent of global fuel use. Biodiesel production, which started from a smaller base, quadrupled (http://www.earth-policy.org/Updates/2006/Update55_data.htm). While corn-based US ethanol and sugarcane-based Brazilian ethanol account for nearly 90 per cent of global production, other countries are rapidly expanding output, using a variety of sugar and grain crops. Europe is currently the leading producer of biodiesel, which is processed from vegetable oils derived from soybeans, oil palm, and canola, among other crops.

In 2005, the US pledged to nearly double ethanol production by 2012, and the European Union announced that biofuels will meet 10 per cent of its transportation fuel needs by 2020. In 2006, world biofuels production rose 28 per cent to 44 billion litres, fuel ethanol was up by 22 per cent, and biodiesel rose by 80 per cent. Some experts feel that over the next 15 to 20 years, biofuels may supply up to 25 per cent of the world's energy needs.

As Table 10.1 indicates, Canada is not a large producer of biofuels in the global context, and the federal government has introduced the ecoAGRICULTURE Biofuels Capital Initiative (ecoABC), a $200-million four-year program ending in 2011 to stimulate production. The program provides repayable funding for the construction or expansion of transportation biofuel production facilities. Funding is conditional on investment in the biofuel projects by agricultural producers and the use of agricultural feedstock to produce the biofuel.

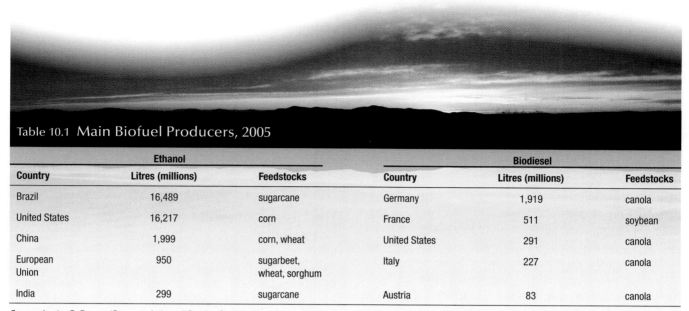

Table 10.1 Main Biofuel Producers, 2005

Ethanol			Biodiesel		
Country	**Litres (millions)**	**Feedstocks**	**Country**	**Litres (millions)**	**Feedstocks**
Brazil	16,489	sugarcane	Germany	1,919	canola
United States	16,217	corn	France	511	soybean
China	1,999	corn, wheat	United States	291	canola
European Union	950	sugarbeet, wheat, sorghum	Italy	227	canola
India	299	sugarcane	Austria	83	canola

Source: Lester R. Brown, 'Supermarkets and Service Stations Now Competing for Grain', Eco-Economy Update *(Washington, DC: Earth Policy Institute, 13 July 2006).*

In view of these advantages, why are there so many critics of biofuels? One question relates to the inputs used to grow many of the crops used for biofuels. Fossil fuels are invested in fertilizers, pesticides, machinery, and processing, and the second law of thermodynamics (Chapter 2) warns us that energy is lost in each of these energy transformations, which contributes to global warming. However, a comparative study in *Science* magazine (Farrell et al. 2006) suggests that overall, using corn ethanol instead of gasoline could reduce GHG emissions by some 13 per cent. For sugarcane, the reduction could be even greater, as high as 87 to 96 per cent, according to some sources (http://www.unica.com.br/i_pages/files/pdf-ingles.pdf) because the crop can be grown at higher yields with fewer fossil-fuel inputs. Furthermore, ethanol derived from cellulosic material, including crop residues and woody plants, offers the potential of reducing GHG emissions by more than 100 per cent, particularly if bio-energy is also used in producing the fuel (http://www.iea.org/dbtw-wpd/Textbase/press/pressdetail.asp?PRESS_REL_ID=127).

Another problem is that with the increased profit to be made from land used for growing biofuel inputs, land is being taken away from food production. The cars of the wealthy may well end up consuming the food of the poor. For example, it took 13 per cent of the US corn harvest in 2005 to displace less than 3 per cent of fuel needs (http://www.ethanolrfa.org/resource/facts/agriculture). Corn prices doubled in 2007, causing social unrest in Mexico, where corn tortillas are a dietary staple, and there was widespread unrest in many countries in 2008 as food prices continued to soar. This is a difficult issue to address, since there are many factors at work in rising prices, but increased competition for land created by biofuels is undoubtedly one of them. However, the relative impacts of the biofuel demand will vary from region to region. Furthermore, biofuel production can boost incomes in rural areas, where three-quarters of the world's poor reside, which may increase their ability to secure food supplies. Deeper analysis is needed to understand the global and local impact of expanded biofuels demand on food prices. However, the International Food Policy Research Institute projects that a drastic increase in biofuel production in sub-Saharan Africa will result in an 8 per cent decline in calorie availability (von Braun 2007).

One answer to both the challenges of energy inputs into biofuel production and competition with food crops is to produce ethanol from non-food sources. Cellulosic ethanol is made from a wide variety of plant materials, including wood wastes, crop residues, and grasses, some of which can be grown on marginal lands not suitable for food production. The process for converting these materials to fuel is often more efficient, because plant material rather than fossil fuels can be used to provide heat and power. As a result, cellulosic ethanol has an energy yield at least four to six times the energy expended during production and can reduce greenhouse gas emissions by 65 to 110 per cent relative to gasoline (Worldwatch Institute 2007b).

With the continued improvement of cellulosic technology, the ethanol feedstock supply is expected to shift from edible crops such as corn and canola oil to fast-growing grasses such as switchgrass and miscanthus, agriculture and forestry residue, and even the organic portion of municipal solid waste. The cellulosic conversion process is still relatively expensive, but the costs of cellulosic biofuel production are declining, and one Canadian company, Iogen Corporation (http://www.iogen.ca), is already selling cellulosic ethanol commercially.

A further challenge for biofuels is that large-scale biofuel production can threaten biodiversity, as seen recently with oil palm plantations in Indonesia and sugarcane plantations in Cambodia that are encroaching on forests. In Brazil, the Cerrado, a vast landscape of biologically rich forests, brush, and pasture just south of the Amazon, is coming under pressure as sugarcane cultivation expands. One response to this threat has been to convince decision-makers that areas rich in biodiversity may yield greater long-term returns by being preserved for ecotourism rather than bulldozed for sugarcane. For example, the government of Uganda had been supporting a proposal for a 7,000-hectare sugarcane concession in the supposedly protected Mabira Forest Reserve near Lake Victoria. However, a study by the conservation group NatureUganda showed that the financial benefits of protecting the forest and encouraging ecotourism vastly outstripped the potential of biofuel crops. The commercial value of tourism and carbon capture in Mabira was estimated at more than $316 million a year, whereas sugarcane production would be worth less than $20 million. A similar rationale has led the Cambodian government to restrict the number of concessions for sugarcane production in the fabled Cardamom Mountains in the southwest of the country to see whether ecotourism can produce sufficient yields to address poverty in the area.

Despite these challenges, it seems that biofuels must be part of a portfolio of options to deal with global warming that also includes dramatic improvements in vehicle fuel economy, investment in public transportation, and better urban planning (see Chapter 13). The long-term potential of biofuels is probably in the use of non-food feedstock, including agricultural and forestry wastes, as well as fast-growing, cellulose-rich energy crops such as perennial grasses and trees. Following the model of Brazil's sugarcane-based biofuels industry,

Lands being cleared for sugarcane in the southern Cardamom Mountains, Cambodia.

cellulosic ethanol could dramatically reduce the carbon dioxide and nitrogen pollution that results from today's biofuel crops.

Modern industrial cropping systems that rely on auxiliary energy flows and other inputs are responsible for the production of a large percentage of the cereals, pulses, oil crops, roots and tubers, fruits, and sugar crops produced and consumed worldwide. However, these crops represent only 60 per cent of the total value of output from the world's agro-ecosystems. Livestock production is responsible for the remaining 40 per cent.

The Livestock Revolution

Hundreds of years ago, livestock (e.g., cattle, sheep, goats) raised for local consumption were permitted to graze on surrounding natural vegetation. Stocking densities were dictated by surrounding environmental conditions—i.e., the availability of water and food supplies on a given unit of land and the ability of the local environment to assimilate animal wastes. As a result, farms were typically small, with less than 100 animals. However, as the industrialized world's appetite for meat continues to grow exponentially, traditional livestock production systems are being replaced. Worldwide, meat consumption has doubled since 1977, and over the past half-century it has increased fivefold. Production of beef, poultry, pork, and other meats has risen to nearly 40 kilograms per person per year, more than twice as much as in 1950.

The **Livestock Revolution** has led to a number of changes with respect to how animals are brought to market. In industrialized countries such as Canada, the desire to supply the phenomenal growth in demand for protein has led to a reliance on industrial feedlots, which now produce more than half of the world's pork and poultry and 43 per cent of the world's beef (Worldwatch Institute 2003b). Industrial systems of livestock production depend on outside supplies of feed, energy, and other inputs. Technology, capital, and infrastructure requirements are based on large economies of scale, and production efficiency is high in terms of output per unit of feed. As the world's main providers of eggs, poultry, beef, and pork at competitive prices, intensive farm operations (or 'factory farms') meet most of the escalating demands for low-cost animal products.

Agricultural production of livestock has grown across Canada, while the number of farmers has declined and the size of the average farm has increased. In Ontario, for example, one quarter of farms account for three-quarters of total farm revenues. New farms are often capital-intensive operations with very large numbers of livestock. Farms with 3,000 or more pigs or 1,200 cattle are increasingly common, and some farms in Ontario and Quebec house in excess of 10,000 animals. For pigs, herd size grew by 150 per cent between 1996 and 2006, and average herd size is now 1,308 pigs per farm. Over the same time period, dairy cattle herd size increased by 44 per cent and beef by 35 per cent (Agriculture and Agri-Food Canada 2008).

As the livestock industry expands and becomes more intensive, health and environmental concerns over livestock manure are growing, particularly when livestock are produced in large numbers under confined conditions such as beef feedlots and intensive hog and poultry barns. The social and environmental impacts associated with intensive farming operations will be discussed in greater detail later.

Agriculture's Impact on the Global Landscape

A variety of impacts are associated with the development of agriculture and more specifically with the development and spread of modern farming systems:

- Natural selection determines the plants and animals to be found in an area, but following the development of agriculture, humans became the major influence on species distributions. In modern agriculture, this has led to large areas of **monoculture cropping**, often not only made up of identical crop species but with each individual having exactly the same genetic code. New species have been created for the purpose of maximizing the output of food for humans, while native species have been displaced. Humans have influenced not only distributions but also the numbers of each species, with domesticated plants and animals greatly increasing in number and range and wild species being drastically reduced. Saumure, Herman, and Titman (2007) provide a good Canadian example of the impact of agricultural practices on a wild species, the wood turtle.
- Energy flows are increasingly directed into agricultural as opposed to natural systems. It is estimated that humans now appropriate some 40 per cent of the net primary productivity of the planet.
- Biogeochemical cycles are interrupted as natural vegetation is replaced by domesticates that are harvested on a regular basis. Auxiliary energy flows in the form of fertilizer are used in an attempt to replenish some of the nutrients extracted through harvesting.
- Auxiliary energy flows used in modern agricultural systems to supplement the natural energy flow from the sun are often in excess of those derived from natural sources.
- In many areas, agriculture involves supplementing rainfall with irrigation to provide adequate water supplies. This has led to large-scale water diversions and to changes in groundwater, soil characteristics, precipitation patterns, and water quality.
- Soils are altered not only chemically but also physically through ploughing. There is no natural process that mimics the disturbance created by ploughing.
- Natural food chains are truncated as humans destroy and replace natural consumers and predators at higher trophic levels.
- Natural successional processes are altered to keep the agricultural systems in an early seral stage; auxiliary energy

Agriculture has had a profound impact on the distribution of species. Large areas of the agricultural landscape are dominated by monocultures—in this case sunflowers—in which each plant has the same genetic makeup.

flows in the form of herbicides and mechanical weeding are often used to accomplish this.
- The stocking densities of domesticated herbivores are often much higher than that of natural herbivores, leading to a reduction in standing biomass and changes in the structure and composition of the primary production system.
- The industrial system of livestock production acts directly on land, water, air, and biodiversity through the emission of animal waste, use of fossil fuels, and substitution of animal genetic resources. It also affects the global land base indirectly through its effect on the arable land needed to satisfy its feed concentrate requirements. The industrial system requires the use of uniform animals of similar genetic composition, contributing to within-breed erosion of domestic animal diversity, as discussed in Box 10.7.

Trends in Canadian Agriculture

Approximately 7 per cent of Canada's total land area (68 million hectares) is agricultural land, of which 46 million hectares are cropland, pasture, or summer fallow. The total area of farmland in Canada has remained relatively constant over the past 50 years. The Prairie provinces contain 81 per cent of the agricultural land base, while Quebec, Ontario, and BC account for 17 per cent, with the remaining 2 per cent in Atlantic Canada. The area of farmland increased by 10 per cent in BC between 2001 and 2006 while declining by 11 per cent in Newfoundland.

The agriculture and agri-food sector is a $86-billion industry (8 per cent of Canada's GDP), exporting more than $28 billion in products annually (Statistics Canada 2007a). Total government support to the agriculture sector rose to a record high level in 2006–7, at approximately $8.4 billion or 40 per cent of total sector GDP. Farmers in Prince Edward Island, Saskatchewan, and Manitoba received the most support as a percentage of agricultural GDP. The number of farms has declined since the early 1970s, with almost 230,000 farms in Canada in 2006,

Box 10.7   Conserving Canada's Breeds at Risk

Modern livestock production systems rely on a few specialized breeds that produce high yields. Consequently, many less productive but genetically valuable traditional breeds are threatened or have disappeared. According to the Canadian Farm Animal Genetic Resources Foundation:

- The genetic base for future breeding of dairy cattle is threatened by reliance on a single breed. Within the Holstein breed, 80 per cent of the cows are bred to 20 sires or their sons.
- The poultry breeding industry is limited to a few large breeding organizations, all of which appear to have similar genetic material. About a dozen such breeding companies provide stock for the world's poultry industry. The farm birds of the 1950s are now seen only as 'fancy' flocks.
- The Canadian swine industry is based predominantly on crossbreds of four breeds.

Loss of farm animal diversity is not just a Canadian problem. FAO estimates that of the 6,500 breeds of domesticated mammals and birds worldwide, 1,350 are at risk of extinction, 119 are officially confirmed as extinct, and 620 are reported to be extinct. The number of mammalian breeds at risk of extinction has risen from 23 per cent to 35 per cent since 1995, while the total percentage of avian breeds at risk of being lost increased from 51 per cent in 1995 to 63 per cent in 1999 (FAO 2000). A recent global assessment suggests that one livestock species per month has become extinct over the past seven years (FAO 2007a).

Since the mid–twentieth century, a few high-performance breeds, usually of European descent—including Holstein-Friesian (by far the most widespread breed, reported in at least 128 countries and in all regions of the world) and Jersey cattle; Large White, Duroc, and Landrace pigs; Saanen goats; and Rhode Island Red and Leghorn chickens—have spread throughout the world, often crowding out traditional breeds.

This progressive narrowing of genetic diversity is largely complete in Europe and North America and is now occurring in many developing countries, which have so far retained a large number of their indigenous breeds. In Vietnam, for example, the percentage of indigenous sows declined from 72 per cent of the total population in 1994 to only 26 per cent in 2002. Of its 14 local breeds, five are vulnerable, two are in a critical state, and three are facing extinction. The loss of genetic diversity in our farm animals may represent a serious food security risk. Modern production systems are very efficient, but they are not well prepared for unexpected challenges, such as new diseases, feed crop failures, or sustained changes in consumer demand (e.g., the desire for low-fat foods and issues of animal welfare and environmental sustainability). Efforts are now underway to raise awareness about the state of the world's animal genetic resources and to encourage improved technical documentation of each country's domestic animal diversity, and some livestock conservancies seek to maintain this diversity. However, at present, there is no agricultural research program to further research activity in the field of farm animal genetic resource conservation in Canada.

a decline of 7 per cent since 2001. At the same time, average farm size has been increasing (Figure 10.6). The largest farms are in Saskatchewan (586 hectares), an increase of 13 per cent since 2001, and the smallest in Newfoundland (65 hectares).

Wheat is still the dominant crop in Canada, although the area in wheat is declining (Figure 10.7). Saskatchewan grows

51 per cent of Canada's wheat. Other traditional grains, such as barley and oats, are also declining. The production of pulses (e.g., dry field peas, lentils, field beans, soybeans) has increased since the late 1970s and now accounts for almost 8 per cent of national crop area. Soybeans, the second largest oilseed crop grown in Canada after canola, are a major field crop in eastern

South Saskatchewan from the Cypress Hills. The Prairie provinces contain 83 per cent of the agricultural land base of the country.

Canada, along with grain corn. The area under soybean production, more than 1.2 million hectares, has grown eightfold since 1976, mostly as a result of breeding that has produced more cold-tolerant varieties with short growing seasons and the strong demand for the crop. The high-protein, high-oil beans are used as food for human consumption, animal rations, and edible oils as well as in many industrial products. They are also nitrogen fixers and help to replenish the soil (Chapter 4).

Western Canada produces most of Canada's second largest crop, hay, which is required to keep up with the increasing numbers of livestock to be fed. The livestock industry is an important part of Canadian agriculture, accounting for $18.9 billion in farm cash receipts in 2001, or 57.9 per cent of the total (Statistics Canada 2003a). The number of cattle and calves on Canadian farms rose by more than 4 per cent between 1996 and 2001 to a record 15.6 million head and was approximately the same in 2006. However, there have been interesting developments since bovine spongiform encephalopathy (BSE) was detected in one cow on a farm in Alberta. Canadian beef exports to the US were immediately banned. Numbers of cattle increased through 2004 until domestic slaughter rates caught up. Then in 2005, the US reopened its border to Canadian cattle less than 30 months old, and numbers started to fall. During this period, the number of beef farms fell by about 10 per cent. Alberta accounts for 43 per cent of the national herd.

The poultry sector experienced significant growth between 1996 and 2001 but little since that time. Most of the 126 million hens are produced in Quebec and Ontario. Abundant supplies of low-cost feed have pushed pig numbers to an all-time high. Between 2001 and 2006, the total number of pigs in Canada increased

by 7.8 per cent to more than 15 million. Quebec and Ontario house more than half of all the pigs in Canada.

Many Canadians have access to a ready supply of fresh produce year-round. Innovations include a thriving greenhouse sub-sector, as well as some of the most advanced storage technologies in the world. Other technologies have increased the availability of fresh food through production techniques that improve yields. Yield-enhancing technologies include mechanization, fertilizers and biocides, and genetic research. Canada is a main producer of food for the global market (Box 10.8). Unfortunately, many of the innovations and technologies currently employed by Canadian farmers have negative implications for ecosystem health. These implications will be discussed in more detail in the next section.

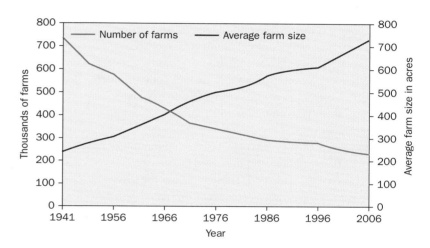

Figure 10.6 Number and size of farms in Canada, 1941–2006. Source: Statistics Canada. *Census of Agriculture,* various years.
Note: It is interesting that although Canada officially uses metric units, Agriculture and Agri-Food Canada continues to use imperial units (acres) even in recent publications such as this graph. In a sense, this can be seen as symbolic of the inertia in Canada's agricultural sector.

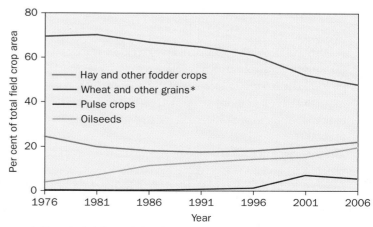

* Other grains include oats, barley, mixed grains, corn for grain, rye, and buckwheat.

Figure 10.7 Crop area allocation, 1976–2006. Source: Statistics Canada. *Census of Agriculture,* various years.

Box 10.8　Food Production and Consumption in Canada

Food Production

- In Canada's early years, agriculture employed more than 80 per cent of the population. Today, only 3 per cent of Canadians are directly occupied in farming.
- Canada exports a wide range of products to more than 200 trading partners around the world. Value-added and processed goods, together with prime-quality meats, live animals, bulk grains, oilseeds, and vegetables are Canada's top agricultural exports. Other important export foods include milk products, fish and seafood, maple syrup and honey, organic, natural, and health foods, and confectionaries and beverages.
- Between 1990 and 2000, canola production in Canada more than doubled and increased by one-third between 2000 and 2006. These increases are driven by consumers' growing awareness of canola's health benefits as well as demand for biodiesel. Saskatchewan and Alberta are the largest producers.
- There are about 16,500 vegetable growers in Canada, producing close to 7 million tonnes of vegetables worth approximately $2 billion annually. The area under glass used to produce vegetables (10.6 million square metres) exceeded that used to produce flowers (9.3 million square metres) for the first time in 2003. This increase in greenhouse area for vegetables offsets the reduction in area for growing the same vegetables, mostly tomatoes and cucumbers, outdoors.
- Potatoes, along with sweet corn and green peas, are the most extensively grown vegetables in Canada. Approximately 4.5 million tonnes of potatoes (valued at more than $700 million) are grown each year. More than half of them are processed, mostly into french fries.
- The total area in fruit climbed by 5.3 per cent between 2001 and 2006 to 110,069 hectares across the country. Fruit production is dominated by blueberries, which accounted for 46.6 per cent of the total acreage in 2006, an increase of 16.7 per cent since 2001. Quebec's area under blueberries increased by 24.5 per cent between 2001 and 2006, and the province is now Canada's leading producer, followed by Nova Scotia and New Brunswick.
- There are more than 12,000 maple syrup producers in Canada. Canada accounts for more than 85 per cent of world production.
- Canada's red meat and meat products industry includes beef, pork, lamb, venison, and bison. From 1990 to 2003, red meat and live animal exports increased in value from $1.9 billion to $5.3 billion.
- Canada's commercial chicken and turkey meat production totals 1.18 billion kilograms, an increase of about 100 million since 2001, and the average Canadian eats almost double the amount of chicken than in 1976.
- Poultry production and processing are among the most highly mechanized sectors in agriculture. One person can operate a unit of 50,000 broiler chickens. Poultry processing plants in Canada are effectively mechanized, which allows for the slaughter and preparation of 25,000 broiler chickens for market per hour.

Food Consumption

- In Canada, food and non-alcoholic beverages cost about 10 per cent of the average person's disposable income, making our food among the least expensive in the world.
- The level of food energy consumed per Canadian, which remained relatively stable from the mid-1970s to the early 1990s, increased by 17 per cent between 1991 and 2001.
- Red meat consumption totalled 27.1 kilograms per person in 2002.
- Consumption of poultry surpassed 13 kilograms per person in 2002, up 23 per cent over the past decade. In 2006, egg consumption amounted to more than 13 dozen eggs per person.

Sources: Agriculture and Agri-Food Canada 2003, 2008; Statistics Canada, Agricultural Division 2003; Statistics Canada 2003a, 2007a.

Environmental Challenges for Canadian Agriculture

Land Degradation

Land degradation includes a number of processes that reduce the capability of agricultural lands to produce food. As agricultural activities have intensified with increased cultivation and addition of agricultural chemicals to produce better yields, so has pressure on the soil resource.

Soil Erosion

Soil erosion is a serious land degradation problem in Canada and is estimated to cause up to $707 million in damage per year in terms of reduced yields and higher costs. In some parts of southwestern Ontario, erosion has caused a loss in corn yields of 30 to 40 per cent. Further costs are incurred off the farm when sedimentation blocks waterways, impairs fish habitat, lowers water quality, increases the costs of water treatment, and contributes to flooding.

Soil erosion is a natural process whereby soil is removed from its place of formation by gravitational, water, and wind processes. Under natural conditions in most ecozones, soil erosion is relatively minimal, since the natural vegetation tends to bind the soil together and keep it in place. Agricultural activities may totally remove this natural vegetation and replace it with intermittent crop plantings, thereby exposing the bare soil to erosive processes, or keep the land under full vegetation for grazing purposes. The latter approach provides much better protection for the soil but may still result in erosion, particularly under conditions of high livestock density.

The rate of soil formation varies as a function of different environmental factors. Because of Canada's latitude, soil formation is slow, and an annual rate of 0.5 to 1.0 tonne per hectare may be considered average. Any soil erosion above this amount will result in some loss of productive capacity. Losses in excess of 5 to 10 tonnes per hectare per year may lead to serious long-term problems. These figures have often been exceeded, and 30 tonnes per hectare has been recorded in BC's Fraser River Valley under row crops, while 20 tonnes per hectare is not uncommon in Prince Edward Island. Wind erosion is more difficult to measure but is estimated as a significant problem in the Prairie provinces in particular, with high wind speeds, dry soils, and cropping practices that often leave the soil unprotected. One study in Saskatchewan detected a net output of soil of 1.5 tonnes per hectare on a near-level field as a result of wind erosion, whereas a field with a greater incline (three degrees) was found to lose 6.6 tonnes per hectare, with water and wind erosion combined.

Soil Compaction

Soil compaction occurs from frequent use of heavy machinery on wet soils or from overstocking with cattle. Compaction serves to break down the soil structure and inhibit the through-flow of water. Crop yields can be reduced by up to 60 per cent

Large areas of tilled soil are particularly vulnerable to erosion.

in such conditions. Soil compaction is a problem mainly in the lower Fraser River Valley in BC and in parts of central and eastern Canada. One estimate puts the annual cost of compaction in Canada at between $68 million and $200 million.

Soil Acidification and Salinization

Acidity in soils can occur naturally but can also be augmented by fallout from acid precipitation (see Chapter 4) and the use of fertilizers. Nitrogen fertilizers undergo chemical changes in the soil that result in production of H+ ions, causing greater acidity. In the Maritime provinces, where significant declines in soil pH have been measured, it is estimated that 60 per cent of the change can be attributed to fertilizer use and 40 per cent to acid precipitation. In the Prairies, concern over acidity is relatively recent because the substrate is generally alkaline. However, increased use of fertilizer has now led to acidification in some areas. Excess acidity reduces crop yields and also leads to nutrient deficiencies and the export of soluble elements such as iron and aluminum into waterways. The yields of crops such as barley and alfalfa fall sharply at soil pH of less than 6. Liming is a common agricultural practice to combat the effects of acidity.

Salinization is a major problem in many areas of the world where irrigation is common. As water evaporates, it leaves behind dissolved salts. Over time, these salts can accumulate in sufficient quantities to render the land unusable. Ancient civilizations that designed complex irrigation systems were unable to counter the effects of salinization, contributing to their eventual decline. Estimates suggest that 50 to 65 per cent of irrigated croplands worldwide are now less productive as a result of salinization.

Alkaline soils occur naturally in areas of western Canada that have high sodium content and shallow water tables. Salinization can also be exacerbated through cropping practices in which natural vegetation is removed and surface evaporation increases to concentrate salts at the surface. Summer fallow has this effect. **Summer fallow** is a practice common on the Prairies in which land is kept bare to minimize moisture losses through evapotranspiration. On the Prairies, crop yields have been reduced by 10 to 75 per cent as a result of salinization. Despite the increased use of fertilizers, it is estimated that in some regions, salinization is increasing in area by 10 per cent every year. However, summer fallow decreased by 25 per cent between 2001 and 2006. Saskatchewan had 9 per cent of its total farmland in summer fallow in 2006, down from 11.9 per cent in 2001. The economic need to keep arable land productive, along with diversified and extended crop rotations, improved seeding and tilling methods, and proper use of herbicides, have all contributed to the reduction in summer fallow.

Organic Matter and Nutrient Losses

Cultivation involves a continuous process of removing plant matter from a field. In so doing, both the organic and nutrient content of the soil are reduced. Organic matter is critical for maintaining the structure of the soil, influencing water filtration, facilitating aeration, and providing the capacity to support machinery. It also helps to maintain water and nutrient levels.

On the Prairies, current organic matter levels are estimated to be 50 to 60 per cent of original levels, representing a probable annual loss of about 112,000 tonnes of nitrogen (Figure 10.8). This nitrogen is replaced by the addition of synthetic fertilizers, which in turn contribute to the problem of acidification. An alternative way to replace the nitrogen is the growth of leguminous crops to enhance biological nitrogen fixation (BNF),

discussed in Chapter 4. It is estimated that before the increased use of fertilizers began in the 1960s, nitrogen exports from prairie grain exceeded fertilizer applications by more than tenfold and phosphate removals exceeded inputs by threefold. Current estimates still show a depletion of the soil but with nitrogen now reduced to double the exports over inputs and phosphorus inputs to about 50 to 60 per cent of the export.

Soil degradation, including erosion and nutrient depletion, is undermining the long-term capacity of many agricultural systems worldwide. According to scientists at the International Food Policy Research Institute, nearly 40 per cent of the world's agricultural land is seriously degraded. The global significance of this degradation remains unclear, but it will likely have serious implications for future generations, since the production of food in sufficient quantity and quality requires a healthy natural resource base.

Biocides

Biocide use affects almost all Canadians—biocides are used to produce and preserve the food we eat, and homeowners use them to control weeds in their lawns, insects in their gardens and homes, and parasites on their pets. Since the publication of Rachel Carson's classic book **Silent Spring** more than 45 years ago, which outlined some of the environmental problems associated with the use of biocides, there has been considerable controversy over the risks of using chemicals to

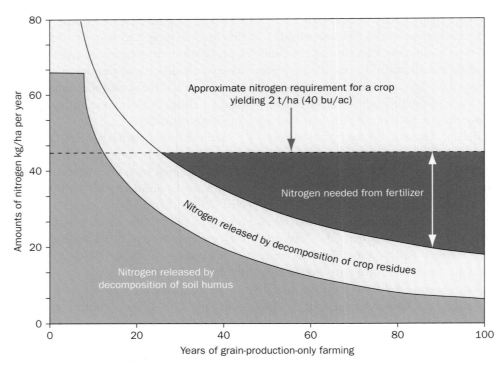

Figure 10.8 Diagrammatic illustration of approximate sources of nitrogen needed to maintain grain yields of about two tonnes per hectare (40 bushels per acre) of barley under a system of continuous grain production in the Prairie region. Note that this diagram illustrates plant requirements, not supply—i.e., the amount of fertilizer nitrogen applied would normally be greater than the plant requirements because of losses due to denitrification and/or leaching. Source: Bentley and Leskiw 1985.

control pests. Much of the controversy stems from the fact that biocides are designed to be toxic and are deliberately released into the environment.

Biocides have helped to boost yields throughout the world to meet the food demands of rising populations (many more people would be starving without their use), and biocides have also saved countless lives throughout the world by assisting in the control of various diseases through attacking vectors, such as malaria-carrying mosquitoes. But scientific evidence indicates that many chemicals have profound negative impacts on ecosystems. The possible environmental and health impacts may be delayed, in some cases for decades. Despite the risk, biocide use continues to grow on a global scale.

Most biocides fall into five main groups (Tables 10.2 and 10.3). Various environmental effects can be attributed to these chemicals, but it should be noted that there is considerable variation among them in terms of these effects.

It was a spring without voices. On the mornings that had once throbbed with the dawn chorus of robins, catbirds, doves, jays, wrens, and scores of other bird voices, there was now no sound; only silence lay over the fields and woods and marsh.

—Rachel Carson. 1962. *Silent Spring.*

Resistance

Part of the scientific debate on crop protection relates to the ability of pests, weeds, and viruses to develop resistance to biocides. When a population of insects, for example, is sprayed with a chemical, individuals within the population will react in various ways. If the biocide is effective, most of the population will be killed, but it is likely that a small number of individuals will have a higher natural resistance and survive the chemical onslaught. This remnant resistant population may then grow rapidly in numbers, a result of the lack of competition from all the dead insects. Seeing a resurgence of the pest insect, the farmer sprays again and is again

Table 10.2 Major Types of Pesticides

Type	Examples
Insecticides	
Chlorinated hydrocarbons	aldrin, chlordane, DDT, dieldrin, endrin, heptachlor, mirex, toxaphene, kepane, methoxychlor
Organophosphates	malathion, parathion, dizainon, TEEP, DDVP
Carbamates	aldicarb, carbaryl (Sevin), carbofuran, propoxur, maneb, zineb
Botanicals	rotenone, nicotine, pyrethrum, camphor extracted from plants
Microbotanicals	bacteria (e.g., Bt), fungi, protozoans
Fungicides	
Various chemicals	captan, pentachorphenol, methyl bromide, carbon bisulphide
Fumigants	
Various chemicals	carbon tetrachloride, ethylene dibromide (EDB), methyl bromide (MIC)
Herbicides	
Contact chemicals	atrazine, paraquat, simazine
Systemic chemicals	2,4-D, 2,4,5-T, daminozide (Alar), alachlor (Lasso), glyphosate (Roundup)

Table 10.3 Characteristics of Common Chlorinated Hydrocarbons

Pesticide	Characteristics
Aldrin	A pesticide applied to soils to kill termites, grasshoppers, corn rootworm, and other insect pests; aldrin can also kill birds, fish, and humans.
Chlordane	Used extensively to control termites and as a broad-spectrum insecticide on a range of agricultural crops; chlordane remains in the soil for a long time.
DDT	An insecticide controlled in many nations but still widely used in tropical countries, particularly to combat malaria; DDT builds up in the food chain and has caused harm to wildlife, including thinning of eggshells.
Dieldrin	Used principally to control termites and textile pests, dieldrin persists in soil; it is highly toxic to fish, aquatic animals, and predatory birds.
Endrin	A chemical used against insects and rodents; endrin is highly toxic to fish.
Heptachlore	Primarily used to kill soil insects and termites, this chemical is believed responsible for the decline of several wild bird populations; heptachlore also kills or causes reproductive problems in test animals.
Mirex	An extremely persistent chemical that was used as a fire retardant and a pesticide; mirex is toxic to a number of plants and animals and is classified as a possible human carcinogen.
Toxaphene	This insecticide is chemically similar to DDT; toxaphene is highly toxic to fish and is classified as a possible human carcinogen.

successful in killing a proportion of the population but not as high a proportion as before, since the natural resistance has been passed on to a larger proportion of the population. As this process repeats itself, the use of the chemical creates a population that will ultimately be quite resistant to it. This results in a constant need to develop new biocide products (or pest-resistant plant varieties) to keep one step ahead of biological adaptation.

The 'biocide treadmill' has led to biological adaptations resistant to most commercially available biocides. One estimate suggests that in Canada and the United States, more than 900 major agricultural pests are now immune to biocides, including some 500 insects and mites, 270 weed species, and 150 plant diseases (UNEP 2002). As a result, more frequent applications are needed today to accomplish the same level of control as in the early 1970s. Across Canada, farmers have experienced increasing difficulty in controlling pests through spraying. In New Brunswick, Colorado potato beetles that used to be killed with one spraying a season now must be sprayed five or six times, and they still cause substantial damage. New approaches to controlling beetle damage have involved hybridization with a naturally resistant wild variety, illustrating the value of maintaining as wide a spectrum of wild species as possible. In British Columbia, the pear psylla, an aphid-like insect that feeds on pears, has become resistant to the five main

synthetic pyrethroids registered for use against it, and another one is being developed.

One of the main beneficial impacts that biocides have had is in controlling disease-bearing organisms such as mosquitoes. Some of the world's deadliest diseases, including

Malaria kills many people a year but receives little funding because it has little impact on the richer countries of the North. This is an educational sign in Sri Lanka showing villagers how to minimize the risks.

malaria, are spread through mosquito bites. More than 1 million people are infected every year, and one person dies from malaria every 30 seconds. The numbers of infections and deaths are rising. Fifty years ago, medical experts predicted that malaria would be eliminated because of the control of mosquitoes with DDT. Unfortunately, the mosquitoes soon developed resistance to DDT, and since that time have become resistant to virtually all control mechanisms. Given global warming trends, it is possible that tropical diseases such as malaria could invade Canada. The advent of West Nile virus may well be a forerunner of what is to come in the future.

Non-selective

Many biocides are popular because they are broad-spectrum poisons. In other words, there is no need to identify the specific pest, because a broad-spectrum poison will kill most insects. Unfortunately, they tend to eliminate not only the pest species but also other, valuable species, including some that may act to control the population of the pest. This may result in a population explosion of the resistant

members of the pest population after spraying as well as the development of new pests previously kept in check by natural predators.

Many biocides are also extremely toxic to species other than those directly targeted, such as soil micro-organisms, insects, plants, mammals, birds, and fish (Box 10.9). For example, in PEI more than 20 instances of fish kills since 1994 have been attributed to pesticides, with up to 35,000 dead fish collected in each incident (Commissioner of the Environment and Sustainable Development 2003). The government instituted new buffer zone regulations as a result of excessive kills, but two further large kills occurred in 2007. Non-target organisms poisoned by biocide use may be beneficial to agriculture or other human economic activities, and they are part of a biodiversity valued by society for recreational, cultural, ethical, or other reasons.

Mobility

The purpose of biocides is to reduce the impact of a particular pest species (or several species) on a particular crop in a particular area. However, the effects of the chemical application

Box 10.9   Carbofuran and Birds

Carbofuran is a carbamate (Table 10.2) that was registered for agricultural use in Canada. It was available either as a liquid or in granular form on particles of grit and was often applied in the latter form during seeding to protect recently germinated seedlings from insects. Considerable amounts of the chemical (up to 30 per cent of that applied) are often exposed on the soil surface following application. These granules are highly attractive to many birds that ingest grit in their gizzards to grind seeds. The chemical is highly toxic to birds, and the consumption of just one grain can be fatal to small seed-eaters. Carbofuran also contaminates invertebrates such as earthworms, which in turn will poison organisms higher on the food chain. Flooded fields are also highly dangerous, since the chemical goes into solution. This is particularly serious in acidic fields, where the breakdown of carbofuran is very slow.

There have been many documented bird kills as a result of the use of carbofuran. Some Canadian examples include:

- December 1973: 50 to 60 pintails and mallards were killed in flooded turnip fields in BC.
- November 1974 to January 1975: 80 ducks were killed in flooded turnip fields in BC.
- October to December 1975: 1,000 green-winged teal were killed in flooded turnip fields in BC.
- May 1984: more than 2,000 Lapland longspurs were killed in a canola field in Saskatchewan.
- September 1986: an estimated 500 to 1,200 seed-eating birds were killed in turnip and radish fields in BC.
- June 1986: 45 gulls died after eating carbofuran-contaminated grasshoppers in Saskatchewan.

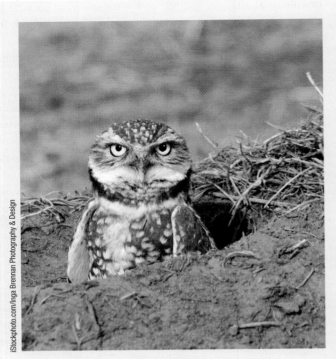

The burrowing owl is an endangered species that has been particularly threatened by the use of agricultural biocides.

Figure 10.9 Estimated pairs of burrowing owls reported by Alberta and Saskatchewan landowners. Source: www.pnr.rpn.ec.gc.ca/nature/endspecies/burrowing/db04s05.en.html.

Of particular concern was the impact of carbofuran on one of Canada's designated threatened species, the burrowing owl. The owl nests on the Prairies in abandoned mammal burrows, feeding on small mammals and insects. Two-thirds of the Canadian breeding population nests in Saskatchewan, where studies suggest that breeding numbers declined by 50 per cent in the south-central region between 1976 and 1987. Carbofuran spraying for grasshopper infestations is suspected to be the main cause. There were significant declines in nesting success and brood size with increasing proximity of carbofuran spraying to the nests. A high percentage of adult owls also disappeared after spraying. There were particularly severe infestations of grasshoppers in the early 1980s, and in Saskatchewan alone more than 3 million hectares was sprayed in 1985, 40 per cent of the area with carbofuran.

The registration of carbofuran was finally cancelled in 1996. The burrowing owl remains on Canada's endangered species list. One evaluation suggested that the voluntary habitat stewardship program (see Chapter 14), Operation Burrowing Owl, is having some success in protecting the remaining habitat (Warnock and Skeel 2004), and 61 pairs were reported in the 2007 annual census. However, this is slightly down from the 68 reported in 2006 and the third consecutive year of such declines.

are often felt over a much wider area, sometimes spanning thousands of square kilometres, because of the mobility of the chemicals in the Earth's natural cycles, particularly the hydrological cycle, and the manner in which chemicals are applied (Figure 10.10). The US Department of Agriculture estimates that aerial spraying of insecticide results in less than 2 per cent reaching the target and for herbicide applications, less than 5 per cent. The remainder finds its way into the ecosystem where it may contaminate local water supplies or be transported by atmospheric processes to more distant sites (Box 10.10).

Places with well-developed agricultural sectors and frequent use of chemicals might find that the entire environment is becoming contaminated. This seems to be the case in PEI, according to studies undertaken by government agencies that were only made available to the public under the Access to Information Act. The province uses more than eight kilograms of biocides per person every year, compared to the average of 1.3 kilograms used in the US. Tests on airborne pollution found every sample to be contaminated. Even tests taken at the end of a wharf, far from any farms, showed chemicals. One of the most heavily used chemicals on the island, chlorothalonil, has been identified by the US government as a potential carcinogen, and its effects can be detected in air samples two hours after spraying has taken place.

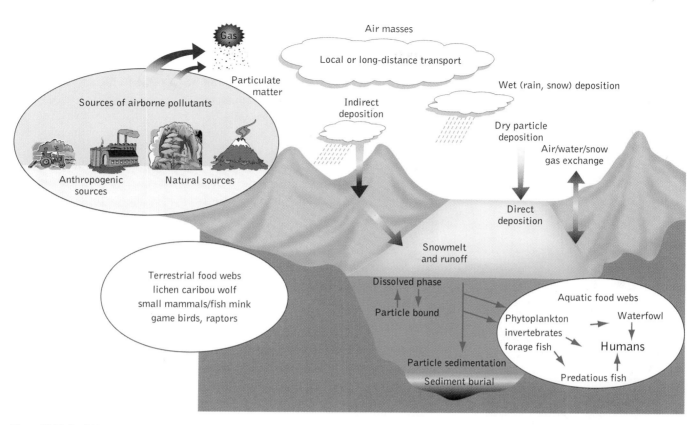

Figure 10.10 Pesticide transportation in the environment. Source: Adapted from Indian and Northern Affairs Canada 1997b. Reproduced with the permission of the Minister of Public Works and Government Services, 2004.

Box 10.10 **The Grasshopper Effect**

Imagine for a moment, if you will, the emotions we now feel: shock, panic, grief—as we discover that the food which for generations nourished us and keeps us whole physically and spiritually is now poisoning us. You go to the supermarket for food. We go out on the land to hunt, fish, trap, and gather. The environment is our supermarket. . . . As we put our babies to our breasts, we feed them a noxious chemical cocktail that foreshadows neurological disorders, cancers, kidney failure, reproductive dysfunction. That Inuit mothers—far from areas where POPs are manufactured and used—have to think twice before breast-feeding their infants is surely a wake-up call to the world.

— Sheila Watt-Cloutier, President,
Inuit Circumpolar Conference (Canada), 2000

We tend to think of the Arctic as 'pristine wilderness'. However, research has indicated that this is far from the truth. At Ice Island, for example, a floating ice-research station 1,900 kilometres above the Arctic Circle, concentrations of a family of pesticides called hexachlorocyclohexanes (HCHs) have been measured that are twice as high as those in agricultural southern Ontario. Yet there is not a single pesticide-dependent product grown in the North! How did the chemicals get there?

The so-called **grasshopper effect** is one reason behind Arctic pollution. Indeed, atmospheric transport and deposition is a major pathway of contamination. After chemicals are introduced into an environment, they are absorbed into the soils and/or plant tissues or deposited into rivers, lakes, and wetlands. Persistent

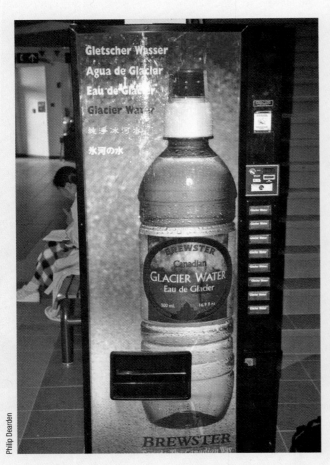

People have the perception of the Arctic and Rockies as pristine wilderness areas. Marketers use this image to sell products such as water. In reality, these cold environments can have high toxic burdens due to the grasshopper effect.

at the top of the food chain and serve to concentrate the chemicals before excreting them on the local landscape, where concentrations can be 60 per cent higher than in the surrounding landscape.

The implications of this long-distance transport are serious. Arctic ecosystems are more vulnerable to toxic chemicals because they last longer in the North. Degradation processes are inhibited by low temperatures and reduced ultraviolet radiation from the sun. The cold also condenses the toxins, keeping them locked up and slowing evaporation rates. There are measurable concentrations of DDT, toxaphene, chlordane, and PCBs in the Arctic, and when fish and other species ingest these chemicals, they travel up the food chain, accumulating in the fatty tissue of animals at the top of the food chain.

Concentrations of several persistent organochlorine pesticides (POPs) remain high in many aquatic food webs in Canada. This has serious implications for the Inuit in particular because of their high consumption of wildlife. More than 80 per cent of Inuit consume caribou, almost 60 per cent consume fish, and almost 40 per cent consume marine mammals. Because of bioconcentration, the consumption of traditional foods places the Inuit at greater risk for developing several ailments related to toxic chemical exposure, including endocrine disruption, reproductive impairment, and cancer. These concerns are not restricted to the Arctic, since large concentrations of toxic chemicals have also been found in the mountains of British Columbia. Fish in alpine lakes have chemical levels that make them toxic to eat in large quantities.

Since POPs travel great distances, a global approach to tackle the issue is required. Various initiatives have been undertaken to control or eliminate POPs, including the Global POPs Protocol, signed in Johannesburg in December 2000. Signatories of the protocol, including Canada, agreed to a global ban of 12 chemicals, including the pesticides aldrin/dieldrin, endrin, DDT/DDE, HCH/lindane, chlordane, heptachlor, chlordecone, mirex, and toxaphene. Canada is also a signatory to the United Nations Economic Commission for Europe (UNECE) POPs Protocol, which lists 16 chemicals for phase-out. But while governments around the globe have a significant role to play in reducing Arctic pollution, consumers also share the responsibility. Chemicals used in consumers' everyday environment can end up polluting some of the most 'pristine' environments on Earth. Consumers must make an effort to phase out domestic use of toxic chemicals.

and volatile pollutants evaporate into the air in warmer climates and travel in the atmosphere towards cooler areas, condensing out again when the temperature drops. The cycle then repeats itself in a series of 'hops' until the pollutants reach climates where they can no longer evaporate. Chemicals released in southern Canada, for example, may go through the grasshopper cycle several times and take 10 years to reach the Arctic. Extremely volatile chemicals will travel farther and recondense in greater concentrations. For example, biocides such as lindane and HCH are quite volatile compared to DDT and therefore usually reach the North in greater quantities.

However, if only air transport were responsible, then the pollutants should have a fairly even distribution across the Arctic. Researchers found this not to be the case, with marked concentrations in some areas, particularly near bird colonies. These birds, such as fulmars, are

Persistence

Not only do biocides spread over vast areas, they also contaminate through time, as many of them are very persistent. DDT is one of the best-known insecticides. First synthesized more than 100 years ago, it was not until the 1940s that DDT became widely used, first in health programs in World War II to control disease vectors and later as an agricultural chemical. Production peaked by 1970 when 175 million kilograms was manufactured. By that time, the environmental effects of DDT were becoming better understood, and its use, but not manufacture, was banned in the US in 1972. It was not until 1985 that registration of all DDT products was discontinued in Canada. More than 7 million kilograms of DDT was sprayed on forests in New Brunswick and Quebec between the early 1950s and the late 1960s. DDT is extremely persistent. Even now, there are still considerable residues of DDT and its main breakdown product, DDE, in the environment. Because it is soluble in fat, DDT may also gradually accumulate over time in the tissues of organisms. This is known as **bioaccumulation** (Figure 10.11).

Organisms with long lifespans are particularly susceptible to bioaccumulation. In British Columbia, for example, an insecticide banned in the late 1970s called hexachlorocyclohexane (HCH), once used as a timber preservative and agricultural spray, is still being found in geoducks, a type of clam, off the west coast of Vancouver Island and in Puget Sound. These large clams are filter-feeders that may live as long as 140 years. They are therefore very susceptible to bioaccumulation, and concentrations have been sufficient to have the clams refused by processing plants. In another example of the persistence of biocides in aquatic environments, researchers have detected high levels of toxaphene in trout in Bow Lake in Banff National Park, AB. The toxaphene had been applied in 1959 to rid the lake of what were then seen as undesirable fish species. DDT and other chemicals have also been detected in trout in many of the lakes in Waterton, Banff, Jasper, and Yoho national parks.

Biomagnification

The Arctic has generally been viewed as one of the few unpolluted regions of the world. But a closer look reveals a different story. High concentrations of persistent organic pollutants (POPs) such as organochlorine biocides have been detected in top predators of the Arctic food chain, including indigenous

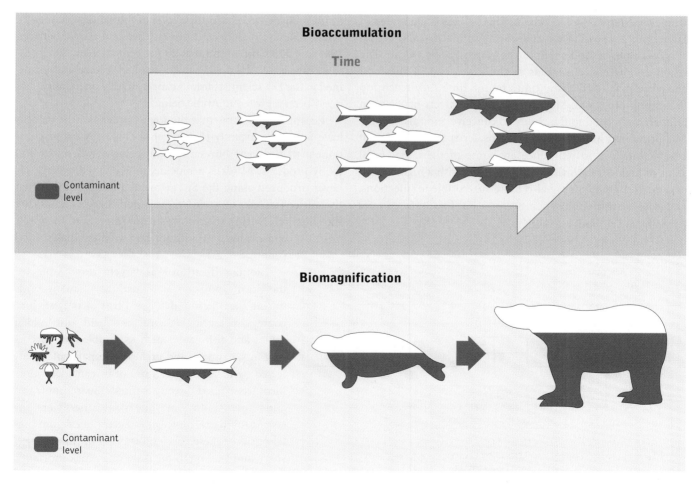

Figure 10.11 Bioaccumulation and biomagnification. Source: Adapted from Indian and Northern Affairs Canada 1997a. Reproduced with the permission of the Minister of Public Works and Government Services, 2004.

Philip Dearden

Although Prince Edward Island is known for its beautiful rural landscape, the intensity of agricultural production has left large biocide residues, even in ocean sediments.

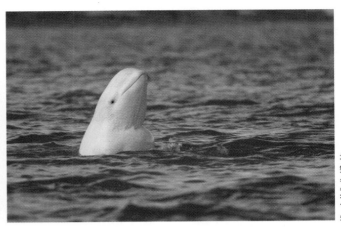

Visuals Unlimited/Corbis

The beluga whale population has declined rapidly in the St Lawrence, and the toxic burden from biocides appears to be one cause.

peoples. The ability of biocides and other toxic chemicals to traverse long distances, combined with the ability of biocides to concentrate and accumulate in the lipids of pelagic marine organisms (e.g., Atlantic cod, whales, seals), is to blame for the introduction of contaminants into the Arctic food web (Box 10.10). Concentrations of POPs multiply five- to tenfold with every step in the food chain. This process is illustrated in Figure 10.12, showing the concentration of DDT and its derivatives along a food chain in the North Pacific Ocean. The relatively low concentrations at the lower end of the food chain are magnified many times by the time they reach top fish-eating predators. Most of the visible effects of POPs on animals are related to the ability to conceive and raise young. Malformations in reproductive organs, fewer young, and even complete failure to reproduce are some of the detrimental signs of high contaminant levels. Animals with a high load of organic contaminants are also more susceptible to infections, and POPs are suspected of being responsible for increased rates of malignant tumours in wildlife.

Some species build up very large concentrations. The beluga whales of the St Lawrence estuary, for example, showed concentrations of 70,000 to 100,000 parts per billion of DDT. Populations fell to less than 10 per cent of the original population in the area, and individual lifespans were about half the normal lifespan for the species, indicating that this level of toxic burden exceeded their level of tolerance (Chapter 2). Scientists examined 73 carcasses that washed ashore between 1983 and 1994 and found that 20 per cent of them had intestinal cancer. Of the 1,800 whales washed ashore and examined in the US, scientists found cancer in only one. Cancer has never been reported in Arctic belugas.

Recent studies show that the toxic burdens of the belugas have fallen significantly (Lebeuf et al. 2007). A major component of the toxic burden for the belugas was mirex and its by-products. Mirex is a biocide, now banned, which was never produced along the St Lawrence. Biologists think that the source was Lake Ontario, where American eels accumulate the chemical. During their downstream migration, these eels constituted a significant part of the whales' food supply.

Biomagnification was largely responsible for the drastic population reductions of many birds during the 1960s and 1970s. Birds of prey such as ospreys, peregrine falcons, and bald and golden eagles and fish eaters such as double-crested cormorants, gannets, and grebes were particularly affected by the widespread use of insecticides. Some birds were killed directly through bioaccumulation and biomagnification, while many others were unsuccessful in breeding. DDT affects the calcium metabolism of these species, resulting in thinner eggshells and leading to breakage and chick mortality. The banning of DDT and similar chemicals has led to a recovery of many of these species in temperate countries (see Chapter 14). However, the continued

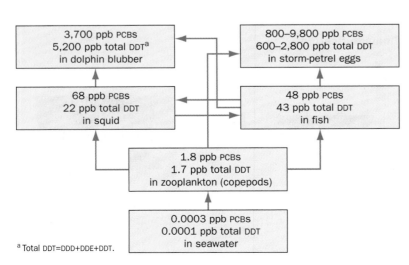

3,700 ppb PCBS
5,200 ppb total DDT[a]
in dolphin blubber

800–9,800 ppb PCBS
600–2,800 ppb total DDT
in storm-petrel eggs

68 ppb PCBS
22 ppb total DDT
in squid

48 ppb PCBS
43 ppb total DDT
in fish

1.8 ppb PCBS
1.7 ppb total DDT
in zooplankton (copepods)

0.0003 ppb PCBS
0.0001 ppb total DDT
in seawater

[a] Total DDT=DDD+DDE+DDT.

Figure 10.12 Organochlorines in a North Pacific food chain. Source: Nobel 1990.

use of the chemicals in some tropical countries still affects populations in these areas as well as the populations of migratory species such as peregrines.

Together, biomagnification and bioaccumulation are often known as **bioconcentration**. Humans are exposed to the harmful effects of biocides through bioconcentration. They may come into contact with chemicals through contaminated water supplies, ingestion of food products, their occupation, and/or domestic use (Box 10.11). Researchers found, for example, that women involved in farming in Ontario's Windsor and Essex counties have nine times the risk of developing breast cancer than non-farm women.

A greater dependence on fish or game birds for food may elevate the risk of biocide exposure, since fish and game birds may have already concentrated significant amounts of toxic matter in their fat deposits. Many indigenous communities are highly dependent on marine fish and mammals and may be particularly at risk. For example, levels of the biocide chlordane are significantly higher in the breast milk of Inuit women in the North than in women in southern Canada (UNEP 2002).

Synergism

When chemicals are tested for their harmful effects, they are tested individually in controlled situations. When applied on farmers' fields, however, the chemicals are free to interact with each other and the environment in myriad ways. A single biocide may contain up to 2,000 chemicals, and as the chemicals break down, new ones are created that may again react with each other in unpredicted ways. The combined effects are often greater than the sum of their individual effects. This is called **synergism**, and it can result in many unanticipated effects.

Box 10.11 **Biocides and You**

Although most biocides in Canada are used by commercial producers, large amounts are also used domestically in homes and on gardens and lawns to control unwanted organisms. How people use, store, and dispose of these chemicals is very important in terms of minimizing environmental damage. Here are a few tips.

- Use chemicals only as a last resort. Ask yourself why you need to kill the organism. If it is just for aesthetic reasons, such as dandelions on your lawn, then maybe you need to change your perceptions rather than automatically reaching for a chemical solution. For each pest or weed, there are usually several other approaches you can take as part of your own integrated pest management strategy. These approaches are too numerous to document here, but you can find out about more specific strategies from government ministries such as the ministry of agriculture in your area.
- Use the safest chemicals available in the minimum quantities. Many plant nurseries now sell products that are less toxic than traditional biocides. Often, they need more skill in application, but they are less environmentally damaging than regular chemicals. Examples include the 'Safer Soap' line of products.

- Apply all chemicals in strict accordance with the manufacturer's instructions.
- Store unused chemicals so that they do not leak and cannot be accidentally upset.
- Dispose of chemicals and containers in a safe manner. Contact your local ministry of environment to see what programs are in place in your province for safe disposal. Some provinces, for example, have specific sites where biocides and other toxic chemicals can be disposed of. If your province does not have such a program or acceptable alternative, start lobbying for one!

You should also protect yourself against the risk of ingesting chemicals that have been applied to food.

- Grow your own food; don't use chemicals.
- Buy organically grown produce whenever possible.
- Fruit and vegetables that look perfect often do so because they have had heavier applications of fertilizers and pesticides. Choose products that show more natural blemishes; this is a sign that chemical use has not been as high.
- Carefully wash all produce in soapy water.
- Remove the outer leaves of vegetables such as cabbage and lettuce, and peel all fruit.

Figure 10.13 Biocides and human health. Source: Cam Cardow/*Ottawa Citizen*.

Biocide Regulation

The concerns tied to the use of biocides—resistance, non-selection, mobility, persistence, bioaccumulation, biomagnification, and synergism—are significant. However, the economic value tied to the use of biocides is also very significant. Because the use of biocides will likely continue to grow in the foreseeable future, there is a pressing need to strengthen the regulatory and enforcement mechanisms governing biocide import, production, and use.

In Canada, an estimated 1,000 new chemicals are introduced annually, adding to the more than 20,000 already in industrial, agricultural, and commercial use. New chemicals cannot be introduced onto the market without undergoing scientific tests regarding their capacity to cause cancer, birth defects, and mutations. The Pest Management Regulatory Agency (PMRA), a branch of Health Canada created in 1995, has the primary responsibility for regulating biocides. Other Health Canada branches and other federal departments and agencies that play important roles in biocide management include Agriculture and Agri-Food Canada, the Canadian Food Inspection Agency, Environment Canada, Fisheries and Oceans Canada, and Natural Resources Canada. The federal government shares the responsibility for managing biocides with provincial, territorial, and in some cases municipal governments (Box 10.12). About 5,000 biocides are currently registered for use in Canada.

Box 10.12 Cosmetic Use of Pesticides

The proportion of households using pesticides on their lawns and gardens dropped slightly from 31 per cent in 1994 to 29 per cent in 2005. Municipalities also use substantial amounts of pesticides. Figures are difficult to obtain, but in Ontario in 1993, the amount was estimated at 1.3 million kilograms, about one-quarter of the amount used in agriculture.

Because of potential negative impacts on human health, the Standing Committee on Environment and Sustainable Development (2000) recommended a ban on the use of biocides for cosmetic purposes—i.e., lawn care—but the federal government refused to endorse the recommendation (Figure 10.13). The government preferred to take a voluntary, educational approach to reducing the cosmetic use of pesticides and launched a 'Healthy Lawns Strategy' to address the issue.

However, the government's educational approach did not satisfy all Canadians, particularly residents of a small community in Hudson, QC. Residents were worried about the health consequences of lawn and park applications of herbicides and insecticides, particularly on children. The community wanted the municipal government to enact a bylaw that would ban the cosmetic use of pesticides, but chemical companies won the first battle, arguing that municipal governments lacked the authority to introduce such bylaws. The community took the matter to court, and in June 2001, Canada's Supreme Court unanimously ruled that towns and cities have the right to enact bylaws banning the purely cosmetic use of pesticides. The Supreme Court ruling grants municipalities across the country the right to impose similar pesticide restrictions. More than 130 communities in Canada now have bylaws banning the use of chemicals for cosmetic lawn purposes, including the entire province of Quebec. Toronto's Board of Health has endorsed a similar bylaw, making it the first large city to begin steps to phase out the use of pesticides on lawns for cosmetic purposes. In early 2008, the Ontario government began consultations with Ontarians on how to shape legislation banning the cosmetic use of pesticides.

To predict the effectiveness of biocides and their risks to human health and the environment, the PMRA relies on the expertise and judgment of the agency's scientists and managers. Evaluation of a new biocide ends with the approval of the biocide label, which describes the biocide's hazards and its proper use. This process, however, has not been foolproof.

The House of Commons Standing Committee on Environment and Sustainable Development reviewed biocide management in Canada and published a report, *Pesticides: Making the Right Choice*, in 2000. The report expressed strong concern over current practices and urged a much stronger regulatory approach based on the precautionary principle. This was followed in 2003 by a report titled *Managing the Safety and Accessibility of Pesticides*, released by the Commissioner of the Environment and Sustainable Development (Box 10.13). This report also documented the adequacy (or lack thereof) of the current regulatory practices used in Canada to approve biocide use, as well as the health and environmental standards relating to compliance, the government's commitment to research, and monitoring. The report identified several instances of poor overall compliance. To predict occupational exposure to pesticides and pesticide residues on food, PMRA evaluators assume that agricultural users will follow good practices— i.e., users will follow label instructions. But evidence suggests otherwise. For example, a 2001 survey of farmers by Statistics Canada concluded that only 14 per cent calibrate their pesticide spraying equipment between applications of different pesticides; as a result, the amounts they actually use on their crops may be higher than the levels specified on the label. In 2001, evaluators collected soil samples from 20 onion growers in Ontario. Of those, 14 (or 70 per cent) had violated the Act by using pesticides not registered in Canada, while four other growers were using pesticides not registered for use on onions. Even though this situation and that described in Box 10.13 has since improved, it is still worth describing, since Canadians place trust in their government to safeguard their health and that of the environment.

Lack of compliance is partly due to problems with pesticide labels. Some agricultural pesticides may have 30 or more pages of directions in fine print, while other label instructions

Box 10.13   A Report Card on Canada's Use of Pesticides

In 2003, the Commissioner of the Environment and Sustainable Development released a report card on Canada's use of pesticides. The document, titled *Managing the Safety and Accessibility of Pesticides*, detailed the shortcomings of the Pest Management Regulatory Agency (PMRA), a branch of Health Canada responsible for the evaluation of new pesticides, monitoring and compliance of existing pesticides, and the re-evaluation of old pesticides to ensure compliance with modern standards. Overall, the commissioner concluded that 'the federal government is not managing pesticides effectively.' A few of the main findings in the report are highlighted here.

Some pesticides are approved based on inadequate information. The temporary use of a pesticide may be granted pending the submission of further studies. Of new pesticide registrations in 2001–2, 58 per cent were temporary. For some temporary registrations, the missing information was to have been included with the original submission. Examples of information gaps at the time of temporary registration include what happens to the pesticide after it is released into the environment, what impact it is likely to have on children's central nervous systems, and how toxic it is to invertebrates and non-target plants. As a result, many pesticides are used before they have been evaluated fully against current health and environmental standards. (See sections 1.36–1.39)

Key assumptions are not tested, and some are not valid. Agency evaluators must make a series of assumptions to link the laboratory studies they receive to the possible impacts of the pesticide's use. Such assumptions include how large a crop area will be treated, how much treated food Canadians will eat, and how the pesticide will be applied. Despite the uncertainties in all of the different assumptions evaluators make, they have not determined how reliable their predictions of the risks are. For example, evaluators have not tried altering their assumptions slightly to see whether that would reverse the decision to approve a pesticide. In addition, agency staff have unrealistic assumptions about user behaviour. They assume that pesticide users will follow label instructions, even though the agency's own compliance reports show that they may not. (See sections 1.40–1.45)

The agency manages a legacy of older pesticides. Many pesticides have been registered in Canada based on evaluations that did not apply the more stringent methods and standards used today. Some of the changes in requirements include considering the impacts on bystanders, the reproductive impacts on later generations, and the greater susceptibility of children. To ensure that older pesticides meet today's standards, the agency has implemented re-evaluation programs. However, many re-evaluations do not consider new information about the pesticide's effectiveness resulting from new research, and as a consequence, opportunities may be missed to reduce the rate of application. In addition, progress on re-evaluations has been very slow. By 2006, the agency planned to re-evaluate all products registered by 1994, which would require reviewing 405 active ingredients, or about 75 per cent of all active ingredients currently registered in Canada. As of March 2003, only 1.5 per cent of the 405 active ingredients had been fully re-evaluated. All pesticides re-evaluated were found to pose unacceptable risks and had to be restricted or removed from the market. (See sections 1.51–1.53)

The agency does not know to what extent users are complying with pesticide labels. The Pest Control Products Act influences how pesticides are produced, distributed, and used. Lack of compliance could cause serious environmental impacts and expose users and bystanders to unnecessary risks. The Food and Drugs Act deals with residues in food; lack of compliance would affect those who eat treated or contaminated food. The agency conducts inspections to determine whether pesticide registrants, distributors, and users are complying with the Pest Control Products Act. In 2002–3, the agency conducted only 510 inspections of users, although in agriculture alone, roughly 216,000 farms across Canada could have used pesticides. Inspection programs check which pesticides are being used but do not determine systematically whether the label requirements are being met. Problems with unclear labels also make it difficult to ensure compliance. Ambiguous labels mean that enforcement action cannot be taken for some possible violations of the Act. (See sections 1.79–1.82)

Methods for measuring pesticide residues on food are not up-to-date. The Canadian Food Inspection Agency (CFIA) conducts an extensive chemical sampling program that includes testing each year for pesticide residues on food. However, methods for measuring pesticide residues are not inclusive or thorough. The CFIA uses a risk-based, multi-residue testing method that screens for 269 different pesticides in various commodities, but the agency has identified more than 190 additional pesticides used in Canada or in countries that export food to Canada. In addition, small samples tested for any given pesticide on one type of food may prevent meaningful conclusions about compliance with the limits; consequently, the CFIA can provide only limited assurance that pesticide residues on food comply with the Food and Drugs Act. Furthermore, some of the residue limits set by the PMRA are based on old assessments and are inconsistent with current standards. (See sections 1.90–1.91)

Critical information on pesticide use and exposure is still missing. The federal government has not yet set up a pesticide sales database despite a commitment to do so in 1994. In addition, the adverse effects of pesticides on human health and the environment are tracked and reported only on an ad hoc basis. Reports of pesticide problems by registrants, doctors, provincial agencies, university researchers, and pesticide users could help the agency understand the impacts of pesticides. In 1994 and again in 2000, the federal government committed to developing a program of mandatory reporting on adverse effects of pesticides, but it has yet to do so. The lack of reliable information on pesticide use, exposure, and impacts is a major hurdle that continues to interfere with the agency's ability to regulate pesticides. (See sections 1.94–1.100)

Federal research on the health impacts of pesticides has not been a priority. Health Canada has very limited dedicated funding for research on human exposure to pesticides or the resulting health effects. Three researchers are working on current pesticides, and they rely primarily on outside funding. (See sections 1.101–1.102)

Source: Report of the Commissioner of the Environment and Sustainable Development—October 2003, Chapter 1, Office of the Auditor General of Canada. *Reproduced with the permission of the Minister of Public Works and Government Services, 2009.*

The Commissioner of the Environment and Sustainable Development published a follow-up audit in 2008 called *Chemicals Management—Pesticide Safety and Accessibility.* The commissioner concluded that the PMRA procedures have improved, as have procedures for re-evaluating chemicals, although there is still a considerable backlog and no action plan to deal with it. The Canadian Food Inspection Agency had also broadened the range of residue tests on fresh fruits and vegetables.

Source: Status Report of the Commissioner of the Environment and Sustainable Development—March 2008, Chapter 2, Office of the Auditor General of Canada. *Reproduced with the permission of the Minister of Public Works and Government Services, 2009.*

are difficult to follow. For example, labels are often ambiguous, and application is therefore dependent on the applicator's interpretation. Ambiguous, vague terms used on pesticide labels include:

- '*Appropriate* buffer zones should be established between: treatment areas and aquatic systems and treatment areas and *significant* habitat.'
- 'Do not apply in areas where soils are *highly* permeable and groundwater is *near* the surface.'
- 'Do not apply *near* buildings inhabited by humans or livestock.'
- 'Do not apply where fish and crustaceans are *important* resources.'

Failure to follow label instructions could increase the risks to consumers and the environment, but it is difficult to apply pesticides appropriately when the directions are unclear. Many poisonings have been attributed to inappropriate application, and this is one reason why farmers in Ontario have welcomed mandatory biocide safety courses dealing with the use, mixing, handling, and transportation of biocides as well as laws governing their use. The certificate from these courses must be renewed every five years, and it must be presented in order to buy agricultural chemicals.

The new federal Pest Control Products Act came into force on 28 June 2006. Its objective is to better protect Canadians and the environment from the risks of pesticide use. The new Act requires that all pesticides be re-evaluated every 15 years against the most current health and environmental standards.

Intensive Livestock Operations

Another aspect of modern agriculture that can have significant and far-reaching environmental implications is intensive livestock farming. The production of livestock manure has both environmental benefits and drawbacks. Although manure is a valuable fertilizer for crop production, it can also become a source of pollution if not managed properly.

Manure consists of a variety of substances, including nitrogen, phosphorus, potassium, calcium, sodium, sulphur, lead, chloride, and carbon. Manure also contains countless micro-organisms, including bacteria, viruses, and parasites. Some of these micro-organisms are pathogenic, and therefore direct consumption or recreational use of water containing these organisms can lead to a variety of illnesses and even death. The contamination of drinking water with E. coli that killed seven residents of Walkerton, ON, in May 2000 was identified as related to livestock manure. Pathogens from manure that have reached watercourses also have the potential to spread disease to livestock. The spread of bovine spongiform encephalopathy (BSE), commonly referred to as 'mad cow disease', is an example of an inter-species disease transmission. BSE is thought to cause Creutzfeldt-Jakob disease among humans.

Other risks to human and ecosystem health arise from air pollution. Odour and air pollution are identified as serious environmental and human health concerns related to **intensive livestock operations** (ILOs). High concentrations of noxious gases such as methane, hydrogen sulphide, carbon dioxide, and ammonia are often found in manure pits and confinement barns. Pigs and poultry, for example, excrete some 65 and 70 per cent of their nitrogen and phosphorus intake, respectively. Nitrogen, under aerobic conditions, can evaporate in the form of ammonia (Chapter 4). Ammonium nitrate and ammonium sulphate emitted to the air from animal housing can be harmful to human and animal health. Foul odours emitted by ILOs are a significant problem for neighbours, and studies have shown an increase in chronic respiratory diseases reported by people who live in close proximity to a large animal farm. Ammonia can also have toxic and acidifying effects on ecosystems. Ammonia in high concentrations in the air can have a direct effect on plant growth by damaging leaf absorption capacities, but its indirect effect on soil chemistry is even more important—ammonia acidifies the soil, interfering with the absorption of other essential plant elements.

While air pollution from intensive livestock operations is a significant issue, larger issues arise from the storage and use of livestock manure. There are two basic manure storage types in operation in Canada: liquid manure storage (concrete enclosures; steel tanks, either open or covered; earthen basins; lined/unlined lagoons) and solid manure storage (e.g., manure stored indoors with bedding or as a pack in the barn; manure stored outside as a pile on the ground). Problems arise when storage systems are inadequately built or when they are sited too close (less than 30 metres) to water supplies. Liquid manure stored in lagoons, for example, may overflow during periods of heavy rainfall, or the lagoons can fail to prevent the leaching of organic and inorganic materials into the surrounding environment. The health and environmental impacts associated with manure spills are similar to those arising from the improper use of manure as fertilizer.

Beef feedlot in southern Alberta.

Guest Statement

Canada Feeding the World?

Peter Schroeder

The myth woven into agricultural policies and subliminally embedded in all agricultural thinking throughout the developed countries is that agricultural output must continually increase, at all costs, or humanity will face the consequences of global starvation. This myth formed the bases for the theology driving the Green Revolution and is the primary reason stated for the global rush to accept and adopt genetically modified organisms in all aspects of modern commercial farming systems. The 'need to feed the world' myth would not be a bad thing if in some form modern agriculture's hyper-productivity actually helped to reduce global hunger, but there is little, if any, evidence of that. There is ample evidence that maximizing production from every hectare has caused more problems than it has solved by ignoring some of history's most important lessons.

Since its inception, agricultural production has been based on the 'natural fertility' of the land on which it is practised. As farmers became more adept at farming, they began to understand what depletes as well as what enriches their soil's inherent fertility. By paying close attention to these factors, they began to manipulate their environment to increase yields and harvest reliability. In a relatively short time, they had a thorough understanding of the direct linkage between sustainable yield and sustainable fertility. Left to themselves, farmers are loath to destabilize this critical balance because they have nothing but the land to sustain them.

The great civilizations throughout antiquity rose and fell according to how well their farming and herding people were able to balance the land's sustainable fertility with the harvest output. The fundamentals of that technology and science have not changed, because now as then, agriculture is all about repackaging solar energy into calories that can be digested by humans. This includes the use of all food animals that convert plant material not digestible by humans. Since so much of this planet is more pasture land than 'farm' land, food animals form an important link in the solar energy reprocessing system that is agriculture.

Modern agriculture, which began in the 1950s and 1960s, is based on cheap petroleum energy and a surplus of nitrogen production left over from Second World War explosives manufacturing. These industrial surpluses were all redirected at agricultural production and marketing. Without fossil fuel in all its forms to power modern agricultural equipment and provide the fertilizer as well as transportation, the contemporary commercial agricultural models could not exist.

It was at this point that the Canadian government turned Prairie agriculture into a full-blown exporting component of the Canadian economy. Grain, specifically wheat, was produced beyond any possible local demand and shipped around the world. Everybody involved in growing grain on the Prairies had a fuzzy warm feeling; they were working hard, making a decent living, and feeding the world. What could be better?

The long-term truth is that the wheat was sold to the highest bidder, often disrupting local markets and bankrupting farmers as far away as India and other developing countries. The cheap Canadian wheat was often used to lower local grain prices, forcing massive disruptions in the local agricultural economies. Canadian wheat was cheap in part because Canadians had learned to use modern farming systems but also because it was subsidized by cheap fuel and transportation.

Intensive livestock production can only happen when the supply of grain in the world is too large for the human population to consume. In the overstimulated production environment as in North American and western Europe, this artificially created surplus of grain is used as livestock feed for everything from poultry to pigs, cattle, and sheep. In addition to producing the cheapest food the world has ever seen, this farming model brings with it environmental and biological challenges that have their own costs.

For almost 100 years, government subsidies have directed agricultural production in western Canada. The best example is how the Canadian taxpayer paid the bulk of the freight costs to get western wheat and barley from the Canadian Prairies to deepwater ports on Canada's west coast or the Great Lakes. The same story is now being repeated with canola, minus the transportation subsidy.

Today's incentives come in the form of tax breaks and risk management programs. A similar scenario is also being played out in Manitoba with its massive export-dependent hog industry and across the Prairies where the beef industry expanded to supply a growing demand in the US.

Today, Canadian and American beef and hog producers are suffering economically as US energy policy diverts corn from feed grain to ethanol production. Because of the feed grain shift to ethanol, Manitoba hog producers are killing newborn piglets and brood sows. Diverting even a small part of the feed grain stocks has also caused food riots in the poorest countries, because it takes only a very small supply shortfall to cause dramatic price increases

From a farmer's perspective, modern agriculture has nothing to do with feeding the world's hungry and everything to do with not going broke. The world's hungry cannot afford to pay the freight for the food it would take to solve their problem, never mind cover the production costs. At the same time that the world's starving poor are rioting, people in developed nations are fighting an epidemic of obesity.

The Green Revolution has been underway for more than 40 years, and one has to wonder how it has changed the world for the better. There are fewer farmers on the land in Canada's Prairies every year. Those who are left earn less per acre and animal unit than their parents did before the Green Revolution started. In their struggle, they have abandoned many of the fundamentals that sustained the generations on the land before them. Many of today's farming practices now contribute to ecological damage, with chemical and nutrient runoff contributing to contaminating waterways and lakes instead of keeping the land productive and healthy.

Peter Schroeder

The Green Revolution has not been kind to farmers. The net beneficiaries of the Green Revolution are the people with shares in the farm service and food processing sector.

There are farmers in North American who have opted to step off the industrial farming treadmill. They distinguish themselves as grass farmers, holistic ranchers, natural and organic farmers. The common link among all of them is that they are striving to achieve a balance between the 'natural' fertility, sustainable yield, and their own economic needs without resorting to the farming stimulants: chemical fertilizers, herbicides, pesticides, and genetically modified seeds. These people are a minority, but their numbers are growing.

In my own farming experience covering 25 years of sheep ranching, we had our best economic return after we adopted the grass farming principles and applied them. Simply put, grass farming is all about using livestock to repackage the solar energy stored in plants for human consumption. Grass farming requires the rancher to put the needs of pasture ahead of those of livestock. It also requires the rancher to forget about maximizing production in favour of sustainable yields. The credo for many grass farmers is the 80:20 rule: 'Take the easy 80 per cent nature is offering for free, and forget about the last 20 per cent.' The net return to the enterprise increased substantially, and there was a marked improvement in lifestyle with less labour and lower stress levels.

The biggest threat facing food production today is that by well-intentioned regulatory means, the traditional subsistence and small-scale commercial farms could be eliminated in the drive to accommodate intensive agri-business operations as they continue to perpetuate the myth that they are feeding the world's hungry.

Peter Schroeder farmed the Canadian Prairies for many years and now shares his experience as a columnist with the Winnipeg Free Press.

Raw, well-rotted manure is often spread onto farm fields as fertilizer, a reasonable environmental practice as long as farmers have sufficient cropland to absorb the manure of their livestock. However, new large-scale farms produce vast quantities of manure and often do not have correspondingly large areas of farmland. In 2001, Canadian livestock produced

an estimated 177.5 million tonnes of manure (Hofmann and Beaulieu 2006). The hogs in Ontario currently produce as much raw sewage as the province's people.

If manure and commercial fertilizers are misused, spilled, or applied in excessive quantities, the result is contamination of soil and water by nitrogen, phosphorus, and bacteria. Although crops take up the bulk of added nutrients, a portion—the nutrient surplus—remains in the field. For all agricultural land in Canada, annual inputs of nitrogen and phosphorus from commercial fertilizers and livestock manure exceed annual outputs. There is a national surplus of approximately 0.3 million tonnes of nitrogen and 56,000 tonnes of phosphorus, or 8.4 kilograms per hectare of nitrogen and 1.6 kilograms per hectare of phosphorus. Fertilizers, whether organic or synthetic, are viewed as essential to maintain crop yield and soil health, but their application in excess of what crops can utilize can have significant implications for human and environmental health.

Nutrient loading can result in runoff to streams, rivers, lakes, and wetlands, spurring additional growth of algae and other aquatic plants. Accelerated eutrophication results in loss of habitat and changes in biodiversity (Chapter 4). For example, long-term exposure to elevated nitrate concentrations has contributed to the recent decline in frog and salamander populations in Canada. Concentrations of nitrate greater than 60 milligrams per litre in water kill the larvae of many amphibians.

Excess aquatic plant growth associated with nutrient loading also reduces oxygen levels in the water, leading to fish kill. Between 1987 and 1997, 353 fish kill events were reported to Environment Canada's Environmental Emergencies Centre. Of these, 22 were caused by discharges of nutrient-containing materials, and almost all of them contained nitrogen compounds. Most of the reported fish kills were associated with agricultural activities, particularly the release of manure through storm or flood runoff, underground tank leaks, overflowing storage facilities, or spraying of fields. From 1988 to 1998, a total of 274 manure spills were reported in Ontario, of which 53 resulted in fish kills, primarily because of the ammonia in liquid manure. Reporting on fish kills from accidental spills/discharges of nutrient-related compounds is currently on a voluntary basis only, so they are believed to be widely under-reported.

The application of fertilizers can also affect the quality of drinking water. Twenty-six per cent of Canadians—approximately 8 million people—rely on groundwater for domestic water supply. In most provinces, 5 to 20 per cent of these wells are contaminated with unsafe levels of nitrate, and up to 60 per cent of wells are contaminated in regions with high-demand crops or intensive livestock operations. A federal-provincial study in PEI found the Wilmot River watershed, just east of Summerside, to be one of the most contaminated areas of the Island. One in five wells tested during the study showed levels of nitrates above Health Canada guidelines, with eight more watersheds in PEI in a similar situation.

Factors influencing the effect of manure and commercial fertilizers on the environment include soil type, climate, precipitation, topography, and the quantities of manure produced. Most of these factors are beyond the farmer's control. However, manure management practices also influence the magnitude and extent of ecological impacts. Unfortunately, the environmental laws governing manure management were created when small operations were the norm, and therefore they fail to address the environmental and health risks that come with more intensive livestock operations. For example, in Ontario there are no legally binding standards for constructing manure storage facilities or for the application of manure. Nor are there any monitoring mechanisms to ensure that farmers use best practices for managing manure.

Sustainable Food Production Systems

While Canadians continue to enjoy a wide variety of high-quality foodstuffs year-round, concern over the vulnerability of the productive capacity of our agro-ecosystems mounts. The future capacity to deliver agricultural outputs depends on the continuing ecological viability of agro-ecosystems, yet the stresses imposed on them by intensification are significant. The challenge is to foster agro-ecosystem management practices that will meet growing food, feed, and fibre needs while providing more environmental protection.

Improving agro-ecosystem management so that all levels of agricultural production can be associated with better environmental performance requires new knowledge and better skills, which can be achieved by improvements in technology, natural resource management systems, and landscape planning, as well as by policies and institutional arrangements that help to integrate environmental values into agricultural investment and management decisions. Examples include integrated pest management, integrated plant nutrient systems, and no-till and conservation agriculture. These approaches seek to meet the dual goals of increased productivity and reduced environmental impact.

Integrated Pest Management

The goal of **integrated pest management** (IPM) is to avoid or reduce yield losses caused by diseases, weeds, insects, mites, nematodes, and other pests while minimizing the negative impacts of pest control (resistance, non-selection, mobility, persistence, bioconcentration, etc.). Originally used to reduce excessive use of pesticides while achieving *zero* pest incidence, the concept has broadened over time. The presence and density of pests and their predators and the degree of pest damage are monitored, and no action is taken as long as the level of pest population is expected to remain within specified limits.

IPM considers the crop and pest as part of a wider agro-ecosystem, promoting biological, cultural, and physical pest management techniques over chemical solutions to pest control. Combinations of approaches are used, including:

- bacteria, viruses, and fungi (pathogens);
- insects such as predators and parasites (biological management);

- disease and insect-resistant plant varieties;
- synthetic hormones that inhibit the normal growth process;
- behaviour-modifying chemicals and chemical ecology products (such as pheromones, kairomones, and allomones).

If pesticide use is deemed essential to pest control, only pesticides with the lowest toxicity to humans and non-target organisms are applied.

The adoption of IPM practices has economic and other benefits for farmers, but its use does require more expertise than simply applying chemicals. For this reason, Ontario has established a formal system of IPM for 21 of the province's 95 agricultural commodities. Producers can obtain expert and current advice on these products, their pests, and optimal courses of action by phone. Ontario has also introduced a formal accreditation program for companies and facilities wishing to be IPM-certified. The program in Ontario cut the use of pesticides by 52 per cent since its introduction in 1987 up to 2004. Overall pesticide use in fruit and vegetable crops decreased by 20 per cent over the past five years alone. Specifically, insecticide use in fruit-growing declined by 57 per cent, while fungicide use for both fruit and vegetables has been reduced by 54 per cent since 1998. Increased adoption of IPM and alternative pest control strategies such as border sprays for migratory pests, mating disruption, alternate row spraying, and pest monitoring are major reasons for these large declines. Internationally, countries such as Indonesia have developed an aggressive IPM program, managing to cut pesticide use by as much as 90 per cent. Sweden has adopted a similar aggressive approach, reducing pesticide use by 50 per cent.

Integrated Plant Nutrient Systems

As discussed earlier in the chapter, agricultural production removes plant nutrients from the soil, reducing its organic and nutrient content. Imbalances in nutrient availability can lead to excessive depletion of nutrients that are in short supply, with corresponding reductions in crop yield. The goal of **integrated plant nutrient systems (IPNSs)** is to maximize nutrient use efficiency by recycling all plant nutrient sources within the farm and by using nitrogen fixation by legumes (Chapter 4) to the extent possible (Box 10.4). Soil productivity is enhanced through a balanced use of local and external nutrient sources, including manufactured fertilizers. Plant nutrient losses arising from nutrients supplied in excess can pollute soils and waters, so IPNSs also seek to minimize the loss of nutrients through the judicious use of external fertilizers. IPNSs aim to optimize the productivity of the flows of nutrients passing through the farming system during a crop rotation (Scialabba 2003). The quantities of nutrients applied are based on estimates of crop nutrient requirements—i.e., knowledge of the quantities of nutrients removed by crops at the desired yield level.

No-Till/Conservation Agriculture

To destroy weeds and loosen topsoil to facilitate water infiltration and crop establishment, agricultural land is ploughed, harrowed, or hoed before every planting. Topsoil disturbance of this magnitude and frequency destabilizes the soil structure, leading to soil erosion and soil compaction, negatively affecting productivity and sustainability. The economic and ecological costs associated with conventional tillage systems are becoming more apparent, leading farmers to search for alternative land preparation techniques, such as **no-till/conservation agriculture (NT/CA)**. NT/CA (or zero, minimum, or low tillage) protects and stimulates the biological functioning of the soil while maintaining and improving crop yields.

Essential features of NT/CA include: minimal soil disturbance restricted to planting and drilling (farmers use special equipment to drill seeds directly into the soil instead of ploughing); direct sowing; maintenance of a permanent cover of live or dead plant material on the soil surface; and crop rotation, combining different plant families (e.g., cereals and legumes). Crops are seeded or planted through soil cover with special equipment or in narrow cleared strips. Soil cover inhibits the germination of many weed seeds, minimizing weed competition and reducing reliance on herbicides. Soil cover also reduces soil mineralization, erosion, and water loss, builds up organic matter, and protects soil micro-organisms. Crop sequences are planned over several seasons to minimize the build-up of pests or diseases and to optimize plant nutrient use by synergy among different crop types, and these sequences involve alternating shallow-rooting crops with deep-rooting ones to utilize nutrients throughout various layers of the soil. Other advantages associated with NT/CA include: increased yields in the order of 20 to 50 per cent higher than with conventional tillage practices; reduction in the variability of yields from season to season; significant reductions in labour costs; and lower input costs, particularly for machinery (e.g., smaller tractors can be used, reducing fuel costs).

Despite these advantages, conventional tillage-based agricultural systems continue to dominate worldwide. There are several reasons for this—a reluctance to change (conventional approaches have been working for decades), a lack of knowledge of the damage to soil systems associated with plough-based agriculture, and the complex management skills required for a successful transition. The short-term economic cost associated with the transition from conventional tillage to NT/CA is also a deterrent. During the transition years, there are extra costs for tools and equipment, higher weed incidence may increase herbicide costs initially, and yields will improve only gradually.

In Canada, farmers appear to recognize the advantages of NT/CA, with more than 63 per cent employing some or all of the elements of NT/CA. No-till increased to 46.4 per cent of the area tilled in 2006 from 29.7 per cent in 2001. The area worked with conventional tillage, which had historically been the most popular tillage method, dropped to 28 per cent in 2006 from

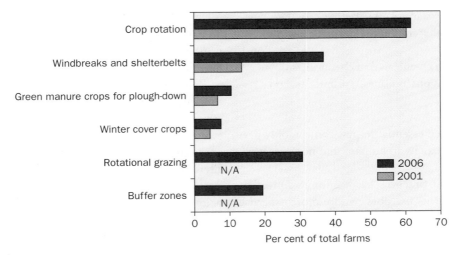

Figure 10.14 Soil conservation practices, 2001 and 2006. Source: Statistics Canada, *Census of Agriculture* 2001 and 2006.

40.5 per cent in 2001. Conservation tillage—the midpoint between conventional and no-till, dropped from 29.8 per cent of tilled area to 25.6 per cent.

Figure 10.14 shows the percentage of farms in Canada using various soil conservation methods. **Crop rotation** is the most common and is an important means of recharging soil nitrogen through use of legumes such as alfalfa and clover. Grassed waterways are used to control overland flow of runoff, thereby controlling the formation of gullies on exposed soil surfaces. **Contour cultivation** involves cultivating the soil parallel to the contour of the slope, which serves to reduce the speed of runoff by catching soil particles in the plough furrows. More than 16 per cent of PEI and Saskatchewan cropland is protected in this way. **Strip cropping** is a similar technique in which different crops may be planted in strips parallel to the slope. While one crop may be harvested, leaving bare soil, the other crop serves to provide some protection. This technique is commonly used against wind erosion and is most prevalent in western Canada. The soil surface can also be protected in winter through growth of a winter cover crop. This is effective not only for wind erosion in winter but also to protect the soil from intense rainfall in the spring.

There is considerable regional variation in the use of these practices. Newfoundland, for example, has the smallest proportion of farmers using erosion control, largely because of the small amount of land devoted to seed crops in the province. In Saskatchewan, however, almost 72 per cent of farmers use some form of erosion control. Overall, 63 per cent of farms in Canada use one or more erosion control practices. During 2000, 24 per cent of farms reported using conservation tillage techniques, and a further 7 per cent used no-tillage, equal to more than 4 million hectares of land. Comparatively, the United States, Brazil, Australia, and Mexico have approximately 19,750,000, 13,470,000, 8,640,000, and 650,000 hectares, respectively, under zero tillage.

Sustainable Agriculture in Action: Organic Farming

Current approaches to sustainable agriculture such as IPM, IPNS, and conservation tillage consider only one aspect of the farming system components—pest ecology, plant ecology, and soil ecology, respectively. Organic agriculture, however, combines these and other management strategies into a single approach, focusing on food web relations and element cycling to maximize agro-ecosystem stability. Organic agriculture is a production management *system* that aims to promote and enhance ecosystem health. It is based on minimizing the use of external inputs and represents a deliberate attempt to make the

Strip farming in Saskatchewan.

best use of local natural resources while minimizing air, soil, and water pollution. Synthetic pesticides, mineral fertilizers, synthetic preservatives, pharmaceuticals, genetically modified organisms, sewage sludge, and irradiation are prohibited in all organic standards (FAO 2001; Canadian General Standards Board 2003).

Organic agriculture encompasses a range of land, crop, and animal management procedures designed to:

- enhance biological diversity within the whole system;
- increase soil biological activity;
- maintain long-term soil fertility;
- recycle wastes of plant and animal origin in order to return nutrients to the land, thus minimizing the use of non-renewable resources;
- rely on renewable resources in locally organized agricultural systems;
- promote healthy use of soil, water, and air as well as minimize all forms of pollution that may result from agricultural practices;
- handle agricultural products with emphasis on careful processing methods in order to maintain the organic integrity and vital qualities of the product at all stages;
- become established on any existing farm through a period of conversion, the appropriate length of which is determined by site-specific factors such as the history of the land and the type of crops and livestock to be produced.

For example, organic practices that encourage soil biological activity and nutrient cycling include manipulation of crop rotations and strip cropping, the use of **green manure** and organic fertilizer (animal manure, compost, crop residues), minimum tillage or zero tillage, and avoidance of pesticide and herbicide use. Organic agriculture significantly increases the density of beneficial invertebrates, earthworms, root symbionts, and other micro-organisms essential to maintaining soil health. For example, the biomass of earthworms in organic systems is 30 to 40 per cent higher than in conventional systems, and earthworm density is 50 to 80 per cent higher.

Enhanced soil fertility is the cornerstone of organic agriculture, but there are other benefits associated with organic production systems. For example, organic fields in excess of 15 hectares contain a greater diversity and abundance of flora and fauna than conventional fields, including endangered varieties (e.g., organic grasslands contain 25 per cent more herb species than conventional grasslands). Organic farming systems are also more energy-efficient per unit crop than conventional farming techniques, since organic systems resemble closed or semi-closed nutrient cycles. Organic land management permits the development of a rich weed flora, and a versatile flora attracts more kinds of beneficial insects. Organic farming systems are also better at controlling erosion, since organic soil management techniques improve soil structure.

Organically grown foods also benefit human health, since they contain fewer pesticide residues than foodstuffs grown under intensive farming methods.

Like most industrial countries, Canada has national organic standards, regulations, and inspection and certification systems that govern the production and sale of foods labelled as 'organic'. In 2007, federal regulations were introduced for organic certification, easing the way for organic trade of produce between provinces and with other countries. Worldwide, an estimated 20 million hectares of land are under certified organic agriculture, with Australia, Argentina, Italy, Canada, and the United States having the largest areas of organic farmland. In 2000, agricultural land under certified organic management averaged 0.25 per cent of total agricultural land in Canada, compared to 2.4 per cent of total agricultural land in western Europe, 1.7 per cent in Australia, and 0.22 per cent in the United States (Scialabba 2003). In most developing countries, agricultural land reported under certified organic production is less than 0.5 per cent, although the extent of non-market, non-certified organic agriculture may be considerable. For example, an estimated one-third of West African agricultural produce is produced organically (Scialabba 2003).

More than 15,000 farm operations in Canada (6.8 per cent) reported at least one type of organic product in 2006. Non-certified organic production is most common, with the most likely application being animal products. Field crops such as buckwheat, rye, and caraway dominate certified Canadian organic production. Canada is among the five top world producers of organic grains and oilseeds. Farm cash receipts from organic production in Canada reached about $1 billion in 2002. The number of certified producers increased by 59 per cent between 2001 and 2006. The domestic organic market is strongest in British Columbia, Alberta, Quebec, and Ontario. However, Saskatchewan has the highest proportion of certified organic producers, with 38 per cent of the national total.

Consumer health and food quality concerns (e.g., concerns about growth-stimulating substances, genetically modified [GM] food, dioxin-contaminated food, and livestock epidemics such as bovine spongiform encephalopathy and foot-and-mouth disease) continue to drive demand for organic products in Canada and around the globe. Organic food production systems are considerably more respectful of the environment than intensive farming practices, but intensive farming systems still dominate. In the absence of governmental support for the expansion of organic production, farmers may be reluctant to convert to organic farming for several reasons. Conversion from conventional, intensive systems to organic production causes a loss in yields, the extent of which varies depending on the biological attributes of the farm, farmer expertise, the extent to which synthetic inputs were used under previous management, and the state of natural resources. Yields can be 10 to 30 per cent lower in organic systems, and it may take several years (e.g., three to five) to restore the ecosystem to the point where organic production becomes economically viable. In addition, production costs per unit of production (e.g., labour, certification and inspection fees) and marketing expenses can be higher with organic produce, but once produce qualifies as *certified* organic, some of these costs can be offset by price premiums. In developed countries, retail organic products can command 10 to 50 per cent more than conventional prices for the same commodity. In Canada, demand for most organic products continues to exceed supply despite the higher prices charged for certified organic produce. Consumers in industrialized countries are willing to pay a premium for organic food because they perceive environmental, health, or other benefits.

In Canada and elsewhere, cases are emerging in which organic production is constrained or no longer feasible because of the advent of GM crops. In North America, the contamination of organic fields by GM crops has been estimated as resulting in an economic loss of more than US $90 million a year (Scialabba and Hattam 2002). Organic farmers in Canada can no longer grow organic canola (i.e., oilseed rape) because of GM canola contamination in Saskatchewan. In 2002, two organic farmers in Saskatchewan filed a class action suit against Monsanto Canada and Aventis Canada in response to the commencement of field trials of genetically engineered (GE) wheat for introduction as a commercial crop in 2004. Wheat is a self-pollinated crop, but there is danger of spread of pollen from the wheat, and the resulting genetic contamination of organic

The case involves legal questions of significant importance to the public, namely liability and rights associated with the development, marketing, sale, and dispersal of GMOs, as well as public access to justice through class certification. The prevalence of open-pollinating GM crops on the landscape is a matter of significant environmental and public interest. These issues transcend provincial or territorial boundaries, as organic farmers in Saskatchewan can no longer grow and sell certified organic canola as a crop.

— Organic Agriculture Protection Fund Class Action Lawsuit against Monsanto and Aventis. http://www.saskorganic.com/oapf/legal-page-2.html

wheat would devastate Saskatchewan's organic sector. The suit calls for compensation for damage caused to certified organic farmers resulting from the introduction of GE canola into the rural environment and an injunction to prevent the introduction of GE-wheat61. The outcome of this precedent-setting legal case will be of major importance to future policies regarding the impact of biotechnology on organic agriculture.

Implications

Agricultural modification is arguably the main impact that humans have had on natural ecosystems. It is, however, also one of the oldest and one that is basically a modification of ecological systems to benefit humans. Over centuries, natural and human-modified agricultural landscapes have existed and transformed from one state to the other with little lasting damage to planetary life-support systems. However, as additional auxiliary energy flows were applied to boost the productivity of agriculture, the differences between these two ecological systems became more distinct, and the impacts of agriculture on natural ecosystems increased. Agricultural production (certainly in Canada's commercial agricultural sector) is now more similar to industrial production than to the natural ecosystems from which agriculture was derived.

This industrialization has led to many environmental challenges for agriculture. Yields are declining in some areas as crops become less responsive to fertilizer input, biocides continue to eliminate many natural enemies of pests, and soils are eroded, salinized, and compacted. In response, researchers are suggesting that a fundamental restructuring is required in how agriculture is undertaken, with the emphasis changing from maximizing productivity to ensuring sustainability.

Sustainability will require greater attention to the agro-ecosystem and particularly to the soil base that sustains agriculture. In addition, the socio-economic and regulatory dimensions will need to be integrated into systemic change. Organic farming will not be successful, however, unless customers are willing to pay for the produce, not only as a benefit to themselves but also to sustain and nourish healthier ecosystems overall. In addition to buying organic produce, we can do a number of things as individuals to ensure that we at least do not exacerbate the challenges now facing agricultural systems (Box 10.14).

Finally, it is important to remember that millions of people still experience hunger on a daily basis. Programs designed to address these problems have always failed to meet their goals. Food prices soared rapidly in 2008, and in countries where more than half of family income (as opposed to less than 10 per cent in Canada) goes to feeding families, civil unrest has already taken place. Hungry families are angry families, and ways must be found to ensure greater food security for the world's poor if this situation is not to deteriorate into widespread violence.

Box 10.14 What You Can Do

Although the challenges facing agriculture at the global and national levels are immense, there are still some ways by which individuals can help.

1. Eat less. This entails finding out about good nutritional habits so that we consume only the food that we really need.

2. Eat lower on the food chain. Most North Americans eat far too much meat. Eating more vegetables will benefit not only the global food situation but your own health as well.

3. Feed your pet lower on the food chain. Dogs and cats will also be healthier if fed on balanced grain pet foods rather than meat.

4. Waste less food. Studies indicate that as much as 25 per cent of food produced in North America is wasted.

5. Grow at least some of your own food. If Canadians were to devote a fraction of the time and resources on growing food that they spend on their lawns, it would allow more food for others elsewhere.

6. Support local food growers and food co-ops. This helps to protect agricultural land in Canada from being transformed to other uses.

7. Learn what foods are in season in your area, and try to build your diet around them.

8. Ask the manager or chef of your favourite restaurant how much of the food on the menu is locally grown, and then encourage him or her to source food locally. Urge that the share be increased. People can do the same at their local supermarket or school cafeteria.

9. Take a trip to a local farm to learn what it produces.

10. Host a harvest party at your home or in your community that features locally available and in-season foods.

11. Produce a local food directory that lists all the local food sources in your area, including farmers markets, food co-ops, restaurants emphasizing seasonal cuisine and local produce, and farmers willing to sell directly to consumers year-round.

12. Buy extra quantities of your favourite fruit or vegetable when it is in season and experiment with drying, canning, jamming, or otherwise preserving it for a later date.

13. Join one of the NGO groups that specializes in rural development in less developed countries.

Summary

1. Agriculture originated at least 9,000 years ago, when societies domesticated both plant and animal species. Societies around the globe flourished by improving their capacity to expand agricultural production.

2. Over much of history, agricultural output has been increased by bringing more land into production. Today, intensification of production—obtaining more output from a given area of agricultural land—has become a key development strategy in most parts of the world to

meet the increased demand for foodstuffs. The future capacity to deliver agricultural outputs depends on the continuing ecological viability of agro-ecosystems.

3. Agriculture is a food chain, with humans as the ultimate consumers. The second law of thermodynamics dictates that the shorter the food chain, the more efficient it will be.

4. The Green Revolution and the Livestock Revolution have led to profound changes to the global land base. Complex natural systems have been replaced by relatively simple control systems in which humans are in command of the species and numbers that exist in a given area.

5. The Green Revolution relies on auxiliary energy flows, such as fertilizers, biocides, fossil fuels, and irrigation systems, to increase yields. Yields are falling in many of the poorest parts of the world because of soil exhaustion and the rising price of fertilizers. Industrial systems of livestock production also depend on outside supplies of feed, energy, and other inputs to satisfy growing worldwide demand for meat.

6. Biofuels have great potential to be part of adaptation to reducing GHGs and creating higher values for agricultural products in many underdeveloped countries, but they also create challenges in terms of competition with food supply and encourage greater transformation of biodiversity-rich forests into agricultural land. Cellulosic production seems to have the greatest potential to address these problems in the future.

7. Agriculture is an important industry in Canada, accounting for 8.5 per cent of gross domestic production, 10 per cent of employment, and about 6 per cent of total merchandise export earnings. Many of the innovations and technologies employed by Canadian farmers to produce agricultural products have extraordinary implications for ecosystems.

8. Large increases in fertilizer inputs have occurred in Canada over the past two decades.

9. Between 1970 and 2000, the value of biocides applied to agricultural crops in Canada increased threefold. More than 30 million hectares were treated with at least one biocide in 2000, compared to less than 10 million hectares in 1970.

10. Land degradation includes a number of processes that reduce the capability of agricultural lands to produce food. One study suggests that such processes cost Canadian farmers more than $1 billion per year.

11. Soil erosion is estimated to cause up to $707 million in damage per year in terms of reduced yields and higher costs. Soil formation in Canada is slow. A rate of 0.5 to 1.0 tonnes per hectare (t/ha) may be considered average. Losses of 5 to 10 t/ha are common in Canada, and figures of 30 t/ha have been recorded in the Fraser River Valley.

12. Increasing acidity as a result of the application of nitrogen fertilizers and acid deposition is also a problem that reduces crop yields. Salinization occurs where there are high sodium levels in the soils and shallow water tables, such as in the Prairies. One estimate suggests that salinization causes economic losses four to five times as great as losses due to erosion, acidification, and loss of nitrogen.

13. Cultivation involves a continual process of removing plant matter from a field. In the process, both the organic content and nutrient content of the soil are reduced. On the Prairies, current organic matter levels are estimated to be 50 to 60 per cent of the original levels.

14. Biocides are applied to crops to kill unwanted plants and insects that may hinder the growth of the crop. They have boosted yields throughout the world and helped feed many hungry mouths. There is also clear scientific evidence that they have serious negative impacts on ecosystem health.

15. Biocides promote the development of resistance among target organisms. Over the past 40 years, more than 1,000 insects have developed such resistant populations. They are non-selective and tend to kill non-target as well as target organisms. They may also be highly mobile and move great distances from their place of application. In addition, they may persist for a long time in the environment and accumulate along food chains. Such biomagnification has resulted in drastic reductions in the populations of some species at higher trophic levels, such as ospreys and bald eagles.

16. Chemicals, and their constituents as they break down, may interact synergistically.

17. The government must register chemicals before they can be used. Registration involves the chemical company providing evidence that the chemical is not carcinogenic or oncogenic. Many chemicals have been registered but subsequently fail these tests.

18. Attention is now being devoted to sustainable food production systems that maintain or enhance environmental quality, generate adequate economic and social returns to all individuals/firms in the production system, and produce a sufficient and accessible food supply. Integrated pest management is now becoming more popular, and several provinces have such programs.

19. Organic farming is growing rapidly but is still a relatively small part of overall agricultural production.

Key Terms

bioaccumulation
bioconcentration
biofuels
biomagnification
contour cultivation
crop rotation
genetically modified
    organisms (GMOs)

grasshopper effect
green manure
Green Revolution
hybridization
integrated pest
    management (IPM)
integrated plant nutrient
    systems (IPNSs)

intensive livestock opera-
    tions (ILOs)
Livestock Revolution
monoculture cropping
no-till/conservation
    agriculture (NT/CA)
salinization

Silent Spring
soil compaction
soil erosion
strip cropping
subsistence farming
summer fallow

Questions for Review and Critical Thinking

1. How do the laws of thermodynamics apply to agriculture?
2. Indicate how the concepts of 'limiting factors' and 'range of tolerance', discussed in Chapter 4, can be applied to agriculture.
3. Considerable interest is being directed towards ecosystem management. What is the relevance of this concept, if any, for agriculture?
4. Do you think that biofuels have an important role to play in agricultural systems in the future? How would you try to mitigate some of the challenges presented by biofuels?
5. What do you think might be the main challenges facing the global agricultural supply in five years' time?
6. In this chapter, mention was made of the resilience of farm systems. In Chapter 6, in relation to adaptive management, attention was also given to resilience. What value do the concepts of resilience and adaptive management have regarding agriculture?

7. How would you go about identifying and assessing the impacts (environmental and social) of agricultural policies and practices in Canada? What ideas from Chapter 6 might be helpful in this exercise?
8. In Chapter 6, attention focused on the ideas of partnerships and stakeholders. What relevance do such concepts have regarding agriculture if partnerships are to be formed among the federal and provincial governments, agribusiness, and family farms?
9. It has been suggested here that conflict or tension can exist between the needs of urban areas and adjacent rural areas. To what extent do you think alternative dispute resolution ideas might be applied to deal with such tensions?
10. If you were a commercial farmer in Canada, to what extent would it be important for you to consider the implications of climate change for your farming operations?

Related Websites

Agriculture and Agri-Food Canada: http://www.agr.gc.ca/site_e.phtml#env

Canadians against Pesticides: http://www.caps.20m.com

Consultative Group on International Agricultural Research (CGIAR): http://www.cgiar.org

Environment Canada: http://www.ec.gc.ca/envhome.html

Food and Agriculture Organization of the United Nations (FAO): http://www.fao.org

Government of Canada, agriculture and the environment: http://www.agr.gc.ca/policy/environment/pubs_sds_e.phtml

Pesticides and wild birds: http://www.hww.ca/hww2.asp?cid=4&id=230

World Hunger: We Feed People: http://sustag.wri.org

World Resources Institute, Earth Trends: http://earthtrends.wri.org

Worldwide Fund for Nature, chemicals and wildlife: http://panda.org/downloads/toxics/causesforconcern.pdf

Further Readings

Note: This list comprises works relevant to the subject of the chapter but not cited in the text. All cited works are listed in the section 'References' at the end of the book.

Environment Canada and US Environmental Protection Agency. 2002. 'State of the Great Lakes 2001'. www .binational.net/sogl2001/download.html.

FAO (Food and Agriculture Organization). 2003. *World Agriculture: Towards 2015/2030. An FAO Perspective.* Rome: FAO.

Miller, G. 2000. 'Intensive farming'. In 'The protection of Ontario's groundwater and intensive farming: Special report to the Legislative Assembly of Ontario'. http://www. eco.on.ca.

Nass, L.L., P.A.A. Pereira, and D. Ellis. 2007. 'Biofuels in Brazil: An overview'. *Crop Science* 47: 2,228–37

Statistics Canada. 2001. 'Urban consumption of agricultural land'. *Rural and Small Town Canada Analysis Bulletin* 3, 2. Cat. no. 21-006-XIE. Ottawa: Statistics Canada.

———. 2003. 'Manure storage in Canada'. In *Farm Environmental Management in Canada* 1, 1. Cat. no. 21-021-MIE. Ottawa: Statistics Canada.

Chapter 11

Water

Learning Objectives

- To appreciate the need for credible science to inform decision-making about water.
- To recognize the water endowment in Canada.
- To understand the hydrological cycle.
- To know the environmental and social impacts associated with water diversions.
- To appreciate various perspectives related to water export from Canada.
- To understand the significance of point and non-point sources of pollution.
- To learn about the concept of 'water security'.
- To appreciate the growing importance of 'water terrorism'.
- To gain an understanding of the concept of a 'multi-barrier approach' to drinking water protection.
- To understand the challenges and opportunities regarding water security on First Nations reserves.
- To appreciate the distinctions among supply management, demand management, and the soft path approach.
- To understand the concept of 'virtual water'.
- To realize that water is both resource and hazard.
- To appreciate the difference between structural and non-structural approaches to flood damage reduction.
- To understand the significance of droughts.
- To appreciate the significance of heritage related to protection of aquatic systems.
- To understand the importance of hydrosolidarity and integrated water resource management (IWRM)
- To appreciate the evolving ideas related to 'water ethics'.

Introduction

Statistics Canada (2003b) reports that in terms of water, although Canada has just 0.5 per cent of the world's population, Canadians have access to almost 20 per cent of the global stock of fresh water, and Canada has 7 per cent of the total flow of renewable water. This apparent natural bounty is often taken for granted, it seems, since Canadians are among the highest consumers of water in terms of per capita water use, second only to citizens of the United States. This high use was characterized some time ago by Foster and Sewell (1981, 7) as due to a 'myth of superabundance'. With or without a myth of superabundance,

Canada has a relatively and absolutely generous endowment of water. Indeed, as O'Neill (2004, xi) has observed, 'There can be no question that Canada's freshwater supply is an immensely valuable national resource. Recent estimates of water's measurable contribution to the Canadian economy range from $7.5 to $23 billion annually, values comparable to the gross figures for agricultural production and other major economic components.' In contrast, countries in the Middle East and Sahelian Africa usually experience significant water deficits that are a major impediment to overcoming poverty and facilitating development. However, even in Canada, water is often not available in the right place or at the right time.

Water Availability in 2025

By 2025, the greater part of the Earth's population will likely live under conditions of low and catastrophically low water supply. Approximately 30–35 per cent of the world population will have catastrophically low freshwater supply (less than 1,000 m³ per year per capita). At the same time . . . high water availability can be found in northern Europe, Canada, and Alaska, almost all of South America, Central Africa, Siberia, the Far East, and Oceania.

—Shiklomanov 2000, 28

The Need to Manage Water in Canada

Why should we act? Canada's rivers, lakes, snow, and ice are part of our national identity. Isn't Canada a country blessed with plenty of water and a small population? Well, we were. But things are changing. Across much of Canada, water tends to be where most of the people aren't. As our population grows and pressure for industrial and agricultural uses of water increases, our water supplies are becoming more and more vulnerable to contamination and overuse. Climate change increases the pressure and the risks. Canada needs to be prepared for the serious water policy challenges that are coming. This flies in the face of our history and image of ourselves, but the dramatic consequences of failing to prepare can be seen in many places around the world.

—Pollution Probe 2007, 5

Hydrological Cycle

Water or aquatic resources are one component of a system that includes the atmosphere, **cryosphere**, biosphere, and terrestrial components. Evaporation from surface water (rivers,

Jonathan Price

Wetlands of the Hudson Bay Lowlands.

lakes, wetlands) and transpiration from plants release water vapour into the atmosphere that condenses and forms clouds while moving upward. The tiny droplets of water in clouds eventually fall to the earth as rain, fog, hail, or snow. After reaching the surface, the water evaporates back into the atmosphere, moves into rivers, lakes, or oceans, or percolates into the soil to become groundwater. Chapter 4 provides a more detailed discussion of the hydrological cycle.

About 12 per cent of Canada (1.2 million square kilometres) is covered by *lakes* and *rivers*, with only 3 per cent of that area located in inhabited regions. There are more than 2 million lakes, with the largest being the Great Lakes shared between Canada and the United States. Other large lakes are Great Bear Lake and Great Slave Lake in the Northwest Territories and Lake Winnipeg in Manitoba. Lake water represents about 98 per cent of the surface water available for human use. Canada has more than 8,500 named rivers, and the Mackenzie River, with an average surface flow of 8,968 cubic metres per second, has the highest volume. There are more than 1,000 named *glaciers*, and they are an important source of fresh water for rivers and lakes.

Various types of **wetlands** exist, all being hybrid aquatic and terrestrial systems. They are a key habitat for waterfowl and also store and gradually release water, thus serving as an important 'sponge' to aid in reducing flooding. Wetlands are found in the greatest number and extent in the Prairie provinces and in northern Ontario. Canada has about 25 per cent of the wetlands in the world, the largest amount of any country.

Groundwater is a key source of water for rivers and lakes and is created by surface water passing into the ground and becoming contained in sand and gravel, as well as in pores and cracks in bedrock. During dry periods, many rivers receive much of their water via base flow from groundwater aquifers.

Sprague (2007, 23–5) has observed that the belief of many Canadians about an abundance of water most likely stems from the apparent large volume of fresh water contained in lakes across the country, totalling about 20 per cent of the water in all the lakes of the world. He emphasizes, however, that water contained in lakes is not the same as what is considered a 'renewable' supply. The latter is based on precipitation that falls, then runs off into rivers, often being held in lakes before draining to the ocean or moving downward into aquifers. The flows associated with precipitation or snowmelt should be identified as the renewable supply. As Sprague (2007, 24) has stated, 'To use a financial analogy, the water sitting in lakes and aquifers is comparable to a capital resource of money that can be spent only once. The rivers running out of the lakes would represent interest and dividends that could be used every year for an indefinite time.'

More specifically, Sprague highlights that the apparent bountiful abundance of water in Canada is due to several characteristics of the landscape and the hydrological cycle, including: (1) a few very large lakes and many shallow, small lakes; (2) a cool climate; and (3) low evaporation of water. Calculations suggest that Canada ranks between third and sixth in terms of renewable water supply. The leaders are Brazil with more than 12 per cent of the global renewable supply, followed by Russia with 10 per cent. After that, Canada (6.5 per cent) is in a virtual tie with Indonesia (6.5 per cent), the United States (6.4 per cent), and China (6.4 per cent). The next five on the list, ranging from 5 to 2 per cent of the global renewable water supply, are Colombia, Peru, India, and the Democratic Republic of Congo.

Canada's **renewable water supplies** amount to some 2,850 cubic kilometres out of the world's total of 43,800 cubic kilometres. Such supplies are based on measurement of surface water flows and groundwater recharge resulting from precipitation within a nation's boundaries. Not counted is water flowing into a country from one or more other nations, water flowing out of a country to one or more other countries, and water lost to evaporation or transpiration.

Some 60 per cent of Canada's water flows northward to Arctic and sub-Arctic areas in which few people live. Such northward-flowing water is generally not available to southern Canada, which is where most people live and work. Consequently, when considering the proportion of renewable water supply available to southern Canada, the proportion drops from 6.5 to 2.6 per cent of the global supply. Sprague argues that it is the 2.6 per cent figure that citizens, managers, and political leaders should keep in mind and remarks that 'notably, the figure [of 2.6 per cent] is tenfold lower than the frequently used and mythical "one-quarter of the world supply".' These data and insights highlight the importance of our understanding the basic hydrological characteristics of Canada.

Human Interventions in the Hydrological Cycle: Water Diversions

Given that water is often not in the right place at the right time, humans modify aquatic systems to store, divert, or modify flows. There are some 600 large dams in Canada and about

Water Diversions in Canada

Interbasin diversion projects are found in almost all provinces, and the total flow of water diverted currently between drainage basins is enormous—approximately 4,500 cubic metres per second. No other country diverts nearly as much water or concentrates so much flow for a single function—hydroelectric power generation.
—Quinn et al. 2004, 3

60 large interbasin diversions. Diversions are completed for one or more of the following reasons:

- To increase water supplies for a community or in a region, as illustrated by the St Mary Irrigation District in Alberta. While diversions for irrigation are important in the southern Prairies, this type of diversion is not as typical of the Canadian experience as it is for countries such as India and the United States.
- To deflect watercourses away from or around areas to be protected, such as the Portage Diversion in Manitoba. Here, the purpose is not to move water to a place of need but to protect a community from flood damage. Other reasons are to drain land to allow agricultural production or drain a mine site.
- To enhance the capacity of a river so that it can be used to support activities such as floating logs or allow passage of ships, disposing of wastes, or sustaining of fish. For example, dams on the Ottawa River were designed partly to facilitate moving logs downriver to sawmills.
- To combine or consolidate water flows from several sources into one channel or route to facilitate hydroelectric generation, such as the James Bay hydroelectric project in northern Quebec. Canada is a global leader in water diversions for hydroelectricity generation, and diversions for hydropower purposes dominate overwhelmingly in both number and scale of diversions in Canada (Day and Quinn 1992, 10–11).

While diversions have a positive capacity for increasing power production, expanding irrigation, enhancing waterborne commerce, and reducing flood damage, they also can cause negative environmental impacts and impose costs on people or regions not benefiting directly from them. The James

Bay Cree and their homeland in northern Quebec represent a case in point.

The James Bay Hydroelectric Project

Governments and private corporations have pursued many **megaprojects** in Canada to meet energy demands, and virtually every region in the country has experienced such megaprojects (Figure 11.1). Perhaps the most massive, and one that has garnered a great deal of national and international attention, has been the **James Bay Project** in Quebec. Other huge hydroelectric developments include Churchill Falls in Labrador, the Nelson-Churchill in Manitoba, and the Columbia and Nechako rivers in British Columbia. Nuclear power plants in Ontario, the development of the Hibernia oilfields off the coast of Newfoundland, the Sable Island natural gas exploration off Nova Scotia, and the exploitation of the oil sands in northern Alberta are among other major Canadian energy projects and are discussed in Chapter 12.

Background

In 1971, Quebec Premier Robert Bourassa proposed hydroelectric development using the rivers on the eastern side of James Bay. The purpose was to satisfy future electricity needs in Quebec. The cost was estimated at $2 billion. After alternative projects were examined, a decision was taken to develop La Grande River basin to double the flow in that river by diverting water from adjacent catchments (Figure 11.2). Other river systems north and south of La Grande were to be developed in later phases.

Two major diversions channelled water into La Grande basin. From the south, 845 cubic metres per second (m³/sec), or 87 per cent of the flow measured at the mouth of the Eastmain River, was redirected into the LG2 reservoir on La Grande. Twenty-seven per cent (790 m³/sec) of the Caniapiscau River was diverted into the LG4 reservoir with the intent that this water would in turn flow through all four power stations to be built on La Grande. These diversions together added an average of 1,635 m³/sec to La Grande, almost doubling the natural flow in that river. Over a 15-year period, the cost increased to $14.6 billion, compared to the $2 billion announced by Premier Bourassa in 1971.

In Phase I of the development, three hydroelectric plants (LG2, 3, 4) with a combined 10,283 megawatt (MW) capacity were built. LG2, with 5,328 MW capacity, became the most powerful underground generating station in the world. The first electricity was generated from LG2 in 1979, and LG4 was completed in 1986. LG1 and other dam construction were deferred to Phase II.

Figure 11.1 Energy megaprojects in Canada. Source: Adapted from Day and Quinn 1992, 16.

Figure 11.2 La Grande River hydroelectric development project, Phase 1. Source: Day and Quinn 1992, 134.

The scope and magnitude of the James Bay development has been described as 'breathtaking'. It produces electricity from rivers flowing in a 350,000-square-kilometre area of Quebec, more than one-fifth of the province or an area equivalent to France. The provincial government and Hydro-Québec justified the James Bay development on the grounds of the jobs to be created, the industrial growth to be attracted to the province, and the stability to be generated.

In the enthusiasm over the perceived benefits from hydroelectricity in the James Bay area, a territory far from the settled part of the province, little regard was given to the fact that it was the homeland for about 10,000 Cree and Inuit whose people had lived and hunted in the region for centuries.

James Bay and Northern Quebec Agreement

The **James Bay and Northern Quebec Agreement** is considered by many to be the first 'modern' Native land claims agreement in Canada. However, when Premier Bourassa first announced the construction of the hydroelectric megaproject, no systematic environmental or social impact assessments had been completed, although such a large-scale development could be expected to have major impacts. The Cree people in northern Quebec, who had not been consulted, soon organized themselves to fight the project. The outcome was the James Bay and Northern Quebec Agreement, signed on 11 November 1975 and subsequently approved by the government of Canada and Quebec's National Assembly.

The agreement is complex and often ambiguous. However, it provided for land rights and guaranteed a process to deal with future hydroelectric developments. The agreement included provisions for environmental and social impact assessment for future developments, monetary compensation, economic and social development, and income security for Cree hunters and trappers.

James Bay II

When Premier Bourassa announced Phase II in 1985, he explained that the development would (1) generate revenue for Quebec through exports of electricity to the United States under long-term contracts and (2) attract energy-intensive industries (such as aluminum and magnesium smelters) as a result of competitively priced electricity. James Bay II involved completion of development in La Grande basin, particularly the building of LG1, as well as new hydroelectric development in the Great Whale and Nottaway-Broadback-Rupert river systems. During 1986, the Cree agreed to the completion of the development in La Grande basin but opposed the projects in the basins to the north and south of La Grande.

The projects in the Great Whale basin would provide just under 3,000 MW of new power by diverting several adjacent rivers. One outcome would be a reduction by 85 per cent in the flow of the Great Whale River at the community of Whapmagoostui (Figure 11.3). The Nottaway-Broadback-Rupert development would produce 8,000 MW of additional power and, as with development on La Grande, would involve inundation of land as a result of dam construction.

During the 10-year construction period from 1974 to 1984, various concerns had emerged. These concerns included the

Figure 11.3 The Great Whale project. Source: Diamond 1990, 32.

relocation of Fort George to a new site at Chisasibi, the quality of the drinking water in the new community, problems in maintaining traditional hunting activity in areas that had become accessible from the new roads built for the construction of dams, and, because of the altered patterns of ice break-up on the lower river and estuary as a result of the release of relatively warmer water from the reservoirs in winter and early spring, the difficulty hunters faced in travelling to the northern coastal area across the river from Chisasibi.

At the community level, other concerns emerged. For example, increased erosion along the banks of La Grande, the result of fluctuating water levels in the river caused by releases from the upstream reservoirs, threatened the site of the new community at one point. The newly built road exposed the community to other people and values, contributing to problems such as alcohol abuse for some individuals.

Following completion of the first three dams on La Grande, the major problem became the very high levels of mercury in fish caught in the reservoirs or connecting rivers. As a result, by the end of 1985, the Cree completely stopped fishing in the LG2 area. Another problem was that hunters from Chisasibi purchased vans to travel to distant inland hunting grounds. However, after construction of LG4 was completed, maintenance of the road network east of LG4 was stopped, and the vans could no longer be used.

Against this changing mix of issues and concerns, we look next at the challenges of estimating impacts regarding some specific issue areas. These changing issues and concerns reinforce the arguments for an adaptive management approach, discussed in Chapter 6, and highlight the presence of uncertainty, complexity, and change in terms of both science and management.

1. *Mercury in reservoirs.* None of the environmental impact assessment studies had predicted the appearance of mercury in reservoir fish. This is a puzzling aspect of the impact assessment work, since evidence about elevated mercury levels in fish was available from earlier hydroelectric projects at the Smallwood Reservoir in Labrador and from Southern Indian Lake in Manitoba. Such impacts apparently were dismissed by investigators as being of only short duration and not significant for La Grande.

As Gorrie (1990, 27–8) explained, mercury is common in rocks throughout the North. It exists in an insoluble form that does not affect air or water. However, when such rocks are inundated by a reservoir, bacteria associated with the decomposition of organic material in the reservoir water transform the insoluble mercury into methyl mercury that vaporizes, is released into the atmosphere, and returns to the water. Once in the water, the mercury enters the food chain and through biomagnification reaches the highest trophic levels in fish species that prey on other fish, as reviewed in Chapter 10. Such predator fish—walleye, pike, lake trout—had been an important source of high-quality protein food for the local people. Berkes (1988) indicated that in most years, about one-quarter of the total community wild food harvest came from fishing, averaging about 60 kilograms per year for every man, woman, and child.

In new reservoirs, a burst of decomposition often accelerates the release of mercury. In the La Grande river system, few trees were removed prior to the flooding of the reservoir area, so there was a lot of organic matter to decompose. Downstream from the dams on La Grande, levels of mercury in fish climbed to six times their normal levels within months of completion of the dams. By the sixth year following the impoundment, concentrations of mercury were four to five times higher in all species sampled. A 1984 survey of the Cree at Chisasibi showed that 64 per cent of the villagers had unsafe levels of mercury in their bodies. In terms of presence in food sources, the highest levels of mercury were in walleye and northern pike, which also ruined any potential for a recreational fishery in the reservoirs behind the three dams.

It was expected that as time passed and the drowned vegetation completely decomposed, the release of mercury would return to normal (and safe) levels. Monitoring, as reported by Chevalier et al. (1997) and Dumont et al. (1998), focused on the species most often consumed by the Cree people, including whitefish, Cisco, lake trout, northern pike, walleye, and longnose sucker. The results indicated that 15 years after the impounding of the LG2 reservoir, the concentrations of methyl mercury were higher than in natural lakes but also that they were decreasing in both predatory (northern pike) and non-predatory (whitefish) species.

Monitoring revealed that mercury concentrations in the hair of the Cree people had decreased and were stabilized and that for most of the population, these concentrations 'do not present a health risk' (Chevalier et al. 1997, 79). Dumont et al. (1998) reached a more qualified conclusion. While mercury levels had dropped in all communities among both sexes and in all age groups between 1988 and 1993–4, in their view, 'the significance of these mercury levels for the health of the Cree population remains unknown.' Their reasoning was that the decrease in mercury concentrations in humans could be due to the Cree having changed the type of fish eaten from contaminated to less contaminated fish or having decreased their total fish consumption, or both. A reduction in other types of bush food (because of game cycles) or new roads providing access to contaminated sites could lead to increases in future fish consumption. As a result, Dumont et al. (1998, 1,444) concluded that 'Communities with historically low [mercury] levels can thus experience increases rapidly. The present low mercury levels in the Cree communities may not be permanent and must not lull health authorities into believing that the mercury problem has been definitively resolved.'

2. *La Grande estuary fish.* The pre-construction impact assessments indicated that the estuarine fishery in La Grande was unlikely to survive the development of the dams and reservoirs. The impact study predicted that as a result of interruption of the water flow in the river during the construction and when the reservoirs began to fill behind the dams and the resulting absence of ice cover to dampen the impact of ebb and flow of the tidal water from James Bay, salt water would move farther up into the river. The consequence would be the elimination of the freshwater overwintering fish habitat for species important for the local fishery. On the other hand, if the river water flow were reduced *after* the formation of ice cover, then saltwater intrusion would be impeded, and a critically important pocket of fresh water could be maintained in the key habitat area.

Partly as a result of pressure exerted by local fishers, the river flow was not cut off until after an ice cover had formed on the river. Monitoring revealed that this action did result in the creation of the necessary freshwater pocket, which remained in place throughout the winter. The freshwater pocket was created even though no minimum flow was provided other than what entered the river from very minor tributaries. The outcome was that the predicted fish kill did not occur, and subsequent fish populations were about the same as in the pre-construction period. In this situation, the impact prediction was incorrect, but the outcome was positive for the fishery.

Limiting Factor Principle

*In Chapter 2, we discussed the **limiting factor** **principle**, which indicates that all factors necessary for growth must be available in certain quantities if an organism is to survive. We also noted that the weakest link is known as the dominant limiting factor. Are these ideas helpful in understanding the impact of interrupting river flow, and changing patterns of ice cover, on the overwintering fish habitat in La Grande estuary?*

Improving Impact Assessment

The James Bay experience reinforces the viewpoint that surprises should be expected in impact assessment and other aspects of environmental management, a point made in Chapters 1 and 6. Some time ago, Berkes (1988) concluded that little mystery surrounds some sources of uncertainty. Impacts do not occur simultaneously but over an extended period and are the result of decisions taken—filling schedules for reservoirs, release of 'excess' water through spillways, and including more or fewer turbines and powerhouses. Furthermore, impacts do not happen in isolation; in many instances, they combine to form *cumulative impacts*. Such a situation is particularly important in instances such as James Bay for which development is contemplated over decades for adjacent catchments on the eastern side of James Bay. The state-of-the-art in impact assessment is still very weak in terms of how to deal with such cumulative impacts and therefore how to prepare mitigation and compensation strategies.

Water Exports, Diversions, and Other Options

Growth and development in the southern United States, especially in the southwest, has the potential to lead to water scarcity in that region. A 'supply management' solution would be to look to Canada as a source and import water in bottles, by ship, or via pipeline. Canada exports oil and natural gas to the United States, so why not water? They are all natural resources, and Canadians have not hesitated to sell oil and natural gas.

As Day and Quinn (1992, 41–2) have explained, those who support the export of water to the United States argue that water is just another resource with value and can be exchanged on the open market, that Canada has more water than it needs to meet its foreseeable needs, that substantial income could be earned from selling water to the US and elsewhere, that jobs would be created through the major construction projects that would be necessary, and that some of the diverted water could be sent to regions in Canada facing shortfalls. These advantages, identified nearly 20 years ago, continue to be put forward by those advocating export of water.

Those who oppose the export of water emphasize that the scale or magnitude of the proposed water diversion projects would be much larger than any previous project, creating significant risks and uncertainty. Once the taps are 'turned on', it would be virtually impossible to turn them off, since receiving areas would have become dependent on the diverted water. In addition, negative environmental and social impacts could be significant and irreversible.

During the 1960s, various southern American state governments considered the feasibility of large-scale water diversions through pipelines from one or both of the Columbia and Mississippi river systems. The northern states that would be the source of these diversions strongly opposed the proposals. One outcome was that some states and private companies began to consider whether water from Canadian river or lake systems could be diverted southward. Two examples illustrate the scope and magnitude of these proposals (Table 11.1; Day and Quinn 1992, 40).

North American Water and Power Alliance (NAWAPA)

The NAWAPA project was the most publicized major diversion project from Canada to the United States. Conceived by the Ralph M. Parsons Company, an engineering firm in California, NAWAPA proposed to store the headwaters of the Yukon, Skeena, Peace, Fraser, and Columbia rivers in the Rocky Mountain Trench in eastern British Columbia. The stored water would then be diverted to both the Canadian Prairie provinces and western states by pipeline. The reaction from Canada was quick and negative. The Canadian federal and provincial governments strongly opposed the possibility of major water diversions from Canada to the US, rejecting Parsons's basic assumption that water in Canada was a 'continental resource'.

Grand Recycling and Northern Development (GRAND) Canal

Thomas Kierans, a Canadian engineer based in Newfoundland, proposed a major diversion that would involve building a dyke across James Bay, creating a reservoir in the bay,

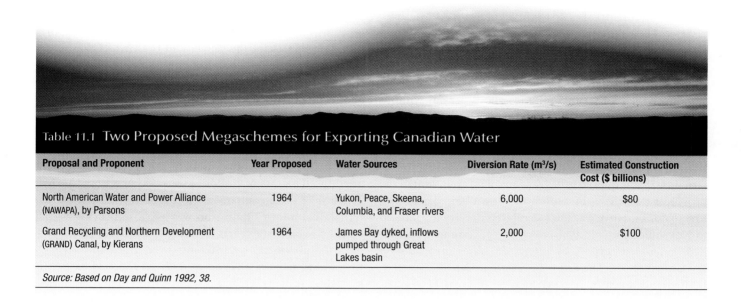

Table 11.1 Two Proposed Megaschemes for Exporting Canadian Water

Proposal and Proponent	Year Proposed	Water Sources	Diversion Rate (m³/s)	Estimated Construction Cost ($ billions)
North American Water and Power Alliance (NAWAPA), by Parsons	1964	Yukon, Peace, Skeena, Columbia, and Fraser rivers	6,000	$80
Grand Recycling and Northern Development (GRAND) Canal, by Kierans	1964	James Bay dyked, inflows pumped through Great Lakes basin	2,000	$100

Source: Based on Day and Quinn 1992, 38.

and then pumping water up 300 metres to move it into the Great Lakes basin from which it could be moved by pipeline to the southwestern states. The eight states (Minnesota, Wisconsin, Michigan, Illinois, Indiana, Ohio, Pennsylvania, New York) and two provinces (Ontario and Quebec) in the Great Lakes basin all opposed this proposal.

Other Options

Exports of water do not have to take place only via pipelines or major diversions. Some exports come about between communities adjacent to one another on either side of the Canadian-US border. In such situations, water supply systems are often shared between communities, such as between St Stephen, NB, and Calais, ME, or Coutts, AB, and Sweetgrass, MT. In these arrangements, water often flows in both directions, and the volumes are modest. As Day and Quinn (1992, 30) observed, in these situations, 'The operating principle appears to be mutual accommodation at the level of local government. National interest is not at stake.'

Another type of diversion involves movement of water between national and boundary waters. For example, in northern Ontario, diversions from Ogoki and Long Lake move water from Ontario rivers into Lake Superior in order to increase power-generating capacity. In the United States, a channel was constructed at Chicago to divert water from Lake Michigan to the Illinois-Mississippi river system to facilitate navigation southward from Chicago. While the amounts of water involved are modest in the Chicago Diversion, there is concern that it could set a precedent that would support arguments for larger-scale diversions out of Lake Michigan for other purposes.

Tanker shipment represents another option. It would involve containerized vessels taking fresh water from coastal rivers to destinations as close as California or as distant as Middle East nations. The arrangement would use ships bringing petroleum to Canada from the Middle East to transport fresh water on the return journey after suitable cleaning of holds. The volumes of water would be relatively small, and the cost of collection and transportation would have to be competitive with alternatives available at the destinations, such as desalinization. However, Canada does not import petroleum from Middle East oil states, so there are no such tanker movements.

Export of water in bottles or similar containers is a possibility. It does not raise concerns on the same scale because the quantities involved would not be significant and such trade could be stopped or modified with minimal consequences, unlike turning off the water passing through a major pipeline.

Reactions and Responses

The situation in the US is evolving. For example, Lasserre (2007, 152) has commented that 'water demand has been stagnating in the United States for the past two decades.' Several reasons account for this pattern. First, although still relatively low, water prices in the US have been steadily rising. Second, water-short cities are relatively wealthy compared to the agricultural sector and have been able to drive reallocation of water from agriculture to urban areas. Third, agriculture in the US is facing growing competition from producers in Mexico and Asia, leading to reduced crop production. As a result, Americans appear more likely to look to other solutions than to Canada for bulk imports.

In Canada and in the Great Lakes states, however, concern has persisted. This concern was reflected during 1985 when the Ontario and Quebec premiers joined the governors of the eight Great Lakes states in signing the Great Lakes Charter. The charter is a non-binding agreement, but those signing agreed to notify and consult each other regarding any possible diversion. In addition, they agreed that no jurisdiction would start a new or increase an existing diversion involving more than 5 million gallons per day without seeking the consent of all Great Lakes states or provinces that would be affected.

The Canadian federal government has taken various actions to curtail water export, drawing on its authority for international and interjurisdictional matters, even though the provinces own the water within their own boundaries. The first major initiative was the federal water policy introduced in November 1987. In that policy, the government prohibited 'large-scale' export by interbasin water diversions to the United States. The rationale was that Canada's water is limited and much of it is not located in areas of population concentration in Canada. In addition, the federal government recognized that large-scale diversions would have major environmental and social costs, especially in the Canadian North.

A different challenge emerged in the late 1980s when 'free trade' negotiations between Canada and the US were at their final stage. Concern was expressed that provisions for free trade would make it impossible for Canada to prohibit export of water to the United States because the free trade agreement would take precedence over Canadian policy or legislation. Reflecting this concern, in June 1988 the government of Ontario introduced a bill entitled the Water Transfer Control Act. After debate in the provincial legislature, the bill was amended to prohibit all forms of water export from Ontario, except for bottled water, and became law in 1989.

During 1988, when serious drought was affecting the US midwest and the Mississippi River was at record low flow levels, southern senators lobbied the US president to triple the diversion of water through the Chicago Diversion. The US Army Corps of Engineers concluded that such action would not be as effective as other alternatives, and as a result nothing was done. Nevertheless, the apparent willingness of some American political leaders to look to the Great Lakes system as a source of water to meet needs in the South caused concern in Canada. Acknowledging the concern, the Canadian government introduced a bill in August 1988 to ban large-scale water exports, defined as those above a daily average rate of one cubic metre per second, or an annual quantity of 20,000 cubic decametres. However, in September 1988, a federal election was called, and the bill did not become law.

Then in 1998, the Ontario government of Premier Mike Harris awarded a licence to the Nova Group to export water from Ontario. The Nova Group proposed to ship water from Sault Ste Marie to Asia. There was an immediate uproar across Ontario and in the Great Lakes states, which led the Ontario government to reverse its decision. The Ontario government's original willingness to approve water exports raised doubts about the capacity of the Great Lakes Charter to prevent such decisions, since it had clearly ignored the provisions of the charter. The incident also raised fears about the potential cumulative effect of many small water diversions from the Great Lakes to meet the growing water needs of cities in the northeastern and midwest US outside the Great Lakes basin.

One outcome was that in 1999, the federal government announced a strategy to address water exports from Canada. There were three components to the strategy. First, the International Joint Commission was asked to provide a legal opinion regarding exports from the Great Lakes basin. Second, the provincial governments were urged to pass laws that would ban water exports. This was viewed as necessary, since provinces own the water within their boundaries. And third, a federal law was to be designed to impose tight restrictions on exports.

With the exception of New Brunswick, all provinces subsequently passed legislation to prevent bulk water diversions or exports outside their borders. The door was still left open for other types of exports, such as bottled water. The risk is that if one province allows what might be viewed as significant water exports, then under the North American Free Trade Agreement (NAFTA), all provinces might be compelled to treat water as a 'tradable good', because under NAFTA, once a good is traded, it cannot be withdrawn from commerce unless all signatories decide to exclude it. The general view is that the US would be unlikely to agree to an exemption for water.

As Lasserre (2007) explained, rather than passing a law, the federal government chose a different approach. The choice was based on a feature of the river basins associated with the international border. Nearly all the large basins are located either entirely in Canada or the US, with only a few systems, such as the Great Lakes and the Columbia River, straddling the border. Given this physical reality, if provinces prohibited transferring water out of their jurisdiction, that would effectively stop water exports to the US. This arrangement could be reinforced by the federal government restricting interbasin transfers because of negative environmental impacts. Thus, Canada did not need to have a direct policy or law prohibiting water exports to the United States, either of which might be challengeable under NAFTA.

The federal government approved an amendment to the International Boundary Waters Treaty Act during December 2002. The amendment prohibits bulk water exports from river basins if they exceed 50 cubic metres per day. It was also intended to deter the diversion of water out of the Great Lakes basin, since the International Joint Commission's legal position in response to Canada's request for guidance was that diversions from boundary waters could only occur if they were authorized by the appropriate governments with jurisdiction for the water.

It also should be appreciated that in the US, some initiatives have been undertaken to restrict large-volume diversions from the Great Lakes. In 2000, the US government amended its Water Resources Development Act to stipulate that any diversion of water from the Great Lakes system by any state or federal agency or private organization for use outside of the Great Lakes basin is prohibited unless the governors of each of the eight Great Lakes states give approval. Further support for this policy came in December 2005 when the eight governors and two premiers, through the Council of Great Lakes Governors, agreed on a set of principles to review proposals for transfers of water from the Great Lakes. Any proposal would be authorized only if:

- no reasonable alternative to the proposed transfer exists, such as conserving existing supplies;
- withdrawals are limited to reasonable volumes for specified uses;
- all withdrawn water, after an allowance has been made for consumptive use, is returned to the Great Lakes basin from which it was removed;
- for 'major' proposals, an explicit conservation plan has been prepared.

The 2005 agreement by the Council of Great Lakes Governors effectively banned major water exports or diversions from the Great Lakes basin but left open the option for low-volume transfers that might be anticipated at a local level between adjacent communities on either side of the border.

Water Quality

Humans can adversely affect water quality in numerous ways, and water quality has been selected as one of the three main Environmental Sustainability Indicators by Environment Canada (Environment Canada, Statistics Canada, and Health Canada 2006). Water quality is assessed according to an index to measure the ability of surface waters to protect aquatic life at selected river and lake locations within Canada. Environment Canada (Environment Canada, Statistics Canada, and Health Canada 2006) has reported three key insights related to its water quality indicator based on information gathered up to 2005:

- Fresh water at the 359 monitoring stations located in southern Canada was rated as 'good' or 'excellent' at 44 per cent of the sites, 'fair' at 33 per cent, and 'marginal' or 'poor' at 23 per cent.

- Freshwater quality at the 36 monitoring sites in northern Canada was rated as 'good' or 'excellent' at 56 per cent of the sites, 'fair' at 31 per cent, and 'marginal' or 'poor' at 14 per cent.
- Phosphorus, derived primarily from human activities, is a primary concern in surface water quality in Canada. Phosphorus levels in southern Canada exceeded limits set under water quality guidelines for aquatic life more than half the time at 127 of 344 monitoring stations.

The most important issue is pollution from various sources but especially from industrial and other urban wastes and from agricultural runoff. The first two are easier to identify because they are usually associated with *point sources*, such as end-of-pipe sources at manufacturing plants or sewage treatment plants. Agricultural runoff is more challenging, since it is usually diffuse pollution from *non-point sources*, such as fertilizers, herbicides, and pesticides from farm fields. Some types of urban runoff, such as oil and salt from road surfaces, are also characterized as non-point, since they cannot be identified with specific places.

Industrial point-source pollution on the Calumet River, Chicago.

Point-source, end-of-pipe type of water pollution, Great Lakes Basin.

Municipal Waste Water

Municipal waste water can result in increased nutrient levels, often leading to algal blooms; depleted dissolved oxygen, sometimes resulting in fish kills; destruction of aquatic habitats with sedimentation, debris, and increased water flow; and acute and chronic toxicity to aquatic life from chemical contaminants, as well as bioaccumulation and biomagnification of chemicals in the food chain.

—Environment Canada 2003b, 34

Point Sources

Urban waste water from households, industries, other commercial operations, and institutions can be treated to varying degrees. Up to three levels of treatment exist: (1) *primary*, which removes only insoluble material; (2) *secondary*, which removes bacterial impurities from water previously having received primary treatment; and (3) *tertiary*, which removes chemical and nutrient contaminants following secondary treatment.

In 1999, 78 per cent of Canadians living in municipalities serviced by sewers had secondary and/or tertiary treatment, an increase from the 56 per cent with such service in 1983. Nine per cent of all Canadians had their wastes go to waste stabilization ponds (also known as sewage lagoons), simple treatment systems that provide the equivalent of secondary treatment. In 1998, 3.46 million Canadians living in larger municipalities were not connected to waste water systems and either used septic tanks or other arrangements to handle their wastes. Examples of ocean disposal and the different approaches taken by Victoria and Halifax are discussed in Chapter 8. Another 4.88 million people lived in smaller municipalities, and almost 57 per cent of them had no waste water treatment facilities. The percentages reported here for the late 1990s remain relevant near the end of the first decade of the twenty-first century.

A growing concern is that many waste water treatment facilities are old and need expensive maintenance or upgrading or replacement. The Canadian Water and Wastewater Association calculated that $5.4 billion in new investment is required *each year* between 1997 and 2012 to modernize and upgrade all existing water and waste water treatment facilities, as well as to provide such facilities to communities currently without them.

Industry is an important source of wastes deposited into water bodies. Tables 11.2 and 11.3 identify the most prevalent chemicals and the water bodies receiving more than 1,000 tonnes of pollutants in 2001. These data are from the National Pollutant Release Inventory (NPRI), which in 2006 tracked the release of more than 300 substances into the environment. Ninety per cent of the 9,000 facilities that reported chemical releases in 2006 were in Alberta (3,717) and Ontario (2,379) (http://www.ec.gc.ca/pdb/npri/npri_factsheet06_e.cfm).

Ammonia and nitrogen represented more than 94 per cent of the total releases to water. Other chemicals such as mercury are released in much smaller amounts but have serious negative impacts on human and aquatic system health. Mercury bioaccumulates and biomagnifies (see Chapter 10) in the liver, kidneys, and muscles of affected organisms, and chronic exposure can result in brain and kidney damage. Monitoring shows that mercury levels in the Canadian environment continue to rise and that the main sources are metal mining and smelting, waste incineration, and coal-fired power plants.

Runoff from urban areas either flows directly into water bodies from roads and other non-point sources or can be channelled by stormwater systems. Stormwater can contain various contaminants such as suspended solids, sediment, and grit; nutrients, including different forms of phosphorus and nitrogen; toxic metals, including copper, lead, and zinc; hydrocarbons, including oil, grease, and polycyclic aromatic

Table 11.2   Top Releases of Chemicals to Water, 2001

Chemical	Releases (tonnes)
Ammonia (total)*	26,106
Nitrate ion in solution at pH equal to or greater than 6.0	22,450
Manganese (and its compounds)	1,157
Methanol	697
Zinc (and its compounds)	308

*Total includes both ammonia (NH_3) and ammonium ion (NH_4^+) in solution.
Source: Statistics Canada 2003, 18.

Table 11.3 Water Bodies Receiving More Than 1,000 Tonnes of Pollutants, 2001

Water Body	Total Release (tonnes)	Dominant Release	Share of Total Release (%)
Fraser River	9,168	Ammonia*	49.2
Lake Ontario	8,877	Ammonia*	41.6
Bow River	8,264	Nitrate ion	90.8
Ottawa River	3,066	Ammonia*	76.6
North Saskatchewan River	2,953	Nitrate ion	61.3
Red River	2,766	Ammonia*	72.7
Hamilton Harbour	1,516	Ammonia*	70.6
South Saskatchewan River	1,275	Nitrate ion	62.4
St Lawrence River	1,086	Nitrate ion	43.6

*Total includes both ammonia (NH_3) and ammonium ion (NH_4^+) in solution.

Source: From Statistics Canada, Human Activity and the Environment: Annual Statistics 2003. *Cat. no. 16-201, 3. Dec. 2003. Ottawa: Statistics Canada, 18.*

hydrocarbons; trace organic contaminants, including pesticides, herbicides, and industrial chemicals; and fecal bacteria. As a result, it is desirable that stormwater be treated in municipal waste water plants along with waste from homes, industries, and other urban institutions. Unfortunately, such treatment does not always occur.

A third important source of wastes into water bodies is from agricultural activity, but that is more appropriately discussed under the category of non-point sources.

Non-point Sources

As discussed in Chapter 10, crop and livestock production has increased significantly as a result of more effective farm machinery, new genetic crops, agrochemicals, and irrigation. The latter two also contribute to environmental impacts, especially through fertilizers, pesticides, and herbicides being carried in runoff from farm fields, which ends up in streams, rivers, and lakes. In this section, we examine the experience with diffuse pollution in the Great Lakes basin.

Diffuse pollution has been recognized as a policy issue in the Great Lakes basin. Since the early 1960s, interest has evolved from concern about sedimentation from soil erosion and eutrophication from phosphorus and nitrate loading to persistent toxic chemicals. As the definition of the problem has evolved, so have ideas regarding appropriate responses.

What has been learned about the strategic implications of how the problem has been defined? One of the key lessons is that diffuse pollution represents a 'layered' problem—that is, it is much like an onion, with many layers. Too often with such problems, attention does not go beyond the first layer, although there is a need to go much deeper. At the first layer, concern about diffuse pollution focuses on *environmental degradation* and the *economic costs* imposed on downstream users. The motivation for defining the problem in this manner appears to be that people will see the connection between diffuse pollution and loss of economic production or increased costs for economic production. A second layer, now given greater attention, is the link between diffuse pollution and negative impacts on *ecosystem health* or *integrity* and especially on *human health*. It is believed that making the connection to human health should create a powerful image in the minds of both policy-makers and residents in any area affected by diffuse pollution. A third layer is the interpretation of diffuse pollution as a problem touching on human *values*, *beliefs*, *attitudes*, and *behaviour*. In other words, from this perspective, the fundamental problem is behaviour by individuals and groups, driven by inappropriate values, beliefs, and attitudes. If attention is focused on this third level, then the prescription to resolve diffuse pollution is certainly different from what it would be if attention were limited to the first level.

What mechanisms have been effective in achieving recognition of diffuse pollution as a policy issue? From experience in Canada in general and the Great Lakes basin in particular, it is clear that there is a need is for *credible science* to document the nature of the problem.

Phosphorus Loadings in Canada

For Canada as a whole, estimated yearly loadings of phosphorus fell by 44 per cent between 1983 and 1999, despite the 24 per cent increase in urban population.

—Environment Canada 2003b, 35

Credible Science and Institutional Commitment

Appreciation of diffuse pollution as a policy issue has been encouraged in Canada and the Great Lakes basin by a combination of science, institutions, and individuals. Several initiatives by the **International Joint Commission**, a bilateral institution created to manage interjurisdictional water issues between Canada and the United States, have been significant.

During the 1960s, the media declared that 'Lake Erie is dying', a reference to the highly eutrophic state of that lake (see Chapter 4). In 1972, the governments of Canada and the United States entered into an agreement to restore and enhance water quality in the Great Lakes. Initial attention focused on reducing phosphorus loading from municipal sewage treatment plants and other point sources. Initiatives in that regard were effective, but it was suspected that non-point sources might also be significant. However, data were not available to indicate the importance of these sources.

Under the 1972 agreement between Canada and the United States, the International Joint Commission was asked 'to conduct a study of pollution of the boundary waters of the Great Lakes System from agricultural, forestry, and other land use activities'. Subsequently, an intensive inquiry was completed by the International Reference Group on Great Lakes Pollution from Land Use Activities, otherwise known as PLUARG. PLUARG examined two major pollution problems in the basin: eutrophication from elevated nutrient inputs, especially in the lower lakes (Erie and Ontario), and increasing contamination by toxic substances. PLUARG studied the pollution potential from various land uses, including agriculture, urbanization, forestry, transportation, and waste disposal, as well as natural processes such as lakeshore and riverbank erosion.

In its final report, PLUARG concluded that depending on the magnitude of point-source loads, the 'combined land drainage and atmospheric [non-point] inputs to individual Great Lakes ranged from 32 per cent (Lake Ontario) to 90 per cent (Lake Superior) of the total phosphorus loads (excluding shoreline erosion). Phosphorus loads in 1976 exceeded the recommended target loads in all lakes' (International Reference Group on Great Lakes Pollution from Land Use Activities 1978, 4–5). The PLUARG study was the first credible science to document the important contribution of non-point sources to phosphorus loading. Produced by a group of international scientists and government officials for a respected bilateral institution, these results were difficult to ignore. The PLUARG report also stated that toxic substances such as PCBs were entering the Great Lakes system 'from diffuse sources, especially through atmospheric deposition. Through land drainage, residues of previously used organochlorine pesticides (e.g., DDT) are still entering the boundary waters in substantial quantities.' In terms of the sources of these loadings, it was reported that 'intensive agricultural operations have been identified as the major diffuse source

contributor of phosphorus.' In addition, 'Erosion from crop production on fine-textured soils and from urbanizing areas, where large-scale land developments have removed natural ground cover, were found to be the main sources of sediment. Urban runoff and atmospheric deposition were identified as the major contributors of toxic substances from non-point sources' (1978, 6).

Regarding necessary actions, PLUARG concluded that

remedying non-point source pollution will be neither simple nor inexpensive. Non-point sources of water pollution are characterized by their wide variety and large numbers of sources, the seemingly insignificant nature of their individual contributions, the damaging effect of their cumulative impact, the intermittent nature of their inputs, the complex set of natural processes acting to modify them, and the variety of social and economic interactions which affect them.

. . . The level of awareness about pollution from non-point sources among Great Lakes Basin residents is inadequate at present. Control of non-point sources will require all basin residents to become involved in reducing the generation of pollutants through conservation practices. Improved planning and technical assistance are prerequisites to long-term solutions of land drainage problems (1978, 10–11).

The PLUARG report, and other analyses, led to renewal of the Great Lakes Water Quality Agreement in 1978 and to the signing in 1987 of a protocol amending the 1978 agreement. The 1987 amendments made the agreement broader in scope than the previous agreements, but diffuse pollution was still recognized as one of the priority problems. For example, in commenting on the 1987 protocol, the International Joint Commission remarked that through it, 'Specific improvements seek to control air-borne toxics, to deal with contaminated sediments, to designate and remedy particular areas of concern, to reduce pollution from non-point sources, to protect groundwater, and to develop water quality objectives for each of the Great Lakes' (Canada and Ontario 1988, 5).

More specifically, Annex 13 of the protocol focused exclusively on 'pollution from non-point sources' and identified 'programs and measures for abatement and reduction of non-point sources of pollution from land-use activities' in order 'to further reduce non-point source inputs of phosphorus, sediments, toxic substances, and microbiological contaminants contained in drainage from urban and rural land, including waste disposal sites, in the Great Lakes System' (Canada and Ontario 1988, 55).

Many observers would agree that the PLUARG study commissioned by the International Joint Commission, along with the commission's prestige and watchdog role, was significant in helping both elected officials and the public to understand the severity of the diffuse pollution problem in the Great Lakes. Without such a credible voice to draw attention to the issue of non-point source pollution, it is unlikely that action would have been forthcoming.

Agricultural Non-point Source Pollution

After the PLUARG studies were completed, the governments of Canada and the United States agreed to deal with the issue of high phosphorus loadings from rural non-point sources. In Canada, the federal and Ontario governments in 1987 created a five-year, cost-shared program—the Soil and Water Environmental Enhancement Program, or SWEEP. The purpose was to meet the target reduction of 200 metric tons per year for Canada for phosphorus loading in Lake Erie by 1990 from non-point sources.

SWEEP consisted of various programs. The first focused on technology evaluation and development and was intended to stimulate adoption of soil management and cropping practices needed to improve water quality and to reduce soil erosion and degradation. A second thrust focused on pilot watershed programs, local demonstrations, and technical assistance at the farm level. The pilot watershed program was particularly innovative, with emphasis on testing the effectiveness of state-of-the-art conservation practices on working farms. The effects of using different practices on all farms in three experimental watersheds were compared with conventional practices in three control watersheds. Control and treatment watersheds were evaluated for impacts on water quality, hydrology, soil quality, crop production, and economics at the farm level. A third component involved information services, with attention to informing the public about the nature and consequences of soil and water quality problems and about the SWEEP objectives. Cressman (1994, 421), whose consulting firm was actively involved in the pilot watershed studies, later remarked that 'It was during this time that interest in, and the practice of, conservation tillage grew significantly among Ontario farmers.'

In parallel with SWEEP, the Ontario Land Stewardship Program was introduced by the Ontario Ministry of Agriculture and Food. It was a three-year, $40-million program that provided financial incentives for first-time adoption of conservation measures on farmland. Financial assistance was provided for practices to protect soil structure, build structures to ameliorate soil erosion, purchase conservation equipment, and obtain technical training. Funds also were dedicated to research projects related to stewardship practices. The five-year SWEEP program (1987–91) overlapped with the National Soil Conservation Program, a three-year, $150-million cost-shared program. When SWEEP and the National Soil Conservation Program both terminated in the early 1990s, a newly elected federal government introduced another program—the Green Plan—which provided funds for soil conservation and diffuse pollution control. However, by that time attention had shifted from soil erosion, sedimentation, and eutrophication to toxic substances. One observer remarked to one of the authors in 1999 that in his view, the three programs had been developed with little consultation among key agencies, resulting in duplication and overlap and a time frame not long enough to allow measurement of program impacts. Indeed, the monitoring for soil erosion, sedimentation, and eutrophication had been cursory rather than systematic—*perceived* results and the provision of support to farmers for production improvements took precedence over ameliorating environmental degradation problems.

Some positive initiatives were undertaken and results achieved. In the Lake Erie and Lake Ontario watersheds, the loads from phosphorus have been reduced significantly as a result of initiatives dealing with both point and non-point sources. The main initiatives for non-point sources helped farmers to modify how their land was cropped, especially by encouraging the use of conservation tillage to reduce erosion and thereby reduce sediment and toxics placed into aquatic systems. The main actions regarding point sources involved upgrading municipal sewage treatment plants, and regulations were established to reduce phosphorus in laundry detergents. The latter action is consistent with the principles of anticipation and prevention—that is, to minimize a problem in the first place rather than to pursue remedial measures after the pollutants enter the environment.

Box 11.1 provides a summary of the environmental status of the Great Lakes, the St Lawrence River, and the St Clair River–Lake St Clair–Detroit River ecosystem in 2006. Environment Canada and the US Environmental Protection Agency (2007, 3) concluded that 'the overall status of the Great Lakes ecosystem was assessed as mixed because some conditions or areas were **good** while others were **poor**. The trends of Great Lakes ecosystem conditions varied: some conditions were ***improving*** and some were ***worsening***' (emphasis in original).

Water Security: Protecting Quantity and Quality

A central concern in water management is to ensure a sufficient quantity of water of adequate quality for human use. At the start of the twenty-first century, it was estimated that 1.1 billion people did not have access to safe water supplies, and two out of five people did not have access to adequate sanitation, notwithstanding substantive efforts during the United Nations International Drinking Water Supply and Sanitation Decade throughout the 1980s to improve conditions.

In terms of current per capita water use, the range extends from as little as 20 litres to more than 500 litres each day. Only 4 per cent of the world's population use water in the range of 300 to 400 litres per person per day, with people in the United States, Canada, and Switzerland being the highest per capita users. In contrast, about two-thirds of

Box 11.1   State of The Great Lakes in 2006

Some of the good features of the ecosystem leading to the *mixed* conclusion by Environment Canada and the US Environmental Protection Agency include:

- Levels of most contaminants in herring gull eggs and predator fish continue to decrease.
- Phosphorus targets have been met in Lake Ontario, Lake Huron, Lake Michigan, and Lake Superior.
- The Great Lakes are a good source of treated drinking water.
- Sustainable forestry programs throughout the Great Lakes basin are contributing to environmentally friendly management practices.
- Lake trout stocks in Lake Superior have remained self-sustaining, and some natural reproduction of lake trout is occurring in Lake Ontario and in Lake Huron.
- Mayfly (*Hexagenia*) populations have partially recovered in western Lake Erie.
- An agreement has been signed between Ontario and the federal government to establish Canada's first park under the National Marine Conservation Act on the north shore of Lake Superior (see Chapter 8). More than 10,000 square kilometres in size, it will be the world's largest freshwater protected area.

Some of the negative features of the ecosystem leading to the *mixed* conclusion include:

- Concentrations of the flame retardant PBDEs are increasing in herring gull eggs.

- Nuisance growth of the green alga *Cladophora* has reappeared along the shoreline in many places.
- Phosphorus levels are still above guidelines in Lake Erie.
- Non-native species (aquatic and terrestrial) are pervasive throughout the Great Lakes basin, and they continue to exert impacts on native species and communities.
- Populations of *Diporeia*, the dominant native bottom-dwelling invertebrate, continue to decline in Lake Michigan, Lake Huron, and Lake Ontario, and they may be extinct in Lake Erie.
- Groundwater withdrawals for municipal water supplies and irrigation, and the increased proportion of impervious surfaces in urban areas, have negatively affected groundwater.
- Long-range atmospheric transport is a continuing source of PCBs and other contaminants to the Great Lakes basin and can be expected to be significant for decades.
- Land-use changes in favour of urbanization along the shoreline continue to threaten natural habitats in the Great Lakes and St Lawrence River ecosystems.
- Some species of amphibians and wetland-dependent birds are showing declines in population numbers, in part because of wetland habitat conditions.

Source: Environment Canada and US Environmental Protection Agency 2007, 3.

the global population get by on less than 50 litres for each person daily.

To place the above information in perspective, most humans become thirsty after losing only 1 per cent of their bodily fluid and are in danger of death once the loss approaches 10 per cent. The minimum water requirement to replace loss of fluid for a normal healthy adult in an average temperate climate is about three litres each day. In tropical or subtropical conditions, the minimum amount becomes about five litres per person per day.

Most Canadians receive their drinking water from the 4,000 municipal water treatment plants across the country, but a significant number depend on private wells or other arrangements. About 9 million Canadians, most living in small towns or rural areas, draw on groundwater for their drinking water.

The relative abundance of water in Canada, the high levels of water use, and the myth of superabundance, all referred to at the beginning of this chapter, made most Canadians complacent about the adequacy and safety of their water supplies.

High Water Use by Canadians

One of the key factors explaining the high residential consumption rates is the lack of financial incentive to Canadian households to use less water. For instance, in 1999, unmetered households, which pay a flat rate for water, used 50 per cent more water than metered households, which pay for water by volume used. About 57 per cent of Canada's municipal population had water meters in 1999, showing a gradual increase since 1991.

—Environment Canada 2003b, 30

Groundwater Quality

Groundwater sources generally provide water that is safe to drink. This is especially true if the well field is protected from pollutants. Aquifers that are close to the surface are more prone to contamination by pollution, which partially explains the poor water quality that is characteristic of many shallow wells in Canada.

—Statistics Canada 2003b, 24

For many, this all changed in mid-May 2000 when the small town of Walkerton in southwestern Ontario, with a population of about 5,000, experienced contamination of its water supply system by deadly bacteria, *Escherichia coli* O157:H7, or **E. coli** as it is often called, and *Campylobacter jejuni*. Children under five years of age and the elderly are at greatest risk. Seven people died, and more than 2,300 became ill. It is not known for certain but is anticipated that some individuals who became sick in Walkerton, especially children, may experience effects for the rest of their lives.

The concern generated by the Walkerton experience was reinforced during March 2001 in North Battleford, SK, a community of 14,000, where thousands of residents suffered from contamination of the municipal water system by the parasite *cryptosporidium*. The parasite got into the water supply system over a period of three weeks following routine maintenance of a chemical filter at the treatment plant. Residents of the city were under a boil-water order for three months. An inquiry

Bruce Mitchell

Walkerton water memorial.

ordered by the provincial premier concluded that the Saskatchewan government had not been effective in safeguarding drinking water in the province.

The Walkerton Inquiry

A public inquiry into the Walkerton tragedy by Justice Dennis O'Connor (2002a; 2002b) established that:

1. The E. coli, contained in manure spread on a farm near one well of the Walkerton water supply system, entered the system through that well.
2. The farmer who spread the manure followed proper practices and was not at fault.
3. The outbreak caused by the contamination would not have occurred if the water had been treated. The water was not treated because the chlorination equipment was being repaired.
4. The provincial government's approvals and monitoring programs were inadequate.
5. In addition to lack of training, the operators of the well system had a history of improper operating practices.
6. When people began to fall ill, the general manager of the water system withheld from the public health unit critical information about adverse water quality test results. This resulted in delay of a boil-water advisory.
7. Budget reductions by the provincial government had led to closure of government laboratory testing services for municipalities, and private laboratories were not required to submit adverse test results to the Ministry of the Environment or to the medical officer of health.

Walkerton: Lessons and Recommendations

In part two of his report, Justice O'Connor offered recommendations to ensure the safety of drinking water across the province. We examine some of them here, since they are relevant to all regions of Canada. Overall, he recommended a **multi-barrier approach** to drinking water safety. In his words, 'Putting in place a series of measures, each independently acting as a barrier to passing water-borne contaminants through the system to consumers, achieves a greater overall level of protection than does relying exclusively on a single barrier (e.g., treatment alone or source protection alone). A failure in any given barrier will not cause a failure of the entire system' (O'Connor 2002b, 5). He argued that in a multi-barrier approach, the first barrier involves selecting and protecting reliable, high-quality drinking water sources. As a result, he recommended 'a source protection system that includes a strong planning component on an ecologically meaningful scale—that is, at the watershed level' and that 'the Province adopt a watershed-based planning process' (6, 3). Within a watershed-based approach, he recommended that:

1. A comprehensive approach for managing all aspects of watersheds was needed.

2. A framework for developing watershed-based source protection plans should be established.
3. To ensure that local considerations are fully taken into account and to create goodwill in and acceptance by local communities, source protection planning should be undertaken as much as possible at a local (watershed) level by those who will be most affected (municipalities and other affected local groups).

The multi-barrier approach has subsequently been endorsed by other federal and provincial governments. For example, at a national level, Environment Canada, Health Canada, the Canadian Council of Ministers of the Environment, and the Committee on Environmental and Occupational Health collaborated to prepare what was labelled a multi-barrier approach. Details about this initiative can be found on the website of the Canadian Council of Ministers of the Environment (http://www.ccme.ca/sourcetotap/) under 'Source to tap—Protecting our water quality'.

First Nations Water Security

The events at Walkerton and North Battleford drew attention to structural and human resource issues related to reliable and safe water supply systems in Canada, systems that most citizens had previously taken for granted. Events in October 2005 at the Cree First Nations community of Kashechewan close to the western shore of James Bay, however, highlighted the 'Third World' conditions regarding water on many First Nations reserves across the country. This situation was verified by the Commissioner of the Environment and Sustainable Development, who observed that 'When it comes to the safety of drinking water, residents of First Nations communities do not benefit from a level of protection comparable to that of people who live off reserves' (Auditor General of Canada 2005a, ch. 5, p. 1).

The commissioner (Auditor General of Canada 2005a, ch. 5, p. 11) reported that information assembled by the Department of Indian and Northern Affairs indicated that in 2004, about 460,000 First Nations people were living on some 600 reserves across the country. In 1995, data showed that 25 per cent of reserves had water systems that 'posed potential health and safety risks to the people they served'. In 2001, when Indian and Northern Affairs completed its next assessment, it found that 75 per cent of the water systems on First Nations reserves posed 'significant risk to the quality of the safety of drinking water'.

This serious situation regarding poor water quality has been attributed to inconsistent testing and inadequate training of equipment operators. Others, while agreeing with this conclusion, have noted that many non-reserve communities in rural Canada also have poor water supply services. Indeed, in December 2006, Health Canada reported to the Senate that 1,174 boil-water advisories were in place at that time, a remarkably high number for a nation that usually perceives itself to be 'developed'. In early November 2006, the CBC News reported that at that time, nearly 100 reserves across Canada were under boil-water advisories and one, on an island off the west coast of Vancouver Island, had had an advisory in effect for nine years.

Persistent First Nations Water Problems

By the end of February 2008, 93 First Nations communities had boil-water advisories or 'do not consume' orders in force. Some, such as the 282 residents of the Neskantaga First Nation in northwestern Ontario that was issued an advisory in 1995, have been operating under them for many years. Furthermore, in January 2008, 85 First Nations communities were flagged as having high-risk drinking water systems.

Source: Eggertson 2008, 985.

Boil-Water Advisories in Canada

In 2008, the Canadian Medical Association reported that 1,760 boil-water advisories were in place across the country, in addition to those in place on 93 First Nations reserves. The number of advisories by province, in descending order, were: Ontario 679; British Columbia 530; Newfoundland and Labrador 228; Saskatchewan 126; Nova Scotia 67; Quebec 61; Manitoba 59; Alberta 13; New Brunswick 2; Northwest Territories 1; Prince Edward Island, Nunavut, and Yukon 0.

The advisories included those for communities, commercial facilities, and trailer parks. Some had been in place for at least five years.

The Federation of Canadian Municipalities estimates that about $31 billion is needed to upgrade water and waste water treatment facilities across the country.

Source: CBC and CTV News, 7 and 8 April 2008.

But Kashechewan, a community of 1,900 people located on the Albany River some 400 kilometres north of Timmins, ON, became a flashpoint for the poor water services in many remote and distant communities. Kashechewan was established by the federal government in 1958 on the flood plain of the Albany River near the shore of James Bay because supply barges were unable to travel through rapids on the Albany River to a site further upstream preferred by the Cree people.

The present water treatment plant was built in 1996, but the intake was located downstream from large sewage lagoons that leach continuously into the creek containing the water supply intake. Furthermore, tidal action from James Bay pushes waste-laden water into the creek, past the water intake. A further challenge was that the plant was run by local operators who were being trained on the job, meaning that they did not have sufficient background or expertise to recognize serious problems.

A contributing factor to the contamination of Kashechewan's drinking water in October 2005 was the failure of a chlorine pump due to an air lock. An emergency back-up system should have taken over but did not because it had not been connected. In most modern plants, an emergency paging system alerts operators to a system malfunction, but such a system had not been installed. Thus, the sequence of events leading to contamination was similar to that at Walkerton, where chlorine equipment was not working and operators did not have the necessary training to realize the implications.

On 12 October 2005, Health Canada discovered unacceptable levels of E. coli in the treatment system but did not alert the band office for two days. Once a qualified contractor arrived at the treatment plant, it took about six hours to have the equipment operating properly, and safe potable water was being provided by 22 October. Nevertheless, nearly 1,000 community members experienced negative side effects—severe skin infections, which frequently occur from exposure to high levels of chlorine necessary to purify the water. All 1,900 residents needed vaccinations for hepatitis A and B. Furthermore, many residents had scabies and impetigo because of the ongoing poor water quality, conditions exacerbated by high chlorine levels.

A further concern was that in 2003, the Ontario Clean Water Agency had alerted the government to problems with the water treatment system at Kashechewan. The next year, the provincial minister of health and the minister of community safety visited the community, but no action was taken. The view of the provincial government was that because First Nations reserves are under the jurisdiction of the federal government, it was the responsibility of federal departments, specifically Health Canada and Indian and Northern Affairs, to take action.

The short-term solution was to evacuate residents needing treatment and care. The evacuation began on 26 October, and by 2 November 815 people had been removed to Ottawa (245), Sudbury (251), Cochrane (206), Timmins (50), Attawapiskat (43), and Moosonee (20). The cost of the evacuation was estimated at about $16 million. The provincial minister of natural resources, David Ramsay, attributed the delay in evacuating people to 'a jurisdictional misunderstanding as to who should have the lead on this'. In addition to the evacuation, other interim steps were taken to ship bottled water to the community, send certified water treatment operators to assess and repair the treatment plant, assess the quantity and quality of the water in the river from which the treatment plant takes water, and assess the state of the sewage treatment lagoon.

Subsequently, the federal government examined various long-term solutions, including a $200-million package over five to seven years to reinforce a dyke to reduce flood damage vulnerability as well as to construct better drainage systems to protect low-lying areas; a $200-million project to relocate the community to Timmins; or a $500-million initiative to move the entire reserve to a new location on higher ground. At the end of July 2007, the federal minister of Indian and northern affairs announced that his government would rebuild and redevelop the low-lying reserve on its present location, since relocation was too expensive. Enhancing a dyke at a cost of $40 million would be the first action taken, followed by improvements to other community infrastructure and housing.

Flooding in Kashechewan in 2008

On 25 April 2008, Kashechewan was again evacuated, this time because of flooding. Hundreds of the most vulnerable residents were evacuated by plane and were housed in motels, hotels, or other types of accommodation in Cochrane, Greenstone, Kapuskasing, Hearst, Sault Ste Marie, and Thunder Bay. Subsequently, about 1,900 people were evacuated from Kashechewan and Fort Albany, with 1,000 sent to Stratford and nearby communities such as St Mary's, Mitchell, and Milverton. The plan was to accommodate them for up to three weeks, but many were able return in less than a week. This was the fourth evacuation caused by flooding for Kashechewan since 2004 and the first for Fort Albany. In mid-May, residents of the Attawapiskat area on the shores of James Bay were evacuated because of flooding and were housed in residences at the University of Waterloo in southern Ontario.

In light of what happened at Kashechewan in 2005, it should be noted that as a result of the Walkerton experience in 2000, the federal government had initiated a First Nations Water Management Strategy in 2003, with a budget of $600 million

Tributary joining the Albany River at Kashechewan, ON.

Dyke and drain control between the Albany River and Kashechewan, ON.

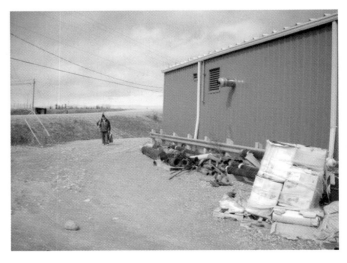

New water treatment plant beside the dyke, Kashechewan, ON. (All photos courtesy Dan McCarthy, 29 May 2008.)

spread over five years. The main purpose was to improve the quality and safety of drinking water on First Nations reserves through developing comprehensive policies, guidelines, and standards; educating on-reserve residents about drinking water issues; clarifying roles and responsibilities; building and upgrading water systems to meet standards; improving operation and maintenance; providing training to operators; and expanding water testing. Given what happened at Kashechewan and continuing conditions on many other First Nations reserves, it appears that much more needs to be done.

Supply Management, Demand Management, and Soft Path

Water security can be achieved through various approaches. The best known are supply management and demand management, but an emerging approach called soft path is also considered below. More details can be found in Gleick (2003), Brooks (2005), and Brandes and Brooks (2005).

Uncertainty and Choices

Water problems (scarcity, flooding, pollution) have no single solution: options to deal with scarcity include supply augmentation through the mobilization of more resources through capital-intensive projects; efforts to conserve water; or redefining allocation to users. All these options have political and financial implications. They all come with risks, costs, and benefits, private or public, which strongly shape what solutions particular stakeholders are likely to push for.

—Molle 2007, 361

Supply Management

Supply management is the traditional approach. When a water shortage is anticipated, the solution is to develop a new source of supply, and this is normally accomplished through either augmenting an existing supply (e.g., raising the height of a dam in order to be able to impound more water) or developing a new supply (e.g., a new dam and reservoir, new wells, a pipeline to a new source [a lake or river], a desalinization plant). The rationale for this approach is that populations will grow and the economy will expand and each requires additional water supplies. Without incremental water, human well-being and development will be impeded.

Supply management has served societies well when it has resulted in making an adequate supply of suitable-quality water available to meet demands. There are also some downsides. If people believe that additional supplies will always be found to meet demands, there is little incentive to avoid wasteful water use practices or to adopt active water conservation measures. One consequence is that a society may invest more money than would otherwise be required to ensure that water needs are met. Such additional costs may be significant, since the most accessible and therefore low-cost sources have usually already been developed, so the new incremental supplies are often more expensive on a per capita basis. Another consequence is that the construction of new dams and reservoirs or pipelines may have significant environmental impacts at local and regional scales.

Given the above mix of benefits and limitations of a supply management approach, it is worthwhile to consider other approaches that can be used in combination with or in place of supply management.

Demand Management

If the basic approach in supply management is to manipulate the natural system to create new sources of supply, the approach in **demand management** is to influence human behaviour so that less water is used. If the latter is achieved, then existing sources may satisfy demand for a longer period of time.

Various methods can be used to influence human behaviour in water use. The most basic is to use pricing as a signal to users that water has a cost and that by using less water, people can save money. There are a number of pricing systems. For example, volume-based pricing can be designed so that as consumers use more water, they pay an increasingly higher per unit charge. For this system to work, all use, whether in homes, offices, institutions, manufacturing plants, or farms, has to be metered. However, significantly higher prices for water could disadvantage the poorest members of a society, who may find it difficult to pay for the minimal amounts of water they require to meet basic needs. To avoid inequities such as this, other social policies must be in place to ensure that vulnerable people are not in jeopardy. This is particularly important with regard to water, because there is no alternative to water for meeting basic needs.

Another incentive is to offer price rebates for the purchase and installation of water-saving devices such as low-flow showerheads or toilets. Low-flow toilets, for example, use on average about 75 per cent less water per flush than a 'regular' toilet. Since showers and baths account for 35 per cent and toilet flushing 30 per cent of water used in Canadian households, reducing volumes of water for showers, baths, or toilet flushing can lead to significant reductions in water used. Further savings can be realized if previously used water, or grey water, is used for waste disposal, rather than using drinking-quality water for toilet flushing.

A third incentive is to restrict outside water use (watering of lawns, flower beds) by regulation during the hottest months when water use peaks. If the high peaks of water use in the hot months can be reduced, there would be no need to invest in additional capacity that would only be required during a few months each year.

A fourth measure in demand management is to inform and educate water users so that, over time, they reduce water use. For example, people can be educated to turn off the faucet when brushing their teeth except when rinsing the toothbrush or to turn off the water in a shower except when initially soaking or rinsing off soap. Other practices include running a dishwasher or washing machine only with a full load.

Promoting a Culture of Water Conservation

To date, progress on water conservation in Canada has been limited and piecemeal. Some municipalities now incorporate water efficiency programs into management strategies, and others, like Calgary and Victoria, are looking to meet all future growth in demand through conservation. At the same time, rapidly growing centres across the nation continue to follow the old, hard path of supply-side management, developing new water sources and investing in engineered solutions and infrastructure such as reservoirs, powerful pumps, and large pipelines.

—Morris et al. 2007, 39

Demand management is not a new concept, but it has often been introduced as a second-level approach after decisions focused on supply management have been taken. An ideal approach would combine the two in a well-integrated system.

Soft Path

The **soft path** approach is an extension of demand management. As Brandes and Brooks (2005, 9) explain, soft path aims to improve water use efficiency but goes further than demand management by challenging basic patterns of consumption. In their view, demand management emphasizes the question of 'how', or how to do the same with less water. In contrast, the soft path approach addresses the question of 'why', or why

water is even used for a function. The 'why' question normally leads to consideration of a broader range of methods.

Brandes and Brooks (2005, 9) provide several examples to illustrate the significance of the 'why' question. In their words,

- Why . . . do we use water to carry away our waste? Demand management would urge low-flow toilets, but waterless systems are available—perhaps not for homes (because of the need for regular maintenance) but certainly for larger buildings.
- Why do we use half the potable water that is piped to a house in the summer for watering lawns and gardens—and sidewalks? Demand management would urge more efficient sprinklers with automatic shut-offs, maybe even water restrictions. The soft path goes further: recycling water from bathtubs and washing machines or, better yet, drought-resistant greenery that requires little or no watering once it is established.

The soft path approach is based on four basic principles, which distinguish it from supply management (Brandes and Brooks 2005, 10–13):

1. *Water is treated as a service rather than as an end in itself.* In the soft path, water is not viewed as the final product, other than for a few human uses (drinking, washing) and for support of ecosystems. Instead, water is viewed as a means to accomplish specific functions, including sanitation, farm production, or yard maintenance. This principle moves managers away from thinking only in terms of providing more water for activities and instead to think about alternative ways that services traditionally supported by water might be achieved. For example, the end is not to flush toilets or irrigate crops but to dispose of wastes or to grow food.

2. *Ecological sustainability is a fundamental criterion.* Ecosystems are viewed as legitimate users of water and also as one foundation of economies. Consequently, it is necessary to include consideration of ecosystem health and ecosystem resilience when calculating the benefit-cost ratio of solutions to meet water demand (Box 11.2). One result is that environmental needs are identified from the outset, and the amount of water required to satisfy such needs is subtracted from what is available in a source of supply to meet human needs.

3. *Quality of delivered water is matched to an end-use requirement.* While high-quality water is necessary for human consumption, the quality required for water may vary significantly for other uses. The soft path seeks to match water quality to that needed to accommodate an end use. One implication is recognition of what is referred to as cascading water systems, which leads to waste water from one use becoming the supply for another use with less

Box 11.2   The Brisbane Declaration

Environmental flows are essential for freshwater ecosystem health and human well-being.

Freshwater ecosystems are the foundation of our social, cultural, and economic well-being. Healthy freshwater ecosystems—rivers, lakes, flood plains, wetlands, and estuaries—provide clean water, food, fibre, energy, and many other benefits that support economies and livelihoods around the world. They are essential to human health and well-being.

Freshwater ecosystems are seriously impaired and continue to degrade at alarming rates. Aquatic species are declining more rapidly than terrestrial and marine species. As freshwater systems degrade, human communities lose important social, cultural, and economic benefits; estuaries lose productivity; invasive plants and animals flourish; and the natural resilience of rivers, lakes, wetlands, and estuaries weakens. The severe cumulative impact is global in scope. . . .

Flow alterations imperil freshwater and estuarine ecosystems. These ecosystems have evolved with, and depend upon, naturally variable flows of high-quality fresh water. Greater attention to environmental flow needs must be exercised when attempting to manage floods; supply water to cities, farms, and industries; generate power; and facilitate navigation, recreation, and drainage.

Environmental flow management provides the water flows needed to sustain freshwater and estuarine ecosystems in coexistence with agriculture, industry, and cities. The goal of environmental flow management is to restore and maintain the socially valued benefits of healthy, resilient freshwater ecosystems through participatory decision-making informed by sound science. Groundwater and flood plain management are integral to environmental flow management.

The above declaration, quoted in part, was developed at the 10th International River Symposium and International Environmental Flows Conference held in Brisbane, Australia, 3 to 6 September 2007. Some 750 scientists, engineers, economists, planners, and resource managers from more than 50 countries attended the conference.

stringent-quality needs. Examples include using water from a washing machine on a garden or shower or bath water for toilet flushing.

4. *Determine the desired future condition, and plan back to the present.* Conventional planning for water takes the present as its starting point. Future needs are projected, and then decisions are taken to meet these needs. The soft path focuses not on the most probable future but instead on the most desirable future, which is normally characterized by ideal sustainability characteristics or criteria. Once the attributes of the desirable future are defined, and reflecting the preceding three principles, decisions are taken about the most appropriate means to meet desired ends—without assuming that the way water is used now will be the same in the anticipated desirable future. This approach is referred to as backcasting, discussed in Chapter 5.

Water security, both quantity and quality, should ideally be achieved through a mix of supply management, demand management, and soft path approaches. The goal should be to use more of the tools of demand management and soft path approaches as part of an integrated strategy that includes supply management. The key is to use creative and innovative ways of moving beyond primary reliance on supply management, which provides little or no incentive for conservation behaviour or attention to the needs of the environment.

Virtual Water

Growing water scarcity, whether shortages of quantity or inadequate quality, has the potential to damage economic development as well as human livelihoods and well-being. An emerging concept in that regard is **virtual water**, which has grown out of recognition of the importance of water for agriculture and food production. For example, Liu, Zehnder, and Yang (2007) have reported that in China, agriculture yields increase on average from 30 to 100 per cent when an area under cultivation is irrigated. The negative consequences of water shortages for food production and other economic activities, and the possibility that such shortages could lead to regional conflict or wars, stimulated thinking about virtual water, characterized as water-scarce countries striving to reduce problems resulting from water shortages by importing foods that require significant water inputs.

During the early 1990s, various writers, such as Starr (1991), Gleick (1993), and de Villiers (1999) commented about the likelihood that water was or would soon be in a crisis state, and terms such as 'water wars' and 'green wars' were used. Others, such as Wolf (1995), noted that water wars had actually been very infrequent and that water shortages normally had to be accompanied by other social problems before armed conflict occurred. Nevertheless, it was pointed out that water security needed to be defined not only in terms of **blue water**, or the water represented by rain falling into lakes, rivers, and aquifers, but also in terms of **green water**, or water created from evapotranspiration by plants and crops.

As Warner and Johnston (2007, 65) noted, the virtual water proposition is basic:

> . . . water-stressed regions of the world can ensure water and food security by actively seeking out a strategy which reduces domestic agricultural production of basic foodstuffs, increases reliance on the global food trade system to secure domestic requirements, and, in turn, increases the availability of water for non-agricultural development needs, reduces environmental stress, and avoids senseless fighting over water.

Continuing, Warner and Johnston (2007, 66) observed that the concept of virtual water is a 'welcome antidote' to the gloomy prospect of water wars. By pursuing a virtual water strategy, national governments in water-stressed or deprived countries can plan to meet food security needs even if their nation has a limited water endowment.

A virtual water strategy does have the potential to ameliorate water shortages at a national level. Nevertheless, as with all concepts or strategies, it has limitations or drawbacks. First, if a nation decided to cut back significantly on domestic production of high-water-demanding crops and import them from other countries, many local farmers would likely lose their livelihoods. One outcome could be further migration of poor rural people into urban areas, where they would very probably become part of a growing marginalized group. Second, a relatively poor nation would not likely have the foreign currency necessary to purchase food products requiring significant water inputs. Third, a nation might be reluctant to become dependent on other countries for food needs. To illustrate, Liu, Zehnder, and Yang (2007, 86) have commented

Professor Tony Allan and Virtual Water

Professor John Anthony Allan of King's College London and the School of Oriental and African Studies received the 2008 Stockholm Water Prize for his pioneering work to develop the concept of 'virtual water'.

The Stockholm Water Prize, established in 1990, is awarded annually by the Stockholm Water Foundation. It is conferred upon an individual, organization, or institution for 'outstanding water-related activities'. The recipient receives US $150,000 and a crystal sculpture. Professor Allan received the prize from HM King Carl XVI Gustaf of Sweden on 21 August 2008.

that with regard to China, 'For food security, the government pays more attention to food self-sufficiency than to water use efficiency. Food self-sufficiency is overwhelmingly favoured by the Chinese government, which regards reliance on international food markets as a threat to domestic security.' And fourth, use of virtual water might mask or obscure the reality of in-country water shortages and lead to delays or inaction regarding policy changes that would have environmental and social as well as economic benefits.

What are your views on 'virtual water'? Does it offer promise as one tool to address water security challenges? Or does it have the potential to disadvantage the people in a country who are already very vulnerable and benefit those who enjoy a relatively high standard of living?

Water Terrorism

In the context of water security, Gleick (2006, 481) has noted that contemporary society relies on complex and sophisticated infrastructure systems to provide safe and reliable supplies and to remove waste water, especially in large cities. He has observed that such infrastructure, while vital for human welfare and economic development, is 'vulnerable to intentional disruption from war, intrastate violence, and, of more recent concern, terrorism'. He further observed that water infrastructure systems are attractive terrorism targets because 'there is no substitute for water.' Communities can also suffer if too much water is present at the wrong time, as when a dam and reservoir is destroyed, leading to flooding immediately downstream.

Terrorists can disrupt water infrastructure or related services by directly attacking the infrastructure or by contaminating water with poison or disease-bearing agents. Damage can come in the form of serious health implications, unusable water, or the destruction of infrastructure facilities and equipment. As Gleick (2006, 482) commented, 'the typical scenario for a terrorist attack on domestic water supplies involves putting a chemical or biological agent into local water supplies or using conventional explosives to damage basic infrastructure such as pipelines, dams, and treatment plants.'

While water infrastructure is vulnerable, the difficulties of disrupting it are considerable. For example, most biological pathogens do not easily survive in water, and most chemicals have to be present in very large volumes to contaminate the supply. Treatment systems normally deal with biological agents or chemicals. In addition, most water systems are designed for redundancy in order to deal with equipment failure or the practical requirement to take systems 'off-line' for routine maintenance. Notwithstanding such redundancy, the bombing of a major pipeline supplying water to Baghdad in 2003 illustrates that serious disruption and hardship can be caused by attacking pipelines, pumping plants, or treatment plants. Another possible threat is using remotely controlled computers to interrupt the functioning of valves, pumps, and chemical processing equipment regulated by computer-based control systems. Such an attack is referred to as cyber-terrorism. For example, during April 2003, a man used a computer along with a radio transmitter to gain control of a district waste water system in Queensland, Australia, and released sewage into parks and rivers and onto property (Gleick 2006, 492).

From a management perspective, two matters must be considered: (1) the probability of a terrorist attack and (2) the consequences of an attack. Both require examination to calculate degree of risk. While vulnerability can never be eliminated, mitigation measures can reduce the probability of an attack as well as the consequences.

To reduce the probability of an attack, various initiatives can be undertaken. The most fundamental is to reduce or remove the basic motivation for a terrorist attack. Such an option is normally well beyond the scope or authority of resource managers, but it should always be kept in mind. However, attention normally focuses on protection and detection initiatives.

Protection involves limiting or denying access to facilities or at least to key or vulnerable parts of the facilities. This often involves 'gates, lights, and guards'. However, such actions are not practical for pipelines or aqueducts that extend over many kilometres, and other options must be considered for them. Detection can involve surveillance cameras or motion detectors, often at significant expense. By themselves, such tools cannot stop intruders, but they act as a deterrent because would-be intruders know that their intrusion will be recorded.

In addition to protection and detection, society can prepare by ensuring that various emergency measures are in place. These measures can include public advisories, temporary shutdown of systems, information about alternative services or supplies, treatment for contamination (e.g., boil-water advisories), health interventions, and emergency response teams. A key element is to have a plan in place for an emergency response, as well as training sessions for key stakeholders, including water utility staff, law enforcement officers, community leaders, and the media.

Water as Hazard

Flooding

Humans settle adjacent to rivers and lakes for good reasons. Proximity to them provides access to potable water, a place to dispose wastes, and sometimes, a source of power and a means of transportation. The relatively flat land beside many rivers and lakes facilitates construction of roads, homes, and places of business. The aesthetic quality—serenity, beauty, natural appeal—of a river or lake view also often means that waterfront lots command a premium price. However, the term **flood plain** exists for a reason. From time to time, rivers and lakes extend beyond their normal limits to cover adjacent areas or flood plains. Flooding is a normal hydrological function. Indeed, many species of flora and fauna depend on flooding to survive and flourish. In addition, humans often benefit from flooding, as when flood plains are enriched by the deposit of silt that then supports agriculture.

When flooding occurs as a result of a river or lake 'reclaiming' adjacent land, the result is often only a minor inconvenience, and this was usually the case when settlements were relatively small. However, as population concentrations on flood plains increase, the potential of flood damage goes up. Examples of major floods and serious associated damages in Canada include the floods in the lower Fraser River Valley in BC (1948); Winnipeg (1950, 1997 [an estimated $300 million in damage]); Toronto (1954 [Hurricane Hazel, causing more than 80 deaths and millions of dollars in damage; see Box 11.3]); Fredericton (1973); Cambridge, ON, Maniwaki, QC, and Montreal (1974); the Saguenay River Valley, QC (1996 [10 deaths and $800 million in damage]); and eastern Ontario and Quebec as well as the St John River in New Brunswick in the spring of 2008. The latter flood was of similar magnitude to the one in 1973 and threatened 1,300 homes.

Humans have various ways of reducing flood damage potential. One relies on *structural approaches* in which the behaviour of the natural system is modified by delaying or redirecting flood waters. Common methods are upstream dams and storage reservoirs, protective dykes or levees, and deepening or straightening river channels to increase their capacity. All these measures provide protection. However, because they are designed and built with a standard in mind, such as the magnitude of flooding that may occur once in 100 years, a flood of greater magnitude will eventually occur (such as a flood that occurs once every 200 or 500 years). If people perceive the structural measures as 'protecting' the flood plain, resulting in more development on it, then when the inevitable flood event greater than the design capacity of the structural measures does happen, potential and actual flood damage will be greater.

A second approach, referred to as *non-structural*, focuses on modifying the behaviour of people. Methods include land-use zoning to restrict or prohibit development in flood-prone areas, relocation of existing flood-prone structures, information and education programs to alert people to the hazard of occupying flood plains, and insurance programs to help people deal with the costs of flood damage. The best strategies to reduce flood damage use a mix of structural and non-structural approaches, since each method has limitations. In the following discussion of the 1997 flood of the Red River in Manitoba, you will see how both structural and non-structural approaches are used.

Red River Flood, 1997

In late April and early May of 1997, the Red River 'experienced a catastrophic regional flood that far surpassed any previous flood in the historical record' (Todhunter 2001, 1,263). As a result, areas in Minnesota, North Dakota, and Manitoba had what the media called the 'Flood of the Century'. Total damages in the US part of the basin were calculated at US $4 billion, nearly US $3.6 billion of which was within the Grand Forks (North Dakota) and East Grand Forks (Minnesota) metropolitan area. Per capita, this event was the most costly flood for a major metropolitan area in the United States. In the Grand Forks metropolitan area, close to 13,000 structures, or 85 per cent of all structures, were damaged, and about 55,000 residents had to evacuate their homes for periods ranging from days to months. Some 1,300 homes were destroyed.

Box 11.3   The Aftermath of Hurricane Hazel, October 1954

The valleys, ravines, and trees of Toronto are some of the city's defining natural features—and we have a hurricane to thank for them.

Half a century ago . . . Hurricane Hazel swept across southern Ontario, unleashing floods that took at least 81 lives and caused millions of dollars in damage. In the years afterward, efforts to prevent a repeat of Hazel resulted in a collection of protected green spaces sprinkled throughout the city and region.

Islands of green amid the concrete and asphalt, places like Riverdale Park, the Toronto Brick Works, the G. Ross Lord Reservoir, and Étienne Brûlé Park were all created or developed in the aftermath of the hurricane.

'If that [hurricane] hadn't happened 50 years ago, I think we probably would have lost the vast majority of the green space in our valleys,' says Gregor Beck, director of conservation science with Ontario Nature (the former Federation of Ontario Naturalists). . . .

The key event in Hazel's immediate wake, Beck says, was the establishment three years later of what is now known as the Toronto and Region Conservation Authority. Its initial role was to mitigate the damage that any future Hazels could cause. But the authority evolved into the most important player in the protection and revitalization of the GTA's natural environment.

In October 1954, when Hazel dropped up to 28 centimetres of rain over just two days onto the already sodden city, most of the deaths and property damage after the hurricane occurred near rivers and creeks. Hazel made it clear that encroaching development in these settings had to be halted, says Brian Denney, chief administrative officer of the conservation authority.

'There was a recognition that the valley systems really belong to the rivers,' Denney says.

People, Beck adds, like to live near rivers and lakes but rarely consider the water's latent power and destructive potential. . . .

The authority began to aggressively acquire land, buying or expropriating homes in watershed areas and taking over open lands and forests abutting rivers and creeks. Places like Raymore Drive in Etobicoke, where a dozen homes were washed away in the Hazel flooding, were transformed from residential streets into riverside parks.

Today, the authority owns more than 14,164 hectares of land, mainly along the . . . Humber, Don, and Rouge rivers and their tributaries, between the Oak Ridges Moraine and Lake Ontario.

Source: Hall 2004, B1–B2.

The Red River Valley

The topography of the Red River Valley is very flat, resulting in slow movement of flood waters and considerable areal extent of flooding when a flood occurs.
—Burn and Goel 2001, 356

Downstream, southern Manitoba experienced $500 million in damage. The Winnipeg area just avoided a catastrophe, with only 54 homes flooded. While there were no deaths, 28,000 rural residents had to evacuate their homes (Figure 11.4).

The Red River originates in the United States, in southern North Dakota/Minnesota, and flows northward, draining into Lake Winnipeg. It is 880 kilometres long, and the watershed is 290,000 square kilometres, including the Assiniboine River basin (163,000 square kilometres), which joins the Red River at the 'Forks' in Winnipeg (Brooks and Nielson 2000). The valley is underlain by Mesozoic bedrock, which has a moderate relief. In the central valley, the bedrock is covered by late Pleistocene glacial sediments, which in turn are capped by glaciolacustrine clay-rich deposits that accumulated within glacial Lake Agassiz, which at one time covered most of Manitoba, the northern part of North Dakota and Minnesota, northwestern Ontario,

Figure 11.4 Extent of Red River flooding, 1997. Source: Brooks and Nielson 2000, 307.

Flooding on the Red River in 1979.

and east-central Saskatchewan in the late Pleistocene and early Holocene. In Manitoba, the modern Red River is on the bed of glacial Lake Agassiz in an eroded valley up to 15 metres deep and 2,500 metres wide.

The river has a long history of flooding. The largest recorded flood occurred in 1826. Records also document major floods in 1852 and 1861. More recently, significant floods happened in 1950, 1979, 1996, and 1997. Various factors combined to cause the major flood of 1997: (1) high precipitation in the autumn of 1996, resulting in the soil being saturated by the time of the winter freeze-up; (2) near-record levels of precipitation during the winter, with twice or more the normal amounts of snowfall; (3) a long and unusually cold winter, leading to a high water content of the snowpack into the spring; and (4) after spring melting had started, a major blizzard over much of the basin on 5 April, with snow accumulations of up to 50 centimetres. These factors combined to produce ideal conditions for flooding.

The flooding of the Red River was typical for that river system but unusual compared to most Canadian rivers. During the flood, an area of about 2,000 square kilometres and up to 40 kilometres wide was covered by water from the Canada-US border to the southern edge of Winnipeg (Figure 11.4). For most rivers, flood waters are confined to the river valley, even when they spread out onto the flood plain. Extensive flooding occurs in the Red River Valley because of the low capacity of the river to accommodate extreme flows as a consequence of the low valley gradient and shallowness of the valley itself. This situation is exacerbated by the fact that the areas adjacent to the river are broad and flat, allowing the flood waters to spread out over many kilometres on each side of the river. The flood

Flooding on the Red River in 1997.

Control gates for the Red River Floodway.

peaked at Emerson, the most southern community along the river in Manitoba, on 27–8 April and then at the Floodway inlet in Winnipeg on 3–4 May. This pattern was typical for the Red River, where floods usually rise and fall slowly over a period of several weeks.

After the devastating flood in 1950, when a large part of Winnipeg was inundated, structural measures were undertaken to protect the city, smaller towns to the south, and transportation routes. These measures included:

1. *Red River Floodway*. Completed in 1968, the **floodway** is an excavated channel (48 kilometres long, 210 to 305 metres wide, and 9.1 metres average depth) that can divert up to 1,700 cubic metres per second of the flow of the Red River around the eastern side of Winnipeg to rejoin the Red River downstream of the city. The capacity is sufficient to divert the river stage to a level 0.6 metres below the top of the primary dykes in the city. During the 1997 flood, use of the Floodway kept flooding in Winnipeg to a minor level, compared to the major damage suffered in the 1950 flood. However, the Floodway operated at maximum capacity (a once-in-160-years flood) during the 1997 flood, indicating that a flood of larger magnitude would have caused serious damage.
2. *Portage diversion*. This **diversion** is located at Portage La Prairie, 84 kilometres west of Winnipeg, and is an excavated channel, 29 kilometres long and 54 to 366 metres wide, that can divert up to 700 cubic metres per second from the Assiniboine River to Lake Manitoba. Completed in 1970, the diversion is designed to reduce the flow of the Assiniboine River into Winnipeg during high water in the Red River.
3. *Shellmouth Dam and Reservoir*. Completed in 1972, this storage reservoir is able to hold water for later release at times of high flows.
4. *Earth dykes*. A system of **dykes** has been built along the Red, Assiniboine, and Seine rivers within Winnipeg. One hundred twenty kilometres of primary dykes have been built within the city, as well as the 32-kilometre West Dyke that extends west and south from the Floodway.
5. *Ring dykes*. South of Winnipeg, large dykes encircle various small towns on the Red River flood plain. Smaller ring dykes surround individual buildings, such as homes and barns. Dugouts also often remained after the earth dykes were constructed and have become small perennial ponds in the valley.
6. *Elevated roads and railway beds*. Roads and railway beds are raised one to two metres above the valley. This elevation protects transportation infrastructure from flooding and also can serve as dykes.

There were both positive and negative outcomes from the 1997 Red River flood (Burn and Goel 2001; Shrubsole 2001; Todhunter 2001). First, the Winnipeg Floodway prevented an estimated $6 to $10 million in damage in Winnipeg. Second, flood-fighting by local and provincial people was supplemented by more than 8,000 military personnel. More than 8 million sandbags were filled and delivered in Winnipeg, more than 600,000 cubic metres of clay were excavated to provide material for new temporary dykes, and 50 temporary dykes were modified to become permanent structures. Third, in addition to filling sandbags, thousands of volunteers looked after children and provided food and other support to their neighbours. Fourth, non-government organizations provided critical assistance, such as the $10,000 grants from the Red Cross to residents whose homes were damaged beyond repair.

Balancing these positive outcomes were some negative aspects. First, the operation of the gates at the Winnipeg Floodway caused the upstream water level to rise, exacerbating flooding by 0.64 metres in some smaller upstream communities (south of Winnipeg). This led some rural upstream residents to conclude that their well-being had been sacrificed to protect the residents of Winnipeg. Second, although there was considerable forewarning of the flood, the peak flow at the Floodway was underestimated by 1.5 to 1.7 metres. Furthermore, the flooding of the upstream community of Ste Agathe was a surprise to many, since the town had never been flooded before and thus a ring dyke had not been built. Eight other small communities on the flood plain but within ring dykes were not flooded. Third, some rural municipalities hesitated to spend their own funds on fighting the flood until arrangements were clarified with the provincial government, since provincial law does not allow municipalities to have operating deficits. Fourth, some Aboriginal communities experienced problems because of confusion regarding which agencies were responsible for them.

The International Joint Commission (2000) conducted an inquiry on the 1997 flood and offered the following conclusions and recommendations: (1) although the 1997 flood was natural but rare, floods of the same or greater magnitude are possible; (2) both people and property in the Red River Valley will remain at risk to flooding until comprehensive, integrated, and binational solutions are developed; (3) a mix of structural and non-structural approaches is needed, since no one approach is adequate; (4) specific communities in both Canada and the US need to take flood damage initiatives; and (5) aspects of the ecosystem need more explicit attention, and in that regard, hazardous material needs to be more carefully controlled, and banned substances need to be removed from flood-prone locations.

Since the 1997 flood, the Canadian and Manitoba governments together have spent more than $130 million on structural measures. Much of this amount ($110 million) has been for protective adjustments for 16 rural communities. In December 2003, the two governments announced an agreement to jointly allocate $240 million for expansion of the Red River Floodway around Winnipeg. This will allow an increase in the capacity of the existing 49-kilometre channel to meet

The inlet control structure during the spring of 2005.

The inlet control structure is situated on the Red River downstream from the Floodway channel entrance, June 2007.

An integral component of the Manitoba Floodway protection system, the outlet structure, located downstream of Lockport, MB, was designed to dissipate energy from flood water as it returns to the Red River. (Photo from 2004.)
(All photos courtesy the Manitoba Floodway Authority.)

a flood event comparable to the one in 1826, calculated as a once-in-280-years occurrence. This will be realized by increasing the width of the diversion channel by about 110 metres and the depth by up to two metres. When the entire expansion is completed, at an estimated cost of $700 million, it will provide protection against a once-in-700-years flood.

The expansion will be complicated and challenging, since it will involve upgrading and improvements to 12 bridge crossings, the Floodway outlet, the West Dyke, utilities, and drainage services. There is also potential to disrupt the groundwater table during the excavation, which will move more than 30 million cubic metres of earth. An environmental impact assessment began in 2004 to determine what adjustments would need to be made to the design. This project, which will take up to a decade to complete, is being managed by the Manitoba Floodway Expansion Authority, and up-to-date details can be found at its website (http://www.floodwayauthority.mb.ca/floodfacts.html).

Canadians will continue to live and work on flood plains for the reasons mentioned at the beginning of this section. Thus, there will continue to be damage from floods. In that context, the comments of Shrubsole (2001, 462) deserve consideration:

> The present practice of flood management in Canada is characterized by at least three realities. First, it is impossible to provide absolute protection to people and communities. Second, a mix of structural and non-structural adjustments that cover the entire range of protection, warning, response, and recovery is needed to effectively protect lives and property. Third, implementation of flood adjustments requires the effective participation of all levels of government and the public.

Droughts

If flooding represents situations in which there is too much water, droughts represent the opposite problem—insufficient water. While flooding is immediate and apparent, the beginning or end of a drought is more difficult to determine, since droughts are a function of a lack of precipitation, temperature, evaporation, evapotranspiration, capacity of soil to retain moisture, and resilience of flora and fauna in dry conditions.

Consequently, as Gabriel and Kreutzwiser (1993) have noted, a significant challenge when studying droughts or seeking to identify 'drought-prone' areas is to define what is meant by a **drought**. As they noted, interpretations are based on different factors: causes and effects. Regarding those based on *causes*, an example is a meteorological drought due to a prolonged deficiency of precipitation, which reduces soil moisture. This type of drought can trigger a second type (hydrological drought, an effect that then becomes a cause), manifested by reduced stream flows and lowered **water table** and/or lake levels. In terms of *effects*, an example is an agricultural drought, which results in reduced crop yields because of a lack of moisture. An urban drought happens when there is insufficient water, because of lower stream flows or water tables, to support all demands in the community. The point here is the importance of understanding the definition of a 'drought' when preparing or examining results or recommendations pertinent to management decisions.

Droughts are of concern because, as Gabriel and Kreutzwiser (1993, 119) explain, they reduce the amount of water for use by depleting soil moisture and groundwater reserves as well as by lowering stream flows and lake levels. These reductions can start a 'depletion cycle': less than normal rainfall leading to low soil moisture, triggering demand for irrigation development, in turn depleting non-recharging surface and groundwater supplies. The high evapotranspiration rate associated with hot, dry periods also contributes to depleting soil moisture and surface water supplies, which are not restored to normal levels without unusually high rainfall.

What has been the experience with drought in Canada, where the Prairies, the interior of British Columbia, and southern Ontario are most vulnerable?

Droughts are most usually associated with the Prairie provinces, especially in that area of southern Alberta and Saskatchewan and extreme southwest Manitoba known as **Palliser's Triangle** (Figure 11.5). This area is named after Captain John Palliser, who was sent by the British government and the Royal Geographical Society to explore the territory between the Laurentian Shield and the Rocky Mountains between 1857 and 1860 and determine the nature of the soil, its capacity for agriculture, the quantity of its timber, and the presence of coal or other minerals. Based on his surveys, Palliser divided the area into two sections—a fertile belt and a semi-arid area. He considered the southern or semi-arid area unfit for settlement. Palliser commented in his journal that this area 'has even early in the season a dry patched look. . . . The grass is very short on these plains, and forms no turf, merely consisting of little wiry tufts. Much of the arid country is occupied by tracts of loose sand, which is constantly on the move before the prevailing winds' (Mackintosh 1934, 11). Palliser concluded that 'There is no doubt that the prevalence of a hard clay soil derived from the cretaceous strata which bakes under the heat of the sun, has a great deal to do with the aridity of these plains, but it is primarily due more to want of moisture in the early spring' (34). Thus, 150 years ago, Palliser identified the drought-prone nature of the southern Prairies. His 'heads-up' was reinforced during both 2001 and 2002 when large areas of the Prairie provinces experienced drought conditions.

Figure 11.6 illustrates that 'in 2002, drought-stricken areas covered over three-quarters of the Prairies (including the northeastern part of British Columbia)' (Statistics Canada 2003b, 13). While there were many impacts from these dry conditions, the most pronounced was inadequate water to support agriculture. As the data in Table 11.4 show, yield of spring wheat, barley, and canola fell significantly during 2002

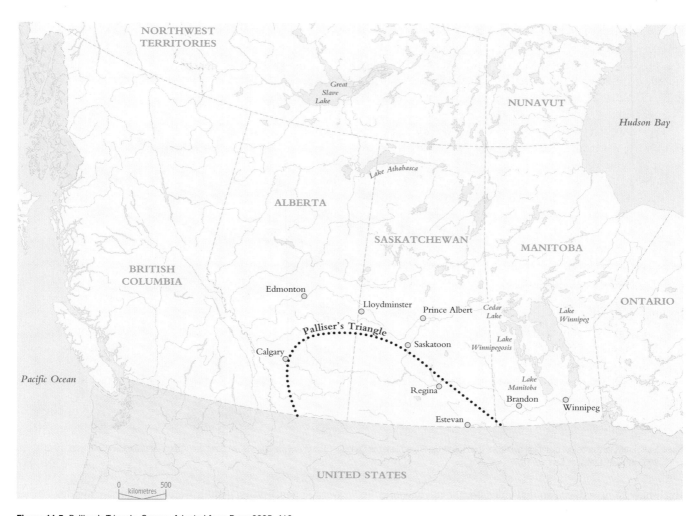

Figure 11.5 Palliser's Triangle. Source: Adapted from Bone 2005, 410.

relative to the average yields between 1991 and 2000, which were non-drought years.

Livestock were also affected negatively by the drought. The greatest impact was in Alberta, where the inventory dropped by 605,000 cattle, a decrease of 10.4 per cent between January 2002 and January 2003. At the same time, declining supplies of cattle feed, a result of the drought conditions, pushed up feed prices and led many ranchers to reduce herds.

Another indicator of the drought conditions was the drying up of many dugouts (small human-made ponds), potholes (small natural ponds), and sloughs. By September 2002, 80 per cent of Prairie farms were in regions in which dugouts were half empty, and 20 per cent reported their dugouts were completely dry. The drying up of potholes and sloughs not only affected agriculture. They are also critically important habitat for migratory wildfowl, which were adversely affected as well (Statistics Canada 2003b, 13, Table 10.4 regarding the effect of drought on wetlands).

In Ontario, drought can occur in any season but is most likely in the summer when demand for water is usually the highest. Southwestern Ontario is most vulnerable, especially in the summer and early autumn. Extended dry periods for more than a month are unusual, but shorter droughts are not uncommon. For example, dry periods of at least seven consecutive days occur at least once a month during the agricultural growing season in southern Ontario, and short-term (10 to 20 days) dry spells occur every year. Longer droughts (more than four weeks) happen once in three years. One of the most severe and extensive droughts occurred in 1966 when during a 41-day period between mid-June and the end of July, most of southern Ontario received less than 25 millimetres of precipitation, or less than a quarter of normal rainfall. Areas from Lake Huron and Georgian Bay to Lake Ontario received less than 15 per cent of normal rainfall in this period. Another significant drought occurred in the summer of 1988 when southwestern Ontario received less than 40 per cent of the average precipitation from early May to mid-July.

Lake levels are affected by dry periods, and the Great Lakes illustrate this effect. The variation between minimum and maximum lake levels is 1.2 metres on Lake Superior, 1.8 metres on Lakes Huron and Erie, and two metres on Lake Ontario. Low levels affect shipping, especially in terms of cargo tonnages,

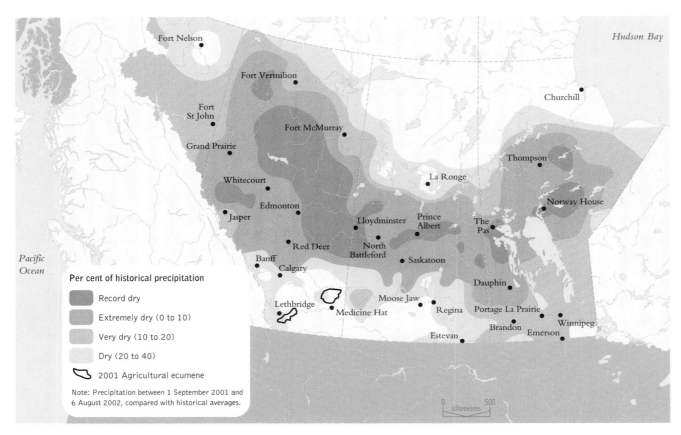

Figure 11.6 Precipitation below historical averages, 2002. Source: From Statistics Canada, *Human Activity and the Environment: Annual Statistics 2003*. Cat. no. 16-201, 3. Dec. 2003. Ottawa: Statistics Canada, 13.

which have to be reduced so that ships do not run aground. To reduce the draft by only 2.5 centimetres requires a reduction of up to 90 tonnes on most ships and more than 180 tonnes on ships between 244 and 305 metres in length. During low levels in the St Clair River in 1964, ships carrying iron ore and grain from Lake Superior to the lower Great Lakes had to reduce their loads by between 725 and 1,360 tonnes to pass through the river.

In Ontario, many streams are almost totally supplied by groundwater discharges during low rainfall periods. In average conditions, groundwater discharge provides 20 per cent of the water for streams and rivers in most of Ontario. For some rivers or streams, the contribution can be up to 60 per cent, extending up to 100 per cent in the summer months. Thus, depletion of groundwater reserves resulting from drought can have a serious impact on surface flows, especially for smaller streams. The lowering of water tables due to drought can also lead to the drying up of wells dependent on shallow aquifers.

Table 11.4 Crop Yields and Insurance Payments, 2002 Percentage Variation from 1991–2000 Average, Prairie Provinces

Province	Spring Wheat	Barley	Canola	Crop Insurance Payment
Alberta	−29.4	−26.8	−13.0	399.1
Saskatchewan	−32.0	−34.1	−21.4	224.1
Manitoba	2.6	7.6	3.9	69.9

Source: Adapted from Statistics Canada 2003b, 13.

Heritage Rivers

The mission of the Canadian Heritage Rivers Program (CHRP) is to 'develop a river conservation program that is nationally valued, internationally recognized, and reflects the significance of rivers in the identity, and history of Canada; and ensure that the natural, cultural, and recreational values for which rivers are designated are managed in a sustainable manner'. The CHRP is overseen by a board, with representatives from the federal, provincial, and territorial governments, under the oversight of appropriate ministers from the three levels of government.

By 2006, 40 rivers had been nominated under this program, and 36 had been formally designated (Figure 11.7). The process leading to designation has several steps, as described in Box 11.4.

For almost a decade after the beginning of the program, designated rivers were located in federal or provincial parks, the territories, or in areas within provinces with relatively few

Rivers As Heritage

Rivers teach valuable lessons about renewal. It is said, and it is true, that you cannot enter a river at the same spot twice; because, of course, the river 'rolls along'. This very character of rivers generates health and well-being. It also connects one part of the waterway to another.

—Harry Collins, chairperson, Canadian Heritage Rivers Board, 2002

Figure 11.7 Heritage rivers in Canada. Source: Canadian Heritage Rivers Board 2003, 5.

Box 11.4   Procedure for Designation of a Heritage River

1. *Submissions and public involvement.* The first step is a community-based initiative to recognize and protect a local river. The initiative can be taken by landowners; recreational, historical, or tourism organizations; Aboriginal peoples; environmental groups; local businesses; interested members of the public; or a combination. This step requires a submission to the board.

2. *Selection.* The governments with responsibility for the river identified in a submission assess the case relative to the natural, cultural, and recreational values of the river and the level of public support for the nomination. The decision is also based on consideration of the number and complexity of conflicting land uses and the capacity for effective management of the river.

3. *Nomination.* If the government(s) determine that the river is a credible candidate, then the relevant government agencies work with local stakeholders to prepare a formal nomination to the board. The nomination must demonstrate that the river is of outstanding Canadian value relative to the CHRP criteria (natural, cultural, and recreational significance) and that sufficient measures will be implemented to ensure maintenance of those values. Once the board has reviewed

the nomination, it forwards a recommendation to the federal minister responsible for Parks Canada and the relevant provincial or territorial minister.

4. *Management plan.* The formal designation process starts when the nominating government submits a management plan to the board. The management plan identifies the policies and practices that will protect the outstanding values of the river.

5. *Designation.* Once the management plan has been approved, the responsible federal and provincial or territorial governments formally designate the river, and a plaque is unveiled at a key location along the river. Designation is intended to ensure that the river will be managed so that the outstanding heritage resources will be protected and the recreational potential will be realized.

6. *Monitoring.* To ensure that designated rivers are managed to protect the outstanding natural, cultural, and recreational values, the board requires that at least every 10 years the managing jurisdictions conduct a systematic assessment to verify the heritage status of each designated river.

people. Designation of such places involved rivers that were primarily on Crown land, avoiding the complication of having to deal with private landowners and municipalities, who often were suspicious of the Heritage Rivers Program, viewing it as possible government intrusion into property or municipal rights.

However, in 1994, the entire Grand River basin, located in southern Ontario and with most of its land in private ownership, was designated, and that experience is explained by Barbara Veale in her guest statement below. Other rivers in highly settled areas have since been designated, such as the Humber River in Toronto and the Thames River and the Detroit River in southwestern Ontario.

Grand River, Cambridge, Ontario. A former factory has become a park setting adjacent to the river, making the river accessible to the public while retaining a sense of heritage.

Guest Statement

How Becoming a Heritage River Can Influence Water Management

Barbara Veale

The Grand River is located in the heart of southern Ontario. Its rich diversity illustrates key elements in the history and development of Canada, and many of the river-related heritage resources remain intact today. These resources include the historical buildings and settlements in the watershed, the land use associated with different ethnic groups, such as Mennonite farmers still using horse-drawn ploughs, and the Six Nations area near Brantford. In addition, the river and its tributaries provide a broad range of excellent recreational opportunities.

In 1987, the Grand River Conservation Authority spearheaded a participatory process to have the Grand River and its major tributaries declared a Canadian heritage river. This was achieved for the Grand River and its major tributaries in 1994.

The designation of the Grand River as a Canadian heritage river marked the beginning of a second generation of heritage rivers. Prior to 1990, almost all nominated rivers either were within protected areas or were short sections of larger rivers. In contrast, the Grand River is located in one of the most densely populated and fastest-growing parts of Canada where almost all lands are privately owned and managed within a complex multi-agency, multi-jurisdictional setting. The management plan presented to the Canadian Heritage Rivers Board—*The Grand Strategy*—deviated from past management plans for heritage rivers in that it provided a framework for an ongoing, community-based watershed approach sustained by consensus, cooperation, and commitment. It was based on a common vision, beliefs, values, principles, and goals, and the designation of the Grand River set a precedent for the Canadian Heritage Rivers Board to accept other rivers in highly settled areas of Canada where river management is complex and shared among all levels of government, First Nations, and non-government entities.

Since 1994, *The Grand Strategy* has evolved into a shared management approach for integrated watershed management. Management partners, including federal and provincial governments, watershed municipalities, First Nations, non-government groups and organizations, and educational institutions, jointly identify critical resources issues, develop creative solutions, pool resources, implement actions, monitor results, and evaluate progress on an ongoing basis.

Under the umbrella of *The Grand Strategy*, the Grand River Conservation Authority works with its partners to address a wide array of existing and emerging resource issues and to identify priorities for action. Progress is documented and celebrated through Grand Actions—a bi-monthly newsletter, special events, an annual Registry of Accomplishments and Commitments, and an annual *Watershed Report*. Opportunities to directly participate in *The Grand Strategy* are offered through various working groups and public forums. The philosophy is that everyone who shares the resources of the Grand River watershed is encouraged to be a part of a collective effort to address key watershed issues, including:

- reducing non-point sources of pollution in rural areas;
- pursuing excellence in waste water treatment;
- slowing increases in water use and advocating the wise use of water;
- protecting groundwater resources;
- developing long-term water quality and water budget/supply plans;
- maintaining the water control system;
- implementing a watershed-wide fisheries management plan;
- developing community-based plans that advance forest management, wildlife management, and natural heritage management;
- developing community riverfront plans;
- developing the watershed's potential for outdoor recreation, cultural education, and eco-tourism;
- building a sense of community around the river and celebrating successes.

Building on the Grand River Basin Water Management Study, a multi-agency initiative completed in 1982 to review the issues of flood protection, water supply, and water quality within the Grand River watershed, *The Grand Strategy* relies on good science and strong technical tools. Within the context of water management, partners are currently evaluating the cumulative impacts of municipal waste water on river water quality, determining what needs to be done to improve water quality, and monitoring and reporting on water quality conditions and long-term trends in the river. Partners are also assessing the cumulative demand for water, identifying where water shortages may occur, and putting drought contingency plans in place. Groundwater sensitivities have been mapped, and priority areas for water protection programs are being identified.

The merits of a collaborative watershed management approach for improving river health were recognized in 2000 when the Grand River Conservation Authority was awarded the international Thiess River Prize for excellence in river management in Brisbane, Australia.

In keeping with Canadian Heritage Rivers System requirements, the first Ten-Year Monitoring Report for the Grand River was completed in 2004. This assessment was undertaken within the participatory structure created by *The Grand Strategy* and provided participants with the opportunity to revisit and reaffirm the vision, values, principles, and goals of *The Grand Strategy*. It revealed that great strides had been made in increasing awareness of the river as a community asset, strengthening heritage conservation, and developing new outdoor recreational opportunities. The general level of concern about the river had increased, resulting in more stewardship activities taking place.

Despite many gains, it was noted that a continuing challenge was nurturing the process of collaborative research, planning, education, action, and monitoring, especially given significant staff turnover in partner agencies and organizations. Participants voiced concern about emerging water quality issues and degradation of ecosystem health.

Today, mounting stresses on water resources place even more emphasis on the need to grow partnerships and strengthen collaborative management efforts. According to the 2006 census, the Regional Municipality of Waterloo was the fourth fastest-growing urban area in the province, with a growth rate of nearly 9 per cent between the years 2001 and 2006. The *Growth Plan for the Greater Golden Horseshoe*, released by the government of Ontario in 2006, anticipates rapid population growth in the watershed's five cities (anywhere from 52 to 60 per cent between 2001 and 2031) and associated land intensification. The new Greenbelt Act, passed in 2005, imposes rigid planning policies limiting growth in the Greater Toronto

Barbara Veale

Area. The greenbelt area runs almost entirely along the eastern boundary of the Grand River watershed. There is concern that the greenbelt will cause 'leapfrogging' of development from the Toronto-Hamilton region into the Grand River watershed, creating even more demands on land and water resources.

The new Clean Water Act passed in 2006 adds another layer of complexity to the governance of water in Ontario. The Clean Water Act creates source water protection authorities charged with developing source water protection plans for source protection areas, which are defined on a watershed basis. In most cases, source protection areas coincide with the boundaries of conservation authorities and are combined to create larger source protection regions. Within each region, one conservation authority is appointed as the source water protection authority. The Grand River Conservation Authority is the source water protection authority for the Lake Erie Source Protection Region. A Source Protection Committee consisting of members from watershed municipalities and various sectors, including agriculture, business, industry, environment, and health, has been selected to guide the development of the source water protection plans for each watershed within the Lake Erie Source Protection Region. Once completed, the source water protection plans will be implemented through municipal planning documents.

The Grand Strategy establishes a firm foundation for collaboration from which new partnerships and approaches can evolve within the Grand River watershed and illustrates that strong science and technical tools must be combined with an ongoing participatory approach that allows people to regard water management within the broadest societal context. In the face of unprecedented growth and increasing governance complexity, this integration is essential for motivating collective actions in a timely fashion to resolve current and anticipated water issues. Focusing on measures that improve water quality and sustain water supply in addition to actions that conserve and interpret the watershed's natural and human heritage will ensure that the Grand River and its major tributaries will continue to deserve their national status as Canadian heritage rivers.

Barbara Veale is co-ordinator of policy planning and partnerships, Grand River Conservation Authority, Cambridge, ON.

In the Fraser River basin in British Columbia, a Charter for Sustainability, which outlines what is needed to achieve sustainability for the Fraser River and its watershed, has been signed by federal, provincial, and municipal governments, as well as by First Nations and other organizations. A Fraser Basin Council, a not-for-profit organization, was created and provides oversight for activities to achieve the future identified by the charter.

These examples illustrate the types of initiatives being taken as part of the management strategies for designated **heritage rivers**. All are oriented to protecting the integrity and health of the ecosystems, ranging from biophysical to cultural components. Shuheng Li provides a perspective from China related to a heritage water system in her country, the Grand Canal. The situation she outlines in China offers an interesting contrast to that of the Grand River in Ontario.

International Guest Statement

The Grand Canal of China
Shuheng Li

The Grand Canal of China (also known as the Jing-Hang Grand Canal) is the longest artificial waterway in the world. Its length is 1,794 kilometres, extending from Beijing in the north to Hangzhou, Zhejiang Province, in the south, via Tianjin, Hebei, Shandong, and Jiangsu provinces. It connects five large rivers: the Haihe, Yellow, Huaihe, Qiantang, and Yangtze. In a country dominated by west-east-flowing rivers, the Grand Canal provides a north-south connection among the river systems.

The canal we see today was built section by section in different areas and dynasties. The oldest part dates back to 486 BC. A canal, called Han Gou, was first cut near Yangzhou, Jiangsu Province, to guide the waters of the Yangtze River north to the Huaihe River by using existing waterways, lakes, and marshes. It was lengthened significantly during the Sui Dynasty. The shift of China's

economic centre from the north to the south prompted Emperor Sui Yang, between AD 605 and 610, to extend the canal to link the productive southern region to the nation's capital, Luoyang, in the north. During the Yuan Dynasty of the thirteenth century, the capital was moved to Beijing. The canal was rebuilt and became what is now known as the Grand Canal.

With a history and culture dating back about 2,500 years, the Grand Canal has served as the major transportation artery between north and south China. In terms of improving communication between the north and south and promoting economic and cultural exchanges, the canal, like the Great Wall, was an astonishingly huge project in the history of Chinese civilization. Once the canal opened, many cities along the waterway developed quickly. Communities along the canal grew into the richest agricultural areas in the country. In addition to its contribution to business and the economy, the canal also helped culture to flourish. It helped different ideas,

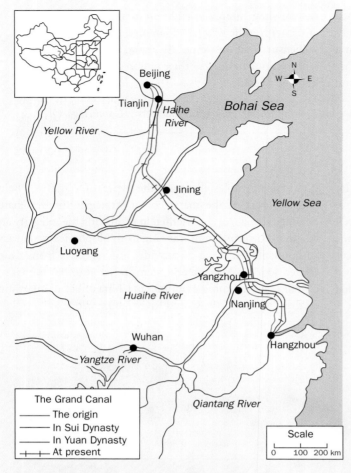

Figure 11.8 Development of the Grand Canal.

However, because of inadequate water quality and the narrow width, most of the canal's northern section (from Beijing to Jining) is inaccessible to boats. Currently, the navigable sector of the Grand Canal is 883 kilometres in length, stretching from Jining to the Sanbao Shiplock in Hangzhou. The accessible part still plays an important role in regional transportation, irrigation, and flood control, despite the decline during the Qing Dynasty and 'Republic of China'. The canal connects major coal deposits in the north and serves as a cheap and convenient channel to transport coal to the fast-developing Yangtze River Delta. More than 100,000 vessels use the canal and carry 260 million tonnes of goods every year, three times as much as the Beijing-Shanghai Railway. It also functions as part of a huge project for water diversion from the south to the north as the main channel of the Eastern Route of the South-North Water Transfer Project. By using the canal, the project can save a great deal of investment, labour, and land. Moreover, the lakes and wetlands it connects also help to reduce flooding.

The historical, cultural, social, and economic significance of the canal has attracted great attention both in China and abroad. Protecting the canal is vitally important not only for maintaining its great historical and cultural values but also for promoting its significant economic and social values. Since 2005, the Chinese government has been making great efforts to have the Grand Canal included on the world heritage list. Led by vice-chairperson Chen Kuiyuan and organized by the History Study Committee, a group of experts toured the Grand Canal in May 2006 for a field study of the canal's preservation and

beliefs, folk customs, and architecture styles to collide and blend. After more than 2,500 years, unique culture and folk customs associated with the canal formed, including countless magnificent cultural relics.

Community of Xitang adjacent to the Grand Canal.

Barges on the North Jiangsu section of the Grand Canal.

utilization. A seminar was held in Hangzhou on preservation of the canal and application for world cultural heritage site recognition for the canal. At the seminar, a declaration was adopted to promote the preservation and application efforts. The State Administration for the Preservation of Cultural Relics has also placed it on a revised list of the country's candidates for world cultural heritage status. The canal was selected as the key cultural relics site under state protection in May 2006.

However, the preservation and application task is challenging, and some thorny problems exist because of inadequate attention paid by the public to the importance of the task. Over the past 10 years, various preservation and maintenance projects have been initiated under different regional administrations, without integrated efforts and results. There has been no unified plan

Shuheng Li

for preservation or a mechanism for overall co-ordination and management. A result is that the canal as a whole has been seriously degrading. The section north of Jining has been dry since the lower part of the Yellow River changed its channel in 1855. Some sections are seriously polluted. Traffic congestion on the canal is frequent along various sections. Some sections of the ancient waterway have been vandalized repeatedly. Some ancient water control projects on the canal and important ancient structures related to the canal have been dismantled to make room for new urban projects. Intangible cultural heritage along the canal, such as folk lifestyles and arts, has been deteriorating.

Thus, China needs to redouble its efforts to protect the Grand Canal and make all the appropriate preparations for successfully getting world cultural heritage status for the canal.

Shuheng Li received her PhD from Nanjing University in 2008. She is a lecturer in the Department of Urban and Resources Science at Northwest University in Xi'an, China. Her research focuses on coastal ocean science and water resource management.

Hydrosolidarity

The term **hydrosolidarity** refers to an approach that recognizes the interconnections among aquatic, terrestrial, and other resource systems, leading to management that is integrated, participative, collaborative, co-ordinated, and shared, whether at local, state, national, or international levels. Understandably, the challenges of achieving hydrosolidarity increase when moving from local to international situations. Notwithstanding such difficulties, hydrosolidarity is most often evoked as necessary in international settings (International Water Resources Association 2000). Hydrosolidarity stands in contrast to more traditional approaches to international rivers or lakes in which sovereign states claim control over the water within their boundaries without regard for the implications for countries that share the resource, especially as downstream users.

Elements of best practice associated with hydrosolidarity reflect the reality of aquatic systems, including that: (1) water flows downhill, leading to upstream and downstream interests in a river basin or catchment; (2) interconnections exist between water and land systems, meaning that land-based activities can have significant implications for water quantity

and quality; and, (3) the multiple uses that water can serve range from drinking, to crop and industrial production, to supporting migratory bird habitats and recreation. As a result, best practice regarding hydrosolidarity includes management by (1) using river basins or watersheds as the spatial unit for planning and management; (2) ensuring attention to upstream-downstream issues; (3) recognizing the interrelationships among water, land, and other resource systems; (4) engaging a range of stakeholders in a collaborative and participatory manner; and (5) acknowledging the needs of biophysical systems as well as humans.

Consistent with hydrosolidarity is the concept of **integrated water resource management (IWRM)**, which the Global Water Partnership (2000, 22) has defined as 'a process which promotes the co-ordinated development and management of water, land, and related resources in order to maximize the resultant economic and social welfare in an equitable manner without compromising the sustainability of vital ecosystems'.

The motivation for IWRM is to overcome the challenges arising from various groups having interest in and organizations being responsible for water and related resources, along with the reality that they each often focus only on their own interests and responsibilities. One result, for example, can be a ministry of agriculture providing support for farmers to drain wetlands in order to expand land in agricultural production, while a ministry of natural resources in the same jurisdiction provides support to farmers to expand wetlands in order to extend migratory bird habitat and enhance capacity to slow the release of flood waters. As Dale and Newman (2007, 59–60) have observed, an integrated approach seeks to overcome the silos, stovepipes, and solitudes that often characterize the approaches of governmental organizations responsible for water, land, and other related resources.

Ideally, IWRM drives managers to consider integration at several levels, including:

1. Integration of various dimensions of water, such as quantity and quality, surface and underground, and upstream and downstream. Use of river basins or watersheds for planning and management is recognized as the most appropriate spatial unit to achieve this.
2. Integration of water considerations with those for terrestrial and other related resources. This involves recognizing that many water problems, such as pollution and flooding, originate from or are exacerbated by land-based activities. Here again, the river basin or watershed is viewed as the most appropriate unit to facilitate attention to a range of interconnected resource systems.
3. Integration of water, as part of the environmental system, with aspects related to economic and social systems. Regional land-use planning, environmental impact

Great Lakes System

The Great Lakes—Superior, Michigan, Huron, Erie, and Ontario—constitute the largest system of fresh surface water on earth, containing about 20 per cent of the world's surface water. Less than one per cent of the system's water is renewed annually by precipitation. In addition to the natural beauty, ecological complexity, and significance of the lakes, the region is home to nearly 40 million Americans and Canadians, supports the culture and life ways of tribes and First Aboriginal Peoples, provides drinking water to millions of people, and is the foundation for billions of dollars in shipping, trade, agriculture, fishing, and recreation.

—Agreement Review Committee 2007, 5.

assessment, and strategic sustainability assessment are used to connect environmental, economic, and social considerations.

The rationale for IWRM is intuitive: water is but one subsystem, and therefore managers need to take an ecosystem approach to ensure linkages and connections are addressed. Such connections occur at scales ranging from a local subcatchment to a major interjurisdictional river basin or catchment. However, the challenges to implementing IWRM can be formidable, because human behaviour is often competitive rather than cooperative, individuals focus on narrow interests without concern for the consequences for others today or in the future, and understanding complex natural and human systems can be very difficult (Biswas 2004; Rahman and Varis 2005).

The following examples illustrate the opportunities and challenges in implementing a hydrosolidarity approach.

Great Lakes Water Quality Agreement

In 1972, the governments of Canada and the United States signed the Great Lakes Water Quality Agreement (GLWQA). The agreement was amended in 1978, 1983, and 1987. It focuses on the Great Lakes ecosystem, defined to include the interacting components of air, land, water, and living organisms, including humans. The purpose is to restore and maintain the chemical, physical, and biological integrity of the waters in the Great Lakes ecosystem.

The objectives of the 1972 agreement were to reduce nuisance conditions and discharge of substances toxic to humans, animals, and aquatic life, as well as to reduce phosphorous loadings in Lakes Erie and Ontario. During 1978, the agreement was amended to specify that an ecosystem approach would be used, with attention directed to both human health and environmental quality. Furthermore, amendments indicated that the intent was to virtually eliminate persistent toxic substances.

In 1983, further amendments stipulated that efforts would be increased to reduce phosphorous inputs into each lake. During 1987, a key change was to identify 'areas of concern', near-shore areas that had been significantly degraded and for which rehabilitation efforts would be pursued. In addition, new initiatives were to be undertaken to address non-point pollution sources and create Lakewide Management Plans, as well as deal with contaminated sediment, airborne toxic substances, and contaminated groundwater.

Despite impressive progress resulting from the combined efforts of national, state/provincial, and municipal governments and First Nations, the Agreement Review Committee (2007, 8) remarked that 'there are still serious threats to the physical, biological, and chemical integrity of the Ecosystem. Many scientists have voiced concern that the Great Lakes are exhibiting symptoms of stress from a variety of sources and impacts, including nutrient loadings, toxic contaminants,

Great Lakes Water Quality Agreement

Today, the Agreement remains one of the most farsighted international agreements and is a model of cooperative environmental research and ecosystem management.

— International Joint Commission 2004, 1

Often cited as one of the most forward-thinking diplomatic achievements for the environment, the Agreement has served as a model for other international agreements to protect and restore environments elsewhere in the world. Its strengths include the establishment of common objectives and commitments for protecting and restoring the waters of the Great Lakes Ecosystem, the facilitation of information sharing, and cooperation on research and monitoring.

— Agreement Review Committee 2007, 6–7.

invasive species, and land use changes.' The drivers behind some of these challenges were viewed by the committee as including climate change, urbanization, long-range transport of toxics, and inadequate approaches to stop invasive species. Regarding the latter, the International Joint Commission (2004, 13) has stated that new aquatic alien species are introduced into the Great Lakes at a rate of one every eight months from ocean-going ships or from bait fish, aquarium fish, aquaculture, or connecting tributaries.

In terms of future needs, the Agreement Review Committee (2007) identified the following:

- clarification related to the ecological scope of the agreement to establish whether the focus is only on water quality or on a full ecosystem approach;
- attention to the geographical scope to clarify whether the focus is on open waters or on a combination of open waters, near-shore waters, inland areas, tributaries, and watersheds within the basin;
- clarification regarding whether groundwater resources are meant to be included as part of the Great Lakes ecosystem (the agreement considers groundwater only as a source of contamination, not as a component of the overall ecosystem);
- clarification regarding whether the entire St Lawrence River system should be included as part of the ecosystem (at the moment, the agreement includes only that part of the river down to the international boundary at Cornwall, ON);

Invasive Species

. . . the flow of new invasive species to the Great Lakes has not been stopped. In 2001, scientists estimated that 162 invasive species had entered the lakes from all pathways. Today, some scientists have raised that estimate to more than 170 non-indigenous fish, invertebrates, plants, algae, protozoa, and parasites. . . .

— International Joint Commission 2004, 15

- determination of how to apply management concepts such as watershed management planning, adaptive management, pollution prevention, biodiversity initiatives, and airshed management;
- determination of how best to deal with chemicals, such as pharmaceuticals, flame retardants, and personal care products, not addressed in the agreement;
- in addition to that of the two national governments, clarification of the critical role and essential participation of other levels of government, such as states and provinces, local governments, and tribes and First Nations.

The above needs capture some of the challenges of achieving a hydrosolidarity approach, in spite of the high regard earned by the Great Lakes Water Quality Agreement through collaboration involving two national governments and numerous state/provincial and municipal governments.

New Pathogens

Some experts believe that the massive and largely unregulated use of antibiotics in agriculture and aquaculture, coupled with the increasing number of antibiotic-resistant pathogens found in nature, may present the greatest risk to the aquatic environment and to public health. Antibiotic-resistant bacteria have been spread in the environment through the indiscriminant use of antibiotics in human and animal health.

— International Joint Commission 2004, 30

Garrison Diversion and Devils Lake Outlet

Two projects in the United States with implications for water systems in Canada, especially in Manitoba, illustrate that best practice regarding hydrosolidarity is not always achieved. Manitoba has been concerned about the projects because, in the words of the provincial government (Manitoba 2001, 1),

> Water resources within the Manitoba portion of the Hudson Bay drainage basin are extremely valuable. More than one million Manitobans live in the basin, with approximately 100,000 relying directly on the Red River, Souris River, and Assiniboine River as sources of drinking water. Lake Winnipeg, the tenth largest freshwater lake in the world, supports a commercial fishery with a direct, landed value estimated to be over $15,000,000 (Canadian) annually to the economies of Canada and Manitoba. . . . About 800 commercial fishing licences are issued to fishers on Lake Winnipeg, many of these to First Nations people. In many cases, fishing is their major source of income. Sport anglers spend another $10 to $15 million (Canadian) annually in the region. The Red River is the single most important sport fishing destination in Manitoba and accounts for nearly 20 per cent of the total value of this $80-million (Canadian) industry to Manitoba's economy.

The Manitoba government also noted that the Boundary Waters Treaty, signed by Canada and the United States in 1909, stated that 'waters flowing across the boundary shall not be polluted on either side to the injury of health or property on the other side.' Under the treaty, the International Joint Commission was created to help governments of both nations to resolve water problems and overcome associated disputes.

Garrison Diversion

The Garrison Dam, with an associated large reservoir, was constructed on the Missouri River in North Dakota during the mid-1950s. Subsequently, various projects were proposed to take water from the reservoir to supply water for municipal and industrial uses and to establish irrigation systems to support agricultural production in the state. The best known is the Garrison Diversion project, which would involve transferring water out of the Missouri River, using a 120-kilometre canal to cross the continental divide to connect with another 71-kilometre canal. One consequence would be that water and biota from the Missouri River system would be moved into the Hudson Bay drainage basin, including Lake Winnipeg with its valuable fishery. As a result, both the Canadian and Manitoba governments objected to the Garrison Diversion project.

The initial concerns from Canada led to an investigation by the International Joint Commission (IJC) during 1977. The conclusion of the IJC was that use of the Missouri River to supply irrigation water for agriculture in the area of North Dakota in the Hudson Bay basin should not occur. The IJC expressed particular concern about the possible negative impacts that might occur from the transfer of non-native species between the two river basins, as well as from the potential for introduction of invasive species, diseases, and pathogens. The IJC recommendation in the late 1970s reflected the best practice elements of hydrosolidarity, especially that activity in an upstream or other basin should not have deleterious consequences for users downstream or in another basin.

In 2000, the North Dakota state government passed the Dakota Water Resources Act to replace the previous Garrison Diversion Reformulation Act of 1986, which had reflected the views of the IJC noted above. The Dakota Water Resources Act will make it much easier for water transfers to occur between the Missouri River and Hudson River systems, because it:

- presumes that by treating the Missouri River water, it is reasonable to transfer it to the Hudson Bay basin (however, there is no agreement among scientists regarding what level of treatment is required);
- makes it possible for interbasin transfer of water in order to meet in-stream flow needs and to augment aquifers, with the same potential problems of interbasin movement of biota (Manitoba Water Stewardship 2007).

The Northwest Area Water Supply Project (NAWS) is one of several proposals to move water that would be pre-treated from the Missouri River across the continental divide (another proposal is termed the Red River Valley Water Supply Project and focuses on water needs in eastern North Dakota). Under NAWS, water would be transferred to Minot in North Dakota, which is in the Hudson Bay basin. This water would receive further treatment and then be distributed to provide drinking water for various communities in western North Dakota, which are all in the Hudson Bay drainage basin. Such an interbasin transfer would be the first of its kind between these two basins, and there is concern that it will establish a precedent that could lead to subsequent and larger interbasin transfers.

For Canada, the concern has been that the state government of North Dakota and the US government bypassed the International Joint Commission, the mechanism established to deal with transborder resource issues. Instead, it was the US Bureau of Reclamation that prepared an environmental impact assessment for the NAWS project under the terms required by the US National Environmental Policy Act. When the assessment was completed in June 2001, the US State Department formally provided a copy of it to the Canadian government. Both Canada and Manitoba disagreed with the findings that there were no potential negative environmental impacts from NAWS. When those concerns were not recognized in the US, Manitoba launched a legal challenge. In February 2005, a US judge ruled in favour of Manitoba, indicating that the environmental impact assessment had not satisfactorily addressed the risks and implications related to interbasin transfer of foreign biota (bacteria, viruses, parasites, and fungi). The US Bureau of Reclamation in March 2006 agreed to conduct a comprehensive environmental assessment, which is what Manitoba had asked for.

Interbasin Water Transfers and Hydrosolidarity

Issues surrounding biota transfer are even more of a concern now than in 1977 [when the IJC examined the Garrison Diversion Project]. The issue of invasive species is recognized as one of the most formidable environmental challenges facing governments in North America. . . . Both the United States and Canada are working cooperatively in many areas to control invasive species through ballast water programs, sharing of scientific information, and regional cooperation to implement the US Aquatic Nuisance Species Act. Economic losses from invasive species can be staggering. A 1999 US study estimates damages and losses caused by invading foreign species adding up to more than $138 billion per year. Zebra mussels cause damage in the US that is estimated at $3 billion annually. The mussels have not yet invaded the Hudson Bay drainage area, but an impact of similar magnitude can be expected should zebra mussels make their way into the Hudson Bay drainage basin.

—Manitoba Water Stewardship 2007, 4

Devils Lake Outlet

Devils Lake is located in North Dakota and is a small 'closed system'. It is a subcatchment within the larger Hudson Bay river basin, but no water flows out of Devils Lake, and hence the designation 'closed'. High precipitation led to water levels reaching record levels, causing damage in the town of Devils Lake. The state government began to lobby for a project that would create an emergency outlet from the lake, allowing water to be pumped into the Sheyenne River, which joins the Red River before it flows into Manitoba. The project also has provision for water to be pumped from the Missouri River into Devils Lake during times of drought.

Since 1999, the governments of Manitoba and Canada, along with Great Lakes state governments and Missouri, have all expressed concern about this project. The main concern

has been that biota from the Missouri River system and Devils Lake would be introduced into the downstream Red River and from there would be introduced into Lake Winnipeg. Canada and Manitoba have argued that the IJC should be asked to investigate the cross-border implications of the Devils Lake Outlet proposal. In the words of the government of Manitoba (2001, 5) with reference to Devils Lake,

> The permanent, large-scale diversion of water between the Missouri River and Hudson Bay basins represents an unacceptable risk of irreversibly damaging the aquatic resources of Manitoba. It is a risk that never can be reduced to zero and a risk that has enormous consequences for Manitoba, given the value of our water resources, including those of Lake Winnipeg.

The state of North Dakota, however, completed the Devils Lake Outlet during 2005. In August of that year, after negotiations between the Canadian and US governments, North Dakota received approval to operate the outlet after additional environmental safeguards had been installed to protect downstream systems. To date, monitoring has shown that the 12 invasive species that have caused problems in other systems are not present in Devils Lake. On the other hand, other biota with potential to cause downstream damage, which are not known to be present in Lake Winnipeg, have been identified.

The spirit of hydrosolidarity is that upstream and downstream jurisdictions work together collaboratively to ensure that initiatives in upstream parts of an aquatic system do not cause significant damage in downstream parts. The International Joint Commission was created for exactly that purpose and is viewed around the world as a model of how transjurisdictional resource issues can be addressed. However, some are concerned that there is a trend in the US, especially on the part of the federal government, to take unilateral actions or to avoid mechanisms created to deal with cross-border issues. Certainly, when a nation agrees to participate in mechanisms and processes to deal with cross-border problems, it acknowledges that it is sacrificing some autonomy in order to reach decisions that benefit all nations. When a country demonstrates that it is not prepared to participate in processes requiring it to share or give up some its authority or power, meeting the spirit of hydrosolidarity becomes very challenging.

Water Ethics

Water has been attracting increasing attention. Some have suggested that in the twenty-first century, it will represent what oil did at the end of the twentieth century as a strategic resource. Some have gone even further, suggesting that regional wars can be expected as a result of water shortages. Whether or not armed hostilities may erupt over water, the discussion in this chapter highlights that water is a necessity for human life. There is no substitute. Furthermore, many needs and interests compete for their share of water. Given this situation, it is puzzling that more attention has not been given to development of **water ethics** to establish principles on which water management decisions could be based. UNESCO has published some relevant reports, such as Selbourne (2000) and Priscoli, Dooge, and Llamas (2004), but study focused on water ethics has been paltry.

Matthews, Gibson, and Mitchell (2007) have addressed the matter of water ethics, and the following comments are based on their work. They remind us that an ethic normally is a statement of principles or values to guide behaviour by individuals or groups. Ethics identify appropriate values and behaviour. At the same time, a set of ethics cannot provide all the answers needed or resolve all dilemmas. For example, different ethics may lead to different outcomes, some positive and some negative. Frequently, uncertainty and complexity make outcomes difficult to predict, contributing to fuzzy understanding about what might happen. Also, different ethical principles, each desirable, can sometimes conflict, making it unclear as to which path to pursue.

Notwithstanding the above difficulties, Matthews, Gibson, and Mitchell (2007, 350–3) offered six 'imperatives' for a new water ethic. They are:

- meet basic human needs to enhance equity today and for the future;
- safeguard ecosystems by allocating sufficient water resources;
- encourage efficiency and conservation of water resources;
- establish open and participative decision-making processes;
- respect system complexity and emphasize precaution; and
- seek multiple sustainability benefits from water-centred initiatives.

Regarding the first imperative (meet human needs to enhance equity today and for the future), the United Nations Human Rights Council has been endeavouring to have water and sanitation recognized as a basic human right and to establish an international monitoring organization to track the actions of nations. The council met for three weeks in March and April 2008 to address this and other matters. The occasion was the third time in six years that the United Nations had attempted to have human rights to water and sanitation recognized.

At the Human Rights Council meeting in Geneva in early April 2008, however, Canada opposed the resolution for water to be recognized as a basic human right, which had been proposed by Germany and Spain. Canada was not the only country to oppose the resolution: Russia and the United Kingdom also did not support it. Nevertheless, it was generally agreed that Canada led the opposition.

Canada has been consistent in its opposition. During a Human Rights Council meeting in 2002, Canada was the only one of 53 nations voting against a motion to appoint a special rapporteur on water. And in October 2006, Canada voted against a resolution to have the Human Rights Council conduct a study on the right to water.

Canada's lack of support appears to be driven by a belief on the part of federal politicians that Canadian sovereignty over its own water is ambiguous under NAFTA. In that trade agreement, water could be viewed as a commodity or service comparable to any other. As a result, if Canada were to support the principle that water is a basic human right, some politicians believe that it could become vulnerable to claims from the United States that since Canada has more water than it needs and places such as Atlanta and the American southwest face water shortages, then Canada has a moral duty to share its water through bulk water transfers or other means. In contrast, others argue that the Human Rights Council resolution explicitly excluded transborder water issues and therefore Canadian sovereignty over its water would not be threatened. This situation highlights that identifying and working from fundamental ethical principles does not automatically simplify and resolve challenges. At the same time, such ethical principles help to clarify the fundamental issues at stake.

We invite you to use the six 'imperatives' or some modification of them as a lens through which to view current water systems planning and management. For example, if the above imperatives were used to guide decisions, what would the implications be for water security on First Nations reserves? Would the outcome of the Devils Lake Outlet project have been different from what actually happened? The imperatives or principles can be used to analyze such situations and to determine other options for consideration.

Implications

Canada is fortunate in having a relative abundance of high-quality water, even though some areas experience scarcity in terms of quantity and/or quality. As a society, however, we have placed stress on aquatic systems, sometimes degrading them significantly. As the discussion in this chapter has shown, considerable scientific understanding can be drawn upon to assist in the management of water systems. And there have also been some significant improvements, confirming that individuals, groups, communities, and societies can reverse degradation and deterioration. Unfortunately, the government is not showing the kind of leadership that might be hoped for, given the significance of water to Canadians. The Canada Water Act requires the tabling of an annual report on water use and management. In early 2008, the most recent report available was for 2001–2. Even if the government complies with its own legislation and produces the required report, do you think it is acceptable for such long delays to occur? If not, what might you do about it?

Each of us has an opportunity to modify our basic values and change behaviour to help protect our water resources and to maintain the integrity and health of aquatic ecosystems. Table 11.5 identifies action that individuals and/or governments can take to further contribute to improved water resources in Canada.

Table 11.5   What You Can Do: Ten Water Conservation Initiatives

Focus	Problem	Solution	Challenges	Savings
1. Educate.	Lack of understanding about need for and potential benefits from water conservation	Introduce outreach and education programs beyond information dissemination to change behaviour.	Engaging community members in meaningful education that changes view of 'water abundance'	Many experts agree 50 to 80 litres of high-quality water per capita per day is needed for a good standard of living.
2. Design communities for conservation.	Municipal-level decisions too often having negative impacts on watersheds	Limit urban sprawl, reduce green lawn syndrome, promote 'green' infrastructure, expect land-use decisions to be judged for their impacts on watersheds.	Belief that water-sensitive urban design is much more expensive than standard approaches	Conservation-oriented urban design can save 50% of outdoor water use.
3. Close urban water loop.	All municipal water treated to drinking-water quality but more than two-thirds used for non-drinking functions	Use reclamation, reuse, and recycling to ensure better match of water quality to end uses.	Possibility that lower water prices may make reuse and recycling less financially attractive; risks associated with reused and recycled water	Up to 50% water savings can be achieved by reusing or recycling water for toilets and outdoor irrigation.
4. Use rainwater.	Rainwater not usually viewed as source of water for homes, businesses, etc.	Use decentralized infrastructure to harvest rainfall and create xeriscaped landscapes that rely on rainfall.	Building and plumbing code restrictions; financial cost of rainwater harvesting infrastructure for homeowners and businesses	Rainwater harvesting and xeriscaping can lead to 50% savings in outdoor water use and up to 40% saving for indoor use for toilet flushing and clothes washing.
5. Plan for sustainability.	Too many conservation programs viewed as short-term solutions until next supply source can be developed	Plan with 10- to 50-year time horizon, involve all stakeholders, and place ecological health in central position.	Looking beyond the three- or four-year electoral cycle and investing in programs that provide long-term returns; engaging the community	Effective water conservation plans can lead to water savings ranging from 20% to 50%.
6. Adopt appropriate pricing.	Normal water pricing rates encouraging wasteful use	Use 'full cost' pricing with volume-based pricing structures.	Need to implement metering and gain political support; need to ensure full accessibility to meet basic needs for water	Effective pricing can lead to 20% reduction in water use over the long term.
7. Link conservation to development.	Present arrangements for funding urban water infrastructure promoting neither conservation nor innovation	Connect conservation to development by requiring water infrastructure funding and development permits to depend on use of demand management.	Local resistance to conditional funding arrangements; capacity for enforcement and follow-up on conditions	Water savings from 20% to 30% can be achieved by using 'off the shelf' technologies and modest water pricing reforms.
8. Make managing demand part of regular business.	Demand management approaches often not comprehensive or part of daily business in most communities	Implement permanent water conservation measures and hire full-time staff with appropriate skills.	Reluctance of utilities to commit financial resources to hire demand management professionals and implement long-term demand management programs	Depending on how aggressive and creative demand management programs are, 'the sky is the limit' in terms of potential savings.
9. Stop flushing the future.	Inefficient fixtures and appliances common in most homes	Install efficient toilets, faucets, and showerheads as well as water-saving dishwashers and washing machines.	Permissive building and plumbing codes and lack of incentives and resources to promote efficient technologies	Efficient fixtures and appliances can achieve 33% to 50% indoor water savings with payback within two years in most instances.
10. Fix leaks and reduce waste.	Significant water loss from leaks, often due to old infrastructure	Detect and repair leaks by regular water audits and maintenance programs.	Financial challenges on utilities of up-front costs for integrated metering, detection, maintenance, and monitoring programs	Fixing leaks can easily result in 5% to 10% water savings, and up to 30% savings is possible with older infrastructure.

Source: Based on Brandes, Maas, and Reynolds 2006, 6–42.

Summary

1. While comprising only 0.5 per cent of the world's population, Canadians have access to almost 20 per cent of the global stock of fresh water and to 7 per cent of the total flow of renewable water. About one-fifth of the Canadian population relies on groundwater for daily water needs.

2. Canada's per capita demands on water resources are the second highest in the world and have been calculated at about 326 litres per person per day at home.

3. Canada is a global leader in terms of water diversions for hydroelectric generation, and diversions for hydropower purposes dominate overwhelmingly in both number and scale of diversions.

4. Canada has used many megaprojects to meet energy demands, and virtually every region in the country has energy megaprojects. One of the most significant is James Bay in Quebec, while others are Churchill Falls in Labrador, the Nelson-Churchill in Manitoba, and the Columbia and Nechako rivers in British Columbia.

5. Various major projects have been proposed, such as NAWAPA and the GRAND scheme, to transfer large volumes of water from Canada to the United States. They all have significant financial costs and environmental implications. At the moment, other options appear to be more realistic.

6. Free trade agreements, such as stipulations under NAFTA, trigger concern that water could become a tradable commodity even if Canada did not want to send bulk exports to the United States.

7. Pollution involves point and non-point sources, with the latter being the most challenging to manage.

8. Progress regarding diffuse pollution has been associated with *credible science* to document the nature of the problem.

9. The governance of the Great Lakes is evolving, and key aspects of the evolution include a shift from an exclusive command-and-control emphasis to voluntary measures and from top-down management to environmental partnerships.

10. At the start of the twenty-first century, it was estimated that 1.1 billion people worldwide did not have access to safe water supplies and that two out of five people did not have access to adequate sanitation.

11. Many Canadians have been complacent about the adequacy and safety of their water supplies. For many, this changed in mid-May 2000 when the small town of Walkerton, ON, experienced contamination of its water supply system by deadly bacteria. Seven people died, and more than 2,300 became ill.

12. The judge who conducted a public inquiry into the Walkerton tragedy recommended a 'multi-barrier approach' to drinking water safety.

13. The water crisis at Kashechewan in 2005 drew attention to the often unsatisfactory water supply infrastructure on First Nations reserves as well as in many other rural communities in remote regions. Water supply systems in such places are often more like those in developing countries.

14. Increasing attention is being given to determining the most appropriate mix of supply management, demand management, and soft path approaches to provide water to communities. The soft path approach shifts attention from questions about *how* to meet needs to questions about *why* needs are met by using water.

15. Environmental flow management is increasingly being used to ensure that needs of biophysical systems are included when water allocation decisions are made.

16. The concept of 'virtual' water is used to encourage nations to determine whether it would be more sensible to import water-intensive crops than to grow them in their own countries. This sometimes leads to conflict between two desirable goals: food self-sufficiency and efficient water use.

17. Awareness is growing about the vulnerability of water infrastructure to terrorism. No infrastructure system can ever be made totally invulnerable, but the risk can be reduced.

18. Floods are a normal component of the behaviour of rivers and lakes. Humans adapt to flooding through some mix of structural and non-structural approaches that aim to modify behaviour of natural systems and humans.

19. Droughts are a function of precipitation (or lack thereof), temperature, evaporation, evapotranspiration, capacity of soil to retain moisture, and resilience of flora and fauna in dry conditions.

20. The most drought-prone areas in Canada are south-central British Columbia, the southern Prairie provinces, and southern Ontario.

21. The Canadian Heritage Rivers Program recognizes the significance of rivers in the identity and history of Canada and is intended to ensure that their natural, cultural, and recreational values are protected.

22. Hydrosolidarity and integrated water resource management are similar concepts that require an ecosystem approach to water management in which attention is given to connections among water, land, and other resources, relationships between upstream and downstream parts of a basin, and linkages between surface and groundwater.

23. The Great Lakes Water Quality Agreement contains many elements of best practice related to hydrosolidarity, while the Garrison Diversion and Devils Lake Outlet violate basic best practice elements.

24. Attention is being drawn towards developing principles related to a 'water ethic' to guide decisions regarding water allocation, development, and use.

Key Terms

blue water
Brisbane
   Declaration
cryosphere
demand
   management
diversion
drought
dykes
E. coli

environmental flow
   management
floodway
green water
heritage rivers
hydrosolidarity
International Joint
   Commission
integrated water resource
   management (IWRM)

James Bay and
   Northern Quebec
   Agreement
James Bay Project
limiting factor principle
megaprojects
multi-barrier approach
Palliser's Triangle
renewable water
   supplies

soft path
supply management
virtual water
water ethics
water table
water terrorism
wetlands

Questions for Review and Critical Thinking

1. Does the 'myth of superabundance' adequately account for the high per capita water use by Canadians?
2. Why do some believe that Canada has 20 per cent of the world's renewable waters supplies while others suggest that this number is 2.6 per cent?
3. What are the main reasons for water diversions in Canada?
4. Why is the likelihood of pressure emerging in the United States to import bulk water supplies from Canada relatively low?
5. Why is the presence of mercury in fresh water a serious concern?
6. Explain the significance of the 'limiting factor principle'.
7. Why are there different levels of treatment for municipal waste water?
8. Explain the significance of bioaccumulation and biomagnification of chemicals in the food chain and what role water plays in it.
9. Why is diffuse pollution from non-point sources a challenge for managers?
10. What is meant by 'water security'?
11. What are the implications of the experience with contaminated drinking water in Walkerton, ON, and North Battleford, SK?
12. What is the significance of the failure of the water supply system in Kashechewan?
13. Explain the distinction among supply management, demand management, and soft path approaches.

14. Explain the intent of environmental flow management.
15. What is the relationship between 'virtual' water and 'green' water?
16. How serious is the threat of water terrorism, and what options are available to reduce the risk from terrorism actions?
17. What is the significance of structural and non-structural approaches for reducing flood damages?
18. What lessons were learned from major floods in Canada in the past five years?
19. How do we know when a drought begins and ends?
20. What are the most significant flood and drought-prone areas in Canada?
21. What is Palliser's Triangle?
22. What criteria and indicators should be used to identify the natural, cultural, and recreational value of rivers?
23. Are there river systems in your region that could be candidates for nomination as heritage rivers?
24. Explain the relationship between the concepts of 'hydrosolidarity' and 'integrated water resource management'.
25. What is the significance of the Great Lakes Water Quality Agreement, the Garrison Diversion, and the Devils Lake project in terms of best practice associated with hydrosolidarity?
26. What principles should underlie a 'water ethic' to guide decisions about allocation and use of water?

Related Websites

Atlantic Coastal Action Program: http://atlantic-web1.ns.ec. gc.ca/community/acap

Canada's Water Quality Guidelines: http://www.ec.gc.ca/ CEQG-RCQE/English/Ceqg/Water

Canadian Council of Ministers of the Environment: http:// www.ccme.ca

Canadian Global Change Program: http://www.globalcentres .org/cgcp

Canadian Pollution Prevention Information Clearinghouse: http://www.ec.gc.ca/cppic

Canadian Water and Wastewater Association: http://www.cwwa.ca

Canadian Water Resources Association: http://www.cwra.org

Clean Annapolis River Project: http://www.annapolisriver.ca

Environment Canada's Water Information Collection: http://www.ec.gc.ca/water/e_main.html

Experimental Lakes Area: http://www.umanitoba.ca/institutes/fisheries

Fisheries and Oceans Canada, Freshwater Institute: http://www.dfo-mpo.gc.ca/regions/CENTRAL/pub/fresh-douces/01_e.htm

Great Lakes Information Network: http://www.great-lakes.net/lakes

International Joint Commission: http://www.ijc.org

National Water Research Institute: http://www.nwri.ca

Northern River Basins Study: http://www3.gov.ab.ca/env/water/nrbs/nrbs.html

Statistics Canada, Human Activity and the Environment: Annual Statistics: http://www.statcan.ca/bsolc/english/bsolc?catno=16-201-X&CHROPG=1

St Lawrence Vision 2000 Action Plan: http://www.slv2000.qc.ca/index_a.htm

UNESCO water portal bi-monthly newsletter: http://www.unesco.org/water/news/newsletter

WaterCan: http://www.watercan.com

Further Readings

Note: This list comprises works relevant to the subject of the chapter but not cited in the text. All cited works are listed in the section 'References' at the end of the book.

Allan, J.A. 1998. 'Virtual water: A strategic resource. Global solution to regional deficits'. *Groundwater* 36: 545–6.

———. 2003. 'Virtual water—the water, food and trade nexus: Useful concept or misleading metaphor?' *Water International* 28: 4–10.

Annan, K.A. 2002. 'The Secretary-General's Message on World Environment Day, June 5, 2002'. New York: United Nations. http://www.escwa.org.lb/information/press/un/2003/june/5/html.

Armstrong, A. 2006. 'Ethical issues in water use and sustainability'. *Area* 38 (1): 9–15.

Bakker, K., Ed. 2007. *Eau Canada*. Vancouver: University of British Columbia Press.

Berger, T.R. 1977. *Northern Frontier; Northern Homeland*, v. 1. Ottawa: Minister of Supply and Services.

Botts, L., and P.K. Muldoon. 2005. *Evolution of the Great Lakes Water Quality Agreement*. East Lansing: Michigan State University Press.

Bourassa, R. 1985. *Power from the North*. Scarborough, ON: Prentice-Hall Canada.

Brelet, C. 2004. *Some Examples of Best Ethical Practices in Water Use*. Paris: UNESCO.

British Columbia Ministry of Environments. 2008. *Living Water Smart*. Victoria: British Columbia Ministry of Environments.

Campbell, K., and Y. Nizami. 2001. *Security or Scarcity? NAFTA, GATT and Canada's Fresh Water*. Vancouver: West Coast Environmental Law.

Canada. 2007. *Canadian Water Sustainability Index: Project Report*. Ottawa: Policy Research Initiative.

Canada. Washington Embassy. 2007. 'Garrison Diversion and the Devils Lake Outlet: The Canadian position'. http://geo.international.gc.ca/can_am/washington/shared_env/garrison-en.asp.

Christensen, R. 2006. *Waterproof 2: Canada's Drinking Water Report Card*. Vancouver: Sierra Legal Defense Fund.

Council of Great Lakes Governors. 2005. 'The Great Lakes–St Lawrence River Basin Water Resources Compact and the Great Lakes–St Lawrence River Basin Sustainable Water Resources.' http://www.cglg.org/projects/water/annex2001Implementing.asp.

Diamond, B. 1985. 'Aboriginal rights: The James Bay experience'. In M. Boldt and J.A. Long, Eds, *The Quest for Justice: Aboriginal People and Aboriginal Rights*, 265–85. Toronto: University of Toronto Press.

Environment Canada. 2003. *Performance Report for the Period Ending March 31, 2003*. Ottawa: Environment Canada.

———. 2004. *Threats to Water Availability in Canada*. NWRI Scientific Assessment Report Series no. 3 and ACSD Science Assessment Series no. 1. Ottawa and Burlington, ON: Environment Canada and National Water Research Institute.

Environment Defense. 2007. *Up to the Gills: Pollution in the Great Lakes*. Toronto: Environment Defense.

Falkenmark, M. 2004. 'Towards integrated catchment management: Opening the paradigm locks between hydrology, ecology and policy-making'. *Water Resource Development* 20: 275–82.

Falkenmark, M., and C. Folke. 2002. 'The ethics of socio-ecohydrological catchment management: Towards hydro-solidarity'. *Hydrology and Earth System Sciences* 6: 1–10.

Falkenmark, M., and J. Rockstrom. 2005. *Balancing Water for Humans and Nature: The New Approach to Ecohydrology*. London: Earthscan.

Griffiths, M., A. Taylor, and D. Woynillowicz. 2006. *Troubling Waters, Troubling Trends*. Calgary: Pembina Institute.

Harremoes, P. 2002. 'Water ethics: A substitute for over-regulation of a scarce resource'. *Water Science and Technology* 45 (8): 113–45.

Heinmiller, B.T. 2003. 'Harmonization through emulation: Canadian federalism and water export policy'. *Canadian Public Administration* 46: 495–513.

Hellegers, P., et al. 2008. 'Interactions between water, energy, food and environment: Evolving perspectives and policy issues'. *Water Policy* 10 (supplement 1): 1–10.

International Joint Commission. 1998. *Ninth Biennial Report on Great Lakes Water Quality*. Windsor, ON: International Joint Commission.

Izquierdo, F. 2004. 'Statement to the Forum Barcelona 2004.' http://www.barcelona2004.oprg/eng/banco_del_conocimiento/documentos/ficha.cfm?IdDoc+749.

McCutcheon, S. 1991. *Electric Rivers: The Story of the James Bay Project*. Montreal: Black Rose Books.

Mainville, R. 1992. 'The James Bay and Northern Quebec Agreement'. In M. Ross and J.O. Saunders, Eds, *Growing Demands on a Shrinking Heritage: Managing Resource Conflicts*, 176–86. Calgary: Canadian Environmental Law Association.

Merrett, S. 2003. 'Virtual water and Occam's razor'. *Water International* 28: 103–15.

Nelson, J.G., B. Veale, and B. Dempster. 2004. *Towards a Grand Sense of Place*. Waterloo, ON: University of Waterloo, Heritage Resources Centre.

Perkel, C.N. 2002. *Well of Lies: The Walkerton Water Tragedy*. Toronto: McClelland and Stewart.

Pollution Probe. 2008. *A New Approach to Water Management in Canada*. Toronto: Pollution Probe.

Schindler, D.W., and W.F. Donahue. 2006. 'An impending water crisis in Canada's western Prairie provinces'. *Proceedings of the National Academy of Sciences of the United States of America* 103: 7,210–16.

Schindler, D.W., and J.P. Smol. 2006. 'Cumulative effects of climate warming and other human activities on freshwaters of Arctic and subarctic North America'. *Ambio* 35: 160–8.

Standing Committee on Agriculture, Fisheries and Forestry. 1984. *Soil at Risk: Canada's Eroding Future*. Ottawa: Senate of Canada, Committees and Private Legislation Branch, Standing Committee on Agriculture, Fisheries and Forestry.

Swain, H., S. Louttit, and S. Hrudey. 2006. 'Report of the Expert Panel on Safe Drinking Water for First Nations,' v. 1. Ottawa: Department of Indian Affairs and Northern Development. http://www.eps-sdw/gc/ca/reprt/index-e.asp.

United Nations World Water Assessment Programme. 2006. *Water: A Shared Responsibility*. The United Nations World Water Development Report. Paris: UNESCO; New York: Berghahan Books.

Walkem, A. 2007. 'The land is dry: Indigenous peoples, water, and environmental justice'. In K. Bakker, Ed., *Eau Canada*, 303–19. Vancouver: University of British Columbia Press.

World Water Commission. 2000. 'A report of the World Water Commission on water for the 21st century'. *Water International* 25: 194–201.

Chapter 12

Minerals and Energy

Learning Objectives

- To understand the characteristics of non-renewable resources relative to the renewable resources discussed in previous chapters.
- To appreciate the significance of minerals and energy for Canada.
- To understand the management issues associated with non-renewable resources in general and minerals and energy in particular.
- To identify the relative importance of different minerals for the Canadian mining industry as well as Canada's importance in global mining trade.
- To discover how science is used in environmental assessments, illustrated by the Ekati Diamond Mine in the Canadian North.
- To appreciate that energy resources can be both renewable and non-renewable.
- To learn of the potential of alternative, renewable energy sources, particularly wind and solar power.
- To understand the significance of non-renewable energy resources, including offshore petroleum and natural gas, the Athabasca oil sands, and nuclear power.
- To know what you can do to have a lighter 'footprint' related to use of minerals and energy.

Introduction

Previous chapters have focused on **renewable** or **flow resources**, usually defined as those that are renewed naturally within a relatively short period of time, such as water, air, animals, and plants (Rees 1985, 14). Other renewable resources are solar radiation, wind power, and tidal energy. Given this mix, a distinction is often made between renewable resources not dependent on human activity (e.g., solar radiation) and those that renew themselves as long as human use allows reproduction or regeneration (e.g., fish).

Figure 12.1 highlights that flow or renewable resources can exist in critical or non-critical zones. Those in the critical zone can be harvested or exploited to exhaustion. The most vulnerable depend on biological reproduction for renewal. Whether through overhunting, overfishing, polluting, or destroying habitats, humans can create conditions such that renewable resources cannot replace or replenish themselves. Indeed, the 'collapse' of the northern cod in the northwest Atlantic (see Chapter 8) is often noted as a classic case of overharvesting of a renewable resource that led to its depletion.

In this chapter, emphasis is mostly on **non-renewable** or **stock resources**, which take millions of years to form. As a result, from a human viewpoint, such resources are for practical purposes fixed in supply and therefore not renewable. However, as with renewable or flow resources, Figure 12.1 indicates that non-renewable or stock resources are not homogeneous. Some are consumed through use, whereas others can be recycled. Those consumed by use are best illustrated by fossil fuels (coal, oil, natural gas). Once used, they are effectively not available to humans, even though you know from earlier chapters that they do not really disappear but are changed into another form, often polluting material. In contrast, stock resources such as metals can be recycled many

STOCK			FLOW	
Consumed by Use	Theoretically Recoverable	Recyclable	Critical Zone	Non-critical Zone
OIL GAS COAL	ALL ELEMENTAL MINERALS	METALLIC MINERALS	FISH FORESTS ANIMALS SOIL WATER IN AQUIFERS	SOLAR ENERGY TIDES WIND WAVES WATER AIR

Flow resources used to extinction

Critical zone resources become stock once regenerative capacity is exceeded

Figure 12.1 A classification of resource types. Source: Rees 1985, 13. Reprinted by permission of Thompson Publishing Services.

times, so the stock of resources in the ground is not the only source. However, recycling often requires significant amounts of energy, so the recycling of one type of stock resource (e.g., aluminum) may hasten the depletion of another (coal, oil, or natural gas) stock resource.

Our attention here focuses on both minerals and energy. And to emphasize that non-renewable resources are not homogeneous, the discussion of energy examines wind and solar power, usually viewed as types of renewable or flow resource. In addition, attention also is given to non-renewable types of energy resources, such as offshore petroleum and natural gas, the Athabasca oil sands, and nuclear power.

Whatever categories are considered, it is important to appreciate that mineral and energy resources are important in and for Canada. For instance, Canada is ranked as the sixth largest user of primary energy (commercially traded fuels) in the world. Such a high level of use is attributed to Canada being a large country with long travel distances, as well as its cold climate, an energy-intensive industrial base, relatively low energy prices, and a high standard of living. Regarding minerals, Canada is one of the major global exporters. More details about energy and minerals will be provided later in the chapter, but the key message here is that both are important for regional and national economies in this country and that in their extraction and use, they can have significant environmental impacts.

Framing Issues and Questions

The challenge for renewable resources is to manage them so that they remain renewable or sustainable. For non-renewable resources, however, extraction usually results in absolute depletion in any time frame other than a geological one. Given the characteristics of non-renewable resources, the management issues are usually different from those associated with

renewable resources (Sanchez 1998, 522; Hilson 2000, 202–3). Specifically, concerns focus on:

1. how to use the proceeds from resource extraction to generate new wealth, which, as new capital, can be used to benefit generations today and in the future;
2. how to conserve mineral or fossil-fuel assets to extend the longevity of reserves and how to identify substitutes for use in the long run;
3. how to minimize negative environmental impacts at each stage in the life cycle of use: exploration, extraction, transformation, consumption, recycling, and final disposal;
4. how to create improved socio-economic relationships with stakeholders, especially the local communities located in the mining area, after the resource has been exhausted;
5. how to manage recyclable non-renewable resources—i.e., many metals and minerals—as a renewable or flow resource.

What would motivate mining and fossil-fuel firms to engage in environmental management, given that their priority is to maximize profits and remain competitive in international markets? As Hilson (2000, 203) noted, if done systematically and correctly, enhanced 'environmental management practices and extended social responsibility almost always generate some kind of economic return on investment for business, although usually over the long term. A documented reduction

Environmental History of the Canadian Mining Sector

Of all the Canadian industries . . . the one in greatest need of sustainable development is mining simply because of its propensity to cause significant environmental damage on numerous fronts and the fact that its operations affect a wider range of stakeholder groups than most industries. . . .

. . . The Canadian mining sector has had a long history of environmental damage, and it has only been in the past 30 years that the industry has received any environmental attention.

—Hilson 2000, 204

in effluent discharges, for example, leads to a reduction in costly government inspections and auditing practices.'

As Sanchez (1998, 522–3) observed, the main environmental issues for the mining and energy sectors include **acid mine drainage**, **sulphur dioxide emissions**, and **metal toxicity**.

1. *Acid mine drainage.* Most non-ferrous metals exist as sulphides and usually are accompanied by iron sulphides. When ore minerals are separated from minerals without economic value, significant quantities of waste rock and tailings are created, and they contain iron sulphides that can readily oxidize to become sulphuric acid. When exposed to precipitation (rain or snow), sulphuric acid can dissolve residual metals, leading to acidic drainage, which can continue for centuries. It has been estimated that liabilities in the Canadian mining industry related to acidic drainage may range between $2 billion and $5 billion.
2. *Sulphur dioxide emissions.* One outcome of smelting sulphide ores is the release of huge quantities of sulphur, mainly in the form of dioxides, into the atmosphere. The Canadian mining sector has been estimated as the main contributor to sulphur dioxide emissions in Canada. The burning of fossil fuels is also a major source of atmospheric emissions of sulphur dioxides, creating pressure for alternative sources of energy that are less damaging to the environment.
3. *Metal toxicity.* As understanding increases through research, the mining industry is being challenged about the toxic effects of metals on human and ecosystem health. For example, many uses of asbestos are now not acceptable because of connections established between it and cancer. Lead is also considered toxic. Emissions from smelting and steelmaking processes can also threaten health, as illustrated by the case study of the Sydney Tar Ponds in Chapter 1.

To these three issues can be added challenges related to energy:

- disruption of remote ecosystems due to exploration, test drilling, and operation of oilfields or gas wells, ranging from habitat degradation to alteration of nesting, denning, and migration patterns of birds and animals;
- disturbance to aquatic ecosystems from escape of waste heat produced from nuclear energy production;
- threat to human and ecosystem health from radioactive waste associated with nuclear energy production over a period of thousands of years;
- alteration to ecosystems from building hydroelectric dams and generating stations.

These issues, individually and collectively, provide a strong rationale for increased attention to environmental aspects and management in the mining and fossil-fuel sectors.

'Best practice' related to environmental management for mining and fossil-fuel firms in Canada should include a combination of basic scientific research to ensure understanding of natural and social systems that can be affected by operations and design of appropriate mitigation measures, environmental impact assessments and reporting, environmental audits, corporate policies that explicitly include environmental aspects, environmental management systems (Hilson and Nayee 2002), and life-cycle assessments.

Non-Renewable Resources in Canada: Basic Information

According to Natural Resources Canada (NRCAN 2006a), Canada is one of the largest exporters of minerals and mineral products in the world. During 2006, minerals and mineral products accounted for almost 17.5 per cent of total exports for the nation and contributed to its trade surplus. About 80 per cent of Canadian mineral and metal production is exported. As an example, Canada is the seventh largest exporter of coal, in 2006 exporting almost 30 million tonnes, which was 42 per cent of total coal production in Canada. Furthermore, as documented by Lemieux (2005), Canada has more large mineral exploration companies than any other country. Canada accounted for 51 per cent of such companies, followed by the Asia-Pacific region with 22 per cent. For these reasons, Canada is considered 'the hub for mining activities worldwide' (NRCAN 2007c).

Mining and mineral processing industries are important to the national and local economies. In 2006, they added $39.9 billion to the Canadian economy, which represented 3.7 per cent of gross domestic product. Furthermore, they directly employed 368,745 Canadians, with 49,173 working in mining, 79,740 in smelting and refining, and 239,832 in manufacturing of mineral and metal products (NRCAN 2007a). At the beginning of 2006, about 200 metal, non-metal, peat, and coal mines existed, more than 3,000 stone quarries and sand/gravel pits, and about 50 non-ferrous smelters, refineries, and steel mills.

In 2006, the value of mineral products totalled a record $33.6 billion, due mainly to the significant increase in prices for most metallic minerals. The top five products in terms of production value in 2006 were nickel ($6.8 billion), copper ($4.6 billion), iron ore ($2.6 billion), gold ($2.24 billion), and cement ($2.21 billion). When coal, petroleum, and natural gas are included, the total for non-renewable resources becomes $77 billion, which highlights the importance of the fossil-fuel industry. Canada ranks first at a global scale as a producer of potash and uranium and third for primary aluminum after China and Russia. It is ranked in the top five for nickel, asbestos, zinc, cadmium, titanium concentrate, platinum group metals, salt, gold, molybdenum, copper, gypsum, cobalt, and lead. Potash is the most important product in the non-metals group. Canadian potash exports totalled 15.8 million tonnes in 2005. The potash industry began in Saskatchewan during the early 1960s and expanded steadily during the 1970s and 1980s to the extent that Canada is now the largest producer and exporter of potash in the world. Furthermore, Saskatchewan's potash industry has been judged the most productive worldwide, accounting for about 25 per cent of world potash production (Saskatchewan 2008). Annual sales for Saskatchewan potash alone total $1 billion.

In 1991, following more than a decade of geological detective work, two Canadian geologists, Chuck Fipke and Stewart Blusson, discovered minable diamonds underneath Lac de Gras in the Northwest Territories, leading to the opening in 1998 of the first diamond mine (BHP's Ekati mine) in Canada, after an investment of $700 million. Ekati, located 200 kilometres south of the Arctic Circle, supplies about 6 per cent of the world's annual $8 billion in rough diamonds. A second diamond mine, Diavik, began producing in 2003 after a $1.3-billion investment by Rio Tinto/Aber Resources. Canada produced almost 13 per cent of the world's diamonds in 2005, making it the third largest producer by value after Botswana and Russia. The output in 2005 was 175.9 million carats, with a value of US $11.3 billion, more than six times the value of production in 2003.

In 2003, the Nunavut Impact Review Board conditionally approved what became Canada's third diamond mine and the first in Nunavut. The mine is operated by Vancouver-based Tahera Corporation, and its Jericho mine, located 420 kilometres northeast of Yellowknife, near Contwoyto Lake, began commercial production on 1 July 2006 (Tahera Diamond Corporation 2008). And De Beers is expected to begin producing diamonds from a fourth mine at Snap Lake, NWT. According to De Beers, 'While the mine is ramping up to full production, construction continues underground, where a crusher and conveyor to bring kimberlite ore to the surface are scheduled for completion around the end of the first quarter of 2008.' In addition to the diamond production and exploration in the NWT and Nunavut, major exploration is underway in Alberta, Saskatchewan, Manitoba, Ontario, and Quebec, and one of the largest bodies of diamond-bearing ore in the world has been found some 50 kilometres northeast of Prince Albert, SK. As of 2008, testing and investigation were ongoing to determine the economic future of diamond mining in Saskatchewan. Some predictions indicate that in 20 years, Canada could be the global leader in terms of the value of diamond production. More details are provided in a later section discussing the development of Ekati.

Recycling Growing in Importance

Canada is not only a producer of metal and non-metal products. Natural Resources Canada (NRCAN 2006a) has estimated that trade in recyclable ferrous and non-ferrous metal, including waste, scrap, ash, and residue, reached 6.7 metric tons, valued at $4.8 billion. Canada imports and exports recycled metals, much of which occurs with the United States. For example, in 2005, Canada exported 87 per cent of its recycled metal to the US, while imports from the US represented nearly 99 per cent. More than 3,000 metal recycling companies operate in Canada, employing some 15,000 people.

Canadian mineral production is not only exported but is also used within the country. For example, coal and uranium are the basis for one-third of electricity production. Alberta, which produces nearly half of the coal mined in Canada, depends on coal for almost 50 per cent of its electrical power as more natural gas

Bruce Mitchell

Shore Gold Inc. test drilling site for diamonds in Saskatchewan.

Bruce Mitchell

generators become operational (Alberta Energy 2008). By the end of 2006, 21 coal-fired power generation plants were operating in Canada, which together represent more than 90 per cent of Canada's coal consumption (NRCAN 2006a).

In the following sections, we look at how science has been incorporated into initiatives to remediate landscape degradation associated with mineral extraction and how it has been used in understanding and mitigating environmental impacts when a new mining venture is being designed. Finally, we consider the role of science in exploring alternatives to fossil-fuel energy sources.

Developing a Diamond Mine: Ekati, NWT

The story of Ekati is now part of Canadian mining lore. In 1980, exploration geologists Chuck Fipke and Stewart Blusson noticed alluvial traces of pyrope garnet, ilmenite, and chrome diopside—all indicator minerals associated with diamonds—while working near the border of Yukon and the Northwest Territories. These indicator minerals had been dispersed during the last ice age 10,000 years ago, and Fipke and Blusson realized that their presence did not mean diamonds were close by. Analysis of paleoglaciation and drainage patterns led them to believe that they would have originated in **kimberlite pipes** somewhere in a 65-million-square-kilometre area of tundra to the east.

Ekati: A First for Canada

Ekati is not just another mine. It is a diamond mine, Canada's first, and it is one of the world's better ones, with roughly 4 per cent of the world production by weight and 6 per cent of world production by value.

—Werniuk 1998a, 9

They began an extensive program of exploration. Usually travelling by helicopter, they collected thousands of alluvial samples and examined each one for traces of indicator minerals. By 1983, they observed a promising pattern. As they moved east, the concentration of indicator minerals increased, and more significantly, the crystals had less alluvial wear. By 1989, the trail of indicator minerals had led them some 640 kilometres to the east, and other geologists also were searching. In that year, although no diamonds had yet been found, Fipke and Blusson staked claims on 1,800 square kilometres of tundra.

Kimberlite

Diamonds are a crystalline type of carbon, stable at depths of 150 km or more beneath the Earth's surface. Kimberlite is a rare igneous rock found at the same or greater depth. Eruptions of kimberlite can transport diamonds to the surface of the Earth. Diamond content can be highly variable in a carrot-shaped kimberlite pipe, ranging from nothing to economic concentrations. Because of glaciation in Canada's North, the top part of the diamond-yielding kimberlite pipes often have been scoured out, leaving a circular depression. Glacial alluvium and water then fill the depression, creating a circular lake over the kimberlite pipe, as was the case at what has become the Ekati mine. There are 136 known kimberlite pipes in the NWT.

—Rylatt 1999, 39; Voynick 1999; Couch 2002, 267

Later that year, travelling again by helicopter, they noticed that Point Lake was circular and much deeper than other lakes in the area. They sampled along its shoreline and discovered a chrome diopside crystal with no alluvial wear, suggesting that a kimberlite pipe was close by, perhaps even underneath the lake. Verification required very expensive core drilling. A partnership was arranged with a major Australian mining company, BHP World Minerals, and in 1991, core samples from Point Lake yielded the first diamonds found in Canada. The subsequent public announcement, required under Canadian law, triggered a huge mineral rush, with 260 companies from eight nations attracted to the Canadian North, staking claims totalling 194,000 square kilometres. Following an environmental impact assessment review during 1994 to 1996, Canada became a major producer of diamonds in October 1998 when the Ekati Diamond Mine started up. By 2001, it was producing $1.7 million worth of diamonds each day. The Etaki mine currently produces 15,000 carats of diamonds daily and is expected to produce a gross value of $9.5 billion over its projected life of 25 years.

The Environmental Context

A claim was staked by BHP for an area of 3,400 square kilometres situated some 300 kilometres northeast of Yellowknife (Figure 12.2). The mining activity is located mainly in the Koala River watershed, which drains into Lac de Gras and then northward into the Coppermine River and on to the Arctic Ocean. The mine is found in the Low Arctic Ecoclimate region in which the average annual temperature is −11.8°C. The temperature range is large, with daily temperatures in

Process of Discovery

Economically viable gem-bearing diamond pipes are very rare. There were only 15 such areas known in Africa, Siberia, Australia, and Brazil. In a seemingly impossible task, Canadian geologists Charles (Chuck) E. Fipke and Dr Stewart Blusson criss-crossed the Arctic by foot and light aircraft for a decade looking for diamonds until they found indicator minerals near Lac de Gras in 1989.

—Couch 2002, 266

Meaning of 'Ekati'

'Ekati' is a Dene word, meaning 'fat lake', but it has been said that it actually refers in this context to the white granite rock outcrops that resemble caribou fat.

—http://www.ccrs.nrcan.gc.ca/resource/ tour/43/index_e.php

Figure 12.2 Location of NWT Diamonds Project. Source: Canadian Environmental Assessment Agency 1996, 6. Reproduced with the permission of the Minister of Public Works and Government Services, Canada, 2005.

Barren-ground caribou.

summer reaching 25°C and winter temperatures often falling below −30°C. Precipitation is low, averaging only 300 millimetres, most of which falls as snow.

The BHP claim area is in the tundra region, 100 kilometres north of the tree line. About one-third of it is covered by some 8,000 lakes, and the landscape has continuous permafrost, with permanently frozen subsoil and rock up to 250 metres deep, and an overlayer of about one metre that thaws during summer. The main vegetation includes stunted shrubs and grass tussocks, with willows and scrub bush appearing in low areas. Wetlands include a combination of water sedges and sedge-willow communities.

Two important mammal species exist in this environment: the Bathurst caribou herd and grizzly bears. The caribou herd has approximately 350,000 animals and moves around a range of about 250,000 square kilometres. They spend the winter south of the tree line, then in the spring start their northward migration to calving grounds near Bathurst Inlet on the Arctic Ocean. The grizzly bear, because of low numbers, density, and reproduction rates, has been designated as vulnerable (see Chapter 14).

Economic and Social Context

The economic aspects of the Ekati Diamond Mine are significant. It has been estimated that the total project capital cost will be $1.2 billion, the contribution to the Canadian gross national product will be $6.2 billion, and the direct, indirect, and induced benefits to the NWT will be $2.5 billion (60 per cent of that would be wages and benefits). In terms of employment, the Ekati mine will employ twice as many people as any other mine in the NWT, a total of 830 over the estimated 25-year lifespan. The mining company's policy has been to first hire NWT Aboriginal people, then non-Aboriginal NWT residents, and finally other Canadians. When Aboriginal people do not have the necessary skills, the mining company provides appropriate education and training related, for example, to literacy and technical skills.

The mining company committed to give preference to businesses owned by Aboriginal people for contracting, to establish scholarship programs, on-the-job training programs for Aboriginal students, and cross-cultural training in the workplace, and to draw upon elders, youth, and organizations to ensure traditional knowledge is included when dealing with

community issues. Furthermore, the workplace is drug- and alcohol-free, and offenders are fired without exception.

The company encountered numerous challenges. The area staked for mining had been used for traditional hunting and fishing by Aboriginal people representing four groups: (1) Treaty 11 Dogrib Dene, (2) Treaty 8 Akaitcho Dene (Chipewyan Dene), (3) North Slave Métis Alliance, and (4) Kitkmeot Inuit Association. Members of these groups speak various languages, including English, several Athapaskan languages, and Inuinnaqtun (a dialect of Inuktitut spoken in Canada's western Arctic), creating communication problems. The situation was further complicated because of two separate land claims agreements under negotiation and a third under consideration. Another issue was the role of *traditional knowledge*. Such knowledge is transmitted verbally from generation to generation, and as a result, there is usually no written record of it. Western scientists have tended to view traditional knowledge as anecdotal and unreliable and therefore inappropriate to use in impact studies. A further difficulty was that because of the land claims negotiations, the Aboriginal people were reluctant to share their traditional knowledge publicly.

You may wish to reread the guest statement by Elise Ho in Chapter 7, which focuses on traditional ecological knowledge, and consider its implications for the development of Ekati. In Chapter 8, a box describes a situation in which traditional knowledge turned out to be more valid than scientific assessments in the North, although it should not be assumed that this is always the case.

The Diamond Mine and Its Impacts

Five kimberlite pipes were initially identified for mining. One set consists of a group of four, while the fifth is about 30 kilometres away on the edge of Lac de Gras. Mining would result in the loss of at least seven lakes covering 890 hectares—five as a result of mining, one for tailings disposal, and one as a source of aggregate material for construction (Figure 12.3). The extracted ore is being processed based on a plan aimed at handling 9,000 tonnes each day; after 10 years, this amount will increase to

Panda Pit.

Figure 12.3 Development plan area, NWT Diamonds Project. Source: Canadian Environmental Assessment Agency 1996, 7. Reproduced with the permission of the Minister of Public Works and Government Services, Canada, 2005.

18,000 tonnes daily. The ore is crushed, and the diamonds are separated by physical rather than chemical processes.

Long Lake was chosen to hold tailings. Dams have been built at the outlet of the lake and around its perimeter to increase holding capacity. Rock dykes have been used to divide the catchment area into five cells. As the tailings held in each cell settle, consolidate, and evolve to permafrost, rocks and soil will be spread over the surface. Revegetation will be started, with the goal of having the entire holding area become a wetland once the mining is completed.

A permanent work camp has been built to accommodate the 400-person workforce, which rotates on the basis of two weeks in/two weeks out with 70-hour workweeks.

No all-weather road has been built. Over the winter months, a 476-kilometre ice road over frozen lakes and rivers is used, including by trucks carrying fuel. Air transport is used to transport workers in and out, as well as to bring food in and take the diamonds out.

Assembling Data Related to Environmental Impacts

The mining company, BHP, collected some baseline data during 1992, with systematic and intensive field sampling begun in 1993. The sampling program addressed biological, cultural, and socio-economic issues. Through the environmental assessment process, various federal government departments provided comments in the fall of 1993 about the field sampling design. The most significant comment focused on the need for BHP to make significant additional effort to incorporate traditional knowledge into the collection of conventional scientific data.

In response, BHP observed that it had faced serious challenges in its efforts to include traditional knowledge into its research program. First, the Treaty 8 and Treaty 11 Dene groups were in the midst of land claims negotiations and as a result were reluctant to release traditional knowledge into the public domain because the traditional knowledge was important for their negotiation strategy. Second, Aboriginal people were expressing concern about using traditional knowledge outside the context of the cultures and broader system of knowledge that gave it meaning and value. Third, there was no one set of traditional knowledge, since the Inuit, Métis, and Dene each have their own traditional knowledge, which do not always coincide. Fourth, traditional knowledge was viewed by Aboriginal people as their intellectual property, and therefore its use and management had to remain within their control. And fifth, there was no documented baseline of traditional knowledge, nor were there any generally accepted standards or methods to guide traditional knowledge research.

Mining Tailings

As observed in the report from the Canadian Environmental Assessment Agency (1996, 24), the management of mining tailings is critically important because of the potential impact on downstream water quality. During the mining operations, 35 to 40 million tonnes of waste rock will be excavated each year.

The ore is being crushed, and diamonds are separated through physical means. The crushed rock or tailings will be placed in the Long Lake tailings impoundment basin for the first 20 years and then in one of the mined-out pits for the final five years.

The capacity of Long Lake has been increased with the building of three perimeter dams. The dams have a frozen core design, meaning that each has a central core of frozen soil saturated with ice and bonded to the natural permafrost. The core is surrounded with granular fill to ensure both stability and thermal protection. The use of a frozen core and permafrost foundation is intended to ensure that no water can escape through the dams as long as the soil remains saturated with ice.

The basin has been divided into five cells, and the cells will be filled sequentially, starting with the cell farthest upstream, leaving time for the tailings to settle and consolidate (Figure 12.3). The last cell, the farthest downstream, will not receive any tailings. It will act as a final settling pond.

The design has the tailings gradually consolidating and becoming permafrost. Once a frozen crust has formed over a tailings cell, it will be covered with waste rock and then topped with fine granular soil. Such a covering will be thick and moist enough to facilitate creation of a new active layer in the new permafrost system. Subsequently, the soil will be revegetated, with the ultimate purpose being to create a wetland. After the mine is closed, the spillway dam at Long Lake will be breached, with water flowing naturally from the fifth cell into a nearby lake and from there into the Coppermine River system.

BHP Approach to Environmental Management at Ekati

The basis of BHP's environmental management approach is an adaptive environmental strategy which involves the establishment of criteria or indicators used to indicate change so that appropriate management actions can be implemented. A sensitive and effective monitoring program is key to the success of this strategy. BHP told the Panel that the environmental management plan is intended to be flexible so that it can be modified in response to changes in the mine development plan, natural environmental or technological advances, research results, and improved understanding of traditional knowledge.

—Canadian Environmental Assessment
Agency 1996, 24

The Panda Pit frozen core dam.

The **frozen core dam** design was chosen because of the lack of impervious fill needed for construction of a conventional dam, because the climate is conducive to a frozen core design, and because previous experience with frozen core designs in Canada and Russia could be drawn upon. As the Environmental Assessment Panel observed, the critical factor for a frozen

core dam is ensuring that the core stays frozen. The design of the dam is based on a requirement that the long-term average temperature of the core not exceed −2°C when the mine is operational. Modelling indicated that this requirement would be satisfied even in future conditions suggested by global warming scenarios (see Chapter 7). The frozen core will be monitored systematically, and various contingency plans are in place in case of an unexpected increase in temperature.

Migratory Caribou

BHP conducted and supported research on the Bathurst caribou herd, the largest in the NWT (Figure 12.4). Baseline data were collected in 1994 and 1995 to determine the numbers of animals using the Lac de Gras area during migrations, the location of migration corridors, and the use of habitat. The migration patterns differed during the two years. This variability was confirmed by the Environmental Assessment Panel, which commented in its report that 'This [variability] was consistent with the observations of several that there is a considerable natural variability in caribou migration and habitat use. For example, the GNWT [Government of the

Figure 12.4 Distribution of Bathurst caribou herds. Source: Canadian Environmental Assessment Agency 1996, 40.

Northwest Territories] agreed that the ability to predict, on an annual basis, the timing and numbers of caribou in the vicinity of the proposed mine was low. A representative of the NWT Barren Ground Caribou Outfitters Association said that over his 18 years of experience, the exact migration route varied from year to year' (Canadian Environmental Assessment Agency 1996, 39).

Since the caribou herd does not follow the same migration route each year and the areas affected by mining activity represent less than 0.01 per cent of the range of the herd, it was believed that mining would have a very small impact. Attention was also given to the possible effects of roads and the new airport landing strip, either through collisions of caribou with vehicles or as a barrier to migration. It was concluded that these developments would not cause problems. Indeed, caribou had been seen using the new habitats adjacent to the roads and landing strip during the exploration period.

Bathurst Caribou Herd

Among environmental issues, the potential effect of the Project on the health, numbers, and migratory patterns of the Bathurst caribou was the most important concern raised. . . . The GNWT told the Panel that the cultural value of the herd could not be estimated but that the dollar value of the harvest, based on meat replacement costs, was $11.2 million annually.

—Canadian Environmental Assessment Agency 1996, 39

However, notwithstanding the above boxed comment by the Canadian Environmental Assessment Agency in 1996, a 2006 survey of breeding females revealed that the Bathurst caribou herd has been declining by about 5 per cent annually during the past decade (Northwest Territories 2006). In 2006–7, the Independent Environmental Monitoring Agency (IEMA) commented on the lack of progress on understanding the regional and cumulative effects on caribou and recommended increased action on the part of territorial governments.

Finally, modelling indicated that the water in the tailings impoundment would be within federal guidelines for protection of livestock (and therefore of wildlife). According to the BHP Billiton Environmental Agreement annual

report in 2001, 'the undisturbed lakes and streams around Ekati are very clean.' Furthermore, environmental monitoring programs for air, water, and wildlife are ongoing. For example, during 2001, vegetation surveys found trace levels of metals (aluminum, iron) in lichens near the mine facilities, attributed primarily to road construction, blasting, and road traffic rather than to direct emissions from the mine. Levels of metals and chemicals are below Canada's accepted levels, and the report notes that 'none of the levels are considered toxic to plants or animals' (BHP Billiton 2001, 15–16). Results from water quality monitoring in the King-Cujo watershed downstream from the mine site indicated balanced levels of zoo- and phytoplankton, indicators of healthy lakes and streams. The slight fluctuations in pH and nitrate levels in the Koala River watershed were not viewed as a threat to fish.

Water Issues

BHP began to collect baseline data on hydrology in the autumn of 1992. Data were collected for lake bathymetry, lake water levels, and surface hydrology. During 1993, preliminary investigations were conducted at eight lakes, followed in 1994 with sampling at 25 lakes and in 1995 with sampling at four lakes and seven stream sites. The water samples were sent to an independent laboratory for analysis.

It was concluded that water flow changes would be caused by draining the lakes to facilitate open-pit mining as well as by diversion of flows around the pits and by the infilling of Long Lake with the tailings. In total, 15 lakes would be affected by the mining activity. It was decided that drainage of the lakes prior to open-pit mining would be managed so that flows would not be greater than 50 per cent of the mean annual flood levels in any downstream water system containing fish. As a result, the main consequence of draining the lakes would be to extend the peak spring flows for a longer period of time. And because the connecting channels between the lakes are both wide and braided, the effects of the extended period of higher flow were judged to be negligible.

The potential impact of mining operations on water quality attracted considerable attention. A primary concern was that contaminants from the mining operation could affect downstream consumers of fish and drinking water in the Coppermine River watershed. The main issue was whether the tailings impoundment in Long Lake would ultimately release water of acceptable quality. Analysis focused on three water quality variables (suspended solids, total nickel, and total aluminum). The holding cells in Long Lake are designed to ensure that water would be held there until it met regulatory standards before release. During the impact assessment process, it was agreed that the design should meet all regulatory standards for water quality.

In terms of follow-up, BHP noted in its 2001 annual report that 'during 2001, technicians took 169 water samples. The samples were sent to independent laboratories to measure water quality. Most of the samples met the water quality guidelines of the licence. This shows that mining activities have not affected the water quality of lakes and streams around the mine. Only two samples exceeded the licence guidelines. Both exceedances were thought to be the result of a laboratory or a field identification error' (BHP Billiton 2001).

Concern also arose about possible contamination from toxicity of kimberlites, acid generation from waste rock, and nitrogen from blasting. Analysis of the toxicity of kimberlites and waste rock and the design of the holding pond led to agreement that such contamination would be controlled satisfactorily. Effects on groundwater also were addressed. Baseline data were collected, and a long-term monitoring program was established so that effects on hydrogeology could be tracked. As for water released downstream from containment in Long Lake, the IEMA commented in 2007: 'While we have yet to receive the monitoring report for 2006, we are less sure than BHPB is about how clean the water is. We have a concern that it may be getting worse for some metals in the water such as molybdenum and copper' (IEMA 2007).

Fish

Fish are present in 12 of the 15 lakes that would be affected by the mining through draining lakes prior to open-pit mining, filling them with tailings, or covering them with waste rock. In addition, 43 connecting streams, outflow streams, and inflow streams would be affected. The main species of fish in the 12 lakes is lake trout, followed by round whitefish, Arctic grayling, and burbot. The federal Department of Fisheries and Oceans (DFO) has a policy of 'no net loss' of productive capacity of fish habitat. That means that whenever fish habitat is degraded or lost, DFO expects there to be a counterbalancing habitat replacement.

BHP has compensated for the lost fish habitat in streams by creating a diversion channel between two of the key lakes, making the channel a quality fish habitat. The cost of this initiative was $1.5 million. The diversion channel is 2.25 times longer than the natural connecting channels that were lost, and DFO stated that this approach was an acceptable way of offsetting the loss of natural stream habitat.

Costs

Couch (2002, 274) reported that the scientific research funded by BHP cost more than $10 million. In addition, the environmental assessment review process cost the Canadian government about $1 million, with another $255,000 for participant funding. These amounts do not include the costs incurred by various federal departments, such as Environment, Fisheries and Oceans, and Indian Affairs and Northern Development, as well as by the government of the Northwest Territories. In Couch's view, 'in comparison with the Project's capital cost, the anticipated profits to BHP Diamonds Inc., and the tax revenue to governments, this outlay was very small' (2002, 274).

Environmental Assessment Process

Beginning in 1992, BHP began research to understand the impact of the proposed mining activity and to develop mitigation measures. Company representatives visited all communities in the project area at least twice. BHP made public presentations, organized field trips, held community meetings and open houses, facilitated cultural exchanges and workshops, and sent a group of Aboriginal people to its mines in New Mexico, where 76 per cent of its employees were Native Americans.

In July 1994, the minister of Indian affairs and northern development referred the mining project for an environmental assessment, and in July 1995 BHP submitted its environmental impact statement. From late January to late February 1996, an Environmental Assessment Panel appointed by the minister of environment held public meetings, and its report was submitted to the federal government in June 1996. In February 1997, the federal government gave its formal approval, and construction started in May 1997. In January 1999, the first diamonds from Ekati were sold in Antwerp, Belgium.

Agreements and Arrangements

Emerging from the process outlined above were a number of agreements and arrangements.

Announcement from the minister of Indian affairs and northern development. In August 1996, the minister announced his acceptance of the Environmental Assessment Panel's report and gave participants only 60 days to work out detailed agreements. Given that land claims negotiations were also underway, there was the potential for much debate and disagreement. The tight time frame put all participants under much pressure to work out detailed agreements.

Environmental Agreement. The Environmental Agreement is legally binding and requires BHP to: (1) prepare a plan for environmental management during the construction and operation of the diamond mine; (2) submit annual reports related to the environmental management plan; (3) prepare an impact report every three years related to the impact of the project; (4) establish a monitoring program for air and water quality and for wildlife; (5) submit a reclamation plan for approval; (6) establish a security deposit ($11+ million) for potential land impacts and a guarantee of $20 million for potential water impacts; and (7) incorporate traditional ecological knowledge into all environmental plans and programs.

In addition, an Independent Environmental Monitoring Agency (IEMA) was to be established as a public watchdog. The IEMA is to: (1) prepare annual reports on the project's environmental implications; (2) review impact reports; and (3) provide a public document repository at its Yellowknife office. This was an innovative feature, since it had not been recommended by the assessment panel.

In 2007, the IEMA (2007, 23) noted that 'Last year [2006], the Government of the Northwest Territories (GNWT) was firm with the company in dealing with a disagreement about the Environmental Impact Report, a report that describes the overall environmental effects of the project every three years. BHPB is taking steps to make sure the GNWT is satisfied with the environmental reporting.' Furthermore, a July 2006 letter from Indian and Northern Affairs Canada to Ekati Diamond Mine indicated that the GNWT found the 2003–5 Environmental Impact Report from BHP unsatisfactory regarding air quality and wildlife. This echoes concerns from the IEMA on environmental reporting, and there appear to be some communication problems and information gathering and reporting issues among the company, territorial government bodies, and the IEMA. Basically, the GNWT and IEMA find that BHP's environmental reporting provides insufficient details and is slower than they would like (http://www.monitoringagency.net/PublicRegistry/EkatiRelatedCorrespondenceAllParties/tabid/89/Default.aspx).

Highlights from the IEMA report (2007) include:

- BHP is doing a satisfactory job in studying and monitoring seepage from waste rock to ensure that water released downstream is of good quality. However, BHP adds calcium chloride to waste water from the processing of kimberlite, which if not controlled properly could make waters downstream salty and could adversely affect fish and wildlife;
- IEMA wants a study on future water quality from the Long Lake Containment Facility as well as a plan for managing processed kimberlite after the mine has closed to ensure proper containment. IEMA also wants BHP to stop using calcium chloride until research shows that water quality will be maintained.
- 'It is our [IEMA] view that BHPB needs to find better ways to record and explain how TK [traditional knowledge] is being used in other operations at the mine site. For example, the 2006 Environmental Impact Report stated that the company is monitoring vegetation based on TK to look for changes in permafrost. A better description of this monitoring and how TK was exactly used should be provided.'
- Good efforts are being made by BHP to avoid attracting wildlife to the mine site by proper garbage containment and by erecting fencing and more *inokhok* (stone piled to resemble human figures) meant to repel wildlife from mine pits and the air strip.

Socio-economic Agreement (SEA). This agreement between BHP and the government of the Northwest Territories focuses on commitments beyond existing statutory requirements. The concern was economic benefits and social impacts related to all NWT residents, not just those who were traditional users of the project area. The agreement covered matters such as preferential hiring of NWT residents (with a target of 62 per cent northerners and 31 per cent Aboriginal peoples), criteria to guide recruitment, employment targets, employment of local contractors, training programs, and employment support. Targets were specified for awarding contracts to and purchases from northern businesses, as well as for employment of Aboriginal and northern residents.

In 2006, total northern employment was just below target at 60 per cent, while employment levels for Aboriginal peoples exceeded targets at 33 per cent. The mine created 1,034 person-years of employment, including 336 and 279 person-years for Aboriginal and non-Aboriginal northern residents, respectively. Mining operations at Ekati provided 20 apprenticeships, 11 of which went to Aboriginals and eight to northerners. Ekati has also been one of Canada's top 100 employers for five straight years. In terms of northern spending targets, BHP bought $326 million in goods and services from northern businesses, representing 78 per cent of total spending and exceeding SEA targets by 70 per cent. Twenty-nine per cent of total spending, or $123 million, was directed to northern Aboriginal-owned businesses.

Although not part of the agreement, a noteworthy initiative has been the establishment of diamond-cutting and polishing businesses in Yellowknife. The traditional centres for polishing are Antwerp, Tel Aviv, New York City, and India. In 1999, a small Vancouver-based diamond-polishing company opened a facility in Yellowknife, recruiting a South African diamond cutter from Antwerp. Shortly afterwards, another company opened, with cutters recruited from Armenia. Other firms have opened facilities as well, and local people are learning the trade under the guidance of cutters from Europe, Israel, and Africa. It was reported that by 2003, almost 200 people were employed by four diamond-polishing companies based in Yellowknife.

Impact and benefit agreements. In 1994, BHP began negotiations with the four Aboriginal groups. Each group was involved in land claim negotiations, and BHP did not want to get entangled in those processes. **Impact and benefit agreements (IBAs)** are one tool for addressing community and industry relations in mining or other extractive resource activities. They are voluntary agreements, beyond formal impact assessment requirements, and are intended to facilitate extraction of resources in a way that contributes to the economic and social well-being of local people and communities. IBAs create opportunities for communities to realize direct economic benefits from natural resource development projects as well as to participate in the management, monitoring, and mitigation of impacts. All of these matters were addressed in the IBA between BHP and the four Aboriginal groups. In the guest statement below, Michael Hitch provides further insight into IBAs.

Guest Statement

Minerals, Mining, and Impact and Benefit Agreements

Michael Hitch

The notion of sustainability as we know it had its modern roots in the well-worn work of Bruntland and the World Commission on Environment and Development (WCED 1987). For most, this report seemed like the first time we thought about managing our current consumption so as not to negatively affect the ability of future generations to meet their needs. It was perhaps Thomas Malthus some 200 years earlier, however, who set the stage with his call for new ways to increase yields while preserving natural habitats and biodiversity. The similarities are striking. The common thread is that we have needs that have to be met today and in the future.

The modern discourse concerning sustainability, addressed in Chapter 1, has since permeated every aspect of our society, whether in the form of multilateral institutions such as the United Nations, corporations that now don the mantle of purveyors of sustainability, consumers and the new 'green' products they buy, or the media and what we teach our children. The forum of this discourse has been driven down to the community level, perhaps where it counts most.

The reality is, however, that sustainability and a sustainable existence is a luxury good that normally can only be afforded by the affluent—even though we all need it. Sustainability is like an environmentally neutral 'Nirvana' that can only be visited once the basic needs of survival are met. An all too common situation in many places on the globe is the subsistence farmer providing the bare necessities for the survival of his nuclear family. The more advanced ideas of sustainability, such as product life cycle, are the furthest things from his mind. Food, shelter, health, and the support of friends and family are currencies of his happiness.

Prosperity presents the opportunity for the consideration of the future and how our consumption patterns today affect those of the future. Prosperity arrives in many forms, manifestations, and origins. Natural resource development, particularly of minerals, allows for

this welcome visitor. The act of taking a mineral from the ground and processing it into a usable product is the creation of wealth for the first time. This wealth is in turn fed into various levels of the economy as wages and purchases of services and goods, as well as savings. At the community level, this wealth provides the platform for building capacity and the transformation of natural capital to more utilitarian forms. This is the development of prosperity. It is at this point that we can consider examining what and how we function as a community and its impact on our common future.

The pace of mineral development in Canada's Arctic has increased dramatically over the past decade with the discovery of diamonds in the western portion of what is now called Nunavut. Three deposits are in production, making Canada a significant participant in the small club of diamond-producing nations. Mineral development anywhere has impacts on the regions where they are located. The western portion of Nunavut is no exception, and perhaps the impacts are more significant because of the fragility of the land and the culture of its people.

The economic history of the Inuit people of Canada's Arctic reads like a Greek epic, full of the trials and tribulations of any hero in search of his or her destiny. The quest of the Inuit has been changed, has evolved from that of a hunter/gatherer society completely dependent on the land for their survival to that of a different kind of dependence—on the federal government in strange new 'white' communities. Today, they are now being challenged with the next step of the journey, entrance into the wage economy.

The settlement of land claims and the conditions for mineral development have set the stage for the aggressive mineral development we have seen and has established the foundation for the Inuit to partake in this new-found wealth and embrace the prosperity that will propel them along the sustainability pathway. One of the conditions placed on all mineral developers is to negotiate an *impact and benefit agreement* with the local regional Inuit association before beginning operations. An impact and benefit agreement, or IBA, is a negotiated business contract that stipulates the roles and responsibilities of both the industrial proponent and the regional Inuit organization. Elements of these agreements include Inuit content in the workforce, development of new businesses to serve the needs of the new mine, education and training opportunities, and community outreach to assist with any negative impacts of the operation. Over the past 15 years, eight such agreements have been concluded in the Kitikmeot region where most of the new mineral development has taken place.

The Kitikmeot Corporation was established to manage the numerous company-Inuit joint ventures springing from the new mines in a manner that will provide for the long-term sustainable

Michael Hitch

growth and prosperity of the beneficiaries of the land claims agreements. These new businesses include heavy equipment maintenance, a new pharmacy in the hamlet of Cambridge Bay, catering services, and home Internet access, among many other independently viable enterprises. The result is a vibrant regional economy no longer dependent on federal government (not shifting from one dependency to another) subsidies and preparing for a time when mining will play a lesser part in their future existence. Nunavut will always need investment from the outside to fuel its long-term sustainability. However, it has embarked upon a pathway of prosperity in order to fuel the long trek of sustainability.

Michael Hitch, a geologist, is an assistant professor in the Norman B. Keevil Institute of Mining at the University of British Columbia in Vancouver. He previously worked for 22 years in the mining industry and in that time was involved in mining projects in 141 countries. He has particular experience with mining operations in the Canadian North and has participated in negotiations to establish the first IBAs.

Energy Resources

Significance of Energy

. . . energy is society's critical master resource: when it's scarce and costly, everything we try to do, including growing our food, obtaining other resources like fresh water, transmitting and processing information, and defending ourselves, becomes far harder.
—Thomas Homer-Dixon 2007, 12

Energy resources are classified as renewable and non-renewable. As noted at the beginning of this chapter, renewable resources are those that can be replenished in a relatively short time period. Figure 12.5 identifies three renewable energy sources, and one of these, gravity, is ongoing and widespread but remains as potential unless associated with significant motion, such as tides or river flow. Geothermal heat, created by nuclear energy, also is persistent and widespread but at great depths

below the surface. Manifestations of geothermal heat at the surface or shallow depths are much more limited and are usually associated with the heat being carried by water or steam, so the renewability of geothermal heat depends on a reliable and ongoing supply of water. Solar supplies come from continuous emission of radiation from the sun, but it arrives discontinuously on the surface of the Earth because of diurnal and seasonal variation as well as cloud cover. As a result, renewable, solar-based energy supplies are intermittent and often cyclic, meaning that they usually must be supplemented by other sources.

Biomass energy sources are frequently used in rural areas in developing countries and can take the form of millions of people and their draft animals doing subsistence work. Metabolic energy (muscle power) is supplemented by heat created from burning firewood, from crop and animal wastes in basic biogas converters, and from direct sunlight used to dry and preserve agricultural or marine products (e.g., dried fish). Biomass energy is renewable as long as the rate of use and capacity to produce biomass are balanced.

The non-renewable sources cannot be replenished over a period of time short enough to support humans. These sources result from geological processes over millions of years, which lead to solid (coal) and liquid (oil) fuels, natural gas, and nuclear fuels. While they all share the characteristic of offering high energy content per unit of weight or volume, they also differ. Solid fuels are mined, which is labour-intensive and requires expensive infrastructure. For efficient

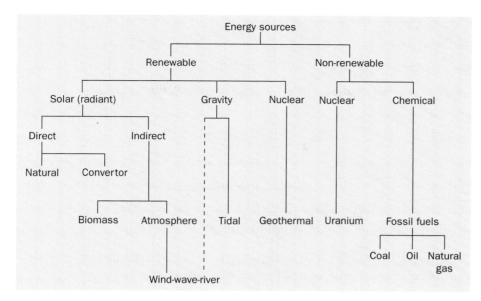

Figure 12.5 Energy sources. Source: Chapman 1989, 4.

refineries or processing plants. Once processed, the product can be transported continuously through pipelines or in batches (trains, ships, trucks). Nuclear fuels contain the highest content per unit of weight but require sophisticated facilities and highly skilled human resources. They are used only to generate electricity and demand careful handling in processing and waste disposal. Given these different attributes, it is important to appreciate that the most appropriate sources of energy will vary depending on circumstances. Box 12.1 highlights the different variables that need to be considered when making a choice.

transport, they must be carried in bulk or batch containers, such as rail cars or ships. When used (burned), solid fuels release gaseous and particulate matter in large quantities. In contrast, oil and natural gas can be produced with facilities requiring relatively little labour but in capital-intensive

Energy Use and Issues in Canada

As noted at the beginning of this chapter, Canada is ranked as the sixth largest user of primary energy in the world. According to Environment Canada (2002, 57), energy consumption in Canada was 9.9 exajoules in 2000, an increase of 10 per cent

Box 12.1   Choosing among Energy Sources

1. *Occurrence*. Many energy sources are confined to specific environments and locations and are only available at other locations when transport systems exist. Even physically present sources may not actually be available because of technical, economic, or other constraints.

2. *Transferability*. The distance over which an energy source may be transported is a function of its physical form, energy content, and transport technology.

3. *Energy content*. This is the amount of usable energy by weight or volume of a given source. Low-energy-content sources are inadequate when demand is large and spatially concentrated.

4. *Reliability*. Uninterrupted availability gives one source an advantage over one that is intermittent.

5. *Storability*. To meet interruptions of supply or peaks of demand, a source that can be stored has an advantage over one that cannot.

6. *Flexibility*. The greater the variety of end uses to which a given source or form may be put, the more desirable it is.

7. *Safety and impact*. Sources that may be produced or used with low risk to human health and the environment will be preferred over less benign sources.

8. *Cleanliness and convenience*. The cleaner and more convenient source will be preferred over the dirty and the cumbersome.

9. *Price*. The less expensive source or form will be preferred over the more expensive.

Source: Chapman 1989, 5.

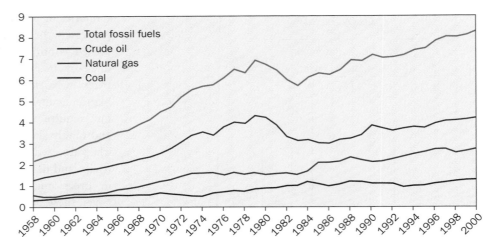

Figure 12.6 Fossil fuel consumption in Canada (exajoules). Source: Environment Canada 2003a. Adapted in part from Statistics Canada Energy Division. Reproduced with the permission of the Minister of Public Works and Government Services, Canada, 2005.

relative to 1990. In the same period, however, energy use per capita decreased, indicating that initiatives to improve energy efficiency may be starting to show results. In terms of fossil fuel, use increased by 20 per cent between 1990 and 2000 (Figure 12.6). In 1999, Canada's fossil-fuel energy use was 2.5 per cent of the global total, as was its total energy use relative to total global energy consumption.

Fossil fuels are the main type of energy consumed by Canadians, and one of the largest consumers of energy is the transportation sector, which depends overwhelmingly on fossil fuels. Between 1990 and 2004, energy use in the transportation sector increased by 31 per cent (NRCAN 2005a). The 2005 Canadian Vehicle Survey reported that there were about 18.6 million light vehicles and medium and heavy trucks on Canadian roads. Together, these vehicles travelled approximately 315.3 billion vehicle-kilometres and 525.7 billion passenger-kilometres in 2005, increasing 2 and 10 per cent respectively since 2000. On-road vehicles consumed some 29.5 billion litres of gasoline and 10 billion litres of diesel. In 2004, passenger vehicles accounted for 54 per cent of all energy use in the transportation sector, an increase of 17 per cent since 1990. Some fuel efficiencies in automobiles occurred between the early 1970s and the early 1980s, but since then there have not been significant improvements. Instead, there has been an increased use of less fuel-efficient vehicles, such as light-duty trucks and sport utility vehicles. For example, the percentage of automobile passenger-kilometres travelled in light-duty trucks almost tripled over 25 years (10 per cent in 1976 to 27 per cent in 2000). A 50 per cent growth occurred in air travel in the same period. These increases in use of automobiles and planes occurred at the same time that there was a drop in use of buses and trains.

A clear message is that Canada is an intensive energy-using nation and that despite growing efficiencies, energy use is climbing. Given our dependency on fossil fuels, it is appropriate to examine what are often referred to as 'alternative energy supplies', which we will do in the next sections. However, it is important to recognize that alternative sources, usually viewed as renewable sources (solar, geothermal, hydro, tides, wind), are not problem-free from an environmental perspective. For example, the impacts of the hydropower development at James Bay, examined in Chapter 11, included increased levels of mercury in aquatic systems, with consequences for both humans and other living species in the area, and changes in fish populations.

Energy and Climate Change

In June 2006, the Round Table released its advice to the federal government on a long-term strategy on energy and climate change. It noted that significant greenhouse gas emission reductions could take place in Canada in mid-century only if energy is used more efficiently and if it is produced while emitting less carbon. It pointed to the need to increase energy efficiency, to perfect carbon capture and storage, and to transform energy generation to clean coal technology, co-generation, and renewable energy, particularly wind power.

—Auditor General of Canada 2006

Wind Power

Natural Resources Canada (NRCAN 2006c) explains that wind energy converts kinetic energy available in wind to forms of energy more useful to humans, such as mechanical energy or electricity. Furthermore, wind energy is 'a pollution-free, sustainable form of energy. It doesn't use fuel; it doesn't produce greenhouse gases, and it doesn't produce toxic or radioactive waste.'

Humans have used wind energy for centuries, beginning with windmills to provide mechanical energy for pumping water and grinding grain. Sailing ships also depended on the power of the wind. Frequent contemporary uses of wind energy are electricity production and water pumping.

Capacity to generate power from wind depends on several variables, the most important being wind speed. The consequence is that wind turbines are located in the windiest areas, and within such areas they are situated on high spots, since wind speed increases with elevation above the surface. On the other hand, when there is little or no wind, wind turbines stay idle and produce no energy.

According to the Global Wind Energy Council (2006, 1),

Against the backdrop of a growing acknowledgement of the twin crises of global climate change and energy security, wind power is the most effective means available now to tackle both issues. At the same time, wind energy development strengthens local economies and insulates countries from the macro-economic shocks of global commodity markets. Wind energy provides a clear solution for policy-makers and a general public increasingly concerned with volatility in the gas market, discussions of 'peak' oil, and growing realization of the true cost of coal.

The expansion of **wind power** has been impressive. The Global Wind Energy Council (2006) noted that between 1996 and 2006, the wind energy industry experienced an annual rate of growth of 30 per cent at a global scale. Europe is the leader in the use of wind energy, followed by North America and Asia. During 2006, slightly more than 15,000 megawatts (MW) of wind energy capacity was installed, bringing the total global installed capacity to more than 74,000 MW, an increase of 9,000 MW compared to 2005.

Germany is the by far the leader in terms of installed wind power capacity at 20,622 MW in 2006, followed by Spain at 11,165, the US at 11,003, India at 6,270, and Denmark at 3,136. As a region, Europe was the leader in 2006, with a total of 48,545 MW of installed capacity, or 65 per cent of the worldwide total. By 2006, 13 countries had more than 1,000 MW of installed wind capacity, including France and Canada, which both crossed this threshold in 2006. Regarding *new* installed capacity, Asia was second to the United States, and within Asia, India represented the strongest market. China more than doubled its capacity during 2006, making it the sixth largest market for wind power. In Latin America, the wind power leaders are Brazil and Mexico.

Looking to the future, it is expected that Europe will remain the leader, but its share will steadily decrease as other regions adopt more wind power. Within Europe, Germany and Spain will most likely remain the biggest users of wind power, but the UK and France are expected to become other major markets, along with Denmark.

In North America, the US is examining the feasibility of a goal of supplying 20 per cent of the nation's electricity from wind power. In 2006, the market share of wind power was less than 1 per cent. The American Wind Energy Association has stated that the target is both feasible and affordable because wind resources are available, technology and manufacturing capacity are in place, and widespread demand exists for more renewable sources of energy. The main barriers are at the policy and regulatory levels, but they can be overcome if there is political will.

Wind Power in Canada, 2006

Canada is on the edge of a wind energy boom as provincial governments are now targeting to have a minimum of 10,000 MW of installed wind energy capacity in place by 2015.

. . . Canada also is predicted to be one of the countries with extraordinary growth rates. It is estimated that by 2010 the cumulative installed capacity in Canada will have reached 6,000 MW. This means that 4,500 MW will be introduced during the period 2007–10, listing Canada in the top five countries for that period.
—Global Wind Energy Council 2006, 6, 12–13

During 2006 in Canada, some 776 MW of new wind power capacity was commissioned, representing an increase of 113 per cent in capacity in one year. At the end of 2006, Canada had 1,459 MW of installed wind capacity (Figure 12.7). In January 2007, the federal government announced a new ecoEnergy Renewable Energy Program, which included an incentive of a one cent per kilowatt-hour production incentive payment over the first 10 years of production by wind farms. The goal of this program is to encourage the installation of an additional 4,000 MW of renewable energy by 2010. Provincial governments also are actively involved, with nearly all provinces having targets for future energy production from wind sources. The Global Wind Energy Council (2006, 33) noted, however, that a major obstacle to reaching such targets is the 'patchwork' of municipal regulations related to renewable energy. A second issue is public opposition to wind farms because of noise and aesthetics. Such opposition was illustrated when citizens organized the Windfarm Action Group to protest the development of a 182-MW project in Bruce County, ON, which would be one of the largest wind farms in the country. The Windfarm Action Group appealed the rezoning decision, but the Ontario Municipal Board did not accept their objections and authorized construction to begin in the summer of 2007, with electricity to be produced during 2008.

Wind Power in Canada

Canada has utility wind turbines installed in Alberta, Saskatchewan, Ontario, Quebec, Prince Edward Island, Nova Scotia, and Yukon. Wind farms have been established in Alberta, Saskatchewan, Ontario, Quebec, and Prince Edward Island.

The opening of the Prince Wind Farm on the shores of Lake Superior near Sault Ste Marie in 2006 made Ontario Canada's leading wind power generator. The project has 126 turbines producing 189 MW of power, or enough to power 40,000 homes. In 2006, Ontario had 413 MW of wind power generation, compared to 384 MW in Alberta. But by 2008, Ontario will have increased provincial wind energy generation threefold to 1,300 MW (Ontario Ministry of Energy 2006).

Canada currently ranks 11th in the world for installed wind energy capacity, producing 1,846 MW in 2007.
      —Canadian Wind Energy Association 2008,
      http://www.canwea.ca/news_releases.cfm?ID=58.

In terms of future use of wind energy, it is useful to consider its advantages and disadvantages.

Advantages of Wind Power

* Wind power is an ideal renewable source of energy because it does not require fuel, does not create greenhouse gases (GHGs), and does not produce toxic or radioactive wastes.
* Wind energy is quiet and not a hazard to birds or other wildlife.

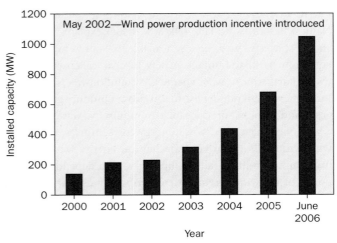

Figure 12.7 Total installed wind power capacity in Canada is growing. Source: Based on information provided by the Canadian Wind Energy Association.

* When large wind farms are established, containing many wind turbines, they only require 2 per cent of the land area, making the balance available for farming, livestock, and other uses.
* Payment is made to landowners, which provides another source of income.

Disadvantages of Wind Power

* Wind is not constant, meaning that there will be times when no power is generated.
* When wind turbines are located on coastlines, conflict often arises because other landowners view them as a negative feature on the landscape.
* With a large wind farm containing many wind turbines, noise from the turbines may be intrusive for nearby landowners.

The advantages and disadvantages noted above can be considered with reference to the 'criteria' for sustainable energy options introduced at the beginning of the chapter.

Environmental Impacts of Wind Turbines

With regard to the significance of wind turbines in terms of birds, noise, safety, and aesthetics, Dillon Consulting Ltd (2000) examined the possible impacts and reached the conclusions outlined below.

Wildlife

Findings from monitoring programs in North America and Europe have shown that bird mortality from wind turbines is very low. For sites with a small number of turbines, it is not unusual to have no recorded fatalities. For sites with many turbines, it is common for there to be no more than one bird fatality per wind turbine per year, and the maximum recorded mortality in North America has been 1.9 birds per turbine per year. Most studies were conducted on variable-speed turbines, which are more dangerous, but even for variable-speed turbines, studies have shown that up to 80 per cent of birds can fly unharmed through rapidly moving blades.

Most birds travel during the day, have good vision, and can readily see and avoid obstacles. Furthermore, most birds fly at altitudes much higher than the tops of wind turbines, so they normally fly over rather than through them. The birds at greatest risk are small nocturnal migrants that fly together in large numbers, but even these types usually fly well above the turbine.

Evidence reveals that other urban structures (tall buildings, communication towers) cause much higher rates of mortality. For example, Dillon Consulting reported that tall communication towers in Canada often each kill more than 1,000 birds annually. A few tall buildings in Toronto are estimated to cause the death of more than 10,000 birds every year. Even normal family homes can be dangerous, with estimates suggesting each home causes from one to 10 bird deaths every year.

A reasonable conclusion from research and monitoring is that wind turbines pose a negligible threat to birds.

Noise

Disturbance from noise is influenced by many variables, including distance from source, nature of background noise, and nature of the source (frequency, time pattern, intensity). All noise levels from wind turbines during operation are lower than what is experienced in a quiet residential area and similar to what is experienced inside an average home. Dillon Consulting concluded that given the normal background noise in an average suburban residence, the noise from a wind turbine would be inaudible at a distance of 260 metres.

Safety

The main safety concern is ice thrown from turbine blades or falling off the tower. There have been no recorded incidents of falling ice and therefore little regulatory experience exists. At the wind turbine towers at Kincardine on the Bruce Peninsula of Ontario, operators have recorded an average of three icing events each winter. Several proactive actions are available. First, setback criteria can be used to ensure that people are kept at a reasonable distance from a wind turbine tower and the rotating blades. Second, temperature sensors as well as sensors to monitor the balance of blades can provide early information about ice build-up. Once ice accumulation is detected, the wind turbine can be shut down and not restarted until operators determine that conditions are safe.

Aesthetics

Some people may view a wind turbine or wind turbine field as an unwelcome visual intrusion on the landscape, especially if they feel that the turbines are not in keeping with an area's historical, cultural, or natural values. On the other hand, some people enjoy the look of a wind turbine, appreciating its modern, futuristic appearance as well as the symbolism and educational role of a visible environmentally benign technology. The challenge, as Dillon Consulting (2000, 44) observed, is that 'Given the conformity of view that windmills are a good thing but that they should be placed "somewhere else" and not "here", and the recognition that everyone's "somewhere else" is someone else's "here", a balanced answer is needed.'

Research regarding wind turbines or windmills in Europe and North America has indicated that prior to their construction, nearby residents usually have some concerns. However, after the wind turbines are operating, their views became either neutral or positive. If the homes receive electricity from the turbines, attitudes are likely to be more positive.

Summary

Evidence indicates that wind turbines have minimal adverse environmental affects. However, compared to conventional fossil-fuel energy sources, they are still relatively expensive, but that could change if the appropriate technology becomes less costly and/or fossil-fuel supplies become more expensive. Their increased use in the future will require governments to be proactive, as is occurring in Germany, and create requirements or incentives for energy suppliers to include renewable sources in their mix of sources. Furthermore, in weighing the cost of alternative sources against that of conventional sources, the total costs of each source should be considered, including the costs entailed in emissions into the atmosphere. If such comprehensive costing were done, the gap between conventional sources and renewable energy sources would not be as large as it seems to be at the moment (Box 12.2).

Solar Power

Solar power is another renewable energy option. As Natural Resources Canada (NRCAN 2005b) explains, energy generated by the sun travels to the Earth as electromagnetic radiation. The solar energy available at any place on the Earth is a function of several variables, the most important of which are how high the sun is in the sky and cloud cover. There are three general categories of use for solar energy: heating/cooling, production of electricity, and chemical processes. The most widespread uses are for heating of space and water.

The International Energy Agency (IEA) has documented its use worldwide and by region (Weiss, Bergmann, and Faninger 2007). The IEA noted that with reference to the annual collector yield of all solar thermal systems operating in 2005, 45 countries produced 66,406 GW_{th}, (the subscript 'th' stands for 'thermal energy'), or the equivalent of 10.7 billion litres of oil and an annual avoidance of 29.3 million tonnes of CO_2 (Weiss, Bergmann, and Faninger 2007, 5). For Canada in 2005, the reported total collector capacity was 175 GW_{th}, and the CO_2 reduction was 78,437 tonnes based on all solar thermal systems. If only swimming-pool heating is considered, the CO_2 saving was calculated at 64,249 tonnes, highlighting the dominant role of solar systems for heating swimming-pool water in Canada.

Box 12.2   Wind Power for Farms

Wind power is not confined to urban areas and can be generated and used by individuals. To illustrate, two Mennonite farmers in a township just north of Waterloo, ON, have each invested more than $35,000 for wind turbines just under 25 metres high, which they expect will generate enough electricity to supply themselves, with enough left over to sell to the local utility company, Waterloo North Hydro. They anticipate that the wind turbines could pay for themselves in 10 to 12 years.

Jacob Martin, one of the farmers, has used two smaller windmills for some time to provide power for his home and barn and has never been connected to the power grid. However, to sell excess power to the grid, he will have to become connected.

As Jacob Martin observed, 'If I could get my power and get a cheque for $5,000 a year, that would be good.'

Source: Based on *Kitchener-Waterloo Record* 2004.

Wind turbine on Jacob Martin's farm, Woolwich Township, Ontario.

The end use for solar power varies from country to country. For example, in China and Taiwan, Europe, and Japan, the main use is for heating water and space heating. In contrast, the dominant use in the US and Canada is for heating swimming pools. Europe has the most diverse and sophisticated market for solar power, with end uses ranging from heating water, space heating for single- and multi-family houses and hotels, and large-scale plants for district scale heating, as well as air conditioning of homes along with cooling and industrial uses.

Based on total installed solar power capacity, the leading countries are China, Turkey, Japan, Germany, Israel, and Greece, with China representing almost 50 per cent of the global market. Canada is ranked 29th. If the measure is 'installed capacity per 1,000 inhabitants', the leading countries become Cyprus, Israel, Austria, and Barbados. China ranks 11th on this measure after Greece, Turkey, Australia,

Germany, Denmark, and Taiwan. Canada is ranked 38th. Regarding heating of swimming pools, the US is the leader, with nine times the capacity of Australia, the second-ranked country, followed by Germany, Canada, Austria, and South Africa.

In considering the prospects for solar energy in the future, Natural Resources Canada (NRCAN 2005b) has compiled a list of its advantages and disadvantages relative to conventional energy sources.

Advantages Relative to Conventional Energy Sources

- Energy from the sun is effectively free after the initial cost of infrastructure is recovered.
- The sun is a virtually unlimited source of solar energy.
- Depending on how the energy is used, the payback time can be very short.

- Solar energy systems can be 'stand alone', making them attractive for isolated areas or facilities, since they do not have to be part of a power or natural gas grid.
- Use of solar energy as an alternative to conventional energy sources normally leads to a proportional decrease in GHG emissions.

Disadvantages Relative to Conventional Energy Sources

- Cloudy conditions significantly reduce capacity to produce energy from solar systems.
- Large surface areas of land are required to install enough solar panels to produce significant quantities of energy, with potential negative disruption to farming as well as concern about aesthetic impact.

Offshore Petroleum

About three-quarters of the surface of the Earth is covered by oceans. As land-based reserves of petroleum and gas become depleted, exploration has moved to offshore locations. The outcome is that about three-fifths of production of petroleum around the world is now from offshore facilities in waters adjacent to more than half of the world's coastal nations. The extraction of offshore petroleum is also increasingly occurring in highly challenging environments, including at greater depths (2,500 metres below the surface in the Gulf of Mexico off the coast of Louisiana) or in extreme climate conditions (storms in the North Sea; hurricanes in the Gulf of Mexico; high winds and waves, cold temperatures, ice and icebergs, and fog in the northwest Atlantic Ocean east of Newfoundland).

With Pacific, Atlantic, and Arctic coastlines, Canada is an offshore producer. Offshore petroleum production started in 1992 from a basin located some 40 kilometres southwest of Sable Island off the coast of Nova Scotia. Production continued there until 1999, with a total output of 44.5 million barrels of light crude. The next major production started in mid-November 1997 at the Hibernia field on the Grand Banks, about 315 kilometres south-southeast of St John's in Newfoundland. The fixed production platform used at Hibernia is the world's largest and is anchored on the seabed at a depth of 80 metres. Because of the prevalence of icebergs, the outer edge of the platform is serrated. Tankers take the petroleum from the production platform to an inshore storage terminal near an oil refinery at Come by Chance. A support vessel is always stationed near the production platform, and one of its tasks is to tow small and medium-sized icebergs away from the platform.

Located 350 kilometres east-southeast of St John's and discovered in 1984, the Terra Nova project is Canada's third field and began production in January 2002. The Terra Nova field is estimated to hold 440 million barrels of recoverable petroleum, making it the second largest on the east coast. Producers use a floating facility with capacity for production, storage, and off-loading. The floating facility design was chosen in light of the harsh environment, and it can be disconnected relatively quickly from its mooring system and moved off-location in case of an emergency. The hull of the facility was designed to withstand the force of an iceberg weighing up to 100,000 tonnes or sea ice covering up to 50 per cent of the ocean surface around the platform. Other protective measures include sub-sea wells within 'glory holes' (excavations on the seabed) to protect the wellheads from icebergs that scour the ocean bottom and flexible pipes to take oil from the wells so that oil can be flushed out of them and replaced by sea water if an approaching iceberg that might damage the pipes is sighted.

Hibernia and Terra Nova together produced 115 million barrels of light crude in 2004. Hibernia had an expected production life of 25 years and Terra Nova 15+ years. In June 2006, the Canada-Newfoundland Offshore Petroleum Board revised its estimate of Hibernia's recoverable reserves, setting it at 1,244 billion barrels, an increase of 379 million barrels from the previous estimate. This upward revision means that Hibernia is now expected to be in production until about 2030 rather than until the early 2020s.

In November 2005, a fourth field containing both petroleum and gas, named White Rose, was brought into production. It is located some 350 kilometres east of St John's and about 50 kilometres from the Hibernia and Terra Nova fields, placing it on the northeastern part of the Grand Banks. Like Terra Nova, it was discovered in 1984. The White Rose field extends over 40 square kilometres at a depth of 120 metres and is estimated to contain 250 million barrels of recoverable oil. A floating production and storage facility is used here, similar to the one at Terra Nova.

White Rose received regulatory approval in June 2001 following an environmental impact assessment from which the federal minister of environment concluded that it was unlikely to cause significant negative environmental effects as long as mitigation measures were used. The environmental assessment report, completed in 1997, concluded that extreme weather and ice regimes at the production site would be the most serious challenges (Canada-Newfoundland Offshore Petroleum Board 1997). The Environmental Assessment Panel believed that use of a floating production system, by allowing avoidance strategies in the face of extreme conditions, reflected the *precautionary approach* that it believed should underlie all aspects of the project. Notwithstanding its confidence in the capacity for avoidance of possible environmental dangers to the production platform, the panel recommended that effort should continue on improving operational forecasting capacity regarding both weather and iceberg trajectories. The panel noted that the threat from ice flows and icebergs was of 'particular concern'.

The panel also observed that the developers of White Rose could not be held responsible for the effects of subsequent development projects on the Grand Banks. The possibility of future projects, however, made it clear to the panel that significant difficulties and uncertainties exist in terms of calculating *cumulative effects* from a number of offshore projects. As viable. By the mid-1990s, improvements in drilling technology and increased prices for natural gas made commercial extraction feasible. As a result, a consortium of oil and gas extraction companies started the process to develop the gas fields in 1996, and production began from the Sable Island Project during 1999. This was the first offshore natural gas project in Canada.

The project has two components. The initial one focused on extracting gas from six fields near Sable Island, building an offshore processing plant (installed in August 1999), and constructing a pipeline to take the product to a plant for further processing near Goldborough, NS. Construction of the Goldborough plant started in January 1998 and was completed in November 1999, while the pipeline was completed in September 1999. The second component involved building what was called a Maritimes and Northeast Pipeline to move processed gas from the Goldborough plant to a transfer point at the border between Canada and the United States.

Given the possible environmental impact of these projects, various federal and provincial departments collaborated on a co-ordinated environmental impact assessment process (Canadian Environmental Assessment Agency 2003). The participating agencies included the Canadian Environmental Assessment Agency, the National Energy Board, Natural Resources Canada, the Nova Scotia ministries of Natural Resources and Environment, and the Canada–Nova Scotia Offshore Petroleum Board. A five-member assessment board was created in September 1996, and its report was made

Iceberg Hazards at White Rose Oil Field

In early April 2008, oil production from the White Rose field was shut down because of a threat from nearby icebergs.

Production was stopped because thick sheets of ice along with large icebergs were close to the offshore production facilities. Six vessels were used to tow icebergs away, in addition to water cannons to deflect the icebergs' trajectories.

The White Rose oil rig, along with its 109 crew members, was towed to an ice-free area some 40 kilometres southwest of the oil field.

one step to respond to this dilemma, the panel recommended development of a systematic and peer-reviewed monitoring system that would become a foundation for *adaptive management* decisions in the future.

Beyond the possibility of a major oil spill, the panel noted that discharges of oil-based drilling mud, various chemicals, and product water into the ocean were the project's biggest environmental hazard. If a major oil spill were to occur, the panel believed that mitigative measures were unlikely to be effective because of the fragile nature of the environment. Consequently, the panel argued that it was 'absolutely essential' that prevention be the top priority. In the view of the panel, the project operators had to 'adopt a zero-tolerance for oil spills of any kind'.

The possible impact of light oil on seabirds was recognized, and the panel concluded that appropriate mitigative measures were feasible if there was a sound monitoring program. Another risk to seabirds would arise when oil is moved from the production site to the shore refinery. An oil spill close to the shoreline could threaten the large seabird colonies on the Avalon Peninsula. The panel noted that a systematic coastal zone management regime was not in place for the shoreline of the Avalon Peninsula and recommended that one be established.

Regarding natural gas, the coastal continental shelf adjacent to Nova Scotia contains significant gas fields. In the 1970s, recoverable reserves were discovered in various locations near Sable Island, some 100 kilometres from the Canadian mainland. During 1979, a drill rig successfully identified a commercial field. At that time, however, the cost of developing the field, combined with low natural gas prices, did not make extraction

Nova Scotia to Use Oil and Gas Revenue to Buy Land for Conservation

In April 2008, the government of Premier Rodney MacDonald announced that it would allocate some of the offshore oil and gas revenue earned by the province to purchase and protect areas with high conservation value in Nova Scotia. Up to $25 million will be dedicated for this purpose. Other revenue from the offshore oil and gas industry will be used to reduce the provincial debt, enhance university infrastructure, and support offshore research. This decision illustrates how revenue earned from a non-renewable resource can be used to enhance the renewable resource base for a jurisdiction.

public in October 1997. Subsequently, all appropriate regulatory agencies gave approval, subject to adoption of the recommendations in the assessment report.

Average daily production of gas is between 400 and 500 million cubic feet. It has been estimated that the Sable Island Project will produce for up to 25 years, and it has also been calculated that royalty payments to the province over that period will range between $1.6 and $2.3 billion.

The above information shows that Atlantic Canada has been the major source of offshore oil and gas production so far. And by 2005, Newfoundland was producing 12 per cent of Canada's crude oil and equivalent. However, significant discoveries have been made in the Beaufort Sea off the coast of the Northwest Territories. Production will likely begin there once a Mackenzie Valley pipeline has been built to move the gas to southern markets. There also are estimates of significant oil and gas reserves off the coast of British Columbia, but a federal moratorium on exploratory drilling has been in place there since 1972.

Estimated Oil and Gas Reserves in BC Offshore Waters

The Geological Survey of Canada has estimated that in the offshore sedimentary basins, there may be as much undiscovered hydrocarbon reserves as 9.8 billion barrels of oil (BBL) and 25.9 trillion cubic feet (TCF) of gas in the Queen Charlotte Basin, 9.4 TCF in the Tofino and Winona Basins, and 6.5 TCF in the Georgia Basin. Such estimates can only be confirmed by drilling, which must be preceded by and based on sophisticated scientific and technical surveys, which in turn will only be carried out if permitted by the governments of British Columbia and Canada.

—Scientific Review Panel 2002, 2–3.

The moratorium affecting the waters off the British Columbia coast reflects at least the following concerns: jurisdictional uncertainty regarding whether the federal or provincial government owns the seabed, Aboriginal land and related ocean claims, and environmental risks. The oil spill from the tanker *Exxon Valdez* when it hit Bligh Reef on 24 March 1989 in Alaskan waters highlighted the vulnerability of BC coastal waters to environmental risk. Some 10.8 million US gallons of unrefined crude oil were released into Prince William Sound from the tanker, the largest oil spill ever to occur in North American

waters. The oil eventually covered more than 1,900 kilometres of rocky shoreline and caused the death of tens of thousands of birds, a thousand sea otters, several hundred seals, and unknown numbers of fish and other sea life. Exxon had 10,000 workers on site in the summer of 1989 for the cleanup work, which ultimately cost US $2.2 billion. Additional costs were a US $1-billion fine payable to the US and Alaskan governments, as well as several billion dollars for damage experienced by fishermen, property owners, and others.

The potential for offshore oil and gas (estimated to be up to 25 per cent of the globe's undiscovered oil and gas) and mineral deposits has led various countries to make political claims to seabed resources in the Arctic. For example, on 2 August 2007, a remote-controlled Russian mini-submarine was used to plant a Russian flag on the seabed at the North Pole. This action was a symbolic gesture by Russia to claim rights to the Lomonosov Ridge, an underwater mountain range extending 1,995 kilometres, as part of Russia's continental shelf. One month later, Russia stated that initial analysis of samples collected by one of its scientific teams during May and June 2007 proved that the mountain range beneath the Arctic Ocean was an extension of its continental shelf. The consequence of Russia asserting this area as part of its economic zone would be to give Russia rights to the resources on the ocean floor in that area. Shortly after that flag-planting event, Danish scientists travelled to the Arctic to search for evidence supporting Denmark's claim that the Lomonosov Ridge is an extension of the continental shelf of Greenland, which would place it under the control of Denmark.

Canada's prime minister declared in April 2007 that the Northwest Passage was 'Canadian internal waters' rather than 'territorial waters'. In international law, foreign ships in general can travel through territorial waters without permission, but foreign warships must obtain permission before sailing in such waters. The complication is that the Northwest Passage consists of a number of linked channels situated between Canadian-owned Arctic islands, and some countries do not recognize Canada's jurisdiction.

Later, in August 2007, the prime minister announced that Canada would spend $7 billion to build seven armed icebreakers, establish a deepwater port at Nanisivik on the northern shore of Baffin Island, and create a military training base at Resolute Bay, all to reinforce Canada's claim that the Northwest Passage is under Canadian jurisdiction. However, the United States has never accepted the Canadian claim that the Northwest Passage is Canadian territorial waters, instead arguing that it is 'international waters', meaning that US warships can use the Northwest Passage without seeking Canada's permission. As Dyer (2007, A9) has commented, 'There is a scramble for the Arctic, but it is not military. It's about laying claim to potentially valuable resources.' As a result, we can expect to see more and more attention given to the petroleum and mineral resources in the Arctic, accompanied by various political manoeuvres as nations jostle to gain access to them.

This situation highlights the need to combine understanding of scientific, technical, management, and policy aspects in resource and environmental management.

Athabasca Oil Sands

Background

Extensive and intensive development is occurring in northeastern Alberta, focused on the oil sands located in an area north of Fort McMurray along both sides of the Athabasca River. The Athabasca oil sands are one of three deposits, the others being in the Peace River and Cold Lake areas (Figure 12.8 and Table 12.1). Together, the three deposits extend under an area of 149,000 square kilometres, which is almost one-quarter the area of Alberta and larger than the state of Florida. The oil sands make Canada second only to Saudi Arabia in terms of global oil reserves and ahead of Iran, Iraq, Kuwait, Abu Dhabi, Venezuela, Russia, Libya, and Nigeria, which have the next largest reserves. Prior to the official recognition of the reserves in the Alberta oils sands in 2003, Canada had been ranked 21st in terms of oil reserves. Given the relatively stable political situation in Canada, the oil sands are attracting billions of dollars in investment.

Figure 12.8 Athabasca oil sands region. Source: Woynillowicz, Severson-Baker, and Raynalds 2005, 1. Reprinted by permission of the publisher.

Table 12.1 Area and Bitumen Resources in the Oil Sand Deposits of Alberta

Deposit	Initial Volume of Crude Bitumen in Place (Barrels)	Land Area (km²)
Athabasca (in situ + surface minable)	1.37 trillion (110 billion is surface minable)	102,610 (2,800 is surface minable)
Cold Lake	201 billion	29,560
Peace River	129 billion	17,250
Total oil sands	**1.7 trillion**	**149,420**

Source: Woynillowicz, Severson-Baker, and Raynalds 2005, 2. Reprinted by permission of the publisher.

The Athabasca oil sands are situated under an area of boreal forest. Development of the oil sands started during the mid-1960s, but it was not until the mid-1990s that it became financially viable as a result of improving technology, preferential financial arrangements (low provincial royalties and federal tax concessions), and strong demand as well as rising prices for oil. Between 1995 and 2004, production from the oil sands more than doubled, reaching an average daily output of 1 million barrels. With oil prices expected to remain high and demand to grow, it is anticipated that between now and 2030, investment in the oil sands will be $100 billion and daily output will reach 5 million barrels. This level of output is bringing major economic benefits to Alberta and to Canada. Nevertheless, the development also has serious implications for the integrity of the northern boreal forest as well as for air quality, water in the Athabasca River, and land-based resources. Each of these aspects is considered here, while the implications for greenhouse gas emissions are addressed in Chapter 7.

Context

In 2001, production of crude bitumen (a thick and heavy oil) surpassed production of conventional crude oil in Alberta. By 2003, slightly more than 50 per cent of Alberta's and 33 per cent of Canada's total oil production was accounted for by crude bitumen. By 2005, production from the oil sands represented about 50 per cent of total crude oil production in Canada. And by 2015, it is expected that the oil sands will account for 75 per cent of total Canadian crude oil production.

Extracting Oil from Oil Sands

The oil sands consist of about 10 to 12 per cent **bitumen**, mixed with sand, silt, clay, and water. The oil removed from the oil sands is referred to as **crude bitumen**, and because it is a thick and heavy oil, it cannot flow towards a well. As a result, two different methods are used to extract bitumen. If the bitumen is not more than 100 metres below the surface, it is removed through surface or strip mining from open pits. Subsequently,

Athabasca oil sands, north of Fort McMurray.

Nancy Groce, Smithsonian Institution

the mined oil sands are mixed with hot water, which washes the bitumen out of the sand. The other method, used at depths greater than 100 metres, is referred to as **in situ recovery**, and the specific technique is called **steam-assisted gravity drainage**, or SAGD. This method is used for more than 90 per cent of the oil sands. The usual approach is to inject high-pressure steam into the oil sands to separate the bitumen from the sand, silt, and clay. Once exposed to the steam, bitumen can flow to a well from which it can be pumped to the surface.

For surface or strip mining, Woynillowicz, Severson-Baker, and Raynalds (2005) have explained that:

- To produce one barrel, or 159 litres, of synthetic crude oil, about four tonnes of material (two tonnes of soil and rock above the deposit, then two tonnes of oil sands) must be removed. On average, every two days the mining operations move oil sands equivalent to what would fill Toronto's Rogers Stadium.
- To extract one barrel of bitumen, two to five barrels of fresh water and 250 cubic metres of natural gas are needed. The amount of natural gas is equivalent to what would be needed to heat an average Canadian home for one and a half days.

The surface mining method results in about 90 per cent of the bitumen being recovered. In contrast, to extract a barrel of bitumen, in situ oil sands extraction requires:

- 2.5 to 4 cubic metres of steam;
- 1,000 cubic metres of natural gas.

Through the SAGD method, 60 to 80 per cent of the bitumen in the oil sands is recovered.

After recovering the bitumen from below the surface, subsequent processing stages involve producing the final synthetic crude oil from the bitumen and then transportation to final destinations in Canada or the US. The removal of the bitumen from the oil sands and the production of synthetic crude oil require significant energy inputs. Woynillowicz, Severson-Baker, and Raynalds (2005, 15) state that 'the oil sands industry consumes about 0.6 billion cubic feet of natural gas per day, enough to heat 3.2 million Canadian homes for a day.' The energy-intensive methods of producing oil from the oil sands has led to exploration of nuclear power as a source of energy to produce the final product.

Pipelines are used to move most of the oil to markets in Ontario and BC and in Washington state as well as in Rocky Mountain and midwest states. A pipeline has also been proposed from Alberta to a new marine terminal in Kitimat, BC, to export the synthetic oil to China and other Asian markets as well as to California. For that to happen, however, the moratorium restricting oil tanker traffic along the BC coast would have to be lifted. Finally, by 2008, the demand in the US for oil from the Alberta oil sands had led several major US oil companies to prepare proposals for building pipelines from Alberta to the Gulf Coast in Texas.

Environmental Impacts

The oil sands operations are having significant impacts on the boreal forest system, water in the Athabasca River, and air quality. We discuss each of these issues here, while the implications for greenhouse gas emissions and climate change are addressed in Chapter 7. The consequences of the development are also significant in relation to *cumulative impact assessment*, discussed in Chapter 6.

Boreal Forest and Wetlands

It is important to appreciate the 'footprint' of oil sands in northeastern Alberta. As reported by Woynillowicz, Severson-Baker, and Raynalds (2005, 41), the footprint on the boreal forest is expanding. Oil sands mines are from 150 to 200 square kilometres in size, and an above-ground system of well pads and pipelines is associated with each mine. More than 1,800 oil sands lease agreements have been approved, and collectively they cover some 32,000 square kilometres. These agreements represent about 20 per cent of the leases that ultimately could be approved for exploration and development.

A major impact of oil sands development is fragmentation of the boreal forests. Fragmentation is significant because boreal ecosystems, involving a mix of forest and wetlands, are the habitat for many species of wildlife and also support the highest diversity of breeding bird species in North America. Furthermore, the boreal forest system is valuable in the context of global change because it influences climate and is a source for storage of carbon. Fragmentation occurs as a result of the removal of forests and wetlands through either surface or strip mining or in situ removal of bitumen, as well as the building of roads and above-ground pipelines. The outcome is the breaking up of what used to be continuous areas of extensive woodland and wetlands into smaller and separated patches. This means reduced habitat for wildlife, as well as constraints on movement of wildlife from patch to patch. Such changes have serious negative implications for birds that nest in interior forest systems and for larger carnivores that range over large territories.

Reclamation programs have been designed, but they have limitations. For example, as noted by Woynillowicz, Severson-Baker, and Raynalds (2005, 40), 'the oil industry is still not required to replant seismic lines and well sites with trees after the soil has been reclaimed.' This limitation is significant, because the amount of boreal forest area removed annually by the oil sands industry for seismic exploration is at least as much as what is removed by the forestry industry. They also note that the eventual reclaimed landscape, as proposed by the oil sands industry, will be different from the mix of forest and wetlands being altered by mining. It will consist mainly of dry, forested hills, more lake area arising from the end-pit lakes used in oil sands production, and absence of peatlands. The latter take

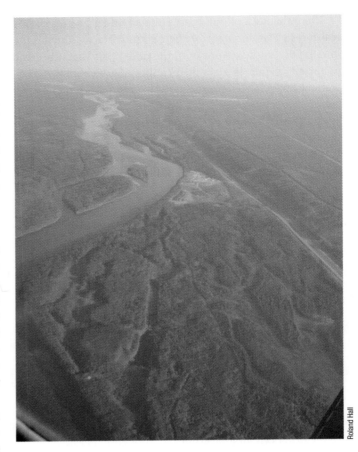

Athabasca River north of Fort McMurray, with the town of Fort McMurray in the background. The road leads to the oils sands operations.

thousands of years to develop and so cannot be replicated by reclamation. It is estimated that after reclamation, about 10 per cent of the wetlands in the original boreal ecosystem in the region will be gone forever. The loss of wetlands will have several consequences because they:

- provide habitat for rare plants and wildlife;
- regulate surface and groundwater flow through retaining snowmelt and summer storm flows;
- recharge aquifers;
- serve as natural filters, removing contaminants from waters that flow through them.

Overall, 'it is likely that the reclaimed landscape will lack the biodiversity of its pre-disturbance state, and it is acknowledged that it will be a major challenge to re-establish self-sustaining ecosystems' (Woynillowicz, Severson-Baker, and Raynalds 2005, 38).

Athabasca River

Both surface mining and in situ extraction of oil sands have significant implications for aquatic systems. Impacts can occur from draining or removing wetlands, dewatering aquifers, withdrawal of water from the Athabasca River, and storing tailings.

The Athabasca River is about 1,540 kilometres in length, starting in Jasper National Park and emptying into Lake Athabasca through the Peace-Athabasca Delta in Wood Buffalo National Park, the largest boreal delta on the planet and a major nesting and staging area for migratory birds. The national park has been designated as a World Heritage Site (discussed in more detail in Chapter 14). Given the huge demand for water for oil sands production, large quantities of water are being removed from the Athabasca River. Such withdrawals pose a potential threat to the Peace-Athabasca Delta, which requires minimum flows from the river as well as natural fluctuations.

In addition, there needs to be sufficient water in the river to support various fish species. The natural flow in the river fluctuates seasonally, with lows occurring in the winter and highs in the spring. Fish, such as northern pike, walleye, and burbot that occupy the river during the winter, are vulnerable if flows drop below their minimum needs, and the natural low flow in winter combined with extraction of water for oil production is a threat to their well-being. Given the low temperatures

Incomplete Understanding of the Causes of River Flow Variability

An interdisciplinary team of scientists has been examining the impacts of climate and river flooding on the ecology of the Peace-Athabasca Delta since AD 1700. Part of the motivation for this research was the argument that regulation of the Peace River for hydroelectric power generation by the Bennett Dam since 1968 and the filling of the Williston Lake reservoir between 1968 and 1971 in BC had created major stress on the Peace-Athabasca Delta. Specifically, it has been suggested that the Bennett Dam has been largely responsible for reduced frequency of spring ice-jam and open-water flooding as well as an extended period of drying, both of which have had significant impacts on the ecology of the delta.

Paleoenvironmental research was used to reconstruct regional climatic variability near the headwaters of the Athabasca River as well as the flood history of the Peace River in the northern section of the Peace-Athabasca Delta. Climate records were reconstructed from carbon and oxygen isotope analyses of a composite tree-ring chronology.

The climate reconstruction indicates the following climate situations: (1) cold and very dry conditions in the 1700s in association with the peak of the Little Ice Age; (2) subsequent warming and wetter conditions from about 1780 until 1940; and (3) progressively drier conditions.

In discussing the implications of the flow regulation of the Peace River for changes in the ecology of the delta, the scientists concluded that 'Although concerns have been focused on observed drying since the Peace River has been regulated since 1968, paleolimnological evidence from Spruce Island Lake indicates this to be part of an extended period of drying that was initiated in the early to mid-1990s. Resolving whether significant river regulation impacts are superimposed upon natural climate-driven hydrological variability over this time interval, however, remains uncertain' (Wolfe et al. 2005, 160).

The research team also observed that 'Recent studies . . . suggest that the absence of high-magnitude flood between 1975 and 1995 may be a result of climatic variation that increased the temperatures during the ice-cover season, reduced the snow-pack depths, and altered the intensity and duration of the pre-melt period. Thus, reduced spring ice-jam flooding due to regulation of the Peace River and climate variability represent competing hypotheses to explain what are perceived to be unusually dry conditions in the PAD [Peace-Athabasca Delta] during much of the past 35 years' (Wolfe et al. 2006, 132).

The findings and conclusions of the interdisciplinary scientific team are a reminder that we often face considerable uncertainty and complexity regarding the behaviour of ecological systems. As a result, scientists are understandably cautious in drawing conclusions about cause-and-effect relationships. In this context, systematic monitoring of the impact of both climate variability and water withdrawals for oil sands production facilities will be needed to determine the long-term effect of the oil sands development on the Athabasca River system.

Source: Wolfe et al. 2005; 2006.

associated with a boreal ecosystem, the low water temperatures result in low reproduction rates and many species not beginning to reproduce until they are six to 10 years old. The challenge then is to determine what is termed 'instream flow needs', or the minimum threshold for water flow needed to sustain the health of an aquatic ecosystem.

Abstraction of Water from the Athabasca River

The ecological integrity of the river is threatened during the winter months in years when low precipitation rates in the Athabasca River basin lead to low flow conditions. Industrial water withdrawals must be limited during these brief periods to protect the health of the river. These withdrawal limits must be based on the precautionary application of current scientific knowledge and must acknowledge the impacts to the Athabasca River's resilience arising from active oil sands mining operations within its basin.

—Woynillowicz and Severson-Baker 2006, 2

Woynillowicz and Severson-Baker (2006, 4) have noted that withdrawal of water from the Athabasca River to support production from the oil sands is further challenging because such operations 'return very little water to the Athabasca River'. They also highlight that oil sands operations are by far the largest withdrawers of water from the river, representing some 65 per cent of withdrawals in 2006, which is about two times the amount needed to meet the annual water demand of Calgary with its population of more than 1 million. Future anticipated oil sands production will increase the withdrawals to about one and a half times the amount now being removed.

Withdrawal of water from the Athabasca River is not the only threat to the integrity of the aquatic system. As Woynillowicz and Severson-Baker (2006, 8) observe, the mining operations also affect quantity and quality of water in the tributaries of the Athabasca River, which in turn affects the main river. Furthermore, changes to the natural habitat in the tributaries resulting from mining activity affect fish in their waters. To illustrate, they point out that intensive mining activity will occur in the Muskeg River catchment, a sub-basin of the Athabasca, and that it 'is unlikely that this river basin can

Oil sands operation in Alberta. The Athabasca River is in the far background, adjacent to the end of the plant. In the foreground are large (soccer-field-sized) cakes of yellow sulphur.

sustain this degree of industrial development and still retain any significant degree of ecological integrity'.

In terms of management response to threats to the Athabasca River system, concern has been expressed that the provincial government's approach 'is not adequately precautionary and protective and relies too much on voluntary actions by companies to protect the river' (Woynillowicz and Severson-Baker 2006, 8). Woynillowicz and Severson-Baker argue that robust monitoring should be done but conclude that no such system is in place to 'detect the effects of low flows on fish habitat and fish population'. The provincial government has considered some monitoring options but has not committed funding to implement a systematic monitoring system. They also argue that to reduce negative impacts from water withdrawal, there should be a systematic examination of options for on-site storage of water, off-stream storage of water upstream of the oil sands area, and enhanced water conservation.

Air Quality

Woynillowicz, Severson-Baker, and Raynalds (2005, 44) comment that development of the oil sands is a major contributor to air pollution emissions in Alberta. By 2003, Alberta had become the major polluter from industrial sources for all of Canada. In that year, industrial sources released more than 1 billion kilograms of air pollutants.

Particular attention has been directed to what are referred to as 'criteria air contaminants', or CACs. These contaminants are the ones most commonly emitted by heavy industry using fossil fuels, and they also negatively affect health. They include nitrogen oxides (NO_x), sulphur dioxide (SO_2), volatile organic compounds (VOCs), and particulate matters ($PM_{2.5}$). All of them are released from oil sands operations.

Oils sands technology has improved, and this has led to a reduction in the volume of pollutants emitted per barrel of oil produced. Nevertheless, the emissions from producing synthetic oil from bitumen are higher than they are in conventional oil production processes. Furthermore, the rapid expansion of oil sands production has meant that overall emissions continue to increase even though emissions per barrel have decreased.

Modelling of air pollution based on approved future oil sands production expansion indicates that maximum emissions of NO_x and SO_2 will exceed provincial, national, and international standards. VOCs are also of concern because in 2002 Alberta was one of the top four states or provinces in North America in terms of emissions, and emissions of VOCs are predicted to go up. Any additional development will make the situation worse. In contrast, forecast emissions related to $PM_{2.5}$ show that although they will increase, they will remain below accepted thresholds.

In early 2008, the federal Conservative government announced a green plan that would allow greenhouse gas emissions from the oil sands to triple from 25 million to 75 million tonnes a year over the next decade, a period when the national goal is to reduce overall emissions by 150 million tonnes. The economic value of the oil sands development appears to override environmental concerns, even issues as serious as global climate change, as discussed in Chapter 7.

Uranium and Nuclear Power

Uranium

Canada is not a stranger to uranium mining. Today, the main use of uranium, once it has been processed, is for fuel in nuclear reactors to generate electricity. When exploration for uranium began in the early 1940s, however, the demand for uranium was not from nuclear reactors, because none existed. Instead, the demand was for military use, specifically for atomic weapons then being developed. During World War II, the federal government established a Crown corporation that later became known as Eldorado Nuclear Ltd. For a short time, it was the only company approved to mine radioactive material. After the end of the war, other firms were given approval to mine and process uranium, and by the late 1950s more than 20 uranium mines were operating. The largest were near Elliot Lake, ON, and Uranium City, SK. When commercial nuclear reactors became available in the 1960s, it stimulated further exploration and led to discovery of major new deposits of uranium in northern Saskatchewan. One outcome was the closure of the lower-grade uranium mines in Elliot Lake and Uranium City.

Uranium mining and production have not been without controversy. Regarding environmental considerations, the mining process results in 'tailings' that need careful storage to keep them from leaking into aquatic systems. After being used in power plants, the nuclear fuel wastes are highly radioactive, generating major challenges in terms of long-term containment and storage. From an economic perspective, however, uranium mines have provided jobs for skilled workers in remote regions of the country where jobs have not been plentiful (see accompanying box).

Uranium Mining in Saskatchewan

Saskatchewan is the world leader in uranium production, and the uranium industry serves to provide many jobs, investment, and economic development in the province. The industry provides all of these benefits in an environmentally and socially responsible manner and is held accountable for its performance. Regular internal and external audits on the environment and safety of operations are ongoing, and thousands of samples of air, water, and vegetation are taken annually. These samples demonstrate, and the government regulatory agencies agree, that the industry is protecting the environment.

—Saskatchewan Mining Association 2007

Canada mines about one third of global uranium, making it the world leader in output. The next largest uranium producers are Australia, Kazakhstan, and Russia. In 2004, production in Canada was 13,676 tonnes of uranium oxide concentrate. It is estimated that Canada has some 524,000 tonnes of uranium oxide (U_3O_8), 9 per cent of the global total. Australia has two and a half times that amount.

Most of the production in Canada is from two mines in northern Saskatchewan: the McArthur River underground mine, the largest uranium mine in the world, and the McClean Lake open-pit mine, to be followed by an underground mine. Both began production in 1999. In 2005, McArthur River contributed more than one-sixth of global uranium production. Another mine, at Rabbit Lake, is at the end of its productive life. Two other mines, Cigar Lake and Midwest, will begin production in about 2010, and some of the ore from Cigar Lake will be processed at the on-site mill at Rabbit Lake. A relatively small deposit at Dawn Lake will likely be mined later. Previous mines at Key Lake and Cluff Lake are no longer in production.

The current provincial government in Saskatchewan supports uranium mining, subject to appropriate safeguards for the environment. This policy contrasts with that of the New Democratic government in the early 1990s, which was to end mining of uranium. Four Saskatchewan uranium mines have been ISO 1401 certified: McClean Lake (2001), Key Lake (2003), McArthur River (2003), and Cluff Lake (2004), indicating that they meet international standards related to environmental considerations. AREVA (2007), operator of the McClean and Cluff Lake mines, has stated that the 'industry's long-term goal is to return all operations, as close as possible, to a natural state suitable for future uses. All uranium mine site operators must post bonds with the federal government to ensure adequate funds are available for proper decommissioning of each site after the reserves have been mined out.' AREVA reported that in 2007, the uranium operations companies spent about $4.8 million on reclamation projects.

As an example of one company's approach to environmental management, AREVA (2007) uses the following environmental policy to guide its activities:

> AREVA recognizes that continued economic and social development depend on a healthy environment and incorporates environmental considerations into all company activities to ensure sustainable development. AREVA is committed to continually improve approaches and technology to minimize the effects of its activities on the environment.
>
> To meet this commitment, AREVA shall:
>
> * comply with all applicable environmental legislation;
> * minimize adverse environmental impacts of its activities by reducing consumption of natural resources, controlling releases, and optimizing waste management;
> * prevent pollution by using processes, practices, material, or products that avoid, reduce, or control pollution;
> * deal proactively with environmental issues by identifying potential impacts and implementing mitigative actions and/or developing effective contingency plans;
> * develop internal objectives and targets to achieve continual improvement;
> * measure performance against established goals;
> * conduct employee training, internal assessments, and periodic reviews to ensure these operations and activities are conducted in compliance with documented procedures;
> * communicate environmental requirements and corporate initiatives to employees and contractors to encourage their participation and compliance; and
> * involve the public, with particular focus in impact communities, in initial planning, ongoing operations, and decommissioning of AREVA activities through an open and transparent public involvement program.

It is, of course, one thing to have a statement of intent and another to follow through with action. Nevertheless, it is instructive to see the approach that one mining company has adopted to address environmental matters.

Nuclear Power

A made-in-Canada experimental nuclear reactor was developed in Chalk River, ON, and began producing power in 1947. That reactor became the forerunner of the CANDU (CANada Deuterium Uranium) pressurized heavy-water reactors that are used around the world.

Canada produces about 12 to 15 per cent of its electricity from **nuclear power**, compared to about 20 per cent in the United States. Some 18 reactors produce more than 12,500 MW of power. Ontario, the province most dependent on nuclear power, has used it since the early 1970s. Quebec (Gentilly) and New Brunswick (Point Lepreau) each has a single-unit CANDU plant.

By 2007, Ontario had 16 commercial nuclear reactors in operation in three locations (Pickering and Darlington, both on the shore of Lake Ontario, and Bruce on the shore of Lake Huron). During the early 1990s, nuclear reactors generated about two-thirds of Ontario's electricity. This had dropped to about 51 per cent by 2005. Part of the reason is that Ontario nuclear reactors have experienced numerous problems, resulting in several units at the Pickering and Bruce plants being taken out of service for long periods for safety reasons. The refurbishing of nuclear units in these plants has been very expensive. For example, the cost to refurbish units at the Bruce plant to give them a further 25 years of life has been estimated at $4.25 billion. In 2005, the New Brunswick government decided to renew its Point Lepreau reactor at an estimated cost of $1.4 billion.

To place such costs into perspective, the Ontario Power Authority conducted a major review of energy requirements and in December 2005 reported that the province would need to spend $83 billion to refurbish its electricity supply system over a 20-year period. The report concluded that nuclear power capacity would have to be increased to maintain its share of production at 50 per cent. Among the arguments in favour of building more nuclear power capacity are that nuclear power has less environmental impact than coal- or petroleum-fuelled power plants and that it operates at a lower cost.

In the summer of 2006, the Ontario government stated that enhanced nuclear power capacity would be a key component of an overall plan to meet emerging shortages of electricity. The government indicated that about $40 billion would be spent on enhancing nuclear capacity, including the addition of two new plants. It also decided to 'fast track' the overall $83-billion plan, meaning that the plan was exempt from a full review under the Ontario Environmental Assessment Act, which most likely would have taken up to five years to complete. Proposals for individual nuclear plants would go through the federal government's assessment process, estimated as a two-year process.

Management of Used Nuclear Fuel

A major concern for many is how to handle used radioactive wastes from nuclear power plants. Recognizing such concern, the federal government created the Nuclear Waste Management Organization (NWMO) in 2002 to identify options for storage and disposal of **nuclear wastes**. In its report, completed in November 2005, the NWMO (2005, 2) noted that Canada has been using nuclear fuel to generate electricity for decades, and as a result, Canada has about 2 million spent fuel bundles, which add up to 36,000 tonnes of uranium. This amount is estimated to increase by a factor of two in the future, assuming that all existing nuclear plants operate for their anticipated lifespans of 40 years.

The used fuel bundles from nuclear plants are now stored in regulated facilities on the sites at which they are produced. This arrangement has always been viewed as a short-term one, and a need exists to determine what should be done in the long term. The NWMO has examined such a long-term solution. In its view (2005, 3), both scientific and technical analysis are critical components in finding a solution, but such analysis 'cannot be the sole basis of our choice. While science can speak to the probability of an occurrence of an event, science cannot speak to the social tolerance for its occurrence.' As a result, the NWMO allocated considerable time to listening to the views of Canadian citizens who were not technical specialists. The outcome was strong agreement between technical and non-technical commentators about two requirements: any approach must be (1) safe and secure for people, communities, and the environment and (2) fair to present and future generations.

The NWMO examined the benefits, costs, and risks of three approaches: (1) deep geological disposal in the Canadian Shield; (2) centralized storage above or below ground; and (3) storage at nuclear reactor sites. In reviewing these options, the following factors were used to guide the assessment: fairness, public health and safety, worker health and safety, community well-being, security, environmental integrity, economic viability, and adaptability. Not surprisingly, no option satisfied all requirements.

Disposal option. Option 1, deep storage in the Canadian Shield, was judged to be better than Options 2 and 3 in the long term, because engineering design and natural barriers would isolate and confine the radioactive waste. The main disadvantage was seen as lack of adaptability in terms of the capacity to respond to changing knowledge or evolving technology.

Storage options. Options 2 and 3 were judged to be appropriate in the short term. A concern, however, was that sites for nuclear power plants had never been selected with regard to their technical suitability as permanent storage sites. In addition, residents of the host communities had a reasonable expectation that the nuclear fuel waste would eventually be moved.

Hybrid option. Given the limitations of the disposal and storage options, the NWMO concluded that a different option should be developed, labelled 'adaptive phased management', which includes a technical method and a management system.

Key features of the hybrid option are: (1) eventual centralized containment and isolation of used reactive fuel in deep underground storage; (2) optional shallow storage at a central site as a contingency option; (3) continuous monitoring; (4) allowance for retrieval; (5) citizen engagement; and (6) phased and adaptive decision-making. It would be implemented in three phases: Phase 1 (30 years), preparing for centralized management of used nuclear fuel; Phase 2 (30 years), demonstrating central storage and technology systems; and finally, Phase 3 (beyond approximately 60 years), long-term containing and isolating of the used nuclear waste combined with monitoring.

Adaptive Phased Management

Adaptive Phased Management tries to find an optimal balance of competing objectives. It embraces the precautionary principle and adaptive management. Societal goals and objectives and successful technology demonstration will determine the pace of implementation. We believe Adaptive Phased Management is the strongest possible foundation for managing the risks and uncertainties that are inherent in the very long time frames over which used nuclear fuel must be managed with care.

—NWMO 2005, 7

The NWMO (2005, 6) was explicit that a key component of the overall strategy would be 'to seek an informed, willing community to host the central facilities'. In the spirit of 'fairness', the NWMO concluded that it would focus on the provinces directly involved in nuclear fuel—Saskatchewan, Ontario, Quebec, and New Brunswick. And given that in August 2007 Alberta announced it would build a nuclear power plant within eight to 10 years, Alberta would likely be added to the list. The Energy Alberta Corporation has chosen a site for two proposed twin-unit CANDU reactors, estimated to cost $6.2 billion, on private land about 30 kilometres west of Lac Cardinal some 480 kilometres northwest of Edmonton. Energy Alberta selected the Lac Cardinal site because of community support, as well as the presence of necessary infrastructure and support services.

To be inclusive, communities and regions in other provinces would be invited to indicate whether they were interested in hosting the central facilities. The NWMO emphasized that the siting process needed to be open, inclusive, and fair to all parties, meaning that anyone with an interest should have an opportunity to make their views known. This would include communities adjacent to routes used for transporting the used nuclear fuel wastes to the central location.

The work of the NWMO is a clear reminder that even if Canada stopped using nuclear plants to produce electricity tomorrow, there would still be a legacy of spent nuclear fuel to be stored and contained for a very long time.

Sustainable Energy Pathways

The Canadian Academy of Engineering (2007) examined various 'energy pathways' to explore how Canada's endowment of energy sources can be developed to meet end-use needs. The academy noted that by 2007, about 75 per cent of the energy demand within Canada was being met by fossil fuels. This percentage is lower than that in many other industrialized countries because of the significant contribution in Canada from both hydro and nuclear power. Nevertheless, the academy (2007, 11) commented that that 'there is clearly a global desire to shift away from fossil fuels.' In that regard, the academy suggested that it would be reasonable for Canada to have a long-term target of satisfying no more than 33 per cent of its energy needs by fossil fuels. For such a target to be achieved, alternative energy pathways need to be considered.

Canada As a Sustainable Energy Superpower?

Energy is one of the cornerstones of civilization and is central to Canada's economic and social well-being, but we lack a compelling national energy vision. With huge, unequalled energy resources, will Canada be able to produce upgraded energy products at reasonable prices with acceptable environmental impacts? New technology will be needed, but successes will also require effective public policy and new concepts of risk-sharing. Transformational changes will not be made through the efforts of individual companies, nor governments working alone.

*It we are able to bring about such transformational change, it has been suggested that Canada would become a '**sustainable energy superpower**'.*

—Canadian Academy of Engineering 2007, 3–4.

To help you consider what alternative pathways might be feasible, 10 of the 15 criteria used by the academy to judge energy pathways are detailed below. The other five are commercial readiness, fitness for the Canadian national initiative, corporate capacity, Canadian competitive advantage, and delivery/infrastructure issues.

1. *Scientific principles.* The scientific principles related to an option are (1) not yet fully understood; (2) well understood, and all relevant information has been disseminated to the scientific community; (3) well understood and have been validated by independent research teams; or (4) such that no significant scientific questions remain to be answered.

2. *Technology validation.* Related technologies have been (1) tested beyond the laboratory bench scale; (2) successfully tested at the first logical scale beyond the laboratory bench; (3) successfully tested on a demonstration scale under commercially relevant conditions; or (4) applied with relevant stakeholders in the demonstration, and results have been accepted or validated.

3. *Sustainability.* Development of an option would (1) neither extend the expected commercial lifespan of an existing energy source nor introduce a new energy source with a lifespan of more than 30 years; (2) measurably (by at least 30 years) extend the expected commercial lifespan of an existing energy source or introduce a new energy source with an equivalent lifespan; (3) extend the expected commercial lifespan of an existing energy source well beyond any current planning horizon (100+ years) or introduce a new source with the same long-term potential; or (4) result in essentially unlimited energy supply.

4. *Enabler for another option.* Commercialization of an option (1) depends on successful commercialization of another option but will not by itself contribute to the successful commercialization of that option; (2) does not depend on the successful commercialization of another option and will not contribute in a measurable way to the successful commercialization of another option; (3) will contribute in a measurable way to the successful commercialization of another option; or (4) is essential to the effective commercialization of another option and will substantially enhance the effectiveness of both.

5. *Social acceptability.* An option would encounter (1) significant societal resistance that would not be readily overcome; (2) societal concerns, and it will be a challenge to address them in a clear and convincing manner; (3) minimal societal concerns that are expected to be relatively easy to address; or (4) no societal concerns.

6. *Economic impact.* Successful commercialization of an option would (1) have a limited or short-term economic impact on the Canadian economy; (2) contribute to the

Canadian economy and possibly have a major impact in a specific geographic region; (3) have a major positive and sustained impact across the Canadian economy; or (4) be equivalent to other major undertakings such as constructing the first gas pipeline across Canada.

7. *Environmental (other than GHG emissions) impact.* Commercialization of an option would have (1) significant negative overall environmental impacts that would be perceived as unacceptable by well-informed citizens; (2) small negative overall impacts on the environment, and well-informed citizens would seek evidence that such impacts would be acceptable; (3) small positive overall environmental impacts in the view of well-informed citizens; or (4) significant positive overall environmental impacts in the view of well-informed citizens.

8. *Greenhouse gas (GHG) emissions impact.* Commercialization of an option would result in (1) significant increases in Canada's total GHG emissions; (2) moderate increases in Canada's total GHG emissions; (3) moderate reductions in Canada's total GHG emissions; or (4) significant reductions in Canada's total GHG emissions.

9. *Energy-efficiency impact.* Commercialization of an option would (1) result in a reduction in the overall efficiency with which energy is produced, converted, delivered, and/or used in Canada; (2) not have a material impact on the level of efficiency with which energy is produced, converted, delivered, and/or used in Canada; (3) increase the overall efficiency with which energy is produced, converted, delivered, and/or used in Canada; or (4) result in major improvements in the efficiency with which energy is produced, converted, delivered and/or used in Canada.

10. *Value-added impact.* Commercialization of an option would (1) not increase the yield or value of products derived from this energy option compared to current commercial practice; (2) moderately increase the yield or value of products derived from this energy option compared to current commercial practice; (3) significantly increase the yield or value of products derived from this energy option compared to current commercial practice; or (4) result in a major transformation in how Canada's resources are recovered and upgraded within our borders.

Figure 12.9 illustrates how over time and around the world, the major sources of primary energy have evolved from muscular and biomass energy, to coal, to oil and natural gas, and to nuclear fuels. Some countries, such as Japan, have moved rapidly from Phase I to Phase IV. In contrast, many developing countries struggle to move beyond Phase I except in their major cities. Canada is now well into Phase IV. A challenging question is to determine what a Phase V might look like.

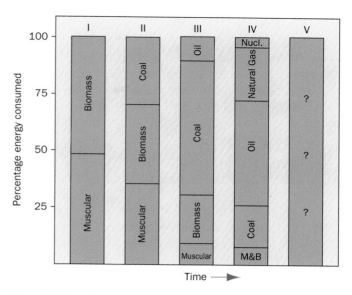

Figure 12.9 The evolution of energy consumption mixes by source. Source: Chapman 1989, 6.

We encourage you to use the above criteria or some other *explicit* set of criteria to assess which options or pathways are most appropriate in the future. When using such criteria, you should decide whether all of them have equal value or whether it would be better to weight them in terms of relative importance. A key consideration is that both we and others should know the basis upon which assessments of energy options are made. This is important, because every option has a different mix of strengths and weaknesses and you need a systematic way of determining how to prioritize them.

Implications

If our ecological footprint is to become lighter in terms of our use of minerals and energy, individuals, institutions, and societies will have to change. Consumers complain that manufacturers do not build 'green cars', while manufacturers say that customer demand does not indicate that green cars are wanted in sufficient quantity to justify producing them. Thus, if change is to occur, adjustments are needed at all levels, with individuals taking initiative to reduce consumption of energy and mineral products, governments providing greater incentives to both individuals and manufacturers to embrace green products, and manufacturers showing leadership to market green products effectively.

In the meantime, what can you do? Box 12.3 identifies some actions or changes in behaviour that will help to extend the life of non-renewable or stock resources, as well as encourage greater use of renewable or flow resources. There is no recipe or formula that will lead readily and easily to a society that is less materialistic and energy-intensive. However, small steps such as the ones identified in Box 12.3 can cumulate to effect significant change. Perhaps most important, thinking about and taking such actions are first steps in shifting basic beliefs and values.

Box 12.3   What You Can Do

1. Individuals can modify behaviour to reduce energy and materials use, waste production, and ecosystem degradation. Examples include:
   - Using a bucket and sponge and a trigger nozzle on a hose saves about 300 litres of water each time a car is washed, as well as saving some of the energy used to pump water through a municipal system.
   - Commuting to work by car pool, public transportation, or bicycle or on foot rather than by driving alone in a car reduces energy consumption and helps to reduce greenhouse emissions and other air pollutants.

2. Individuals can use more efficient technology or use products with lower environmental impact throughout their life cycle. Examples include:
   - Use smaller, more energy-efficient automobiles and major appliances with the lowest energy ratings. This too will reduce greenhouse gas emissions and other air pollutants.
   - Install water-saving devices in the home, such as low-flow showerheads and toilet dams. Energy will be saved by heating less water and by moving less clean and grey water through the water supply system (discussed in Chapter 11).

   - Replace incandescent light bulbs with fluorescent bulbs, which use about 75 per cent less energy and last 10 times longer.

3. As part of a larger society, individuals can ask for appropriate information and insist that products, services, and planning explicitly address environmental implications. Examples include:
   - Individuals in their communities can promote better planning of urban transit and bicycle routes and reduced dependency on passenger vehicles (discussed in Chapter 13).
   - Individuals can lobby their local government and others to show leadership in educating the community about types of behaviour and products that are environmentally benign.
   - Individuals can support manufacturers committing to including environmental considerations into their production processes via life-cycle management and environmental management systems.

Source: Based on Environment Canada 2002, 68.

Summary

1. Non-renewable or stock resources take millions of years to form. Consequently, from a human viewpoint, they are for all practical purposes fixed in supply and therefore not renewable.

2. If done systematically and correctly, enhanced environmental management practices and extended social responsibility with regard to non-renewable resources almost always generate some kind of economic return on investment for business, although usually over the long term.

3. The main environmental issues for the mining and energy sectors include: acid mine drainage; sulphur

dioxide emissions; metal toxicity; disruption of remote ecosystems as a result of exploration, test drilling, and operation of oil fields or gas wells; disturbance to aquatic ecosystems from escape of waste heat produced by nuclear energy production; and threats to human and ecosystem health from radioactive waste associated with nuclear energy production over thousands of years.

4. 'Best practice' for environmental management in mining and fossil-fuel firms in Canada should include a combination of basic scientific research to ensure understanding of natural and social systems that can be affected by operations, design of appropriate mitigation measures, environmental impact assessments and reporting, environmental audits, corporate policies that explicitly include environmental aspects, environmental management systems, and life-cycle assessments.

5. Canada is one of the largest exporters of minerals and mineral products in the world, and there are more large mineral exploration companies based in Canada than in any other country. Canada accounts for about one-half of all such companies.

6. Canada's top five minerals in terms of production value in 2006 were nickel, copper, iron ore, gold, and cement.

7. In 1998, BHP's Ekati mine, located 200 kilometres south of the Arctic Circle, became the first diamond operation in Canada. A second diamond mine, Diavik, began producing in 2003. Jericho was the third and began production in July 2006.

8. In October 1998, Canada became a major producer of diamonds when the Ekati Diamond Mine started up. Today, it produces $1.7 millionworth of diamonds each day, and Canada is the third largest diamond producer in the world.

9. Mining tailings will be held in a lake at Ekati. As the tailings settle, consolidate, and evolve to permafrost, rocks and soil will be spread over the surface. Revegetation will be started, with the goal of having the entire holding area become a wetland once the mining is completed.

10. Serious challenges were encountered by Ekati in incorporating traditional ecological knowledge into environmental research: (1) two Aboriginal groups were in the midst of land claims negotiations and as a result were reluctant to release traditional knowledge into the public domain because this knowledge was important for their negotiation strategy; (2) concern was expressed by Aboriginal people about using traditional knowledge outside of the context of the cultures and broader system of knowledge that give it meaning and value; (3) there was no one set of traditional knowledge, since the Inuit, Métis, and Dene each have their own traditional knowledge, which do not always coincide; (4)

traditional knowledge was viewed by Aboriginal people as their intellectual property, and therefore its use and management had to remain within their control; and (5) there was no documented baseline of traditional knowledge.

11. The Bathurst caribou herd is the largest one in the NWT and includes about 350,000 animals. Since the caribou herd does not follow the same migration route each year and the areas affected by the Ekati mine represent less than 0.01 per cent of the range of the herd, it was believed that the mining activity would have a very small impact.

12. Beginning in 1992, the company responsible for Ekati initiated scientific research to understand the impact of the proposed mining activity and to develop mitigation measures. In July 1994, the minister of Indian affairs and northern development referred the mining project for an environmental assessment, and in July 1995 the company submitted its environmental impact statement. From late January to late February 1996, an Environmental Assessment Panel held public meetings and in June submitted its report to the federal government. In February 1997, the federal government gave its formal approval, and construction started in May. In January 1999, the first diamonds from Ekati were sold in Antwerp.

13. Impact and benefit agreements were pioneered in Canada and are intended to ensure that Aboriginal communities benefit from mining projects and that if they contain compensation provisions, the communities are compensated for the negative impact on their communities, their land, and their traditional way of life. Targets for employment of Aboriginal people at the Ekati mine have been met.

14. Nearly 200 diamond cutters work in Yellowknife for four different companies.

15. Canada is ranked as the sixth largest user of primary energy in the world. Fossil fuels are the main type of energy consumed by Canadians, and one of the largest consumers of energy is the transportation sector.

16. The combustion of fossil fuels emits greenhouse gases, such as carbon dioxide and nitrous oxide, which accumulate in the atmosphere and contribute to climate change.

17. Alternative energy sources are solar, geothermal, hydro, tides, and wind.

18. Wind power is the fastest-growing sector in the world's energy market. Wind turbines have minimal adverse environmental affects. However, compared to conventional fossil-fuel energy sources, they are expensive.

19. Europe is the leader in use of wind energy. Canada is now one of 13 countries with more than 1,000 MW of installed wind capacity.

20. Wind turbines have minimal impact on flying birds and generally create minimal noise problems or human health issues.
21. China is the leading country in terms of total installed capacity of solar power; Canada is ranked 29th.
22. Europe has the most diverse end uses of solar power. In Canada and the US, the main use of solar power is to heat water in swimming pools.
23. In Atlantic Canada, oil production is based in the Hibernia, Terra Nova, and White Rose fields, and natural gas production occurs near Sable Island.
24. Discoveries of petroleum below the Beaufort Sea off the coast of the Northwest Territories and off the coast of British Columbia offer potential for development. However, there is a prohibition on transporting petroleum by ship along the coast of British Columbia because of the *Exxon Valdez* spill in 1989.
25. Various nations are challenging Canada's claims to Arctic sovereignty, not only for strategic military reasons but also to establish ownership over fossil fuels on the sea floor.
26. The Athabasca oil sands is the second largest reserve of petroleum in the world, outranked only by the reserves in Saudi Arabia.

27. Extraction of the bitumen from the oil sands is having negative environmental impacts on the boreal forest and wetland systems, the Athabasca River, and air quality.
28. Because of production from the Athabasca oil sands, Alberta is now Canada's leading industrial polluting province.
29. Canada produces about one-third of the world's uranium, and most of that production is from two mines in northern Saskatchewan.
30. Twelve to 15 per cent of electricity in Canada is supplied from 18 nuclear power plants, 16 of which are in Ontario.
31. Regarding nuclear energy, much of the concern focuses on how to dispose of used nuclear fuel.
32. Three strategies for disposing of used nuclear fuel have been considered: deep underground geological storage, centralized storage above or below ground, and storage at nuclear power plant sites. A hybrid approach, labelled 'adaptive phased management', has been recommended.
33. The Canadian Academy of Engineering has examined 'sustainable energy pathways' for Canada and has concluded that Canada has the potential to become a 'sustainable energy superpower'.

Key Terms

acid mine drainage
bitumen
crude bitumen
frozen core dam
impact and benefit
    agreements (IBAs)

in situ recovery
kimberlite pipes
metal toxicity
non-renewable or
    stock resources
nuclear power

nuclear wastes
renewable or flow
    resources
steam-assisted gravity
    drainage

sulphur dioxide
    emissions
sustainable energy
    superpower
wind power

Questions for Review and Critical Thinking

1. What are the implications of non-renewable or stock resources for strategies related to 'sustainable development' or for 'sustainable resource management'?
2. How important are non-renewable resources for the Canadian economy?
3. What have been elements of 'best practice' related to the opening of diamond mines in the Canadian North?
4. What was learned from the environmental assessment for the Ekati mine regarding incorporation of local knowledge into scientific understanding of impacts?
5. Why is Canada so dependent on fossil fuels? What would have to change for there to be less dependence?
6. What are the main uses of primary energy in Canada?
7. What are the advantages and disadvantages of alternative energy sources?

8. Which European country is the global leader in terms of wind energy?
9. What is the significance of the Prince Wind Farm on the shore of Lake Superior?
10. What country is the global leader in terms of installed capacity of solar power?
11. What is the main use of solar power in Canada?
12. What are the greatest environmental risks associated with extracting fossil fuels from the seabeds of the Atlantic, Pacific, and Arctic oceans?
13. What is the significance of the *Exxon Valdez* incident off the coast of Alaska in 1989 in terms of the development of offshore oil resources?
14. Why did the Russians plant a flag on the seabed at the North Pole in August 2007?

15. How much water and energy is required to produce one barrel of crude oil from the Athabasca oil sands?

16. Why is fragmentation of the boreal forests and wetlands in the area of the Athabasca oil sands of concern in terms of biodiversity?

17. Why does abstraction of water from the Athabasca oil sands to support oil production pose a threat to the Peace-Athabasca Delta?

18. What are the best strategies for facilitating the economic benefits related to oil production from the Athabasca oil sands while minimizing negative environmental impacts?

19. What are the main controversies associated with mining of uranium?

20. Why are there reservations about storing used nuclear fuel on the sites of nuclear power plants?

21. What are the characteristics of 'adaptive phased management' in relation to spent nuclear fuel?

22. Do you agree or disagree that Canada has the potential to be a 'sustainable energy superpower'?

23. What changes should be made by individuals, organizations, businesses, and governments to reduce energy and mineral use?

Related Websites

Canadian Wind Energy Association (Canwea): http://www.canwea.ca/home.html?phpLang'en

Danish Windpower Organization: http://www.windpower.org

Environment Canada, Wind Atlas: http://www.cmc.ec.gc.ca/rpn/modcom/eole/CanadianAtlas.html

European Wind Energy Association (Eurwea): http://www.ewea.org

Independent Environmental Monitoring Association, A Public Watchdog for Environmental Monitoring of the Ekati Diamond Mine: http://www.monitoringagency.net/Home/tabid/36/Default.aspx

Mining Association of Canada: http://www.mining.ca/www/index2.php

Mining Watch Canada: http://www.miningwatch.ca/index.php

Natural Resources Canada, Canadian Minerals Yearbook: http://www.nrcan.gc.ca/mms/cmy/pref_e.htm

Natural Resources Canada, Office of Energy Efficiency: http://oee.nrcan.gc.ca/english/index.cfm?attr=0

Natural Resources Canada Wind Energy Information Site: http://www.canren.gc.ca/wind/index.asp

Pollution Probe, Energy Program: http://www.pollutionprobe.org/whatwedo/Energy.htm

Windpower Monthly: http://www.wpm.co.nz

Further Readings

Note: This list comprises works relevant to the subject of the chapter but not cited in the text. All cited works are listed in the section 'References' at the end of the book.

Bramley, M., D. Neabel, and D. Woynillowicz. 2005. *The Climate Implications of Canada's Oil Sands Development*. Drayton Valley, AB: Pembina Institute.

British Petroleum. 2007. 'Statistical review of world energy 2007'. http://www.bp.com/productlanding.do?categoryId=6848&contentId=7033471.

Canadian Aboriginal Minerals Association. 2001. *Tapping Aboriginal Resources: Opportunities for Aboriginal Community Development, Mining Metals and Diamonds, Oil and Gas*. Whitehorse, Yukon: Canadian Aboriginal Minerals Association.

Diamonds North Resources Ltd. 2006. 'Targeting Canada's next diamond mine: Shareholders' report 2005'. http://www.diamondsnorthresources.com/i/pdf/DDN_SAR_2006.pdf.

Ebner, D., and B. McKenna. 2008. 'The pipes that bind'. *The Globe and Mail* 29 February: B1.

Hanks, C., and S. Williams. 2002. 'Perceptions of reality: Cumulative effects and the Lac de Gras diamond field'. In A.J. Kennedy, Ed., *Cumulative Environmental Effects Management*, 411–24. Edmonton: Alberta Society of Professional Biologists.

Harding, J. 2007. *Canada's Deadly Secret: Saskatchewan Uranium and the Global Nuclear System*. Halifax: Fernwood.

Hoos, R.A.W., and W.S. Williams. 1999. 'Environmental management of BHP's Ekati diamond mine in the western Arctic'. In J.E. Udd and A. Keen, Eds, *Proceedings of the International Symposium on Mining in the Arctic*, v. 5, 63–70. Rotterdam: Balkema.

Indigenous Environmental Network. 2002. 'Indigenous Mining Campaign Project'. http://www.ienearth.org/mining_campaign.html#project.

International Energy Agency, Photovoltaic Power Systems Programme. 2006. 'PVPS annual report 2006: Implementing Agreement on Photovoltaic Power Systems'. Paris: International Energy Agency. http://www.iea-pvps.org/products/download.rep_ar06.pdf.

Kennett, S. 1999. *A Guide to Impact and Benefit Agreements*. Calgary: Canadian Institute of Resources Law.

McAllister, M.L., and C.J. Alexander. 1997. *A Stake in the Future: Redefining the Canadian Mineral Industry*. Vancouver: University of British Columbia Press.

Mining Watch Canada. 2002. 'Aims and objectives'. http://www.miningwatch.ca/mwC_profile_short.html#anchor28160192.

Noble, B.F, and K. Storey. 2001. 'Towards a structured approach to strategic environmental assessment'. *Journal of Environmental Assessment Policy and Management* 3: 483–508.

Northern News Service. 2003. 'Diamonds double manufacturing figure'. http://www.nnsl.com/ops/man.html.

Northwest Territories. 2006. Press release: 'Bathurst caribou survey completed'. http://www.nwtwildlife.com/NWTwildlife/caribou/bathursttop.htm.

NRCAN (Natural Resources Canada). 2000. *Energy in Canada, 2000*. Ottawa: Canada Communication Group.

———. 2001. *Focus 2006: A Strategic Vision for 2001–2006*. Ottawa: Minister of Public Works and Government Services.

———. 2001. 'Canada's ranking in world mining'. In *Canadian Minerals Yearbook 2000*. Ottawa: Minister of Public Works and Government Services.

———. 2007. Information Bulletin: 'High metal prices spur mineral production to a record $34 billion in 2006'. http://www.nrcan.gc.ca/mms/pdf/minprod-07_e.pdf.

NRCAN, Minerals and Metals Sector. 1998. *From Mineral Resources to Manufactured Products*. Ottawa: Minister of Public Works and Government Services.

O'Reilly, K., and E. Eacott. 1998. *Aboriginal Peoples and Impact and Benefit Agreements: Report of a National Workshop*. Northern Minerals Program Working Paper no. 7. Ottawa: Canadian Arctic Resources Committee.

Stanley, A. 2008. 'Citizenship and the production of landscape and knowledge in contemporary Canadian nuclear fuel waste management'. *Canadian Geographer* 52: 64–82

Storey, K., and L.C. Hamilton. 2003. 'Planning for the impacts of megaprojects'. In R.O. Rasmussen and N.E. Doroleva, Eds, *Social and Environmental Impacts in the North*, 281–302. Amsterdam: Kluwer Academic Publishers.

Strong, M. 1995. 'Planning for the future'. In J.M. Gunn, Ed., *Restoration and Recovery of an Industrial Region: Progress in Restoring the Smelter-Damaged Landscape near Sudbury, Ontario*, 109–22. New York: Springer-Verlag.

Whiteman, G., and K. Maman. 2001. 'Community consultation in mining: A tool for community empowerment or for public relations'. *Cultural Survival Quarterly* 25: 30–5.

Witteman, J., L.M. Davis, and C. Hanks. 1999. 'Regulatory approval process for BHP's Ekati Diamond Mine, Northwest Territories, Canada'. In J.E. Udd and A. Keen, Eds, *Proceedings of the International Symposium on Mining in the Arctic*, v. 5, 7–11. Rotterdam: Balkema.

Wolfe, B.C., et al. 2006. 'From isotopes to TK interviews: Towards interdisciplinary research in Fort Resolution and the Slave River Delta, Northwest Territories'. *Arctic* 60: 75–87.

Chapter 13

Urban Environmental Management

Learning Objectives

- To understand the nature and significance of urbanization.
- To understand the quality of environmental conditions in Canadian cities.
- To become aware of the impacts of urban areas on the environment.
- To understand the vulnerability of urban areas to natural and human-induced events.
- To become aware of 'best practices' related to urban environmental management.
- To identify strategies for governments, the private sector, civil societies, and individuals to reduce the impact of urban areas on the environment.

Introduction

According to the United Nations Population Fund (2007a, 1), the world passed a significant milestone during 2008, with more than half the global population, some 3.3 billion people, living in urban areas for the first time in the history of the planet. Looking forward to 2030, it is estimated that almost 5 billion people will be urban residents. The most dramatic changes will occur in Africa and Asia, where urban populations are expected to double between 2000 and 2030.

Regarding such dramatic change in the growth of urban areas, the United Nations Population Fund (2007a, 1) observed:

> Cities also embody the environmental damage done by modern civilization; yet experts and policy-makers increasingly recognize the potential value of cities to long-term sustainability. If cities create environmental problems, they also contain the solutions. The potential benefits of urbanization far outweigh the disadvantages: The challenge is in learning how to exploit its possibilities.

Canada is a highly urbanized society, as the box on page 482 illustrates, with four of five Canadians living in cities. As a result, both the environmental challenges and the opportunities identified by the UN Population Fund apply to Canada,

making it important to be aware of the issues and alternative ways to address them.

In this chapter, we examine the nature of **sustainable urban development** at global and Canadian scales, determine what impacts cities have had on the environment, examine how urban areas become vulnerable to environmental variability, and consider strategies to enable cities to become part of the solution rather than the problem regarding environmental quality and sustainable development. In other words, we want to learn not only 'what is the right thing to do' but also 'how to do the thing right' with regard to urban environmental management.

Sustainable Urban Development

The National Round Table on the Environment and the Economy (NRTEE 2003c, 3) observed that it is difficult to define a 'sustainable city' or 'urban sustainability'. The NRTEE noted, however, that preserving or protecting the environment for future generations, and giving equal consideration to economic, social, and environmental issues, are normally accepted as key features of 'sustainability'. Building on those ideas, the NRTEE (2003c, 3) defined urban sustainability as 'The enhanced well-being of cities or urban regions, including integrated economic, ecological, and social components, which will maintain the quality of life for future generations'. You may wish

The State of the Canadian Urban Environment

The 2001 Canadian census revealed that 80 per cent of Canadians live in urban centres and that over half of them live in the four largest urban regions—the extended Golden Horseshoe [Greater Toronto Area plus], the Montréal region, the Lower Mainland of British Columbia, and the Calgary-Edmonton corridor—where virtually all of Canada's population growth occurred in the five preceding years.

Yet the recent environmental performance of Canada's cities has been patchy at best. Despite improvements in areas such as the fuel efficiency of passenger vehicles (with the notable exception of SUVs and light trucks), most key indicators suggest negative trends: the use of cars is on the rise, urban transit ridership is down, and cities are using land less efficiently. Concentrations of ground-level ozone—which is linked to childhood asthma, respiratory illnesses, and a range of other health issues—are also increasing.

—NRTEE 2003c, xiii

Land-Use Patterns in Cities

Average urban densities (that is, the number of inhabitants per square kilometre of built-up area) have been declining for the past two centuries. As transportation continues to improve, the tendency is for cities to use up more and more land per person.

Sprawl, at least in its current form, is not conducive to sustainable development.

—United Nations Population Fund 2007a, 47, 51

to compare this definition with the ideas about sustainable development introduced in Chapter 1.

Achieving urban sustainable development requires attention to at least four key factors: *urban form, transportation, energy*, and *waste management*. Each is considered below.

Urban Form

Urban form refers to the type and distribution of infrastructure (e.g., buildings, roads) in communities and is a key factor influencing environmental quality in cities. For example, the configuration of roads and other transportation networks has a major impact on energy use for travel within cities (see quotation box below). Furthermore, regulations related to buildings influence their energy efficiency. Energy use by both transportation and buildings becomes a major contributor to greenhouse gas (GHG) emissions. Finally, **urban sprawl** contributes to loss and disruption to or degradation of adjacent agricultural land, environmentally sensitive areas, natural habitats, and water and air quality.

The External Advisory Committee on Cities and Communities (2006, 50) summarized research findings related to the implications of urban form for environmental quality. Many have concluded that a compact urban form is more environmentally desirable than the 'sprawl' typical of North American cities. But even though leaders in many Canadian cities have been working

hard to revitalize their core areas, the trend continues to be towards sprawl. As the External Advisory Committee reported:

- The average home in Canada is farther away from a city centre than it was a decade ago.
- The proportion of low-rise, low-density homes, except in major cities, is expanding steadily and frequently accounts for more than two out of every three homes constructed.
- While house sizes have increased, the number of people in households has decreased, resulting in space and energy use per person going up significantly.
- Commuting times have increased, with traffic congestion costs estimated at $2.3 to $3.7 billion each year, with obvious negative consequences for productivity.
- Sprawl causes higher servicing and infrastructure costs and less effective public transit service, displaces large tracts of habitat and prime agricultural land, and contributes to water quality degradation.

On the basis of the above characteristics, the External Advisory Committee (2006, 52) concluded that 'the principal land use challenge for the immediate present and the foreseeable

Urban sprawl.

future is to reduce sprawl in our growing places.' It has been observed that urban areas with a high population density in their cores lead to more efficient and effective land use than in lower-density areas. They are also much more likely to be able to provide effective public transit. In addition, it is normally cheaper to provide services such as water supply and waste removal in higher-density areas. Later in this chapter, we examine alternative ways of reducing or minimizing urban sprawl and other important aspects of urban sprawl.

To illustrate the importance of urban form, the National Round Table (NRTEE 2003c, 30) noted that part of the federal government's strategy to address climate change is based on calculations indicating that improved energy efficiency in federal government buildings would reduce GHG emissions by 0.2 megatons. It argued, however, that in the medium to long term, such a reduction in GHG emissions would be achieved or bettered by locating or relocating federal buildings to minimize travel time for workers and visitors, as well as supplying the buildings with goods and services delivered by van or truck rather than by car. We examine considerations related to transportation in the next subsection.

Transportation

The negative consequences of low-density urban development are at the heart of many serious critiques of automobile-dependent cities and the environmental impacts of cities. The strong relationship revealed by many studies between more compact, mixed-use urban form and reduced car use is reflected in efforts around the world to reduce urban sprawl and create more transit-oriented communities (Kenworthy 2006, 69).

Several variables affect energy used for transportation in cities: distance travelled, vehicle loading, and vehicle mode. Each is significantly affected by urban form. Other influential variables are density of the urban area, urban structure, mixes of land use, and street patterns. All affect the number, length, and type of trips (that is, walking, cycling, using transit, or driving a car). The more spread-out a city, the farther people have to travel between places of work and home. The lower the population density, the more challenging it is to provide high-quality public transit services. The usual outcome is higher reliance on automobiles as the preferred way of commuting between home and work, resulting in greater energy use.

Transportation and Energy Use in Toronto

In the City of Toronto, transportation is the largest source of emissions, responsible for an estimated 90% of carbon monoxide emissions, 83% of emissions of nitrogen oxides, and 60% of sulphur dioxide emissions.

—NRTEE 2003c, 19

The reason that transportation is a major contributor to GHG emissions is that most vehicles are powered by fossil fuels. In contrast, for commercial and industrial buildings, the source of energy for heating, cooling, and lighting is often cleaner, such as natural gas or hydroelectricity. In Canada, almost 60 per cent of the energy used in transportation is for moving people, with automobiles accounting for the largest share. The National Round Table (NRTEE 2003c, 13) has stated that 'transit is a more environmentally sustainable form of urban transportation than the automobile' (see box on page 484). As shown in Figure 13.1, however, even though transit ridership in Canada has been increasing steadily since 1960, the average number of transit rides per person has been decreasing. Indeed, there was a notable drop in both total and per capita transit ridership between 1960 and 2000. In contrast, total travel by automobiles, including SUVs and light trucks, has been increasing rapidly and certainly at a rate faster than population growth (Figure 13.2).

Various strategies can be used to reduce energy use by transportation within cities. They include: (1) facilitating tele-working and tele-services to reduce travel time; (2) ensuring that parking arrangements encourage reduced car travel (providing ample parking adjacent to public transit departure

Commuter traffic.

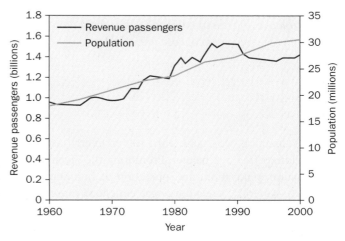

Figure 13.1 Urban transit ridership and population growth, Canada, 1960–2000.
Source: NRTEE 2006.

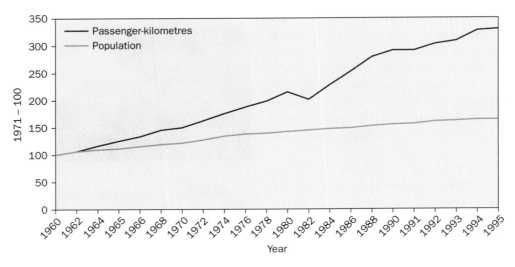

Figure 13.2 Total passenger-kilometres travelled and population growth, Canada, 1960–95. Note: Vehicles included are automobiles. Source: NRTEE 2003c, 13 (from Environment Canada. 1998. *Canadian Passenger Transportation*. SOE Technical Supplement no. 98–5: *How Canadians Travel*, Figure 3; and Statistics Canada CANSIM II Series, v. 1, Table 510005).

nodes; setting appropriate [higher] charges for parking cars near workplaces); (3) encouraging development of ride-sharing programs; (4) initiating transit pass programs to provide a seamless public transit system, such as the systems in Hong Kong and various European cities that allow a single pass to be used on buses, trains, and ferries; and (5) facilitating the use of bicycles and other means with a small **ecological footprint**.

Enhancing Use of Urban Transit

Shifting from automobile travel to transit will likely have the single greatest impact on the environmental quality of Canadian cities and their affect on the global environment. Bringing about this shift involves making transit attractive and competitive relative to the automobile in terms of convenience, cost, and comfort.

— NRTEE 2003c, 33

Energy Use

In this subsection, we examine the role of energy use in urban buildings (see box below). Promoting the green design, construction, renovation, and operation of buildings could cut North American greenhouse gas emissions more deeply, quickly, and cheaply than any other available measure (Commission for Environmental Cooperation 2008). Buildings in North America release more than 2,200 megatons of CO_2 into the atmosphere annually, about 35 per cent of the continent's

total. Rapid market uptake of currently available and emerging advanced energy-saving technologies could result in a reduction of more than 1,700 megatons by 2030 from the emissions projected for that year under a business-as-usual approach. A cut of that size would come close to equalling the CO_2 emitted by the entire US transportation sector in 2000.

In terms of end use, residential, commercial, and industrial buildings accounted for 63 per cent of GHG emissions in Canada during 2004, and most of it occurs in urban areas. For the residential and commercial sectors, energy in buildings is used mostly for heating water and space and for cooling space. Indeed, for residential buildings, these three end uses are responsible for 80 per cent of energy use.

Energy use in residential buildings is influenced by the construction materials, shape, and orientation of the building, internal temperature settings, internal use activity, and climate conditions. Urban form also is an important influence. For example,

Fossil Fuels and Sustainability

Canada's cities and communities are not environmentally sustainable on their current path. Use of fossil fuels is the main source of the problem. Canadian cities are also characterized by low densities of population, and Canada is still sprawling more than ever before.

— External Advisory Committee on Cities and Communities 2006, xiii

This residential house in Waterloo, ON, features a BIPV (Building Integrated PV 7.6kW system) installed by the ARISE Technologies Corporation. The system uses roof vents and pipes to draw warm air down through the ground in order to cool it before it re-enters the house, thus avoiding the need for air conditioning. Also shown is a clothesline on the deck, another example of energy saving.

townhouses and apartments are usually more energy-efficient than single detached houses. As the National Round Table (NRTEE 2003c, 20) observed, 'evidence suggests that overall energy use is inversely related to the density of development: more compact, mixed-use cities, which support greater use of sustainable forms of transportation and less energy-intensive building types, tend to use less energy.' It is common now for more advanced green buildings to routinely reduce energy usage by 30, 40, or even 50 per cent over conventional buildings, with the most efficient buildings now performing more than 70 per cent better than conventional properties. Despite proven environmental, economic, and health benefits, however, green building today accounts for a only small fraction of new home and commercial building construction, just 2 per cent of the new non-residential building market, less than 0.5 per cent of the residential market in the US and Canada, and less than that in Mexico (Commission for Environmental Cooperation 2008).

Waste Management

Various factors affect total per capita amounts of waste generated. For residential areas, key factors include demographic characteristics such as household size, age structure, and annual income, as well as type of dwelling unit, geographical location, and time of year (Maclaren 2004, 374). Not all of these factors are affected by urban form, but some are. For example, presence or absence of yard wastes influences the amounts and composition of wastes produced by households. In that regard, apartment dwellers generate lower per capita wastes than single-family dwellings because the former do not have yards and instead often share a common area adjacent to the apartment building.

An integrated approach to waste management strives to divert as much waste as possible away from disposal through the 3Rs: source **reduction**, followed by **reuse** and **recycling**. Energy **recovery** is sometimes considered a fourth R, but it is challenging because it involves burning waste in incinerators.

Paper-recycling depot.

Establishing an incinerator facility is usually controversial, since most people are not enthusiastic about having one nearby. The other well-known option is disposal of waste in a landfill site. Finding such sites can also be controversial, since they, along with incinerators, are viewed as LULUs (locally unwanted land uses) that trigger NIMBY (not in my backyard) reactions. As Maclaren (2004, 377) has observed, 'The precedence of a 3R option within the hierarchy is determined by

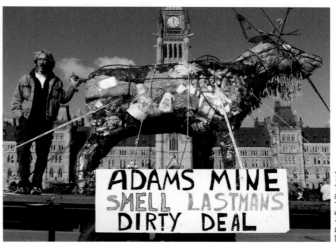

A garbage-filled moose on display at Parliament Hill as part of a protest against the dumping of Toronto's garbage in Adams Mine in Northern Ontario.

its ability to generate less waste, conserve more raw material resources, save more energy, and create fewer environmental impacts than an option lying below it in the hierarchy.'

Based on a survey in 2006, Statistics Canada (2007b) reported that urban Canadians are recycling and composting more than in the past but with significant variation across the country. For example, in Montreal and Calgary, less than one-third of waste is diverted from landfill sites. Toronto diverts just over 40 per cent. In contrast, in Halifax about 55 per cent of waste is diverted and in Markham, ON, about 70 per cent. Markham officials attribute this success to a practice of

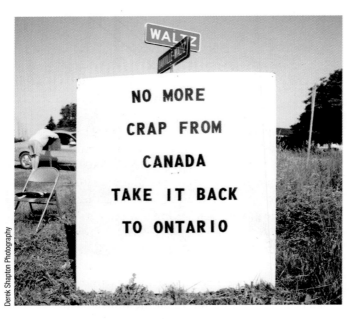

NO MORE
CRAP FROM
CANADA
TAKE IT BACK
TO ONTARIO

Derek Shapton Photography

Sign in Michigan protesting imports of Toronto garbage.

collecting recyclable (blue boxes) and organic waste (green bins) twice as frequently as garbage.

While we have discussed urban form, transportation, energy, and wastes separately, they are closely interrelated. As Kenworthy (2006, 67) has noted, 'Not only do urban form, transportation systems, and water, waste and energy technologies have to change, but the value systems and underlying processes for urban governance and planning need to be reformed to reflect a sustainability agenda.'

Summary

The External Advisory Committee on Cities and Communities (2006, 46) concluded that:

- The way Canadians are living now in their cities and communities uses too much natural capital [the environment and natural resources] and is outstripping the capacity of the environment to recover from pressures placed on it.
- Decisions and actions that enhance or degrade the environment also affect the Canadian economy, society, and culture.
- Investment in new environmental technologies could create major new opportunities for long-term prosperity and improved competitiveness.
- Cities and communities in Canada 'clearly lie at the heart of the problems and solutions for the sustainability of the Canadian environment'.

The ideas of the National Round Table (NRTEE 2003c, 31) related to a sustainability checklist for location and site design of buildings offer useful ideas (Box 13.1 on page 488).

Guest Statement

Urban Waste Management
Virginia W. Maclaren

Urban waste management practices in Canada have come a long way since the 1980s when governments first started thinking about how to reduce, reuse, and recycle waste rather than just how to dispose of it. Residential curbside collection programs for recyclables are thriving across the country, and new organic waste curbside collection

programs are spreading quickly. Many provinces have implemented or are forging ahead with innovative **extended producer responsibility** (EPR) and product stewardship programs. EPR extends the producer's responsibility for a product to the post-consumer stage of the product's life cycle, while product stewardship involves shared responsibility among producers, retailers, users, and disposers for reducing a product's environmental impact across its life cycle. Both involve diversion of products from landfill by providing infrastructure for collection of the used product and incentives for its recycling or reuse. EPR or product stewardship programs now exist in several provinces for used pharmaceuticals, tires, paint, solvents, lead acid batteries, and e-waste (electronic waste). All provinces have some form of beverage container deposit system, which promotes reuse or recycling of the returned containers.

Even landfills are receiving attention. More and more municipalities are turning their old and existing landfills into energy sources by extracting methane, which is produced by the anaerobic decomposition of organic waste. Burning methane and turning it into electricity or heat not only provides energy but also reduces greenhouse gas emissions from landfills.

Where do we need to turn our attention to now? Many challenges exist, but I will address just three of them. The first and perhaps the most difficult is the issue of consumption. We live in a consumer society in which consumption is often associated with social status and personal happiness. For many people, the more they consume, the more status they attain and the happier they are. One consequence of excessive consumption is excessive waste production. Sustainable consumption is a relatively new concept that focuses on how to break the link between consumption and waste production. One way to do this is by simply purchasing less, an approach that is part of a lifestyle known as 'voluntary simplicity'. Another way to break the link is to purchase services rather than products. The rationale here is that if a company leases its products (i.e., provides product services) rather than selling them, it has an incentive to optimize the product's operation, increase its service life, and gain value from the old product once the lease is over by recycling or refurbishing it rather than sending it to landfill. A third way to break the link is to purchase products that produce less waste because they are more durable, they have less packaging, or they require fewer material inputs and are less pollution-intensive in their production.

The second challenge is what to do about packaging waste. About one-third of the waste produced in Canada comes from packaging, yet except for beverage containers, no national or provincial programs currently focus on reducing packaging waste. This is in sharp contrast to efforts in Europe, where the European Union requires its member countries to meet national packaging waste reduction targets for many different types of packaging. For example, member countries were expected to be recycling between 55 and 80 per cent of all packaging waste by 2008. Targets for specific materials varied, depending on the availability of recycling markets and technologies for recycling.[1]

The third challenge is to improve residential waste diversion. Diversion of waste from landfills is working well in single-family households but not so well in apartments or multi-family dwellings (MFDs).[2] For example, in Toronto in 2007, residents of single-family dwellings (SFDs) diverted 59 per cent of their waste, while residents of MFDs diverted only 13 per cent. Part of the reason for the wide difference in diversion rates was that SFDs had access to curbside collection of both recyclable and organic waste, while residents of MFDs only had recycling programs. Even accounting for the lack of an organics collection program in MFDs, waste diversion was still much lower than in SFDs. What makes waste diversion in MFDs so difficult? There are many reasons. It starts with the diversity of structures that comprise MFDs. Some have garbage chutes and recycling/organic waste rooms on individual floors, while others do not. These structural differences are important because they affect the convenience of participating in diversion programs. The less convenient a program is, the lower its participation rate. In MFDs with garbage chutes on each floor, garbage disposal is easy, but residents cannot use the same chute for recyclables or organics and have to take them down to bins on the ground floor or in the basement. This arrangement clearly makes it less convenient to divert waste than to throw it out. Another factor is the difficulty of finding storage space for recyclables and organic waste in apartment units. Some municipalities provide residents of MFDs with small containers, such as those shown in the photograph, to help with storage. A third factor affecting participation is that residents of MFDs do not face the same level of public scrutiny of their recycling hab

Toronto pilot project in-unit containers for recyclables in MFDS.

its by their neighbours as do residents of SFDs, who see their neighbours participating (or not) every time they set their recyclables and organics out at the curbside. Finally, many MFDs consist of rental units, and since renters tend to be more transient than owners, they may not receive the same amount of education about why, where, and how to divert waste locally or within their buildings.

Virginia Maclaren

Given the enormous progress that has been made in waste reduction and waste diversion since 1981, when the first **blue box** curbside recycling program was launched in Kitchener, ON, I am optimistic that municipalities will be successful in meeting the above challenges and in further transforming the face of modern urban waste management practices.

Virginia Maclaren is an associate professor and chair of the Department of Geography and Program in Planning at the University of Toronto. Her primary research interests are in the field of waste management, with a regional focus on Canada and Southeast Asia.

[1] European Environment Agency. 2005. *Effectiveness of Packaging Waste Management in Selected Countries: An EEA Pilot Study.* Copenhagen: European Environment Agency. http://reports.eea.europa.eu/eea_report/2005_3/en.

[2] City of Toronto. 2007. *Waste Collected and Waste Diverted Rates.* http://www.toronto.ca/garbage/residential-diversion.htm.

Box 13.1   Sustainability Checklist for Location and Site Design of Buildings

Location

- use of existing buildings in already urbanized areas before new construction on **greenfields** (undeveloped land, as opposed to **brownfields**, or previously developed properties);
- easy access to good transit service;
- potential for walking and cycling access by employees and visitors;
- proximity (walking distance) to amenities and services for workers (e.g., restaurants, personal services, and daycare);
- potential to link to a community energy system;
- potential to contribute to the regeneration of economically depressed urban areas.

Site Design

- maximized building density;
- integration with transit facilities (e.g., covered walkways connecting transit to the facility);
- facilities for bicycle riders, such as racks and showers;
- minimization and appropriate treatment of parking (e.g., creation of underground or structured parking lots, landscaping lots to maintain street frontages);
- maximized site permeability;
- easy pedestrian access to the facility;
- integration of other uses into the facility (e.g., restaurants, services, amenities, and residences).

Environmental Issues in Cities

Air Pollutants

Concentrations of some common air pollutants in Canadian cities have been decreasing (Figure 13.3 [from NRTEE 2003c, 14]). The main reason for this drop is stricter regulation of emissions from automobiles combined with enhanced regulations regarding industrial emissions. One anomaly to this trend is ground-level ozone (also referred to as photochemical smog or summer **smog**) created when nitrogen oxides and volatile organic compounds combine in sunlight. There has been a striking increase in the number of 'smog advisory days' in many of Canada's major cities, especially in southern Ontario. The principal explanation is more ground-level ozone.

Environment Canada (Environment Canada, Statistics Canada, and Health Canada 2006) and other federal agencies monitor air quality through two measures: ground-level ozone and fine particulate matter ($PM_{2.5}$). Respiration and heart rates can be increased by ground-level ozone. Other health problems can be asthma attacks, bronchitis, and emphysema. Children are usually most vulnerable.

Environment Canada (Environment Canada, Statistics Canada, and Health Canada 2006, 5) reports that at a national level during the period 1990 to 2004, there was an average annual increase of 0.9 per cent for ozone as well as year-to-year variation. In urban areas, the triggers for ozone, such as nitric oxide and volatile organic compounds (VOCs), which are

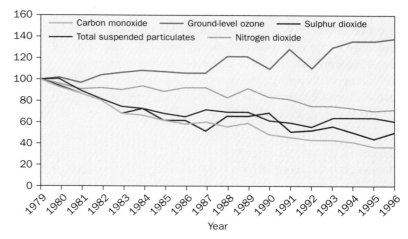

Figure 13.3 Concentration of common air pollutants, Canada, 1979–96. Source: Centre for Sustainable Transportation. 2002. *Sustainable Transportation Indicators, Report on Phase 3*, 35 (data from Environment Canada).

Box 13.2 Human Deaths from Air Pollution

Researchers at the University of British Columbia and the University of Alberta have concluded that up to 25,000 premature deaths annually in Canada are due to exposure to air pollution, hazardous chemicals, and pesticides. The associated costs to the health care system are estimated at up to $9.1 billion related to illnesses such as cancers, respiratory diseases, heart problems, and congenital problems associated with pollutants, all of which have demonstrated connections to environmental contaminants.

They also concluded that such pollutants cause major disabilities and estimated that the types of pollution noted above cause between 1.1 million and 1.8 million 'restricted activity days' annually for people with asthma

and result in Canadians every year spending between 600,000 and 1.5 million additional days in hospital.

David Boyd, one of the researchers, was quoted in *The Globe and Mail* (Mittelstaedt 2007) as commenting that 'In our cultural DNA, we think of Canada as a pristine nation, but this is at odds with our track record on the environment.' Boyd argued that such deaths, impairments, and costs could be reduced if Canada introduced and enforced more stringent standards regarding air quality, drinking water, food, and consumer products. He suggested that Sweden represents best practice in this regard and would be a good model for Canada to strive to match.

Source: Mittelstaedt 2007.

generated from local emissions, have dropped. The main explanation for such a decrease is improved quality of fuels and better emission control technology in vehicles. Nevertheless, over the same period there was an increasing trend in southern Ontario, which had 'the highest concentrations and fastest rise of all regions monitored' (Environment Canada, Statistics Canada, and Health Canada 2006, 6) as well as about 30 per cent of Canada's population. Specifically, southern Ontario experienced an average annual increase of 1.3 per cent in ozone levels. Part of the explanation for this increase is proximity to the industrialized northeastern US, along with prevailing winds that result in long-range transport of ozone and its trigger chemicals. Environment Canada also reported that southern Ontario had the highest average levels of PM$_{2.5}$ in Canada during 2004.

Urban Heat Island Effect

The **urban heat island** effect occurs as a result of increased temperatures in core urban areas relative to surrounding areas. It is not uncommon for the temperature of city centres to be from 2° to 6°C higher than that of nearby rural areas. Such an effect can be reduced or countered by skilful use of green areas, given their cooling effect within urban areas. A secondary benefit of reducing the urban heat island effect is a decrease in the need for air conditioning buildings in city cores, thereby reducing use of electricity. Higher temperatures created by the urban heat island effect can also generate smog and enhance ground-level ozone, neither of which is confined to the urban area. The smog and ozone can drift to nearby rural areas and in some cases can reduce agricultural productivity, increase health risks, and even contribute to triggering tornadoes and thunderstorms (United Nations Population Fund 2007a, 59).

The urban heat island effect is one factor that has encouraged development of 'green roof technology', which involves creating a new or retrofitting an existing roof with a growing medium allowing plants, shrubs, or trees to grow on it. Leaders in this technology include countries in Europe and Japan. It has been estimated that a one-degree reduction in the

Urban Heat Island Effect in Canada

Regina and Saskatoon: The night temperatures in the city centres on average are three to four degrees warmer than those in the adjacent countryside. The heat island effect is greatest on calm and cloudless nights, when conditions are best for rural areas to cool more quickly than city surfaces.

—http://esask.uregina.ca/entry/ urban_heat_islands.html

Winnipeg: The average low temperature at the Forks, in the centre of the city, is 2.73 degrees warmer than at the airport on the edge of the city. The comparable difference for the average warm high temperature is 1.57 degrees.

—http://www.fcpp.org/main/publication_detail .php?PubID=1884

Toronto: The average summer temperatures are four to 10 degrees higher in the city centre compared to temperatures in nearby rural communities.

—http://www.team.gc.ca/english/dbProjects; viewProuect.asp?id=5335&typ=comm

urban heat island effect leads to significantly reduced demand for electricity for air conditioning and refrigeration, which in turn results in lower GHG releases.

In addition to monitoring for energy savings, the proponents of green roofs have also tracked its impact on stormwater retention. To deal with water runoff, Portland, OR, has used green roofs for some time. It has learned that most of the rain that lands on a green roof is absorbed by plants and soil and eventually evaporates or transpires back into the atmosphere. Almost all the summer rain in Portland is held by green roofs, and in the fall and spring, retention is between 40 and 50 per cent. In the winter months, retention drops to between 10 and 20 per cent. This retention significantly reduces peak runoff flows following rain events, which reduces the possibility of localized flooding (Dawson 2002).

Murphy and Martin (2001, 69–70) observed that 'solutions to heat island effects ultimately depend on reductions in energy consumption, pollution, and urban sprawl, but there are mitigation efforts that can work.' In that regard, they identify the following:

1. Design buildings and neighbourhoods to balance building structures with the geometric shapes and characters

Green roofs, Portland, OR.

Green Roofs for Healthy Cities/Walker Macy Landscape Architects of the areas (or canyons) between buildings to reduce the amount of energy hitting the surface of buildings and roadways and thus the amount of energy to be re-radiated. North-south street orientations reduce the amount of energy that will reach roads.

2. Use light-coloured surfaces and less thermally absorptive exterior facing on buildings. However, this initiative reduces the potential for solar-based heating systems.

3. Provide vegetation surfaces in place of or to shade heat-absorbing surfaces. This can reduce the urban heat island effect by between 25 and 80 per cent. However, Murphy and Martin also note that there can be negative effects in terms of creating obstructions to walking and hiding places for assailants, injuries from falling branches, and an increase in pollen and mould that trigger allergies.

In assessing the third option, Murphy and Martin (2001, 70) provide sound advice in observing that 'Ultimately, the question is about the relative short and long-term risks, costs, and benefits of an ecological mitigation of urban heat island effects versus doing nothing. We believe the cost of doing nothing is greater, especially since the benefits of ecological mitigation go beyond the urban heat island effect.'

Hydrological Cycle

Urban areas affect the hydrological cycle in several ways in terms of both quantity and quality of water.

Regarding quantity, one of the most obvious impacts is due to urban infrastructure creating an impervious rather than a pervious surface. The implications are twofold. First, expansion of roads and construction of parking lots and buildings results in precipitation running off the surface more quickly, since it is less likely to soak into the soil because of the human-made impervious surface. The consequences are: (1) surface flooding and (2) reduced recharge of aquifers. Both of these outcomes have led to initiatives to build retention ponds in urban areas so that water can be collected during rainfalls and allowed to either percolate into the ground or be released more slowly than would otherwise be the case. These outcomes have also reminded urban residents that urban sprawl, especially when suburbs are built on the surface of aquifer recharge areas such as moraines, can significantly affect the amount of water available in aquifers for human use.

Water quality is negatively affected because pollutants such as oil and gas from vehicles and salt from winter applications (to make driving safer) get washed into surface streams and groundwater systems. The result is degraded water quality, with negative health consequences for humans and other living species. Such outcomes have been summarized well by the United Nations Population Fund (2007a, 58), which commented that 'Urban areas can affect water resources and

Retention pond.

the hydrological cycle . . . through the expansion of roads, parking lots, and other impervious surfaces, which pollute runoff and reduce the absorption of rainwater and aquifer replenishment.'

The United Nations Population Fund (2007a, 58) continued its observation by noting another impact, which was 'through large-scale hydroelectric installations that help supply urban energy needs'. This perspective provides a counterview to those who argue that hydroelectric installations are 'clean' in terms of GHG emissions. Although that conclusion is valid, it is important to recognize that hydroelectric dams and

Road Salt Challenges in Toronto

The GTA (Greater Toronto Area) is one of North America's fastest-growing regions, and serious questions are being raised regarding the environmental sustainability of the anticipated urban growth and the potential long-term impacts on the quality and quantity of ground and surface water resources. Degradation of groundwater quality by NaCl de-icing salt is the primary concern since there are no cost-effective alternatives to NaCl de-icing salt for large-scale use and there is little evidence that salt loadings to the subsurface can be significantly reduced. According to Environment Canada, almost 5 million tonnes of NaCl are released to the environment every year, with the City of Toronto alone receiving over 100,000 tonnes at a rate averaging 200 grams for every square metre of land surface.

—Howard and Maier 2007, 147

reservoirs can inundate significant amounts of habitat and can modify downstream flow regimes, which can affect both flora and fauna dependent on natural fluctuations in the hydrological cycle, fluctuations that are evened out when dams are introduced.

Brownfield Sites

Many cities have a legacy of abandoned or active industrial sites. On the surface or underground are soils contaminated through disposal practices that were accepted in earlier times before people appreciated the long-term consequences. Such contaminated sites are often referred to as 'brownfields'.

One specific problem results from LUST, or 'leaking underground storage tanks'. Industrial sites and garages usually have underground tanks in which materials, including gasoline and chemicals, have been or are stored. While the storage tanks are in active use, it is not uncommon for them to develop leaks, resulting in some of the contents being gradually released into the surrounding soil. Alternatively, when an industrial plant or garage is closed, it was not uncommon in the past for the contents of storage tanks to be left in the tanks or for the tanks to be filled with unwanted liquids. Eventually, many of these tanks would develop leaks, thereby releasing the contents into the soil. Once out of the tanks, the liquids may move through the soil, often ending up in aquifers. When such aquifers are the source of wells to provide water, there is a risk to human health. Or if the contaminants are underneath fields on which crops are grown, the contaminants can be drawn into the plants by their root systems and eventually absorbed by the plant material, which is later consumed by humans or animals.

A similar contamination problem can occur through the burying of uncontained wastes under the property of a factory or production facility. For example, in Ontario it was normal practice for industries or cities to bury wastes on their properties. The provincial government estimated in the 1980s that a minimum of 41 coal tar–contaminated sites existed in 36 communities in the province.

The case study of the Sydney Tar Ponds, reviewed in Chapter 1, highlights the huge cost of remediating large sites. As another example, the municipal government in Kitchener, ON, spent about $19 million to remove coal tar from beneath one square block in the centre of the city. The city had operated a coal gasification plant there from 1883 until 1958. Coal was heated in ovens, and the flammable gas created was used to provide heat and lighting for businesses and residences in the central part of the city. During the process of gasification, some coal became an oily tar, commonly called 'coal tar'. This coal tar, a waste by-product, was placed in underground tanks or into open pits and then buried under the property.

After the gasification plant was closed, a Canada Post building was constructed over the coal tar site, and roads were built and paved on each side of the old site. Over decades, the buried coal tar gradually seeped laterally through the fine-grained silt beneath the property. Some was discovered under an adjacent property, leading to a $5-million lawsuit against the city. The suit was eventually dropped when the city agreed to remove the coal tar from under the old gasification plant site and purchase the property.

The City of Kitchener ended up paying almost $15 million more than was initially estimated for the remediation of the gasification site, an experience that highlights the uncertainty involved in such contaminants. The first estimate for the cleanup was $5.6 million. This amount was later increased to $9 million and then finally to $19.5 million. The reason for the extra cost was that there was much more coal tar and much more of it was hazardous waste than had originally been estimated.

Initial estimates of the amount and kind of soil contaminated by coal tar were based on test drilling of boreholes between 1986 and 2004. The conclusion was that the contaminated soil was not more than 5.5 metres below the surface. The test drilling also indicated that none of the contaminated soil was hazardous. In estimating remediation costs, the consultants included an allowance of 1,500 tonnes of hazardous waste to be removed in case some was present.

During May 2006, after work began on building a right-of-way to the site, it was discovered that the coal tar–contaminated soil extended at least seven metres below the surface. As a result, new boreholes were drilled between July and November 2006. The results verified that the contaminated soils extended to seven metres but also that there were 13 times more contaminated soils that were hazardous than estimated at the outset, as well as almost twice as much non-hazardous contaminated soils (Table 13.1).

The cost of removing hazardous soils ($156 per tonne) was about three times as much as that of disposing of non-hazardous

Table 13.1 Estimated Coal Tar–Contaminated Soil at Coal Gasification Plant Site, Kitchener, ON, 2004 and 2006

Year	Hazardous soils	Non-hazardous soils
2004	1,500 tonnes	35,700 tonnes
2006	20,000 tonnes	60,000 tonnes

Source: http://www.kitchener.ca/gaukelstreetproject.html.

soils ($48.50 per tonne). The hazardous material was trucked to a site in Quebec with the capacity to handle hazardous soil. The City of Kitchener subsequently reported that $9.37 million was required to deal with the higher than initially estimated remediation costs. The remediation was completed during the summer of 2007, and by the time it was finished, about 2,500 truckloads of contaminated soil had been removed.

The Kitchener situation is not unique. In nearby Cambridge, ON, a law firm filed a $110-million class action lawsuit in 2006 on behalf of a woman against a firm allegedly responsible for contaminating groundwater beneath at least 100 homes. The lawsuit stated that the firm had allowed trichloroethylene (TCE) to escape into the subsoil, contaminating several blocks around its plant. TCE is a toxic substance used primarily as an industrial degreasing solvent (Environment Canada 2002b). The lawsuit was filed on behalf of homeowners whose property was within a 4.8-kilometre radius of the plant. After the discovery of the TCE in 2005, the claimant and her family had moved into a motel, where they had lived for three months. Five other families had also vacated their homes because of the presence of TCE vapour. The class action lawsuit argued that the TCE contamination had reduced property values and prevented the families from obtaining mortgages and insurance coverage.

The firm observed that it had been working closely with the ministries of the Environment and Public Health, local officials, and residents to map, assess, and resolve the situation, which it had inherited after decades of improper disposal of chemicals on the property. By the summer of 2007, the company had installed soil vapour extraction units in more than 60 homes to pump contaminated TCE vapour from the ground under them. The residents acknowledged this action by the company but still had concerns that neither the firm nor the provincial Environment Ministry was making arrangements for monitoring air quality in homes.

Scott Lackenbauer 2007

Coal tar cleanup in Kitchener, ON.

These examples illustrate that the presence of brownfield sites within urban areas can be a major challenge for city officials, private companies, residents, and provincial government regulatory agencies. They also highlight the considerable uncertainty and complexity involved in determining the nature of the problem and identifying solutions.

Vulnerability of Urban Areas to Natural and Human-Induced Hazards

Many cities are vulnerable to hazards, whether triggered by natural or human-induced actions, because of their high concentration of people and, in some situations, the low-quality construction of buildings. The United Nations Population Fund (2007a, 59) states that natural disasters have become both more frequent and more severe since 1990. Furthermore, the United Nations Environment Programme has reported that 75 per cent of the global population lived in areas affected by natural disasters between 1980 and 2000.

Environment Canada (2003c) has reported that Canadians experience many natural disasters. Geophysical hazards such as earthquakes have stayed constant over the past 50 years, but weather-associated hazards have increased dramatically. Environment Canada (2003c, 2) has noted that the following factors make us vulnerable: population growth, urbanization, environmental degradation (e.g., removing timber from hillsides, leading to landslides), urban sprawl in hazard-prone areas, loss of collective memory about hazardous events because of increased mobility, aging infrastructure, and historical over-dependence on technological solutions. Specifically, Environment Canada (2003c, 2) observed that 'Higher concentrations of people living in urban areas means that if disasters hit, they affect a larger number of individuals. Urban sprawl has led to more development in high-risk areas, such as flood plains.'

McBean and Henstra (2003) provide further information and insight on why 'urban Canada' is susceptible to hazards. One of the core factors is that about 60 per cent of Canadians live in urban areas of 100,000 or more people and about 80 per cent in cities of 10,000 or more. Earthquakes are a significant hazard, with the most vulnerable areas in British Columbia and the St Lawrence Valley. However, in their view, 'about 80 per cent of the impacts are due to weather and weather-related hazards.' Weather hazards include tornadoes, hailstorms, winter storms, and heat waves, while what are termed 'weather-related hazards' are drought, storm surges, floods, and moving ice. In terms of specific weather and weather-related hazards, the probability of tornadoes and hailstorms is highest on the Prairies and in southern Ontario, storm surges are most frequent along the Atlantic coast, and winter storms are ubiquitous throughout the country.

McBean and Henstra (2003, 2) have also observed that the viability of Canadian urban areas depends on the continuity of basic infrastructure such as sewer and water, transportation, and electrical power systems. In their words:

> Past experience in Canada has demonstrated that this urban machinery is highly susceptible to disaster damage. For example, the 1998 ice storm which struck Ontario, Quebec, and New Brunswick crippled power grids, caused structural damages to buildings, and brought transportation to a halt, becoming Canada's most expensive natural disaster, exceeding $5 billion in losses, and causing at least 28 deaths. Similarly, the British Columbia blizzard of 1996 shut down the cities of Vancouver and Victoria for days, with a major economic impact.

The following two subsections illustrate in more detail the variety of hazards that can be experienced by urban dwellers.

Hurricane Katrina and New Orleans

About 65 per cent of urban areas with populations of 5 million or more are located in low-elevation coastal areas throughout the world. A striking example of a low-elevation coastal area being damaged by a natural disaster occurred on 29 August 2005 when Hurricane Katrina hit the Gulf Coast of the United States. More than 2,800 people died across the various states, thousands of homes were destroyed, and hundreds of thousands were left homeless. The city of New Orleans received the greatest damage, but almost 10 million people living in the Gulf Coast states of Alabama, Louisiana, and Mississippi were affected by the winds from Katrina. In southeast Louisiana alone, 90 per cent of residents were evacuated.

In New Orleans, slightly less than 50 per cent of which is below sea level, about 80 per cent of the city had been flooded two days after the hurricane struck, with some areas under five metres of water. Much of the flooding occurred because the levees built over four decades to provide flood protection had been breached. Construction of the levees had started in 1965 and was scheduled to be completed in 2015. The Superdome stadium became a 'refuge of last resort', and by the evening of 28 August, some 20,000 to 25,000 people were staying there.

As with many natural disasters, human behaviour contributed to the problems. Notwithstanding the mandatory evacuation order, many chose to stay in New Orleans, citing various reasons: belief that their homes were adequately protected, lack of access to transportation or sufficient money to cover the cost of leaving, and conviction that it was necessary to 'guard' their homes and possessions. Unfortunately, the third reason was valid. Within a day of Hurricane Katrina making landfall, violence and looting had spread throughout many areas of New Orleans. Sometimes such behaviour occurred in daylight while police officers watched on. Some looting for non-essential goods involved armed robbery. One incident involved looters with guns, using boats, who attempted to break into the Tulane University Medical Center to steal drugs. In contrast, many other looting incidents involved residents removing food, water, and other basic consumable goods from closed grocery stores.

Those who remained in their homes often became stranded when Hurricane Katrina had passed by but flood waters had risen. To escape, some people cut holes in their roofs in order to climb out. Clean water was not available from the municipal water system, and power outages were common. Significant time and resources were needed to remove stranded people.

While Hurricane Katrina was an extreme natural hazard, it highlights that humans often make themselves vulnerable because of where they choose to live and work. Individuals living or working adjacent to rivers or shorelines place themselves at risk of floods, and those living in snow-prone areas are vulnerable to the risks of severe winter storms. The key message is that we do have choices and often have not been attentive enough to the risks posed by natural hazards.

Earthquake and Peru

The experience in New Orleans highlighted how vulnerable an urban area can be to a natural disaster and how challenging it can be to respond when many basic infrastructure services become inoperable. Another dramatic example was the deadliest earthquake to hit Peru in 35 years. With a magnitude of 8.0 on the **Richter scale** (Box 13.3), the earthquake struck in the early evening on 16 August 2007. The epicentre of the quake was in the south of the country, and two cities, Ica and Pisco, were badly damaged. The United Nations estimated at least 450 people died and 1,500 were injured. In Pisco, a city of 116,000, it was reported that up to 60 per cent of the buildings were damaged and that more than 200 people were buried when a church collapsed during a special service to mark a Roman Catholic holy day. In Ica, with a population of 120,000, one-quarter of the buildings in the city collapsed. Telephone and mobile phone services were knocked out between the area affected by the quake and Lima, and electricity and water systems also were seriously disrupted.

Regarding earthquakes in Canada, Environment Canada (2003c, 3–4) has observed that

> Although the only significant earthquake in Canada occurred off the East Coast in 1929, triggering a tsunami that killed 28 people, scientists predict that an earthquake in the Vancouver area is the most likely major disaster on our horizon. Since quakes occur where tectonic plates converge, only certain regions of the country are at risk: the West Coast, the St Lawrence and Ottawa valleys, off the coast of Nova Scotia and Newfoundland, and certain parts of the Arctic.

What is the nature of risk from earthquakes in your community? What about other natural hazards? In making your assessment, do you differentiate between high probability, low risk and low probability, high risk hazards? To which type does

(a)

(b)

(c)

(d)

(e)

Impact of Hurricane Katrina on New Orleans (a) The Lakeview neighborhood of New Orleans stands under two and a half to four metres of water in the days following Hurricane Katrina. The levee breach on the Seventeenth Street Canal, which separates the Orleans and Jefferson parishes, caused thousands of homes to flood and millions of dollars in damage to the area's infrastructure. (b) Rescuers worked for weeks following Hurricane Katrina to save people and recover bodies. In many areas, workers were forced to wait until flood waters receded before they could access homes in outlying regions. (c) The many levels of destruction caused by Hurricane Katrina may be seen on this vehicle and home. St Bernard Parish, where this picture was taken, lies just to the east of New Orleans and was hit the hardest by the storm. No property, public or private, was spared from destruction in this area. (d) A traditional camel-back home sits in flood waters in the days following Hurricane Katrina. A combination of the storm's surge and poor design and upkeep of local levees led to flooding across more than 80 per cent of New Orleans. (e) Many conspiracy theorists say that New Orleans officials blew the levees to protect the richer communities from flood damage, however, this Lakefront mansion is evidence that even affluent neighborhoods were not spared when the levees broke. (All photos courtesy of Cpl Frans E. Labranche.)

Box 13.3   Richter Scale

Earthquake severity is measured related to *magnitude* (amount of energy released at the hypocentre of an earthquake, measured through the amplitude of earthquake waves) and *intensity* (observed effect of movement of the ground on people, buildings, and natural landscape features). The Richter scale, developed in 1935, is a measure of magnitude.

The Richter scale is logarithmic, which means that a whole number increase (e.g., from 5 to 6) represents a tenfold increase in magnitude. Expressed in terms of energy released, the difference between two whole numbers is 31 times more energy. Earthquakes with a Richter scale number of up to 2.0 are not normally noticed by people and are only recorded on local seismographs. Quakes with a magnitude of 4.5 or higher are detected by seismographs around the world, and several thousand such events occur annually. Exceptionally high-magnitude earthquakes measure 8.0 or

higher, and on average one such event happens each year. Examples are the 1906 earthquake in San Francisco, with a Richter number of 8.3, the 'Good Friday' earthquake centred in Alaska in 1964, which measured 8.6 and affected the west coast of British Columbia and Vancouver Island, and the 7.9 quake in Mindanao, Philippines, in 1976. Damage from an 8.0 earthquake usually extends over at least 300 kilometres. Theoretically, there is no upper limit for the Richter scale, but the largest measured events have been between 8.8 and 8.9. The movement of the earth due to a high-magnitude earthquake can cause serious damage and disruption, but often more damage is caused by fires triggered from ruptured gasoline lines, etc. Most of the property damage in San Francisco after the 1906 earthquake was from fire.

Source: United States Geological Survey. *The Severity of an Earthquake*. http://pubs.usgs.gov/gip/earthq4/severitygip.html.

our society devote the most resources? Is this the most appropriate strategy? To what extent is there awareness of natural hazards on the part of residents in your community? What are the main mitigation measures in place? What might 'next steps' to protect residents, buildings, and infrastructure be?

Rain, Wind, and Sedimentation and the Greater Vancouver Area

In mid-November 2006, the Lower Mainland and Vancouver Island in British Columbia experienced heavy rain from storms over one week. This led to boil-water advisories for some 2 million people in Greater Vancouver, Canada's third largest metropolitan area. The advisory applied to water for drinking and food preparation. It was issued because of mudslides on the mountains inland from North and West Vancouver that affected the reservoirs that provide much of the water to residents of the Greater Vancouver Area.

The mudslides increased turbidity levels in the reservoirs. At one time, turbidity in water was considered an aesthetic rather than a health issue. That view changed, however, once researchers learned that bacteria and parasites attach themselves to dirt suspended in water, with the result that once attached, they are less likely to be killed by chlorine.

During the boil-water advisory period, restaurants, seniors' care homes, and daycare facilities used bottled water. Many homeowners also turned to bottled water. Rough jostling broke out in stores as supplies started to run out, indicating how quickly civil order can start to break down in the face of stress caused by hazards, as was also seen in New Orleans after Katrina.

The agencies responsible for the reservoirs monitored turbidity by computer every 30 seconds. As turbidity levels dropped, the advisory was partially removed on 17 November for people receiving water from the Coquitlam reservoir, which had not been as severely affected. Nevertheless, at that stage the advisory continued for Vancouver, Burnaby, North Vancouver, and West Vancouver, affecting 900,000 people. The advisory remained in place for those cities because their water is drawn from the Seymour and Capilano reservoirs in the nearby Coast Mountains, which continued to have high turbidity levels.

This event highlights that basic infrastructure and related services in urban areas are vulnerable to natural events, even if they do not lead to the loss of life and destruction of buildings that occurred with Hurricane Katrina along the Gulf Coast and the earthquake in southern Peru. Such incidents are a reminder that environmental managers responsible for urban systems must understand and have the capacity to adapt to variations in natural systems. It is also a reminder that urban areas are often constructed with little attention to their vulnerability to hazards. What are the most significant lessons from the three experiences outlined above?

Urban Sustainability

In this section, we turn our attention to cities or communities taking action to establish an urban sustainability trajectory.

Whistler, BC

The Resort Municipality of Whistler, popularly known as Whistler and one of Canada's best-known four-season destination resorts, is located 120 kilometres from Vancouver and will be a key venue during the 2010 Winter Olympic Games (Figure 13.4). With a permanent population of about 11,000 people, its economy is based primarily on tourism. Some 2 million people a year visit for both winter and summer activities.

Kelly and Williams (2007, 68) have commented that 'Energy use in tourism destinations is normally disproportionately greater than what is typically associated with other similar-sized communities. This is largely due to the extensive use of energy-intensive technologies that deliver tourism amenities.' The largest portion of energy use is related to travel from other regions to tourist destinations, an aspect over which destination communities have little influence. On the other hand, options exist to achieve energy efficiency within their communities. Kelly and Williams (2007) outline what Whistler has done in that regard by highlighting five strategies in Whistler's 2003 *Integrated Energy, Air Quality and Greenhouse Gas Management Plan*. These strategies have the potential for significantly reducing internal energy consumption and related GHG emissions.

1. *Comprehensive transportation strategy*. Efficiencies in energy use related to transportation in Whistler start with land-use policies and plans to create compact forms of development, which reduce travel distances as well encourage walking and cycling, provide attractive public transit service, manage parking spaces to encourage modes of transport other than automobiles, and improve road systems to make neighbourhoods more accessible. The expectation is that by 2011, the vehicle-kilometres travelled will be 20 per cent lower than if these measures had not been introduced.

2. *Municipal vehicle fleet efficiency*. The Regional Municipality of Whistler has a fleet of gasoline and diesel vehicles. The intent is to gradually change the fleet passenger vehicles to hybrids, use vehicles with smaller engines, use fuel additives to improve fuel economy, and use exhaust scrubbers for trucks and large engines. It is anticipated that by 2020, these measures will enhance energy efficiency of the gasoline-powered fleet by 50 per cent and of the diesel-powered fleet by 15 per cent.

3. *New and redeveloped building energy efficiencies*. Whistler will implement energy conservation programs (LEED, CBIP) for contractors as well as homeowners (EnerGuide) (Box 13.4). It is expected that such programs will increase energy efficiency in buildings by at least 25 per cent.

4. *Natural gas as a primary energy supply*. Whistler is supplied by a piped propane system, the capacity of which is expected to be exceeded. One option for the future is a natural gas system. The demand for energy would not be affected by changing to natural gas, but there would be substantial reduction in GHGS and other air pollutants because natural gas has lower emission intensities compared to propane.

5. *Small-scale and localized renewable energy sources*. Various localized and small-scale projects, such as micro-hydro and geothermal, are being considered. As with natural gas, they would not affect demand for energy but should reduce GHG emissions, since they would reduce the draw on the provincial electricity grid.

Figure 13.4 Whistler, BC. Based on BritishColumbia.com.

Box 13.4   LEED and CBIP

LEED stands for the Leadership in Energy and Environmental Design Green Building Rating System. It is a national standard, established in the US, related to the design, construction, and operation of 'high-performance green buildings'.

LEED provides benchmarks for performance related to five aspects regarding human and environmental health: (1) sustainable site development, (2) water savings, (3) energy efficiency, (4) materials selection, and (5) indoor environmental quality.

CBIP, or the Commercial Building Initiative Program of Natural Resources Canada, is based on experience showing that new buildings can be designed to reduce overall energy consumption in a major way. Performance better than 25 per cent relative to minimum requirements specified in the *Model National Energy Code for Buildings* (MNECB) is a common target in the marketplace. More than 1,000 building design models validated by Natural Resources Canada's Office of Energy Efficiency have a higher average design efficiency—35 per cent better than the MNECB.

Whistler has moved ahead with implementing actions consistent with the initiatives outlined above. For example, an inventory of GHG emissions was completed in the spring of 2006. While the inventory revealed that passenger vehicle transportation and space heating for residential and commercial buildings were the main contributors to GHG emissions, GHG emissions had decreased from the municipality's landfill site because of new processes to cap and capture gases. In the spring of 2007, an audit of appliances and heating systems was completed, and construction of a 50-kilometre natural gas pipeline from Squamish was completed along the Sea-to-Sky Highway. The pipeline provides the municipality with a reliable supply of natural gas. During the autumn of 2008, all existing propane appliances and heating systems were converted to natural gas. It has been estimated that the change from propane to natural gas will result in a 15 per cent decrease in GHG emissions from homes and businesses. A complementary step will be to introduce natural gas vehicle programs for the municipal fleet of vehicles.

To publicize sustainability issues, the municipality has introduced other activities. For example, Whistler has an annual 'dumpster dive' during which staff from the Public Works Department dive into trash bins to determine which items could have been recycled. In addition, the municipality provides biodegradable 'doggy bags' and makes them available at many parks and areas designated for dog walking.

Whistler's actions have won several awards. At the end of 2005, Whistler was recognized as the first municipality in Canada to have completed all five milestones in the Partners for Climate Change Protection Program created by the Federation of Canadian Municipalities. Intended to help

municipalities reduce GHG emissions through both municipal operations and community-wide activities, the five milestones include: (1) creating a GHG inventory and forecast; (2) setting an emissions reduction target; (3) developing a local action plan; (4) implementing the local action plan or a set of activities; and (5) monitoring progress and reporting results. Whistler has completed all of them. The community won a second award when it was one of seven receiving the inaugural Green City Award from the BC government in September 2007. Whistler was recognized for exemplary planning and development in balancing protection of the natural environment with promotion of its tourism economy.

Sudbury, ON: Remediating Mined Landscapes

Located about 400 kilometres north of Toronto, Sudbury at one time was 'notorious across the country for the air pollution and the barren, blackened landscape created by its smelters' (Richardson, Savan, and Bodnar 1989, 4). Starting in the 1970s, however, some remarkable initiatives were undertaken to rehabilitate the landscape and restructure the economy of Sudbury.

Sudbury was established in the early 1880s as a construction camp for the Canadian Pacific Railway. During the building of the railway, copper and nickel deposits were discovered only a few kilometres north of the construction camp. The construction camp soon evolved into a mining community, and Sudbury became the second largest producer of nickel in the world. In addition to Sudbury, other mining communities such as Falconbridge were established in the 30 × 60-kilometre Sudbury basin (Figure 13.5).

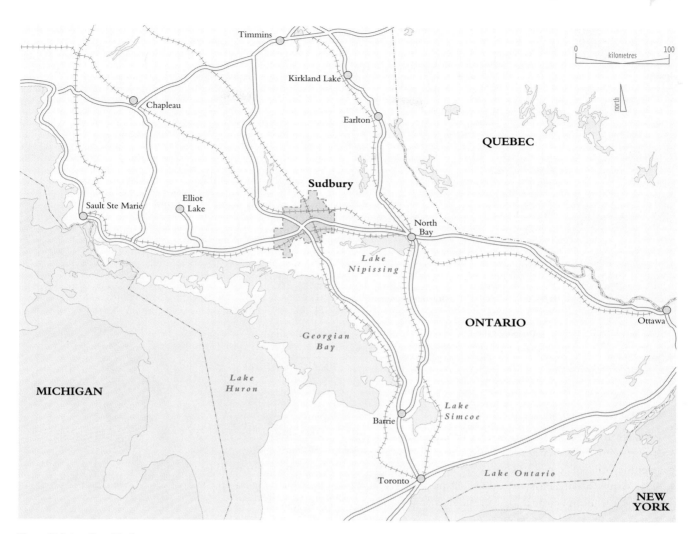

Figure 13.5 Location of Sudbury.

As with most mining-based communities, Sudbury experienced boom-and-bust cycles. Peak production occurred in 1970, when nickel output from the Sudbury basin reached an annual high of 200,000 tonnes at the same time that prices for nickel had climbed to their highest levels. At that time, Sudbury had one of the highest average household incomes in Canada. In addition, the economy had begun to diversify as cultural and advanced educational facilities were built and good transportation links to southern Ontario were put in place. The economic prospects of the Sudbury basin at the start of the 1970s appeared bright.

On the negative side, however, two issues were of concern. First, the community had some serious social divisions as a result of bitter and prolonged labour-management and inter-union conflicts within the mining industry. As a result, Sudbury was a deeply divided community. Second, the strong economic growth, based on the mining industry, was offset by a dramatically degraded physical environment as a result of previous mining activities. Tens of thousands of hectares of the Sudbury area were devoid of significant vegetation because of air pollution and past mine practices. Many lakes in the area had become highly acidic and degraded by metal

Reclamation in Sudbury

At times the environmental devastation humankind has wrought on our planet seems overwhelming. But when individuals come together and bring their time, energy, money, quick minds, strong backs, and collective will to a task, they not only can move mountains, they can recreate a landscape. Sudbury is living, thriving proof.

. . . The Sudbury story is a beacon of environmental hope. If our experience can teach the world anything, it is that we are all responsible for our natural environment, we can learn from our mistakes, and nature is forgiving. Each of us has a role to play, and collectively we can make a difference.

—Ross 2001, 8

Figure 13.6 Sudbury: Extent of barren and semi-barren landscape. Source: Lautenbach 1985, 4.

contaminants, partly as a result of acid deposition associated with the smelter, as described in Chapter 4. The city periodically experienced episodes of choking air pollution from the mining smelters.

While Sudbury's economic growth had been strong and impressive, the reversal of its economic fortunes during the 1970s was equally dramatic. Increasing pressure from international and domestic nickel-producing competitors caused a steady drop in nickel prices, which in turn led to a significant drop in mining employment. Over a 12-year period (1971 to 1983), mining employment in the Sudbury basin fell from 47 per cent to 11 per cent of the labour force. This resulted in significant out-migration, but even with that, the unemployment rate in 1983 had climbed to 17 per cent. By 1986, mining employment had fallen further to 7,700, and between 1971 and 1986 the population of Sudbury had dropped by 10 per cent, from 169,200 to 152,000. The prospects for the area appeared bleak.

Several factors contributed to the subsequent economic restructuring in the Sudbury basin. First, in 1973 the Regional Municipality of Sudbury was formed, creating for the first time a federation of the 15 smaller communities, which allowed a regional perspective on planning and development. However, it is important to note that questions have been raised about the value of regional government for Sudbury. Critics are calling for de-amalgamation because of what they view as very high costs for services as well as poor delivery of services to outlying communities. Earlier in this chapter, we highlighted the importance of increasing population density as one foundation to help achieve sustainability. In that context, the reality is that the Regional Municipality of Sudbury is a collection of relatively small and dispersed mining towns. It has one-sixteenth the population of Toronto but extends over an area five times as large. Such a relatively low population density is one explanation for the relatively high cost and poor delivery of services.

Sudbury's daunting economic prospects led to much improved labour-management and inter-union relationships as people began to realize that it was imperative for them to work together. The provincial and federal governments relocated some of their offices to Sudbury, and regional hospital services were established. Machinery manufacturing also started to grow. In addition, significant growth occurred in service-oriented businesses.

Another important initiative was the development of a community-led strategy to help rejuvenate and diversify the local economy, with the ultimate goal being to create a sustainable community (Katary 1982). The *economic goal* was to help Sudbury become a self-sustaining metropolitan area based on a diversified economic structure. The *social goal* was to allow individuals to discover their role and potential in the community, and one thrust was to develop self-help and mutual assistance programs. For the *physical dimension*, the goal was to enhance the quality of the physical environment, including the ecosphere. Particular attention was given to minimizing adverse environmental impacts and to maximizing the aesthetic ambience within the community.

On the environmental side, some initiatives had begun before the negative consequences of the economic downturn had become so evident. During the 1970s, Inco and Falconbridge Ltd had shut down part of their smelter capacity and had significantly reduced emissions of sulphur dioxide. Inco also built its 381-metre 'Superstack'—the tallest smokestack in

Inco Superstack in Sudbury, ON.

the world—to disperse the emissions farther afield. Other ore processing measures were adopted in Sudbury's smelters, including 'removing sulphur from ore before smelting, increasing the efficiency of the roasting and smelting processes, and containing sulphur through the production of sulphuric acid, which is then sold' (Clean Air Sudbury 2005, 8). The result was that the average level of sulphur dioxide in the air at Sudbury dropped from 54 to five parts per billion between 1971 and 2002, while total emissions fell by 88 per cent between 1960 and 2002. The outcomes were tangible. The city was no longer periodically subjected to air pollution fumigations. Vegetation could grow and began to return. At the same time, some of these positive outcomes were offset by the fact that the Superstack had simply shifted a portion of the air pollution problem from the Sudbury basin to areas downwind, sometimes as far away as parts of Quebec.

With local air quality significantly improved, opportunities for restoration presented themselves. Beginning in 1971, scientists established small plots to test various combinations of soils and plant species. Based on this experience, techniques were developed at Inco and in the Department of Biology at Laurentian University to grow grass, clover, and then tree seedlings on formerly degraded land. Experience through trial and error led to procedures that resulted in healthy vegetation. The main lesson was that applying crushed limestone resulted in neutralization of acidic soil, inhibition of the uptake of metals, and enhanced bacterial activity, all of which allowed some vegetation types to do well. The context for this work has been highlighted by Lautenbach and his colleagues (1995, 112) who commented that 'There was little scientific information to guide the design and implementation of an effective reclamation program for this type of landscape. Restoration ecology was and still is a very new and developing field of study. Therefore, testing and monitoring were essential for achieving the objectives of the reclamation program.' The objectives and outcomes of this work are highlighted in Box 13.5.

The regional municipality became involved in the reclamation program with the establishment of VETAC—a Vegetation Enhancement Technical Advisory Committee to Regional

Box 13.5   Objectives of the Grassing and Reforestation Activities in Sudbury

The objectives were to:

- create a self-sustaining ecosystem with minimal maintenance;
- use plant species tolerant of acidic soils and low nutrient concentrations;
- use seed application rates that allow for natural colonization and thus increase species diversity;

- give preference to the use of native species;
- restore nutrient cycles and pools by the use of species that fix nitrogen (legumes);
- use species that attract and provide cover for wildlife;
- undertake initiatives that speed up natural successional changes.

Council. This committee includes botanists, ecologists, landscape architects, horticulturalists, agriculturalists, planners, fisheries experts, gardeners, and interested citizens from the mining companies, university and college, provincial ministries, Hydro One, municipalities, and the general public. Together, this group addressed Sudbury's vegetation restoration. After surveying the degraded areas, they identified a target to rehabilitate about 30,000 hectares of barren and semi-barren land.

During the summer of 1978, 174 students worked to apply crushed limestone, fertilizer, and grass seed on an area adjacent to the highway close to the airport and on another area beside the Trans-Canada Highway. At the end of the summer, these students had planted grass on 115 hectares of barren land, removed debris from a further 206 hectares, planted 6,000 trees, shrubs, and plants, collected some 30,000 samples for pH and nutrient testing, created 122 new test plots, and gathered 365 kilograms of native seeds. The next summer, the number of students increased to 325, and from that time on, 150 to 300 students were employed each summer on the reclamation project.

By 1982, nearly 400,000 trees had been planted, and by 1984 more than $12 million had been spent, with 2,636 hectares planted. In 1982, the regional municipality began to hire workers laid-off from Inco and Falconbridge to plant trees in the spring before students could start to work. The work continued during the 1980s and 1990s, and in the 1990s, because of improved air quality, some tree species were colonizing previously barren areas. As Ross (2001, 60) reported, one VETAC member described how amazed team members felt about the natural regeneration: 'One day we wandered around up behind the smelter, and our mouths dropped open. We saw all these birch seedlings coming back, and we hadn't planted them.'

Over time, emphasis shifted to planting trees rather than grass, since trees could grow in some places without needing

limestone applied first. Thus, while at the outset it was usual for hundreds of hectares or more to be limed, fertilized, and seeded, the extent was reduced to fewer than 50 hectares annually during the late 1980s and 1990s. In contrast, tree plantings grew to more than 100,000 annually for red and Jack pine, red oak, and tamarack between 1983 and 2000. The one-millionth tree was planted in 1990, and in 1998 alone 985,574 trees were planted.

The results have been impressive. While slag heaps and bare, blackened rock are still present, by the end of 2007 3,392 hectares had been limed, 3,194 had been fertilized, and 3,121 had been seeded, and 8.83 million trees as well as 43,427 shrubs had been planted, at a cost of just under $24 million. In summer, this formerly barren land is now green with grass and trees. The Land Reclamation Program annual report for 2007 highlighted that 95 per cent of all planted trees have been coniferous, less than 5 per cent have been hardwood, and less than 1 per cent have been shrubs. Of the coniferous trees, some 75 per cent are pines (Jack, red, and white), while red oak trees represent about 50 per cent of the hardwoods. The aim has been to plant species believed to be the main components of the original forests in the Sudbury region.

All of this reclamation activity provided short-term employment for students and unemployed individuals, with much of the labour cost borne by employment programs funded by the federal government. In addition, Inco and Falconbridge have restored degraded company-owned land and between them have planted 3.5 million trees on their properties. Volunteers have also made a significant contribution through the Land Reclamation Program. By the end of 2007, some 8,502 volunteers had planted 289,924 seedlings (VETAC 2007).

As noted above, the area totally reclaimed (the area limed, fertilized, and seeded) amounts to 3,100 hectares, or 3.6 per cent of about 85,000 hectares of barren and semi-barren lands

(VETAC 2005). Another 3,392 hectares had only been limed as of 2007. A 1,300-hectare untreated site of blackened rock without any vegetation has been left just south of one smelter as a reminder of the landscape before the work started and as a benchmark for measuring future progress.

The program has faced some special challenges. For example, in early June 2006, liming was started on a barren area to the west of the city's landfill site, and as a result, bears are frequent visitors. A 'bear awareness' training session was initiated, and the program's annual report for 2006 reported that 'Numerous sightings were reported over the season, fortunately all without incident' (Land Reclamation Program 2006, 5).

Given that the remediation initiative has been in part a research project, care has been taken to monitor results. Monitoring was started during the early years regarding the survival and growth of tree seedlings, as was tracking the biodiversity/forest floor transplant plots. Identification of future planting sites was also systematically developed. Newer monitoring initiatives have included establishment of plots to evaluate the health of shrubs, long-term survival, and spring versus fall planting success, as well as wetland assessment and selection of future potential seed collection sites (Land Reclamation Program 2006, 8).

Other benefits have emerged. Almost a century after elk disappeared from the local landscape, more than 150 have been reintroduced and appear to be thriving in their traditional habitat south of Sudbury. Following an absence of 30 years, pairs of peregrine falcons now nest in Sudbury, and during the spring of 2000, a pair of trumpeter swans hatched cygnets.

The positive results outlined above have attracted recognition for Sudbury both within and outside Canada. Among the many awards Sudbury has received from within Canada are the Lieutenant Governor's Conservation Award from the Council of Ontario and an award from Canadian Land Reclamation. Outside Canada, Sudbury has been recognized through a United Nations Local Government Honours Award and awards from the Society for Ecological Restoration and from Germany for Global Responsibilities through Local Action.

An intriguing question, posed by Richardson, Savan, and Bodnar (1989), is the extent to which the environmental enhancement contributed to the economic restructuring and recovery of Sudbury. They concluded that a definitive answer could not be provided. Nevertheless, they did conclude that the positive changes in environmental quality and economic structure had a common cause—modernization of the production facilities at Inco and Falconbridge. The technological changes in the smelting process allowed the two companies to substantially reduce emissions of sulphur dioxide into the atmosphere. Once air quality started to improve, other improvements became possible: revegetation, reduction of acidity in lakes, and re-establishment of fisheries.

Richardson and his colleagues (1989, 24–6) concluded that

environmental improvement was probably not a direct cause of economic diversification but rather a necessary

condition without which the economy might have languished. Thus, while government policies and other factors certainly changed Sudbury's economic structure, we suggest that these policies might not have achieved their intended result without the improvements in environmental quality that took place. . . .

In other words, while environmental improvement may not have generated economic benefits directly, it contributed to creating the conditions necessary for the positive economic changes in Sudbury to take root and flourish. One example includes the long-term prospect of restoring renewable resources such as timber to help diversify the Sudbury economy, which has long focused on a non-renewable mining industry (Bullock and Watelet 2006, 45). In ecological terms, environmental quality could be a factor that does not directly *produce* economic benefits but that, when lacking, can *limit* economic development. As a result, the experience in Sudbury reinforces the argument about sustainable development in Chapter 1 that economic development and environmental management should be viewed as complementary, that each can support the other, and that new decision-making processes that consider a broader mix of factors and include the public are necessary.

Best Practice for Urban Environmental Management

In the following subsections, we highlight key factors deserving attention as best practice related to urban environmental management. To begin, we present the ideas that the Canada Mortgage and Housing Corporation (2005, 1) has developed regarding what it calls 'smart growth' for cities, or in its words, 'land use and development practices that limit costly urban sprawl, use tax dollars more efficiently, and create more livable communities'. In particular, it argues that 'in order to be sustainable, cities should alter their development patterns so as to be more compact, diverse in local/district land uses, with well-defined urban boundaries and clear internal structures.' The corporation (2005, 2) uses the following 'indicators' to assess whether a municipality is using smart growth principles:

- promote denser, mixed-use development in greenfield areas;
- intensify the existing fabric to moderate greenfield development;
- take advantage of specific intensification opportunities;
- increase transportation choice and reduce car usage;
- increase the supply of new affordable housing;
- improve the range of housing types;
- preserve agricultural lands;
- preserve lands essential to maintaining regional ecosystem functions;

- direct employment to strengthen the core and designated sub-centres;
- provide infrastructure to reduce ecological impacts of development.

In the following five subsections, we consider spatial scale, parks and pathways, air quality, waste management, and transportation. Earlier in the chapter, we examined urban form or structure in some detail. In the final subsection, we review approaches being taken in Victoria, BC, Toronto, and Sydney, Australia, in that regard.

Spatial Scale

The United Nations Population Fund (2007a, 53) has noted that

fragmentation of the urban territory brings both administrative inefficiency and environmental setbacks. The boundaries of the city's administration rarely coincide with its actual area of influence. . . . Without some sort of regional entity, the administration of key services, such as water and transport, that cut across different boundaries, is very difficult. By the same token, fragmentation breaks up the contiguity that natural processes require. Fragmentation also makes it difficult to protect ecologically fragile areas or regulate for environmental integrity.

The challenges outlined by the United Nations Population Fund have been recognized for many years but are difficult to overcome. The boundaries of cities or urban municipalities usually reflect administrative or political, not ecological, considerations. For example, the cities of Ottawa in Ontario and Gatineau in Quebec are located across from each other on opposite sides of the Ottawa River. Their urban boundaries do not reflect the reality that both of them are within the Ottawa River catchment or basin. Furthermore, many cities are located in the upper portion of river basins and dispose of much of their waste in the river, which then carries the waste-laden water to downstream communities that have to bear the cost of treating the water if they use the river as their source of supply.

The most common way to address the problem of what is an appropriate scale for an ecosystem approach is to create a 'regional authority' based on landscape features in order to assist managers in considering the larger ecosystem within which a city is located. This is not a perfect solution, but it is a first step. For instance, many cities within a region are often reluctant to give up their authority or autonomy. This can subsequently lead to conflicts and requires a high level of capacity to facilitate cooperation and negotiation.

Parks and Pathways

Public parks and pathways in urban areas contribute to physical fitness and the well-being of residents. In addition, urban

Community garden.

trees and other vegetation can filter air pollutants, moderate the urban heat island effect, and help to enhance water quality and reduce flood damage potential. Thus, parks or informal 'green spaces' offer considerable ecological value and should not be viewed as a frill or luxury.

In addition to creating new parks or enhancing existing parks, municipal governments can facilitate community gardens, ban the use of pesticides for 'cosmetic' reasons, and protect or restore wetlands and other natural areas that filter air pollutants. At the household level, residents can plant more trees and shrubs, which take up rainwater and help to keep buildings cooler. Using native species for home gardens instead of lawns and imported plants is beneficial because native species are well adapted to local climate conditions and therefore need less watering or fertilizing. Another choice is to have a vegetable garden or grow vegetables in an allotment garden, since locally grown vegetables do not have to be shipped from afar and thus do not contribute as much to energy use and emissions. Finally, there are benign alternatives to pesticides, toxic cleaning products, and other chemicals.

Air Quality

The discussion earlier in the chapter about urban form, transportation, and energy use indicated that municipal governments have various options to enhance air quality. Local governments can use more energy-efficient vehicles in their municipal fleets, which will reduce fuel consumption and GHG emissions. Furthermore, in terms of energy for buildings or other purposes, options exist to develop renewable energy sources, such as wind or solar power. In terms of urban form, sprawl can be constrained through land-use regulations and by providing user-friendly and convenient public transit. Finally, building standards can be modified to improve the energy efficiency of new or renovated buildings. With regard to the latter, the city government in Chicago requires all new buildings to meet the silver standard in LEED (Leadership in

Guest Statement

Feng Shui Landscapes in the Context of Urban Environments in Hong Kong and South China
Lawal Marafa

Patches of forest landscapes located in urban areas play important ecological roles in the overall context of city structures. As urban areas experience numerous problems such as deterioration of air quality, rising temperature, and psychological stress, urban forests and green spaces filter air and water, trap sunlight, and provide shade and recreation areas for the resident population. In fact, such forest areas can help to reduce the urban heat island effect that characterizes most urban areas. Although many urban forests are remnants of wilderness that either purposely or accidentally escaped development processes, globally their existence reflects value attributed to them regarding both cultural purposes and ecological functions.

In past civilizations, forests attracted people for various reasons that include shelter, wood for homes and fires, meditation and worship, or even physical support. Forests and culture have therefore indeed been intertwined throughout human history. But these relationships have often been altered by the growth of urbanized landscapes and the extension of related values. In Hong Kong and South China, the forested landscapes, often at the periphery of urban areas, have been modified as a result of adherence to the *feng shui*, or the science of spatial placement and traditional geomancy. While geomancy in the Eastern tradition involves the interpretation of topography of space and place, it is also seen as a system of human propensity for pattern recognition. Consequently, the *feng shui* has shaped much of the human landscape of Hong Kong and South China through constituting an established system of land cover and landmarks in the region. Characteristically, the *feng shui* woodland is preceded in the front by agricultural fields, followed by the residential area at the bottom of the hills adjacent to hilly woodland areas. But as agriculture has been reduced or has ceased to exist in most urban areas, these open fields at the bottom of the *feng shui* **landscapes** are assuming other functions,

such as becoming flat open spaces for recreation or other urban infrastructure.

The *feng shui* woods are a common feature associated with most settlements of South China. These wood groves primarily consist of broad-leaved trees that are often juxtaposed along the residential areas. In fact in historical terms, *feng shui* explicitly states that human beings should live in harmony with nature, leading to retaining trees in settled areas. Reflecting this historical notion, the art and science of *feng shui* are gradually becoming incorporated into various aspects of city planning that is now called an 'ecocity' approach. As societies develop and urbanization escalates in South China, patches of these forests have often been retained in urban areas and in some cases at the periphery of the urban environment. One of the main reasons for this retention is the cultural belief of the East that there is a connection between trees and good fortune. It is partly as a result of this that the *feng shui* woods and forests have been preserved, despite the influence of urbanization and human activities that continuously encroach or exert pressure on the forest woodlands.

Although the *feng shui* woodlands are limited in size, ranging in Hong Kong from 600 square metres to about

Luk Keng is a village in the North East New Territories of Hong Kong. The classical layout includes an agricultural field in the lower topography, a village at the foot of the hill, and an area of *feng shui* woodland on the elevated area adjacent to the village.

six hectares and numbering almost 350 patches across the Special Administrative Region, they are repositories of a well-maintained biodiversity in the urban structure and periphery and can provide well-nurtured shelter and habitat for wildlife such as birds, insects, and mammals. In addition, in an urban environment like Hong Kong, they also provide resources that are used for education and academic and scientific research, as well as the promotion of nature conservation.

Given that the existence of *feng shui* woodlands in a locality is often rooted in the residents' belief of geomancy, the *feng shui* woodlands can be of immense cultural and ecological value. Socially, the trees can represent symbols, meeting points, and sacred places. While their spatial location and characteristics signify the local belief in harmony between humans and nature, their ecological structure has immense beneficial influences on the microclimate. When *feng shui* woods are well managed and preserved, they also can act to

Lawal Marafa

prevent soil erosion and landslides. They can also help to mitigate the impact of typhoons that often accompany the subtropical monsoon as well as modify the local temperature. With growing scientific understanding of the science and myth upon which *feng shui* is based, scholars, researchers, and others schooled in the Western traditions of natural resource management and environmental sciences are beginning to emphasize principles of planning natural place and space in urban areas that reflect the centuries-old *feng shui* beliefs. The basic aspects of the natural environment that are considered in such planning tasks include climate, physiography, hydrology, vegetation and ecology, landscapes, and habitats, all of which *feng shui* requires attention to for sites in cities to determine the desirability of building layouts and locations. Accordingly, the principles related to *feng shui* are now accepted by most urban planners and natural resource managers as a comprehensive environmental evaluation system that reflects the relevance of the physical nature of any given landscape poised for development.

Lawal Marafa was born and educated in Nigeria and completed graduate studies in Russia and Hong Kong. He is currently an associate professor in the Department of Geography and Resource Management at the Chinese University of Hong Kong, and his main research interests are ecotourism, tourism and environment, natural resource management, and sustainable development related to the Millennium Development Goals.

Energy and Efficiency and Design). Chicago also leads all cities in the world in the construction of green roofs on both public and private buildings, which Chicago encourages through subsidies of $5,000 to homeowners and small businesses.

Individual urban residents also have choices. They can walk, cycle, or use public transit, any of which contribute to reduced GHG emissions. Insulating or draft-proofing homes and switching to energy-efficient lighting and appliances will reduce the draw on energy sources, thereby contributing to lower GHG emissions. Setting the home furnace thermostat at a lower temperature and wearing a sweater in winter or setting the air conditioner to a higher temperature and wearing lighter clothing in summer reduces energy consumption and GHGs. Finally, purchasing food grown in the local area year-round contributes to lower energy consumption than buying food transported from distant locations. These ideas highlight the importance of disseminating information to sensitize individuals to the impact of their behaviour, thus facilitating longer-term shifts in

basic values, attitudes, and behaviour. Opportunities for individual actions are addressed in detail in the final chapter.

Waste Management

Municipal governments can enhance reduction of the waste stream by expanding the types of items that can be placed in blue boxes and by providing blue box services to apartments and office buildings. Reduction also can be improved by requiring manufacturers to reduce packaging for products and by establishing deposit refunds for glass bottles. Recycling can be enhanced by searching for and developing more markets for recycled materials. More aggressive options include the banning of certain products, as San Francisco did in banning plastic bags in larger supermarkets and pharmacies in 2006. Oakland, CA, has banned polystyrene or styrofoam containers, which has led to restaurants using paper, cardboard, or recyclable plastic containers.

Box 13.6   Leaf Rapids, MB: The First Canadian City to Ban Plastic Bags

The community of Leaf Rapids is located about 980 kilometres northeast of Winnipeg adjacent to the Burntwood River. In 1969, a copper and zinc ore deposit was discovered in the area, and the first residents arrived in the new mining town of Leaf Rapids in 1971. The population soon grew to 2,000.

In early April 2007, the municipal council of Leaf Rapids banned plastic shopping bags. This action was preceded by a decision in May 2006 to charge a three-cent levy on every plastic shopping bag given out by local merchants.

The bylaw prohibits the sale or distribution of single-use bags. Violations can lead to a fine of $1,000 a day. To encourage compliance, the town distributed cloth shopping bags to residents. Retail stores provide paper shopping bags.

The manager of the local co-op grocery store stated that prior to the ban, his store used at least 2,000 bags a week.

No one has been charged with violating the bylaw.

Source: http://www.townofleafrapids.ca/aboutus.html.

Individuals can become more disciplined in their use of blue and **green boxes** and use reusable coffee containers, water bottles, and shopping bags. Other options include purchasing products with minimal packaging, buying reusable instead of disposable products, and when appropriate, reusing paper, bottles, and other material.

Transportation

If people are to change their behaviour, incentives must be provided, and people need to be educated in order to understand why behavioural change is necessary. If it is clear that many people will continue to use automobiles, then they should be encouraged to use more fuel-efficient vehicles, which also would reduce GHG emissions. Ontario has considered issuing green-toned licence plates to car owners who purchase verified 'environmentally friendly, low-emission' vehicles such as hybrids or 'smart cars' because of their low fuel consumption.

Various technical matters would have to be resolved before cars could be assessed and designated for that purpose. For example, what thresholds for fuel efficiency or GHG emissions would be appropriate, and how could authorities verify that the vehicles were maintained well enough to continue meeting the thresholds? Would one person travelling in a smart car achieve the same fuel efficiency as four people car-pooling in a conventional vehicle?

What incentives might be provided to encourage people to purchase more fuel-efficient vehicles? Suggestions include charging a lower licence fee for such vehicles, providing free parking in municipal parking lots, and access to commuter lanes on highways that are normally reserved for buses or other vehicles with a minimum number of passengers. Another option is to encourage people not to use cars but rather other transportation means. For example, it could be made more expensive to drive a car into the centre of a city, as was done in London, UK, where drivers pay a fee for driving in the central area during the day. And in Paris, people can rent a bicycle for only a few euros at one of the many docking stations throughout the city. When they are finished with the bicycle, they can leave it at any other docking station. Paris intends to provide a similar service for electric cars, making 2,000 of them available on the same basis as bicycles.

Bike rental station in Paris, France.

AP Photo/Michel Spingler

What information and education programs do you think should be developed to help individuals understand the impact of travel behaviour on the environment and why it is in everyone's interest to modify travel behaviour with the environment in mind?

Best Practice: Dockside Green (Victoria), Toronto, and Sydney, Australia

Award-winning urban development in Victoria, BC, Toronto, and Sydney, Australia, offer insight into best practice in environmental management.

Dockside Green, Victoria, BC

The Dockside Green development in Victoria received the Excellence in Urban Sustainability Globe Award in 2008, an international award presented to a local government, private sector company, or consortium that has developed and applied outstanding urban sustainability principles. It is not the only award that the project has won; Dockside Green also received the international Brilliant Development award at the first Discover Brilliant Conference as well as several others. A showpiece of sustainable design and technology, Dockside Green is a multi-million-dollar project designed for residential, home office, retail, office, and light industrial uses and including extensive public amenities. It includes 26 buildings (all aiming for LEED platinum certification) totalling 117,000 square metres and is being built over the next six to eight years. Superior building practices and extensive community planning are transforming the seven-hectare waterfront site from a contaminated industrial wasteland into a healthy and lively community that 2,200 residents will call home. As you will see from the following list of attributes, one thing that characterizes the development is the addition of innovative features, such as training for First Nations and a car-pooling program, that are rarely seen in conventional developments. Some of its most outstanding attributes include:

- a wood waste gasification plant providing renewable heating to the development, making it 'greenhouse gas positive' and allowing sale of surplus heat;
- sewage treated on-site using technology by Zenon, a Canadian company recently purchased by General Electric;
- heat recovered from a sewage treatment plant and a city sewage trunk line;
- treated sewage water used in toilets, irrigation, and water features;
- projected potable water savings estimated at 70,318 million litres per year, equivalent to the entire region's use on the driest day of the year;
- green roofs on two residential and one commercial building;
- construction training program for First Nations;
- housing affordability strategy offering qualified buyers one of 26 units at 25 per cent below market value;
- another 45 units promised as rental units;

- built on a brownfield—a former contaminated industrial site;
- no potable water used for landscaping;
- LED lights in corridors and compact fluorescents in suites and common areas;
- motion-sensor lights in closets, storage areas, and bathrooms;
- heat-recovery technology that captures heat from ventilated air to pre-warm incoming air;
- energy metering to individual suites, allowing individuals to control their energy use;
- every appliance Energy Star rated;
- solar trash compactor;
- 1,000 trees planted throughout the development;
- every resident a member of a car-share program with 10 cars;
- more than 93 per cent of construction waste diverted from the landfill.

Toronto

Toronto and Region Conservation is one of 36 conservation authorities in Ontario with a mandate to protect and enhance the environment at a river basin or catchment scale and in that regard is responsible for the watersheds in the Greater Toronto Area. In that context, Toronto and Region Conservation (2006, 4) states:

Sustainability means living respectfully within the natural system. A sustainable society lives within its environmental means. It mimics nature by using only renewable energy and resources and by returning 'resources', not 'garbage' or 'pollutants', to the environment. It seeks creative synergies among the needs of the environment, human society, and the economy and understands their interconnections. . . .

How do we achieve this difficult and necessary goal? By viewing everything we do through a 'sustainability lens' to ensure that we continually contribute to The Living City rather than work against it.

To achieve sustainability, Toronto and Region Conservation has identified a range of needs and opportunities in an urban environment:

- *Water*: protect and conserve it.
- *Biodiversity*: protect and restore local natural habitats.
- *Land use*: confine an increasing proportion of city building to already urbanized or serviced areas.
- *Transportation*: choose clean, fuel-efficient cars and reduce automobile use in favour of transit, bicycling, and walking.
- *Energy*: conserve energy, and move from fossil fuels to alternative and renewable energy sources.
- *Green buildings*: design buildings that contribute to a healthy environment rather than produce waste and harmful emissions.
- *Waste*: work toward zero garbage through reuse, recycling, digesting, and composting.

- *Education*: spread ideas about sustainability and positive change in the community.
- *Human health*: contribute to clean air and water and other conditions essential for good health.
- *Social equity*: help to eliminate poverty and include everyone in community amenities, responsibilities, and career opportunities.
- *Green economy*: encourage sustainable practices, products, and services in business and industry through purchasing, partnership, and investment decisions.

Sydney, Australia

Newman (2006, 287–8) has reported on the eight criteria used by Sydney, Australia, related to sustainable development. The criteria focused on natural resources or environment are presented below, and we encourage you to consider whether they extend the ideas offered by the Dockside Green development in Victoria and by Toronto and Region Conservation.

Overall goal: To reduce the city's ecological footprint (water, energy, land, materials, waste . . .) and enhance the environment while simultaneously improving quality of life (health, housing, employment, community . . .) within the capacity constraints of the city and bioregion. [A **bioregion** is not defined by surveyor's lines but instead reflects a place that makes sense for its inhabitants related to its natural and cultural characteristics. A bioregion is usually delineated to reflect the natural and cultural diversity and richness of a place.]

Natural resources: Live within natural resource limits and minimize the ecological footprint.

- Water: Manage the total water cycle to keep water extraction levels within sustainable yields.
- Land: Minimize the urban footprint and disruption.
- Energy/greenhouse: Use energy efficiently and reduce greenhouse gases.
- Materials: Use appropriate materials and recycle waste.
- Waste: Minimize, reduce, and recycle wastes.

Environmental protection: Protect and enhance biodiversity, coasts, air, water, and agricultural land.

- Biodiversity and ecosystem function: Preserve core biodiversity values and enhance natural ecosystem of the bioregion with corridors and natural areas retained.
- Coastal protection: Protect and enhance the character of the coast and access to it and ensure coastal hazards are recognized and avoided.
- Air quality: Improve air quality.
- Water quality: Maintain and improve waterway health.
- Agricultural land: Ensure important agricultural land is conserved.

Places of high quality: Provide high-quality places to live and play.

- Parks: Preserve open space corridors and ensure local parks are provided.
- Heritage: Protect and preserve regionally significant cultural landscapes and places, including places of relevance to indigenous peoples.
- Sense of place: Protect and enhance the character and identify of the area.
- Scenic places: Protect and enhance scenic areas.
- Walkability: Provide easy accessibility for walking and cycling in local areas.

Jobs-economy and access: Provide employment opportunities through increasing the city's role in the global economy and in regionally based jobs. Provide sustainable accessibility among homes, jobs, services, and recreation.

- Offices: Provide quality office space in centres and along corridors serviced by public transit.
- Public transport infrastructure: Ensure all knowledge-intensive centres and corridors have quality public transport at their core.
- Local access: Facilitate short trips by sustainable modes for local accessibility.
- Density and mix: Create appropriate zonings and opportunities for density and mix of uses in centres and corridors to reduce car dependencies and create efficient land use.

Using the ideas for best practice from Victoria, Toronto, and Sydney, we challenge you to complete an inventory and assessment of sustainability initiatives in your community. How many of these best practices are in place in your community? Of those being used, how effectively have they been developed and implemented? If one new initiative might be undertaken, which one do you recommend?

To motivate you to think imaginatively, we encourage you to check out the aspirations and accomplishments of Copenhagen (Copenhagen 2007). Copenhagen, a city of some 1.5 million people, is aiming to become the world's Eco-Metropole by 2015, or the capital city with the best urban environment in the world.

Today, 36 per cent of its residents commute to work by bicycle along some 300 kilometres of dedicated bike lanes. The intent is to increase that to 50 per cent by 2015 and also to reduce the number of seriously injured cyclists by more than half. To achieve that ambition, about $40 million will be spent between 2008 and 2011 on enhancing road arrangements for cyclists. And since 1995, Copenhagen has provided bicycles for use in the city. In 2007, for a deposit of 20 Danish krone (a bit less than $5), people can use one of the 2,000 city-provided bicycles between April and November. The bicycles can be picked up and dropped off at one of the 110 city bicycle stations.

When the 50 per cent target of Copenhagers bicycling to work or school each day is achieved, CO_2 emissions will have been reduced by 80,000 tonnes. The overall goal is to reduce CO_2 emissions by at least 20 per cent relative to 2005.

Today, Copenhagen reuses about 90 per cent of all building waste and incinerates about 75 per cent of household waste. The energy from burning waste is used for both electricity generation and district heating. Another ambition is to clean up the harbour waters so that people can swim in it and catch fish.

What initiatives could or should be taken in your community?

Implications

If cities persist in the uncontrolled expansion of urban perimeters, indiscriminate use of resources, and unfettered consumption, without regard to ecological damage, the environmental problems associated with cities will continue to worsen.

—UN Population Fund 2007a, 67

In this chapter, our goal has been to help you understand the implications of current values and behaviour with regard to urban environmental management. Understanding the implications of urbanization for the environment is important in Canada because four of five Canadians live in urban areas. We have choices regarding urban form or structure and design (sprawl or compact), transportation (private vehicles or public transit), energy consumption (fossil-fuel or hybrid engines for vehicles), and green space. These choices have consequences for matters such as air quality, greenhouse gas emissions, and the health of humans and other species.

We invite you to develop a vision for urban sustainability or urban sustainable development for the community in which you live. If your community were on a trajectory toward sustainability, what would be different from what happens today? What changes would you have to make as an individual? What choices would have to be made by the community as a whole? What would the costs of making such changes be? What would the costs of not making changes be? In the short term, however, there are actions you can take as an individual, and they are noted in Box 13.7.

Box 13.7   What You Can Do

1. Support municipal and regional elected officials in efforts to promote and introduce land-use planning practices that minimize urban sprawl.
2. Regarding transportation, consider public transit, car-sharing, and bicycling as alternatives to travelling alone in your automobile.
3. Regarding household or work-related waste, find opportunities to reuse, reduce, and recycle.
4. Become familiar with green building technology, and determine how it could be incorporated in a new home or when renovating an existing home.
5. Regarding energy and water conservation at home or work, when buying new or replacing older

appliances, ensure that you purchase energy-efficient units.
6. Replace low energy-efficient light bulbs with energy-efficient bulbs.
7. Consider setting your furnace thermostat at a lower temperature during winter and your air conditioner at a higher setting during summer.
8. Consider native trees and shrubs for your gardens and yards.
9. Volunteer to help create and maintain trails or walking paths in your community.

Summary

1. Fifty per cent of the world's population live in urban areas, and four out of five Canadians live in urban areas.
2. Urban sustainable development is based on four considerations: urban form, transportation, energy, and waste management.
3. In Canada, the proportion of low-rise, low-density homes is steadily increasing, and commuting times are increasing because of reliance on private automobiles. A major challenge is to reduce 'urban sprawl'.
4. Urban sprawl can be reduced by compact, mixed-use urban form and reduced private car use.
5. Use of private automobiles is a major contributor to greenhouse gas emissions. Shifting from cars to public transit is the one action likely to have the greatest single impact on improving environmental quality in Canadian cities.
6. Residential, commercial, and industrial buildings contribute more than 60 per cent of GHG emissions in Canada.
7. Advanced 'green' buildings can save up to 50 per cent in energy use relative to conventional buildings.
8. Regarding waste management, key activities are reduction, reuse, and recycling, followed by energy recovery.
9. About one-third of waste in Canada is from packaging, but except for beverage containers, there are no national or provincial programs focused on reducing packaging waste. In contrast, the European Union has aggressive programs to reduce packaging waste.
10. Residential waste diversion is most challenging for apartments and multi-family dwellings.
11. To enhance sustainability of urban buildings, attention should be given to location and site design.
12. Major environmental issues in cities include air pollution, urban heat island effect, poor water quantity and quality, and brownfield sites.
13. Many cities are vulnerable to natural or human-induced hazards, and in Canada notable hazards include earthquakes, storm surges, floods, droughts, and snow or ice storms.
14. The impact of Hurricane Katrina on New Orleans in 2005, rain, wind, and sedimentation in the Greater Vancouver Area in 2006, and an earthquake in Peru in 2007 highlight the vulnerability of urban infrastructure and residents to hazards.
15. Dockside Green in Victoria, Whistler, Toronto and Region Conservation, Sydney, Australia, and Copenhagen offer excellent examples of urban sustainability initiatives.
16. Leaf Rapids, MB, was the first Canadian community to ban use of plastic bags when it did so in April 2007.

Key Terms

bioregion
blue box
brownfields
ecological footprint
green box
greenfields

LEED
LULU
LUST
NIMBY
recovery

recycling
reduction
reuse
Richter scale
smog

sustainable urban
     development
urban form
urban heat island
urban sprawl

Questions for Review and Critical Thinking

1. Why is Canada so urbanized? Why is urbanization a global phenomenon?
2. What are the key attributes of urban sustainability or of urban sustainable development?
3. What is the significance of urban form, transportation, energy use, and waste management for urban sustainability, and what are the key connections among them?
4. What are the 3Rs?
5. How can the locating and siting of urban buildings help to achieve urban sustainability?
6. Why is air pollution generally increasing in highly urbanized areas?
7. How can the urban heat island effect be reduced?
8. What impact does urbanization have on the hydrological cycle?
9. What is the significance of LULUs, NIMBY, and LUST for urban sustainability?
10. What are the general lessons from Hurricane Katrina related to the vulnerability of urban areas to natural and human-induced hazards?
11. What are the lessons from Whistler and Sudbury regarding how to evolve towards urban sustainability?
12. How could you use the goals, objectives, and principles of urban sustainability developed in Victoria, Toronto, and Sydney to develop an urban sustainability vision and strategy for your community?

Related Websites

Canada Urban Institute: http://www.canurb.com

Environment Canada, Urban Environmental Management: Resource Library: http//www.ec.gc.ca/cppic/en'refView .cfm?refId=1186

Great Northern Way Campus, Centre for the Advancement of Green Roof Technology: http://www/gnwu.ca/cagrt

International Green Roof Association, Greenroofs Project Database: http://www.greenroofs.com

National Round Table on Environment and Economy: http://www.nrtee.com

Transport Canada, Environmental Affairs: http://www.tc.gc .ca/programs/environment/menu.html

Toronto and Region Conservation: http://www.trca.on.ca

United Nations Population Fund: http://www.unfpa.org

World Health Organization, Health and Environment Linkages Initiative, The Urban Environment—A General Directory of Resources: http://www.who.int/heli/risks/ urbanvdirectory/en/index2.html

Further Readings

Note: This list comprises works relevant to the subject of the chapter but not cited in the text. All cited works are listed in the section 'References' at the end of the book.

Alberti, M., and J.M. Marzluff. 2004. 'Ecological resilience in urban ecosystems: Linking urban patterns to human and ecological functions'. *Urban Ecosystems* 7: 241–65.

Allen, A. 2003. 'Environmental planning and management of the peri-urban interface: Perspectives on an emerging field'. *Environment and Urbanization* 15 (1): 135–48.

Arenstam Gibbons, S.J., and R. Nicholls. 2006. 'Island abandonment and sea-level rise: An historical analog from the Chesapeake Bay, USA'. *Global Environmental Change* 16 (1): 40–7.

Beatley, T. 2000. *Green Urbanization: Learning from European Cities*. Washington: Island Press.

Beckett, P.J. 2000. *A Reflection of Two Landscapes—The Copper and Nickel Mining Region of Sudbury before and after 25 Years of Land Reclamation Activity*. Sudbury, ON: Laurentian University Library.

Campbell, S. 1996. 'Green cities, growing cities, just cities? Urban planning and the contradictions of sustainable development'. *Journal of the American Planning Association* 62: 296–312.

De Sherbinin, A., A. Schiller, and A. Pulsipher. 2007. 'The vulnerability of global cities to climate hazards'. *Environment and Urbanization* 19 (1): 39–64.

Federation of Canadian Municipalities. 2002. *Municipal Governments and Sustainable Communities: A Best Practice Guide*. Ottawa: Federation of Canadian Municipalities and C2HMHill.

Fung, M., and C.A. Kennedy. 2005. 'An integrated model for assessing urban sustainability'. *Environment and Planning B* 32: 639–56.

Gunn, J.M., Ed. 1995. *Restoration and Recovery of an Industrial Region: Progress in Restoring the Smelter-Damaged Landscape near Sudbury, Ontario*. New York: Springer-Verlag.

Hardoy, J., D. Mitlin, and D. Satterthwaite. 2001. *Environmental Problems in an Urbanizing World*. London: Earthscan.

Hostovsky, C. 2006, 'The paradox of the rational comprehensive model of planning: Tales from waste management planning in Ontario, Canada'. *Journal of Planning Education and Research* 25: 382–95.

Kennedy, C., J. Cuddihy, and J. Engel-Yan. 2007. 'The changing metabolism of cities'. *Journal of Industrial Ecology* 11 (2): 43–59.

Land Reclamation Program. 1999, 2000, 2001, 2002. *Annual Report*. Sudbury, ON: Economic Development and Planning Services.

———. 2006. *Land Reclamation Program Annual Report, 2006*. Sudbury, ON: Environmental Planning Initiatives.

Lautenbach, W.W. 1987. 'The greening of Sudbury'. *Journal of Soil and Water Conservation* 42: 228–31.

Lees, D. 2000. 'Green rebirth: How three decades of grassroots determination cleaned up Sudbury's industrial gloom'. *Canadian Geographic* 120: 60–70.

Lo, C.P., and D.A. Quattrochi. 2003. 'Land-use and land-cover change, urban heat island phenomenon, and health implications: A remote sensing approach'. *Photogrammetric Engineering and Remote Sensing* 69 (9): 1,053–63.

Marafa, L.M. 2003. 'Identifying wilderness in the landscapes of Hong Kong urban periphery'. *International Journal of Wilderness* 9 (3): 39–42.

———. 2003. 'Integrating natural and cultural heritage: The advantage of *feng shui* landscape resources'. *International Journal of Heritage Studies* 9 (4): 307–23.

Mazmanian, D., and M. Kraft, Eds. 1999. *Toward Sustainable Communities: Transition and Transformation in Environmental Policy*. Cambridge, MA: MIT Press.

Muller, M. 2007. 'Adapting to climate change: Water management for urban resilience'. *Environment and Urbanization* 19: 99–113.

Newman, P.W.G., and I. Jennings. 2006. *Cities As Sustainable Ecosystems*. Kobe, Japan: UNEP-IETC.

Newman, P.G., and J.R. Kenworthy. 1999. *Sustainability and Cities: Overcoming Automobile Dependence*. Washington: Island Press.

Pierce, J., and A. Dale, Eds. 1999. *Communities, Development and Sustainability across Canada*. Vancouver: University of British Columbia Press.

Regional Municipality of Sudbury. 1990. 'Five-year land reclamation plan 1990–1994 (Draft)'. Sudbury, ON: Regional Municipality of Sudbury, Planning and Development Department, Vegetation Enhancement Technical Advisory Committee.

Register, R. 2003. *Ecocities: Building Cities in Balance with Nature*. Berkeley, CA: Berkeley Hills Books.

Richardson, N.H. 1991. 'Reshaping a mining town: Economic and community development in Sudbury, Ontario'. In J. Fox-Przeworski, J. Goddard, and M. de Jong, Eds, *Urban Regeneration in a Changing Economy: An International Perspective*, 164–84. Oxford: Clarendon Press.

Saarinen, O. 1992. 'Creating a sustainable community: The Sudbury case study'. In M. Bray and A. Thomson, Eds, *At the End of the Shift: Mines and Single Industry Towns in Northern Ontario*, 165–86. Toronto: Dundurn Press.

Stewart, I. 2006. 'Influence of meteorological conditions on the intensity and form of the urban heat island effect in Regina'. *Canadian Geographer* 44: 271–85.

Winterhalder, K. 2002. 'The effects of the mining and smelting industry on Sudbury's landscape'. In D.H. Rousell and K.J. Jansons, Eds, *The Physical Environment of the City of Greater Sudbury*, 145–73. Ontario Geological Survey Special Volume no. 6. Toronto: Ontario Ministry of Northern Development and Mines, Mines and Minerals Information Centre.

Worldwatch Institute. 2007. *State of the World 2007: Our Urban Future*. New York and London: W.W. Norton.

Chapter 14

Endangered Species and Protected Areas

Learning Objectives

- To understand why endangered species are important and the factors leading to endangerment.
- To become aware of the extrinsic and intrinsic values of nature.
- To learn why some species are more vulnerable to extinction than others.
- To be able to discuss the main responses to endangerment at the international and national levels.
- To appreciate the many roles played by protected areas.
- To gain an international and a Canadian perspective on protected areas.
- To know some of the main management challenges faced by protected areas in Canada.

Introduction

Most people are aware that many more species are becoming endangered than is natural—that we have a 'biodiversity crisis'. A common perception, however, is that this is a problem more for the tropics than for countries such as Canada. Although it is true that threat levels and the numbers of endangered species are higher in the tropics, Canada also has plenty of challenges in this regard.

Canada has a long-established national parks system designed to protect species and their habitats, but parks do not necessarily afford adequate protection to all species that need it. The decline in turtle populations in Point Pelee National Park, ON, is a good example of some of the problems in our national parks and the challenges park managers face.

Turtles have evolved for hundreds of millions of years. Historically, their adaptations—terrestrial nesting, low adult mortality, late maturation, and longevity—have served them well. But these attributes are no longer adequate, and throughout the world many turtle species are experiencing dramatic declines. Historical records show that Point Pelee, the southern tip of Canada's mainland, at one time had seven different species of indigenous turtles. Research by Browne and Hecnar (2003) of Lakehead University revealed that several species, including the stinkpot, map, and Blanding's turtles, now exist only in

small populations, while other species may be headed towards extirpation—no spotted turtles were found in the park, and only one individual of the threatened spiny softshell turtle was recorded. Reasonably large populations now exist for only one species, the painted turtle (Browne and Hecnar 2007). The authors examined the age structures of the populations and found a preponderance of older animals for most species, indicating aging populations, especially for Blanding's and snapping turtles.

The sand spit at the Lake Erie end of Point Pelee National Park in Southwestern Ontario marks the southernmost tip of mainland Canada.

Thomas Kitchin and Victoria Hurst/All Canada Photos

Although the populations are 'protected' within the park, at least three problems threaten the long-term viability of turtle populations. Roads are implicated in two of the three and illustrate the need to minimize development within parks (discussed later in the chapter). Roads are a source of direct mortality. Some species are attracted by the soft shoulders of roads for nesting, and while migrating across roads in search of suitable sites, many turtles are killed. In addition, roadside nesting sites were found to be very vulnerable to nest predation (100 per cent loss), compared to the 62 to 64 per cent loss in more remote areas of the park. Roadsides are a favourite scavenging area for predators such as raccoons. These and other predators, such as striped skunks and opossums, are reportedly at higher levels in the park than previously, and raccoons may be the dominant limiting factor (Chapter 2) on population growth.

Contaminants are also implicated in turtle population declines. There are still elevated levels of DDT and DDE (see Chapter 10) in areas of the park from past agricultural practices, illustrating the vulnerability of parks to threats from surrounding land uses.

This example illustrates the plight of many species to which we give relatively little attention. Plant and animal populations are often assumed to be 'healthy', especially if they are within the boundaries of a national park. However, *no* national park—anywhere on the planet—is big or remote enough to exclude the impacts of modern society. This chapter focuses on endangered species and the factors behind endangerment. One of the main responses to endangerment is to protect habitats and

Extinction is a natural process that has been occurring since life first evolved on Earth. However, it is the speed of current extinction rates across many different forms of life that concerns scientists.

Philip Dearden

species in park systems. The designation and management of park systems constitute the second main topic discussed.

But why should we be concerned about endangerment? Extinction is a natural process that has been taking place since life first evolved on this planet more than four billion years ago (Chapter 3). Consequently, concern for the **extinction** of species does not focus on the process itself but on the increasing rates of extinction over time—i.e., what humans are doing to speed up the process. Before looking at some of the pressures responsible for this increase, we need to have a clear understanding of why high extinction rates are undesirable.

Valuing Biodiversity

Changes in biodiversity due to human activities were more rapid in the past 50 years than at any time in human history, and the drivers of change that cause biodiversity loss and lead to changes in ecosystem services are either steady, show no evidence of declining over time, or are increasing in intensity. Under the four plausible future scenarios developed by the Millennium Ecosystem Assessment, these rates of change in biodiversity are projected to continue or to accelerate (Millennium Ecosystem Assessment 2005).

Extrinsic and Intrinsic Values

Extrinsic values are values that humans derive from other species. These values can be *consumptive* (i.e., the organism is harvested) or *non-consumptive* (i.e., the organism is not harvested or the resource is not destroyed). There is no universally accepted framework for assigning value to biological diversity, but various approaches have been proposed. For example, the value of biodiversity can be calculated by examining import and export statistics for products bought and sold in markets. However, it is often difficult to assign an economic value to

External Influences on Parks and Species

. . . it is a common fallacy that protected areas are unimpaired swaths of wilderness with pristine natural conditions. In reality, Ontario's provincial parks and conservation reserves are threatened by numerous ecological stresses and threats, some of which originate beyond their boundaries. Indeed, in many cases, the boundaries of protected areas are political constructs that do not reflect natural boundaries. As such, there may be an issue of concern outside of a protected area that affects its management.

—Wilkinson 2008, 185

biodiversity. How do you put a price tag on environmental services provided by biological communities, such as photosynthesis, protection of watersheds, or climate regulation? These services are not directly consumed by humans, but they are vital for survival.

While important, the extrinsic reasons for species protection should not be allowed to dominate our thinking. Such thinking could lead to the protection of a selection of species believed to be of higher value, while species with less use value are afforded little or no protection. Thus, arguments for biological conservation focus on the **intrinsic value** of nature—nature has value in and of itself, apart from its value to humanity. The following discussion outlines some of the key ecological, economic, and ethical reasons for conservation.

Art Wolfe/Stone/Getty Images

The great hornbill is found throughout the tropical forests of Southeast Asia where it has been extirpated from many areas by hunting. Even where it exists in small numbers, it is often considered ecologically extinct as there are no longer sufficient numbers to crack and distribute the seeds of many of the tree species. Ultimately, this will also cause changes in the tree species composition of the forests.

If the land mechanism as a whole is good, then every part is good, whether we understand it or not. If the biota, in the course of aeons, has built something we like but do not understand, then who but a fool would discard seemingly useless parts? To keep every cog and wheel is the first precaution of intelligent tinkering.
—Aldo Leopold. 1953. *Round River.*

Ecological Values

The elimination of species affects ecosystem functioning, which can lead to unfortunate consequences, such as the impact on coastal marine ecosystems on the Pacific coast when the sea otter was extirpated (discussed in Chapter 3). Species become **extirpated** when they have been eliminated from one part of their range but still exist somewhere else. A species is considered **ecologically extinct** in an area when it exists in such low numbers that it can no longer fulfill its ecological role in the ecosystem. For example, the eastern mountain lion may still exist in the Maritime provinces and has not yet been declared extinct, but if this species does persist, it exists in such low numbers that it no longer acts as a significant control for species in the preceding trophic level. The eastern mountain lion may therefore be considered ecologically extinct in this region.

All species in a community combine to maintain the vital ecosystem processes that make human life possible on this planet—oxygen to breathe, water to drink, and food to

eat. Humans are part of this web of life, but if we continue to eliminate components of the web, its strength will be compromised, which will in turn have a significant impact on the ability of humans to survive. The elimination of species from an ecosystem has been likened to the removal of rivets from an airplane—the system may continue to function after losing a few components, but sooner or later the system will crash. One of the roles of science is to understand how these systems work, but it is difficult to achieve this understanding if components are missing as a result of extinction.

In addition to the value of species in ecosystem functioning, species should be protected for their evolutionary value, their value to future generations. Species evolve over time, as discussed in Chapter 3, and as more species become extinct, genetic variation in the ecosphere on which to base future adaptability is reduced. Fewer species means a more impoverished biosphere on which to base evolutionary adaptability for future generations. In short, the need for preservation of species exemplifies the precautionary principle on a grand scale. In a fanciful manner, the movie *Star Trek: The Voyage Home* showed Captain Kirk, Spock, and his colleagues returning back in time to the present to rescue a humpback whale and transport it to the future, where it had become extinct, in order to save the planet. This is science fiction, to be sure, but it makes the point well—we cannot presume to know what importance or value a particular species might have for future generations.

Economic Values

The preservation of ecosystem components and functions is also economically beneficial. Countless products used in agriculture and industry originate in the natural world. Naturally occurring plants in the tropics, for example, are the source of 90 per cent of the world's food supply. Corn, or maize, feeds millions of people and is now estimated to be worth at least $50 billion annually worldwide. Corn was first domesticated by indigenous people in Central America some 7,000 years ago.

More than 99.8 per cent of the world's plants have never been tested for human food potential. Some of these plants may become important food staples in the future, so preservation of their habitat is important. Furthermore, wild animals still provide an important source of food for millions of people worldwide, particularly indigenous peoples, including Canada's First Nations communities.

Other products besides food are also of economic importance. Many plant and animal products are used extensively in a variety of industries. Rubber, for example, is an important commodity in the automotive sector. It is just one example of a chemical that tropical plants produce to prevent insect damage. Many other chemicals produced by plants are used in the pharmaceutical industry. Fifty-six per cent of the top 150 prescribed drugs in the United States contain ingredients from wild species, with an economic value of US $80 billion a year, according to the UN. In 2002–3, 80 per cent of drugs introduced could be traced to natural ingredients. One drug, a compound derived from a sea sponge to treat herpes, is valued at up to $100 million annually.

However, fewer than 1 per cent of the world's tropical plants have been screened for potential pharmaceutical application. Next time you take an aspirin tablet, thank the white willow, the species in which the active ingredient was first discovered. Taxol, found in the bark of the western yew—a small understorey tree in the forests of the Pacific northwest—was recently discovered as a treatment for cancer. Prior to the discovery, the species was of little commercial value, and it was routinely cut down in clear-cuts and left to rot.

It is difficult to assess the economic value of many of the products we derive from nature. For example, natural gene pools provide a source of material to aid in the development of new genetic strains of crops needed to feed the world's burgeoning human population. Wheat, the mainstay of the western Canadian agricultural economy, originated in Mediterranean countries, where most of its wild forebears have disappeared. But preservation of such wild strains is necessary to allow selective breeding based on the widest range of genetic material to continue.

Ecosystems also provide humans with a wide array of economically important services. Natural pollinators, for example, provide an essential service to commercial crops. Pollination of flowers by diverse species of wild bees, wasps, butterflies, and other insects—not just managed honeybees—accounts for more than 30 per cent of all food production that humans depend upon. Environment Canada (2003b) has conservatively estimated the value of pollination services to crops in Canada at $1.2 billion annually. When New Brunswick switched from spraying its forests with DDT to fenitrothion in 1970 to control the spruce budworm (see Chapter 9), there was a devastating impact on pollination of the blueberry crop because fenitrothion is highly toxic to bees. The commercial crop fell by 665 tonnes per year, and growers successfully sued the government.

Natural predator-prey relationships also aid in food production. It has been shown, for example, that woodpeckers provide an economically important service in the control of pests such as coddling moths in the orchards of Nova Scotia. Such predator-prey relationships can significantly reduce the need to apply biocides to control pests. Similarly, the natural toxicity found in some species can occasionally be refined into a natural biocide for use in agriculture. For example, a

Whether sold in a modern or a traditional pharmacy, many of our medicinal products are based on products found in nature.

Philip Dearden

It is impossible to fully assess the value of natural processes such as pollination.

powerful insect repellent—trans-pulegol—was recently discovered in an endangered member of the mint family.

Other Extrinsic Values

The ecological and economic benefits mentioned above do not encompass all the values associated with protecting other species. How many of us are permanently enriched and emotionally uplifted by a wildlife encounter at some point in our lives? Wildlife contributes to the joy of life, but this joy often translates into a contribution to the economic values attached to wildlife. It is estimated that more than 20 million Canadians spend more than $11 billion participating in nature-related activities, creating approximately 245,000 jobs (NRTEE 2003b). In fact, viewing wildlife is a major reason for travel to some areas, both in Canada and elsewhere. The economic value attached to nature-based tourism can provide a significant impetus to conservation when managed so as to enhance biodiversity and educational values. This form of tourism can also provide a sustainable livelihood for local residents.

Chris Cheadle/All Canada Photos

How much is it worth to see Orca whales surface near your boat?

Ethical Values

Ethical arguments can be made for preserving all species, regardless of their use value to humans. Arguments based on the intrinsic value (value unrelated to human needs or desires) of nature suggest that humans have no right to destroy any species. In fact, humans have a moral responsibility to actively protect species from going extinct as a result of our activities. This philosophy reflects an *ecocentric view*—humans are part of the larger biotic community in which all species' rights to exist are respected. In stark contrast, an *anthropocentric view* of life favours protection of species with an economic and/or ecological use to humans (Chapter 5).

In the past, extinction has been viewed simply as a biological problem. The points raised above emphasize the need to make links among the biological process of extinction and the ethical and economic reasons why extinction is undesirable. However, decisions to protect species and communities often come down to arguments over money—how much will it cost, and how much is it worth? All too often, governments demonstrate a willingness to protect biodiversity only when its loss is perceived to cost money. Unfortunately, standard economic systems tend to undervalue natural resources, and as a consequence, the cost of environmental damage has been ignored and the depletion of natural resources has been disregarded. Ecosystems are being destroyed and species are now being driven to extinction at a rate greater than at any time in the past. An economic system that undervalues natural resources is one of the main underlying causes of extinction. The Millennium Ecosystem Assessment (2005) suggests that the amount of biodiversity remaining on the planet in another 100 years will reflect society's ability to understand and take into account the different values associated with biodiversity conservation (Figure 14.1). The next section reviews some of the main causes of biodiversity loss.

Main Pressures Causing Extinction

Chapter 3 described some human activities that have contributed to the increasing rates of extinction over the past 200 years. This section discusses some of the main pressures on biodiversity in greater detail. Rarely do these pressures act alone; they must therefore be seen as part of the overall stress that human demands are placing on the biosphere.

Humans appropriate more than 40 per cent of the net primary productivity (NPP) of the planet (Chapter 2). With global populations predicted to increase by 70 million people each year over the next 20 years, this figure will only increase. As the amount of NPP increases to support one species—*Homo sapiens*—the amount available to support all other species decreases. The extinction vortex (Figure 14.2) is driven by these human pressures, manifested as habitat loss, overharvesting, pollution, and the introduction of exotic species (Chapter 3). These pressures

The outer circle in the figure represents the present level of global biodiversity. Each inner circle represents the level of biodiversity under different value frameworks. Question marks indicate the uncertainties over where the boundaries exist and therefore the appropriate size of each circle under different value frameworks.

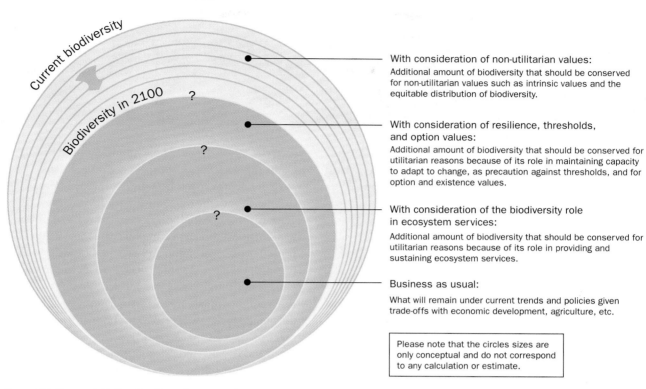

With consideration of non-utilitarian values:
Additional amount of biodiversity that should be conserved for non-utilitarian values such as intrinsic values and the equitable distribution of biodiversity.

With consideration of resilience, thresholds, and option values:
Additional amount of biodiversity that should be conserved for utilitarian reasons because of its role in maintaining capacity to adapt to change, as precaution against thresholds, and for option and existence values.

With consideration of the biodiversity role in ecosystem services:
Additional amount of biodiversity that should be conserved for utilitarian reasons because of its role in providing and sustaining ecosystem services.

Business as usual:
What will remain under current trends and policies given trade-offs with economic development, agriculture, etc.

Please note that the circles sizes are only conceptual and do not correspond to any calculation or estimate.

Figure 14.1 How much biodiversity will remain a century from now under different value frameworks? Source: Millennium Ecosystem Assessment 2005.

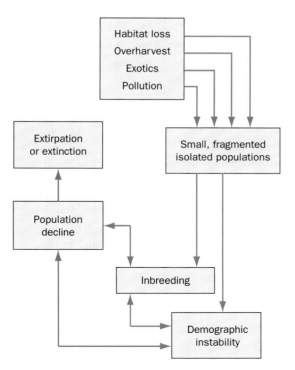

Figure 14.2 Extinction vortex.

result in small, isolated populations that are then vulnerable to inbreeding and demographic instability. Unfortunately, there is a positive feedback loop (Chapter 3) between these factors and population decline. The more the population declines, the greater the impact of these factors becomes, and ultimately this leads to extinction. The fragmented populations of the endangered mountain caribou in BC are a good example of this. As the populations have become progressively reduced and isolated, mainly because of logging, these populations become more vulnerable to extirpation through such factors as bad weather and increased predation, and there are fewer and fewer sites from which repopulation can take place (Apps and McClellan 2006). The vortex eventually spins down to extinction.

Most attention regarding extinction has been devoted to the tropical countries, mainly because they are 'hotspots' for biodiversity and because they are experiencing many of the pressures responsible for increasing rates of extinction. Of the 10 to 15 million terrestrial species thought to be on Earth, up to 90 per cent are estimated to exist in the tropics, particularly in the tropical rain forests. Tropical ecosystems are being degraded and/or destroyed at alarming rates (Box 14.1). Estimates suggest that perhaps 50 per cent of tropical rain forests have already disappeared, and at current rates of destruction, only a few forest fragments will remain in 30 years.

Box 14.1   The Global Toll

Through its Species Survival Commission, the World Conservation Union (IUCN) evaluates and categorizes species according to their degree of endangerment:

- extinct
- extinct in the wild
- critically endangered
- endangered
- vulnerable
- near-threatened
- least concern
- data-deficient
- not evaluated

This so-called **Red List** is produced by thousands of scientific experts and is the best source of knowledge on the status of global biodiversity.

Some highlights from recent lists:

- The Red List includes 16,306 species threatened with extinction, falling into the critically endangered (CR), endangered (EN), or vulnerable (VU) categories.
- Close to half of all primate species (296 out of 683) appear somewhere on the list.
- 8,447 plants were assessed as threatened in 2007, an increase of 2,733 since 2002. However, with only approximately 4 per cent of the world's described plants evaluated, the true percentage of threatened plant species is likely to be much higher. Most of the plant species listed are trees, since they have been relatively thoroughly assessed.
- There are now 776 plant and animal species recorded as extinct and a total of 850 if species that no longer exist in the wild are included. In 2000, the number of species assessed as extinct and extinct in the wild was 811.
- Countries with the most threatened mammals and birds are Indonesia, India, Brazil, China, and Peru. Countries with the most threatened plant species

are Ecuador, Malaysia, China, Indonesia, and Brazil. Countries with the most threatened amphibians are Colombia, Mexico, Ecuador, China, and Peru. Countries with the most threatened fish are the US, Tanzania, Mexico, Indonesia, and Malawi. Countries with the most threatened molluscs are the US, Australia, Portugal, Ecuador, and France. Countries with the most threatened species overall are Ecuador, the US, Malaysia, Indonesia, and Mexico.

- Birds and mammals are increasingly moving towards the higher-threat categories (i.e., more bird and mammal species are entering the critically endangered and endangered categories).
- It is more difficult to assess the trends for reptiles, amphibians, fish, and invertebrates because there are still many information gaps. Within these taxa, however, declines appear to be occurring. For example, a number of amphibian species have shown rapid and unexplained disappearances in Australia, Costa Rica, Panama, and Puerto Rico. There are similar indications that freshwater fish species are suffering serious deterioration, particularly river-dwelling species.
- Habitat loss and degradation affect 89 per cent of all threatened birds, 83 per cent of threatened mammals, and 91 per cent of threatened plants assessed. Habitats with the highest number of threatened mammals and birds are lowland and mountain tropical rain forests. Freshwater habitats are extremely vulnerable, with many threatened fish, reptile, amphibian, and invertebrate species.
- Invasive alien species—a major threat to global biodiversity—are responsible for the loss of island populations of native plants and animals. Hundreds of plant assessments in 2003 from Hawaii, the Falkland Islands (Malvinas), the British Virgin Islands, the Seychelles, Tristan da Cunha, St Helena, and Ascension point to a bleak outlook.

Sources: World Conservation Union 2002; 2003; 2007.

About 7.4 million hectares of tropical forest is destroyed every year. Causes of tropical deforestation include:

- rapidly growing population levels—more people equals less biological diversity, since people use natural resources;
- overconsumption of resources—the rise of industrial capitalism and materialistic modern societies has greatly accelerated demands for natural resources, particularly in developing countries;

- inequality in the distribution of wealth—poor rural people with no land or resources of their own are forced to destroy biological communities and hunt endangered species just to stay alive.

Recent global and national endangered species statistics reveal another major area of concern—aquatic ecosystems (Box 14.2). In both the oceans and freshwater ecosystems, our knowledge of species abundance is very limited. As knowledge

Box 14.2   Aquatic Ecosystems in Trouble

Aquatic ecosystems are not only among our least known ecosystems but also among those most inadequately protected in park systems. You will also see later that many of Canada's most recent extinctions have been from aquatic ecosystems. The following global examples highlight this plight.

- Fifty-six percent of the 252 endemic freshwater Mediterranean fish are threatened with extinction, the highest proportion in any regional freshwater fish assessment so far. Seven species, including carp relatives *Alburnus akili* in Turkey and *Telestes ukliva* in Croatia, are now extinct. Of the 564 dragonfly and damselfly species so far assessed, nearly one in three (174) are threatened, including nearly 40 per cent of endemic Sri Lankan dragonflies.
- In East Africa, human impacts on the freshwater environment threaten more than one in four (28 per cent) freshwater fish species. This could have major commercial and dietary consequences for the region. For example, in Malawi, 70 per cent of animal protein consumed comes from freshwater fish. The lake trout, or Mpasa (*Opsaridium microlepis*), from Lake Malawi is fished heavily during its spawning runs upriver but has suffered a 50 per cent decline in the past 10 years as a result of siltation of its spawning grounds and reduced flows because of water abstraction. It is now listed as endangered.

  Larger freshwater species, such as the common hippopotamus (*Hippopotamus amphibius*), are also in difficulty. One of Africa's best known aquatic icons, it has been listed as threatened for the first time and is now classified as vulnerable, primarily because of

A female scarlet basker dragon fly in Sri Lanka.

Philip Dearden

a catastrophic decline in the Democratic Republic of the Congo (DRC). In 1994, the DRC had the second largest population in Africa—30,000, after Zambia's 40,000—but numbers have plummeted by 95 per cent. The decline is due to unregulated hunting for meat and the ivory of their teeth.
- A key addition to the 2006 Red List of Threatened Species was the first comprehensive regional assessment of selected marine groups. Sharks and rays were among the first marine groups to be systematically assessed, and of the 547 species listed, 20 per cent are threatened with extinction. This confirms suspicions that these mainly slow-growing species are exceptionally susceptible to overfishing and are disappearing at an unprecedented rate across the globe.

Source: http://www.iucn.org/en/news/archive/2006/05/02_pr_red_list_en.htm.

improves, many species have been found to be threatened, and these numbers will continue to increase. Freshwater fish and amphibians now hold the top two spots for animal species threatened worldwide (Table 14.1).

Extinctions also occur in developed and temperate countries. Since the European colonization of North America, more than 500 species and subspecies of native plants and animals have become extinct. Some comparative figures among Canada, North America, and the world are shown in Table 14.1. The Committee on the Status of Endangered Wildlife in Canada (COSEWIC) has identified 13 extinctions and 22 extirpations in Canada since the arrival of Europeans (Table 14.2). Most of the following examples have been chosen to illustrate extinction pressures in Canada.

Overharvesting

There are many examples in Canada of species under pressure due to overharvesting. A couple of historical examples that took place at least partly in Canadian territory are well known.

The great auk. The great auk, a large flightless bird, inhabited the rocky islets of the North Atlantic. For many years, fishers in these waters used the great auk as a source of meat, eggs, and oil. It was reasonably easy to catch and club these birds to death, and once the feathers became an important commodity for stuffing mattresses in the mid-1700s, extinction soon followed. On Funk Island off the east coast of Newfoundland, the species was extirpated by the early 1800s. The

Table 14.1 Number of Threatened Species in Each Major Group of Organisms in Canada, North America (Not Including Mexico), and Worldwide

	Canada	North America	Worldwide
Mammals	17* (14**)	58 (51)	1,094 (1,137)
Birds	18 (8)	93 (64)	1217 (1,192)
Reptiles	3 (2)	35 (29)	422 (293)
Amphibians	1(1)	54 (26)	1808 (157)
Fish	26 (16)	193 (146)	1201 (742)
Molluscs	2 (0)	275 (256)	978 (939)
Other invertebrates	10 (10)	308 (311)	1130 (993)
Plants	1 (1)	243 (170)	8,447 (5,714)
Total	**78 (52)**	**1,259**	**16,306 (11,167)**

Source: World Conservation Union 2002**; 2007*.

Table 14.2 Summary of COSEWIC's Assessment Results for the Risk Categories

	Extinct	Extirpated	Endangered	Threatened	Special concern	Total
Mammals	2	3	21	17	26	69 (64)
Birds	3	2	27	16	23	71 (56)
Reptiles		4	12	14	8	38 (31)
Amphibians		1	7	5	7	20 (19)
Fish	6	4	34	26	41	111 (80)
Lepidopterans		3	5	3	2	13
Arthropods		3		6	4	28
Molluscs	1	2	16	3	4	26 (18)
Plants		2	84	50	32	168 (136)
Mosses	1	1	7	3	4	16 (8)
Lichens			2	1	6	9 (6)
Total	**13 (12)**	**22 (21)**	**225 (153)**	**141 (102)**	**155 (143)**	**556 (431)**

Source: COSEWIC 2003; 2007. http://www.cosewic.gc.ca/rpts/Full_List_Species.html. Note: Figures shown are for 2007, with 2003 figures in parentheses.

Metropolitan Toronto Reference Library

Great auks, painted by John James Audubon.

Metropolitan Toronto Reference Library

A pair of passenger pigeons, painted by John James Audubon.

from their breeding grounds in southeastern Canada and the American northeast to their wintering grounds in the southeastern US. So great were their numbers that tree limbs would break under the pigeons' weight and trees would die as a result of the large amount of guano (bird excrement) deposited. Many eyewitness accounts tell of flocks so huge that they blotted out the sun for hours and sometimes days. Flocks of passenger pigeons were an easy target for hunters, and they were slaughtered in great numbers for food to feed growing urban populations. It is estimated that more than 1 billion birds were killed in Michigan alone in 1869. They were shot, netted, and clubbed into extinction. Hunting took place concurrently with a reduction in their breeding grounds as habitat was converted into agricultural lands. The last passenger pigeon sighted in Canada was at Penetanguishene, ON, in 1902; the last pigeon died in a zoo in Cincinnati in 1914. The world will never again experience the sound and sight of millions of passenger pigeons darkening the heavens.

The Passenger Pigeon

The noise they made, even though still distant, reminded me of a hard gale at sea, passing through the rigging of a close-reefed vessel. As the birds arrived and passed over me, I felt a current of air that surprised me. Thousands of the Pigeons were soon knocked down by the pole-men, while more continued to pour in. . . . The Pigeons, arriving by the thousands, alighted everywhere, one above another, until solid masses were formed on the branches all around. Here and there the perches gave way with a crash under the weight and fell to the ground, destroying hundreds of birds, beneath, and forcing down the loaded. The scene was one of uproar and confusion. I found it quite useless to speak, or even to shout, to those persons nearest me. Even the gun reports were seldom heard, and I was made aware of the firing only by seeing the shooters reloading.

—John James Audubon

last two great auks were clubbed to death off the shores of Iceland in 1844.

The passenger pigeon. There are claims that the passenger pigeon was the most abundant land bird on Earth, totalling up to five billion birds. These great flocks used to migrate annually

This astonishing tale of a species going from such abundance to extinction is not that unusual. Three fish species of the Great Lakes (blue walleye, deepwater cisco, longjaw cisco) were at one time all very abundant, and millions of kilograms of the fish were harvested commercially. By 1950, they were fished into extinction. Similarly, the northern cod, once one of the most abundant fish on the planet, was fished into commercial extinction by the early 1990s. Two populations of Atlantic cod were designated as threatened and endangered by COSEWIC in 2003 (see Chapter 8). Overharvesting is the major cause of endangerment for marine species currently listed by COSEWIC and gives further support for Canada to fulfill its international commitments to establish a network of marine protected areas by 2012, as discussed in Chapter 8.

Perhaps extinction of the passenger pigeon and the great auk are sufficiently in the past that we can excuse their demise on the grounds of a lack of knowledge. However, such a case cannot be made for more recent extinctions. They stand as the ultimate symbol of the failure of resource managers and decision-makers to understand the natural dynamics of the species they are supposed to be managing.

One aspect of overharvesting generally given little consideration is the demand for captive species. In the past, the zeal of zoo collectors to exhibit various species, particularly rare species such as pandas and orangutangs that visitors would pay to see, was of serious concern. International regulations, such as the Convention on International Trade in Endangered Species of Wild Fauna and Flora (CITES), of which Canada is a signatory, now make it difficult for this kind of trade to occur. However, despite international regulations, trade in rare and endangered species still occurs because of demand from private collectors. Exotic species such as tigers, monkeys, parrots,

and tropical fish belong in their native habitats, not in people's homes. Some of the most sought-after Canadian species are falcons, particularly the gyr and peregrine falcons, and they can fetch thousands of dollars each on the international market. Although there is an allowable harvest of wild falcons in some provinces and territories, poaching is a problem because of the high price tags attached to these birds.

Hunting/fishing and the harvesting of live specimens for captivity can have a significant impact on populations. There are, however, more subtle instances of 'non-consumptive' activities having detrimental impacts on species by causing displacement from valuable habitat or even death. For example, research is underway on both the Atlantic and Pacific coasts of Canada to assess the potential impact of whale-watching vessels on the well-being of whales. Such research requires detailed knowledge of a species' natural distribution and behaviour before it can be ascertained whether changes have occurred as a result of disturbance. Impacts associated with the non-consumptive use of natural resources are usually much more difficult to document than the more direct effects of consumptive use. However, researchers have detected costs to killer whales in terms of energy use—for example, from boat traffic in BC (Williams, Lusseau, and Hammond 2006).

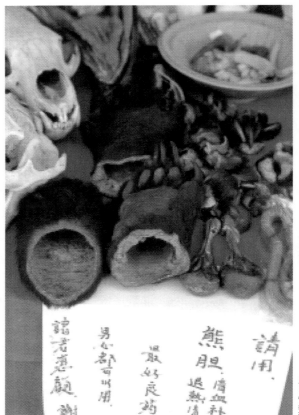

Although habitat destruction is a main cause of biodiversity declines in tropical countries, so is hunting. Species are shot to put on display, as in this hotel foyer in Vietnam, and are also killed for various superstitious and so-called 'medicinal' purposes, such as these bear parts offered for sale in a hill tribe village visited by Taiwanese tourists in northern Thailand.

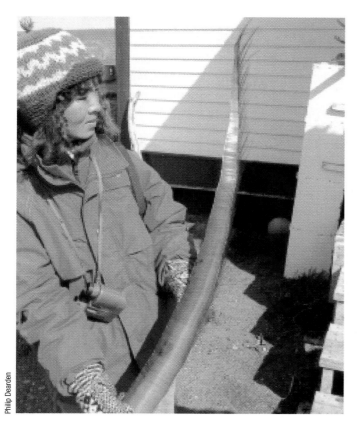

Whales were hunted to the point of extinction around the world because of their commercial value. Baleen, such as what is shown here, was used for such 'indispensable' things as umbrellas, buggy whips, and ladies' corsets.

Predator Control

Several species have been targeted for elimination by humans because they compete directly with humans for consumption of the same resource. One North American example is the Carolina parakeet, the only member of the parrot family native to North America. It was exterminated in the early part of the previous century because of its fondness for fruit crops. Another example is the prairie dog, which was extensively poisoned because of the mortality of horses and cattle after they stepped into prairie dog burrows and broke their legs. This extermination program has been highly successful—prairie dog populations have declined by 99 per cent. Once such a decline occurs, repercussions occur elsewhere in the food chain. In this case, the drastic decline in prairie dogs led to the collapse of their main predator, the black-footed ferret, for which prairie dogs made up more than 90 per cent of their diet. Fortunately, the ferrets have been successfully bred in captivity, an example of ex situ conservation. They have been reintroduced to the wild in places where prairie dogs are protected. The swift fox is another example of predator control and ex situ conservation (Box 14.3).

Predator control is not a thing of the past. For example, deer are sometimes culled to prevent or stop over-browsing of vegetation. Across parts of their North American range, white-tailed deer populations are literally eating themselves out of house and home—deer populations are in excess of what can be supported by the natural resource base. This is due in large part to reductions in range and population declines of their

Box 14.3   Ex Situ Conservation at Work

Not all species subject to heavy pressures are pushed to extinction. Some, such as the beaver, may recover in numbers and start to repopulate their old range. The beaver was able to repopulate with relatively little help. The swift fox, on the other hand, was the target of a 20-year, $20-million reintroduction program, which emphasizes the difficulties and costs associated with trying to reverse extinction trends.

The swift fox is so called because of its ability to run down rabbits and other prey in its home terrain, the dry, short-grass prairie. Speeds of more than 60 kilometres per hour have been recorded. The swift fox is a small fox (about

half the size of a red fox) that at one time roamed all the way from Central America to the southern prairies of Canada. Unlike most other members of the dog family, the swift fox uses dens throughout the year, which are preferably located on well-drained slopes close to a permanent water body. This may be for protection because of their small size. Natural enemies include coyotes and birds of prey such as eagles and red-tailed and rough-legged hawks.

The last swift fox in the wild was spotted in Alberta in 1938. A combination of factors led to its demise, including habitat degradation, overhunting, and predator

Philip Dearden

Hunting of coyotes is allowed in most parts of Canada as part of predator control programs. However, numbers continue to increase in most areas.

control programs. The short-grass prairie came under heavy pressure from cultivation, leading to a loss of habitat for the swift fox and many other species. In addition, the fox was heavily trapped in the mid- and late 1800s for its soft, attractive pelt. The Hudson's Bay Company sold an average of 4,681 pelts per year between 1853 and 1877; by the 1920s, the take had declined to just 500 pelts per year. However, predator control programs against the coyote and wolf finally removed the swift fox from the Canadian Prairies. Predator control programs often relied on extermination methods that were not species-specific, such as poisoned bait and leg traps. As with many species, more than one factor typically drives a species to extinction, and these factors often interact synergistically.

Since 1978, efforts have been made to return the fox to the prairie. Foxes were bred in captivity, and wild populations from the United States were relocated to Alberta. The captive breeding program was initiated by two private citizens, which illustrates the positive impact that individuals can have on environmental issues. There are now more than 600 swift foxes living and breeding in the wild on the Canadian Prairies.

The best strategy for the long-term protection of wildlife species is the preservation of populations in the wild—only in natural communities are species able to continue their process of evolutionary adaptation to a changing environment. Conservation strategies that focus on the organism within its natural habitat are referred to as **in situ preservation**. The Thelon Game Sanctuary in the Northwest Territories, for example, was established in 1927 to help protect the remaining population of muskoxen. Since that time, much of the mainland habitat of the animal has been recolonized by out-migration from this sanctuary. However, in situ preservation may not be a viable option for many rare species, including the swift fox. If remnant populations are too small, on-site preservation strategies will be ineffective. In such cases, the only way to prevent extirpation or extinction is to temporarily maintain individuals in artificial conditions under human supervision. This strategy is known as **ex situ preservation**. Zoos, game farms, aquariums, private breeders, and botanical gardens are all examples of ex situ facilities. The swift fox is an example of ex situ conservation.

Given the scale of change now evident in many species groups, some scientists believe that the only chance for survival for some is through the activities of zoos. One example is Amphibian Ark, which aims to prevent the world's more than 6,000 species of frogs, salamanders, and caecilians from disappearing through ex situ conservation. Scientists estimate that up to 170 species of frogs have become extinct over the past decade through fungal attack and other causes and that an additional 1,900 species are threatened. Amphibian Ark wants zoos, botanical gardens, and aquariums in each country to take in at least 500 frogs from a threatened species to protect them from the killer fungus, which is thought to have originated in Africa. The fungus prevents amphibians from breathing through their pores and has wiped out frog populations from Australia to Costa Rica and the US. However, this is only a stopgap measure to buy time and prevent more species from going extinct while researchers figure out how to keep amphibians from dying off in the wild. Unfortunately, given the spread of alien species described in Chapter 3 combined with the effects of global change, these kinds of catastrophic measures are going to become much more common in the future.

predators, particularly wolves. In stark contrast, other parts of the country, such as Vancouver Island, do not have enough deer, and provincial governments have considered shooting wolves in order to increase the number of deer that humans can kill for sport (Box 14.4). A similar debate is also underway in Newfoundland, where coyotes, which first invaded the island in the 1980s, are now taking significant numbers of caribou.

Box 14.4   Predator Control

The wolf was vilified as a rapacious killer and enemy of humans for centuries. It was shot, poisoned, and extirpated throughout large areas of its range, particularly in parts of the United States, where it became listed as an endangered species. Canada has some of the healthiest wolf populations in the world, numbering around 58,000, and they are being used for reintroductions, such as the recent transfer of wolves from Alberta to Yellowstone National Park. However, wolves are still being shot and poisoned in Canada for predator control programs. Between 2002 and 2007, for example, 41 wolves were killed in the Quesnel region of BC, partly to save the caribou and partly to safeguard the ranching industry. Trappers officially took a further 143 wolves in the winter of 2004–5.

Despite the estimated number of wolves in Canada, local populations are clearly at risk. In Banff National Park, for example, the wolf population is estimated at about 60. Since 1981, 52 wolves have been killed on roads in Banff, Kootenay, and Yoho national parks, while many others have been shot or trapped legally outside park boundaries. At least 13 wolves were shot or trapped legally in close proximity to the boundary of Banff National Park in 1999–2000. A new provincial regulation now requires all hunters and trappers north of the Bow Valley to register their kills. A 10-kilometre buffer zone around the park where wolves cannot be killed has been suggested. This is also the approach taken in Ontario, where in the spring of 2004 the government announced that there would be a permanent moratorium on wolf hunting and trapping in the 39 townships that surround Algonquin Provincial Park. A moratorium was first enacted in 2001, designed to protect the largest remaining population of the eastern wolf. However, a loophole allowed traps to be set if coyotes were the intended target species, which led to the death of several wolves.

The grey wolf has now been removed from the endangered species list in the US, and an unlikely beneficiary will probably be the pronghorn antelope. In areas where wolves became re-established in Montana, Wyoming, and Idaho, pronghorn antelope numbers grew by more than 50 per cent. Research shows that the increase was due to increased calf survival, since wolves find a pronghorn calf too small to be worth the effort. The same is not true for coyotes, which take a high number of the calves. As the wolves move into an area, they displace the coyotes, and the pronghorns thrive. This relationship illustrates how complex predator control can be, as will be seen in the next example.

Another predator cull was carried out on Vancouver Island, the home of Canada's only endemic, endangered mammal, the Vancouver Island marmot. At one time, numbers of the marmot were as low as 30 before ex situ breeding programs were established and began reintroducing animals back into their mountain homelands. Of the 96 marmots released to the wild, 51 are still surviving. Released marmots have also successfully bred in the wild, resulting in five additional litters (14 pups) born in the wild to mixed (wild-born/captive-born) and captive-born parents. There are now potential breeding pairs on 13 mountains—up from three mountains in 2006. However, this highly vulnerable population is threatened by predation. In a study conducted in 2002, six of 18 fitted with radio-transmitter collars were killed by predators—wolves killed four, an eagle killed one, and a cougar killed the other. In 2003, despite government estimates that pointed to a decline in wolf and cougar populations on the island, a cull of up to 30 wolves and 20 cougars was approved. Some environmentalists suggest that if the government was truly concerned about recovering marmot populations, it would advocate an end to clear-cut logging at higher elevations. Non-lethal predator management techniques are now being tested, such as the use of human shepherds, which has shown some success in deterring cougars and wolves but not golden eagles. New release sites are also being tested to see if predation rates vary from site to site.

Do you think that it is acceptable to cull one endangered species if it is threatening the survival of an even more endangered species? The marmot case is not unique; the same problem arises on the west coast with the reintroduced sea otter that is feasting on another endangered species, abalone. What would you do?

The seal hunt off Canada's eastern seaboard is also regaining momentum, partly as a predator control program to protect cod in hopes of reviving the fishing industry (Chapter 8), and a quota of 275,000 was set for the 2009 hunt. Recently, there have been calls in Prince Edward Island and Ontario to kill double-crested cormorants, accused of taking too many fish. As the human population grows and as our demands increase, conflicts over who will consume another organism will escalate. So far, other species appear to be losing the battle.

Habitat Change

Habitat change is *the* most important factor causing biodiversity loss at national and international scales. For **endangered** species in Canada, habitat degradation is responsible for 100 per cent of the listed reptiles, amphibians, invertebrates, and lichens, 99 per cent of the listed plants, 90 per cent of the listed birds, 85 per cent of the listed fish, and 67 per cent of the listed mammals. Habitat loss is the most prevalent threat overall, accounting for 84 per cent of the listings (Venter et al. 2006).

Human demands are causing both physical and chemical changes to the environment. Physical changes such as deforestation *remove* important habitat components, while chemical pollution may *degrade* habitats to the point that they are no longer able to support wildlife even if the physical structure of the habitat remains. Humans place further pressure on species by the introduction of alien species (Chapter 3).

Physical Changes

Some impacts arising from physical changes in the natural environment have already been discussed in the context of forestry and agricultural practices (Chapters 9 and 10). It is difficult historically to separate these influences from the more general impact of colonization in North America. Large areas of forests in central and eastern Canada were cleared to make way for agriculture. Species dependent on such forests, such as the eastern cougar and wolverine, suffered accordingly. Forests are still being replaced by agriculture in some areas. In Saskatchewan, for example, Hobson, Bayne, and Van Wilgenburg (2002) calculated that conversion of the boreal forest between 1966 and 1994 occurred at an annual rate of 0.89 per cent, three times the global average. Venter et al. (2006) found that agriculture, followed by urbanization, was the largest cause of endangerment in Canada, although 70 per cent of listed species are under pressure from more than one source.

Before the mid-1850s, there were some 101 million hectares of long-grass prairie in central North America; now less than 1 per cent remains. Other prairie ecozones have not fared much better, with only 13 per cent of short-grass prairie, 19 per cent of mixed-grass prairie, and 16 per cent of aspen parkland remaining. Millions of bison and antelope once grazed these regions, and the land trembled with their migrations. The bison and the antelope have now been replaced by cattle. Not surprisingly, one-half of Canada's

The Prairies have been virtually completely transformed into an agricultural landscape, with the result that many species native to this habitat are now endangered.

Another form of habitat degradation caused the demise of this pilot whale off the east coast of Newfoundland. It drowned after getting caught in a fish trap.

endangered and **threatened** mammal and bird species are prairie dwellers.

Accompanying the transformation of the natural prairie grassland for agricultural purposes, thousands of hectares of wetlands have been drained to create more agricultural land. Until the early 1990s, the Canadian Wheat Board Act made it financially attractive for farmers to expand cropland instead of managing their land more effectively, and as a result much marginal land was brought under the plough. It is now estimated that more than 70 per cent of prairie wetlands have been drained. Of the remaining wetlands, 60 to 80 per cent of the habitat surrounding the basins is affected by farming practices. Such changes have been a major factor behind declines in the numbers of waterfowl breeding on the Prairies, a trend that is only just being reversed by wildlife management practices (Box 14.5).

Box 14.5   Is Canada's 'Duck Factory' Disappearing?

Millions of ducks, geese, and swans darken the skies every year as they migrate across the length of the continent and back again. This annual migration evokes a sense of wonder and mystery in the more than 60 million North Americans who watch migratory birds each year. But for some, wonder and mystery is accompanied by anxiety over the future status of the 35 species of waterfowl that spend part of each year in Canada. Waterfowl depend on a complex and increasingly vulnerable chain of habitats that extend across international borders. Accelerated conversion and degradation of habitat caused by human activities have led to a series of record low populations of most duck species, as can be seen in the graph below.

Many of the most productive wetlands in Canada have been drained in an effort to bring more land under cultivation. Wetlands in the Prairie provinces are particularly productive. The retreat of glaciers that at one time covered all of Canada left behind significant nutrient deposits, which have formed the basis of highly productive ecosystems. Waterfowl such as mallards and pintails feed on the plants and invertebrates that feed on the nutrients. But as farm intensification has increased, prairie wetlands have diminished in number and extent, making it difficult for ducks to secure adequate food supplies and nesting sites along their long migratory routes.

Unfortunately, habitat loss and degradation are not the only pressures on migrating waterfowl. Duck mortality rates also vary in response to weather, climate, competition for resources, environmental contamination, and hunting. In the Canadian and US prairies, weather has a particularly strong influence on waterfowl breeding habitat conditions and consequently on the abundance of waterfowl populations. According to Environment Canada (2002a), drought in the late 1980s and early 1990s created particularly difficult breeding conditions for ducks. Spring habitat conditions, measured by the number of ponds in May, improved into the late 1990s from the low levels during the drought of the 1980s but have been in decline since 1997. Severe drought in 2002 in the Canadian Prairies resulted in a 48 per cent decline in the estimated number of May ponds compared to 2001. In 2002, pond numbers were 58 per cent below the 10-year and the long-term (1961–2002) averages. As a result of the drought, total duck populations for the Canadian Prairies declined by 33 per cent (to 7.2 million ducks), illustrating the dramatic impact that weather can have on the reproductive potential of waterfowl. However, approximately 13 per cent more ponds were recorded in 2006 compared to 2005, 35 per cent above the long-term (1955–2005) average. The number of breeding ducks observed in 2006

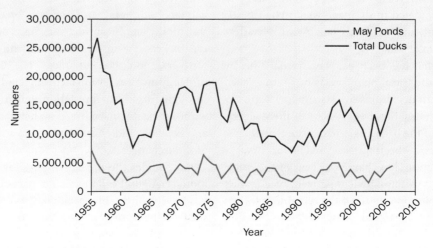

Figure 14.3 Estimates of total ducks and May ponds in southern Prairie Canada, 1955–2006. Source: Environment Canada 2007a.

increased by 19 per cent from 2005 estimates but still remains 1 per cent below the long-term average.

With more than 3 million Canadians and Americans shooting migratory waterfowl each year, hunting also has a significant impact on waterfowl populations. For example, in 2001 more than 5.5 million mallard ducks were killed by hunters. Hunting regulations were introduced decades ago by the Canadian and American governments to protect waterfowl populations, but governments were slow to recognize the impact of land-use practices on waterfowl habitat and therefore abundance. The issue was not formally addressed until 1986, when Canada and the US signed the North American Waterfowl Management Plan (NAWMP) (Mexico joined in 1994). A distinctive feature of the plan was the focus on public-private stewardship initiatives, and it remains one of the most successful examples of this kind of stewardship approach to conservation. The priority goals of the plan were to:

- sustain average waterfowl populations of the 1970s;
- stop further wetland loss;
- stop further loss of native lands, especially native grasslands;
- restore lost wetlands, especially small basins;
- restore the function of upland habitats in landscapes conducive for maintenance of bird populations.

To date, the partnerships have secured 5.4 million acres and invested $641 million. Conservation efforts under the NAWMP include involvement from various stakeholder groups—government agencies at all levels, industry, conservation groups, hunters, farmers, and other landowners. Duck conservation practices include maintaining nesting areas on land close to shallow water for land breeders such as mallards, pintails, teal, gadwalls, wigeons, and shovellors and ensuring that water levels are managed for diving ducks such as redheads and canvasbacks. Encouraging better cropping practices is also important. Since the early 1990s, winter wheat production has increased from less than 81,000 hectares to more than 486,000 hectares in 2006. More than 17,000 landowners deliver habitat conservation programs on their lands. A cooperative program has been set up with the University of Saskatchewan, where researchers are working towards developing hardier winter wheat varieties.

Draining of wetlands is not restricted to the Prairies. Eighty per cent of the wetlands of the Fraser River Delta have been converted to other uses, as have 68 per cent of the wetlands in southern Ontario and 65 per cent of the Atlantic coastal marshes. Drainage not only has a negative impact on marsh-dwelling species but also serves to increase pollutant loads and sediment inputs accumulated in drainage water. High pesticide, fertilizer, sediment, and salt levels may negatively affect organisms further downstream.

Another ecozone that has been particularly hard-hit by habitat destruction is the Carolinian forests of southwestern Ontario. These southern deciduous forests support a greater variety of wildlife than any other ecosystem in the country, including 40 per cent of the breeding birds. More than 90 per cent of this habitat has now been transformed by forestry, agriculture, and urbanization; less than 5 per cent of the original woodland remains. An estimated 40 per cent of Canada's species at risk are in this zone. Most of the remaining forests are in tracts belonging to regional conservation authorities or in privately owned woodlots that are potentially open to logging. Landscapes such as these are particularly suited to stewardship initiatives, discussed in greater detail later in the chapter.

Concern about the loss of habitat for Canadian species extends beyond the Canadian border. The harsh winters and productive summers that characterize much of Canada mean that many species in Canada, especially birds, are migratory. Over the past few decades, significant population declines have occurred for species that spend most of the year in tropical habitats but migrate to Canada to breed. In BC, for example, statistically significant declines have occurred among northern flickers, Swainson's thrushes, chipping sparrows, yellow warblers, and dark-eyed juncos. The reason for these declines probably involves several factors, including loss of winter range through tropical deforestation and increased fragmentation within their northern breeding habitat. More long-term data and detailed studies are required to sort out the complexities of these changes.

Sometimes the impact of physical habitat change can be indirect. A good example is the parasitic habit of the brown-headed cowbird. This species lays its eggs in the nests of other species. The unsuspecting parents often lavish more attention on this large interloper and neglect their own young, leading to their death. The cowbird is an indigenous grassland species, but its distribution has expanded dramatically with human disturbance, since it prefers fragmented habitats and is adept at interloping on the forest-edge nesting sites of other birds. Some of its favourite targets are endangered species, such as the Kirtland's warbler, for which rates of up to 70 per cent parasitism have been recorded.

Physical habitat change has usually focused on aspects of the land or water that a species inhabits. An integral component of this habitat is the climate of the area. Ecological theory, as outlined in Chapter 2, tells us that vegetation growth is primarily controlled by climate in most areas and that animal communities depend on the vegetation for their sustenance. Global climate change will have a dramatic effect on these relationships.

The climate change scenarios for Canada show that we will be one of the most affected countries in the world because of our high latitudes, as discussed in Chapter 7. The northern regions are already showing large-scale reductions in snow and ice cover, reductions in permafrost, coastal inundation, and stressed populations of northern species such as polar bears. Ironically, the US government has suggested listing the polar bear under its Endangered Species Act as a result of thinning ice sheets due to global warming—at the same time that the Bush administration was still denying that global change is happening and refusing to join global treaties to combat it.

Scientists predict that each 1°C rise in temperature will cause biomes to migrate northwards some 300 kilometres. Given the predicted minimum increase of 2 to 5C° in 70 to 100 years, this will translate into 600 to 1,500 metres in elevation and 300 to 750 kilometres in distance. Species must either be able to migrate fast enough to keep up with these changes, evolve to deal with them, or go extinct. Certain biomes, such as Arctic-alpine and the boreal forest, will be very vulnerable to these changes. Parmesan (2006) reviewed 866 scientific studies on the effects of climate change on biodiversity and found that at least 70 species of frogs, mostly mountain dwellers that had nowhere to escape to, have become extinct. A further 100 to 200 cold-dependent species, such as the polar bear, are showing signs of increasing stress. Emperor penguin numbers, for example, dropped from 300 breeding pairs to nine in the western Antarctic Peninsula.

In Canada, many of the great caribou herds have plummeted in numbers. In the High Arctic, the most northerly caribou, Peary's, numbered more than 50,000 in the 1970s and are now down to about 8,000. Unusually warm weather has caused an increase in freezing rains, creating a surface that the caribou cannot penetrate to access the tundra vegetation beneath it, and they starve to death. Although scientists had expected these impacts, they had not anticipated that they would be occurring so rapidly or to such an extent.

Some of the effects might be quite subtle. For example, hybridization is now becoming much more common among species. Climate change has allowed the blue-winged warbler to move further north in Quebec, Ontario, and Manitoba where it not only competes with the endangered golden-winged warbler but has also started to interbreed with it. A similar phenomenon is occurring in the North where a trophy hunter shot what he

thought was a polar bear but turned out to be a hybrid between a polar and a grizzly bear. Scientists theorize that the grizzly father might have wandered farther afield because of warmer temperatures and mated with the female polar bear because there were no female grizzly bears available. Similar cross-overs are happening with bobcats and lynx in New Brunswick.

Hybrids occur in nature, but what happens when hybrids occur much more frequently because of human interference, such as through climate change? Are they to be celebrated as a natural evolutionary adjustment to climate change or despised as the ultimate symbol of human interference with the greatest biological process on Earth, evolution? However, it should be noted that animal hybrids are often infertile. Even if not, they might have trouble finding a mate. Species with small populations can rarely afford the luxury of wasted reproductive effort, so this is particularly troubling for endangered species. Furthermore, interbreeding may lead to genetic swamping of a rare species by a more common one. There are fears that the endangered red wolf of eastern Canada may become genetically swamped by interbreeding with the more common coyote, for example.

Chemical Changes

As the number of chemicals introduced into the environment continues to increase, concern over chemical degradation of habitats intensifies. Pollution is the second most important cause of endangerment for freshwater species in Canada. More than 20,000 chemicals are in use in Canada, and more than 1,000 new chemicals are introduced every year. The effects of chemical pollution are often more difficult to assess than those of physical destruction. Unless there is a catastrophic chemical spill, the signs of declining populations often go unnoticed for several years, even decades. Even after population declines have been documented, it may take many years of careful analysis before a conclusive link to chemical pollution can be established. This was the case with the decline in the numbers of birds at the top trophic level (Box 14.6) after pesticide biomagnification (Chapter 10) led to thinner eggshells and ultimately lower breeding success. Bald eagles, for example, had been persecuted for a long time around the Great Lakes, but it was the total breeding failure due to high chemical levels that led to their extirpation from the Ontario side of Lake Erie by 1980. In 1980, there were only seven nests along the entire Canadian shoreline of the Great Lakes, including Lake Superior, and not one healthy chick was produced.

Bald eagles have now recolonized many areas in the Great Lakes region where they were extirpated as a result of chemical use. There are now at least 136 nests, and the birds are producing strong, healthy chicks. But the birds are often dying young—at 13 to 15 years, less than half their normal natural lifespan. Autopsies completed on dead birds show high levels of lead and mercury contamination. The former is likely persisting in the environment from the time when lead was used in the manufacture of bullets and fishing lures; it should decrease over time. The source of mercury, which is highly toxic,

Box 14.6   Raptors As Indicators of Chemical Degradation of Habitat

Sitting as they do at the top of the food chain, birds of prey or raptors are recognized as powerful indicator species of ecosystem health. Raptors were discovered to be useful indicators of environmental health during the 1960s when research into drastic population declines in bird- and fish-eating species revealed that eggshell-thinning and reproductive failure were caused by organochlorine pesticides. The decline of peregrine falcon populations is particularly well documented.

In Canada, there are three subspecies of peregrine falcon: the continental *anatum* subspecies that breeds south of the tree line from the Atlantic to the Pacific Ocean; the northern *tundrius* subspecies that nests along Arctic rivers, lakes, coastline, and inland escarpments; and the western *pealei* subspecies that occupies coastal islands and areas of adjacent mainland BC. Peregrine falcons are extremely powerful birds of prey, catching other birds in flight while attaining speeds as great as 300 kilometres per hour. Prey is killed by a direct blow of the closed fist delivered at great speed. Favourite prey include songbirds, waterfowl, pigeons, shorebirds and seabirds, and, especially among the Arctic peregrines, small mammals such as lemmings. Falcons nest on cliffs or in trees where they can look down over water bodies. Tall buildings in cities may serve as a substitute, in which case urban pigeons are the main prey. Falcons are quite territorial during the breeding season, with nests seldom closer than one kilometre apart.

The peregrine falcon once bred all across Canada. Populations appeared remarkably stable until the 1940s when populations started to crash, linked to the bioaccumulation of pesticides (Chapter 10) such as DDT, DHC, dieldrin, and heptachlor epoxide (Rowell, Holroyd, and Banasch 2003). Surveys in the 1970s documented the continuing downfall of the peregrine, and by that time the species had been extirpated from large areas of its previous range. In 1978, COSEWIC classified *anatum* peregrines as endangered, *tundrius* as threatened, and *pealei* as rare.

In the late 1980s, urban populations were established in southern Canada through the reintroduction of captive-raised young (an example of ex situ conservation). The program was expanded, and to date more than 700 birds have been released to the wild at more than 20 sites from the Bay of Fundy to southern Alberta and the Okanagan Valley in BC.

Peregrine falcons appear to be on the road to recovery. Nevertheless, chemical habitat degradation is not the only stress threatening the long-term viability of peregrine falcons across Canada. In BC, for example, falcon populations are threatened by declines in the raptor's supply of colonial seabirds due to habitat loss and competition from alien predators such as rats. Alien species are discussed in more detail in a later subsection.

In 1996, the captive breeding station at Wainwright, AB, was closed in the belief that it was no longer needed, and in 1999 COSEWIC downlisted *anatum* peregrines because extinction was no longer a threat. Following a 2005 country-wide survey, none of the three subspecies is now listed by COSEWIC.

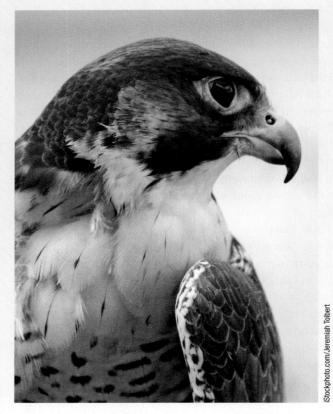

The peregrine falcon suffered badly as a result of the biomagnification of agricultural chemicals but is now recovering in numbers in many areas.

is a mystery. It has now been eliminated from most products in which it was once used, and discharges from human sources are down about 80 per cent. However, the metal is emitted as a by-product of burning coal to produce electricity. Mercury also tends to build up in fish that live in the reservoirs behind power dams, as discussed in Chapter 11 regarding the James Bay Power Project. Scientists are now investigating these sources and their links to bald eagles.

Alien Species

Introduced species can also have a significant impact by outcompeting native species for necessary resources or by direct predation on native species (Chapter 3). The introduction of new species to insular habitats provides some graphic examples of the destruction that can be wrought. On the Queen Charlotte Islands, for example, the introduction of both raccoons and Norway rats is having a catastrophic impact on ground-nesting seabirds. The breeding population of ancient murrelets on Langara Island off the north coast of the Queen Charlottes declined by approximately 40 per cent between 1988 and 1993 to less than 10 per cent of the original size of the population. Further south in the new national park reserve of Gwaii Hanaas, the main ancient murrelet colony on Kunghit Island decreased in size by approximately one-third between 1986 and 1993. In both cases, predation by Norway rats appears to be mainly responsible for the declines.

Aside from direct predation, alien species can affect native species in other ways. For example, populations of the Newfoundland crossbill have declined significantly; competition for food (pine cones) with the introduced red squirrel is believed responsible. However, in contrast to the US, Venter et al. (2006) found that currently, alien species are the least influential causes of endangerment in Canada.

Vulnerability to Extinction

The effects of overhunting and habitat degradation differ among species, since not all species are equally vulnerable to extinction. Ecologists have identified a set of extinction-prone characteristics that help to identify which species are most vulnerable to extinction. Through the identification of such characteristics, conservationists are better able to anticipate the need for protection, and as a result, governments are able to allocate limited resources more effectively. Species with one or more of the following characteristics are more **vulnerable** to extinction.

- *Species with specialized habitats for feeding or breeding.* Once a habitat is altered, the environment may no longer be suitable for specialized species. A good example is the northern spotted owl, discussed in Chapter 9. Northern spotted owls require old-growth habitat for survival and reproduction.
- *Migratory species.* Many songbirds are experiencing population declines in Canada as a result of their long and hazardous migrations to South and Central America.

Species that migrate seasonally depend on two or more distinct habitat types, and if either one of these habitats is damaged, the species may be unable to persist.

- *Species with insular and local distributions.* Dawson's caribou, endemic to the Queen Charlotte Islands, became extinct because of the ease with which it could be hunted in such a restricted habitat, with no hope of an emigrating population for replacement.
- *Valuable species.* Many organisms are overharvested to the point of extinction because of their high economic value. The American ginseng was once abundant in the forests of eastern North America but is now rare because of the demand in Asian countries for dried ginseng roots for medicinal purposes. Similarly, populations of Asian bears have been all but eliminated across their range. Their gall bladders are highly valued in Asian markets, and North American bears are now coming under pressure from the same markets. A single gall bladder can be worth more than $5,000. One illegal dealer in BC was found with 1,125 gall bladders in his possession. BC has now passed a law making possession of endangered animal contraband an offence.
- *Animal species with large body size.* Large animals tend to have large home ranges, require more food, and are more

Species that have high economic values are targeted by hunters and poachers. Here a leopard has fallen victim to a poacher's snare in a park in Sri Lanka, and these cobras in Laos are now part of a local drink.

easily hunted by humans. Top carnivores, for example, depend on the abundance of many different species lower in the food chain. If the numbers of prey species are disrupted, the impacts are felt at the top of the food chain. Furthermore, animals higher up the food chain are more vulnerable to the concentration of toxic materials (Chapter 10). Killer whales are a good example of this vulnerability.

- *Species that need a large home range.* Species that need to forage over a wide area are prone to extinction when part of their range is damaged or fragmented. Grizzly bears, for example, are very sensitive to fragmentation caused by logging and agricultural clearance.
- *Species with only one or a few populations and species in which population size is small.* Any one population may 'blink out' as a result of chance factors (e.g., earthquakes, fire, disease), increasing the species' vulnerability to extinction. Small populations are also more likely to become extinct locally because of their greater vulnerability to demographic and environmental variation. This is why one of the main goals of the Vancouver Island marmot reintroduction has been to start a number of different colonies that are geographically dispersed.
- *Species that are not effective dispersers.* Species unable to adapt to changing environments must migrate to a more suitable habitat or face extinction. Species that cannot migrate quickly have a greater chance of extinction. This

factor will become much more important as the impact of climate change increases.

- *Species with behavioural traits that leave them susceptible.* Some species have behavioural traits that make them particularly vulnerable to clashes with human activities. For example, the red-headed woodpecker flies in front of cars, and the Florida manatee appears to be attracted by motorboats, which are a main cause of death for the animal. Other species, although not attracted by human activities, may be too slow to get out of the way. Tragically, this is the case with the few remaining right whales off the east coast (Box 14.7).

Although some species are more vulnerable to extinction than others, some species also naturally occur at lower population densities than others. The cougar, for example, was once found all across North America but at very low densities because of its need for an area large enough for each individual to secure sufficient food. How do scientists determine whether a species is just naturally rare or declining and in danger of extinction? How small does a population have to be before it is considered endangered? These and similar questions are now answered by scientists on the basis of standardized, quantitative criteria. In Canada, COSEWIC uses criteria based on those suggested by IUCN and used in the global Red List. Table 14.3 shows the criteria for species to be judged endangered or threatened.

Box 14.7 Northern Atlantic Right Whales

At one time, the right whale ranged throughout the northern hemisphere but were extirpated from most areas as a result of hunting this 'right' whale for oil. The largest remaining population summers off the east coast of Canada and winters along the US coast and into the eastern Caribbean. The Roseway Basin, located approximately 30 kilometres south of Cape Sable Island, NS, is one of only two known areas where large numbers of North Atlantic right whales gather on a seasonal basis in Canadian waters. In fact, Roseway Basin, along with the lower Bay of Fundy and three critical habitat areas in United States waters, are the only areas in the western North Atlantic where right whales are known to repeatedly gather on a seasonal basis for several months at a time. The Roseway Basin is an important feeding and socializing habitat for this critically endangered species,

which scientists estimate number only about 400 individuals today. Unfortunately, each of these important habitats is intersected by or located near major shipping routes.

New shipping lanes in the Bay of Fundy, designed to protect the endangered North Atlantic right whale population from ship strikes, were put into operation in 2003. Unfortunately, right whales were still being struck, leading Canada to propose the establishment of a recommended seasonal Area to be Avoided (ATBA) for ships of 300 gross tonnage and upwards during the seven-month period from June 1 to December 31 when the largest percentage of right whales is known to be in the area and when the risk of ship strikes is greatest. The proposal to the International Maritime Organization was adopted and came into effect in the spring of 2008.

Table 14.3 COSEWIC Quantitative Criteria and Guidelines for the Status Assessment

	Endangered	Threatened

A. Declining Total Population

Reduction in population size based on any of the following four options and specifying a–e as appropriate:

	Endangered	Threatened
	≥ 70 %	≥ 50 %

(1) population size reduction that is observed, estimated, inferred, or suspected in the past 10 years or three generations, whichever is longer, where the causes of the reduction are clearly reversible AND understood AND ceased, based on (and specifying) one or more of a–e below.

	Endangered	Threatened
	≥ 50 %	≥ 30 %

(2) population size reduction that is observed, estimated, inferred, or suspected over the past 10 years or three generations, whichever is longer, where the reduction or its causes may not have ceased OR may not be understood OR may not be reversible, based on (and specifying) one or more of a–e below.

(3) population size reduction that is projected or suspected to be met within in the next 10 years or three generations, whichever is longer (up to a maximum of 100 years), based on (and specifying) one or more of b–e below.

(4) population size reduction that is observed, estimated, inferred, projected, or suspected over any 10-year or three-generation period, whichever is longer (up to a maximum of 100 years in the future), where the time period includes both the past and the future AND where the reduction or its causes may not have ceased OR may not be understood OR may not be reversible, based on (and specifying) one or more of a–e below.

   a. direct observation
   b. an index of abundance appropriate for the taxon
   c. a decline in area of occupancy, extent of occurrence, and/or quality of habitat
   d. actual or potential levels of exploitation
   e. the effects of introduced taxa, hybridization, pathogens, pollutants, competitors, or parasites

	Endangered	Threatened

B. Small Distribution and Decline or Fluctuation

	Endangered	Threatened
(1) extent of occurrence	< 5,000 km²	< 20,000 km²

 Or

	Endangered	Threatened
(2) area of occupancy	< 500 km²	< 2,000 km²

For either of the above, specify at least two of a–c.

	Endangered	Threatened
(a) either severely fragmented or known to exist at # locations	≤ 5	≤ 10

(b) continuing decline observed, inferred, or projected in one or more of the following:

   i. extent of occurrence
   ii. area of occupancy
   iii. area, extent, and/or quality of habitat
   iv. number of locations or populations
   v. number of mature individuals

	Endangered	Threatened
(c) extreme fluctuations in one or more of the following:	> 1 order of magnitude	> 1 order of magnitude

   i. extent of occurrence
   ii. area of occupancy
   iii. number of locations or populations
   iv. number of mature individuals

C. Small Total Population Size and Decline

	Endangered	Threatened
Number of mature individuals and one of the following two:	< 2,500	< 10,000
(1) an estimated continuing decline rate of at least:	20% in five years or two generations, whichever is longer (up to a maximum of 100 years in the future)	10% in 10 years or three generations, whichever is longer (up to a maximum of 100 years in the future)

(continued)

	Endangered	Threatened
(2) continuing decline observed, projected, or inferred in numbers of mature individuals and at least one of the following (a–b):		
(a) population structure in the form of one of the following:	(i) no population estimated to contain >250 mature individuals	(i) no population estimated to contain >1,000 mature individuals
	(ii) at least 95% of mature individuals in one population	(ii) all mature individuals are in one population
(b) extreme fluctuations in the number of mature individuals		
D. Very Small Population or Restricted Distribution		
(1) number of mature individuals estimated to be	< 250	< 1,000
Or		
(2) applies only to threatened	Population with a very restricted area of occupancy (area of occupancy typically < 20 km²) or number of locations (typically five or fewer) such that it is prone to the effects of human activities or stochastic events within a very short time period in an uncertain future and thus is capable of becoming highly endangered or even extinct in a very short time period	
E. Quantitative Analysis		
Indicating the probability of extinction in the wild to be at least:	20% in 20 years or five generations, whichever is longer (up to a maximum of 100 years)	10% in 100 years

Source: COSEWIC 2007.

Alive or dead, exotic species do not belong in your home. Do not be tempted to buy such souvenirs because by doing so, you only create a market for their death. These dead hawksbill turtles are hanging outside a souvenir ship in Hanoi. Other common tourist offerings in many locations are seashells. Although some may have washed up on beaches, many species are killed just to supply the tourist trade and will be forbidden, through CITES regulations, from entering another country.

Philip Dearden

Philip Dearden

The extinction of species may now be occurring roughly 100 to 1,000 times faster than the natural rate of extinction. This rate of extinction is much faster than the evolution of new species, so we are in a period in which the world's biological diversity is in decline. What are nations doing to arrest this decline in biodiversity? What are the best strategies for the long-term preservation of biological diversity? In the sections that follow, we will discuss the international and Canadian responses to our biodiversity crisis.

Responses to the Loss of Biodiversity

The International Response

Awareness that we are living in an unprecedented period of mass extinction has led to the formulation of several international conventions and programs. Some programs have a regional orientation, such as the North American Waterfowl Management Plan with the United States and Mexico (Box 14.5), while others include many different nations. Some conventions have a fairly long history. Established in 1973, the **Convention on International Trade in Endangered Species of Wild Fauna and Flora (CITES)** is one of the longest-standing treaties. Ratified by more than 120 countries (including Canada), this treaty establishes lists of species for which international trade is to be controlled or monitored (e.g., orchids, cacti, parrots, large cat species, sea turtles, rhinos, primates). International treaties such as CITES are implemented when a country passes laws to enforce them. Once CITES laws are passed within a country, police, customs inspectors, wildlife officers, and other government agents can arrest and prosecute individuals possessing or trading in CITES-listed species and seize the products or organisms involved. Countries that ratify CITES also vote on the species that should be protected by the treaty.

Several well-known Canadian species are listed under Appendix II of the convention, such as the lynx, bobcat, cougar, polar bear, river otter, and burrowing owl. These species may only be traded with a valid permit from the Canadian Wildlife Service. When travelling through international airports in Canada, you will commonly see information on CITES to warn travellers about trying to import listed plants or animals.

CITES has been instrumental in restricting trade in certain endangered wildlife species. Its most notable success has been a global ban on the ivory trade, instituted in 1989. Without this ban, it is unlikely that any elephants would be left in East Africa today. Despite the treaty's success in protecting some species, Canada's support has been disappointing. According to Le Prestre and Stoett (2001, 197), the Canadian voting record on whether species should be listed has reflected 'domestic economic and/or cultural interests at every turn', and Canada has consistently failed to pay its dues to support the convention.

Other international treaties with a focus on conserving biodiversity include:

- Convention on Conservation of Migratory Species of Wild Animals (the Bonn Convention, 1979);
- Convention on Conservation of Antarctic Marine Living Resources (1982);
- International Convention for the Regulation of Whaling, which established the International Whaling Commission (1946);
- International Convention for the Protection of Birds (1950);
- Benelux Convention on the Hunting and Protection of Birds (1970).

Unfortunately, participation in all of these treaties is voluntary, and countries can withdraw at any time to pursue their own interests when they find the conditions of compliance too arduous. Canada, for example, withdrew from the whaling convention in order to unilaterally permit indigenous whaling in the Arctic. Although Japan still participates, it acts against the spirit of the convention by continuing to hunt whales under scientific pretences.

One of the most important international agreements to protect biodiversity—the **Convention on Biological Diversity (CBD)**—emerged from the World Summit on Sustainable Development held in Rio de Janeiro in 1992 (Chapter 1). Canada was the first industrialized nation to sign the convention and provided extra funding to house the secretariat in Montreal. The CBD requires signatories to develop biodiversity strategies, identify and monitor important components of biodiversity, develop endangered species legislation–protected areas systems, and promote environmentally sound and sustainable development in areas adjacent to protected areas. Canada has been implementing many of the strategies required by the convention over the past decade. The *Canadian Biodiversity Strategy* (Canada 1995) was developed as a direct response to the requirements of the CBD, and other developments, such as the new Species at Risk Act (described below) are also consistent with these requirements.

The Canadian Response

In 1973, the United States became the first country to pass endangered species legislation. Australia followed some 20 years later, and a host of other countries, including the European Union and Japan, have developed similar legislation. In Canada, six provinces have endangered species legislation (New Brunswick, 1974; Quebec, 1989; Manitoba, 1990; Nova Scotia, 1998; Newfoundland and Labrador, 2001; Ontario, 2007). However, this provincial legislation is fairly weak in that many

Box 14.8   Conservation Can Work

Thanks to conservation action, the status of certain species that were red-listed has improved. Some examples include:

- Following large recoveries in many European countries, the numbers of white-tailed eagles doubled in the 1990s, and it has been downlisted from *near threatened* to *least concern*. Enforcement of legislation to protect the species from being killed and protective measures to address threats from habitat changes and pollution have resulted in increasing populations.
- On Australia's Christmas Island, the seabird Abbott's booby was declining as a result of habitat clearance and an introduced invasive alien species, the yellow crazy ant, which had a major impact on the island's ecology. The booby, listed as *critically endangered* in 2004, is recovering thanks to conservation measures and has now moved down a category to *endangered*.
- Other plants and animals highlighted in previous Red List announcements are now the focus of concerted conservation actions, which should lead to an improvement in their conservation status in the near future.
- The 300-kilogram Mekong catfish of Southeast Asia is one of the largest freshwater fish in the world and was listed as *critically endangered* in 2003. Adopted as one of four flagship species by the Mekong Wetlands Biodiversity and Sustainable Use Program, it is the focus of regional cooperation on fisheries management issues and conservation activities, although improvements in status are yet to be seen.
- Swift action since the dramatic 97 per cent population crash of the Indian vulture, listed as *critically endangered* in 2002, means that the future for this and related species is more secure. The veterinary drug that unintentionally poisoned them, diclofenac, is now banned in India . A promising substitute has been found, and captive breeding assurance colonies will be used for a reintroduction program.

Source: http://www.iucn.org/en/news/archive/2006/05/02_pr_red_list_en.htm.

of the provisions to protect endangered species are discretionary. It has taken many years for the federal government to respond to the legislative challenge of protecting endangered species. The **Species at Risk Act (SARA)** was finally passed in 2002. The federal government had little choice; as a signatory to the CBD, it was required to enact such legislation.

One of the reasons that the government was so reluctant to introduce and pass federal endangered species legislation had to do with the Canadian Constitution. Unlike the situation in most other countries, most of the land in Canada is publicly owned, with 71 per cent held by the provinces and 23 per cent held by the federal government (see Chapter 1). Most responsibilities are shared by these two levels of government, with the federal government responsible for oceans and freshwater ecosystems, migratory birds, and the management of federal lands, including the Northwest Territories Nunavut. Yukon now has responsibilities for its own land a similar devolution is underway for the NWT and he federal government also has responsibility for Aboriginal lands south of the 60th parallel, although this will change as land claim negotiations are settled. Therefore, most of Canada's public land, and the resources it contains, are under provincial jurisdiction.

The provinces have numerous individual programs related to nature conservation, and they also participate in joint programs with the federal government. Ontario has passed a comprehensive package that it hopes will help protect endangered species:

- The Endangered Species Act, 16 May 2007. Compared to Ontario's previous legislation, the new Act provides broader protection provisions for species at risk and their habitats, greater support for volunteer stewardship from private landowners and partners, a stronger commitment to recovery of species, and more effective enforcement provisions. Significantly, the Ontario approach avoids many of the criticisms levelled at the new federal legislation by allowing science-based assessments and listings of endangered species rather than allowing politicians to make the final decisions.

- Programs and policies to implement the new legislation.
- Enhanced stewardship programs that support volunteer efforts to protect and restore species at risk and their habitats. The $18-million, four-year Species at Risk Stewardship Fund is part of Ontario's stewardship-first approach to species protection. The fund is open to landowners, farmers, Aboriginal peoples, academic institutions, industries, municipalities, conservation organizations, stewardship councils, and others across the province for eligible protection and recovery activities. To be eligible for funding, a proposal must aim to do one or more of the following:
  - improve the status of species at risk and their habitats through stewardship and recovery activities;
  - encourage involvement in stewardship activities through outreach, education, or youth employment;
  - increase stewardship-related knowledge and skills of interested landowners or groups.

The Canadian Endangered Species Conservation Council (CESCC) comprises the three federal ministers responsible for environment, Canadian heritage, and fisheries and oceans, as well as provincial and territorial ministers responsible for the conservation and management of wildlife. The council coordinates federal, provincial, and territorial government activities related to the protection of species at risk and provides general direction on the activities of COSEWIC and the preparation of recovery strategies and action plans.

Since 1976, COSEWIC has been responsible for determining the status of endangered species. The committee—which includes representatives from relevant federal agencies, provincial and territorial wildlife agencies, and the Aboriginal community, as well eight scientific subcommittees that are species specialist groups—meets annually to consider status reports on candidate species and to assign them to various categories (Table 14.4). By 2007, COSEWIC had assigned 552 species to five risk categories (Table 14.2).

The committee's assessment is the first step in the process for protecting a proposed species at risk under SARA (Box 14.9). However, even if COSEWIC lists a species, this does not guarantee that the species will receive any form of protection. The ultimate decision is in the hands of the politicians who make up the CESCC, and for this reason SARA has been strongly criticized by many who feel that the process should be scientific, not political. The danger of allowing politicians to make such decisions was illustrated in Alberta in 2003. According to the accepted international criterion, a species having less than 1,000 mature breeding pairs should be listed as threatened, which would mean a total of about 2,000. Following this criterion, members of the Endangered Species Conservation Committee in Alberta (appointed by the government) recommended that Alberta's grizzly bear population be listed as a

Table 14.4 Definitions Used by the Committee on the Status of Endangered Wildlife in Canada (COSEWIC)

Not at risk (NAR)	A species that has been evaluated and found to be not at risk
Data-deficient (DD)	A species for which there is insufficient scientific information to support status designation
Delisted	A species not at risk (no designation required)
Special concern (SC)	A species of special concern because of characteristics that make it particularly sensitive to human activities or natural events
Vulnerable (V)	A species particularly at risk because of low or declining numbers, small range, or for some other reason but not a threatened species
Threatened (T)	A species likely to become endangered in Canada if factors threatening its vulnerability are not reversed
Endangered (E)	A species threatened with imminent extinction or extirpation throughout all or a significant portion of its range in Canada
Extirpated (XT)	A species no longer existing in the wild in Canada but occurring elsewhere
Extinct (X)	A species that no longer exists

threatened population. Nonetheless, the minister responsible refused to act on the recommendation, despite the fact that a total population of only 300 to 650 grizzlies remains, with the most recent information suggesting the first number as the most likely.

The influence of politics was clear even when the first new species listings under SARA were made in April 2004, when the fisheries minister decided to delay by nine months a decision on whether to list 12 aquatic species recommended by COSEWIC. The Atlantic cod, for example, is estimated to have a population of less than 1 per cent of historic levels and shows no signs of recovery (Chapter 8), but its listing has not been confirmed. By 2006, 11 per cent of the species suggested for listing under SARA on biological grounds had been rejected on other grounds.

An analysis of the kinds of species that were assessed for inclusion on the list between 2004 and 2006 under SARA found that only five of 29 species of harvested fish and mammals were listed as opposed to 27 out of 29 non-harvested species. Furthermore, species in the North were significantly less likely to

Box 14.9   The Process for Protecting a Species at Risk

1. COSEWIC assesses and classifies a wildlife species as extinct, extirpated, endangered, threatened, of special concern, data-deficient, or not at risk (see Table 14.4). COSEWIC provides its report to the minister of the environment and the Canadian Endangered Species Conservation Council, and a copy is deposited in the Public Registry.

2. Within 90 days, the minister of the environment indicates how he or she intends to respond to a COSEWIC assessment. Within nine months, the government makes a decision about whether or not to add the species to the List of Wildlife Species at Risk. If no government action is taken, the species is automatically added.

3. When a species is on or added to the List of Wildlife Species at Risk, extirpated, endangered, or threatened species and their residences have:
   - immediate protection on federal lands (except for those species in the territories that go through the safety net process described below);
   - immediate protection if it is an aquatic species;
   - immediate protection if it is a migratory bird;
   - protection through a safety net process if it is any other species in a province or territory.

4. For all species included on the List of Wildlife Species at Risk on 5 June 2003:
   - a recovery strategy must be prepared within three years for endangered species and within four

years for threatened species or extirpated species (progress regarding such strategies is discussed below);
   - a management plan must be prepared within five years for a special concern species.

   For all species added to the List of Wildlife Species at Risk after 5 June 2003:
   - a recovery strategy must be prepared within one year for endangered species and within two years for threatened or extirpated species;
   - a management plan must be prepared within three years for a special-concern species.

5. Recovery strategies and action plans, which must include the identification of critical habitat for the species if possible, and management plans are published in the Public Registry. The public has 60 days to comment on these documents.

Five years after a recovery strategy, action plan, or management plan comes into effect, the minister must report on the implementation and the progress towards meeting objectives.

Source: Canada. 2004. 'Species at Risk Act'. www.sararegistry.gc.ca/background/process_e.cfm.

---

be listed than species in southern Canada. Mooers et al. (2007) suggest that these patterns result from the lack of capacity or willingness on the part of certain agencies, particularly the Department of Fisheries and Oceans and wildlife management boards in the North, to accept the additional stewardship responsibilities required by SARA. Delays for further con-'tation with management boards may lead to the extinction ˥e species, such as the eastern beluga. A second reason is ˋ suggested species undergoes a cost-benefit analysis ˥y little attention is given to the benefits whereas

the costs are studied in detail. Furthermore, these analyses are not open to peer review. In 2005, the Commissioner of the Environment and Sustainable Development undertook an audit and recommended that Environment Canada develop a decision-making process by 2006 on the listing of species at risk that would be consistent, systematic, and transparent. A subsequent audit in 2008 found that such a listing process had yet to be established (Auditor General of Canada 2008).

For species listed under SARA, recovery and management plans must be developed and implemented, unless the minister responsible feels that recovery is not 'feasible', a caveat that provides another political opportunity to block action. Under the Act, proposed recovery strategies allow for a 60-day comment period during which any person may file written comments with the minister responsible. Within 30 days of the closing of the public comment period, the proposed recovery strategy must then be finalized. Recovery strategies are evaluated every five years and updated as necessary.

In 2007, COSEWIC had 366 species listed as endangered or threatened. 'Recovery strategies' are to be devised for these species, complete with 'action plans' that will be subject to a cost-benefit analysis. Sixty-one recovery strategies had been approved by 2008, while 147 are officially in violation of SARA because they have exceeded the time periods noted above. These delays appear to be a direct consequence of a budget freeze on the Canadian Wildlife Service applied by the Conservative government in 2007.

The Act has been further criticized on the grounds that even when a species is listed, it receives automatic protection only on federal lands. In southern Canada, where many endangered species live, a significant proportion of the federal lands are national parks in which the species are already protected. Therefore, there is no incremental gain in protection unless provincial jurisdictions agree to provide it. There is a so-called 'safety net' whereby the federal government can invoke powers to act if a provincial government refuses to do so and if the case is seen as critical, but many observers feel that the federal government is unlikely to take this course of action, given the history of federal-provincial relations.

The Species at Risk Act is quite different from the American approach to endangered species protection in that it lays out a framework for cooperation on the protection of endangered species and relies primarily on volunteerism. The federal government has adopted the same approach for habitat protection, establishing the Habitat Stewardship Program to provide information to landowners on how best to manage their lands to protect endangered species. Landowners may receive compensation for any economic losses they might incur. Warnock and Skeel (2004), in their evaluation of Operation Burrowing Owl in Saskatchewan, present evidence that some voluntary habitat stewardship programs can be effective.

Only time will tell whether the Canadian approach will be successful in protecting endangered species in Canada. If there is political interference at any of the numerous opportunities available, if the provinces disagree with the federal government, or if cost-benefit analysis indicates that it is less expensive to let a species become extinct, we will have failed in this experiment. The evidence already suggests that all three are occurring. The costs of such failure will be extremely high. Fortunately, endangered species legislation is not the only means of protecting biodiversity (Box 14.10). However, in its review of progress made by the federal government on protecting endangered species, the status report of the Commissioner of the Environment and Sustainable Development,

released in March 2008, concluded that 'the federal government as a whole has made unsatisfactory progress in responding to our 2001 recommendations relating to the development of a comprehensive inventory of species at risk and of recovery strategies.' This further underlines one of the key messages

Environment Canada and Fisheries and Oceans Canada have made unsatisfactory progress in responding to our 2001 recommendation relating to the development of a comprehensive inventory of species at risk, while Parks Canada has made satisfactory progress on this recommendation.

- *The three organizations have made unsatisfactory progress in responding to our 2001 recommendation relating to the development of recovery strategies and have not complied with specific deadline requirements established by the Species at Risk Act. As of June 2007, recovery strategies should have been completed for 228 species at risk, but recovery strategies completed at that date address only 55 of those species.*
- *Departments and organizations are also required under the Act to identify, to the extent possible, critical habitat necessary for the survival or recovery of species at risk. As of June 2007, critical habitat had been identified for 16 of the 228 species at risk for which recovery strategies were due.*
- *Despite the progress noted at Parks Canada, the federal government as a whole has made unsatisfactory progress in responding to our 2001 recommendations relating to the development of a comprehensive inventory of species at risk and of recovery strategies. While work is underway to develop appropriate data-sharing agreements with third parties, such as provincial and territorial governments, and non-governmental organizations such as Nature Serve, inventory data collections vary across Canada. Ongoing improvements to data quality and data consistency are needed.*

—Auditor General of Canada 2008

contained in the Millennium Ecosystem Assessment (2005) that 'Science can help ensure that decisions are made with the best available information, but ultimately the future of biodiversity will be determined by society.'

Protected Areas

Protected areas have emerged as one of the key strategies to combat the erosion of biodiversity both internationally and in Canada (Box 14.11). **Protected areas** play different roles in society (Box 14.12), and in many cases their conservation role has been recognized only recently as the dominant one. This is especially true in Canada, where Banff, our first national park (1885), was set aside mainly to promote tourism and generate income rather than to protect species and ecosystems. However, since that time, the crucial role in species and ecosystem protection played by protected areas in Canada has led to both legislation and policy directives that make it clear that biodiversity protection is the prime mandate for the national park system.

Overall, Canada's national park system is central to the protection of rare and endangered species. Although the 42 parks cover only about 3 per cent of the land base, they contain more than 70 per cent of the native terrestrial and freshwater vascular plants and 80 per cent of the native vertebrate species. More than 50 per cent of the endangered vascular plant species and almost 50 per cent of the endangered vertebrate species are found in national parks. Of the species considered at risk in

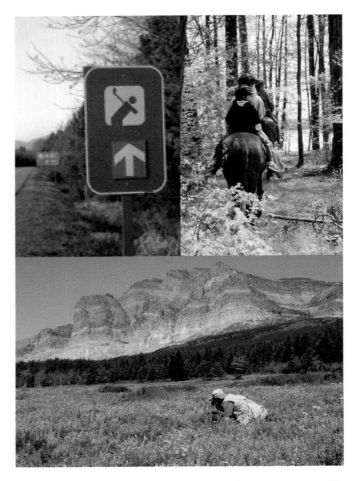

The 'playground' role is one of the most controversial issues for park managers. This Parks Canada golf course sign in Pacific Rim National Park, for example, is advertising a facility that is not actually located on park lands. Nonetheless, by advertising it, a link is created in the public mind between national parks and certain recreational activities. What about traditional activities in some parks, such as horse-back riding, which can have a significant impact on the environment, or are national parks places where we should rely on our own feet to explore the beauty of the landscape? (Top left: Philip Dearden; Top right: iStockphoto.com/Cathleen Clapper; Bottom: iStockphoto.com/Peter Blanchard)

Box 14.10 What You Can Do for Endangered Species

Although the challenges created by endangered species can seem quite daunting to the individual, you can do several things.

1. If you own land, even your own backyard, try to encourage the growth of native species and promote high diversity among these species. Provide the three staples—food, water, and shelter. Plant perennials ~ch as fruit and nut trees, nectar-producing flowers, ̣berry bushes. Don't use chemicals!

2. Write letters to politicians at all levels encouraging them to adopt specific measures. For example, write to local politicians urging protection for a natural habitat in your area.

3. Join and support an environmental group that takes a special interest in endangered species.

4. Take part in an active biodiversity monitoring project such as the Christmas Bird Count, the Canadian Lakes Loon Survey, Frogwatch, Project FeederWatch, or one of the other many organized activities that

take place across the country. Details on these projects are available from local NGOs and university and college departments.

5. Don't keep exotic pets.

6. If you have a pet, try to make sure that it does not injure or harass wildlife. Put a bell on your cat. Domestic cats kill large numbers of songbirds every year.

7. Don't buy products made of endangered animals or plants.

8. Vote for political candidates who share your views on conservation matters.

9. Keep informed of biodiversity issues by watching nature programs on television, reading books, attending public lectures, and having discussions with local conservationists.

10. Actively try to learn more about wildlife, not just by reading and watching television but also by becoming more aware of the wildlife in your region through field observation. Encourage others, especially children, to do likewise.

Box 14.11 Protected Areas: A Global Perspective

From the origins of the conservation movement in the US, the idea of having areas protected by government for conservation and public benefit, education, and enjoyment has spread throughout the world. Implementation has differed to reflect local conditions, but one ubiquitous concern for managers from Canada to England to Southeast Asia is the relationship between protected areas and local populations. As human populations grow, so do pressures for increased use of protected areas. In the UK, this overuse might be mainly recreational, and significant biophysical impacts may result just from the sheer numbers of people enjoying the parks. In many tropical areas, conflicts arise as local people, often driven by poverty and land-use pressures, encroach on the parks in large numbers, hunting wild animals and cutting down trees to make way

for agriculture, obtain firewood, and/or sell on international markets. Estimates suggest that more than half a million people are illegally occupying national parks and wildlife sanctuaries in Thailand, with almost half the area of some parks suffering environmental degradation.

These kinds of management problems are challenging. There is little point in trying to manage the area of land officially designated as a protected area if, in fact, it is not protected from resource use. In the past in Thailand, management activities focused on a preventive approach, with armed guards patrolling boundaries. Since most large remaining areas of forest and most wild animal populations are within the protected area system, there have been some benefits to this approach. Nevertheless, it is clear that large-scale poaching continues in many areas, and shoot-outs between poachers and

Philip Dearden

Park wardens in Thailand receive little pay and risk their lives to protect what remains of the wildlife. Every year lives are lost in battles with poachers.

park guards are not an ideal management tool. Attention therefore has also spread to trying to address some of the underlying motives behind poaching, such as poverty, although here too there are substantial challenges. Economic development programs initiated in some villages have triggered an increase in land prices, leading some villagers to sell their lands and encroach further into the park lands. Unscrupulous local leaders may also encourage villagers to sell so that they can gain control over more land.

As with many environmental management problems, the answer does not lie in one single solution. Each case is different, and an adaptive management approach (see Chapter 6) to the protected area ecosystem is essential. In the long term, education must play a lead role. Many people are unaware of the vital functions played by protected areas. It is better to achieve voluntary compliance with more flexible management regimes than to have armed standoffs and mass non-compliance, as has often occurred in the past.

Box 14.12 The Many Roles of Protected Areas

Art gallery: Many parks were designated for their scenic beauty, and this is still a major reason why people visit parks.

Zoo: As one component of the art gallery, parks are usually places to view wildlife easily in relatively natural surroundings. Because it is protected from hunting in most parks, the wildlife is not as shy of humans as wildlife outside parks.

Playground: Parks provide excellent recreational settings for many outdoor pursuits.

Movie theatre: Just like a movie, parks can lift us into a setting different from that of our everyday life.

Cathedral: Many people derive spiritual fulfillment from communing with nature, just as others go to human-built places of worship—churches, temples, and mosques.

Factory: The first national parks in Canada were designated with the idea of generating income through tourism. Since these early beginnings, the economic role of parks has been recognized, although it is a controversial one because of potential conflict with most of the other roles.

Museum: In the absence of development, parks serve as museums, reminding us of how landscapes might have looked to early settlers. These museums also perform a valuable ecological function, since they encompass important areas against which ecological change in the rest of the landscape can be measured.

Bank: Parks are places in which we store and protect our ecological capital, including threatened and endangered species. We can use the 'interest' from these 'accounts' to repopulate areas with species that have disappeared.

Hospital: Ecosystems are not static and isolated phenomena but are linked to support processes all over the planet. Protected areas constitute one of the few places where such processes still operate in a relatively natural manner. As such, they may be considered ecosystem 'hospitals' where air is purified, carbon stored, oxygen produced, and ecosystems 'recreated'.

Laboratory: As relatively natural landscapes, parks represent outdoor laboratories for scientists to use in unravelling the mysteries of nature. Killarney Provincial Park in Ontario, for example, was an important laboratory for early research on acidic precipitation in Canada.

Schoolroom: Parks can play a major role in education as outdoor classrooms.

Source: Dearden 1995.

Table 14.5 Examples of Reintroductions of Endangered Species into Canadian National Parks

Species	Park
American beaver	Cape Breton Highlands National Park, Prince Edward Island National Park
American bison	Prince Albert National Park, Riding Mountain National Park
Plains bison	Elk Island National Park
Wood bison	Nahanni National Park Reserve, Jasper National Park, Waterton Lakes National Park, Elk Island National Park
Fisher	Georgian Bay Islands National Park, Riding Mountain National Park, Elk Island National Park
American marten	Fundy National Park, Kejimkujik National Park, Terra Nova National Park, Riding Mountain National Park
Moose	Cape Breton Highlands National Park
Muskox	Ivvavik National Park
Trumpeter swan	Elk Island National Park
Caribou	Cape Breton Highlands National Park
Swift fox	Grassland National Park

Source: Dearden 2001.

Canada, 135 can be found in national parks. Canada's national parks have also played a critical and increasingly important role as sites for reintroduction of endangered species (Table 14.5).

Protected Areas: A Global Perspective

The World Conservation Union (IUCN) is an international body that draws together governments, non-governmental organizations, and scientists concerned with nature conservation. The IUCN helps to provide leadership and set global standards for conservation. A protected area, as defined by the IUCN, is 'a clearly defined geographical space, recognised, dedicated, and managed, through legal or other effective means, to achieve the long-term conservation of nature with associated ecosystem services and cultural values'. There are many different kinds of protected areas, such as national and provincial parks, wilderness areas, and biosphere reserves. They all offer some form of protection but with differing degrees of stringency. To help bring some order and understanding to the different types of protected areas, the IUCN has developed a system of classification that ranges from minimal to more intensive use of the habitat by humans (Table 14.6).

The growth of protected areas has been strong, especially over the past decade (Figure 14.4). In 1994, approximately 8,600 protected areas had been designated worldwide, covering some 8 million square kilometres; by 2007, 107,034 protected areas had been designated, covering more than 19.5 million square kilometres. This is equivalent to the combined area of Southeast and South Asia and China and represents almost 13 per cent of the Earth's land surface. Several factors explain the rapid growth in protected area establishment:

1. *Increased realization of the rate of biodiversity loss and the severity of the issue.* In 1990, for example, statistics on endangered wildlife in Canada listed 194 species at risk, compared to 552 in 2007.
2. *Growing awareness at the political level of the links between environmental and societal health.* When ecosystems collapse, livelihoods and economies collapse too. This interdependency between ecosystem protection and poverty is now recognized by many international development agencies and has helped to spur the worldwide interest in protected area establishment. For example, the Global Environmental Facility (GEF), an international agency formed in 1991 to assist developing countries in undertaking activities that benefit the global environment, provided approximately $2.2 billion in grants and leveraged about $5.17 billion in co-financing in support of more than 750 biodiversity projects in 155 countries between 1991 and 2006.
3. *Realization of the value of ecosystem services.* Increased methodological sophistication has also allowed monetary values to be placed on ecosystem values. One team of researchers put an average price tag of US $33 trillion a year on fundamental ecosystem services such as nutrient cycling, soil formation, and climate regulation. This figure is nearly twice the annual global GDP of US $18 trillion and demonstrates, in economically understandable terms, the value of the so-called 'free' services of functioning environments (Costanza et al. 1997).
4. *Growing evidence of the effectiveness of protected areas in helping to combat environmental degradation.* A study by Bruner et al. (2001, 126), based on a survey of tropical countries, found that parks were 'surprisingly effective at protecting ecosystems and species within their borders in the context of chronic underfunding and significant land use pressures'. Protected areas are a main cornerstone for biodiversity protection under international treaties, such as the Convention on Biological Diversity, and a main recommendation of the Millennium Ecosystem Assessment (2005).

Although major gains have been seen in terms of terrestrial park systems, greater progress is required in the marine realm. Less than 0.5 per cent of freshwater and oceanic ecosystems enjoy any effective protection (Chapter 8).

Protected Areas: A Canadian Perspective

Canada has a large variety of protected areas, ranging from small ecological reserves to vast multiple-use areas, and the growth in protected areas has been strong (Figure 14.4). These

Table 14.6 IUCN Classification of Protected Areas

Category	Description
Ia: Strict Nature Reserve (protected area managed mainly for science)	Ecosystems, geological or physiological features, and/or species, available primarily for scientific research and/or environmental monitoring
Ib: Wilderness Area (protected area managed mainly for wilderness protection)	Large area of unmodified or slightly modified land and/or sea, retaining its natural character and influence, without permanent or significant human habitation, which is protected and managed so as to preserve its natural condition
II: National Park (protected area managed mainly for ecosystem protection and recreation)	Natural area of land and/or sea, designated to (a) protect the ecological integrity of one or more ecosystems for present and future generations, (b) exclude exploitation or occupation inimical to the purposes of designation of the area, and (c) provide a foundation for spiritual, scientific, educational, recreational, and visitor opportunities, all of which must be environmentally and culturally compatible
III: Natural Monument (protected area managed mainly for conservation of specific natural features)	Area containing one or more specific natural or natural/cultural feature, which is of outstanding or unique value because of its inherent rarity, representative or aesthetic qualities, or cultural significance
IV: Habitat/Species Management Area (protected area managed mainly for conservation through management intervention)	Area of land and/or sea subject to active intervention for management purposes so as to ensure the maintenance of habitats and/or to meet the requirements of specific species
V: Protected Landscape/Seascape (protected area managed primarily for landscape/seascape conservation and recreation)	Areas of land, with coast and sea as appropriate, where the interaction of people and nature over time has produced an area of distinct character with significant aesthetic, ecological, and/or cultural value and often with high biological diversity. Safeguarding the integrity of this traditional interaction is vital to the protection, maintenance, and evolution of such an area
VI: Managed Resource Protected Area (protected area managed mainly for the sustainable use of natural ecosystems)	Area containing predominately unmodified natural systems, managed to ensure long-term protection and maintenance of biological diversity, while providing at the same time a sustainable flow of natural products and services to meet community needs

Source: World Conservation Union (IUCN—International Union for the Conservation of Nature and Natural Resources). 2001. Protected Area Management Categories. http://www.unep-wemc.org/protected_areas/categories. Reprinted by permission of the publisher.

*Excludes 43,500 sites with no year of establishment.

Figure 14.4 Growth in nationally designated protected areas, 1872–2006. Source: UNEP World Conservation Monitoring Centre. http://www.unep_wmc.org/wdpa/PA_growth_chart_2007.gif.

Figure 14.5 Canadian protected area growth, 1885–2005. Source: Environment Canada 2006b.

areas are protected by a wide range of authorities, from municipal to federal and even international agencies. The most important protected areas are in our national and provincial park systems, and the total area of land protected overall by these two levels of government is about equal. The Parks Canada Agency is the main federal agency in charge of federal protected areas, such as national parks, although Environment Canada also has jurisdiction over large areas, such as migratory bird sanctuaries and national wildlife areas, that are set aside primarily for wildlife. The location of national parks is strongly influenced by Parks Canada's national park **system plan**, which divides the country into 39 physiographic regions that are representative of Canada's natural heritage. The goal of the national park system is to have at least one national park in each of these regions (Figure 14.5). All provinces have now adopted similar system plans, and the overall total amounts to 486 natural regions across the country. Malcolm (2008) discusses the various provincial park systems in more detail.

In 1969, the federal minister in charge of parks, Jean Chrétien, pronounced a goal of achieving system completion by 1985. However, by that date the system was less than half complete. In 1992, Canada's federal, provincial, and territorial ministers of environment, parks, and wildlife signed a Statement of Commitment to Complete Canada's Network of Protected Areas. Terrestrial systems were to be completed by 2000, whereas marine designation was to be 'accelerated'. The goal was again not met, and in 2002 then–prime minister Jean Chrétien announced a five-year plan to establish 10 new terrestrial parks, increasing representation to 35 of the 39 regions, and five marine protected areas under the 29-region marine system plan.

Canada is still far from meeting these commitments. Parks Canada has fallen behind in reaching the goal (Figure 14.6), with only 28 terrestrial and two marine regions represented. However, there has been major progress at the provincial level. In 1968,

Ontario had 90 regulated protected areas totalling 1.6 per cent of the province by area. Currently, Ontario has 631 protected areas (329 provincial parks, 292 conservation reserves, 10 wilderness areas) totalling more than 9.4 million hectares, or 8.7 per cent of the province. In BC, the area of parkland doubled between 1977 and 2005 and now totals more than 12 million hectares (Figure 14.6). BC is the only jurisdiction to accomplish the 12 per cent target set by the World Commission on Environment and Development (WCED 1987). However, in terms of overall protection of Canada's 177 ecoregions, only 29 per cent enjoy a high level of protection (i.e., more than 12 per cent of their area), 12.4 per cent have moderate protection (6 to 12 per cent) and 41.9 per cent low protection (<6 per cent), and 16.6 per cent have no protected areas (Environment Canada 2006b). Less than 0.5 per cent of Canada's marine area is set aside in protective designation, and Canada ranks 70th globally in terms of the percentage of oceans protected (Environment Canada 2006b). Projections suggest that Canada will achieve perhaps 33 per cent of its international requirement for marine protection by 2012.

Environment Canada also has responsibilities for protected areas; specifically, the agency manages 51 national wildlife areas and 92 migratory bird sanctuaries. These areas were established to protect significant habitat for wildlife, including species at risk and migratory birds. The performance of the agency was audited by the auditor general in 2001, with a follow-up audit in 2008. The audit found that:

> Environment Canada has made unsatisfactory progress in responding to our recommendations on national wildlife areas and migratory bird sanctuaries. These areas are at risk.

- Environment Canada has identified specific threats to each of its protected areas, but the Department

Figure 14.6 Canada's national park system plan and state of completion as of 2008. Source: Parks Canada Agency. 2008. http://www.pc.gc.ca/progs/np-pn/cnpn-cnpn/itm1-/index_e.asp.

has not assessed whether conditions are improving or deteriorating at the sites, nor used the information collected to address threats on a priority basis.

- Environment Canada has developed a national strategy to guide the management of sites in its protected areas network, but the strategy is not being fully implemented. For example, most protected areas still lack up-to-date management plans.
- Environment Canada has not established explicit performance expectations against which progress can be assessed and does not comprehensively monitor or regularly report on the condition and management of its network of protected areas.
- According to its own analyses, Environment Canada has allocated insufficient human and financial resources to address urgent needs or activities related to the maintenance of sites and enforcement of regulations in protected areas (Auditor General of Canada 2008).

About 10 per cent of Canada's terrestrial area has been awarded protective designation, short of the international goal of 12 per cent first suggested by the WCED (1987) and well short of the average 14.6 per cent protected by OECD countries (Environment Canada 2006b). Canada manages 5.1 per cent of the world's terrestrial protected area estate. However, 95 per cent of Canada's terrestrial protected areas fall within IUCN categories I–IV (Table 14.6) and hence have a strong protective mandate. Among OECD countries, Canada ranks 16th out of 30 in terms of the proportion of land protected. The US protects almost 25 per cent compared to our 10 per cent yet ranks fourth in terms of proportion of land with strong protection (IUCN categories I–IV). Furthermore, two-thirds of Canada's protected area is situated within a small number of sites that are at least 300,000 hectares in size. Few countries have the opportunity to preserve such large intact landscapes.

Although Canada still has a distance to go to attain the 12 per cent protection goal, the quality of Canadian protected

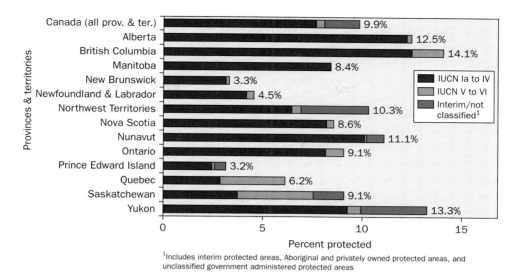

Figure 14.7 Percentage of land in protected areas in each province and territory. Source: Environment Canada 2006b.

areas is very high, with several national parks appearing on the World Heritage list. These sites are judged to be of national and global significance. High international accolades depend on first-rate park management practices. The next section identifies some of the management challenges facing protected areas in Canada. Most of the section is devoted to national parks, but many of the challenges also apply to provincial parks.

Park Management Challenges

Park management is guided by legislation and relevant policies. Park management plans articulate how these requirements will be translated into on-the-ground activities in different parks. However, a national survey by Environment Canada (2006b) on the status of Canada's protected areas found that only 25 per cent had management plans in place. Furthermore, although most park systems in Canada now recognize ecological integrity as their main purpose, only two jurisdictions (Parks Canada and Ontario) have measures in place to monitor changing conditions. Ecosystems are considered to have integrity when they have their native components and processes in place. Several factors make park management a challenging process.

Development within the Parks

Management in national parks is determined by the National Parks Act. Although the first national park was created in Banff, AB, in 1885 (Rocky Mountain National Park, as it was then known), the first National Parks Act was not passed until 1930. Both this Act and earlier legislation dedicated the parks to 'the people of Canada for their benefit, education and enjoyment . . . such Parks shall be maintained and made use of so as to leave them unimpaired for the enjoyment of future generations.' Reconciling the balance between 'making use of' and maintaining the parks 'unimpaired' has been a major topic of debate ever since.

Significance of Ecological Integrity in Protected Areas

The most significant change to the government of Ontario's protected areas is that ecological integrity is now the guiding purpose for planning and management. Subsection 3(1) of the statute [Provincial Parks and Conservation Reserves Act, 2006] states that the 'maintenance of ecological integrity shall be the first priority and the restoration of ecological integrity shall be considered' for all provincial parks and conservation reserves.

—Wilkinson 2008, 182

Tourism and income generation were the main reasons behind the establishment of many parks, including Banff. Thus, catering to the demands of tourists was the most important management priority, and those involved with providing services for tourists were very influential. This is identified as the 'entrepreneurial' period in Figure 14.8. However, by the early 1920s, development voices were not the only ones being heard. James Harkin, the chief commissioner for national parks, was well aware of the economic value of tourism, but he pressed management for a greater sense of nature protection.

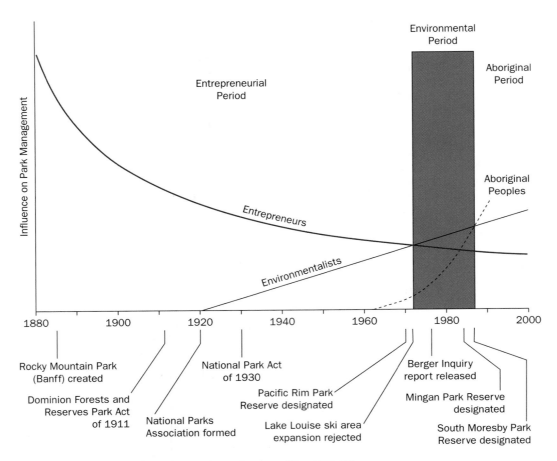

Figure 14.8 Administrative penetration model. Source: Dearden and Berg 1993, 198.

By the 1960s, the number of visitors to Canada's national parks had increased tremendously, as had developments to serve them, including ski hills, golf courses, roads, and hotels. A massive proposal to expand the Lake Louise ski area was rejected in the early 1970s, signifying that the environmental movement was at last starting to be heard in the parks (Figure 14.8). The next 40 years witnessed many debates on the issue of controlling development in protected areas. Although the Lake Louise expansion was thwarted, smaller developments permeated the parks. In 1988, the balance between development and protection was clarified in amendments to the National Parks Act. Protection of ecological integrity became the prime mandate of Canada's national parks.

Despite this legislative mandate, development pressures continued. For example, the level of development at Banff townsite escalated rapidly throughout the 1980s. Shopping space almost doubled between 1986 and 1994, compared to increases of between 10 and 15 per cent for the rest of the country. The amount of shopping space per person in Banff is three times larger than it is in Toronto. The development in Banff is greater than all the development in Yellowstone, Grand Canyon, Yosemite, and Great Smoky Mountain national parks in the US combined. Banff National Park is also the core of the four mountain parks (Banff, Yoho, Jasper, and Kootenay) that

were accorded World Heritage status in 1985 because of their importance for wildlife protection on a continental basis.

Although the park itself is large, most of it is rock and ice; the town of Banff is situated in one of the most productive and rare habitats found in the Rocky Mountains. The 'montane' zone, which covers between 2 and 5 per cent of the park, is characterized by frequent chinook winds, low snow

Gros Morne National Park in western Newfoundland is a World Heritage Site.

accumulation, and warm winter temperatures. These conditions permit a dense concentration of wildlife and critical winter refuge areas.

I do not suppose in any portion of the world there can be found a spot, taken all together, which combines so many attractions and which promises in as great a degree not only large pecuniary advantage to the Dominion, but much prestige to the whole country by attracting the population, not only on this continent, but of Europe to this place. It has all the qualifications necessary to make it a place of great resort. . . . There is beautiful scenery, there are curative properties of the water, there is a genial climate, there is prairie sport, and there is mountain sport; and I have no doubt that it will become a great watering-place.

—Sir John A. Macdonald on Banff, 1887

Although the environmental lobby was successful in its fight against the construction of an upper and lower village at Lake Louise funded by Imperial Oil and supported by Parks Canada in the early 1970s, many would argue that the subsequent incremental developments have achieved almost the same result. This photograph shows the enlarged Château Lake Louise in front of what is advertised as the largest ski hill in Canada. Is this a national park landscape?

The montane zone is also very attractive for development purposes. More than 70 per cent of the zone has already been developed into highways, golf courses, towns, and resorts, and more development is planned. Within Banff National Park, the four-kilometre-wide Bow Valley contains the Trans-Canada Highway, the 1A highway, a national railway, an airstrip, a 27-hole golf course, three ski resorts, the village of Lake Louise, and the town of Banff with a population 7,500! Between 1989 and 1994, 176 animals were killed on the railway tracks alone, and the annual toll on the roads was more than 200. These road-kill figures were obviously unacceptable in a national park, and management responded by erecting fences along 44 kilometres of road to direct animals to a series of underpasses and overpasses. The deer and elk adapted to them immediately; it took the carnivores a little longer. In a further mitigation measure, it was recently announced that CP Rail will refurbish 6,300 grain hopper cars owned by the federal government so that grain will no longer spill onto the tracks and attract animals, particularly bears.

Banff is also of major economic importance. More than 3.5 million people visit every year, and Banff tourism operators employ thousands of workers and generate hundreds of millions of dollars in annual revenue. Tour operators argue that they have to compete on an international basis and that if further developments are not permitted, tourism will dry up.

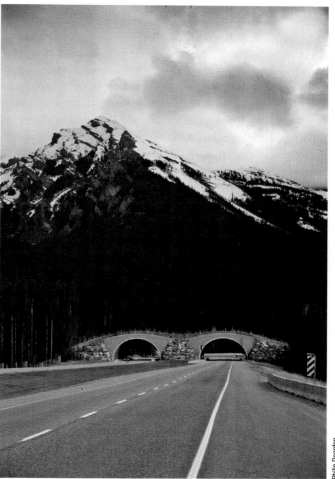

Wildlife overpass built over the Trans-Canada Highway in Banff National Park.

In response to these pressures, the minister of Canadian heritage created the Banff–Bow Valley Task Force in 1994 to advise on the management of Banff and the surrounding areas of the Bow Valley. Swinnerton (2002) provides a detailed overview of the task force, including the stakeholder participation process discussed in Chapter 6. The task force found that grizzly bear populations were declining rapidly and that aquatic ecosystems were compromised as a result of exotic introductions and dams. The task force predicted that current rates of growth would cause 'serious, and irreversible, harm to Banff National Park's ecological integrity' (Banff–Bow Valley Study 1996, 4).

The federal minister committed to implementing the recommendations of the task force and in 1998 created another inquiry—the Ecological Integrity Panel—to look at similar issues in all of Canada's national parks. The Ecological Integrity Panel's report concurred with the earlier study on Banff and made strong recommendations for a more adaptive approach to park management, with greater attention to ecosystem-based management and greater consultation with stakeholders (Chapter 6). The panel prepared 127 recommendations and delivered one central message: *ecological integrity in all the national parks is in peril.* The minister accepted the panel's findings and began implementing the recommendations. The proclamation of a new National Parks Act in 2000, which further strengthens the ecological mandate of the parks, was one response to the panel's recommendations.

The findings of the panel and subsequent recommendations emphasized what had already been well known to biogeographers. The theory of **island biogeography** suggests that small islands are unable to support as many species as large islands of similar habitat. Given that many terrestrial parks are separated from colonizing sources from outside the park, they are analogous to islands. Development inside the parks essentially makes them smaller, and smaller parks are more likely to experience extinctions, as demonstrated by Newmark (1995) (Figure 14.9).

Over time, there has been a progressive change in attitudes towards development in Canada's parks. Initially, development for tourism and recreation was encouraged. Parks were created for the enjoyment of Canadians, and protection of wildlife was secondary—Canada still had large intact wilderness areas, so there seemed little need to protect wildlife within the parks. There was, however, a need to generate revenue, and parks were regarded as playing an important role in generating income from wilderness. But as the Canadian landscape became increasingly developed, this perception changed. Particularly in southern Ontario, parks became one of the few areas that were not developed.

Research has quite clearly shown that development within parks is detrimental to many species (Box 14.13). Fortunately, the federal government has acted accordingly. Over the past 15 years, it has revised the National Parks Act twice in favour of a mandate that favours wildlife protection over recreational opportunities. But while these changes are positive, there has been a growing realization that development is not the only threat to ecological integrity within our parks. Many management challenges arise from threats originating beyond park boundaries.

External Threats

Parks do not exist in isolation—they are intimately linked to surrounding and global ecosystems. It is therefore necessary to be aware of any influences from outside park boundaries that may have a detrimental impact on wildlife resources within the park (Box 14.14). This awareness is relatively recent (Figure 14.10).

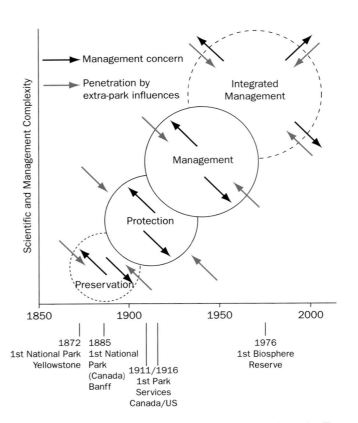

Figure 14.10 Protected area: evolving relationships from isolation to integration. The circles represent the growing size of the protected area system over time. Boundaries (circle circumferences) were initially of little importance but assumed greater significance in the protection and management phases. It is now realized that for park management (arrows) to be effective, it must pay equal attention to environmental changes outside park boundaries. Source: Dearden and Rollins 1993, 10.

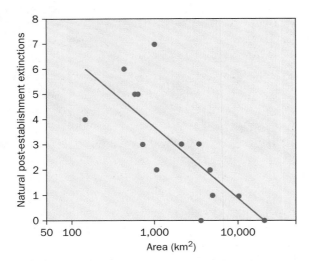

Figure 14.9 Relationship between number of natural post-establishment extinctions and park area in 14 western North American national parks. Area is plotted on a logarithmic scale. The straight line shows the relationship $Y = 11.95 - 2.76 \times$ (log area). Source: Newmark 1995.

Box 14.13 The Banff Longnose Dace: Extinction in a National Park

The Banff longnose dace was a small minnow found in only one location on the planet, a marsh downstream from the famous Cave and Basin hot springs in Banff National Park. Discovered in 1892, the minnow was declared extinct by COSEWIC in 1987. As has been the case with many extinctions, a combination of factors likely led to the animal's demise. Tropical fish were introduced into the warm waters of the marsh, where they flourished and were able to outcompete the dace. The dace may also have interbred with other species of dace, and collecting for scientific purposes may also have reduced numbers. However, it seems that the most important factor in its extinction was the decision to allow a hotel to chlorinate hot spring water and discharge it into the marsh. The chlorine reacted to form chlorinated hydrocarbons, which are very toxic to fish even at low concentrations. Shortly thereafter, the Banff longnose dace disappeared forever, and thus it was pressure from tourism that ultimately led to Banff's first extinction.

Box 14.14 The Role of Parks in Endangered Species Protection: Wood Buffalo National Park

Straddling the Alberta–Northwest Territories boundary, Wood Buffalo covers 44,807 square kilometres. The park is recognized both as a World Heritage Site and as a Ramsar Site. Ramsar Sites are wetlands of global significance, and the Ramsar Convention on Wetlands is named for Ramsar, Iran, where it was signed in 1971. Wood Buffalo contains critical habitat for two endangered species: North America's largest terrestrial mammal, the bison, and the tallest bird, the whooping crane.

> We came to places where, as far as the eye could see, untold thousands [of buffalo] were in sight; the country being fairly black with them . . . these immense herds were moving north and there seemed no end to them.
>
> —Cecil Denny in Saskatchewan, 1874

Bison: The image of vast herds of bison ranging back and forth along the Great Plains of North America is one that will never be seen again. With up to 60 million animals, bison herds probably constituted the greatest large-mammal congregations that ever existed on Earth and were important in the subsistence lifestyles of many Aboriginal peoples in western Canada. But by the 1860s, the bison had been extirpated from the plains of Manitoba. Trainloads of meat and hides were sent east from the slaughter. As American Indians flooded into Canada to seek the protection of the Great White Mother (Queen Victoria), the pressure on the remaining herds increased dramatically, and the wild bison herds were extirpated from the Canadian Prairies.

> All through today's journey, piled up at the leading stations along the road, were vast heaps of bones of the earliest owners of the prairie—the buffalo. Giant heads and ribs and thigh bones, without one pick of meat on them, clean as a well-washed plate, white as driven snow, there they lay, a giant sacrifice on the altar of trade and civilization.
>
> —Traveller on the Canadian Pacific Railway, 1888

Several remnants remained, however. A small number had been protected by the earlier establishment of Yellowstone National Park. Yellowstone is the only place where wild, free-ranging plains bison have survived since colonial times. Banff also had a growing population kept as a tourist attraction in an animal compound. In addition, two remnants had been brought together by an American rancher. The herd was bought by the Canadian government, and the 703 animals were transported to a national park (created for that purpose) adjacent to the railway near Wainwright, AB. In the mid-1920s, the herd, then numbering 6,673, was relocated to Wood Buffalo National Park.

These were plains bison (*Bison bison*). Less well known are their non-migratory, taller, and darker cousins, the wood bison (*Bison bison athabascae*). These bison were at one time widely distributed from the aspen parklands of Saskatchewan and Alberta to the eastern slopes of the Rockies and British Columbia and north to the coniferous forests of the Mackenzie Valley. They are endemic to Canada. Estimates suggest that there were once more than 168,000 wood bison in Canada. The wood bison was never as abundant as the plains bison, and by 1891 fewer than 300 of them remained. Wood Buffalo National Park was established at least partly to protect this remnant, and by 1922 the herd had grown from 1,500 to 2,000 animals. Shortly thereafter, the herd of plains bison was imported from Wainwright, and interbreeding led to the disappearance of the distinctive wood bison characteristics. Wood bison were believed to have become extinct.

In 1957, however, an isolated group of wood bison was located in a remote area of the vast park. This herd was relocated to guard against further interbreeding within Wood Buffalo. Some animals were removed to the Mackenzie Bison Sanctuary in the North; the herd now numbers more than 200, and individuals have expanded their range outside the sanctuary. Other animals were removed to Elk Island National Park near Edmonton, where their numbers have to be closely controlled because of the small area available. This herd has provided animals for satellite herds in the Yukon, the Northwest Territories, northwestern Alberta, and Manitoba. Some have even been sent to repopulate parks in Russia. As of 2006, there were an estimated 4,188 wood bison in seven free-ranging, disease-free herds, 6,216 animals in four diseased, free-ranging herds, and 1,029 animals in captive conservation (public and private) and research herds. Two wild herds exceed the minimum viable population of 400 individuals. In 1988, COSEWIC downlisted the wood bison from endangered to threatened. In May 2000, the species was reassessed; its status did not change.

The dangers for the bison are not yet over, however. When the plains bison were imported from Wainwright, they brought with them bovine diseases such as brucellosis and tuberculosis. These diseases have already taken a toll on bison populations—from highs of more than 12,000 animals, they had dropped to a quarter of this number by the early 1990s. Diseases afflicting the bison have raised concerns from the agricultural sector. Bison represent the last focus for both diseases in Canada, and as agriculture has impinged on the western boundary of the park, farmers are concerned that domestic stock will become infected. This has led to calls from the agricultural lobby for elimination of the herd.

In 1995, the Canadian heritage minister announced that the herd would not be slaughtered. In addition, the minister introduced a management plan calling for more research on the impact of habitat change on bison ecology, a buffer zone between the diseased animals and the Mackenzie Bison Sanctuary to the north, and more investigation into the impact of the diseases on the park's ecosystem. These directions were followed, but considerable uncertainty remains, even among conservation scientists, about whether to proceed with 'depopulation' (slaughter) and then repopulate with disease-free calves.

Several factors besides disease also threaten bison populations. The Peace-Athabasca Delta, for example, supported the highest concentrations of bison. However, since the construction of the Bennett Dam upstream in British Columbia, water levels on the delta have fallen considerably, causing habitat changes that have negatively affected many animal species, including the bison. A more recent impact on the delta has been changes in the

Parks Canada/W.Lynch/09.90.10.01(06)

The vast herds of plains bison had been extirpated from Canada until efforts were made to reintroduce them from the US and eventually transport them to Wood Buffalo National Park, where they mixed with the wood bison population.

Athabasca River system as a result of extractions of water for oil production at the oil sands in Alberta, discussed in Chapter 12. This impact from outside the park again emphasizes the need to take an ecosystem-based perspective on park management (Chapter 6).

Whooping crane: Unlike the bison, whooping cranes (*Grus americana*) were never very numerous. Historical accounts suggest a population of 1,500. What they lacked in numbers, they made up for in presence. More than 1.5 metres high and pure white except for black wing tips, black legs, and a red crown, with wing spans in excess of two metres, these majestic birds migrate annually from wintering grounds on the Gulf of Mexico coast of Texas to the Northwest Territories. These wintering grounds are all that remain of a breeding summer range that at one time stretched from New Jersey in the east to Salt Lake City in the west and as far north as the Mackenzie Delta and a winter range that included marshes from southern Louisiana into central Mexico (see Figure 14.11). Requiring undisturbed breeding habitat, the cranes soon declined under the expansion of agriculture. Unrestricted hunting along their long migration routes also contributed to the decline. By 1941, there were only 22 whooping cranes left.

The governments of the US and Canada agreed to a joint program to try to save the species from extinction. They used the 1916 Migratory Bird Treaty between the US and Canada to stop legal hunting. In 1937, the US government bought the Arkansas National Wildlife Refuge to protect the wintering habitat on the Gulf coast. In 1954, the only known nesting area was discovered in the northern part of another protected area, Wood Buffalo. Finding the breeding grounds enabled direct human interventions, such as artificial incubation of eggs. Whooping cranes generally lay two eggs, but usually only one chick survives. A captive propagation program in the 1960s and 1970s moved one of the eggs for incubation. By 2003, the world population of whooping cranes had risen to more than 400, including captive populations, with more than 185 in Wood Buffalo National Park. The species is currently listed as endangered by COSEWIC.

If you turn back to the discussion on factors that influence the vulnerability of species to extinction earlier in the chapter, you will see that the whooping crane possesses many of the characteristics that make species vulnerable. They are especially at risk because of their long migration and the vulnerability of their wintering grounds to both natural (e.g., hurricanes) and human-caused destruction.

An interesting dilemma arose with their recovery program. As mentioned above, the cranes lay two eggs, but for various reasons, usually only one chick survives. As part of the recovery program, second eggs were removed from the nests and hatched separately. However, Parks Canada, in an effort to maintain natural processes in Wood Buffalo, disallowed the removal of the second egg. This has slowed down the population growth rate (Boyce, Lele, and Johns 2005). What would you do—maintain natural processes or speed up recovery?

Figure 14.11 The original range of the whooping crane in recent times.

The whooping crane, the tallest North American bird and one of the rarest, makes its habitat in muskeg, prairie pools, and marshes.

In the early days, boundaries were easily penetrated as society adjusted to the idea of preserving nature. But as time passed, park boundaries became less permeable, and protection of wildlife resources was more assured. However, as development surrounding parks intensified, park managers began to realize that development pressures outside park boundaries were affecting resources within parks. In response, they began to develop integrated management plans that took external threats into account.

Although the creation of integrated management plans is a step in the right direction, external threats to parks are often difficult to eliminate or even control. Invasions by exotic species constitute one of the most challenging problems, and alien species make up to 50 per cent of the flora in some national parks. External threats may originate from private landowners surrounding park boundaries, and it may be difficult or impossible to restrict activities on private land. External threats also come in all shapes and sizes—to tackle them all would require significant human and economic resources. Examples of external threats include mining, logging, agriculture, urbanization, water projects, hunting, exotic species, tourism, acid precipitation, and chemical pollution.

Sometimes external threats can be readily identified and even managed, such as forestry activities along a park boundary. In 2003, for example, Pukaskwa National Park in Ontario entered into a dispute resolution process with a logging company over the development of forestry roads within five kilometres of its boundary through an otherwise road-less wilderness. The road would fragment the entire 25-kilometre length of the northern park boundary, have a potential impact on the wolf population, impede caribou movements, increase illegal access to the park in isolated areas, increase poaching, encourage encroachment of invasive species, result in the eventual loss of connectivity to other protected areas, and drive wildlife populations to move in and out of the park because of greater harvesting pressure (Theberge 2004). The company subsequently agreed to mitigation measures, and a zone of cooperation was established, extending five kilometres from the park boundary into the provincial lands, in which future forestry operations and prescribed burns would be jointly determined.

However, sometimes the influences of external factors are too distant and diffuse for park managers to control. Global climate change (see Chapter 7) is a good example, and it will obviously have serious implications for protected areas. On the one hand, protected areas will have a huge role to play in terms of their *hospital role* (Box 14.12) in helping sequester carbon from the atmosphere. On the other hand, the *bank role*, providing refuge for natural populations, will be very vulnerable to the changes. Protected area networks must be made as resilient as possible against these changes. One of the main mechanisms for doing this is through the kind of large-scale bioregional planning illustrated by the Yellowstone to Yukon initiative discussed below that emphasizes connectivity, especially north-south connectivity, among protected areas. We will also require new protected areas that help to facilitate

migration, provide source populations, and offer suitable habitat for incoming populations. Private lands will also be important. The biodiversity implications of climate change are likely to be especially severe in the oceans, yet Canada has created virtually no marine protected areas, let alone functioning networks of such areas, as discussed in Chapter 8.

Suffling and Scott (2002) have used climate change models to indicate some of the possible effects on the national parks system over the next century. The researchers conclude that all regions and parks will be dramatically affected and suggest that Parks Canada is going to have to intervene in park ecosystems to a greater degree in an effort to maximize the capacities of ecosystems and species to adapt to climate change. However, a survey of protected area jurisdictions in Canada found that 80 per cent had not completed a comprehensive assessment of the potential impacts and implications of climate change on policy and management and had no adaptation strategy or action plan. Furthermore, 86 per cent felt that they did not have the capacity necessary to deal with climate change issues (Lemieux, Beechey, and Scott 2007). Hannah (2003) suggests that one of the most important steps in dealing with global change in protected areas is to close this management gap as soon as possible before it becomes too wide to bridge. Unfortunately, in almost all provincial jurisdictions in Canada, politicians have been driving things the other way by consistently cutting the funding available to park agencies. One often overlooked role for protected areas in addressing climate change is their *schoolroom role* (Box 14.12), making visitors more aware of the challenges of global change and of the things they can do to help. One of these things is to help persuade politicians to restore funding to park programs.

The idea of park managers actively intervening in park ecosystems rather than leaving change to the vagaries of nature is known as **active management**. It recognizes that the human forces of change are now so prevalent throughout the land-

Building greater awareness among visitors of the role of national parks in society is a central facet of sound management. Parks Canada has developed some of the best interpretive facilities in the world. This is the visitor centre at Greenwich in PEI National Park.

to protect ecosystems. In practice, stewardship takes many forms, including:

- landowners voluntarily restricting damaging use of land, planting native species rather than exotic ones, and placing protective covenants on their land;
- community members contributing to wildlife monitoring programs, providing passive education for tourists and visitors, and participating in collective restoration;
- park visitors voluntarily choosing to avoid hikes along sensitive trails or participating in park host programs;
- corporations introducing sustainable land practices that reduce damage to wildlife habitat.

A diversity of land trust and other conservation organizations is emerging across Canada, using a variety of tools to conserve lands under private ownership. From 1986 to 2003, $170 million in federal funds was leveraged by environmental organizations to secure or steward 1.8 million hectares of lands valued at $3.2 billion.

The Nature Conservancy of Canada—one of many conservation organizations—has protected approximately 2.1 million hectares since 1962. In total, Ducks Unlimited has been responsible for the protection of more than 10 million hectares of Canadian wetlands since 1938. Smaller, provincially based and local land trusts are also increasing in number. In the southern Vancouver Island–Gulf Islands area, where natural heritage is fading quickly and land prices are increasing, at least eight land trusts are operating to protect endangered spaces, most of which have been established since 1998. Government agencies such as Environment Canada and various provincial ministries are also embracing stewardship. For example, numerous funding programs exist for community stewardship, and some ministries publish guides or maintain websites (e.g., see Environment Canada's 'A guide to conservation programs and funding sources for agro-Manitoba') to educate and to support local initiatives. Legislative changes, particularly tax deductions, have also recently been implemented, resulting in incentives and encouragement for ecological gifts and donations.

These initiatives are critically important for the future of conservation in Canada and will play a significant role in restoring connectivity among other protected areas. Whitelaw and Eagles (2007) provide an excellent example of how this might work in southern Ontario. Private lands will never replace the role played by strictly protected areas, but they do play an essential role in 'gluing together' the larger wilderness areas set aside in government parks. The international Biosphere Reserve Program is one of the best-known initiatives promoting greater stewardship surrounding protected areas (Box 14.15).

Stakeholder Interests

Balancing the interests of the range of stakeholders that may be affected by the establishment and/or management of parks is a formidable challenge. Private landowners, local communities, Aboriginal peoples, industry, tourists, conservation

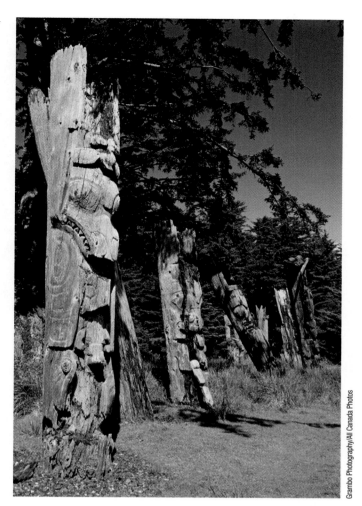

The totem poles of Ninstints, an abandoned Haida village on Gwaii Hanaas (Moresby Island in the Queen Charlotte Islands), give some impression of the Haida's spiritual connection with the environment. The village is now part of Gwaii Hanaas National Park Reserve and is co-managed by Parks Canada and the Haida.

Grambo Photography/All Canada Photos

organizations, and government agencies all have an influence over how parks are managed. Environment Canada (2006b) reports that 13 out of 15 jurisdictions provide opportunities for community involvement in most or all of their protected areas. Community participation is enshrined in the legislation governing eight protected areas agencies.

Aboriginal peoples have a particularly powerful role in the designation and management of many parks and should be considered another level of government rather than another stakeholder. An amendment to the National Parks Act in 1972 created a special category of park, the *national park reserve*, which does not prejudice future land-claim negotiations. Several large parks have been created in the Arctic as a result. In the south, on the other hand, where the provincial governments have jurisdiction over the land base, progress has been slow. It was not until after protests and several court cases that the legitimacy of Native claims over land and resources in southern Canada was taken seriously. Figure 14.8 depicts the rise in influence of Aboriginal peoples on park designation and management in southern Canada following the successful interventions by the Haida Nation regarding how their

Porcupine caribou.

homeland in Haida Gwaii would be managed. More than 50 per cent of the land area in Canada's national park system has now been protected as a result of Aboriginal peoples' support for conservation of their lands, and this proportion will only increase in the future as Native land claims are settled.

The goals of Aboriginal peoples and conservationists, however, are not always identical. Aboriginal peoples, for example, retain the right to hunt in many parks. At the moment, there are few restrictions on the size or means of harvest, a source of concern for some conservation scientists. Several parks in both northern and

Guest Statement

Canada's Great Bear Rainforest
Ian McAllister

Environmental campaigns do not originate in boardrooms or around coffee tables. They begin when one person or a small group of people fall head-over-heels in love with a mountain, a river, an animal, or some combination of all three. You can trace the roots of any successful environmental campaign back to the simple emotion of love, and that is why there is such passion in the environmental movement today.

I fell in love with the central coast of British Columbia about 18 years ago. My father had chartered a sailboat to visit the Koeye River Valley, a mysterious estuary halfway up the BC coast where grizzly bears were rumoured to walk white-sand beaches and to fish in waters teeming with salmon. I had managed to squeeze aboard that boat, and by the end of the trip I was hooked. It turned out that the rumours had been true.

At that time, none of the river valleys on the coast had been extensively explored or protected by legislation, yet the majority had already been licensed out to logging companies. In other words, British Columbia didn't know what it was about to lose! Six months after that fateful voyage, along with my wife Karen, my father Peter, and some friends, we launched the campaign to protect the Great Bear Rainforest.

Now, 18 years later, the campaign to save the Great Bear Rainforest (GBR) is one of the largest environmental movements in Canadian history. It has had a huge impact on the international marketplace for timber and has become a keystone example of how a group of normal individuals can combine their strengths to bring about change. From scientists to celebrities, from letter-writers to lawyers, people from all walks of life joined together in a mutual love of nature and the commitment to help preserve it. I guess it makes sense that we are such a diverse group. After all, diversity is what we strive to protect.

In 2008, 22 per cent of the GBR is poised for legislated protection, with an additional 10 per cent designated off-limits to logging. Scientists continue to warn that more needs to be protected to ensure survival over time of all species, such as Spirit bears (an endemic 'white' black bear), grizzly bears, salmon, and wolves. However, that 22 per cent is a huge step forward from the 12 per cent cap that previous governments have placed on environmental protection in British Columbia—and maybe the greatest achievement of the campaign to protect this wildlife-rich ecosystem to date.

Ian McAllister

There's no question that it has been a long time coming, but there's also no doubt that it has been worth it. Over the years, we have been called heroes, and we have been called traitors (we've even been called colonialists for renaming the coast the Great Bear Rainforest!). But now, governments and businesses listen to our voice and recognize the strength in numbers behind it. And these numbers continue to grow.

Environmental work is one of the most exciting and creative occupations I can think of. It is a constant challenge filled with unrivalled spiritual and physical rewards. Perseverance is a prerequisite, as is a sensitivity to other people's points of view, and the positive result of this is that environmentalism attracts fun-loving, hard-working, passionate individuals who believe very strongly in the value of teamwork.

I didn't know what I was getting into when I hopped aboard that boat 18 years ago. But that's the best thing about environmental work: you can learn everything you need to know along the way, and I wouldn't trade it for anything.

Ian McAllister is a conservationist and founder of PacificWild.org.

southern Canada have some form of **co-management** arrangement between Parks Canada and Aboriginal peoples. In general, they have worked to the satisfaction of both groups, but the issue of shared responsibility remains an ongoing challenge.

Implications

There is no debate about whether rates of extinction have increased as a result of human activities over recent times. There is still some uncertainty, however, regarding the impact of the extinctions. Is there ecological redundancy such that the Earth can afford to lose some species without major impacts? Is the loss of any species as a result of human activities ethically and morally acceptable? There are many unanswered questions regarding the implications of reduced biodiversity.

Extinction is commonly viewed as simply a biological problem. Yet there is a need to link the biological process of extinction with the economic and ethical reasons why extinction is undesirable, the reasons why human-caused extinctions are increasing, and the kinds of measures that have to be taken to prevent this from happening. In other words, extinction is not just the domain of biologists but involves consideration from a broad range of perspectives, including all the social sciences, geography, law, and ethics.

A Biosphere Reserve surrounding Waterton Lakes National Park has been set aside to allow for joint decision-making between park authorities and local landowners. The agreement has been particularly useful in reducing conflicts between grizzly bears and local ranchers.

Wave/All Canada Photos

Box 14.16 What You Can Do for Protected Areas

1. Visit parks and other protected areas often through-out the year. Enjoy yourself. Tell others that you have enjoyed yourself, and encourage them to visit.
2. Always follow park regulations regarding use. Feeding wildlife, for example, may seem kind or harmless, but it can lead to the death of the animal.
3. If you have questions regarding the park's management or features, do not be afraid to ask. A questioning public is a concerned public.
4. Many park agencies have public consultation strategies relating to topics ranging from park policy to the management of individual parks. Let them know your interests so that you can be placed on the mailing list to receive more information.
5. Join a non-governmental organization, such as the Canadian Parks and Wilderness Society or Nature Canada, that has a strong interest in parks issues.
6. Many parks now have cooperating associations in which volunteers can help with various tasks. Find out whether a park near you has such an organization.
7. Write to politicians to let them know of your park-related concerns.

Summary

1. Extinction levels have reached unprecedented levels. There are several reasons why we should be concerned. Life-supporting ecosystem processes depend on ecosystem components. As we lose components through extinction, these processes become more impaired. We also derive many useful and valuable products from natural biota, including medicines. In addition to these utilitarian reasons, there are ethical and moral reasons why we should be concerned about species extinction.
2. Many factors are behind current declines. The underlying factor is human demand as population and consumption levels grow. Much attention has concentrated on the tropics because of the high biodiversity levels and high rates of destruction there. However, Canada has experienced 13 extinctions and 22 extirpations since European colonization.
3. Some of the main pressures causing extinction include overharvesting, predator control, and habitat change. Habitat change includes not only physical changes (e.g., conversion of habitat into agricultural land) but also those caused by chemicals and the introduction of alien species.
4. Not all species are equally vulnerable to extinction. Species that have specialized habitat requirements, migratory species, species with insular and local distributions, valuable species, animal species with a large body size, species that need a large home range, species that are not effective dispersers, and species with low reproductive potential tend to be the most vulnerable.
5. Canada is party to several international treaties for the protection of wildlife and was the first industrialized nation to sign the Convention on Biological Diversity at the Earth Summit in 1992. As a signatory to the convention, Canada was required to introduce legislation to protect endangered species. In 2002, the federal government passed the Species at Risk Act (SARA).

6. The Committee on the Status of Endangered Wildlife in Canada (COSEWIC) is responsible for determining the status of rare species and categorizing them as extinct, extirpated, endangered, threatened, or vulnerable. As of May 2007, 556 species had been classified. The committee's assessment is the first step in the process for protecting a proposed species at risk under SARA.

7. For species listed under SARA, recovery and management plans must be developed and implemented, unless the minister responsible feels that recovery is not 'feasible'. By 2007, only 61 out of 366 plans had been completed.

8. Protected areas have emerged as one of the key strategies to combat the erosion of biodiversity, both internationally and in Canada. Protected areas fulfill many roles in society, including species and ecosystem protection, maintenance of ecological processes, and as places for recreation and spiritual renewal, aesthetic appreciation, tourism, and science and education in natural outdoor settings.

9. There are many different kinds of protected areas in Canada, including national and provincial parks, wilderness areas, First Nations parks, wildlife refuges, ecological reserves, and regional and municipal parks.

The amount of protection given to ecosystem components varies among these different types.

10. National parks are outstanding natural areas protected by the federal government because of their ecological importance and aesthetic significance. There are 42 national parks in Canada. The goal is to have at least one national park in each of the 39 regions of the national system plan. At present, only 28 of these regions are represented.

11. Banff, the first national park in Canada, was established in 1885. Since that time, the national parks fulfilled a dual mandate that required protection of park resources in an unimpaired state but also permitted their use. This conflicting mandate was clarified in an amendment to the National Parks Act in 1988 and further clarified in the National Parks Act of 2000, giving first priority to protecting the ecological integrity of the parks.

12. Current management challenges to the national parks system include external threats and fragmentation, which require more of an ecosystem approach. The involvement of First Nations peoples is also essential. Banff National Park exemplifies some of the difficulties encountered in trying to control tourism development.

Key Terms

active management
co-management
Convention on Biological Diversity
Convention on International Trade in Endangered Species of Wild Fauna and Flora (CITES)

ecologically extinct
endangered
ex situ preservation
extinction
extirpated
extrinsic values
fragmentation

in situ preservation
intrinsic value
island biogeography
minimum viable population (MVP)
population viability analysis (PVA)

protected areas
Red List
Species at Risk Act (SARA)
stewardship
system plan
threatened
vulnerable

Questions for Review and Critical Thinking

1. What are the main reasons that we should be concerned about species extinctions?

2. Why are some species more vulnerable to extinction than others?

3. What is being done to protect endangered species in your province?

4. What are some of the strengths and weaknesses of Canada's Species at Risk Act?

5. What do you think should be the relative importance of the various roles played by protected areas?

6. What different classifications of protected areas exist in your province, and what kinds of protection are offered by these different systems?

7. What is your province doing to achieve the 12 per cent protected area that all jurisdictions in Canada have committed to establishing?

8. What ecozone is Banff townsite situated in, and what is the importance of this zone?

Related Websites

Canadian Biodiversity Information Network: http://www.cbin.ec.gc.ca

Canadian Environmental Assessment Agency: http://www.ceaa.gc.ca

Canadian Parks and Wilderness Society: http://www.cpaws.ca

Canadian Wildlife Service, Ontario Region: http://www.on.ec.gc.ca

Clayoquot Sound: http://www.forests.org

Committee on the Status of Endangered Wildlife in Canada: http://www.cosewic.gc.ca

Ducks Unlimited: http://www.ducks.org

Ducks Unlimited Canada: http://www.ducks.ca

Environment Canada: http://www.ec.gc.ca

Global Environment Facility: http://www.gefweb.org

Nature Canada: http://www.naturecanada.ca

North American Waterfowl Management Plan: http://www.nawmp.ca

Northwest Territories Wildlife Division: http://www.nwtwildlife.com

Ontario Ministry of Natural Resources: http://www.mnr.gov.on.ca

Parks Canada: http://www.pc.gc.ca

Science and Technology for Canadians: http://www.science.gc.ca

Sierra Youth Coalition: http://www.syc-cjs.grg

Species at Risk Act Public Registry: http://www.sararegistry.gc.ca

Species at Risk, Environment Canada: http://www.speciesatrisk.gc.ca

Transport Canada: http://www.tc.gc.ca

United Nations Environment Programme, World Conservation Monitoring Centre: http://www.unep-wcmc.org

Vancouver Island Marmot Recovery Foundation: http://www.marmots.org

Western Canada Wilderness Committee: http://www.wildernesscommittee.org

Wildlife Collision Prevention Program: http://www.wildlifecollisions.ca

World Commission on Protected Areas: http://www.iucn.org/themes/wcpa

World Conservation Union—Red List: http://www.iucnredlist.org

World Wildlife Fund Canada: http://www.wwf.ca

Further Readings

Note: This list comprises works relevant to the subject of the chapter but not cited in the text. All cited works are listed in the section 'References' at the end of the book.

Abbey, E. 1968. *Desert Solitaire: A Season in the Wilderness*. New York: Simon and Schuster.

Beazley, K., and R. Boardman, Eds. 2001. *Politics of the Wild: Canada and Endangered Species*. Toronto: Oxford University Press.

Clevenger, A.P., and N. Waltho. 2000. 'Factors influencing the effectiveness of wildlife underpasses in Banff National Park, Alberta, Canada'. *Conservation Biology* 14: 47–56.

Dearden, P. 2004. 'Parks and protected areas'. In B. Mitchell, Ed., *Resource and Environmental Management in Canada: Addressing Conflict and Uncertainty*, 3rd edn, 314–41. Toronto: Oxford University Press.

Dempsey, J., and P. Dearden. 2008. 'Stewardship: Expanding ecosystem protection'. In P. Dearden and R. Rollins, Eds, *Parks and Protected Areas in Canada: Planning and Management*, 3rd edn, 432–54. Toronto: Oxford University Press.

Dudley, N. *Guidelines for Applying Protected Area Management Categories*. Gland, Switzerland: IUCN.

Langdon, S., and P. Dearden. 2008. 'Aboriginal people and national parks'. In P. Dearden and R. Rollins, Eds, *Parks and Protected Areas in Canada: Planning and Management*, 3rd edn, 373–402. Toronto: Oxford University Press.

McNamee, K. 2008. 'From wild places to endangered spaces'. In P. Dearden and R. Rollins, Eds, *Parks and Protected Areas in Canada: Planning and Management*, 3rd edn, 24–55. Toronto: Oxford University Press.

Parks Canada Agency. 2000. *Unimpaired for Future Generations? Protecting Ecological Integrity with Canada's National Parks*, v. 2, *Setting a New Direction for Canada's National Parks*. Report of the Panel on the Ecological Integrity of Canada's National Parks. Ottawa: Minister of Public Works and Government Services.

———. 2003. *State of Protected Heritage Areas, 2002 Report*. Ottawa: Minister of Supply and Services.

———. n.d. *National Park System Plan*. Ottawa: Minister of Supply and Services.

Rivard, D.H., et al. 2000. 'Changing species richness and composition in Canadian national parks'. *Conservation Biology* 14: 1,099–109.

Searle, R. 2000. *Phantom Parks: The Struggle to Save Canada's National Parks*. Toronto: Key Porter Books.

Statistics Canada. 2000. *The Importance of Nature to Canadians: The Economic Significance of Nature-Related Activities*. Ottawa: Statistics Canada.

Theberge, J.C., and M.T. Theberge. 2004. *The Wolves of Algonquin Park: A 12-Year Ecological Study*. Department of Geography Publication Series no. 56. Waterloo, ON: University of Waterloo.

Willcox, L., and P. Aengst. 1999. 'Yellowstone to Yukon: Romantic dream or realistic vision of the future?' *Parks* 9: 17–24.

Woodley, S. 2002. 'Planning and managing for ecological integrity'. In P. Dearden and R. Rollins, Eds., *Parks and Protected Areas in Canada: Planning and Management*, 2nd edn, 97–114. Toronto: Oxford University Press.

Wilkinson, C.J.A., and P.F.J. Eagles. 2002. 'Strengthening the conservation of biodiversity: Reforming Ontario's Provincial Parks Act'. *Natural Areas Journal* 21: 330–7.

World Conservation Union. 2001. 'Protected area management categories'. http://www.unep-wcmc.org/protected_areas/categories.

Part E
Environmental Change and Challenge Revisited

*'The only person who likes change is a wet baby,'
observes educator Roy Blitzer. 'Two basic rules of life are: 1)
change is inevitable; and 2) everybody resists change. Much of
the world has its defences up to keep out new ideas.'*
—von Oech 1990, 180

Change is ubiquitous. This book provides an overview of environmental change and challenge in Canada. Change occurs as a result of both natural and human-induced pressures, and it is often difficult to determine the balance between them. However, over time it seems that human-induced changes are becoming the dominant driving factor for many aspects of environmental change. In many cases, these changes, such as extinction and climate change, are irreversible. Current activities are serving to impoverish the planet for future generations.

The preceding chapters have emphasized the need to understand the ecological aspects of environmental change, along with the various management approaches that may be useful. Reading the book should enable you to understand the background to many of the environmental problems we face and also to appreciate the different management approaches to their resolution. In each chapter, we have attempted to make you aware not only of the nature of the challenges being faced but also of some of the solutions that are being tried.

This final part contains one chapter, and its main focus is on solutions. We provide an assessment of progress at the global and national levels in coming to terms with environmental change. Perhaps more important, though, we end with some suggestions on the kinds of actions that you can take to help contribute to the positive changes that need to be made to ensure that future generations will inherit a planet every bit as beautiful and productive as have the generations before.

We hope that you will not only read the chapter carefully but also take action to try to improve your balance sheet with or your footprint on the environment! Your efforts, combined with those of thousands of others acting individually, can make substantial changes in the environment of tomorrow.

Chapter 15

Making It Happen

Learning Objectives

- To identify some of the main global responses to environmental degradation.
- To understand some of the main Canadian responses to environmental degradation.
- To place Canada within the global context for environmental response.
- To assess how important environment is to the administration of your university.
- To make better decisions to minimize your impact on the environment.
- To use your influence more effectively to benefit the environment.
- To clarify what 'the good life' means for you.

When I call to mind my earliest impressions, I wonder whether the process ordinarily referred to as growing up is not actually a process of growing down; whether experience, so much touted among adults as the thing children lack, is not actually a progressive dilution of the essentials by the trivialities of life.

—Aldo Leopold. 1949. *A Sand County Almanac.*

Introduction

Aldo Leopold, one of the greatest conservation thinkers and writers, points out that as we get older and our lives get busier, we often get distracted from the important things in life, like protecting the environment. Everyone says that environmental protection is important, but most devote minimal effort to actually doing anything about it. We are all members of NATO: No Action, Talk Only. And the same is true of our country. On paper, Canada has an impressive raft of legislation, policies, strategies, and action plans regarding the environment. Sadly, the translation of these into 'on-the-ground improvements' is often chronically under-resourced. Many examples have been cited in this book, ranging from lack of resources to implement Canada's Oceans Strategy as mandated under the Oceans Act (Chapter 8) through to lack of a detailed plan on how to meet our obligations under the Kyoto Protocol (Chapter 7).

In this chapter, we will provide a brief overview of global and Canadian responses to environmental change. But governments are only part of the answer. This final chapter rests on the firm conviction that individuals can make a *significant difference* in how the environmental challenges presented in this book will develop over the next decade, if we are aware of the problems and are willing to do something about them. This chapter provides some ideas about how *you* can become involved in creating change.

Global Perspectives

The previous century witnessed many changes. It may be characterized as an age of diminishing imperial powers, ongoing wars, atomic bombs, the harnessing of the entire globe into an interconnected economic system, rising consumer demands, and an exploding human population. This century will witness the continuation of some of these trends, but many scientists seem convinced that global climatic change, water shortages, biological impoverishment, declining food yields per capita, desertification, pollution, and overpopulation will constitute the backdrop for the events of the next century.

Many of these trends are driven by consumption of material goods, which emerged over the last couple of decades of the past century as the dominant international ideology. From its

Resist the temptation to give in to the consumer binge.

heartland in Europe, North America, and Japan, the globalization of consumption will be one of the main developments of the next couple of decades, if not the entire century. Although population growth is still a concern, convincing signs point to falling rates of increase and the stabilization of populations, probably within the next 50 years. In contrast, consumption knows no bounds. Indeed, our whole global economic system is focused on increasing consumption levels. At the individual level, our psyches are dominated by images of the consumer goods we hanker for. The shopping mall has become the new place of worship.

Roughly one-quarter of humanity is now within this consumer class, and this number is divided more or less equally between those in developed countries and the rapidly increasing numbers of consumers in developing countries such as China and India. In the developed nations, demands continue to grow insatiably. In the US, for example, the average size of new houses was 38 per cent larger in 2002 than in 1975, yet the number of people per household is falling. These large houses also have big appetites for furniture, electronics, and all manner of consumer goods. They are also expensive to heat and maintain.

The world's auto manufacturers produced a record 67 million vehicles in 2006, putting more cars on the road than ever before (Worldwatch Institute 2007b). Global production grew by 4 per cent, but China increased its production by nearly 30 per cent, overtaking Germany to become the third largest producer. China's rise represents the most dramatic change in the world auto industry, with production there more than quintupling over the past decade. Sales within China surpassed the 3 million mark in 2005, with nearly 9 million passenger cars on the country's roads. While this is still a comparatively small fleet, it is likely to grow rapidly in coming years, and China is expected to become a major exporter within the next four years. Traffic congestion, road accidents, air pollution, climate change, and peak oil are all challenges the world faces from car-centred transportation. On average, urban car travel uses nearly twice as much energy as urban bus travel.

The impacts of growing consumer demands are far-reaching. The Millennium Ecosystem Assessment (2005), introduced in Chapter 1, provided a detailed review of the state of planetary ecosystems, calculating that 15 of the 24 major ecosystem services that support humanity, such as climate regulation, water provision, and soil production, have been pushed beyond their limits and are in a degraded state. Estimated values (UNEP 2007) of some of these services are:

Value of
- annual world fish catch: US $58 billion (provisioning service);
- anti-cancer agents from marine organisms: up to US $1 billion per year (provisioning service);
- global herbal medicine market: roughly US $43 billion in 2001 (provisioning service);
- honeybees as pollinators for agriculture crops: US $2–8 billion per year (regulating service);
- coral reefs for fisheries and tourism: US $30 billion per year (see Box 5.5) (cultural service).

Cost of
- mangrove degradation in Pakistan: US $20 million in fishing losses, US $500,000 in timber losses, US $1.5 million in feed and pasture losses (regulating provisioning services);
- Newfoundland cod fishery collapse: US $2 billion and tens of thousands of jobs (provisioning service).

Until we find an effective way to incorporate these values into the decision-making process, the values will continue to be eroded. There is clear scientific evidence of the deterioration of global ecosystems, but degradation cannot be seen outside the human context that drives the activities causing degradation. The Millennium Ecosystem Assessment's framework shown in Figure 15.1 illustrates these connections. The box in the bottom left, the ecosystem services provided by functioning ecosystems, is influenced by the direct drivers of change promoting indirect drivers that affect ecosystems and reduce human well-being. In turn, the desire for human well-being is the force behind the indirect drivers. The pincers on the arrows show links that are amenable to strategic interventions.

The **Millennium Development Goals** (MDG) introduced in Chapter 1 are globally accepted goals for development. The target date for meeting the goals is 2015, and in 2007 the UN issued a mid-point review outlining some of the positive developments:

- The proportion of people living in extreme poverty fell from nearly a third to less than one-fifth between 1990 and 2004. If the trend is sustained, the MDG poverty reduction target will be met for the world as a whole and for most regions.
- The number of extremely poor people in sub-Saharan Africa has levelled off, and the poverty rate has declined by nearly six percentage points since 2000. Nevertheless, the region is not on track to reach the goal of reducing poverty by half by 2015.

- Progress has been made in getting more children into school in the developing world. Enrolment in primary education grew from 80 per cent in 1991 to 88 per cent in 2005. Most of this progress has taken place since 1999.
- Women's political participation has been growing, albeit slowly. Even in countries where only men were allowed to stand for political election in the past, women now have seats in the legislature.

- Child mortality has declined globally, and it is becoming clear that the right life-saving interventions are proving effective in reducing the number of deaths caused by the main child killers, such as measles.
- Key interventions to control malaria have been expanded.
- The tuberculosis epidemic finally appears on the verge of decline, although progress is not fast enough to halve prevalence and death rates by 2015.

Changes in drivers that indirectly affect biodiversity, such as population, technology, and lifestyle (upper right corner of figure), can lead to changes in drivers directly affecting biodiversity, such as the catch of fish or the application of fertilizers (lower right corner). These result in changes to ecosystems and the services they provide (lower left corner), thereby affecting human well-being. These interactions can take place at more than one scale and can cross scales. For example, an international demand for timber may lead to a regional loss of forest cover, which increases flood magnitude along a local stretch of a river. Similarly, the interactions can take place across different time scales. Different strategies and interventions can be applied at many points in this framework to enhance human well-being and conserve ecosystems.

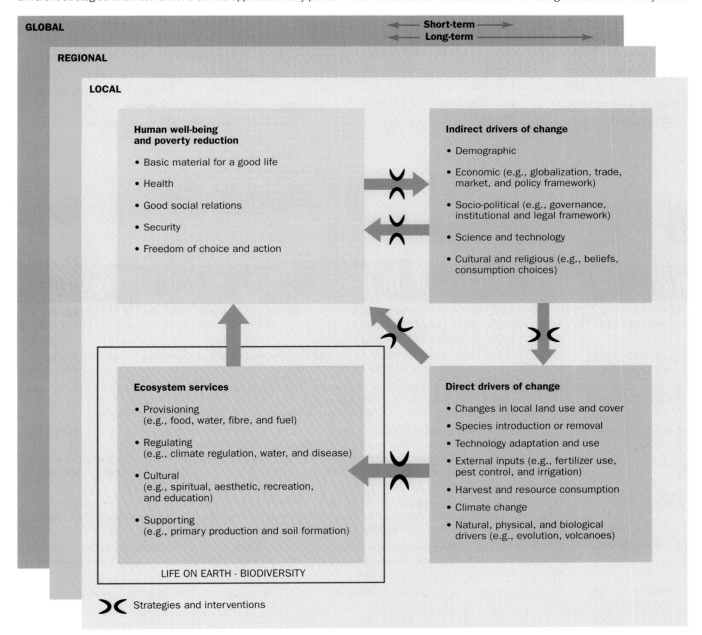

Figure 15.1 Conceptual framework of the Millennium Ecosystem Assessment. Source: Millennium Ecosystem Assessment 2005.

However, of the goals and indicators shown in Table 1.1 in the first chapter, only one of the eight regional groups is on track to achieving all the Millennium Development Goals by 2015, with severe shortfalls in many areas such as sub-Saharan Africa. There are close links between the MDGs and the environment (Table 15.1), and achieving the goals will not be possible without giving greater attention to environmental protection.

Although some landmark international agreements have been signed in the past 15 years, such as the Kyoto Protocol, the Convention on Biological Diversity, and the Stockholm Convention on Persistent Organic Pesticides, progress on many of them has been slow. For example, the Convention on Biodiversity requires signatory nations to establish networks of marine protected areas by 2012. However, at the current rate this will not be achieved until 2085, or three decades *after* the predicted collapse of the global fisheries that marine protected areas are intended to forestall (UNEP 2007). In contrast, global change is continuing apace. In 2007, the Arctic ice retreated to a record-setting minimum area of about 4.3 million square kilometres. This is a 10.4-million-kilometre loss from the mid-March maximum, and scientists are predicting that for the first time in history, the North Pole could be ice-free by as early as 2010.

The degradation of ecosystem services is already a significant barrier to achieving the Millennium Development Goals agreed to by the international community in September 2000, and the harmful consequences of this degradation could grow significantly worse in the next 50 years. The consumption of ecosystem services, which is unsustainable in many cases, will continue to grow as a consequence of a likely three to sixfold increase in global GDP by 2050 even while global population growth is expected to slow and level off in mid-century. Most of the important direct drivers of ecosystem change are unlikely to diminish in the first half of the century, and two drivers—climate change and excessive nutrient loading—will become more severe.

—Millennium Ecosystem Assessment 2005

Table 15.1 Links between the MDGs and the Environment

1. Eradicate extreme poverty and hunger	Livelihood strategies and food security of the poor often depend directly on healthy ecosystems and the diversity of goods and ecological services they provide. Natural capital accounts for 26 per cent of the wealth of low-income countries. Climate change affects agricultural productivity. Ground-level ozone damages crops.
2. Achieve universal primary education	Cleaner air will decrease the illnesses of children due to exposure to harmful air pollutants. As a result, they will miss fewer days of school. Water-related diseases such as diarrhoeal infections cost about 443 million school days each year and diminish learning potential.
3. Promote gender equality and empower women	Indoor and outdoor air pollution is responsible for more than 2 million premature deaths annually. Poor women are particularly vulnerable to respiratory infections, since they have high levels of exposure to indoor air pollution. Women and girls bear the brunt of collecting water and fuelwood, tasks made harder by environmental degradation, such as water contamination and deforestation.
4. Reduce child mortality	Acute respiratory infections are the leading cause of death in children. Pneumonia kills more children under the age of five than any other illness. Environmental factors such as indoor air pollution may increase children's susceptibility to pneumonia. Water-related diseases such as diarrhoea and cholera kill an estimated 3 million people per year in developing countries, the majority of whom are children under the age of five. Diarrhoea has become the second biggest killer of children, with 1.8 million children dying every year (almost 5,000 per day).
5. Improve maternal health	Indoor air pollution and carrying heavy loads of water and fuelwood adversely affect women's health and can make women less fit for childbirth and at greater risk of complications during pregnancy. Provision of clean water reduces the incidence of diseases that undermine maternal health and contribute to maternal mortality.
6. Combat major diseases	Up to 20 per cent of the total burden of disease in developing countries may be associated with environmental risk factors. Preventive environmental health measures are as important and at times more cost-effective than health treatment. New biodiversity-derived medicines hold promises for fighting major diseases.
7. Ensure environmental sustainability	Current trends in environmental degradation must be reversed in order to sustain the health and productivity of the world's ecosystems.
8. Develop a global partnership for development	Poor countries and regions are forced to exploit their natural resources to generate revenue and make huge debt repayments. Unfair globalization practices export their harmful side-effects to countries that often do not have effective governance.

Source: UNDP 2007.

However, progress is being made in some areas, such as certification. The Forest Stewardship Council (FSC) (Chapter 9) has now certified a total of more than 80 million hectares of commercial forest in 80 countries. A Marine Stewardship Council, modelled on the FSC, has been initiated, and to date it has certified 22 fisheries offering 500 certified seafood products and covering 7 per cent of the world's marine catch. However, significant barriers to international and national efforts to enact environmentally sound practices are increasingly being challenged as 'protectionist' (Box 15.1).

Advances have also been made in alternative energy sources. Global production of solar photovoltaic (PV) cells, which turn sunlight directly into electricity, has risen sixfold since 2000 and grew by 41 per cent in 2006 alone. Although grid-connected solar capacity still provides less than 1 per cent of the world's electricity, it increased by nearly 50 per cent in 2006 to 5,000 megawatts, propelled by booming markets in Germany and Japan. There has been a dramatic growth in PV production

in China. In 2006, China surpassed the United States to become the world's third largest producer, trailing only Germany and Japan. Global wind power capacity increased by almost 26 per cent in 2006, exceeding 74,200 megawatts by year's end. Global investment in wind power was roughly $22 billion in 2006, and in Europe and North America, the power industry added more capacity in wind than it did in coal and nuclear combined. The global market for wind equipment rose 74 per cent in 2005 and 2006. Germany, Spain, and the United States generate nearly 60 per cent of the world's wind power, but again, production in Asia is growing rapidly. In 2006, India was the third largest wind turbine installer, and China took the fifth spot, thanks to a 170 per cent increase in new wind power installations over the previous year. More than 50 nations now tap the wind to produce electricity, and 13 have more than 1,000 megawatts of wind capacity installed (Worldwatch Institute 2007b).

Obviously, given the very different conditions prevailing in different parts of the world, different solutions are needed to

Box 15.1   Trade and the Environment in Conflict?

Meetings of the World Trade Organization (WTO) have attracted increasing attention since the organization was first formed in 1995. Environment is a central concern. One of the goals of the WTO is to break down unfair trade barriers that prevent countries from gaining access to the markets of others. This is particularly important to many developing countries that want to market and sell their goods to the rich consumers of the developed world. Over the years, many developed nations have erected trade barriers to protect their own industries from being undercut by such overseas competition.

However, many countries have also established laws and policies that relate to minimum standards, including environmental standards, that must be met by locally produced and imported goods sold within the country. Less developed countries argue that such laws and policies discriminate against their products, since it is too costly for them to meet these standards. Therefore, setting such standards is protectionist and violates international trade agreements.

This difference of perspective has set the stage for some high-profile international disputes relating to the conflict between environmental protection and trade barriers. The

US, for example, placed an embargo on tuna from Mexico following declaration of the domestic Marine Mammal Protection Act. Mexican tuna fishing led to the deaths of thousands of dolphins through the practice of setting tuna nets around the highly visible dolphins that school with the tuna. Mexico claimed, successfully, that the embargo discriminated against their fishers. Several Asian countries recently won a similar case involving the shrimp fishery. US law requires shrimp fishers to have turtle-excluding devices on their nets to prevent the drowning of turtles. All species of marine turtle are listed as endangered.

International trade and the rulings of the WTO obviously have major implications for future environmental goals. How they will be played out is far from certain. On the one hand, there is enormous potential for international trade law to actively encourage a transition to a more sustainable society. Some of the most damaging environmental activities, for example, are the perverse subsidies that many governments use to support unsustainable fishing, agricultural, and forestry activities. On the other hand, past rulings do not contribute to much optimism that this proactive environmental role will be played in the near future.

Treaties, Agreements, and Commitment

The number of international environmental accords has exploded as countries awaken to the seriousness of transboundary and global ecological threats. The UN Environment Programme (UNEP) estimates that there are now more than 500 international treaties and other agreements related to the environment, more than 300 of them negotiated in the last 30 years.

But reaching agreements is only the first step. The larger challenge is seeing that the ideals expressed in them become reality. What is needed is not more agreements but a commitment to breathe life into the hundreds of existing accords by implementing and enforcing them.

—Mastny and French 2002, 13

Extreme poverty forces people in some parts of the world to consume the resources today that they and later generations will need to survive in the future. Here a poor fisher from Sri Lanka's east coast wheels dynamited coral reef fragments to the brick factory for processing. He knows he is destroying the reef that should feed his children in the future, but the reef has been so overfished that it no longer provides for the needs of today. Intact, the reef could have been a protective barrier against the Asian tsunamis of 26 December 2004.

Global challenges require different solutions in different parts of the world. While most of us need to cut our consumption, this peasant family in Myanmar has nothing to cut. They need to increase their levels of consumption just to meet what many of us would classify as basic needs.

find the most effective way to achieve sustainability. For some countries, increased consumption is required if people are to meet their basic needs. For others, drastic reductions in consumption are necessary. Such reduction, however, does not necessarily imply a reduction in quality of life. The Human Development Index (HDI) of the United Nations, for example, takes into consideration education, longevity, and living standards rather than just GNP (gross national product) as measures of development. For very poor people, even a small increase in energy consumption can make a major difference to their living standard. It is difficult to spend long hours studying at night, for example, if there is no electricity available. This retards educational levels, which in turn holds back economic development. The benefits of additional energy availability and income increase up to a certain point. After that point, there is no relationship between consumption and the HDI.

Other research points to similar conclusions regarding the alleged need for energy consumption. Robert Prescott-Allen (2001) measured 87 indicators ranging from life expectancy and school enrolment to deforestation in 180 countries to develop his Wellbeing Index. He found no relationship between energy consumption and the well-being of a country. In fact, the United Arab Emirates, with the world's second highest per capita energy consumption, was ranked 173rd in terms of well-being. Austria was ranked fifth in well-being and 26th in energy consumption. Sweden and the Netherlands are ranked very closely in terms of human well-being, yet the Netherlands has a much lower environmental rating. This illustrates that a high degree of human well-being can be achieved independent of environmental costs.

Imagine that this was not the case but rather that there was a direct correlation between energy consumption and societal well-being. This would trigger an imperative to raise the level of energy consumption throughout the world at least to North American levels, which would entail a fivefold increase in energy use and a probable fivefold increase in all the attendant problems, including acid rain generation (Chapter 4) and global climatic change (Chapter 7). The scale of disruption to planetary biogeochemical cycles (Chapter 4) would be catastrophic.

Thankfully, this is not the case. The studies cited above confirm what many people already know intuitively, even though the truism is something we struggle to address in terms of national policies or individual choices: *The best things in life aren't things.* Consuming more will not, after a certain threshold, improve the overall well-being of society or individuals. The challenge at the international level is to enact policies and programs that will see consumption levels raised in needy countries but reduced in overconsuming countries such as Canada. Reduction in material consumption in wealthier nations will have to be in the order of 90 per cent for some semblance of sustainability and equity to emerge. In short, people in developed countries need to rediscover the importance of the *quality* of life as opposed to the current emphasis on the *quantity* of goods that can be acquired.

The Good Life

Rethinking what 'constitutes the good life' is overdue in a world on a fast track to self-inflicted ill-health and planet-wide damage to forests, oceans, biodiversity, and other natural resources. By redefining prosperity to emphasize a higher quality of life rather than the mere accumulation of goods, individuals, communities, and governments can focus on what people most desire. Indeed, a new understanding of the good life can be built not around wealth but around well-being: having basic survival needs met, along with freedom, health, security, and satisfying social relations. Consumption would still be important, to be sure, but only to the extent that it boosts quality of life. Indeed, a well-being society might strive to minimize the consumption required to support a dignified and satisfying life.

—Gardner and Assadourian 2004, 165

The **Happy Planet Index (HPI)** is one index that attempts to provide a different perspective on human well-being and environmental impact and focus on achieving sustainability. The HPI is based on the idea that most people want to live long and fulfilling lives and that the country doing the best is one that allows its citizens to do so while avoiding infringing on the ability of people in the future and in other countries to do

Imagine

Imagine no possessions
I wonder if you can
No need for greed or hunger
A brotherhood of man.
Imagine all the people
Sharing all the world . . .

—John Lennon

the same. Human well-being is assessed as Happy Life Years and impacts by measuring ecological footprints per capita, as discussed in Chapter 1. The HPI is not a measure of the 'happiest' countries in the world. Countries with relatively high levels of life satisfaction are found throughout the index, but the index provides an assessment of the environmental efficiency of supporting well-being in a given country. Such efficiency could emerge in a country with a medium environmental impact and very high well-being, but it could also emerge in a country with only mediocre well-being but very low environmental impact (such as Vietnam). Each country's HPI value is a function of its average subjective life satisfaction, life expectancy at birth, and ecological footprint per capita. The exact function is a little more complex, but conceptually it approximates multiplying life satisfaction and life expectancy and dividing that by the ecological footprint. In 2006, the island state of Vanuatu was the highest scoring, followed by Colombia and Costa Rica, while Burundi, Swaziland, and Zimbabwe ranked at the bottom of the list. Canada ranked 111th.

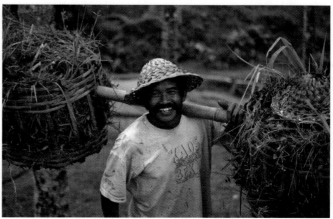

'The best things in life aren't things.' This Bali rice farmer has none of the mechanical aids of our modern farmers, none of the household appliances that you and I have, but he is a happy man.

Philip Dearden

These kinds of indices are useful, even if controversial, because they make us think of different ways of looking at progress and what we are trying to achieve. The traditional way of looking at global development along the consumer-based lines that Western societies have 'progressed' is simply untenable for the rest of the world. All evidence clearly shows that planetary ecosystems are collapsing under current burdens and can no longer provide the services required for future generations. Adding further to these stresses is an option only for the most short-sighted and self-centred of societies. Unfortunately, as Jared Diamond points out in his book *Collapse* (2005), some societies took exactly that route in the past. It usually occurred where there was a ruling class that benefited greatly from the existing state of affairs and preferred to see it continue even over the short term rather than risking their extravagant lifestyle. Diamond suggests that there could be a lesson for current society in his analysis.

The challenge then is to reduce the level of stress below current levels while raising the standard of living of the world's poor to acceptable levels. Although international agencies such as the World Bank do define a 'poverty level', ongoing analysis shows that this level varies remarkably among societies and that a broader range of values has to be taken into account. The Happy Planet Index is one example of such an approach, and it applies not only to the so-called underdeveloped nations but equally to developed nations such as Canada. The end goal is to produce as much human well-being as possible for society with minimal environmental impact. The next section discusses some national perspectives on this challenge.

National Perspectives

Canada is a remarkable country. It is the world's second largest country, with some of the largest remaining wilderness areas in the world. Canadians enjoy one of the highest standards of living in the world. No country has topped the global Human Development Index rankings more often than Canada, although it has not done so since 2000 (UNDP 2007). We seem to be doing fine, and this is exactly the main challenge we face. We are few in number with a big country and a high standard of living. Most Canadians have no idea of the scope of the global challenges because of our isolation from these global pressures. As a result, our performance in most areas of environmental management, from failing to comply with global treaties or to pass and enforce effective legislation to the everyday statistics that document our excessive per capita global impact, is largely overlooked. Throughout this book, we have cited many examples. The audits undertaken on environmental and sustainability issues by the government's own Office of the Auditor General also provide many examples. The audits repeatedly show failures by government departments to comply with their own policies and legislation. We make commitments at international and national levels that we consistently fail to fulfill. To emphasize the seriousness of these shortcomings, we refer you to Box 15.2 in which we reproduce the main findings of the Commissioner of the Environment and Sustainable Development for 2008. The 2008 report is particularly significant because the commissioner re-examined past audits to see whether appropriate changes had been made.

Box 15.2   **What Did We Find?**

We found mixed progress by departments and agencies in addressing and resolving the recommendations and findings included in this Status Report. Of the 14 chapters in the report, five show satisfactory progress and nine show unsatisfactory progress. Where satisfactory progress was made, four success factors were present—realistic objectives, strong commitment at senior levels, clear direction, and adequate resources. Where progress was unsatisfactory, some or all of these factors were absent.

We report satisfactory progress in addressing recommendations and findings in our three chapters on chemicals management, our chapter on insurance for nuclear operators, and our chapter on military dumpsites. In each of these five chapters, the success factors were present. For example, Chapter 3 notes that in dealing with abandoned mines in the North, the government identified the major ones, created a central management capability to work with individual departments and agencies, established a clear work plan with timelines, and provided significant funding.

Unfortunately, we report unsatisfactory progress in addressing recommendations and findings on the other topics covered in this Status Report. In each case, we observed lack of commitment at senior levels and often inadequate funding. Our chapters dealing with protection of species at risk and their habitat, control of aquatic invasive species, and the restoration of heavily polluted areas of concern in the Great Lakes provide ample evidence of this.

Of particular concern is the poor performance by departments and agencies in conducting strategic environmental assessments when developing policy and program proposals. These assessments are required when proposals that are submitted to Cabinet have an environmental impact. Public reporting is required whenever assessments are done. This is similar to the situation we found last October when examining sustainable development strategies where performance was also poor. Once again, the lack of commitment at senior levels is a root cause of these problems. Strategic environmental assessments and sustainable development strategies need to be revitalized on a priority basis in order to help the government address environmental issues of the past, avoid similar issues in the future, and achieve sustainable development over the longer term.

Source: Auditor General of Canada 2008.

Violations of our commitments under the Kyoto Protocol, discussed in Chapter 7, are well known, but there are many others. In terms of marine conservation, projections suggest that we will manage less than a third of our commitments under the Convention on Biological Diversity to establish a network of marine protected areas. Perhaps most astonishing is the lack of government response to many of these failures. On global change, we are charting a course diametrically opposed to that of the rest of the world. The 2008 proposal by the federal Conservative government allows Alberta's oil sands to more than triple their annual emissions of greenhouse gases over the next decade and exempts more than 20 per cent of emissions from smaller operations from any control. Oil sands operations would be allowed to offset (i.e., pay for) their entire pollution through 'pre-certified investments'. Estimates suggest that annual emissions from the oil sands would rise from 25 to 75 million tonnes per annum over the next 10 years. The Commissioner of the Environment and Sustainable Development, Johanne Gélinas, was suddenly relieved of her duties by the Conservative government in 2007 shortly after she released her independent report raising concerns about the continued expansion of the oil sands.

Pollution does not come without costs, not only to the environment but also to people. A recent study (Boyd and Genuis 2008) suggests that pollution could be causing up to 25,000 premature deaths in Canada each year and burdening the health care system with up to $9.1 billion annually in extra costs. Pollution is responsible for 8,000 to 24,000 new cases of cancer and 500 to 2,500 low–birth weight babies annually. The study estimated that pollution causes about 1.1 million to 1.8 million restricted activity days for asthma sufferers each year and that Canadians as a group spend from 600,000 to 1.5 million extra days in hospital annually.

The availability of information to interested citizens also seems increasingly restricted in Canada, in contrast to many other countries where growing Internet use has promoted greater citizen access to government data. The quotation box

The government has made unsatisfactory progress toward providing a complete and understandable picture of the results expected from Canada's international environmental agreements. While Environment Canada, Fisheries and Oceans Canada, Transport Canada, and Foreign Affairs and International Trade Canada generally make information available on Canada's obligations under the agreements, they provide less information on the programs and means in place to meet the obligations. In addition, the departments do not generally make complete and understandable information available on the results the government both expected to achieve and has achieved toward fulfilling obligations under the agreements.

The government has made unsatisfactory progress in planning, monitoring, and reporting the extent to which Canada is meeting its commitments from the World Summit on Sustainable Development in 2002. While it has followed the United Nations Commission on Sustainable Development approach to monitoring and reporting, it still has no longer-term plan for ensuring that it will be in a position to report significant progress on its commitments, while taking into account the review timetable established by the UN.

—Auditor General of Canada 2008

on page 577 contains the findings of the Office of the Auditor General in its 2008 follow-up on the flow of information from government regarding Canada's international environmental obligations. Environment Canada's *Environmental Signals: Canada's National Environmental Indicator Series* (2003b), which was produced in 2003 and provided an excellent and comprehensive database that was used extensively in the previous edition of this book, has been discontinued. The *State of the Forests* reports cited in Chapter 9 have become much less comprehensive in their data coverage. Even government scientists are finding that their efforts to report their work now have to be cleared to ensure that they respond along 'approved lines' (Munro 2008).

Reports required by legislation are frequently delayed for long periods. Under the Canada Water Act, the government is required to produce an annual report on the state of Canada's water resources. The first one was produced in 2001. The second, the report for the 2004 fiscal year, finally appeared in 2007. A *State of Protected Heritage Areas* report is required by legislation every two years; the last one came out in 2005.

These reports are required by legislation. Yet in many cases, little is done to ensure that the law is followed because of a lack of public and hence political interest.

This is beginning to change. The attention focused on global climate change at the international level is so intense that it is even creating change within Canada. By 2007, more than two-thirds of Canadians ranked climate change as a 'very serious problem' (Box 15.3). Politicians who were firmly sceptical of global climate change are now recognizing that the public is becoming increasingly well-informed on the issue

Box 15.3   Poll: Increasing Majority Call Canada's Pollution Laws Inadequate; Global Warming Is Top Environmental Issue

Vancouver, 4 September 2007—A newly-released poll by the Environmental Monitor research program shows that seven in 10 (70%) Canadians now call this country's pollution laws inadequate, up sharply from 11 years ago when just fewer than half (41%) felt this way. The Environmental Monitor poll also found that when Canadians are asked to spontaneously name their top environmental concern, the most frequent mention is global warming (29%), up nine points nationally from July 2006. When asked to rate the seriousness of climate change, 66 per cent of Canadians view climate change as a 'very serious' problem, up from 57 per cent in 2006 and 44 per cent in 2000.

Comparing this data to other countries polled, Canadians rank among the top five countries in the world for their views on the seriousness of climate change—84 per cent of Brazilians rate climate change a very serious problem, followed by Chile (70%), India (68%), and Canada, tied with Mexico and France. Interestingly, a bare majority of Americans (50%) view climate change as a very serious problem.

More Canadians than ever now favour stricter regulation of industrial pollution, sentiments varying according to industrial sector. Asked about stricter regulations for various industries to combat global warming, clear majorities strongly favour these for the chemical industry (67%), oil and gas industry (62%), and the automobile industry (58%). Just 43 per cent strongly favour stricter regulations for the forestry, manufacturing, and mining industries. Canadians, especially younger adults under 25, are least likely to favour stricter regulations for the airline industry (39%), electrical power generators (39%), and the agriculture industry (34%).

'Canadians are growing increasingly concerned about the threat of climate change,' says Angus McAllister, president of McAllister Opinion Research. 'The public are initially open to less onerous alternatives to regulation like voluntary measures and tax incentives. But when the performance of a particular sector is perceived to be lagging, the public is quick to demand regulation and are slow to back off.'

Drawing from his firm's 20-nation polling over 15 years, GlobeScan president Doug Miller observes, 'Canadians are among the most concerned citizens in the world today about the seriousness of climate change—and that's saying a lot given the worldwide surge in concern over the last four years.'

Concern about global warming is by far the highest in Canada's North, while it is lowest in Quebec. Global warming is the top concern in all other regions, including Alberta.

University graduates and young adults under 25 are the most likely to name global warming as a concern, while high school graduates and women aged 55+ are least likely to do so.

Asked how much can be done to reduce the amount of future climate change or global warming, nearly eight in 10 (79%) Canadians express the view that a great deal or a good amount can be done. An identical question asked in a Washington Post/Stanford poll in April 2007 shows that just 63 per cent of Americans say a great deal or good amount can be done. McAllister says 'Canadians expect Canada to be a world leader when it comes to solving environmental problems.' The pollster cautions that this means higher expectations for environmental performance and greater potential for disapproval when these expectations are not met.

Source: 2007 Environmental Monitor, McAllister Opinion Research, http://www.mcallister-research.com/EMMediaRelease SEPT_4_2007.pdf.

and that their political futures are at risk if they continue to ignore it. Whether this will translate into meaningful action at the federal level remains to be seen, as discussed in Chapter 7. The proposal by the federal Liberal party in 2008 to follow in the footsteps of their provincial cousins in BC and impose a national carbon tax (Box 15.4) was ridiculed by the ruling Conservative party and was quickly dropped.

However, the rationale for such a tax is simple: people should be rewarded for adopting behaviours that are less likely to exacerbate the effects of global warming. This means taxing carbon-based activities, such as those using fossil fuels. In BC, for example, people pay extra at the gas pump every time they fill up. For 40 litres, they pay an additional $1; 100 fill-ups would cost an extra $100. However, the tax is designed to be revenue-neutral, with all income given back to taxpayers. In BC, everyone received a $100 cheque in June 2008. For anyone who does not participate in activities based on carbon, the $100 is free and clear; however, for anyone continuing to use fuel at their previous rate, it will soon be eaten up in fill-up costs.

Interestingly, BC politicians are already calling the tax a success even though polls show that more than half of the population oppose it. The politicians judge that success by the amount of attention that the tax has attracted and the increased awareness of the challenges of climate change as a result. The tax has also spawned a movement centred on people discussing what they could do with the $100 if they wanted it to be even more effective against climate change. Suggestions include investing it in a home energy audit to pinpoint sources of energy loss, replacing incandescent light bulbs, installing programmable thermostats, buying a clothesline, buying a bicycle and leaving the car at home, buying a bus pass, buying seeds and growing food, and donating the money to an organization that addresses global change (see www.bcsea .org). Perhaps most remarkable is the fact that BC's political leadership instituted such an innovative scheme knowing that most voters do not support it. In today's world, politicians often choose to do what the polls tell them and not necessarily what is best for society over the long run.

Other provinces are also moving forward with innovative ideas. Nova Scotia, for example, is hoping to harness the power of the world's largest tides at the Bay of Fundy to develop an electricity supply that could meet up to 15 per cent of the province's needs. The idea is not new. In the 1970s, a facility producing up to 20 megawatts daily was built, which acted much like a giant hydropower plant with a barrage facility. In contrast, the new projects involve using fixed or floating turbines within the tides. The province is also exploring an equally innovative project to see whether the old coal mines in Cape Breton, some 3,200 kilometres in length, can be used to produce geothermal energy. The mines are flooded with water at a temperature of 9 to 15°C, and the idea is to turn the mines that once supplied half of Canada's coal into a new energy resource.

The examples above illustrate two key principles for moving forward. First, enlightened political leadership can play a major role in visioning and implementing change at the societal level, and second, human ingenuity is vast but needs to be applied to the key challenges facing society.

The bottom line is that Canada and Canadians are not rising to meet environmental challenges and in many cases are falling sadly behind. We are followers and laggards rather than leaders. A recent example is the Canadian response to the International Chemical Management Agreement signed in Dubai in 2006. The agreement has prompted many governments to institute new measures to control chemical use. Leading the way is the European Union, which has started a program that requires industries to prove the safety of their products before they reach the market. Companies must pay for studies, reveal the chemical composition and toxicity of their consumer product, and have the data entered in a public registry. In Canada, under what was boasted as flagship legislation, the 1999 Canadian Environmental Protection Act, the taxpayer foots the bill for the assessment of new chemicals in a restricted number of tests (that were highly criticized by the Office of the Attorney General in 2008), companies can sell their products before tests are complete, and there is no obligation for the government to ban chemicals found

Box 15.4   Addressing Climate Change in BC

British Columbia has set a target to reduce greenhouse gases by at least 33 per cent below current levels by 2020. This is 10 per cent less than the 1990 level, bettering California's goal of returning to 1990 levels by 2020. Some initiatives include:

- Interim targets will be set for 2012 and 2016 through a Climate Action Team that will determine the most credible, aggressive, and economically viable targets.
- A longer-term emissions reduction target will be set for 2050.
- The Climate Action Team will also be asked to identify practicable options and actions for making the government of BC carbon-neutral by 2010.
- All electricity produced in BC will be required to have net zero greenhouse gas emissions by 2016.
- Greenhouse gas emissions from the oil and gas industry will be reduced to 2000 levels by 2016, including a zero-flaring requirement at producing wells and production facilities.
- A new $25-million Innovative Clean Energy Fund will be established to encourage the commercialization of alternative energy solutions such as bioenergy, geothermal energy, tidal, run-of-the-river, solar, and wind power.
- Tailpipe emission standards for all new vehicles sold in BC will be phased in between 2009 and 2016, reducing carbon dioxide emissions from autos by 30 per cent.
- A low-carbon fuel standard will be established that will reduce carbon intensity of all passenger vehicles by at least 10 per cent by 2020.
- The energy plan will require that 90 per cent of BC's electricity come from clean, renewable sources.
- Effective immediately, BC will become the first jurisdiction in North America, if not the world, to require 100 per cent carbon sequestration for any coal-fired electricity project.
- The $2,000 sales tax exemption on new hybrid vehicles will be extended.
- All new cars leased or purchased by the Province will be hybrids.
- A new unified BC Green Building Code will be developed with industry and communities.

- Legislation will be developed to phase in requirements for methane capture at landfills, the source of about nine per cent of BC's greenhouse gas emissions.
- New incentives to retrofit existing homes and buildings to make them energy-efficient will be introduced.
- New measures will help homeowners undertake 'energy audits' to identify possible energy savings.
- Real-time, in-home smart metering will help homeowners measure and reduce energy consumption.
- New strategies will be launched to promote Pacific Green universities, colleges, hospitals, schools, prisons, ferries, and airports.
- The Province will substantially increase its tree planting efforts.
- The Province will ensure that school curricula inform students about how they can reduce individual impacts on the environment at home and at work.
- Beehive burners will be eliminated.
- Trees infested by the mountain pine beetle will be used to create new, clean energy.
- A federal-provincial partnership will invest $89 million in hydrogen fuelling stations and the world's first fleet of 20 fuel cell buses. The new fuelling stations are part of the initial phase of the hydrogen highway from Whistler to Vancouver, Surrey, and Victoria.
- The Province will work with Pacific states to encourage a hydrogen highway from Whistler to San Diego by 2020. It would be the longest hydrogen highway in the world.
- BC will work with its neighbours to create electrified truck stops to reduce idling.
- The Premier will meet with governors to assess and address the impact of climate change on our oceans and establish common standards for Pacific ports.
- The Province will seek federal cooperation to electrify ports and reduce container ship carbon emissions in all Canadian ports.
- The Province will work with the federal government and Pacific partners to develop a sensible, efficient system to register, trade, and purchase carbon offsets and credits.

Source: British Columbia 2008.

dangerous. In the EU, if a company cannot prove the safety of a product, it cannot be sold. In Canada, the product can be sold before testing, while the taxpayer must pay for the tests and can only hope that the product is withdrawn if it is found to be unsafe.

Sustainable Development in Canada

The government's progress toward providing departments preparing sustainable development strategies with guidance on greening their operations is unsatisfactory. The government-wide targets it produced for departments to use in preparing their strategies were non-specific, reiterated previous objectives, or were open to interpretation, and departments were not required to use them. As a result, departments did not incorporate the targets consistently in their strategies, and the government is therefore not in a position to know what progress it is making in greening its operations.
—Auditor General of Canada 2008

Unfortunately, the longer we delay action to address environmental degradation, the more dramatic (and often more costly) the remedial actions will be when they do occur. In some cases, remedial action is delayed so long that the desired environmental conditions cannot be restored. Extinction is the ultimate example, but there are many others. Scientists have found that although politicians have met agreed-upon targets for reduced sulphate depositions, large areas are still receiving excessive sulphates, pH levels are not improving in many lakes, and biological indicators of recovery are even further behind. Reductions were made, but they were too little and too late to halt the acidification of many lakes in eastern Canada. The real challenge is to identify and mitigate these problems before they develop so far that they become either irreversible or so costly to reverse that they cannot be resolved.

Despite our poor response as a nation in many areas of environmental management, there are many actions that individual Canadians can take that will have a significant impact not only on the Canadian environment but also globally. If an average Cambodian were to halve his or her consumption of global resources, the incremental gains would be small. But because of the scale of our personal consumption, even individual Canadians can make significant changes. If 34 million Canadians made changes, the cumulative impact would

be staggering. The next section discusses some perspectives on environmental change at the personal level.

Personal Perspectives

Pick up That Degree
The future is increasingly a race between education and catastrophe.
— H.G. Wells

The difficulty with many of the environmental challenges discussed above is the vastness of their scale. They are so widespread that most Canadians do not realize they are out there, especially because we are more sheltered from their effects than people in smaller, more densely populated countries. Most of these problems also have long lag times (the period between the time when the processes are set in motion and when the effects are felt), especially when the effects may have different impacts in different parts of the globe. Global warming, for example, may lead to cooling and increased precipitation in some areas, while the effect may be just the opposite in other areas. Changes of this complexity and magnitude require long-term study before they can be understood. The scale of change also suggests that within the human lifespan, many of these trends may be irreversible.

Is our educational system preparing you and the students who will follow you to understand and deal with these changes? Urgent actions are required if we are to help diffuse these trends, but actions do not occur in a vacuum. They require some understanding of the road we are on, where it goes, and how we can get on other, more desirable roads. Educational systems should help to bring about this understanding. Currently, many schools, colleges, and universities graduate students who have little or no idea about how the ecosphere functions and how human activities impair those functions. They shop, travel, eat, drink, work, and play in blissful ignorance of the impact they may be having on life support systems.

Colleges and universities are frequently accused of not being part of the 'real' world. By 'real' world, people usually mean the economic realities of today's society. However, that is not the real 'real' world; it is essentially a game that humans invented to facilitate barter and exchange. Important? Yes! But is it the real world? Only partly! The real world is composed of the air we breathe, the water we drink, the organisms that keep the life support systems going, and the ground we stand on. Without these things, there can be no invented world, however 'real' it might seem. We concentrate on balancing these 'play' budgets when in reality, the more significant budgets of energy throughflow and material balance determine the future of society. The deficit? Yes, Canadians have a huge deficit, but it relates to the 60 million bison that no longer roam the Prairies, the skies cleared of all but a fraction of the birds that used to flock in such large numbers that day turned to night, and the seas that were once home to the largest animals ever to evolve on the planet but where now not even the smallest fish are safe from sonar detection, vacuum trawling, and human consumption. Yes, we have a deficit, most of which remains unrecognized in the routine financial and economic accounting procedures that help to shape national policies.

One major challenge contributing to lack of understanding and action on environmental matters is **nature deficit disorder**. The term was coined by author Richard Louv (2006) to describe what he saw as an increasing gap in understanding of the 'real world' on the part of the younger generation. Instead of playing outdoors in fields, woods, streams, lakes, or the ocean, an increasing proportion of the children of today are glued to their computer or TV screens. They seldom visit the outside world, especially areas dominated by nature rather than human activities. The decline in the number of visitors to national parks and similar areas across North America bears testimony to this trend. A feedback loop develops. The less exposure that young people have to the 'real world', the less they understand it and the less comfortable they feel outdoors. As a result, they tend to avoid encounters with nature and become even further estranged.

At a time when it is critical that more people become involved in significant environmental action, we seem to be producing a new generation that is further than ever from developing any attachment to the environment. This does not bode well for society in the future as this generation matures and becomes the main economic drivers and decision-makers. Concern over this situation has prompted movements across North America to provide opportunities and facilities to encourage outdoor re-engagement by younger people (e.g., see The Kesho Trust, http://www.naturechildreunion.ca). BC, for example, is looking at a program (Kesho Trust 2007) to reconnect children and families with nature and the outdoors, with these four objectives:

1. Children have outdoor playtime included in their school and out-of-school lives and have opportunities to freely access the outdoors and/or wild nature with friends and significant adults.
2. We design and build neighbourhoods and environments that allow and recognize nature to be nearby, accessible, and attractive to children and their families and seen as friendly, safe, joyful, and beneficial.
3. Children are seen as independent and competent and able to handle and benefit from the challenges of being outdoors and/or in wild nature.
4. We are seen as part of nature and are comfortable with the environment in which we live.

A primary function of our educational system should be to give students a general level of understanding about the nature of our environment and resources. We have requirements for general levels of language and mathematical competence but require nothing from our students in terms of this most fundamental challenge of the future. Indeed, there are now pressures, especially on universities, to put greater and greater emphasis on meeting the short-term economic demands of society. Business schools and faculties of commerce flourish, yet few additional resources are allocated for programs dealing with the environment.

Even in such programs, colleges and universities have seldom done a good job of instilling in students an appreciation for and love of the planet as discussed above. Increasingly, science programs have become a process of learning more and more about less and less. They have produced technically competent scientists, but these programs have often missed the mark considerably in terms of maintaining students' wonder about the natural world and combining the rigour of scientific inquiry with deep moral questioning. Science programs often mistake the laboratory for the 'real' world and cut students off from a more comprehensive understanding of and passion for their environment.

On Education

. . . *without significant precautions, education can equip people merely to be more effective vandals of the Earth. If one listens closely, it may even be possible to hear the Creation groan every year in May when another batch of smart, degree-holding, but ecologically illiterate* Homo sapiens *who are eager to succeed are launched into the biosphere.*

—Orr 1994

You can create change on your campus. Are there sufficient courses on the environment? Do these courses cover a wide spectrum from the technical to the philosophical, and more important, are students encouraged or even required to select from courses all along this spectrum? You should also remember that campuses are large consumers and processors of matter and energy. How efficient are they? Has anyone undertaken an environmental audit of your campus? How are wastes disposed of? How much recycling occurs? Are chemicals used for landscaping? Does the faculty pension fund invest in businesses with unsound environmental practices? There are many questions that you can investigate through course work, in environmental clubs, or as an individual. If you are interested in pursuing these ideas, you may wish to draw on such experiences elsewhere (e.g., see the guest statement in this chapter and Bardati 1995; 2006) and start some activities on your own campus.

So pick up that degree and encourage others to increase their understanding of environmental challenges. Do not be intimidated by people who think that interest in the environment and higher learning is not the 'real' world. Challenge your teachers to inspire you. Be interested. Apply what you learn to your life. Do not be misled, however, into thinking that the formal education system is the only source of learning. Keep on reading. Many of the most inspiring works on the environment do not make their way onto college or university reading lists, and most of the rewarding environmental experiences are certainly not part of the curriculum. Challenge society to change, and seek a new kind of relationship between humanity and our home, planet Earth.

Figure 15.2 One of the main ways to fight environmental degradation is to become aware of the implications of your own actions. Source: Paul Zanetti.

Guest Statement

Sustainability in Higher Education: Learning It, Teaching It, and Doing It
Darren Bardati

Our greatest challenge lies in rethinking what kind of education is appropriate for a species whose standards of success threaten its ecological foundations.

—Orr 1992

I first read the above quotation in the summer of 1993. I was trying to come up with a topic for a senior undergraduate independent study project and had stumbled across a special issue of the journal *New Directions for Higher Education* called 'The campus and environmental responsibility'. Dr Orr's opening chapter, in particular, impressed upon me a new perspective on my university education. He wrote of the irony of so-called 'higher education' graduates who were ignorant about their individual impacts on the earth's ecosystems. I asked myself: Would I also become an ignorant graduate?

The special issue contained 11 chapters on how university and college campus communities had begun to respond to the environmental challenges of the times. The authors observed that for decades, academics had been 'pointing the finger' at corporations and industrial managers for poor environmental practices that damage the environment, while turning a blind eye to the impacts of

their own campus activities. Noting that the campus is a major natural resource user and a microcosm of environmental problems linked to the larger community in a myriad of ways, these chapters discussed practical experiences towards 'greening' campuses through improvements such as energy efficiency, waste reduction, purchasing and investment decisions, and raising environmental awareness. I asked myself: What was the state of the environment on my university campus?

Thus began my 15-year (and counting) journey into the fascinating world of sustainability in higher education. That summer's independent study turned into an undergraduate thesis the following year, where I performed the first sustainability audit at Bishop's University and crafted an environmental action plan with specific recommendations based on my findings.

When I began teaching at Bishop's in 1996, I incorporated the periodic campus sustainability audit into a senior undergraduate environmental planning course. It was a natural fit to involve students in a 'hands-on' project that would apply the theory from the class lectures. Thanks to the collaboration of every sector of the campus community—including other students, faculty, staff, administrators, and members of the Board of Governors, a slow and gradual but systemic transformation took place toward a more sustainable campus. Major barriers to progress were identified and resources mobilized to overcome them. Students played the leading role in this transformation by bringing the environmental awareness issues to the forefront of every campus user's mind and by voluntarily doing a lot of the groundwork that paid university employees did not have time to do. The sustainability audit was performed every three years, and the number of people involved grew from a small handful to more than 100. Several actions were taken every semester to raise environmental awareness and to reduce the university's ecological footprint.

By 2006, when the fifth edition of the student-led sustainability audit was published, it had the full support of the university's Board of Governors and its administration. Now sustainability assessments, policy statements, measurable targets, and much improved practices are firmly established in strategic planning documents, in daily operations, and in the overall culture of the university.

The experience at Bishop's University was certainly not unique, nor should Bishop's be seen as a leader in sustainability. While my students and I struggled to integrate sustainability theory with practical experience on our own campus, the scope of the sustainability assessments grew, drawing from experiences in other parts of North America and Europe, where a 'greening the campus' movement had taken hold. In the past decade or so, dozens of books have been published on the topic of campus sustainability, ranging from the fundamental reasons for such a transformation to practical 'how to' manuals. A refereed journal—the *International Journal of Sustainability in Higher Education*—was launched, and numerous associations sprang to life or grew in stature, such as the Association of University Leaders for a Sustainable Future, the Sierra Youth Coalition, and the Association for the Advancement of Sustainability in Higher Education. A quick glance at their websites will convince anyone that the expectation is fully established for institutions of higher education to lead by example.

One lesson learned from this experience is that progress towards a

UPEI Photography

Darren Bardati

more sustainable campus happens best when enthusiasm and patience walk hand in hand. The ability to create, maintain, and harness the enthusiasm required to make positive changes must be matched with sufficient patience, communicative skill, and logic to enable these desired changes to happen in a manner that has long-term significance, or 'staying power'. No one group, whether students, faculty, or administrators, can do it on its own or should be held responsible for doing it all. Although 'paper changes' such as adopting internal policy statements and signing international declarations like the Talloires and Halifax Declarations are useful tools in the transformation, as are standardized criteria to compare sustainability progress among universities, each campus community must create its own individual path towards sustainability.

Thankfully, nowadays, sustainability assessments are well entrenched on most university campuses, and students have many opportunities to get involved in practical sustainability work on their campuses. It seems obvious to repeat it: Everyone has a part to play! So join the local environmental club and get involved! If none exists, start one! And please, please do not be afraid to contact your university's administration and ask them about their environmental commitments and targets and how you can help the university to achieve them.

Despite so much progress in the past 15 years in the area of campus sustainability, I still wonder about Dr Orr's challenge of rethinking our university education. Are university graduates more environmentally literate than they were back in the early 1990s? Are we moving from promoting campus sustainability to achieving more in-depth knowledge and concern for sustainability in higher education? Regardless of what degree our students earn, are they more personally responsible for reducing their own ecological footprint? I hope so.

After serving in the Department of Environmental Studies and Geography at Bishop's University for many years, Darren Bardati is now the director of Environmental Studies at the University of Prince Edward Island.

Table 15.2 Annual Expenditure on Luxury Items Compared with Funding Needed to Meet Selected Basic Needs			
Product	**Annual Expenditure**	**Social or Economic Goal**	**Additional Annual Investment Needed to Achieve Goal**
Makeup	$18 billion	Reproductive health care for all women	$12 billion
Pet food in Europe and the US	$17 billion	Elimination of hunger and malnutrition	$19 billion
Perfumes	$15 billion	Universal literacy	$5 billion
Ocean cruises	$14 billion	Clean drinking water for all	$10 billion
Ice cream in Europe	$11 billion	Immunizing every child	$1.3 billion

Source: Worldwatch Institute, State of the World, 2004, *www.worldwatch.org.*

Light Living

'Light living' expresses the need to tread as lightly as possible, to minimize our own ecological footprints (Chapter 1). What follows is a brief selection of ideas that you may wish to try. Given the statistics presented elsewhere in the text regarding per capita energy consumption, waste production, and water consumption levels in Canada, you can be assured that Canadians contribute greatly to overconsumption. **Light living** is often characterized by the four R's: refuse, reduce, reuse, and recycle.

Refuse

We live in a society geared towards making consumption easy. Our newspapers are full of advertisements regarding the best buys. Turn on the radio, and you hear from the sponsor, or you are bombarded with commercials on TV. Estimates suggest that the average American will see 35,000 television ads every year. Most of us are surrounded by shopping opportunities on a daily basis. We can drive to one of several mega-malls in most Canadian cities, park with ease and at little expense, and consume from a wide variety of stores, pay by credit card, ATM, cheque, or even cash. We frequently shop not to fulfill basic needs but to indulge frivolous and petty whims (Table 15.2). Clothes are discarded when no longer fashionable rather than when they lose their durability. Gadgets are discarded in favour of newer and shinier models. Christmas as a religious holiday is now the time when we pay homage to our greatest god, **consumerism** (Box 15.5).

Resist and *refuse* to buy anything that you do not really, really *need*. If the purchase is necessary, shop carefully. Buy items that are less harmful to the environment during all stages of the product life cycle, from manufacture to consumption and

disposal. There are products certified by the government for their low impacts, as discussed later. Buy one quality item rather than a succession of several shoddy ones to fulfill your needs. Buy organic produce wherever available, or better still and if possible, refuse to buy any and grow your own. It is difficult, because consuming is easy. All the messages that we receive from society extol the virtues of buying things.

Reduce

Can you reduce your consumption of certain items? Energy is a good place to start and a major contributor to greenhouse gas emissions. More than a quarter of Canada's GHG emissions occur as a result of the everyday activities of Canadians. The previous Liberal government asked every Canadian to reduce their emissions by one tonne per year, or about 20 per cent of current consumption, to contribute to the solution and provided plenty of examples on how to do this. Unfortunately, the program was discontinued, but it is still an excellent goal for you to adopt personally. A lot of energy is used in space heating in Canada, but must you set that thermostat so high? Canadians tend to keep their houses much warmer inside than northern Europeans do, for example. These cultures (New Zealanders, too) are accustomed to setting a low thermostat and wearing warmer clothing in the house in winter. They would not expect to be comfortable wearing just a T-shirt. Turn down the thermostat, wear warmer clothes, and turn the thermostat down further when you go out or go to bed.

Reduce lighting costs by replacing burnt-out bulbs with long-life bulbs. They may be more expensive to buy, but they last 10 years or longer and use 75 per cent less electricity. An incandescent light bulb converts only about 5 per cent of the

Box 15.5   Dreaming of a 'Green' Christmas?

Christmas heralds the biggest consumer bash of the year, although merchants are also trying to persuade us to be equally excessive at other times. Take control of your consumer lifestyle at Christmas, and the battle is half won. Consider the following gifts:

1. Arrange an event, an outing, or a personal service rather than giving a material item.
2. Increase 'green' education by giving a book or subscription to a magazine. You could also buy someone a membership in a 'green' organization, such as Pollution Probe, the Canadian Parks and Wilderness Society, or Greenpeace.
3. Give a houseplant, a backyard composting kit, or an unbreakable coffee mug to replace the use of disposable ones. The World Wide Fund for Nature also enables you to protect an acre of rain forest by making a donation to fund a project that was started by unemployed students.
4. Give, in your gift recipient's name, a goat, a llama, a water buffalo, or any number of other animals through Heifer International, a 60-year-old NGO with a mission of alleviating hunger, poverty, and environmental degradation by providing food-producing animals and sustainable agriculture education to families in need.
5. Give something that conserves energy, such as a bus pass or an energy-saving showerhead.
6. Give second-hand items.
7. Make your own gifts, such as a sweater, dried flowers, or jam.
8. Give items that display the EcoLogo of three doves.
9. Choose gifts that require little wrapping. Reuse old wrapping paper or use reusable fabric gift bags instead.
10. Put 'Planet Earth' at the top of your list. If we all gave the planet an offering for Christmas that would make it feel better, the Earth would be a little more loved and a little less stressed.

energy it produces into light; the rest is wasted heat energy, as you will recall from the discussion of the laws of thermodynamics in Chapter 2. In 2007, Australia became the first country in the world to ban incandescent light bulbs to prevent wasted energy and cut greenhouse gas emissions. Calculations suggest that phasing out incandescent bulbs over the next three years could save about 800,000 tonnes of carbon dioxide emissions a year by 2012 and as much as 4 million tonnes by 2015. If you want a simple and easy way to make your own contribution, changing your light bulbs is a good place to start.

When replacing electrical appliances, energy efficiency should be a major factor in your choice. Before you turn on a light or an appliance, think about whether you really need it. Every time you flick the switch or plug in an appliance, you are sending out the message of demand. Electricity-producing utilities and governments will react to your message by building new production facilities with all their attendant environmental costs. If you do not want those costs, try not to send along as many messages signalling your demand.

Transportation is also a big energy consumer, accounting for one-quarter of all energy used in Canada. Road vehicles are responsible for 83 per cent of that share. Canadians' average per capita gasoline consumption is 1,100 litres per year, compared with 350 to 500 litres in European countries. It is also estimated that each kilometre of road or highway takes up about 6.5 hectares of land. In Ontario, which has 155,000 kilometres of highways, roads, and streets, this would add up to one million hectares for motorized vehicles. Whenever feasible, walk or ride a bicycle. If you have to use motorized transport, use public transport such as buses and trains. If you must have a car, get a small economical one with a standard transmission, use it sparingly, and try to car-pool (Box 15.6). Most people are aware that larger cars (and SUVs) consume more gasoline. However, since it takes 18 litres of water to produce one litre of gasoline, they also contribute to water deficiencies (Sawin 2004).

Box 15.6   The Environment-Friendly Driver

How you drive is as important as what you drive in terms of minimizing negative environmental impacts. Here are some tips:

1. Slow down. Cutting speed from 112 kilometres per hour to 80 reduces fuel consumption by 30 per cent.
2. Keep your tires inflated. By cutting tire drag, radials give you a 6 to 8 per cent fuel saving.

3. Keep your car tuned. This will ensure maximum efficiency and minimum pollution.
4. Minimize idling times. Even −20°C requires only a couple of minutes of warm-up. Many municipalities in Canada have passed bylaws that forbid idling for longer than three minutes. Has yours?

Philip Dearden

In Europe, few students have cars. Cities are designed for people, and it's easy to get to campus by bicycle or mass transit. Even when people buy cars, they prefer small ones.

Food choices are second only to transportation in terms of their environmental impacts. Agriculture covers more than 25 per cent of the world's surface and profoundly affects the health of natural ecosystems. A meat-rich meal with ingredients imported from afar generates as much as nine times the carbon emissions of a vegetarian meal made from local produce and requires two to four times the land to produce (FAO 2003a). It is not only what we choose to eat but how much we eat as well. In Canada, as in other Western countries and in the consumer classes of developing countries, growing prevalence of obesity is one of the main challenges not only to personal health but also to global health.

In fact, the number of clinically obese people now exceeds the number of chronically hungry people.

In 2008, researchers calculated that obese people consume 18 per cent more calories than the average, which overall results in more fuel being consumed to produce and transport food. Overeating drives up the cost of food by increasing demand. It also contributes to global warming by boosting food production. Edwards and Roberts (2008) call for policies to promote walking and cycling that would help to reduce obesity in the population as well as reducing fuel consumption by the transportation sector. Reducing the prevalence of obesity would, they say, reduce the global demand for both fuel and food. They conclude that decreased car use would reduce greenhouse gas emissions and thus the need for biofuels produced from vegetative matter. It would also mean more physical activity and reduce the risk of injury from traffic accidents as well as air pollution, thereby improving population health.

The excess consumption by the world's obese costs $20 billion annually to which must be added indirect costs of $100 billion resulting from premature death and related diseases.

—Jacques Diouf, director general, FAO,
World Food Summit, 2008

Eat only as much as your body needs to function in an effective and healthy manner. Become part of the food democracy movement. This means using your purchasing power to help establish connections between consumers and producers in the local area and making sure that you eat the kinds of food you wish to eat, not the kinds promoted by government subsidies and industrial agriculture. See how close you can come to the **100-mile diet**. The term was introduced in Vancouver in 2005 and refers to buying and eating food that has been grown, manufactured, or produced entirely within a 100-mile radius of one's residence (Smith and MacKinnon 2007) The average distance in the modern North American industrial food system is 1,500 miles from farm to plate (Halweil 2002), which obviously involves significant environmental costs.

You can also think about reducing the waste associated with the things you buy. Many products are overpackaged. They may look good on the store shelf but will only add to the amount of waste sent to the landfill. When you can, buy groceries in bulk to help cut down on packaging. Reduce your waste by starting a compost heap for kitchen wastes. Reduce and if possible eliminate your use of toxic materials. Products that may seem quite innocuous (paint, solvents, and cleaning agents, for example) become hazardous wastes when disposed of. Try to find alternatives to these products. Do not buy more products than you need, and dispose of them in full accordance with the instructions from your local municipality.

Some countries, such as Bangladesh and China, have banned plastic bags while others, such as Ireland, South Africa, and Taiwan, have legislation to discourage plastic bag use, with authorities either taxing shoppers who use them or imposing fees on companies that distribute them. Canada has no such law, but Leaf Rapids, MB, was the first municipality in North America to propose, pass, and adopt a law forbidding shops to use plastic bags.

Reuse

Buy products that can be reused, such as rechargeable batteries. Try to find another use for something no longer useful in its original state. Use plastic food containers to store items in your fridge or workshop. Return to the store with the same plastic bags that you used the previous time and use them for your next load of groceries, or better still, use cotton bags. When you are finished with something, it may still be useful to someone else. Organize a garage sale or donate the items to charity rather than throwing them out.

Recycle

Recycling facilities have sprung up all across the country over the past decade. Recyclable materials include newspaper, cardboard, mixed paper, glass, various metals, some plastics, car batteries, tires, and oil. These materials can be reprocessed into new goods. It takes 30 to 55 per cent less energy, for example, to make new paper from old paper than it does to start fresh from a new tree. Estimates suggest that if we recycled all the paper we use in Canada, we would save 80 million trees. Canadians consume more than 50 kilograms of newsprint per person per year—enough to account for one whole mature tree. Similar efficiencies can be obtained by recycling other materials that require less energy for remanufacture. Oil, for example, fuels many industrial processes. Estimates suggest that if 1 per cent of the Canadian population recycled instead of trashing one aluminum can a day, the oil saved by remanufacturing would produce 21 million litres of gasoline. Recycling one aluminium can saves enough energy to run your TV for three hours. Making steel from recycled material uses only one-quarter of the energy it takes to make steel from virgin ore, while recycling copper results in energy savings of up to 85 per cent.

The vast majority of Canadian households that have access to recycling programs make use of them, regardless of household income, the occupants' education levels, or the type of dwelling (Statistics Canada 2007c). Canada has one of the highest levels of recycling on the planet. This is one of the ratings you can find on the 'Greendex' developed by the National Geographic Society (2008). The index was created to measure and monitor consumer progress towards sustainability. Based on surveys in 14 countries, it focuses on actual behaviour and material lifestyles, including measures such as the relative penetration of green products versus traditional products, household footprint, energy use, transportation habits, and food consumption. Canadians led in recycling and in their ability to answer a knowledge question regarding the water demands of different types of food. Unfortunately, they fared much worse in the other categories, and Canada ended up ranking second from the bottom, surpassing only the US. Brazil, India, and China topped the list. Among developed countries, average consumers in the UK, Germany, and Australia received the highest (and virtually equal) scores.

Consumers can adopt more sustainable consumption habits and help to raise Canada's Greendex score by doing any of the following:

- eating less meat, more locally produced foods, and more fruits and vegetables and drinking less bottled water;
- improving the energy efficiency of their homes by sealing drafts, upgrading windows, and installing more efficient water heaters and other appliances;
- keeping air heating at lower and cooling at higher settings;
- using only cold water to wash laundry and minimizing water use overall;
- driving alone less often (e.g., car pooling);
- driving less overall;
- driving smaller and/or more fuel-efficient vehicles;
- having fewer vehicles for their household;
- walking or riding a bicycle when distance allows;
- maximizing the lifespan of household items and minimizing disposal;
- avoiding environmentally harmful products and packaging and seeking out environment-friendly alternatives;
- recycling whenever possible;
- using reusable shopping bags rather than accepting new disposable ones;
- having fewer televisions sets and personal computers in their household.

Unfortunately, the bottom line remains that while it is better to recycle than not to, reducing consumption levels in the first place is still the preferred option, and significant progress in this area has yet to be achieved.

The Law of Everybody

There are many other ways by which you can reduce the pressure on the environment. We have offered just a few suggestions to get you started. The key point to remember is that the accumulated actions of many concerned individuals acting together will make a difference. In this book, we have introduced you to many scientific laws and principles related to understanding and managing the environment. However, if everyone knew and enacted the following law, that would go a long way towards ameliorating our environmental challenges: it is, simply, the **law of everybody**.

'Everybody's Got to Do Something'

If we all did a myriad of small things that could be done easily, without having much negative or even noticeable influence on our lifestyle (e.g., drive 10 per cent less a year, buy fewer toys, shower with a friend), we would find that many of our environmental challenges would be greatly reduced in scale. It is not important that all people do everything, but everyone has to do something!

Several websites can help you to find out more about your own environmental impact and suggest concrete ways in which your impact can be reduced. For example, try the *personal challenges* on the website of the famous Canadian environmentalist, David Suzuki (www.davidsuzuki.org), and explore some of the websites at the end of this and other chapters. One site you will enjoy is 3rd Whale. It is a 'green facebook' run out of Simon Fraser University in Vancouver that lets you explore everything from finding a green partner to finding a green job. Every year, they run a 'greenest person on the planet' competition and in 2008 received more than 600 entries from more than 20 countries. Nearly every contestant eats locally grown organic foods, recycles, and walks, cycles, or takes transit everywhere. A few go out of their way to power-down devices that sip away at electricity even when they are turned off, like TVs, radios, DVD players, and computers.

Consider these contestants the triathletes of the green world. We applaud their inventiveness and dedication, but unlike 3rd Whale, we are not sure that everyone is ready to compete in the triathlon. However, this does not mean that people who, comparatively speaking, are just out for a quiet stroll cannot help in achieving sustainability. Everyone can help, but not everyone can do the triathlon. We feel that engaging as broad a segment of society as possible in as great a number as possible, with each making a contribution, is a better goal than either doing the triathlon—or doing nothing.

Influence

One of the best ways to influence business is through your purchasing power as a consumer. If consumers band together and refuse to buy certain products because of their impact on the environment, the manufacturer will either have to respond to these concerns or go out of business. There are many successful examples of these kinds of actions. During the 1970s and 1980s, for example, conservationists were able to exert pressure on hamburger chains to change their source of beef supply to ensure that tropical rain forests were not being cut down and replaced by grass to feed cattle for hamburgers. Following boycotts, all the major chains were persuaded to ensure that their sources did not contribute to the destruction of the rain forests. Similar campaigns have been directed at tuna canners to ensure that tuna is not caught by methods that kill dolphins, which often swim with schools of tuna.

Managing Planet Earth

A second myth is that with enough knowledge and technology, we can, in the words of Scientific American (1989), 'manage planet Earth'. Higher education has largely been shaped by the drive to extend human domination to its fullest. In this mission, human intelligence may have taken the wrong road. Nonetheless, managing the planet has a nice ring to it. It appeals to our fascination with digital readouts, computers, buttons, and dials. But the complexity of Earth and its life systems can never be safely managed. The ecology of the top inch of topsoil is still largely unknown, as is its relationship to the larger systems of the biosphere. What might be managed, however, is us: human desires, economies, politics, and communities. But our attention is caught by those things that avoid the hard choices implied by politics, morality, ethics, and common sense. It makes far better sense to reshape ourselves to fit a finite planet than to attempt to reshape the planet to fit our infinite wants.

—Orr 1994, 9

Consumer boycotts can be a very effective way of influencing business practices and reducing environmental impacts. When choosing fish to eat, for example, check out the websites at the end of Chapter 8. Download Canada's Seafood Guide (www.Seachoice.org). This wallet-sized and regularly updated guide lists three categories of seafood—best choices, concerns, and avoid—based on independent assessments of the sustainability of that particular fishery. If consumers followed these guidelines, it would alleviate much of the pressure on fishery resources.

Not only should we be sending messages to producers about our environmental standards through purchasing power, we should also let non-conforming producers know why we are not buying their products. If you believe in eating wild rather than farmed salmon, always ask the origin of the salmon before you buy it. If you believe in fair-trade coffee or certified wood products, always let retailers know that is what you are shopping for. Home Depot, for example, is the largest retailer of wood products in the world. Following a consumer boycott organized by a non-governmental organization, Home Depot finally made a commitment to phase in sale of only certified wood products in its stores. This decision, rapidly echoed by

two other major retailers, probably did more to protect old-growth forests in BC than all the protests of the past 20 years. However, the change was slow in coming, and there is still opportunity for improvement in Home Depot's purchasing practices. Home Depot argues that one of the main reasons behind its tardy behaviour is simply that customers do not ask for certified wood products and often do not buy them preferentially (Mastny 2004).

As a responsible consumer, you should let your preferences be known. This also raises the question of **corporate responsibility** (Box 15.7). Is it the responsibility of the retailer to take part in, or even lead, the switch to a more sustainable way of doing things by informing its customers about the environmental implications of different products? Or is it sufficient merely to react to customer demands once they are manifest in the marketplace?

Many countries have enacted what are known as **extended producer responsibility** laws that require manufacturers and importers to accept responsibility for their products at the end of their useful lifespan. These laws provide an incentive for companies to design their products so that they can be recycled or reused and to eliminate toxic materials, since they would have to dispose of them. Ultimately, products should be designed so that they are either biodegradable or can be disassembled into their components for reuse in the future. Canada has no such laws, but some companies have taken the initiative, as discussed in Box 15.7.

Life-cycle assessments (**LCAs**) are now gaining greater support from both government and industry. LCAs identify inputs, outputs, and potential environmental impacts of a product or service throughout its lifetime. The 2002 **World Summit on Sustainable Development** called for greater investment in LCAs, and the United Nations Environment Programme is developing a program to assist in the development and dissemination of practical tools to help in life-cycle assessment. One industrial example is Volvo, which provides LCAs for the various components involved in vehicle manufacture. Some NGOs, such as Green Seal, provide standards for various products that manufacturers must meet to gain their endorsement. Others have excellent programs to assist consumers, industry, and government in making wise procurement decisions, such as the US-based Center for a New American Dream. Among useful websites are Fair Trade, Ethiquette, and Climate Counts, all listed at the end of the chapter. Climate Counts produces a scorecard rating various companies on their performance in measuring, assessing, and reducing the carbon print of their products and on how supportive of climate change initiatives and policies they are. In 2008, 84 per cent of the companies had improved their scores reported on the 2007 scorecard.

In Canada, the Canadian Environmental Choice product-labelling program aims to influence business practices by shifting consumer and institutional purchasing to products and services that conform to environmental guidelines. A board

Box 15.7   Corporate Contributions

Large corporations and retailers are highly visible targets for environmental action and have attracted a lot of attention. However, many corporations do much better than we do personally, or than our governments do, in taking a systematic look at some of their environmental impacts and moving to address them. For example, the ubiquitous global fast-food corporation McDonald's has the Earth Effort program, which embraces the reduce, reuse, and recycle philosophy. Every year, McDonald's is committed to buying at least $100 million worth of recycled products for building, operating, and equipping its facilities. Carry-out bags are made from recycled corrugated boxes and newsprint; take-out drink trays are made from recycled newspapers. New restaurants have been constructed with concrete blocks made from recycled photographic film and roofs made from computer casings. McDonald's has also cut down on the amount of waste produced; for example, sandwich packaging has been reduced by more than 90 per cent by switching from foam packaging to paper wraps. It also uses compostable food packaging made from reclaimed potato starch and other materials.

McDonald's also has a strict policy that they will not buy the beef of cattle raised on land that was converted from rain forests for that purpose. In addition, it has programs to reduce energy consumption and to take part in a wide range of local initiatives ranging from tree planting to local litter drives. For its efforts, McDonald's in the US has won White House awards and the National Recycling Coalition's Award for Outstanding Corporate Leadership.

Another company that environmentalists love to hate, Wal-Mart, also has tremendous potential to influence sustainability. The chain has 4,100 stores in the US and more than 300 in Canada. It operates the continent's largest trucking company, is the largest consumer of electricity, and with 1.3 million employees in the US and 75,000 in Canada stands as the biggest employer. More than 100 million customers per week visit Wal-Mart stores. With that kind of size, even a small incremental change towards sustainability could have a tremendous impact. However, Wal-Mart has introduced some of the most far-reaching programs and demands that its suppliers also achieve sustainability thresholds. It has launched a $30-million Personal Sustainability Program that raises awareness among employees of measures that they can take as individuals to live more sustainable lifestyles. The combined impact of these kinds of programs can be many times that of government agencies.

Many other well-known corporations are making contributions. Eddie Bauer, for example, has joined forces with an NGO, American Forests, in a tree-planting effort on damaged forest ecosystems in eight reforestation sites in the US and Canada. Consumers can opt to add a dollar to the price of their purchase at the store, and in turn the store will plant a tree for each dollar and donate another tree. Nike, the shoe, clothing, and sports equipment giant, has not only boosted the proportion of organic cotton in its own products but also helped to launch Organic Exchange, a network of 55 businesses all committed to significantly expanding the use of organic cotton. Nike has programs to achieve zero toxics, zero waste, and 100 per cent recovery, recycling, and reuse of the products they manufacture. They are designing a running shoe with a biodegradable sole and an upper portion that can be endlessly remanufactured into a new shoe. Other companies, such as Texas Instruments, Levi Strauss, and the Ford Motor Company, have banded their purchasing power together to buy recycled paper.

Also consider Valhalla Pure Outfitters of Vernon, BC, makers of high-quality outdoor clothing. They now make all their polyester fleece jackets from recycled materials that look, feel, wear, and last the same as new fabric, even though they are made from recycled pop bottles. Each pop bottle diverted from a landfill site for recycling is sent to a factory in South Carolina, where they are separated by colour and then reprocessed before being sent to finishing plants in the US and Quebec. The final cost is similar to that of making new material but has added environmental advantages. Each jacket is equivalent to about 25 pop bottles.

These are not the only ways by which the corporate and business world is contributing positively towards addressing environmental problems. Some of them also have substantial foundations to which environmental groups can apply for funding for specific problems. As government coffers are drained, more pressure is put on these sources. The Laidlaw Foundation, for example, used to give $50,000 a year for environmental projects, and requests rarely exceeded this amount. The Foundation now disburses $350,000 but has requests exceeding $1 million. The Richard Ivey Foundation in London, ON, provides up to 80 per cent of its $2-million budget for environmental projects each year but still manages to meet only 5 per cent of the amount requested. With such pressure on resources, it is essential to ensure that the money is used effectively to address the most pressing problems.

If there are environmental problems in your area that you think need to be addressed, do not be afraid to approach local businesses for support. They may well agree with you and be happy to make a contribution to a carefully crafted solution.

establishes the guidelines that each category of product must achieve to obtain recognition, signified by the three-doves logo. These guidelines consider factors such as recyclability, design of packaging material, use of recyclate in manufacturing, effluent toxicity, and energy consumption. However, the government requires the board to take into account the economic impacts of the guidelines to ensure that they do not impinge on the economic competitiveness of Canadian products. There are now more than 7,000 EcoLogo-certified products in more than 120 categories (http://www.ecologo.org/en).

Governments make many decisions that can promote or retard sustainability. Public transport can be subsidized instead of private cars, land-use planning and building codes can be designed to minimize energy and material use, and governments can facilitate shared public consumption, such as provision of libraries and swimming pools, rather than encouraging private acquisition. Many municipalities have inherited what has been called the 'infrastructure of consumption' from previous generations (OECD 2002). However, it is necessary now to redesign the way we live in accordance with our increased understanding of planetary limits. Box 15.7 offers several examples of this kind of approach.

What can you do to encourage governments to act? Election times offer a major opportunity. Make a point of asking your local candidates at all levels of government for their views on environmental or sustainable development issues. Help to publicize these views. It is much more difficult for politicians to change their positions if their views are well known. Make sure that politicians follow through on their pre-election promises on environmental matters. Make sure that you express your views on the environment to politicians between elections. Write them letters, call them, go and see them, and organize demonstrations. The position of the Commissioner of the Environment and Sustainable Development was specifically created by the government within the Auditor General's office to provide a place where *any Canadian* can go with a question about the environmental performance of a particular ministry or program and ask to have it audited. If you have such a question, go to the commissioner's website listed at the end of this chapter and review the simple process on how to proceed. Canadians need to make much greater use of this mechanism.

Governments are often more than willing to adopt environment-friendly, voter-popular measures—as long as they do not involve any significant economic cost. Issues are often very complex, too complex for the understanding of many individuals who nonetheless care about acting in an environmentally responsible fashion. For this reason, concerned people may band together in **non-governmental organizations** (**NGOs**). Such organizations represent the collective concern and resources of many people and consequently are in a

The symbol of three doves in the shape of a maple leaf identifies products certified under Canada's Ecologo Program. The three doves represent Canada's consumers, industry, and government working together to improve the environment. (*Printed with permission from TerraChoice Environmental Marketing, administer of the EcoLogo Program.*)

and its life support systems. So great are our capabilities to affect these systems that human-controlled influences now dominate many natural processes, resulting in the many critical environmental problems described throughout this book. Government programs at the international and national levels continue to emerge, but actions that make a difference on the ground are few and far between. Still, we are not powerless to change the direction of society. As individuals and as concerned individuals banding together, we can effect many of the changes that need to be made, and we can strive for greater well-being, not wealth, in the future. So despite the severity of the challenges, it is important to remain optimistic.

A wind farm in Pincher Creek, Alberta.

much better position to attack a problem than an individual is. One of the best ways to spend your conservation dollar is to support such a group. They have had significant impacts on government policies in Canada. The appendix lists of some of these organizations.

Implications

The overall implications raised in this book should be brutally clear by now. Humanity is facing major challenges over the next decade regarding our relationship with the Earth

Do Less

We should at least entertain the idea of doing less, concentrating on just being in the world. How foolish to laud a work ethic that keeps people nose-to-the-grindstone, busy making and doing far beyond their needs. Here is a monumental mistake. We ought to do less, simply because much of so-called 'productive labour' is destructive: it consumes resources, encourages overpopulation, creates garbage, and weakens the Earth-source. By working less, we free up time for the more worthy task of harmonizing with the surrounding reality from which we came and will soon return.

—Rowe 1993, 8

On Well-Being and Accumulation

Societies focused on well-being involve more interaction with family, friends, and neighbours, a more direct experience of nature, and more attention to finding fulfillment and creative expression than in accumulating goods. They emphasize lifestyles that avoid abusing your own health, other people, or the natural world. In short, they yield a deeper sense of satisfaction with life than many people report experiencing today.

—Gardner and Assadourian 2003, 16

Above all, stay cheerful, stay active, look after yourself, look after others, love this planet, and do not give up! We are on the most beautiful planet we know about. Canada has some of the most breathtaking wonders in the universe. We have a responsibility. We can think of no better advice than that offered by Edward Abbey (1977):

One final paragraph of advice: *Do not burn yourselves out. Be as I am—a reluctant enthusiast . . . a part-time crusader, a half-hearted fanatic. Save the other half of yourselves and your lives for pleasure and adventure. It is not enough to fight for the land; it is even more important to enjoy it. While you can. While it's still here. So get out there and . . . ramble out yonder and explore the forests, encounter the grizz, climb the mountains, bag the peaks, run the rivers, breathe deep of that yet sweet and lucid air, sit quietly for a while and contemplate the precious stillness, that lovely, mysterious, and awesome space. Enjoy yourselves, keep your brain in your head and your head firmly attached to the body, the body active and alive, and*

I promise you this much: I promise you this one sweet victory over our enemies, over those desk-bound people with their ears in a safe deposit box and their eyes hypnotized by desk calculators. I promise you this: you will outlive the bastards.

Summary

1. The previous century witnessed the start of a consumer-dominated society in several parts of the world that is rapidly spreading and threatens to engulf most of the world in the early part of this century.

2. Population growth rates are showing declines, but consumerism continues to expand. Environmental conditions continue to deteriorate and are likely to do so even more as consumerism spreads.

3. Today, most emphasis is on meeting the Millennium Development Goals, but overall progress is slow. There is potential for significant conflict between the movements to improve global environmental conditions and those seeking reductions in international trade barriers.

4. Important global agreements have been forged, including the Kyoto Protocol and the Convention on Biological Diversity, that set frameworks for global actions, but progress on their implementation, by and large, is very slow.

5. Progress has been made in some areas. The Forest Stewardship Council (FSC) has now certified more than 80 million hectares of commercial forest in 80 countries. A Marine Stewardship Council, modeled on the FSC, has been initiated and has certified 22 fisheries offering 500 certified seafood products covering 7 per cent of the world's marine catch.

6. Research examining the relationship between consumerism and well-being shows that a positive relationship does exist when incomes are small. However, after a certain point is reached, there is no relationship. Most of the developed world has already attained triple this income level.

7. Only one of the eight regions is on track to achieve all of the Millennium Development Goals by 2015, with severe shortfalls in many areas such as sub-Saharan Africa.

8. The Happy Planet Index is a measure of human well-being in various countries against the impacts caused. Canada ranks 111th on the scale.

9. Polls indicate that the environment is important to Canadians, yet our response to environmental degradation at both international and national scales continues to be very slow despite clear signals regarding the state of the environment. We are in violation of many international environmental agreements.

10. Canadians rank among the top five countries in the world for their views on the seriousness of climate change, and although the federal government has yet to demonstrate any leadership, there are innovative programs at the provincial level.

11. BC has introduced the most ambitious plan to address global climate change in North America, including carbon taxes. BC has set a target of reducing greenhouse gases by at least 33 per cent below current levels by 2020. This would be 10 per cent lower than the 1990 level, exceeding California's goal of returning to 1990 levels by 2020.

12. The availability of environmental information to interested citizens seems increasingly restricted in Canada, in contrast to the situation in many other countries where growing Internet use has promoted greater citizen access to government data.

13. Ordinary citizens can have a positive impact on the environmental challenges facing society in many ways. A first step is building awareness of these problems. Universities and colleges should be intimately involved in this process by ensuring that all students have a measure of environmental literacy before they graduate. Universities and colleges are also large consumers of matter and energy and should lead by example in reducing their impact on the environment.

14. There is growing evidence of 'nature deficit disorder' among younger people as they spend more and more time on their computers and in front of TV screens and less and less time interacting with nature. This makes building awareness of environmental change and getting people emotionally engaged with change even more challenging.

15. Individuals can help by living in accordance with the four Rs: *refuse* to be goaded into overconsumption, *reduce* consumption of matter and energy whenever possible, *reuse* materials whenever possible, and *recycle* those that you cannot reuse.

16. Individuals can also wield influence by banding together and taking collective action—for example, through consumer boycotts of products and companies that engage in environmentally destructive practices. One of the most effective ways of doing this is to join a non-governmental organization (NGO) composed of and supported by like-minded individuals. As an individual, you will probably not be able to afford to support an environmental lobbyist in Ottawa. However, if thousands of people contribute, it can become a reality.

17. Big corporations have a lot of opportunity to influence sustainability, and many of them have accepted that responsibility.

18. In spite of the serious nature of many of the environmental problems described in this book, it is important to remain optimistic that they can be solved and to take time to get out and enjoy the beauty and challenge of one of the most splendid parts of the planet, Canada, our home.

Key Terms

100-mile diet
consumerism
corporate responsibility
extended producer responsibility
Happy Planet Index (HPI)

law of everybody
life-cycle assessments (LCAs)
light living
Millennium Development Goals
nature deficit disorder

non-governmental organizations (NGOs)
World Summit on Sustainable Development (WSSD)

Questions for Review and Critical Thinking

1. What do you think are the main factors driving global environmental degradation?
2. Which factors do you think are the most serious, and why?
3. Discuss what you think should be the top priorities for global action.
4. Outline some of the main responses to environmental degradation at both the international and national levels.
5. Discuss Canada's current and potential role at the international level.
6. What do you think are the main barriers to more effective government response in Canada?
7. Discuss your institution's present and potential role in raising environmental awareness.
8. What are three concrete steps towards 'light living' that you are willing to take over the next month?
9. What are some initiatives that you could take as an individual or as part of your community to reduce negative environmental consequences from human activities?
10. Find out the names and mandates of the environmental NGOs in your area. Is there anything you can do to help them?
11. Who is your local MP? What is his or her view on environmental issues? Exercise your democratic right—phone and find out.

Related Websites

Auditor General of Canada: www.oag-bvg.gc.ca

BC Sustainable Energy Association: www.bcsea.org

Center for a New American Dream: http://www.newdream.org

City Green: http://www.citygreen.ca

Climate Counts: www.ClimateCounts.org

Commissioner of the Environment and Sustainable Development: http://www.oag-bvg.gc.ca

EcoLogo: www.EcoLogo.org

David Suzuki Foundation: http://www.davidsuzuki.org/consumer

Environment Canada, What You Can Do: http://www.ec.gc.ca/eco/main_e.htm

Environmentally Sound Transportation: http://www.best.bc.ca

Ethiquette: www.ethiquette.ca

Fair Trade Certified: www.transfair.ca

Happy Planet Index: http://www.happyplanetindex.org

Heifer International: http://www.heifer.org; http://www.catalog.heifer.org/index.cfm

Kesho Trust: http://www.naturechildreunion.ca

Marine Stewardship Council: http://www.msc.org

Natural Resources Canada, Office of Energy Efficiency: http://oee.nrcan.gc.ca/english

Nature Serve: http://www.natureserve.ca

Nature Watch: http://www.naturewatch.ca/english

Sierra Youth Coalition: http://www.syc-cjs.grg

3rd Whale: http://www.3rdwhale.com

United Nations Environment Programme: http://www.unep.org

United Nations Human Development Index: http://hdp.undp.org/reports/global/2003

University of British Columbia, Sustainability Project: http://www.sustain.ubc.ca

University of Victoria, Sustainability Project: http://uvsp.uvic.ca/main.php

University of Waterloo, Sustainability Project: http://watservl.uwaterloo.ca/~uwsp

World Summit on Sustainable Development: http://www.joannesburgsummit.org

Further Readings

Note: This list comprises works relevant to the subject of the chapter but not cited in the text. All cited works are listed in the section 'References' at the end of the book.

Barber, J. 2003. *Production, Consumption and the World Summit for Sustainable Development*. Rockville, MD: Integrative Strategies Forum.

Cross, G. 2002. *An All-Consuming Century: Why Commercialism Won in Modern America*. New York: Columbia University Press.

French, H. 2004. 'Linking globalization, consumption and governance'. In Worldwatch Institute, *State of the World 2004*, 144–61. New York: W.W. Norton.

McAllister Opinion Research. 2007. *Environmental Monitor 2007*. Vancouver: McAllister Opinion Research.

United Nations. 2003. *Plan of Implementation of the World Summit on Sustainable Development*. New York: United Nations.

———. 2007. *The Millennium Development Goals Report 2007*. New York: United Nations.

Appendix: Conservation Organizations

INTERNATIONAL ORGANIZATIONS

Alliance for the Wild Rockies
P.O. Box 8731
Missoula, MT 59807
Phone: 406 721-5420
Fax: 406 721-9917
E-mail: awr@wildrockiesalliance.com
Website: http://www.wildrockiesalliance.com

Antarctic and Southern Ocean Coalition
1630 Connecticut Avenue NW, 3rd Floor
Washington, DC 20009
Phone: 202 234-2480
Fax: 202 387-4823
Website: http://www.asoc.org

Center for Marine Conservation
1725 DeSales Street NW
Washington, DC 20036
Phone: 202 429-5609
Fax: 202 872-0619
E-mail: dccmc@ix.netcom.com

Center for Plant Conservation
P.O. Box 299
St Louis, MO 63166-0299
Phone: 314 577-9450
Fax: 314 577-9465

Clean Water Action Project
4455 Connecticut Avenue NW, Suite A300
Washington, DC 20008
Phone: 202 895-0420
Fax: 202 895-0438

Conservation International
1919 M Street NW, Suite 600
Washington, DC 20036
Phone: 202 912-1000

Cousteau Society
710 Settlers Land Road
Hampton, VA 23669
Phone: 757 722-9300
Fax: 757 722-8185
E-mail: cousteau@cousteausociety.org

Earth Council
Phone: 416 498-3150
Fax: 416 498-7296
E-mail: ecsondra@web.ca

Earth Island Institute
300 Broadway, Suite 28
San Francisco, CA 94133-3312
Phone: 415 788-3666
Fax: 415 788-7324
E-mail: earthisland@earthisland.org

EarthKind
2100 L Street NW
Washington, DC 20037
Phone: 202 778-6149
Fax: 202 778-6134
E-mail: earthknd@ix.netcom.com

Earthwatch
3 Clock Tower Place, Suite 100 Box 75
Maynard, MA 01754
Phone: 978 461-0018
Toll-free: 800 776-0188
Fax: 978 461-2332

Ecological Agricultural Projects
Macdonald Campus
McGill University
Ste-Anne-de-Bellevue, QC H9X 3V9
Phone: 514 398-7771
Fax: 514 398-7621
E-mail: ecological.agriculture@mcgill.ca

Environmental Action Foundation
333 John Carlyle Street,
    Suite 200
Alexandria, VA 22314
Phone: 703 548-3118
Fax: 703 548-3119
E-mail: info@agc.org

Friends of Animals
777 Post Road, Suite 205
Darien, CT 06820
Phone: 203 656-1522
Fax: 203 656-0267

Friends of the Earth International
International Secretariat
P.O. Box 19199 1000 GD
Amsterdam
The Netherlands
Phone: 31-20-622-1369
Fax: 31-20-639-2181

Greenpeace International
Keizersgracht 176 1016 DW
Amsterdam
The Netherlands
Phone: 31-20-523-6222
Fax: 31-20-523-6000
E-mail: supporter.services@int.greenpeace.org
Website: http://www.greenpeace.org

International Water Resources Association
University of New Mexico
1915 Roma NE
Albuquerque, NM 87131-1436
Phone: 505 277-9400
Fax: 505 277-9405

International Wildlife Coalition
70 East Falmouth Highway
East Falmouth, MA 02536
Phone: 508 548-8328
Fax: 508 548-8542

IUCN (World Conservation Union)
Rue Mauverney 28 CH 1196
Gland, Switzerland
Phone: 41-22-999-0155
Fax: 41-22-999-0015
Website: http://www.iucn.org

Jane Goodall Institute for Wildlife Research, Education, and Conservation
8700 Georgia Avenue, Suite 500
Silver Spring, MD 20910
Phone: 240 645-4000
Fax: 301 565-3188
Website: http://www.janegoodall.org

Jersey Wildlife Preservation Trust
Les Augres Manor
Trinity, Jersey
Channel Islands JE3 5BP
Phone: 15-34-864-666
Fax: 15-34-865-161
E-mail: ldexec@itl.net

National Audubon Society
700 Broadway
New York, NY 10003
Phone: 212 979-3000
E-mail: webmaster@audubon.org
Website: http://www.audubon.org

Ocean Alliance and Whale Conservation Institute
191 Weston Road
Lincoln, MA 01773
Phone: 781 259-0423
Toll-free: 800 969-4253
Fax: 781 259-0288
E-mail: question@oceanalliance.org

Ocean Voice International
P.O. Box 37026
3332 McCarthy Road
Ottawa, ON K1V 0W0
Phone: 613 264-8986
Fax: 613 521-4205
E-mail: ah194@freenet.carleton.ca

Rainforest Action Network
450 Sansome Street, Suite 700
San Francisco, CA 94111
Phone: 415 398-4404
Fax: 415 398-2732
E-mail: rainforest@ran.org

Sierra Club
85 Second Street, 2nd Floor
San Francisco, CA 94105
Phone: 415 977-5500
Fax: 415 977-5799
E-mail: information@sierraclub.org
Website: http://www.sierraclub.org

Soil and Water Conservation Society
7515 North Ankeny Road
Ankeny, IA 50021-9764
Phone: 515 289-2331
Toll-free: 800 843-7645
Fax: 515 289-1227
E-mail: swcs@swcs.org

United Nations Environment Programme
Information Officer Regional Office for
    North America
2 United Nations Plaza, Room DC 2-803
New York, NY 10017
Phone: 212 963-8210
Fax: 212 963-7341
E-mail: uneprona@un.org
Website: http://www.unep.ch/unep.html
Convention on Biodiversity: http://www.unep.ch/
    biodiv.html

**World Commission on Forests and
Sustainable Development**
CP 51
CH-1219 Châtelaine
Geneva, Switzerland
Phone: 22-979-9165
Fax: 22-979-9060
E-mail: dameena@1prolink.ch
Website: http://www.iisd.org/wcfsd

World Resources Institute
1709 New York Avenue NW, Suite 700
Washington, DC 20006
Phone: 202 638-6300
Fax: 202 638-0036
Website: http://www.wri.org

World Society for the Protection of Animals
29 Perkins Street
P.O. Box 190
Boston, MA 02130
Phone: 617 522-7000
Fax: 617 522-7077
E-mail: wspa@world.std.com

Worldwatch Institute
1776 Massachusetts Avenue NW
Washington, DC 20036
Phone: 202 452-1999
Fax: 202 296-7365
E-mail: worldwatch@worldwatch.org
Website: http://www.worldwatch.org

CANADIAN NATIONAL ORGANIZATIONS

Animal Alliance of Canada
221 Broadview Avenue, Suite 201
Toronto, ON M4M 2G3
Phone: 416 462-9541
Fax: 416 462-9647
E-mail: aac@inforamp.net

Assembly of First Nations
1 Nicholas Street, Suite 1002
Ottawa, ON K1N 7B7
Phone: 613 241-6789
Fax: 613 241-5808

**Association for the Protection of Fur-Bearing
Animals**
3727 Renfrew Street
Vancouver, BC V5M 3L7
Phone: 604 435-1850
Fax: 604 435-1840

Canadian Arctic Resources Committee
1276 Wellington Street, 2nd Floor
Ottawa, ON K1Y 3A7
Phone: 613 759-4284
Toll-free: 866 949-9006
Fax: 613 759-4581

Canadian Council on Ecological Areas
Secretariat/Manager
c/o Leigh Warren
4507 Torbolton Ridge Road
R.R. 2
Woodlawn, ON K0A 3M0
Phone: 613 832-1995
E-mail: leigh.warren@ec.gc.ca

Canadian Earth Energy Association
130 Slater Street, #1050
Ottawa, ON K1P 6E2
Phone: 613 230-2332
Fax: 613 237-1480

Canadian Environmental Law Association
130 Spadina Avenue, Suite 301
Toronto, ON M5V 2L4
Phone: 416 960-2284
Fax: 416 960-9392

Canadian Global Change Program
University of Victoria
P.O. Box 1700 STN CSC
Victoria, BC V8W 2Y2
Phone: 250 472-4337
Fax: 250 472-4830
E-mail: cgcp@uvic.ca
Website: http://www.globalcentres.org/cgcp

Canadian Nature Federation
1 Nicholas Street, Suite 520
Ottawa, ON K1N 7B7
Phone: 613 562-3447
Fax: 613 562-3371

**Canadian Parks and Wilderness
Society (CPAWS)**
National Office
880 Wellington Street, Suite 506
Ottawa, ON K1R 6K7
Phone: 613 569-7226
Toll-free: 800 333-WILD (9453)
Fax: 613 569-7098

**Canadian Water Resources
Association**
P.O. Box 1329
Cambridge, ON N1R 7G6
Phone: 519 622-4764
Fax: 519 621-4844
E-mail: cwranat@grandriver.ca
Website: http://www.cwra.org

**City Farmer—Canada's Office
of Urban Agriculture**
Box 74561, Kitsilano RPO
Vancouver, BC V6K 4P4
Phone: 604 685-5832
Fax: 604 685-0431
E-mail: cityfarm@unixg.ubc.ca
Website: http://www.cityfarmer.org

Earth Day Canada
111 Peter Street, Suite 503
Toronto, ON M5V 2H1
Phone: 416 599-1991, ext. 107
Fax: 416 599-3100

Energy Probe
225 Brunswick Avenue
Toronto, ON M5S 2M6
Phone: 416 964-9223
Fax: 416 964-8239
E-mail: EnergyProbe@nextcity.com

Environment Bureau
Agriculture and Agri-Food Canada
Co-operatives Secretariat
1525 Carling Avenue, Room 367
Ottawa, ON K1A 0C5
Phone: 613 759-7309

Environment Canada
Inquiry Centre
70 Crémazie Street
Gatineau, QC K1A 0H3
Fax: 819 994-1412
TTY: 819 994-0736 (Teletype for the hearing
    impaired)
E-mail: enviroinfo@ec.gc.ca
A Guide to a Green Government: http://www.ec.gc.
    ca/sd-dd_consult/factsheetGreengov1_e.htm
Site of the Environment InfoBase: http://www.ec.gc.
    ca/soer-ree
Sustainable Development Strategy: http://www.
    ec.gc.ca/sd-dd_consult/SDS2004/index_e.cfm

Fisheries and Oceans Canada
Communications Branch
200 Kent Street, 13th Floor, Station 13228
Ottawa, ON K1A 0E6
Phone: 613 993-0999
Fax: 613 990-1866
TDD: 613 941-6517
E-mail: info@www.ncr.dfo.ca
Website: http://www.ncr.dfo.ca/ home_e.htm

Friends of the Earth Canada
206-260 St Patrick Street
Ottawa, ON K1N 5K5
Phone: 613 241-0085

Greenpeace Canada
33 Cecil Street
Toronto, ON M5T 1N1
Phone: 416 597-8408
Toll-free: 800 320-7183
Fax: 416 597-8422

**International Fund for
Animal Welfare**
1 Nicholas Street
Ottawa, ON K1N 7B7
Phone: 613 241-8996
Toll-free: 888 500-4329
Fax: 613 241-0641

**International Institute for Sustainable
Development**
161 Portage Avenue East, 6th Floor
Winnipeg, MB R3B 0Y4
Phone: 204 958-7700
Fax: 204 958-7710
E-mail: info@iisd.ca
Website: http://www.iisd.org

Marine Protected Areas Network
Commission for Environmental
    Co-operation
393, rue Saint-Jacques ouest, bureau 200
Montreal, QC H2Y 1N9
Phone: 514 350-4300
Fax: 514 350-4314

National Parks Directorate, Parks Canada
25 Eddy Street, 4th Floor
Gatineau, QC K1A 0M5
Phone: 888 773-8888
E-mail: National-ParksWebmaster@pch.gc.ca
Website: http://parkscanada.pch.gc.ca

Nature Canada
1 Nicholas Street, Suite 606
Ottawa, ON K1N 7B7
Toll-free: 800 267-4088

Nature Conservancy of Canada
110 Eglinton Avenue West, Suite 400
Toronto, ON M4R 1A3
Phone: 416 932-3202
Toll-free: 800 465-0029
Fax: 416 932-3208
E-mail: nature@natureconservancy.ca

Probe International
225 Brunswick Avenue
Toronto, ON M5S 2M6
Phone: 416 964-9223
Fax: 416 964-8239

Sea Shepherd Conservation Society
P.O. Box 48446
Vancouver, BC V7X 1A2
Phone: 604 688-7325

Sierra Club of Canada
1 Nicholas Street, Suite 412
Ottawa, ON K1N 7B7
Phone: 613 241-4611
Fax: 613 241-2292

Wildlife Habitat Canada
1750 Courtwood Crescent, Suite 310
Ottawa, ON K2C 2B5
Phone: 613 722-2090
Toll-free: 800 669-7919
Fax: 613 722-3318
E-mail: reception@whc.org

Wildlife Preservation Canada
RR 5, 5420 Highway 6 North
Guelph, ON N1H 6J2
Phone: 519 836-9314
Toll-free: 800 956-6608
Fax: 519 836-8840
E-mail: admin@wildlifepreservation.ca
Website: www.wildlifepreservation.ca

World Wildlife Fund Canada
245 Eglinton Avenue East, Suite 410
Toronto, ON M4P 3J1
Phone: 416 489-8800
Toll-free: 800 267-2632
Fax: 416 489-8055

Zoocheck Canada Inc.
2646 St Clair Avenue East
Toronto, ON M4B 3M1
Phone: 416 285-1744
Fax: 416 285-4670
E-mail: zoocheck@zoocheck.com

BRITISH COLUMBIA

BC Spaces for Nature
Box 673
Gibsons, BC V0N 1V0
Phone: 604 886-8605
Fax: 604 886-3768

BC Wild
Box 2241, Main Post Office
Vancouver, BC V6B 3W2
Phone: 604 669-4802
Fax: 604 669-6833

CPAWS, BC Chapter
610-555 West Georgia Street
Vancouver, BC V6B 1Z6
Phone: 604 685-7445
Fax: 604 685-6449
E-mail: info@cpawsbc.org

Friends of Clayoquot Sound
P.O. Box 489
Tofino, BC V0R 2Z0
Phone: 604 725-4218
Fax: 604 725-2527

Ministry of Sustainable Resource Management
P.O. Box 9054, Stn Prov Govt
Victoria, BC V8W 9E2
Phone: 250 356-9076
Fax: 250 356-8273

Ministry of Water, Land, and Air Protection
P.O. Box 9047, Stn Prov Govt
Room 112, Parliament Buildings
Victoria, BC V8W 9E2
Phone: 250 387-1187
Fax: 250 387-1356

Nature Trust of British Columbia
260-1000 Roosevelt Crescent
North Vancouver, BC V7P 1M3
Phone: 604 924-9771
Toll-free: 866 288-7878
Fax: 604 924-9772
E-mail: info@naturetrust.bc.ca

Ocean Resource Conservation Alliance
P.O. Box 1189
Sechelt, BC V0N 3A0
Phone: 604 885-7518
Fax: 604 885-2518

Sierra Club of Canada, BC Chapter
302-733 Johnson Street
Victoria, BC V8W 3C7
Phone: 250 386-5255
Fax: 250 386-4453

Western Canada Wilderness Committee
227 Abbott Street
Vancouver, BC V6B 2K7
Phone: 604 683-8220
Toll-free: 800 661-9453
Fax: 604 683-8229
E-mail: info@wildernesscommittee.org
Website: http://www.wildernesscommittee.org

Wildlife Rescue Association of British Columbia
5216 Glencairn Drive
Burnaby, BC V5B 3C1
Phone: 604 526-7275
Fax: 604 524-2890

ALBERTA

Alberta Environmental Protection
Communications
9915-108 Street, 9th Floor
Edmonton, AB T5K 2G8
Phone: 780 427-6267 (communications)
General inquiries:
Phone: 780 427-2700 (Toll-free by first dialing 310-0000)
Fax: 780 422-4086

Alberta Sport, Recreation, Parks, and Wildlife Foundation
Communications Branch
7th Floor, Standard Life Centre
10405 Jasper Avenue
Edmonton, AB T5J 4R7
Phone: 780 427-6530
E-mail: Comdev.Communications@gov.ab.ca

Alberta Wilderness Association
Box 6398, Station D
Calgary, AB T2P 2E1
Phone: 403 283-2025
Fax: 403 270-2743

Bow Valley Naturalists
Box 1693
Banff, AB T0L 0C0
Phone/Fax: 403 762-4160

CPAWS, Calgary/Banff Chapter
1120-1202 Centre Street SE
Calgary, AB T2G 5A5
Phone: 403 232-6686
Fax: 403 232-6988
E-mail: info@cpawscalgary.org

CPAWS, Edmonton Chapter
P.O. Box 52031
8540-109 Street, Suite 202
Edmonton, AB T6G 2T5
Phone: 780 432-0967
Fax: 780 439-4913
E-mail: info@cpaws-edmonton.org

Federation of Alberta Naturalists
11759 Groat Road
Edmonton, AB T5M 3K6
Phone: 780 427-8124
Fax: 780 422-2663 (c/o FAN)
E-mail: info@fanweb.ca

SASKATCHEWAN

CPAWS, Saskatchewan Chapter
203-115 2nd Avenue
North Saskatoon, SK S7K 2B1
Phone: 306 955-6197
Fax: 306 665-2128
E-mail: info@cpaws-sask.org

Environment and Resource Management
Education and Communications
3211 Albert Street
Regina, SK S4S 5W6
Toll-free: 800 667-2757

Nature Saskatchewan
1860 Lorne Street, Suite 206
Regina, SK S4P 3W6
Phone: 306 780-9273
Toll-free: 800 667-4668
Fax: 306 780-9263

Saskatchewan Environmental Society
P.O. Box 1372
Saskatoon, SK S7K 3N9
Phone: 306 665-1915
Fax: 306 665-2128

MANITOBA

CPAWS, Manitoba Chapter
Box 344
Winnipeg, MB R3C 2H5
Phone: 204 949-0782
Fax: 204 949-0783

Department of Conservation
Phone: 204 945-3744
Fax: 204 945-4261
E-mail: mgi@gov.mb.ca

Department of Environment
123 Main Street, Suite 160
Winnipeg, MB R3C 1A5
Phone: 204 945-7100

Manitoba Naturalists Society
63 Albert Street, Suite 401
Winnipeg, MB R3B 1G4
Phone/Fax: 204 943-9029

ONTARIO

Citizens Network on Waste Management
17 Major Street
Kitchener, ON N2H 4R1
Phone: 519 744-7503
Fax: 519 744-1546
E-mail: jjackson@web.net

CPAWS, Wildlands League Chapter
401 Richmond Street West, Suite 380
Toronto, ON M5V 3A8
Phone: 416 971-9453
E-mail: cpaws@web.net

Earthroots Coalition
401 Richmond Street West, Suite 410
Toronto, ON M5V 3A8
Phone: 416 599-0152
Fax: 416 340-2429

Elora Centre for Environmental Excellence
82 Metcalfe Street
Elora, ON N0B 1S0
Phone: 519 846-0841
Fax: 519 846-2642
E-mail: info@ecee.on.ca

Energy Action Council of Toronto
401 Richmond Street West, Suite 401
Toronto, ON M5V 3A8
Phone: 416 488-3966
Fax: 416 977-2157

Environment North
704 Holly Crescent
Thunder Bay, ON P7E 2T2
Phone: 807 475-5267
Fax: 807 577-6433

Federation of Ontario Naturalists
355 Lesmill Road
Don Mills, ON M3B 2W8
Phone: 416 444-8419
Fax: 416 444-9866

Ministry of Environment and Energy
Communications
135 St Clair Avenue West, 2nd Floor
Toronto, ON M4V 1P5
Phone: 416 323-4321
Toll-free: 800 565-4923

Water Environment Association of Ontario
P.O. Box 176
Milton, ON L9T 4N9
Phone: 416 410-6933
Fax: 416 410-1626

QUEBEC

La Fondation pour la sauvegarde des espèces menacées
Édifice Marie-Guyart, rez-de-chaussée
675 boulevard René-Lévesque est
Québec, QC G1R 5V7
Phone: 418 521-3830
Toll-free: 800 561-1616
Fax: 418 646-5974
E-mail: info@menv.gouv.qc.ca

La Fondation québécoise en environnement
1255 carré Phillips, bureau 706
Montréal, QC H3B 3G1
Phone: 514 849-3323
Toll-free: 800 361-2503
Fax: 514 849-0028
E-mail: info@fqe.qc.ca

L'Union québécoise pour la conservation de la nature (UQCN)
1085 avenue de Salaberry, bureau 300
Québec, QC G1R 2V7
Phone: 418 648-2104
Fax: 418 648-0991
E-mail: courrier@uqcn.qc.ca

Ministry of Environment and Wildlife
Institutional Affairs
675 boulevard René-Lévesque est, 8ième étage
Québec, QC G1R 5V7
Phone: 418 643-1853
Toll-free: 800 561-1616

Ministry of Natural Resources
Direction des communications
5700 4th avenue ouest, B 302
Charlesbourg, QC G1H 6R1
Phone: 418 627-8609
Fax: 418 643-0720
E-mail: communications@mrnfp.gouv.qc.ca

NEW BRUNSWICK

Conservation Council of New Brunswick
180 St John Street
Fredericton, NB E3B 4A9
Phone: 506 458-8747
Fax: 506 458-1047

Department of the Environment
Communications and Environmental
  Education
P.O. Box 6000
Fredericton, NB E3B 5H1
Phone: 506 453-3700

Nature Trust of New Brunswick
P.O. Box 603, Station A
Fredericton, NB E3B 5A6
Phone: 506 457-2398
Fax: 506 450-2137
E-mail: ntnb@nbnet.nb.ca

New Brunswick Federation of Naturalists
924 Prospect Street, Suite 110
Fredericton, NB E3B 2T9
Phone: 506 459-4209
E-mail: nbfn@nb.aibn.com

New Brunswick Protected Natural Areas Coalition
180 St John Street
Fredericton, NB E3B 4A9
Phone: 506 451-9902
Fax: 506 458-1047

PRINCE EDWARD ISLAND

Environment, Energy, and Forestry
Jones Building, 4th and 5th Floors
11 Kent Street
P.O. Box 2000
Charlottetown, PEI C1A 7N8
Phone: 902 368-5000
Fax: 902 368-5830

Island Nature Trust
P.O. Box 265
Charlottetown, PEI C1A 7K4
Phone: 902 566-9150
Fax: 902 628-6331
E-mail: intrust@isn.net

NOVA SCOTIA

CPAWS, Nova Scotia Chapter
1099 Marginal Road, Suite 201
Halifax, NS B3H 4P7
Phone: 902 446-4155
E-mail: coordinator@cpawsns.org
Website: http://cpawsns.org

Department of Environment and Labour
5151 Terminal Road
P.O. Box 697
Halifax, NS B3J 2T8
Phone: 902 424-5300
Fax: 902 424-0503

Ecology Action Centre
1568 Argyle Street, Suite 31
Halifax, NS B3J 2B3
Phone: 902 429-2202

Federation of Nova Scotia Naturalists
6360 Young Street
Halifax, NS B3L 2A1

NEWFOUNDLAND AND LABRADOR

Department of Environment and Conservation
4th Floor, West Block Confederation Building
P.O. Box 8700
St John's, NL A1B 4J6
Phone: 709 729-2664
Fax: 709 729-6639

Protected Areas Association of Newfoundland and Labrador
Box 1027, Station C
St John's, NL A1C 5M5
Phone/Fax: 709 726-2603

Wilderness and Ecological Reserves Advisory Council
c/o Parks and Natural Areas Division, Department
of Tourism, Culture and Recreation
P.O. Box 8700
St John's, NL A1B 4J6
Phone: 709 729-2421
Fax: 709 729-1100

YUKON

CPAWS, Yukon Chapter
P.O. Box 31095
211 Main Street
Whitehorse, YT Y1A 5P7
Phone: 867 393-8080
Fax: 867 393-8081

Department of Environment
Box 2703
Whitehorse, YT Y1A 2C6
Phone: 867 667-5652
Toll-free (in Yukon): 800 661-0408, local 5652
Fax: 867 393-6213
E-mail: environmentyukon@gov.yk.ca

Renewable Resources Communications
P.O. Box 2703
Whitehorse, YT Y1A 2C6
Phone: 403 667-5237

Yukon Conservation Society
302 Hawkins Street
Whitehorse, YT Y1A 1X6
Phone: 867 668-5678
Fax: 867 668-6637
E-mail: ycs@ycs.yk.ca

NORTHWEST TERRITORIES

Department of Resources, Wildlife, and Economic Development
Government Information Office
P.O. Box 1320
Yellowknife, NWT X1A 2L9
Phone: 403 669-2302

Ecology North
5013 51st Street
Yellowknife, NWT X1A 1S5
Phone: 867 873-6019
Fax: 867 873-9195
E-mail: econorth@ssimicro.com

NUNAVUT

Department of Sustainable Development
Government of Nunavut Communications
Iqaluit, NU X0A 0H0
Phone: 867 975-5925
Fax: 867 975-5980
Website: http://www.gov.nu.ca/sd.htm

Nunavut Parks
Government of Nunavut Communications
Iqaluit, NU X0A 0H0
Phone: 867 975-5900
Fax: 867 975-5990
E-mail: Parks@gov.nu.ca
Website: http://www.nunavutparks.com/visi-
torscentre/links.cfm

Glossary

100-mile diet: A term introduced in 2005 referring to buying and eating food grown, manufactured, or produced entirely within a 100-mile radius of one's residence.

abiotic components: Non-living parts of the ecosystem, including chemical and physical factors, such as light, temperature, wind, water, and soil characteristics.

acid deposition: Rain or snow that has a lower pH than precipitation from unpolluted skies; also includes dry forms of deposition, such as nitrate and sulphate particles.

acid mine drainage: Acidic drainage from waste rock and mine tailings caused by the oxidization of iron sulphides to create sulphuric acid, which in turn dissolves residual metals.

acid shock: The build-up of acids in water bodies and standing water over the winter, resulting in higher acidity than experienced through the rest of the year.

active management: Purposeful interference by resource and environmental managers in ecosystems, which recognizes that the human forces of change are now so ubiquitous that even protected areas are affected; includes habitat restoration, creation of wildlife corridors, reintroduction of extirpated species, prescribed burning, and management of hyper-abundant species.

adaptation: Adjustment to different or changing circumstances, such as when insurance companies modify their claims forecasting and setting of premiums with regard to future climate change conditions. The largest challenge for adaptation strategies will occur in the future when the most significant consequences from climate change will appear.

adaptive co-management: Management concept including such key attributes as learning-by-doing, integrating different knowledge systems, collaborating and power-sharing among community, regional, and national levels, and managing for flexibility.

adaptive environmental management: An approach that develops policies and practices to deal with the uncertain, the unexpected, and the unknown; approaches management as an experiment from which we learn by trial and error.

aerobic: Requiring oxygen.

albedo: The extent to which the surface of the Earth reflects rather than absorbs incoming radiation from the sun. Snow has a high albedo, but as temperatures rise, the area covered in snow will be replaced by areas free of snow, uncovering rocks and vegetation with lower albedo values that absorb radiation and thus add to warming.

alien (also known as exotic, introduced, invader, or non-native) species: Any organism, such as zebra mussels, purple loosestrife, and Eurasian water milfoil in Canada, that enters an ecosystem beyond its normal range through deliberate or inadvertent introduction by humans.

alternative dispute resolution: A non-judicial approach to resolving disputes that uses negotiation, mediation, or arbitration, with a focus on reparation for harm done and on improving future conduct.

anadromous: Aquatic life, such as salmon, that spend part of their lives in salt water and part in fresh water.

anaerobic: Lacking oxygen.

annual allowable cut (AAC): The amount of timber that is allowed to be cut annually from a specified area.

anthropocentric view: Human-centred, in which values are defined relative to human interests, wants, and needs.

aquaculture: Seafood farming, the fastest-growing food production sector in the world.

aquifer: A formation of permeable rocks or loose materials that contains usable sources of groundwater and may extend from a few square kilometres to several thousand square kilometres.

arbitration: A procedure for dispute resolution in which a third party is selected to listen to the views and interests of the parties in dispute and develop a solution to be accepted by the participants.

artisanal: Small-scale fisheries.

aspirational approach: With reference to climate change, an approach emphasizing long-term term but unspecific and non-binding targets for reducing greenhouse gas emissions. Advocated by developed countries.

Atlantic Maritime: Ecozone stretching from the mouth of the St Lawrence River across New Brunswick, Nova Scotia, and Prince Edward Island that is heavily influenced by the Atlantic Ocean, which creates a cool, moist maritime climate but with quite variable conditions between the upland masses, such as the Cape Breton and New Brunswick highlands, and the coastal lowlands that support most of the population.

atmosphere: Layer of air surrounding the Earth.

autotrophs: Organisms, such as plants, that produce their own food, generally via photosynthesis.

Bali Conference, 2007: A UN-sponsored climate change conference held in the first two weeks of December 2007 in Bali, Indonesia, to start a process to create a new framework to replace the Kyoto Protocol, which ends in 2012. There was agreement that both developed and developing countries must participate in reducing greenhouse gas emissions but reluctance from key developed countries to commit to binding targets.

benthic: Of or living on or at the bottom of a water body.

bioaccumulation: The storage of chemicals in an organism in higher concentrations than are normally found in the environment.

biocapacity: The amount of biologically productive area—cropland, pasture, forest, and fisheries—available to meet humanity's needs.

biocentric (ecocentric) view: A view that values aspects of the environment simply because they exist and accepts that they have the right to exist.

biocides: Chemicals that kill many different kinds of living things; also called pesticides.

bioconcentration: The combined effect of bioaccumulation and biomagnification.

biodiversity: The variety of life forms that inhabit the Earth. Biodiversity includes the genetic diversity among members of a population or species as well as the diversity of species and ecosystems.

biofuels: Solid, liquid, or gas fuel derived from relatively recently dead biological material and distinguished from fossil fuels, which are derived from long-dead biological material.

biogeochemical cycles: Series of biological, chemical, and geological processes by which materials cycle through ecosystems.

biological oxygen demand (BOD): The amount of dissolved oxygen required for the bacterial decomposition of organic waste in water.

biomagnification: Build-up of chemical elements or substances in organisms in successively higher trophic levels.

biomass: The sum of all living material, or of all living material of particular species, in a given environment.

biomass pyramid: Related to the fact that in terrestrial ecosystems, greater biomass generally exists at the level of primary consumers, with the least total biomass at the highest trophic levels; in marine ecosystems, the reverse is true, and the pyramid is inverted—greater biomass is at the highest trophic level, while the primary consumers, phytoplankton, at any given time comprise much less biomass but reproduce rapidly.

biome: A major ecological community of organisms, both plant and animal, that is usually characterized by the dominant vegetation type; for example, a tundra biome and a tropical rain forest biome.

bioregion: Not defined by surveyors' lines but rather a place that makes sense to its inhabitants based on natural and cultural attributes. Bioregions are usually identified to reflect the natural and cultural diversity of a place.

biotic components: Those parts of ecosystems that are living; organisms.

biotic potential: The ability of species to reproduce regardless of the level that an environment can support; see **carrying capacity**.

bitumen: Any of various minerals that will burn, such as asphalt or petroleum.

blue box: The boxes introduced for curbside collection programs in urban neighbourhoods to divert selected household waste material from landfill sites to be recycled.

blue water: The water represented by rain falling into lakes, rivers, and aquifers.

Boreal Cordillera: Ecozone, to the south and west of the Tundra Cordillera, in northern BC and southern Yukon made up of mountains in the west and east, separated by intermontane plains, with a wet climate that gives rise to tree growth.

boreal forest: One of the largest forest belts in the world, extending all across North America and Eurasia, encompassing roughly a third of the Earth's forest land and 14 per cent of the world's forest biomass and separating the treeless tundra regions to the north from the temperate deciduous forests or grasslands to the south.

Boreal Plains: Ecozone extending from the southern part of the Yukon in a wide sweeping band down into southeastern Manitoba, consisting of a generally flat to undulating surface similar to the Prairie zone to the south.

Boreal Shield: The largest ecozone in Canada, stretching along the Canadian Shield from Saskatchewan to Newfoundland.

bottom trawling: One of the most destructive means of fishing in which heavy nets are dragged along the sea floor scooping up everything in their path.

Brisbane Declaration: Specified in September 2007 that environmental flows are essential for freshwater ecosystem health and human well-being.

brownfields: Abandoned or active industrial sites. On the surface or underground are soils contaminated through disposal practices accepted in earlier times before people appreciated the long-term consequences.

bycatch: Non-target organisms caught or captured in the course of catching a target species, as in the fisheries, where estimates suggest that 25 per cent of the world's catch is dumped because it is not the right species or size.

calorie: A unit of heat energy, the amount of heat required to raise one gram of water by 1°C.

carbon balance: A balance between the amount of CO_2 in the atmosphere and bicarbonate in the water.

carbon offset: A measure of greenhouse gas reduction, measured in metric tons of carbon dioxide equivalent. One carbon offset equals a reduction in one metric ton of carbon dioxide equivalent of greenhouse gas emission. Companies or governments can arrange to receive credit against emission caps from carbon offsets. At a personal level, some travellers pay for a carbon offset related to greenhouse gas emissions produced when they travel by air. The amount paid is used by airlines on planting trees or other initiatives that reduce greenhouse gas emissions.

carbon sequestration: Reforestation and afforestation to ameliorate carbon dioxide loadings in the atmosphere because trees and shrubs use the excess CO_2.

carbon tax: An approach in which greenhouse gas emissions by individuals or companies are taxed. The purpose is to change human behaviour towards activities that produce fewer greenhouse gas emissions.

carnivore: An organism that consumes only animals.

carrying capacity: Maximum population size that a given ecosystem can support for an indefinite period or on a sustainable basis.

certification: The confirmation of certain characteristics of an object, person, or organization.

chemoautotroph: A producer organism that converts inorganic chemical compounds into energy.

chlorophylls: Pigments of plant cells that absorb sunlight, thus enabling plants to capture solar energy.

clear-cutting: A forest harvesting technique in which an entire stand of trees is felled and removed.

climate: The long-term weather pattern of a particular region.

climate change: A long-term alteration in the climate of a particular location or region or for the entire planet.

climate modelling: Various mathematical and computerized approaches for determining past climate trends in an effort to build scenarios predicting future climate, which use any or all of the following factors in measurement: incoming and outgoing radiation; energy dynamics or flows around the globe; surface processes affecting climate, such as snow cover and vegetation; chemical composition of the atmosphere; and time step or resolution (time over which the model runs and the spatial scale to which it applies).

climax community: Last stage of succession; a relatively stable, long-lasting, complex, and interrelated community of organisms.

coevolution: Process whereby two species evolve adaptations as a result of extensive interactions with each other.

collaboration: The art of working together.

co-management: An arrangement in which a government agency shares or delegates some of its legal authority regarding a resource or environmental management issue with local inhabitants of an area.

commensalism: An interaction between two species that benefits one species and neither harms nor benefits the other.

competitive exclusion principle: The principle that competition between two species with similar requirements will result in the exclusion of one of the species.

compound: The coming together of two different atoms to form a different substance, such as water (H_2O), a compound made up of two hydrogen atoms (H) and one oxygen atom (O).

consumerism: Wasteful consumption of resources to satisfy wants rather than needs.

consumers: Organisms that cannot produce their own food and must get it by eating or decomposing other organisms; in economics, those who use goods and services.

contemporary evolution: Evolution that occurs on short time scales.

context: Specific characteristics of a time and place.

contour cultivation: The cultivation and seeding of fields parallel to the contour of the slope, which serves to reduce the speed of runoff by catching soil particles in the plough furrows.

Convention on Biological Diversity (CBD): International treaty that emerged from the World Summit on Sustainable Development in Rio de Janeiro in 1992 that requires signatories, including Canada, to develop biodiversity strategies, identify and monitor important components of biodiversity, develop endangered species legislation/protected areas systems, and promote environmentally sound and sustainable development in areas adjacent to protected areas.

Convention on International Trade in Endangered Species of Wild Fauna and Flora (CITES): A 1973 treaty currently ratified by more than 120 countries (including Canada) that establishes lists of species for which international trade is to be controlled or monitored (e.g., orchids, cacti, parrots, large cat species, sea turtles, rhinos, primates).

co-ordination: The effective or harmonious working together of different departments, groups, and individuals.

coral bleaching: Death of corals caused by water temperatures becoming too warm.

coral polyps: Individual biotic members of a coral reef.

corporate responsibility: Occurs when corporations systematically examine the environmental impact of their activities, then take action to reduce the negative impacts. Initiatives range from reducing, reusing, and recycling to specifying environment-friendly production practices for suppliers to providing funds to environmental groups.

crop rotation: Alternating crops in fields to help restore soil fertility and also control pests.

crude birth rate (CBR): Number of births per 1,000 individuals per year.

crude bitumen: A thick and heavy oil.

crude death rate (CDR): Number of deaths per 1,000 individuals per year.

crude growth rate (CGR): Produced by subtracting the crude death rate (CDR) from the crude birth rate (CBR).

cryosphere: Based on the Greek word *kryos* meaning 'cold', refers to those parts of the Earth's surface where water is in solid form as ice or snow. The cryosphere includes sea, lake, and river ice, snow cover, glaciers, ice caps and ice sheets, as well as frozen ground (permafrost).

culmination age: The age of economic maturity of a tree crop, which varies widely but usually falls within the 60- to 120-year range in Canada.

custom-designed solutions: Management approach in which the specific conditions of a place and time are recognized and the attempt to ameliorate or resolve a problem takes these specifics into account.

DDT (dichlorodiphenyltrichloroethane): An organochlorine insecticide used first to control malaria-carrying mosquitoes and lice and later to control a variety of insect pests but now banned in Canada because of its persistence in the environment and ability to bioaccumulate.

decomposer food chain: A specific nutrient and energy pathway in an ecosystem in which decomposer organisms (bacteria and fungi) consume dead plants and animals as well as animal wastes; essential for the return of nutrients to soil and carbon dioxide to the atmosphere. Also called detritus food chain.

demand management: Emphasizes influencing human behaviour so that less water or energy is used.

demographic transition: Transition of a human population from high birth rate and high death rate to low birth rate and low death rate.

denitrification: The conversion of nitrate to molecular nitrogen by bacteria in the nitrogen cycle.

detritus: Organic waste, such as fallen leaves.

disturbance: A natural or human-induced event or process that interrupts ecological succession.

diversion: Movement of water from one water system to another in order to enhance water security, reduce flood vulnerability, or generate hydroelectricity.

double-loop learning: Relates to situations for which there is a mismatch between intention and outcome and happens when such a mismatch is addressed by challenging underlying values and behaviour rather than assuming that the prevailing values and behaviour are appropriate.

drought: Condition in which a combination of lack of precipitation, temperature, evaporation, evapotranspiration, and the inability of soil to retain moisture leads to a loss of resilience among flora and fauna in dry conditions.

dykes: Walls or earth embankments along a watercourse to control flooding, or encircling a town or a property to protect it from flooding (ring dyke), or across a stream so that the flow of water is stopped from going upstream by a sluice gate (cross dyke).

dynamic equilibrium: Occurs when two opposing processes proceed at the same rate.

E. coli: *Escherichia coli*, a bacterium present in fecal matter that can get into a water supply and pollute it, as happened in Walkerton, ON, in May 2000.

ecocentric (biocentric) view/values: The view that a natural order governs relationships between living things and that a harmony and balance reflect this natural order, which humankind tends to disrupt.

ecological footprint: The land area a community needs to provide its consumptive requirements for food, water, and other products and to dispose of the wastes from this consumption.

ecological restoration: Renewing a degraded, damaged, or destroyed ecosystem through active human intervention.

ecological succession: The gradual replacement of one assemblage of species by another as conditions change over time.

ecologically extinct: A species that exists in such low numbers that it can no longer fulfill its ecological role in the ecosystem.

ecosphere: Refers to the entire global ecosystem, which comprises atmosphere, lithosphere, hydrosphere, and biosphere as inseparable components.

ecosystem: Short for ecological system; a community of organisms occupying a given region within a biome, including the physical and chemical environment of that community and all the interactions among and between organisms and their environment.

ecosystem-based management (EBM): Holistic management that takes into account the entire ecosystem and emphasizes biodiversity and ecosystem integrity, as opposed to focusing primarily or solely on a resource or resources, such as water or timber, within an ecosystem.

ecotone: The transitional zone of intense competition for resources and space between two communities.

El Niño: A marked warming of the waters in the eastern and central portions of the tropical Pacific that triggers weather changes and events in two-thirds of the world.

emission credits: Can be earned by a nation based on land-use or forestry (afforestation, reforestation) initiatives that reduce measurable greenhouse gas emissions.

emission trading: Under the Kyoto Protocol, a system whereby one country that will exceed its allotted limit of greenhouse gas emissions can buy an amount of greenhouse gas emissions from another country that will not reach its own established emissions limit.

endangered: An official designation assigned by the Committee on the Status of Endangered Wildlife in Canada to any indigenous species or subspecies or geographically separate population of fauna or flora that is threatened with imminent extinction or extirpation throughout all or a significant portion of its Canadian range.

endemic species: A plant or animal species confined to or exclusive to a specific area.

endocrine disruption: The interference of normal bodily processes such as sex, metabolism, and growth by chemicals in such products as soaps and detergents that are released into an ecosystem, as happens among aquatic species, often causing feminization.

energy: The capacity to do work; found in many forms, including heat, light, sound, electricity, coal, oil, and gasoline.

entropy: A measure of disorder. The second law of thermodynamics applied to matter says that all systems proceed to maximum disorder (maximum entropy).

environment: The combination of the atmosphere, hydrosphere, cryosphere, lithosphere, and biosphere in which humans, other living species, and non-animate phenomena exist.

environmental flow management: Managing water in recognition that ecosystems are legitimate users of water. Consequently, environmental needs are identified from the outset, and the amount of water required to satisfy such needs is subtracted from what is available in a source of supply to meet human needs.

environmental impact assessment: Part of impact assessment that identifies and predicts the impacts from development proposals on both the biophysical environment and on human health and well-being.

epidemiological transition: Change in mortality rates from high to low in a human population.

estuary: Coastal regions, such as inlets or mouths of rivers, where salt water and fresh water mix.

euphotic zone: Zone of the ocean to which light from the sun reaches.

eutrophic: Pertaining to a body of water rich in nutrients.

eutrophication (also known as nutrient enrichment): The over-fertilization of a body of water by nutrients that produce more organic matter than the water body's self-purification processes can overcome.

evapotranspiration: Evaporation of water from soil and transpiration of water from plants.

evolution: A long-term process of change in organisms caused by random genetic changes that favour the survival and reproduction of those organisms possessing the genetic change; organisms become better adapted to their environment through evolution.

ex situ preservation: The conservation of species outside their natural habitat, including breeding in captivity, so that they can be reintroduced to their natural habitat, as has been done, for example, with the black-footed ferret and swift fox.

exclusive economic zones (EEZs): Areas off the coasts of a nation that are claimed by that nation for its sole responsibility and exploitation, as permitted by the UN Convention on the Law of the Sea.

exponential growth: The growth when a population increases by a certain percentage rather than an absolute amount, producing a J-shaped curve.

extended producer responsibility: The concept underlying laws or regulations that require manufacturers and importers to accept responsibility for their products at the end of their useful life span. They provide an incentive for companies to design their products so that they can be recycled or reused and to eliminate toxic materials, since they would have to dispose of them.

extinction: The elimination of all the individuals of a species.

extirpated: An official designation assigned by the Committee on the Status of Endangered Wildlife in Canada to any indigenous species or subspecies or geographically separate population of fauna or flora no longer known to exist in the wild in Canada but occurring elsewhere.

extrinsic values: Values that humans derive from other species, including consumptive and non-consumptive values.

falldown effect: The lower volume of harvestable timber at the culmination age for second growth on sites where old-growth forest was previously harvested.

***feng shui* landscapes:** Reflect the East Asian belief that humans should live in harmony with nature and that trees bring good fortune. Thus, *feng shui* landscapes involve the inclusion of trees and other vegetation within urbanized or settled areas.

fishing down the food chain: Harvesting at progressively lower trophic levels as higher trophic levels become depleted.

flood plain: Low-lying land along a river, stream, or creek or around a lake that under normal conditions is flooded from time to time.

floodway or diversion: An excavated channel to divert flood waters away from a population centre.

food chain: A specific nutrient and energy pathway in ecosystems proceeding from producer to consumer; along the pathway, organisms in higher trophic levels gain energy and nutrients by consuming organisms at lower trophic levels.

food web: Complex intermeshing of individual food chains in an ecosystem.

forest tenure: The conditions that govern forest ownership and use.

fossil fuels: Organic fuels (coal, natural gas, oil, tar sands, and oil shale) derived from once-living plants or animals.

fragmentation: The division of an ecosystem or species habitat into small parcels as a result of human activity, such as agriculture, highways, pipelines, and population settlements.

frozen core dam: Type of dam used in mining in the North, as at the Ekati diamond mine in the Northwest Territories, with a central core of frozen soil saturated with ice and bonded to the natural permafrost and surrounded by granular fill to ensure both stability and thermal protection.

gaseous cycles: Cycles of elements that have most of their matter in the atmosphere.

general circulation models (GCMs): The most prominent and most complex type of climate modelling, which takes into account the three-dimensional nature of the Earth's atmosphere and oceans or both.

generalist species: Species, like the black bear and coyote, with a very broad niche where few things organic are not considered a potential food item.

genetically modified organisms (GMOs): Organisms created by humans through genetic manipulation combining genes from different and often totally unrelated species to create a different organism that is economically more productive and/or has greater resistance to pathogens.

global climate change: Impacts of accumulation of greenhouse gases on the Earth's climate.

global warming: Changes in average temperatures of the Earth's surface, although these changes are not uniform (i.e., some regions experience significantly higher temperatures, others only slight changes upward, and still others might experience somewhat cooler temperatures).

governance: The processes used to determine how policy decisions are taken and by whom. Governance arrangements identify how disputes will be resolved and require capacity to identify trade-offs and compromises.

government: The formal rules or authority over a country, state, or other jurisdiction, facilitated by people, structures, and processes and designed to provide for transparency and accountability for decisions taken.

grasshopper effect: Atmospheric transport and deposition of persistent and volatile chemical pollutants whereby the pollutants evaporate into the air in warmer climates and travel in the atmosphere towards cooler areas, condensing out again when the temperature drops. The cycle then repeats itself in a series of 'hops' until the pollutants reach climates where they can no longer evaporate.

green box: The second stage of urban curbside collection programs involving residents placing organic and similar wastes into green boxes with the contents then composted and sold or made available at nominal cost for residents to use as soil.

green manure: Growing plants that are ploughed into the soil as fertilizer.

Green Revolution: Development in plant genetics (hybridization) in the late 1950s and early 1960s resulting in high-yield varieties producing three to five times more grain than previous plants but requiring intensive irrigation and fertilizer use.

green water: Water created from evapotranspiration by plants and crops.

greenfields: Areas not yet developed and known not to have surface or underground soil contamination.

greenhouse effect: A warming of the Earth's atmosphere caused by the presence of certain gases (e.g., water vapour, carbon dioxide, methane) that absorb radiation emitted by the Earth, thereby retarding the loss of energy to space.

greenhouse gas (GHG): A gas that contributes to the greenhouse effect, such as carbon dioxide.

gross national product (GNP): The total value of all goods and services produced for final consumption in an economy, used by economists as an index or indicator to compare national economies or periods of time within a single national economy.

gross primary productivity (GPP): The total amount of energy produced by autotrophs over a given period of time.

groundwater: Water below the Earth's surface in the saturated zone.

habitat: The environment in which a population or individual lives.

Happy Planet Index (HPI): An index that attempts to provide a perspective on human well-being and environmental impact and to focus on achieving sustainability. Each country's HPI value is a function of its average subjective life satisfaction, life expectancy at birth, and ecological footprint per capita—it approximates multiplying life satisfaction and life expectancy and dividing that by the ecological footprint.

heat: The total energy of all moving atoms.

herbivores: Animals that eat plants—that is, primary consumers.

heritage rivers: Rivers designated for special protection by the Canadian Heritage Rivers Board because of their historical, cultural, ecological, and recreational significance.

heterotroph: An organism that feeds on other organisms.

high-quality energy: Energy that is easy to use, such as a hot fire or coal or gasoline, but that disperses quickly.

hybridization: The crossbreeding of two varieties or species of plants or animals.

hydrochlorofluorocarbons (HCFCs): Compound containing hydrogen, chlorine, fluorine, and carbon. HCFC is one type of chemical being used to replace chlorofluorcarbons (CFCs) because HCFCs have less impact on reducing ozone in the stratosphere. CFCs are formed by chlorine, fluorine, and carbon and are broken down by ultraviolet light in the stratosphere. When broken down, CFCs release chlorine atoms, which deplete the ozone layer. CFCs are used as refrigerants, solvents, and foam-blowing agents.

hydrological cycle: The circulation of water through bodies of water, the atmosphere, and land.

hydrosolidarity: An approach that recognizes the interconnections among aquatic, terrestrial, and other resource systems, leading to management that is integrated, participative, collaborative, co-ordinated, and shared, whether at local, provincial, national, or international levels.

hydrosphere: One of three main layers of the ecosphere, containing all the water on Earth.

hypoxic: A hypoxic system with low DO concentration—in the range between 1 and 30 per cent DO saturation.

impact and benefit agreements (IBAs): Voluntary agreements between extractive industries and communities that go beyond formal impact assessment requirements and are intended to facilitate extraction of resources in a way that contributes to the economic and social well-being of local people and communities.

impact assessment: Thorough consideration of the effects of a project that takes into account its potential and probable impacts on the environment and on society or a community and that assesses the technology proposed for the project as well as the technology available for dealing with any negative impacts.

in situ preservation: Conservation strategies that focus on a species within its natural habitat.

in situ recovery: Refers to the general practice used at depths greater than 100 metres to remove crude bitumen from oil sands. The specific technique is called steam-assisted gravity drainage, or SAGD. The usual approach is to inject high-pressure steam into the oil sands to separate the bitumen from the sand, silt, and clay. Once exposed to the steam, bitumen can flow to a well from which it can be pumped to the surface.

indicators: Specific facets of a particular system, such as the population of a key species within an ecosystem, that tell us something about the current state of the system but do not help us understand why the system is in that state.

indigenous knowledge: Understanding of climate, animals and animal behaviour, soil, waters, and/or plants within an ecosystem based on experiential knowledge from living or working in a particular area for a long period of time; also referred to as traditional ecological knowledge (TEK) or local knowledge.

inertia: The tendency of a natural system to resist change.

ingenuity gap: Refers to the critical gap between the need for ideas to fix complex problems resulting from resource scarcity and the actual supply of such ideas. This concept highlights the importance of being able to generate and disseminate ingenuity to ensure that scarcity of resources does not negatively affect well-being for humans and other living organisms.

integrated pest management (IPM): The avoidance or reduction of yield losses caused by diseases, weeds, insects, etc., while minimizing the negative impacts of chemical pest control.

integrated plant nutrient systems (IPNSs): Maximization of the efficiency of nutrient use by recycling all plant nutrient sources within the farm and by using nitrogen fixation by legumes.

integrated water resource management (IWRM): An approach that promotes the co-ordinated development and management of water, land, and related resources in order to maximize the resultant economic and social welfare in an equitable manner without compromising the sustainability of vital ecosystems.

intensive livestock operations (ILOs): Factory farms, feedlots, etc. where large quantities of external energy inputs are required to raise for market larger numbers of animals than the area in which they are raised can support, which can result in problems of disease and dealing with animal waste.

International Joint Commission: A bilateral institution, consisting of three Canadian commissioners and three American commissioners, established by the 1909 Boundary Waters Treaty to manage interjurisdictional resource issues between Canada and the United States.

interspecific competition: Competition between members of different species for limited resources such as food, water, or space.

intraspecific competition: Competition between members of the same species for limited resources such as food, water, or space.

intrinsic value: A belief that nature has value in and of itself apart from its value to humanity; a central focus for the preservation of species.

invasive: An introduced species that spreads out and causes harmful effects on other species and ecosystems.

island biogeography: A field within biogeography that attempts to establish and explain the factors that affect the species richness of natural communities.

James Bay and Northern Quebec Agreement: Treaty signed in 1975 by the James Bay Cree and Quebec Inuit with the Quebec government permitting the continuance of the James Bay Project and granting to the Natives, among other things, $232.5 million in compensation and outright ownership of 5,543 square kilometres, as well as exclusive hunting, fishing, and trapping rights to an additional 62,160 square kilometres; often considered as the first modern Aboriginal land claims settlement in Canada.

James Bay Project: A hydroelectric mega-project in northern Quebec, begun in the 1970s, that has involved extensive dams on the La Grande and other rivers and has flooded thousands of square kilometres of the James Bay Cree homeland.

keystone species: Critical species in an ecosystem whose loss profoundly affects several or many others.

kimberlite pipes: Rare, carrot-shaped igneous rock formations sometimes containing diamonds and found in parts of northern Canada.

kinetic energy: The energy of objects in motion.

Kyoto Protocol: An international agreement reached in Kyoto, Japan, in 1997 that targets 38 developed nations as well as the European Community to ensure that 'their aggregate anthropocentric carbon dioxide equivalent emissions of the greenhouse gases [e.g., carbon dioxide, methane, nitrous oxide, hydrofluorocarbons, perfluorocarbons, sulphur hexafluoride] . . . do not exceed their assigned amounts'. The protocol came into effect in 2004 when 55 countries accounting for 55 per cent of 1990 global carbon dioxide emissions had ratified it.

law of conservation of energy: Law stating that energy cannot be created or destroyed; it is merely changed from one form to another. Also known as the first law of thermodynamics.

law of conservation of matter: Law that tells us that matter cannot be created or destroyed, but merely transformed from one form into another.

law of everybody: The understanding that if everyone did many small things of a conserving and environmentally aware nature, major environmental problems, threats, and dangers would be ameliorated or alleviated.

LEED: Acronym for the Leadership in Energy and Environmental Design Green Building Rating System. It is a national standard, established in the US, related to the design, construction, and operation of 'high-performance green buildings'. LEED provides benchmarks for performance related to five issues regarding human and environmental health: sustainable site development, water savings, energy efficiency, materials selection, and indoor environmental quality.

life-cycle assessments (LCAs): Identification of inputs, outputs, and potential environmental impacts of a product or service throughout its lifetime, from manufacture to use and ultimate disposal.

light living: The need to tread as lightly as possible, to minimize our ecological footprints, often characterized by the four Rs: refuse, reduce, reuse, and recycle.

limiting factor: A chemical or physical factor that determines whether an organism can survive in a given ecosystem. In most ecosystems, rainfall is the limiting factor.

limiting factor principle: Stipulates that all factors necessary for growth must be available in certain quantities if an organism is to survive.

lithosphere: The Earth's crust.

Livestock Revolution: The shift in production units from family farms to factory farms and feedlots that depend on outside supplies of feed, energy, and other inputs to produce vastly more livestock, a shift that has fuelled the growth in meat consumption worldwide, which has doubled since 1977.

Living Planet Index: An index that quantifies the overall state of planetary ecosystems.

longline: Type of commercial fishing using lines with many baited hooks.

long-run sustained yield: The yield for an area that is equal to the culmination of mean annual increment weighted by area for all productive and utilizable forest land types in that area.

low-quality energy: Energy that is diffuse, dispersed, at low temperatures, and difficult to gather; most of the energy available to us.

LULU: Acronym for 'locally unwanted land use', often the source of a NIMBY ('not in my backyard') reaction.

LUST: Acronym for 'leaking underground storage tanks', which results in contaminated aquifers.

macronutrient: A chemical substance needed by living organisms in large quantities (for example, carbon, oxygen, hydrogen, and nitrogen).

marine protected areas (MPAs): Underwater reserves set aside and protected from normal human exploitation because of the fragility, rarity, or valued biodiversity of their ecosystems.

matter: What things are made of—92 natural and 17 synthesized chemical elements such as carbon, oxygen, hydrogen, and calcium.

mediation: A negotiation process guided by a facilitator (mediator).

megaprojects: Large-scale engineering or resource development projects that cost at least $1 billion and take several years to complete.

mesosphere: Layer of the atmosphere extending from the stratosphere, from about 50 to about 80 kilometres above Earth.

metal toxicity: The poisonous or harmful nature of metals and minerals, such as asbestos and lead, both to humans and to ecosystems.

micronutrient: An element needed by organisms but only in small quantities, such as copper, iron, and zinc.

Millennium Development Goals: Globally accepted goals for development agreed to by member states of the United Nations. The target date for meeting the goals is 2015. One of the goals is to 'ensure environmental sustainability'.

Millennium Ecosystem Assessment: Assessment carried out by the UN to assess the consequences of ecosystem change for human well-being and to establish the scientific basis for actions needed to enhance the conservation and sustainable use of ecosystems and their contributions to human well-being.

mineralization: The process by which biomass is converted back to ammonia (NH_3) and ammonium salts (NH_4) by bacterial action and returned to the soil when plants die.

minimum viable population (MVP): The smallest population size of a species that can be predicted to have a very high chance of persisting for the foreseeable future.

mitigation: Strategies to reduce or minimize the negative consequences from a hazard such as climate change. Mitigation requires action today in order for initiatives to be able to reduce the most serious negative impacts in the future.

Mixed Wood Plains: The most urbanized ecozone in Canada, spreading from the lower Great Lakes north and east through the St Lawrence Valley, with gently rolling topography and a continental climate characterized by warm, humid summers and cool winters.

monitoring: Explicit and systematic checking of outputs and outcomes related to a management initiative in order to understand what works and what does not work and to determine needed modifications to enhance effectiveness.

monoculture cropping: Cultivation of one plant species (such as corn) over a large area, which leaves the crop highly susceptible to disease and insects, especially when all of the individual plants are genetically identical.

Montane Cordillera: Ecozone in the BC Interior with considerable contrast between the summits of the snowbound peaks and high montane valleys, rolling plateaus, and deeply entrenched desert-like conditions and a climate generally characterized by long, cold winters and short, warm summers.

Montreal Protocol: Signed in 1987 by 32 nations, established a schedule for reducing use of chlorofluorocarbons and halons to reduce the rate of depletion of the ozone layer.

multi-barrier approach: A method of ensuring the quality of a water supply by using a series of measures (e.g., system security, source protection through pollution regulations within a watershed, water treatment and filtration, testing), each independently acting as a barrier to water-borne contaminants through the system.

mutualism: Relationship between two organisms having to do with food supplies, protection, or transport that is beneficial to both.

nature deficit disorder: The increasing gap in understanding of the 'real world' on the part of the younger generation. Instead of playing outdoors in fields, woods, streams, lakes, or the ocean, an increasing proportion of the children of today are glued to their computer or TV screens. They seldom visit the outside world, especially areas dominated by nature rather than human activities.

negative feedback: Control mechanism present in the ecosystem and in all organisms—information in the form of chemical, physical, and biological agents influences processes, causing them to shut down or reduce their activity.

negotiation: One of the two main types of alternative dispute resolution when two or more parties involved in a dispute join in a voluntary, joint exploration of issues with the goal of reaching a mutually acceptable agreement.

neo-liberalism: A political or policy perspective that places high value on the role of free markets to allocate resources efficiently, leading to a belief that it is best to allow markets to function with minimum intervention by government regulations.

net primary productivity (NPP): Gross primary productivity (the total amount of energy that plants produce) minus the energy plants use during cellular respiration.

new forestry: A silvicultural approach that mimics natural processes more closely through emphasizing long-term site productivity by maintaining ecological diversity.

niche: An organism's place in the ecosystem: where it lives, what it consumes, and how it interacts with all biotic and abiotic factors.

NIMBY: 'Not in my backyard', a phrase used to describe local people's reactions when a noxious or undesired facility—for example, a landfill site, a sand and gravel pit, or an expressway— is proposed in an area adjacent to or near their property.

nitrogen fixation: Conversion of gaseous (atmospheric) nitrogen (N_2) into ammonia (NH_3) by bacteria, such as those that grow on the root nodules of legumes.

non-governmental organizations (NGOs): Organizations outside of the government and private sectors, usually established to address a specific societal issue or need. Also referred to as 'not-for-profit' or 'social profit' organizations. They are one element of 'civil society' and when focused on environmental matters, can be referred to as ENGOs (environmental non-governmental organizations).

non-point sources: Sources of pollution from which pollutants are discharged over a widespread area or from a number of small inputs rather than from distinct, identifiable sources.

non-renewable or stock resources: Resources, such as oil, coal, and minerals, that take millions of years to form and thus, for practical purposes, are fixed in supply and therefore not renewable.

non-timber forest products (NTFPs): Forest resources of economic value but not related to the lumber and pulp and paper industries, such as wild rice, mushrooms and berries, maple syrup, edible nuts, furs and hides, medicines, and ornamental cuttings, which currently contribute an estimated $442 million annually to the Canadian economy.

no-till/conservation agriculture (NT/CA): Zero, minimum, or low tillage to protect and stimulate the biological functioning of the soil while maintaining and improving crop yields, which includes direct sowing or drilling of seeds instead of ploughing, maintenance of permanent cover of plant material on the soil, and crop rotation.

nuclear power: Refers to power, usually electric power, produced by atomic energy. Atoms contain atomic energy, which can be released slowly in reactors or rapidly in bombs through alteration of the nuclei of atoms.

nuclear wastes: The radioactive wastes remaining from the uranium used to fuel nuclear fuel reactors. Such wastes have an extremely long life and are life-threatening, thereby creating significant storage and containment challenges.

nutrients: Elements or compounds that an organism must take in from its environment because it cannot produce them or cannot produce them as quickly as needed.

old-growth forests: Forests that generally have a significant number of huge, long-lived trees; many large standing dead trees; numerous logs lying about the forest floor; and multiple layers of canopy created by the crowns of trees of various ages and species.

oligotrophic: Nutrient poor.

omnivores: Organisms that eat both plants and animals.

optimal foraging theory: The relationship between the benefit of making a kill and feeding and the cost of the energy expended to make the kill.

oxygen sag curve: The drop in oxygen levels in a body of water when organic wastes are added and the number of bacteria rises to help break down the waste.

ozone: An atmospheric gas (O_3) that when present in the stratosphere helps to protect the Earth from ultraviolet rays. However, when present near the Earth's surface, it is a primary component of urban smog and has detrimental effects on both vegetation and human respiratory systems.

ozone layer: Thin layer of ozone molecules in the stratosphere that absorbs ultraviolet light and converts it into infrared radiation, effectively screening out 99 per cent of the ultraviolet light.

Pacific Maritime: Ecozone characterized by the influence of the Pacific Ocean, with the highest rainfall figures in Canada (up to 3,000 mm) and the warmest average temperatures.

Palliser's Triangle: Semi-arid area of southern Alberta and Saskatchewan and extreme southwest Manitoba first identified by Captain John Palliser during an expedition to the Canadian West in 1857–60 sponsored by the Royal Geographical Society and the British Colonial Office.

parasitism: Relationship in which one species lives in or on another that acts as its host.

partnerships: A sharing of responsibility and power between two or more groups, especially a government agency and a second party, regarding a resource or environmental issue; co-management is an example of a partnership.

photosynthesis: A two-part process in plants and algae involving (1) the capture of sunlight and its conversion into cellular energy and (2) the production of organic molecules, such as glucose and amino acids from carbon dioxide, water, and energy from the sun.

phototrophs: Organisms that produce complex chemicals through photosynthesis.

phytoplankton: Single-celled algae and other free-floating photosynthetic organisms.

point sources: Easily discernible 'end-of-pipe' sources of pollution, such as a factory or a town sewage system.

policy target value: A target set as a result of compromise among scientific, social, economic, cultural, and political objectives.

polynyas: Ice-free areas of permanent open water in the Arctic surrounded by ice, created by tides, currents, ocean-bottom upwellings, and winds, that are biologically productive and vary greatly in size from 60 to 90 metres in diameter to areas as large as the North Water polynya between Ellesmere Island and Greenland that may cover as much as 130,000 square kilometres.

population: A group of organisms of the same species living within a specified region.

population age structure: The relative distribution of age cohorts in the population.

population viability analysis (PVA): A process that determines the probability that a population will go extinct within a given number of years.

positive feedback loop: A situation in which a change in a system in one direction provides the conditions to cause the system to change further in the same direction.

potential energy: Stored energy that is available for later use.

precautionary principle: A guideline stating that when there is a possibility of serious or irreversible environmental damage resulting from a course of action, such as a development project, lack of scientific certainty is not an acceptable reason for postponing a measure to prevent environmental degradation or for assuming that damage in the future can be rectified by some kind of technological fix.

predator: An organism that actively hunts its prey.

prescribed burning: Burning purposely initiated, usually to achieve ecological goals of restoring natural fire regimes.

prey: An organism (e.g., deer) that is attacked and killed by a predator.

prey switching: A familiar foraging behaviour whereby a predator shifts from its target species after it is depleted or not available in an area to the next most preferred or profitable species until that too is depleted and then continuing to move down the food chain, as wolves do in moving from caribou to Arctic hare to small rodents or as humans have done in fishing down the food chain in commercial fisheries.

primary consumers: The first consuming organisms in a given food chain, such as a grazer in grazer food chains or a decomposer organism or insect in decomposer food chains; primary consumers belong to the second trophic level.

primary succession: The development of a biotic community in an area previously devoid of organisms.

producers: Autotrophs capable of synthesizing organic material, thus forming the basis of the food web.

protected areas: Areas such as national and provincial parks, wildlife sanctuaries, and game preserves established to protect species and ecosystems.

rainshadow effect: The decrease in precipitation levels as the air warms up in its descent from the mountains and can hold more moisture (i.e., there is considerably less precipitation on the leeward side of a mountain or a mountain range than on the windward side).

range of tolerance: Range of abiotic factors within which an organism can survive, from the minimum amount of a limiting factor that the organism requires to the maximum amount that it can withstand.

reclamation: The process of bringing an area back to a useful, good condition—similar to rehabilitation.

recovery: Involves burning waste in incinerators and then recovering the generated energy to provide heating for homes, offices, and other buildings.

recycling: The third stage in waste management, involving the return of used products (glass, plastic, or metal containers; newspapers) to be processed so that the glass, plastic, metal, or newsprint can be used for other products.

Red List: An annual listing of species at risk prepared by the IUCN and produced by thousands of scientific experts; the best source of knowledge on the status of global biodiversity.

REDD: A mechanism for compensating countries for reducing emissions from deforestation and forest degradation.

reduction: The first stage in waste management, involving using less material or products to meet needs.

renewable or flow resources: Resources that are renewed naturally within a relatively short period of time, such as water, air, animals, and plants, as well as solar radiation, wind power, and tidal energy.

renewable water supplies: Supplies based on precipitation that falls, then runs off into rivers, often being held in lakes before draining to the ocean or moving downward into aquifers. The flows associated with precipitation or snowmelt should be identified as the renewable supply.

replacement level fertility: Fertility rate that will sustain a population.

resilience: Ability of an ecosystem to return to normal after a disturbance.

resource partitioning: A situation in which resources are used at different times or in different ways by species with an overlap of fundamental niches, such as owls and hawks, which seek the same prey but at different times during the day.

resources: Such things as forests, wildlife, oceans, rivers and lakes, minerals, and petroleum.

reuse: The second stage in waste management, involving the reuse of a product rather than discarding and replacing it.

Richter scale: Measures the magnitude of earthquake severity on a logarithmic scale. Earthquakes of exceptional magnitude measure 8.0 or higher.

Rio Declaration: The 27 principles related to sustainable development developed at the Earth Summit in Rio de Janeiro during June 1992.

risk assessment: Determining the probability or likelihood of an environmentally or socially negative event of some specified magnitude.

rock cycle: The relationship among three rock-forming processes and how each rock type can be transformed into another type.

salinization: Deposition of salts in irrigated soils, making soil unfit for most crops; caused by a rising water table due to inadequate drainage of irrigated soils.

science target value: A target set on the basis of scientific information.

second growth: A second forest that develops after harvest of the original forest.

second law of thermodynamics: Law stating that when energy is converted from one form to another, it is degraded—that is, it is converted from a concentrated to a less concentrated form. The amount of useful energy decreases during such conversions.

secondary consumers: Second consuming organisms in a food chain and belonging to the third trophic level.

secondary succession: The sequential development of biotic communities after the complete or partial destruction of an existing community by natural or anthropogenic forces.

sedimentary cycles: Those cycles of elements, such as the phosphorus and sulphur cycles, that hold most of their matter in the lithosphere.

seed bank: Locations where plant seeds accumulate.

seral: A stage in a successional process.

serial depletion: When one stock after another becomes progressively depleted as a result of prey switching, even if the total catch remains the same.

serotiny: Behaviour of some plant species that retain their non-dormant seeds in a cone or woody fruit for up to several years but release them after exposure to fire.

shifting baseline: When scientists have no other option than to take the current or recent degraded state as the baseline for stock biomass rather than the historical ecological abundance.

Silent Spring: Rachel Carson book published in 1962 that detailed the disastrous effects of biocides on the environment.

silviculture: The practice of directing the establishment, composition, growth, and quality of forest stands through a variety of activities, including harvesting, reforestation, and site preparation.

single-loop learning: Learning that emphasizes ensuring a match between intent and outcome.

smog: As classic smog, a mixture of smoke and fog in urban areas, mainly due to burning of coal to heat homes and the subsequent mixing of smoke with the released sulphur dioxide. Now more usually due to photochemical reactions of sunlight with hydrocarbons and nitrogen oxide emitted into the atmosphere from vehicles and industries.

social learning: Learning applied not only to individuals but also to social collectives, such as organizations and communities. The implication is that resource and environmental management processes should be designed so that both individuals and organizations are able to learn from their experience and thereby become more knowledgeable and effective in the future.

soft path: Aims to improve water use efficiency by challenging basic patterns of consumption. While demand management emphasizes the question of 'how', or how to do the same with less water, the soft path asks why water is even used for a function. The 'why' question normally leads to consideration of a broader range of methods.

soil compaction: The compression of soil as a result of frequent heavy machinery use on wet soils or the overstocking of cattle on the land.

soil erosion: A natural process whereby soil is removed from its place of formation by gravitational, water, and wind processes.

soil horizons: Layers found in most soils.

specialist: Organism that has a narrow niche, usually feeding on one or a few food materials and adapted to a particular habitat.

speciation: Formation of new species.

Species at Risk Act (SARA): Canadian legislation passed in 2002 that mandates the Committee on the Status of Endangered Wildlife in Canada to maintain lists of species at risk and to recommend to the minister responsible that particular species be given special protection in their environment.

species: A group of individuals that share certain identical physical characteristics and are capable of producing fertile offspring.

species-area curves: A graph showing the numbers of species found in areas of different size.

stakeholders: Persons or groups with a legal responsibility relative to a problem or issue, or likely to be affected by decisions or actions regarding the problem or issue, or able to pose an obstacle to a solution of the problem or issue.

steam-assisted gravity drainage: See **in situ recovery**.

stewardship: Activities undertaken towards caring for the Earth.

strategic environmental assessment (SEA): Focuses on policies, plans, and programs (PPPs) in order to integrate environmental considerations at the earliest possible stage of decision-making. SEA occurs before development decisions are made and when alternative futures and options for development are still open. Emphasis is on opportunities, regions, and sectors as opposed to projects. The objective is to integrate environmental considerations in the development of PPPs and to identify preferred futures and the means to achieve them, rather than focusing on mitigating the most likely outcomes of an already taken development decision.

stratosphere: The layer of the atmosphere (about 10 to 50 kilometres above the Earth's surface) in which temperatures rise with increasing altitude.

strip cropping: A technique similar to contour cultivation in which different crops are planted in strips parallel to the slope.

subsidiarity: A policy and management approach stipulating that decisions should be taken at the level closest to where consequences are most noticeable or have the most direct impact.

subsistence farming: The production of food and other necessities to satisfy the needs of the farm household.

sulphur dioxide emissions: Release into the atmosphere of huge quantities of sulphur, mainly in the form of dioxides, as a result of smelting sulphide ores and burning fossil fuels, which causes air pollution and climate change.

summer fallow: A practice common on the Prairies in which land is ploughed and kept bare to minimize moisture losses through evapotranspiration but which leads to increased salinization.

supply management: Approach based on manipulating the natural system to create new sources of supply, normally through either augmenting an existing supply or developing a new supply.

sustainability assessment: Focuses on efforts to apply some form of sustainability analysis, appraisal, or assessment to adopt sustainability objectives as core guides for evaluations and decisions.

sustainable development: Economic development that meets current needs without compromising the ability of future generations to meet their needs.

sustainable energy superpower: A view that Canada, with its huge energy resources, should be able to produce upgraded energy products at reasonable prices with acceptable environmental impacts, based on new technology, effective public policy, new concepts of risk-sharing, and individual companies and governments working together.

sustainable livelihoods: A human-centred approach to broad environmental management directed towards ways for local people to meet basic needs (food, housing), as well as other needs related to security and dignity, through meaningful work, at the same time minimizing environmental degradation, rehabilitating damaged environments, and addressing concerns about social justice.

sustainable urban development: The enhanced well-being of cities or urban regions, including integrated economic, ecological, and social components, which will maintain the quality of life for future generations.

sustained yield: The amount of harvestable material that can be removed from an ecosystem over a long period of time with no apparent deleterious effects on the system.

synergism: An interaction between two substances that produces a greater effect than the effect of either one alone; an interaction between two relatively harmless components in the environment.

system plan: An idealized blueprint of the distribution of protected areas within a given jurisdiction.

taiga: The portion of the boreal forest lying between the southern boundary of the tundra and the closed-crown coniferous forest to the south, characterized by coniferous forests, soil that thaws during the summer months, abundant precipitation, and high species diversity.

Taiga Plains: Gently rolling northern ecozone with a high proportion of surface water storage, wetlands, and organic soils, a cold and relatively dry climate, and on better-drained localities and uplands, mixed coniferous-deciduous forests.

technocentric perspective: The assumption that humankind is able to understand, control, and manipulate nature to suit its purposes and that nature and other living and non-living things exist to meet human needs and wants.

tertiary consumers: In a food chain, organisms at the top that consume other organisms.

thermocline: Sharp transition in temperature between the warmer surface waters of the ocean and the cooler waters underneath, generally occurring at a depth of 120 to 240 metres.

thermohaline circulation: The movement of carbon-saturated water around the globe, mainly as a result of differing water densities.

thermosphere: Uppermost layer of the atmosphere, beyond the mesosphere.

threatened species: A species designated by the Committee on the Status of Endangered Wildlife in Canada as likely to become endangered in Canada if factors threatening its vulnerability are not reversed.

threshold: A point or limit beyond which something is unsatisfactory relative to a consideration, such as health, welfare, or ecological integrity.

total allowable catch (TAC): The amount, in tonnage, of a particular aquatic species that the federal Department of Fisheries and Oceans, for example, determines can be landed within a particular fishery in a given year.

total fertility rates: Average number of children each woman has over her lifetime.

traditional ecological knowledge (TEK): Different systems of belief, knowledge, and practice gained through experience, normally shared and transmitted verbally. Often referred to as 'indignous knowledge' (see also **indigenous knowledge**).

transpiration: The loss of water vapour through the pores of a plant.

trophic level: Functional classification of organisms in a community according to feeding relationships: the first trophic level includes green plants, the second level includes herbivores, and so on.

troposphere: Innermost layer of the atmosphere that contains 99 per cent of the water vapour and up to 90 per cent of the Earth's air and is responsible for our weather, extending about 6 to 17 kilometres up from the Earth, depending on latitude and season.

uncertainty: A situation in which the probability or odds of a future event are not known and therefore indicates the presence of doubt.

urban form: The type and distribution of infrastructure (e.g., buildings, roads) in communities and a key factor influencing environmental quality in cities.

urban heat island: Occurs as a result of increased temperatures in core urban areas relative to surrounding areas. It is not uncommon for the temperature of city centres to be from 2°C to 6°C higher than that of nearby rural areas.

urban sprawl: Urban areas characterized by low population densities and significant travel costs because of the high priority the majority of residents place on living in single-family, detached homes with large properties. Sprawl contributes to loss of, disruption to, or degradation of adjacent agricultural land, environmentally sensitive areas, natural habitats, and water and air quality.

virtual water: A concept recognizing that because significant amounts of water are required to grow some foodstuffs, nations can reduce pressure on their water resource by importing such products, allowing use of water for other, higher-value products.

vision: A view for the future for a region, community, or group that is realistic, credible, attractive, and attainable.

vulnerable species: An official designation assigned by the Committee on the Status of Endangered Wildlife in Canada to any indigenous species or subspecies or geographically separate population of fauna or flora that is particularly at risk because of low or declining numbers, because it occurs at the fringe of its range or in restricted areas, or for some other reason but which is not a threatened species.

water ethics: A statement of principles or values to guide behaviour by individuals or groups with regard to water.

water table: The top of the zone of saturation.

water terrorism: Involves deliberate disruption of water infrastructure or related services by directly attacking the infrastructure or by contaminating water with poison or disease-bearing agents.

weather: The sum total of atmospheric conditions (temperature, pressure, winds, moisture, and precipitation) in a particular place for a short period of time.

wetlands: Areas that are hybrid aquatic and terrestrial systems, such as swamps and marshes, where the ground is saturated with water much or all of the time.

wind power: The fastest-growing sector in the world's energy market, which uses wind turbines that have minimal adverse environmental affects to generate electricity.

windthrow: Uprooting and blowing down of trees by wind.

World Summit on Sustainable Development (WSSD): Conference held in Johannesburg in 2002, 10 years after the Earth Summit in Rio, that made various commitments, such as halving the proportion of people without access to adequate sanitation by the year 2015.

zone of physiological stress: Upper and lower limits of the range of tolerance in which organisms have difficulty surviving.

zooplankton: Non-photosynthetic, single-celled aquatic organisms.

zooxanthellae: Unicellular algae.

References

Abbey, E. 1977. *The Journey Home: Some Words in Defense of the American West.* New York: Dutton.

Agreement Review Committee. 2007. *Review of the Great Lakes Water Quality Agreement, Volume 1, Final Report to the Great Lakes Binational Executive Committee.* Windsor, ON: Environment Canada–Ontario, Great Lakes Office.

Agriculture and Agri-Food Canada. 2003. *Canada's Agriculture, Food and Beverage Industry: Organic Industry.* http://ats-sea.agr.ca/supply/factsheet-e.htm.

———. 2008. *An Overview of the Canadian Agriculture and Agri-Food System 2007.* http://www4.agr.gc.ca/AAFC-AAC/display-afficher.do?id=1204750285466&lang=e.

Alberta Energy. 2008. 'Electricity frequently asked questions'. http://www.energy.gov.ab.ca/Electricity/683.asp#where.

Alberta Environment. 2005. *Report on the Implementation Progress of Water for Life: Alberta's Strategy for Sustainability.* Edmonton: Alberta Environment.

Alberta Food and Rural Development. 2001. 'Woodlot harvesting'. http://www.agric.gov.ab.ca/sustain/woodlot.

Alter, S.E., E. Rynes, and S.R. Palumbi. 2007. 'DNA evidence for historic population size and past ecosystem impacts of gray whales'. *Proceedings of the National Academy of Sciences,* 10.1073/pnas.0706056104.

Andrey, J., and L. Mortsch. 2000. 'Communicating about climate change: Challenges and opportunities'. In D. Scott et al., *Climate Change Communication: Proceedings of an International Conference.* Waterloo, ON: University of Waterloo and Environment Canada, Adaptation and Impacts Research Group, wp1–wp11.

Anielski, M., and S. Wilson. 2005. *Assessing the Real Value of Canada's Boreal Ecosystem.* Edmonton: Pembina Institute.

Apps, C.D., and B.N McClellan. 2006. 'Factors influencing the dispersion and fragmentation of endangered mountain caribou populations'. *Biological Conservation* 130: 84–97.

AREVA. 2007. 'Uranium in Saskatchewan'. http://www.arevaresources.com/publications/uranium_in_sask_01/environment and safety.html

Armitage, D., F. Berkes, and N. Doubleday. 2007. 'Introduction: Moving beyond co-management'. In D. Armitage, F. Berkes, and N. Doubleday, Eds, *Adaptive Co-management: Collaboration, Learning and Multi-Level Governance,* 1–15. Vancouver: University of British Columbia Press.

Arnstein, S. 1969. 'A ladder of citizen participation'. *Journal of the American Institute of Planners* 35 (4): 216–24.

Auditor General of Canada. 2002. *2002 Report of the Commissioner of the Environment and Sustainable Development.* Ottawa: Minister of Supply and Services.

———. 2005a. *2005 Report of the Commissioner of the Environment and Sustainable Development.* Chapter 5: 'Drinking water in First Nations communities'. http://www.oag.bvg.gc.ca/domino.reports .nsf.html.c2005menu_e.html.

———. 2005b. *2005 September Report of the Commissioner of the Environment and Sustainable Development.* Ottawa: Minister of Supply and Services.

———. 2006. *2006 Report of the Commissioner of the Environment and Sustainable Development.* http://www.\.0ag-bvg.gc.ca/domino/reports.nsf/html/ c20060903ce. html).

———. 2008. *The March 2008 Status Report of the Commissioner of the Environment and Sustainable Development.* Ottawa: Minister of Supply and Services.

Bakker, K. 2007. 'Introduction'. In K. Bakker, Ed., *Eau Canada: The Future of Canada's Water,* 1–20. Vancouver: University of British Columbia Press.

Banff–Bow Valley Study. 1996. *Banff–Bow Valley: At the Crossroads. Summary Report of the Banff–Bow Valley Task Force.* R. Page et al., Eds. Ottawa: Ministry of Canadian Heritage.

Bardati, D.R. 1995. 'An environmental action plan for Bishop's University'. *Journal of Eastern Townships Studies* 6: 19–38.

———. 2006. 'The integrative role of the campus environmental audit: Experiences at Bishop's University, Canada'. *International Journal of Sustainability in Higher Education* 7 (1): 57–68.

Barlow, M., and E. May. 2000. *Frederick Street: Life and Death on Canada's Love Canal.* Toronto: HarperCollins. Copyright © 2000 by Maude Barlow and Elizabeth May. All Rights Reserved.

Barnosky, A.D., et al. 2004. 'Assessing the causes of late Pleistocene extinctions on the continents'. *Science* 306: 70–5.

Beanlands, G.E., and P.N. Duinker. 1983. *An Ecological Framework for Environmental Impact Assessment in Canada.* Halifax: Institute for Resource and Environmental Studies, Dalhousie University.

Bentley, C.F., and L.A. Lewkiw. 1985. *Sustainability of Farmed Lands: Current Trends and Thinking,* p. 22. Ottawa: Canadian Environmental Advisory Council.

Bergeron, Y., et al. 'Natural fire frequency for the eastern Canadian boreal forest: Consequences for sustainable forestry'. *Canadian Journal of Forest Research* 31: 384–91.

Berkes, F. 1988. 'The intrinsic difficulty of predicting impacts: Lessons from the James Bay hydro project'. *Environmental Impact Assessment Review* 8: 201–20.

Berkes, F., D. Armitage, and N. Doubleday. 2007. 'Synthesis: Adapting, innovating, evolving'. In D. Armitage, F. Berkes, and N. Doubleday, Eds, *Adaptive Co-management: Collaboration, Learning and Multi-Level Governance,* 308–27. Vancouver: University of British Columbia Press.

BHP Billiton. 2001. 'Environmental Agreement annual report 2001'. http://ekati.bhpbilliton.com/common/downloads/2001env.pdf.

Birkedal, T. 1993. 'Ancient hunters in the Alaskan wilderness: Human predators and their role and effect on wildlife populations and the implications for resource management'. In W.E. Brown and S.D. Veirs Jr, Eds, *Partners in Stewardship: Proceedings of the 7th Conference on Research and Resource Management in Parks and on Public Lands,* 228–34. Hancock, MI: The George Wright Society.

Biswas, A.K. 2004. 'Integrated water resource management: A re-assessment'. *Water International* 29: 248–56.

Bone, R.M. 2005. *The Regional Geography of Canada,* 3rd edn. Toronto: Oxford University Press.

Boyce, M.S., S.R. Lele, and B.W. Johns. 2005. 'Whooping crane recruitment enhanced by egg removal'. *Biological Conservation* 126: 395–401.

Boychuk, Rick. 2008. 'Seeing the Light'. *Canadian Geographic.* http://www. canadiangeographic.ca/Magazine/jun08/ednotebook.asp.

Boyd, D.R., and S.J. Genuis. 2008. 'The environmental burden of disease in Canada: Respiratory disease, cardiovascular disease, cancer, and congenital affliction'. *Environmental Research* 106: 240–9

Branch, T.A. 2008. 'Not all fisheries will be collapsed in 2048'. *Marine Policy* 32 (1): 38–9.

Brandes, O.M., and D.B. Brooks. 2006. *The Soft Path for Water in a Nutshell.* Ottawa: Friends of the Earth; Victoria: University of Victoria, POLIS Project.

Brandes, O.M., R. Maas, and E. Reynolds. 2006. *Thinking beyond Pipes and Pumps: Top 10 Ways Communities Can Save Water and Money.* Victoria: University of Victoria, POLIS Project.

Bregha, F., et al. 1990. *The Integration of Environmental Considerations into Government Policy.* Prepared for the Canadian Environmental Assessment Research Council. Ottawa: Minister of Supply and Services.

Briggs, D., et al. 1993. *Fundamentals of Physical Geography,* 2nd Canadian edn. Toronto: Copp Clark Pitman.

British Columbia. 2008. *Budget Speech: Turning to the Future, Meeting the Challenge.* 19 February. Victoria: Ministry of Finance. Reprinted by permission of the publisher.

British Columbia Ministry of Environment, Lands and Parks. 1992. *New Approaches to Environmental Protection in British Columbia: A Legislation Discussion Paper.* Victoria: Ministry of Environment, Lands and Parks.

British Columbia Ministry of Environments. 2006. *BC Seafood Industry Year in Review.* http://www.env.gov.bc.ca/omfd/reports/YIR-2006.pdf.

Brook, B.W., and D.M.J.S. Bowman. 2004. 'The uncertain blitzkrieg of Pleistocene megafauna'. *Journal of Biogeography* 31: 517–23.

Brooks, D.B. 2005. 'Beyond greater efficiency: The concept of water soft paths'. *Canadian Water Resources Journal* 30: 83–92.

Brooks, G.R., and E. Nielson. 2000. 'Red River, Red River Valley, Manitoba'. *Canadian Geographer* 44: 306–11.

Browne, C.L., and S.J. Hecnar. 2003. 'Dwindling turtle populations'. *National Park International Bulletin* no. 9 (May).

———. 2007. 'Species loss and shifting population structure in freshwater turtles despite habitat protection'. *Biological Conservation* 138: 421–9.

Bruner, A.G., et al. 2001. 'Effectiveness of parks in protecting tropical biodiversity'. *Science* 91 (5,501): 125–8.

Bruno, J.F., and E.R. Selig. 2007. 'Regional decline of coral cover in the Indo-Pacific: Timing, extent, and subregional comparisons'. PLoS ONE 2 (8): e711. doi:10.1371/journal.pone.0000711.

Bullock, R., and K. Hanna. 2008. 'Community forestry in British Columbia: Mitigating or creating conflict?' *Society and Natural Resources* 2: 77–85.

Bullock, R., and A. Watelet. 2006. 'Exploring conservation authority operations in Sudbury, Northern Ontario: Constraints and opportunities'. *Environments* 34 (2): 29–50.

Burn, D.H., and N.K. Goel. 2001. 'Flood frequency analysis for the Red River at Winnipeg'. *Canadian Journal of Civil Engineering* 28: 355–62.

Burt, B. 2007a. 'Land claim stalls wind project'. *The Record* 12 August: B1–B2.

———. 2007b. 'Wind farm claim riles Tory'. *The Record* 16 August: A1–A2.

Cameron, R.P. 2006. 'Protected area–working forest interface: Ecological concerns for protected area management in Canada. *Natural Areas Journal* 26: 403–7.

Cameron, S.D. 1990. 'Net losses: The sorry state of our Atlantic fishery'. *Canadian Geographic* 110 (2): 28–37.

Canada. 1995. *Canadian Biodiversity Strategy: Canada's Response to the Convention on Biological Diversity*. Ottawa: Minister of Supply and Services.

Canada and Ontario. 1988. *First Report of Canada under the 1987 Protocol to the 1978 Great Lakes Water Quality Agreement*. Toronto: Environment Canada, Communications Directorate.

Canada Mortgage and Housing Corporation. 2005. *Smart Growth in Canada: A Report Card*. Research Highlight Socio-economic Series 05-036. Ottawa: Canada Mortgage and Housing Corporation.

Canada-Newfoundland Offshore Petroleum Board, Terra Nova Development Project Environmental Assessment Panel. 1997. *Terra Nova Development: An Offshore Petroleum Project*. Ottawa: Minister of Public Works and Government Services.

Canadian Academy of Engineering. 2007. *Energy Pathways Task Force Phase 1: Final Report*. Ottawa: Canadian Academy of Engineering.

Canadian Council of Forest Ministers. 2008. *A Vision for Canada's Forests: 2008 and Beyond*. Ottawa: NRCAN.

Canadian Council of Ministers of the Environment. 2003. *Climate, Nature, People: Indicators of Canada's Changing Climate*. Winnipeg: Canadian Council of Ministers of the Environment. Reprinted by permission of the publisher.

———. 2005. *Five-Year Review of the Canada-Wide Acid Rain Strategy for Post-2000*. http://www.ccme.ca/assets/pdf/5_year_review_acid_rain_strategy_e1.0_web.pdf.

———. 2006. '2004–2005 progress report on the Canada-Wide Acid Rain Strategy for Post-2000'. http://www.ccme.ca/assets/pdf/2004_2005_ar_progrrpt_1.0_e_web.pdf.

Canadian Environmental Assessment Agency. 1996. *NWT Diamonds Project: Report of the Environmental Assessment Panel*. Cat. En105-53 1996.

———. 2003. 'Federal environmental assessment making a difference: Sable Island offshore gas project'. http://www.ceaa.gc.ca/017/0003/fea03_e.htm.

Canadian Environmental Assessment Research Council. 1988. *Evaluating Environmental Impact Assessment: An Action Perspective*. Ottawa: Minister of Supply and Services.

Canadian Food Inspection Agency. 2004. *Action Plan for Invasive Alien Terrestrial Plants and Plant Pests*. Ottawa: Canadian Food Inspection Agency.

Canadian General Standards Board. 2003. *Draft National Standard: Organic Agriculture*. Ottawa: Canadian General Standards Board, Committee on Organic Agriculture.

Canadian Heritage Rivers Board. 2002. *The Canadian Heritage Rivers System: Annual Report 2001–2002*. Ottawa: Minister of Public Works and Government Services.

———. 2003. *The Canadian Heritage Rivers System: Annual Report 2002–2003*. Ottawa: Minister of Public Works and Government Services.

———. 2006. *The Canadian Heritage Rivers System: Annual Report 2005–2006*. Ottawa: Minister of Public Works and Government Services.

Canadian Institute of Forestry. 2003. 'Silviculture Systems'. http://www.cif-ifc.org/practices.

Canadian Press. 2007. 'No end in sight to standoff'. *The Record* 27 February: C11.

———. 2008. 'Walkerton victims waiting for help'. *The Record* 10 March: A1, A3.

Canadian Wildlife Service. 'Contaminants levels in double-crested cormorant eggs, 1970–2000'. http://www.ec.gc.ca/soer-ree/English/indicator_series/techs.cfm?tech_id=9$issue_id=2&supp=4.

Canadian Wind Energy Association. 2008. 'Wind energy sets global growth record in 2007'. http://www.canwea.ca/news_releases.cfm?ID=58.

Carpenter, R.A. 1995. 'Communicating environmental science uncertainties'. *Environmental Professional* 17: 127–36.

Carson, R. 1962. *Silent Spring*. Boston: Houghton Mifflin.

CCFM (Canadian Council of Forest Ministers). 2006. *Facing the Challenge of Forest Industry Restructuring*. Ottawa: Canadian Council of Forest Ministers.

———. 2007. 'Compendium of Canadian Forestry Statistics'. http://www.nfdp.ccfm.org.

———. 2008. *A Vision for Canada's Forests: 2008 and Beyond*. Ottawa: Canadian Council of Forest Ministers.

Chalecki, E.L. 2000. 'Same planet, different worlds: The climate change information gap'. In D.N. Scott et al., *Climate Change Communication: Proceedings of an International Conference*, A2, 15–22. Waterloo, ON: University of Waterloo and Environment Canada, Adaptation and Impacts Research Group.

Chambers, P., et al. 2001. 'Agricultural and forestry land use impacts'. In Environment Canada, *Threats to Sources of Drinking Water and Aquatic Ecosystem Health in Canada*, 57–62. Ottawa: Environment Canada.

Chapman, J.D. 1989. *Geography and Energy: Commercial Energy Systems and National Policies*. Harlow, UK: Longman Scientific and Technical.

Chestnut, L.G., and D.M. Mills. 2005. 'A fresh look at the benefits and costs of the US acid rain program'. *Journal of Environmental Management* 77: 252–66.

Chevalier, G., et al. 1997. 'Mercury in northern Québec: Role of the Mercury Agreement and the status of research and monitoring'. *Water, Air and Soil Pollution* 97: 75–84.

Chomitz, K.M., et al. 2007. *At Loggerheads? Agricultural Expansion, Poverty Reduction, and Environment in the Tropical Forests*. Jakarta, Indonesia: World Bank.

Clarke, A., and C.M. Harris. 2003. 'Polar marine ecosystems: Major threats and future change'. *Environmental Conservation* 30: 1–25.

Clean Air Sudbury. 2005. *Clearing the Air: Air Quality Trends in Sudbury*. Sudbury: Clean Air Sudbury.

Cline, W.R. 2007. *Global Warming and Agriculture: Impact Estimates by Country*. Washington: Center for Global Development.

Commission for Environmental Cooperation. 2008. *Green Building in North America: Opportunities and Challenges*. Montreal: Commission for Environmental Cooperation.

Commissioner of the Environment and Sustainable Development. 2003. 'Managing the safety and accessibility of pesticides'. www.oag-bvg.gc.ca/domino/reports.nsf/html/c20031001ce.html/$file/c20031001ce.pdf.

———. 2008. 'Selected aspects of managing the safety and accessibility of pesticides'. http://www.oag-bvg.gc.ca/internet/English/aud_ch_cesd_200803_02_e_30128.html#ch2hd3a

Conference Board of Canada. 2007. *Mission Possible*, v. II. Ottawa: Conference Board of Canada.

Conservation Authorities of Ontario. 1993. *Restructuring Resource Management in Ontario: A Blueprint for Success*. Mississauga, ON: Credit Valley Conservation Authority.

Copenhagen. 2007. *Eco-Metropole: Our Vision for Copenhagen 2015*. Municipality of Copenhagen: Technical and Environmental Centre.

COSEWIC (Committee on the Status of Endangered Wildlife in Canada). 2003. 'Canadian Species at Risk'. http://www.cosewic.gc.ca.

———. 2007. 'Canadian Species at Risk'. http://www.cosewic.gc.ca.

Costanza, R., et al. 1997. 'The value of the world's ecosystem services and natural capital'. *Nature* 387: 253–60.

Couch, W.J. 2002. 'Strategic resolution of policy, environmental and socio-economic impacts in Canadian Arctic diamond mining: BHP's NWT diamond project'. *Impact Assessment and Project Appraisal* 20: 265–78.

CPAWS (Canadian Parks and Wilderness Society). 2003. 'Boreal forests: For the birds'. Boreal Forest Factsheet Series. www.cpaws.org/boreal/english/aboutboreal/boreal-bird-factsheet.pdf.

Cressman, D.R. 1994. 'Remedial action programs for soil and water degradation problems in Canada'. In T.L. Napier, S.M. Camboni, and S.A. El-Swaify, Eds, *Adopting Conservation on the Farm*, 413–34. Ankeny, IA: Soil and Water Conservation Society.

Dale, A., and L. Newman. 2007. 'Governance for integrated resource management'. In K.S. Hanna and D.S. Slocombe, Eds, *Integrated Resource and Environmental Management: Concepts and Practice*, 56–71 Toronto: Oxford University Press.

Darwin, C.R. 1859. *On the Origin of Species*. London: John Murray.

David Suzuki Foundation. 2007. 'What is carbon offset?' http//www.davidsuzuki.org/Climate_Change/What_You_Can_Do/carbon_offsets.asp

Davis, A., and J. Wagner. 2006. 'A right to fish for a living? The case for coastal peoples' determination of access and participation'. *Ocean and Coastal Management* 49: 476–97.

Dawson, D. 2002. 'Plant-covered roofs ease urban heat'. *National Geographic News* 15 November. http://news.nationalgeographic.com/news/2002/11/1115_021115_GreenRoofs.html.

Day, J.C., and F. Quinn. 1992. *Water Diversion and Export: Learning from Canadian Experience*. Department of Geography Publication Series no. 36. Waterloo, ON: University of Waterloo.

de Loe, R.C., S. Di Giantomasso, and R.D. Kreutzwiser. 2002. 'Local capacity for groundwater protection in Ontario'. *Environmental Management* 29 (2): 217–33.

de Villiers, M. 1999. *Water Wars: Is the World Running out of Water?* London: Weidenfeld and Nicolson.

Dearden, P. 1995. 'Park literacy and conservation'. *Conservation Biology* 9: 1,654–6.

———. 2001. 'Endangered species and terrestrial national parks'. In K. Beazley and R. Boardman, Eds, *Politics of the Wild: Canada and Endangered Species*, 75–93. Toronto: Oxford University Press. Article originally published by Blackwell Publishing.

Dearden, P., and L. Berg. 1993. 'Canada's national parks: A model of administrative penetration'. *Canadian Geographer* 37: 194–211. Article originally published by Blackwell Publishing.

Dearden, P., and R. Rollins, Eds. 1993. *Parks and Protected Areas in Canada: Planning and Management*. Toronto: Oxford University Press.

Desjardins, R.L., et al. 2008. 'Moving Canadian agricultural landscapes from GHG source to sink'. In Hengeveld et al., *Enhancement of Greenhouse Gas Sinks: A Canadian Science Assessment*, 19–37. Ottawa: Ministry of Environment.

Diamond, B. 1990. 'Villages of the dammed'. *Arctic Circle* November/December: 24–34.

Diamond, J. 2005. *Collapse: How Societies Choose to Fail or Succeed*. New York: Viking.

Diduck, A. 2004. 'Incorporating participatory approaches and social learning'. In B. Mitchell, Ed., *Resource and Environmental Management in Canada: Addressing Conflict and Uncertainty*, 3rd edn, 497–527. Toronto: Oxford University Press.

Diduck, A., and B. Mitchell. 2003. 'Learning, public involvement and environmental assessment: A Canadian case study'. *Journal of Environmental Assessment Policy and Management* 5: 339–64.

Dillon Consulting Ltd. 2000. *Wind Turbine Environmental Assessment: Draft Screening Document*. Toronto: Dillon Consulting Ltd.

Doubleday, W.G. 2000. 'Seals & cod'. *Isuma* 1 (1). http://www.isuma.net/v01n01/doubleda/doubleda_e.shtml.

Duinker, P.N., et al. 1991. 'Community forestry and its implications for northern Ontario'. *Forestry Chronicle* 67: 131–5.

———. 1994. 'Community forests in Canada: An overview'. *Forestry Chronicle* 70: 711–20.

Dumont, C., et al. 1998. 'Mercury levels in the Cree population of James Bay, Quebec, from 1988 to 1993/94'. *Canadian Medical Association Journal* 158: 1439–45.

Dunbar, D., and I. Blackburn. 1994. *Management Options for the Northern Spotted Owl in British Columbia*. Victoria: Ministry of Environment.

Dwivedi, O.P., and R. Khator. 2006. 'Sustaining development: The road from Stockholm to Johannesburg'. In G.M. Mudacumura, D. Mebratu, and M.S. Haque, Eds., *Sustainable Development Policy and Administration*, 114–33. Boca Raton, FL: Taylor and Francis.

Dyer, G. 2007. 'Arctic scramble is about seabed resources'. *The Record* 14 August: A9.

Earth Policy Institute. *Earthtrends*. 2007. http://www.earth-policy.org/Updates/2006/Update55_data.htm.

Easterling, W.E., et al. 2007. 'Food, fibre and forest'. In Intergovernmental Panel on Climate Change, *Climate Change 2007: Impacts, Adaptations and Vulnerability*. Cambridge, UK: International Panel on Climate Change.

Ebeling, J. 2006. *Tropical Deforestation and Climate Change: Towards an International Mitigation Strategy*. Oxford, UK: Oxford University.

Edwards, P., and I. Roberts. 2008. 'Transport policy is food policy'. *The Lancet* 371: 1,661.

ÉEM inc. 2007. 'Environmental paper procurement: A review of forest certification schemes in Canada'. http://marketsinitiative.org/uploads/MI-EEMcert-2.pdf.

Eggertson, L. 2008. 'Despite federal promises, First Nations' water problems persist'. *Canadian Medical Association Journal* 178: 985.

Ehrlich, P., and A. Ehrlich. 1996. *Betrayal of Science and Reason: How Anti-Environmental Rhetoric Threatens Our Future*. Washington: Island Press.

El Lakany, H., M. Jenkins, and M. Richards. 2007. 'Background paper on means of implementation'. Contribution by PROFOR to discussions at UNFF-7, April. Program on Forests (PROFOR).

Environment Canada. 1988. 'Acid rain national sensitivity assessment'. Environment Fact Sheet 88-1. Ottawa: Environment Canada.

———. 1991. *The State of Canada's Environment*. Ottawa: Minister of Supply and Services Canada.

———. 2000. 'Ecological assessment of the Boreal Shield ecozone'. http://manitobawildlands.org/pdfs/BorealShield_Ecozone.pdf.

———. 2002a. *Population Status of Migratory Game Birds in Canada*. CWS Migratory Birds Regulatory Report no. 7. Ottawa: Canadian Wildlife Service Waterfowl Committee.

———. 2002b. 'Trichloroethylene and tetrachloroethylene in solvent degreasing'. http://www.ec.gc.ca/nopp/docs/rpt/degrease/en/summary.cfm.

———. 2003a. 'Atmospheric science: Acid rain'. http://www.ns.ec.gc.ca/msc/as/acidfaq.html.

———. 2003b. 'Environment Signals: Canada's National Environmental Indicator Series 2003'. Ottawa: Environment Canada. http://www.ec.gc.ca/soer-ree/English/Indicator_series/default.cfm.

———. 2003c. 'Natural disasters on the rise'. *The Science and Environment Bulletin* March/April.

———. 2003d. 'NPRI: Substance information, dioxins and furans'. http://www.ec.gc.ca/pdb/npri/npri_dioxins_e.cfm.

———. 2004. 'Canadian acid deposition science assessment'. http://www.msc-smc.ec.gc.ca/saib/acid/assessment2004/summary/summary_e.pdf.

———. 2006a. 'Canada-US Air Quality Agreement progress report'. http://www.ec.gc.ca/cleanair-airpur/caol/canus/report/2006canus/toc_e.cfm.

———. 2006b. *Canadian Protected Areas Status Report 2000–2005*. Warsaw, ON: Canadian Parks Council.

———. 2006c. *National Inventory Report*. Ottawa: Environment Canada.

———. 2007a. *2006 Prairie Waterfowl Status Report*. Ottawa: Environment Canada.

———. 2007b. *Canadian Environmental Sustainability Indicators*. Ottawa: Environment Canada.

Environment Canada, Statistics Canada, and Health Canada. 2006. *Canadian Environmental Sustainability Indicators*. Ottawa: Statistics Canada, Environment Accounts and Statistics Division.

Environment Canada and US Environmental Protection Agency. 2003. *State of the Great Lakes, 2003*. Ottawa: Environment Canada; Washington: US Environmental Protection Agency.

———. 2007. *State of the Great Lakes 2007*. Ottawa: Environment Canada.

Environmental Assessment Panel. 1991. *Rafferty-Alameda Project: Report of the Environmental Assessment Panel*. Ottawa: Federal Environmental Assessment Review Office.

Environmental Commissioner of Ontario. 2003. *Environmental Commissioner of Ontario 2002–2003 Annual Report*. Toronto: Environmental Commissioner of Ontario.

Estes, J.A., D.O. Duggins, and G.B. Rathbun. 1989. 'The ecology of extinctions in kelp forest communities'. *Conservation Biology* 3: 252–64.

External Advisory Committee on Cities and Communities. 2006. *From Restless Communities to Resilient Places: Building a Stronger Future for All Canadians*. Ottawa: Infrastructure Canada.

Fa, J.E., C.A. Peres, and J. Meeuwig. 2002. 'Bushmeat exploitation in tropical forests: An intercontinental comparison'. *Conservation Biology* 16: 232–7.

FAO (Food and Agriculture Organization). 2001. 'Conservation agriculture'. *FAO Magazine Spotlight*. www.fao.org/ag/magazine.

———. 2003a. 'FAOSTAT statistical data base'. http://www.apps.fao.org.

———. 2003b. 'Water management: Towards 2030'. www.fao.org/ag/magazine.

———. 2006. *The State of World Fisheries and Aquaculture*. Rome: FAO.

———. 2007a. *The State of the World's Animal Genetic Resources for Food and Agriculture*. Rome: FAO.

———. 2007b. *World Food Outlook*. Rome: FAO.

Farrell, A.E., et al. 2006. 'Ethanol can contribute to energy and environmental goals'. *Science* 311: 506–8.

Festa-Bianchet, M. 2003. 'Exploitative wildlife management as a selective pressure for the life-history evolution of large mammals'. In M. Festa-Bianchet and M. Apollonio, Eds, *Animal Behavior and Wildlife Conservation*, 191–207. Washington: Island Press.

Fischer, G., M. Shah, F. Tulbello, and H. van Velhulzen. 2005. 'Socio-economic and climate change impacts on agriculture: An integrated assessment'. *Philosophical Transactions of the Royal Society B* 360: 2,067–83.

Fisheries and Oceans Canada. 2008. 'Canadian fisheries statistics 2005'. http://www.dfo-mpo.gc.ca/communic/Statistics/publications/commercial/index_e.htm.

Flanagan, K., and S. Rasheed. 2002. 'Population viability analysis applied to woodland caribou in Jasper National Park'. *Research Links* 10: 16–18.

Forest Products Industry Competitiveness Task Force. 2007. *Industry at a Crossroads: Choosing the Path to Renewal*. Ottawa: Forest Products Association of Canada.

Foster, H.D., and W.R.D. Sewell. 1981. *Water: The Emerging Crisis in Canada*. Ottawa: Canadian Institute for Economic Policy.

Fraser Basin Council. 1997. *Charter for Sustainability*. Vancouver: Fraser River Basin Council.

Fraser Institute. 1999. *Environmental Indicators for Canada and the United States*. Vancouver: Fraser Institute.

Freedman, B. 1981. *Intensive Forest Harvest: A Review of Nutrient Budget Considerations*. Information Report M-X-121. Fredericton: Maritimes Forest Research Centre.

Freedman, B., P.N. Duinker, and R. Morash. 1986. 'Biomass and nutrients in Nova Scotia forests, and implications of intensive harvesting for future site productivity'. *Forest Ecology and Management* 15: 103–27.

Freiwald, A., et al. 2004. 'Cold-water coral reefs'. Cambridge, UK: UNEP-WCMC. www.unepwcmc.org.

Gabriel, A.O., and R.D. Kreutzwiser. 1993. 'Drought hazard in Ontario: A review of impacts, 1960–1989, and management implications'. *Canadian Water Resources Journal* 18: 117–32.

Gardner, G., and A. Assadourian. 2004. 'Rethinking the good life'. In Worldwatch Institute, *State of the World 2004*, 164–80. New York: W.W. Norton.

Garshelis, D.L. 1997. 'Sea otter mortality estimated from carcasses collected after the *Exxon Valdez* oil spill'. *Conservation Biology* 11 (4): 905–16.

Gauthier, G., et al. 2001. 'Seasonal survival of greater snow geese and effect of hunting under dependence in sighting probability'. *Ecology* 82: 3,105–19.

Gibson, R.B. 2007. 'Integration through sustainability assessment: Emerging possibilities at the leading of edge of environmental assessment'. In K.S. Hanna and D.S. Slocombe, Eds, *Integrated Resource and Environmental Management: Concepts and Practice*, 72–96. Toronto: Oxford University Press.

Gleick, P.H. 1993. *Water in Crisis: A Guide to the World's Freshwater Resources*. New York: Oxford University Press.

———. 2003. 'Global freshwater resources: Soft-path solutions for the 21st century'. *Science* 302: 1,524–8.

———. 2006. 'Water and terrorism'. *Water Policy* 8: 481–503.

Global Water Partnership. 2000. *Integrated Water Resources Management*. Technical Advisory Committee Background Paper no. 4. Stockholm: Global Water Partnership.

Global Wind Energy Council. 2006. *Global Wind 2006 Report*. Brussels: Global Wind Energy Council.

Goodarzi, F., and M. Mukhopadhyay. 2000. 'Metals and polyaromatic hydrocarbons in the drinking water of the Sydney Basin, Nova Scotia, Canada: A preliminary assessment of their source'. *International Journal of Coal Geology* 43: 357–72.

Gorrie, P. 1990. 'The James Bay power project: The environmental cost of reshaping the geography of northern Quebec'. *Canadian Geographic* 110 (1): 21–31.

———. 2008. 'Green giant'. *Toronto Star* 8 March: ID1–ID2.

Gould, S.J. 1989. *Wonderful Life: The Burgess Shale and the Nature of History*. New York: W.W. Norton.

———. 1994. 'The evolution of life on the earth'. *Scientific American* 271: 84–91.

Gray, B. 1989. *Collaborating: Finding Common Ground for Multiparty Problems*. San Francisco: Jossey-Bass.

Greer, A., V. Ng, and D. Fisman. 2008. 'Climate change and infectious diseases in North America: The road ahead'. *Canadian Medical Association Journal* 178: 715–22.

Gruber, N., and J.N. Galloway. 2008. 'An Earth-system perspective of the global nitrogen cycle'. *Nature* 451: 293–6.

Hall, J. 2004. 'Hazel's gift wrapped in green'. *Toronto Star* 15 October: B1–B2. Reprinted with permission—Torstar Syndication Services.

Halweil, B. 2002 'Home grown: The case for local food in a global market'. Worldwatch Paper 163. Washington: Worldwatch Institute.

Hangs, R.D., J.D. Knight, and K.C.J. Van Rees. 2003. 'Nitrogen uptake characteristics for roots of conifer seedlings and common boreal forest competitor species'. *Canadian Journal of Forest Research* 33: 156–63.

Hannah, L. 2003. 'Protected areas management in a changing climate'. In N.W.P. Munro et al., Eds, *Making Ecosystem-Based Management Work*. Proceedings of the Fifth International Conference on Science and Management of Protected Areas, Victoria, BC, May 2003. http://www.sampaa.org/publications.htm. Wolfville, NS: Science and Management of Protected Areas Association.

Harris, Michael. 1998. *Lament for an Ocean: The Collapse of the Atlantic Cod Fishery, A True Crime Story*. Toronto: McClelland and Stewart.

Health Canada. 2007. 'Canada's Health Concerns from Climate Change and Variability'. Ottawa: Health Canada. www.hc-sc.gc.ca/ewh-semt/alt_format/hecs-sesc/pdf/climat/health_table-tableau_sante_e.pdf.

Hearnden, K.W., et al. 1992. *A Report on the Status of Forest Regeneration*. Toronto: Ontario Independent Forest Audit Committee.

Heimann, M., and M. Reichstein. 2008. 'Terrestrial carbon ecosystem dynamics and climate feedbacks'. *Nature* 451: 289–92.

Hengeveld, H. 1991. *Understanding Atmospheric Change: A Survey of the Background Science and Implications of Climate Change and Ozone Depletion*. SOE Report no. 91-2. Ottawa: Minister of Supply and Services.

———. 2006. *CO_2/Climate Report*. Toronto: Environment Canada, Atmospheric Science Assessment and Integration Branch.

Hengeveld, H., E. Bush, and P. Edwards. 2002. *Frequently Asked Questions about Climate Change Science*. Ottawa: Minister of Supply and Services.

Hengeveld, H., B. Whitewood, and A. Fergusson. 2005. *An Introduction to Climate Change: A Canadian Perspective*. Toronto: Environment Canada, Atmospheric Science Assessment and Integration Branch.

Hengeveld, H., et al. 2008. *Enhancement of Greenhouse Gas Sinks: A Canadian Science Assessment*. Ottawa: Ministry of Environment.

Hilson, G. 2000. 'Sustainable development policies in Canada's mining sector: An overview of government and industry efforts'. *Environmental Science and Policy* 3: 201–11.

Hilson, G., and V. Nayee. 2002. 'Environmental management system implementation in the mining industry: A key to achieving cleaner production'. *International Journal of Minerals and Processes* 64: 19–41.

Himmelman, A.T. 1996. 'On the theory and practice of transformational collaboration: From social service to social justice'. in C. Huxham, Ed., *Creating Collaborative Advantage*, 19–43. Thousand Oaks, CA: Sage.

Hites, R.A., et al. 2004. 'Global assessment of organic contaminants in farmed salmon'. *Science* 303: 226–9

Hobson, K.A., E.M. Bayne, and S.L. Van Wilgenburg. 2002. 'Large-scale conversion of forest to agriculture in the boreal plains of Saskatchewan'. *Conservation Biology* 16 (6): 1530–41.

Hofmann, N., and M.S. Beaulieu. 2006. *A Geographical Profile of Manure Production in Canada, 2001*. Report 21-601-MIE-N.07. Ottawa: Statistics Canada.

Hofmann N., G. Filoso, and M. Schofield. 2005. *The loss of dependable agricultural land in Canada*. Rural and Small Town Canada Analysis Bulletin 6. Cat. no 21-006-XIE. Ottawa: Statistics Canada.

Holling, C.S., Ed. 1978. *Adaptive Environmental Assessment and Management*. Chichester: John Wiley.

———. 1986. 'The resilience of terrestrial ecosystems: Local surprise and global change'. In W.C. Clark and R.E. Munn, Eds, *Sustainable Development in the Biosphere*, 292–317. Cambridge: Cambridge University Press.

Homer-Dixon, T. 2000. *The Ingenuity Gap*. New York: Alfred A. Knopf.

———. 2007. *The Upside of Down: Catastrophe, Creativity and the Renewal of Civilization*. Toronto: Vintage Canada.

Howard, K.W.F., and H. Maier. 2007. 'Road de-icing salt as a potential constraint on urban growth in the Greater Toronto Area, Canada'. *Journal of Contaminant Hydrology* 91 (1–2): 146–70.

Human Development Resource Office. 2007. *Human Development Report 2007/2008: Fighting Climate Change*. Houndmills, UK: Palgrave Macmillan.

Hummel, M., Ed. 2000. *Protecting Canada's Endangered Spaces: An Owner's Manual*. Toronto: Key Porter Books.

Huntington, H.P., et al. 2005. 'The changing Arctic: Indigenous perspectives'. In L. Arris, Ed., *Arctic Climate Impact Assessment (ACIA)*, 61–98. Cambridge: Cambridge University Press.

Hutchings, J.A., and R.A. Myers. 1994. 'What can be learned from the collapse of a renewable resource? Atlantic cod, *Gadus morhua*, of Newfoundland and Labrador'. *Canadian Journal of Fisheries and Aquatic Sciences* 51: 2,126–46.

IEMA (Independent Environmental Monitoring Agency). 2007. 'Plain language annual report 2006-07'. http://www.monitoringagency. net/Portals/0/ pdf/key_documents/annual%20reports/2006-07%20annual%20report/pe_07_part_a.pdf.

Indian and Northern Affairs Canada. 1997a. *Highlights of the Canadian Arctic Contaminants Assessment Report: A Community Reference Manual*. Ottawa: Minister of Public Works and Government Services.

———. 1997b. 'Pathways of transportation of persistent organics and metals to Arctic freshwater and marine ecosystems'. In *Canadian Arctic Contaminants Assessment Report*. Ottawa: Minister of Public Works and Government Services.

International Joint Commission. 2000. *Living with the Red*. Ottawa and Washington: International Joint Commission.

———. 2004. *Twelfth Biennial Report on Great Lakes Water Quality*. Windsor, ON: International Joint Commission.

International Reference Group on Great Lakes Pollution from Land Use Activities. 1978. *Environmental Management Strategy for the Great Lakes System: Final Report to the International Joint Commission*. Windsor, ON: International Joint Commission. Reprinted by permission of the publisher.

International Water Resources Association. 2000. 'Towards hydrosolidarity'. *Water International* 25 (2: special issue).

IPCC (Intergovernmental Panel on Climate Change). 2001a. *Climate Change 2001: Impacts, Adaptation, and Vulnerability. Contribution of Working Group II to the Third Assessment Report of the Intergovernmental Panel on Climate Change*. J.J. McCarthy, et al., Eds. Cambridge: Cambridge University Press.

———. 2001b. *Climate Change 2001: The Scientific Basis. Contributions of Working Group I to the Third Assessment Report of the Intergovernmental Panel on Climate Change*. J.T. Houghton, et al., Eds. Cambridge: Cambridge University Press.

———. 2001c. *Climate Change 2001: Synthesis Report. A Contribution of Working Groups I, II and III to the Third Assessment Report of the Intergovernmental Panel on Climate Change*. R.T. Watson and the Core Writing Team, Eds. Cambridge: Cambridge University Press.

———. 2007a. *Climate Change 2007: Impacts, Adaptation and Vulnerability*. Paris and Geneva: IPCC Secretariat, c/o World Meteorological Organization.

———. 2007b. *Climate Change 2007: The Physical Science Basis. Contribution of Working Group I to the Fourth Assessment Report of the Intergovernmental Panel on Climate Change*. Paris and Geneva: IPCC Secretariat, c/o World Meteorological Organization.

———. 2007c. 'Summary for policy makers of the synthesis report of the IPCC fourth assessment report'. http://www.ipcc.ch.

Ivey, J.L., R.C. de Loe, and R.D. Kreutzwiser. 2002. 'Groundwater management by watershed agencies: An evaluation of the capacity of Ontario's conservation authorities'. *Journal of Environmental Management* 64: 311–31.

Jachmann, H., P.S.M. Berry, and H. Imae. 1995. 'Tusklessness in African elephants: A future trend'. *African Journal of Ecology* 33: 230–5.

Jackson, J.B.C., et al. 2001. 'Historical overfishing and the recent collapse of coastal ecosystems'. *Science* 293: 629–38.

Jacques Whitford Engineering, Scientific, Planning and Management Consultants. 2006. *Final Report: 100% Design Report, Demolition and Disposal of the Old SYSCO Cooling Pond*. Prepared for the Sydney Tar Ponds Agency. Dartmouth, NS: Jacques Whitford.

James, C. 2006. *Global Status of Commercialized Biotech/GM Crops: 2006*. ISAAA Brief no. 35. Ithaca, NY: International Service for the Acquisition of Agri-biotech Applications.

Jørgensen, C., et al. 2007. 'Managing evolving fish stocks'. *Science* 318: 1,247–8.

Kanninen, M., et al. *Do Trees Grow on Money? The Implications of Deforestation Research for Policies to Promote REDD*. Bogor, Indonesia: Center for International Forestry Research (CIFOR).

Katary, N. 1982. *Origins and Evolution of 2001: A Developmental Odyssey*. Sudbury, ON: Department of Planning and Development, Regional Municipality of Sudbury.

Kay, C.E. 1994. 'Aboriginal overkill: The role of Native Americans in structuring western ecosystems'. *Human Nature* 5: 359–98.

KBM Forestry Consultants. 2007. *National Forest Strategy (2003-2008) Evaluation: Final Report*. Ottawa: Canadian Council of Forest Ministers

Keith, L.B., et al. 1984. 'Demography and ecology of a declining snowshoe hare population'. *Wildlife Monographs* 90: 1–43.

Kelly, J., and P.W. Williams. 2007. 'Modelling tourism destination energy consumption and greenhouse gas emissions: Whistler, British Columbia, Canada'. *Journal of Sustainable Tourism* 15 (1): 67–90.

Kenworthy, J.R. 2006. 'The eco-city: Ten key transport and planning dimensions for sustainable city development'. *Environment and Urbanization* 18 (1): 67–85.

Kesho Trust, The. 2007. 'Reconnecting children and families with nature'. Report of workshops sponsored by ActNow BC, 28 June and 17 July, Royal Roads University, Victoria, BC.

Kimmins, J.P. 1977. 'Evaluation of the consequences for the future tree productivity of the loss of nutrients in whole-tree harvesting'. *Forest Ecology and Management* 1: 169–83.

Kitchener-Waterloo Record. 2004. 'Harvesting the wind: Local farmers say they can provide power—if regulations let them'. 9 August: A1–A2.

———. 2007. 'Six Nations can't dictate new laws'. 15 September: A16.

Krebs, C.J., et al. 2001. 'What drives the 10-year cycle of snowshoe hares?' *Bioscience* 51: 25–35.

Kreutzwiser, R. 1998. 'Water resources management: The changing landscape in Ontario'. In R.D. Needham, Ed., *Coping with the World around Us: Changing Approaches to Land Use, Resources and Environment*, 135–48. Department of Geography Publication no. 50. Waterloo, ON: University of Waterloo.

Kreutzwiser, R., and R. de Loe. 2004. 'Water security: From exports to contamination of local water supplies'. in B. Mitchell, Ed., *Resource and Environmental Management in Canada: Addressing Conflict and Uncertainty*, 166–94. Toronto: Oxford University Press.

Krkoek, M., et al. 2007. 'Declining wild salmon populations in relation to parasites from farm salmon'. *Science* 318 (5,857): 1,772–5.

Kupchella, C.E., and M.C. Hyland. 1989. *Environmental Science: Living within the System of Nature*, p. 45. Toronto: Allyn and Bacon.

Kyoto Protocol to the United Nations Framework Convention on Climate Change. 1997. http://unfcc.int/resource/docs.convkp/html.

Lahey, A. 1998. 'Black lagoons: They're ugly. They stink. But are the tar ponds really killing the people of Sydney, Nova Scotia?' *Saturday Night* 113 (8): 37–40.

Land Reclamation Program. 2007. *Land Reclamation Program Annual Report, 2007*. Sudbury, ON: Environmental Planning Initiatives.

Landry, M., V.G. Thomas, and T.D. Nudds. 2001. 'Sizes of Canadian national parks and the viability of large mammal populations: Policy implications'. *The George Wright Forum* 18 (1): 13–23.

Laroque, C.P., and D.J. Smith. 1999. 'Tree-ring analysis of yellow-cedar (*Chamaecyparis nootkatensis*) on Vancouver Island, British Columbia'. *Canadian Journal of Forest Research*. 28: 115–23.

Lasserre, F. 2007. 'Drawers of water: Water diversions in Canada and beyond'. In K. Bakker, Ed., *Eau Canada*, 143–62. Vancouver: University of British Columbia Press.

Lautenbach, W.W. 1985. *Land Reclamation Program 1978–1984*. Sudbury, ON: Regional Municipality of Sudbury, Vegetation Enhancement Technical Advisory Committee.

Lautenbach, W.W., et al. 1995. 'Municipal land restoration program: The regreening process'. In J.M. Gunn, Ed., *Restoration and Recovery of an Industrial Region: Progress in Restoring the Smelter-Damaged Landscape near Sudbury, Ontario*, 109–22. New York: Springer-Verlag.

Lautenschlager, R.A., and T.P. Sullivan. 2002. 'Effects of herbicide treatments on biotic components in regenerating northern forests'. *Forestry Chronicle* 78: 695–731.

Lawrence, D., Ed. 2003. *Environmental Impact Assessment: Practical Solutions to Recurrent Problems*. Toronto: John Wiley.

Lebeuf, M., et al. 2007. 'Temporal trends (1987–2002) of persistent, bioaccumulative and toxic (PBT) chemicals in beluga whales (*Delphinapterus leucas*) from the St Lawrence estuary, Canada'. *The Science of the Total Environment* 383 (1–3): 216–31.

LeDrew, E. 2002. *A Uniquely Canadian Dimension to Climate Change: The Role of the Cryosphere in the Everyday Lives of Canadians*. Research Lecture 2002. Waterloo, ON: University of Waterloo, Faculty of Environmental Studies.

Lee, K.N. 1993. *Compass and Gyroscope: Integrating Science and Politics for the Environment*. Washington: Island Press.

Lemieux, A. 2005. 'Canada's global mining presence'. *Canadian Minerals Yearbook, 2005*. Ottawa: Minister of Public Works and Government Services.

Lemieux, C., T. Beechey, and D. Scott. 2007. 'A survey on protected areas and climate change (PACC) in Canada: Survey update'. Canadian Council on Ecological Areas. *ECO* 16: 2–3.

Lennon, John. 1971. 'Imagine'. Words and music by John Lennon © 1971 (renewed 1999) LENONO MUSIC. All rights controlled and administered by EMI BLACKWOOD MUSIC INC. All rights reserved. International Copyright Secured. Used by permission.

Leopold, A. 1949. *A Sand County Almanac*. Oxford: Oxford University Press.

Le Prestre, P.G., and P. Stoett. 2001. 'International initiatives, commitments, and disappointments: Canada, CITES, and the CBD'. In K. Beazley and R. Boardman, Eds, *Politics of the Wild: Canada and Endangered Species*, 190–216. Toronto: Oxford University Press.

Lesieur, D., S. Gauthier, and Y. Bergeron. 2002. 'Fire frequency and vegetation dynamics for the south-central boreal forest of Quebec, Canada'. *Canadian Journal of Forest Research* 32: 1,996–2,002.

Liu, J., A.J.B. Zehnder, and H. Yang. 2007. 'Historical trends in China's virtual water trade'. *Water International* 32: 78–90.

Longhurst, A. 2002. 'Murphy's law revisited: Longevity as a factor in recruitment to fish populations'. *Fisheries Research* 56: 125–31

Lord, G.R. 1974. 'Introduction'. In A.H. Richardson, *Conservation by the People: A History of the Conservation Authority Movement to 1970*, ix–xi. Toronto: University of Toronto Press.

Louv, R. 2006. *Last Child in the Woods*. Toronto: Algonquin Books.

Lovelock, J.E. 1988. *The Ages of* Gaia. New York: W.W. Norton.

McBean, G., and D. Henstra. 2003. *Climate Change, Natural Hazards, and Cities*. Paper Series no. 31. Toronto: Institute for Catastrophic Loss Reduction.

McGinnity, P., et al. 2003. 'Fitness reduction and potential extinction of wild populations of Atlantic salmon (*Salmo salar*) as a result of interactions with escaped farm salmon'. *Proceedings of the Royal Society: Biological Sciences* 270 (1,532): 2,443–50.

McIntosh, R.P. 1980. 'The relationship between succession and the recovery process in ecosystems'. In J. Cairns, Ed., *The Recovery Process in Damaged Ecosystems*, 11–62. Ann Arbor, MI: Ann Arbor Science Publishers.

McIver, J.D., A.R. Moldenke, and G.L. Parsons. 1990. 'Litter spiders as bio-indicators of recovery after clear-cutting in a western coniferous forest'. *Northwest Environmental Journal* 6: 410–12.

Mackintosh, W.A. 1934. *Prairie Settlement: The Geographic Background*. Toronto: Macmillan.

Maclaren, V.W. 2004. 'Waste management: Integrated approaches'. In B. Mitchell, Ed., *Resource and Environmental Management in Canada: Addressing Conflict and Uncertainty*, 3rd edn, 371–97. Toronto: Oxford University Press.

McLeman, R., and B. Smit. 2003. 'Climate change, migration and security'. Commentary no. 86. Ottawa: Canadian Security Intelligence Service.

Malcolm, C. 2003. 'The current state and future prospects of whale-watching management with special emphasis on whale-watching in BC, Canada'. PhD dissertation, University of Victoria.

———. 2008. 'Provincial parks'. In P. Dearden and R. Rollins, Eds., *Parks and Protected Areas in Canada: Planning and Management*, 3rd edn, 56–82. Toronto: Oxford University Press.

Malcolm, J.R., et al. 'Global warming and extinctions of endemic species from biodiversity hotspots'. *Conservation Biology* 20: 538–48.

Manitoba. 2001. 'Manitoba's concerns: Garrison Projects and Devils Lake Outlet'. November. http://www.gov.mb.ca/waterstewardship/transboundary/positions/man-position/backgr.html. Permission to reproduce this text is provided by the Queen's Printer for Manitoba. The Queen's Printer does not warrant the accuracy or currency of the reproduction of this information.

Manitoba Round Table on Environment and Economy. 1992. *Sustainable Development: Towards Institutional Change in the Manitoba Public Sector.* Winnipeg: Manitoba Round Table on Environment and Economy.

Manitoba Water Stewardship. 2006. 'About us, and minister's message'. http://www.gov.mb.ca/waterstewardship/misc.about.html.

———. 2007. 'Potential transboundary water projects'. http://www.gov.mb.ca/waterstewardship/water_info/transboundary/potential. html.

Martin, P.S. 1967. 'Prehistoric overkill'. In P.S. Martin and H.E. Wright Jr, Eds, *Pleistocene Extinctions: The Search for a Cause,* 75–120. New Haven: Yale University Press.

Mascarenhas, M. 2007. 'Where the waters divide: First Nations, tainted water and environmental justice in Canada'. *Local Environment* 12: 565–77.

Mastny, L. 2004. 'Purchasing for people and the planet'. In Worldwatch Institute, *State of the World 2004,* 122–42. New York: W.W. Norton.

Mastney, L., and H. French. 2002. 'Crimes of (a) global nature'. *Worldwatch* (September-October): 12–23.

Matthews, C., R.B. Gibson, and B. Mitchell. 2007. 'Rising waves, old charts, nervous passengers: Navigating toward a new water ethic'. In K. Bakker, Ed, *Eau Canada,* 335–58. Vancouver: University of British Columbia Press.

May, E., and M. Barlow. 2001. 'The tar ponds'. *Alternatives Journal* 27 (1): 7–11.

Mayhew, P.J., G.B. Jenkins, and T.G. Benton. 2008. 'A long-term association between global temperature and biodiversity, origination and extinction in the fossil record'. *Proceedings of the Royal Society of London [B]* 275 (1,630): 47–53.

Meadows, D.H., et al. 1972. *The Limits to Growth.* London. Earth Island.

Meloche, C., and S.D. Murphy. 2006. 'Managing tree-of-heaven (*Ailanthus altissima*) in parks and protected areas: A case study of Rondeau Provincial Park (Ontario, Canada). *Environmental Management* 37: 764–72.

Mercredi, O., and M.E. Turpel. 1993. *In the Rapids: Navigating the Future of First Nations.* Toronto: Viking.

Millennium Ecosystem Assessment. 2003. *Ecosystems and Human Well-being: A Framework for Assessment.* Washington: Island Press.

———. 2005. *Ecosystems and Human Well-being: Synthesis.* Washington: Island Press.

Miller, G. 2002. *Climate Change: Is the Science Sound?* Special Report to the Legislative Assembly of Ontario. Toronto: Environmental Commissioner of Ontario.

Mills, T.J., T.M. Quigley, and F.J. Everest. 2001. 'Science-based natural resource management decisions: What are they?' *Renewable Resources Journal* 19 (2): 10–15.

Mitchell, B., Ed. 2004. *Resource and Environmental Management in Canada: Addressing Conflict and Uncertainty,* 3rd edn. Toronto: Oxford University Press.

Mittelstaedt, M. 2007. 'Pollution causing premature deaths'. *The Globe and Mail* 3 October: A1. http://www.theglobeandmail.com/servlet/story/LAC.20071003. TOXIC03/TPStory/TPNational/BritishColum.

Molle, F. 2007. 'Scales and power in river basin management: The Chao Phraya River in Thailand'. *Geographical Journal* 173: 358–73.

Mooers, A.O., et al. 2007. 'Biases in legal listing under Canadian endangered species legislation'. *Conservation Biology* 21: 572–5.

Morris, T.J., et al. 2007. *Changing the Flow: A Blueprint for Federal Action on Freshwater.* Toronto: The Gordon Water Group of Concerned Scientists and Friends.

Mosquin, T. 1994. 'A conceptual framework for the ecological functions of biodiversity'. *Global Biodiversity* 4: 2–16.

Mosquin, T., P.G. Whiting, and D.E. McAllister. 1995. *Canada's Biodiversity: The Variety of Life, Its Status, Economic Benefits, Conservation Costs and Unmet Needs,* p. 15. Ottawa: Canadian Museum of Nature.

Mulkins, L.M., et al. 'Carbon isotope composition of mysids at a terrestrial-marine ecotone, Clayoquot Sound, British Columbia, Canada'. *Estuarine, Coastal and Shelf Science* 54: 669–75.

Munro, M. 2008. 'Environment Canada "muzzles" scientists' dealings with media; Policy aims to stop Baird from seeing "surprises" in news. *Ottawa Citizen* 1 February.

Murphy, S.D., and L.R.G. Martin. 2001. 'Urban ecology in Ontario, Canada: Moving beyond the limits of city and ideology'. *Environments* 29 (1): 67–83.

Myers, R.A., and B. Worm. 2003. 'Rapid worldwide depletion of predatory fish communities'. *Nature* 423: 280–3.

Nakashima, D.J. 1990. *Application of Native Knowledge in EIA: Inuit, Eiders and Hudson Bay Oil.* Prepared for the Canadian Environmental Assessment Research Council. Ottawa: Minister of Supply and Services.

Nanus, B. 1992. *Visionary Leadership.* San Francisco: Jossey-Bass.

National Geographic Society. 2008. *Greendex 2008: Consumer Choice and the Environment: A Worldwide Tracking Survey.* Washington: National Geographic Society.

National Research Council, Division of Earth and Life Studies, Committee on the Science of Climate Change. 2001. *Climate Change Science: An Analysis of Some Key Questions.* Washington: National Academy Press.

Neave, D., et al. 2002. 'Forest biodiversity in Canada'. *Forestry Chronicle* 78 (6): 779–83.

Neilson, E.T., et al. 2007. 'Spatial distribution of carbon in natural and managed stands in an industrial forest in New Brunswick, Canada'. *Forest Ecology and Management* 253: 148–60.

Nellemann, C., S. Hain, and J. Alder, Eds. 2008. *In Dead Water: Merging of Climate Change with Pollution, Over-harvest, and Infestations in the World's Fishing Grounds.* Arendal, Norway: UNEP/GRID-Arendal. www.grida.no.

Newman, P.W.G. 2006. 'The environmental impact of cities'. *Environment and Urbanization* 18 (2): 275–95. Reproduced by permission of SAGE Publications Ltd, London, Los Angeles, New Delhi, Singapore, and Washington, DC.

Newmark, W.D. 1995. 'Extinction of mammal populations in western North American national parks'. *Conservation Biology* 9: 512–26.

NFSC (National Forest Strategy Coalition). 2005. *Highlights of Accomplishments: Advancing the National Forest Strategy.* Ottawa: Canadian Council of Forest Ministers.

Nobel, P.G. 1990. *Contaminants in Canadian Seabirds.* SOE Report no. 90-2. Ottawa: Environment Canada.

Nova Scotia Department of Environment and Labour. 2002. *A Drinking Water Strategy for Nova Scotia: A Comprehensive Approach to the Management of Drinking Water.* Halifax: Nova Scotia Environment and Labour.

Nova Scotia Tourism Partnership and Nova Scotia Tourism, Culture and Heritage. 2007. *Nova Scotia Strategy for Sustainable Coastal Tourism Development.* Halifax: Nova Scotia Tourism, Culture and Heritage.

NRCAN (Natural Resources Canada). 1995. 'Silviculture terms in Canada'. http://www.nfdp.ccfm.org/silviterm.

———. 2003. 'Competitiveness of the forest industry'. http://www.cfl.scf.rncan.gc.ca.

———. 2005a. 'Canadian vehicle survey 2005'. http://oee.nrcan.gc.ca/Publications/statistics/cvs05/index.cfm?attr=0.

———. 2005b. *The State of Canada's Forests: Annual Report 2004–5: The Boreal Forest.* Ottawa: Canadian Forest Service, Natural Resources Canada.

———. 2005c. 'Technologies and applications: About solar energy'. Ottawa: Natural Resources Canada. http://www.canren.gc.ca/tech_appl/index. asp?Cald+5&PgId+121.

———. 2006a. 'Canadian minerals yearbook'. Minerals and Mining Statistics On-line. http://mmsd1.nrcan.gc.ca/mmsd/minstatistics_e.asp.

———. 2006b. 'Energy efficiency trends in Canada 1990–2004'. http://oee.nrcan. gc.ca/publications/statistics/trends06/chapter6.cfm?attr=0.

———. 2006c. 'Technologies and applications: About wind energy'. Ottawa: Natural Resources Canada. http://www.caren.gc.ca/tech_applindex. asp?Cald+6&PgId=232.

———. 2007a. Information Bulletin: 'Employment in Canada's mining and mineral processing industries remains robust in 2006'. http://www.nrcan.gc.ca/mms/pdf/employment-2007_e.pdf.

———. 2007b. *The State of Canada's Forests: Annual Report 2007.* Ottawa: Canadian Forest Service, Natural Resources Canada.

———. 2007c. 'Strong Growth Expected for Canada's Mining Sector'. Natural Resources Canada's News Room. http://www.rncan.gc.ca/media/newsreleases/2007/200789a_e.htm.

NRTEE (National Round Table on Environment and Economy). 2003a. *2003 Environment and Sustainable Development Indicators for Canada.* Ottawa: NRTEE.

———. 2003b. *Securing Canada's Natural Capital: A Vision for Nature Conservation in the 21st Century.* Ottawa: NRTEE.

———. 2003c. *State of the Debate: Environmental Quality in Canadian Cities: The Federal Role.* Ottawa: NRTEE.

———. 2006. *The State of the Debate on the Environment and the Economy: Environmental Quality in Canadian Cities: The Federal Role.* Ottawa: NRTEE.

NWMO (Nuclear Waste Management Organization). 2005. *Choosing a Way Forward: The Future Management of Canada's Used Nuclear Fuel: A Summary.* Toronto: NWMO.

O'Connor, D.R. 2002a. *Report of the Walkerton Inquiry: Part One, The Events of May 2000 and Related Issues.* Toronto: Ontario Ministry of the Attorney General.

———. 2002b. *Report of the Walkerton Inquiry: Part Two, A Strategy for Safe Drinking Water.* Toronto: Ontario Ministry of the Attorney General.

OECD (Organisation for Economic Co-operation and Development). 2002. *Towards Sustainable Consumption: An Economic Conceptual Framework.* Paris: OECD Environment Directorate.

Olsson, P. 2007. 'The role of vision in framing adaptive co-management processes: Lessons from Kistianstads Vattenrike, southern Sweden'. In D. Armitage, F. Berkes, and N. Doubleday, Eds, *Adaptive Co-management: Collaboration, Learning and Multi-level Governance*, 168–85. Vancouver: University of British Columbia Press.

O'Neill, D. 2004. 'Threats to water availability in Canada: A perspective'. In *Threats to Water Availability in Canada*, xi–xvi. NWRI Scientific Assessment Report Series no. 3 and ACSD Science Assessment Series no. 1. Ottawa and Burlington, ON: Environment Canada and National Water Research Institute.

Ontario Advisory Committee on Watershed-Based Source Protection Planning. 2003. *Protecting Ontario's Drinking Water: Toward a Watershed-Based Source Protection Planning Framework: Final Report*. Toronto: Ministry of the Environment.

Ontario Ministry of Energy. 2006. Press release: 'Ontario leads Canada in wind power generation'. http://www.energy.gov.on.ca/index.cfm?fuseaction=english.news&body=yes&news_id=141D.

O'Riordan, T. 1976. *Environmentalism*. London: Pion.

Orr, D.W. 1992. 'The problem of education'. *New Directions for Higher Education* 77 (spring): 3–8.

———. 1994. *Earth in Mind: On Education, Environment and the Human Prospect*. Covelo, CA: Island Press.

Pahl-Wostl, C., and M. Hare. 2004. 'Processes of social learning in integrated resources management'. *Journal of Community and Applied Social Psychology* 14: 193–206.

Parmesan, C. 2006. 'Ecological and evolutionary responses to recent climate change'. *Annual Review of Ecology and Systematics* 37: 637–69.

Parson, E.A. 2000. 'Environmental trends and environmental governance in Canada'. *Canadian Public Policy* 26 (supplement): S123–S143.

Partidario, M.F., and R. Clark, Eds. 2000. *Perspectives on Strategic Environmental Assessment*. Boca Raton, FL: Lewis Publishers.

Pauly, D. 2006. 'Major trends in small-scale fisheries, with emphasis on developing countries, and some implications for the social sciences'. *Maritime Studies* 4/2: 7–22.

Peterson, C.H., et al. 2003. 'Long-term ecosystem response to the *Exxon Valdez* oil spill'. *Science* 302 (5,653): 2,082–6.

Plummer, R., and J. FitzGibbon. 2007. 'Connecting adaptive co-management, social learning, and social capital through theory and practice'. In D. Armitage, F. Berkes, and N. Doubleday, Eds, *Adaptive Co-management: Collaboration, Learning and Multi-Level Governance*, 38–61. Vancouver: University of British Columbia Press.

Pollution Probe. 2007. *Towards a Vision and Strategy for Water Management in Canada*. Toronto: Pollution Probe.

Population Reference Bureau. 2007. 'World population highlights'. *Population Bulletin* 62: 2.

Prescott-Allen, R. 2001. *The Well-being of Nations: A Country by Country Index of Quality of Life and the Environment*. Washington and Ottawa: Island Press and IDRC.

Priscoli, J.D., J. Dooge, and M.R. Llamas. 2004. *Water and Ethics: An Overview*. Series on Water and Ethics, Essay 1. Paris: UNESCO.

Quinn, F.J., et al. 2004. 'Water allocation, diversion and export'. In *Threats to Water Availability in Canada*, 1–8. NWRI Scientific Assessment Report Series no. 3 and ACSD Science Assessment Series no. 1. Ottawa and Burlington, ON: Environment Canada and National Water Research Institute.

Rahman, M.M., and O. Varis. 2005. 'Integrated water resources management: Evolution, prospects and future challenges'. *Sustainability: Science, Practice and Policy* 1: 15–21.

Rainham, D. 2002. 'Risk communication and public response to industrial chemical contamination in Sydney, Nova Scotia: A case study'. *Journal of Environmental Health* 65 (5): 26–32.

Rees, J. 1985. *Natural Resources: Allocation, Economics and Policy*. London: Methuen.

Regehr, E.V., et al. 2007. 'Effects of earlier sea ice breakup on survival and population size of polar bears in western Hudson Bay'. *Journal of Wildlife Management* 71 (8): 2,673–83.

Richardson, N.H., B.I. Savan, and L. Bodnar. 1989. *Economic Benefits of a Clean Environment: Sudbury Case Study*. Prepared for the Department of Environment, Canada. Toronto: N.H. Richardson Consulting.

Rivers, N., and D. Sawyer. 2008. *Pricing Carbon: Saving Green—A Carbon Price to Lower Emissions, Taxes and Barriers to Green Technology*. Vancouver: David Suzuki Foundation.

Roberts, C.M. 2003. 'Our shifting perspectives on the oceans'. *Oryx* 37: 166–77.

Rochette, P., et al. 2008. 'Estimation of N$_2$O emissions from agricultural soils in Canada, II: 1990–2005 Inventory'. *Canadian Journal of Soil Science* 88 (5): 655–69.

Rohde, K. 1992. 'Latitudinal gradients in species diversity: The search for the primary cause'. *Oikos* 65: 514–27.

Ross, N. 2001. *Healing the Landscape: Celebrating Sudbury's Reclamation Success*. Sudbury, ON: Vegetation Enhancement Technical Advisory Committee.

Ross, P.S., et al. 2000. 'High PCB concentrations in free-ranging Pacific killer whales, *Orcinus orca*: Effects of age, sex and dietary preference'. *Marine Pollution Bulletin* 40: 504–15.

Rowe, S. 1993. 'In search of the holy grass: How to bond with the wilderness in nature and ourselves'. *Environment Views* (winter): 7–11.

Rowell, P., G.L. Holroyd, and U. Banasch. 2003. 'Summary of the 2000 Canadian peregrine falcon survey'. In J. Kennedy, Ed., *Bird Trends: A Report on Results of National Ornithological Surveys in Canada*, 52–6. Ottawa: Canadian Wildlife Service.

Royal Commission on the Future of the Toronto Waterfront. 1992. *Regeneration: Toronto's Waterfront and the Sustainable City, Final Report*. Ottawa and Toronto: Minister of Supply and Services Canada and Queen's Printer of Ontario.

Ruddiman, W.F. 2003. 'The anthropogenic greenhouse era began thousands of years ago'. *Climatic Change* 61: 261–93.

Rylatt, M.G. 1999. 'Ekati diamond mine background and development'. *Mining Engineer* 51: 37–43.

Sanchez, L.E. 1998. 'Industry response to the challenge of sustainability: The case of the Canadian nonferrous mining sector'. *Environmental Management* 22: 521–31.

———. 2001. 'The climate change: Soil fertility–food security nexus'. Address at the Sustainable Food Security for All by 2020 Conference, Bonn, September.

Saskatchewan. 2008. 'Mineral resources of Saskatchewan'. http://www.ir.gov.sk.ca/Default.aspx?DN=3558,3541,3538,3385,2936,Documents.

Saskatchewan Mining Association. 2007. 'Commodities mined in Saskatchewan—Uranium. http://www.saskmining.ca/commodities.php?COMM=1&SEC=Introduction.html.

Saskatchewan Round Table on Environment and Economy. 1992. *Conservation Strategy for Sustainable Development in Saskatchewan*. Regina: Saskatchewan Round Table on Environment and Economy.

Saumure, R.A., T.B. Herman, and R.D. Titman. 2007. 'Effects of haying and agricultural practices on a declining species: The North American wood turtle (*Glyptemis insculpta*). *Biological Conservation* 135: 565–75.

Sawin, J.L. 2004. 'Making better energy choices'. In Worldwatch Institute, *State of the World 2004*, 24–43. New York: W.W. Norton.

Schueler, F.W., and D.E. McAllister. 1991. 'Maps of the number of tree species in Canada: A pilot project GIS study of tree diversity'. *Canadian Biodiversity* 1: 22–9.

Scialabba, N.E. 2003. *Organic Agriculture: The Challenge of Sustaining Food Production While Enhancing Biodiversity*. Rome: Food and Agriculture Organization.

Scialabba, N.E., and C. Hattam, Eds. 2002. *Organic Agriculture, Environment and Food Security*. Rome: Food and Agriculture Organization.

Scientific Review Panel. 2002. *British Columbia Offshore Hydrocarbon Development: Report of the Scientific Review Panel*. Submitted to the BC Minister of Energy and Mines, January 15.

Seaweb. 2003. 'Danger at sea: Our changing ocean'. http://www.seaweb.org.

Selbourne, Earl of. 2000. *The Ethics of Freshwater Use: A Survey*. Paris: UNESCO.

Selin, S., and D. Chavez. 1995. 'Developing a collaborative model for environmental planning and management'. *Environmental Management* 19: 189–95.

Shaftoe, D., Ed. 1993. *Responding to Changing Times: Environmental Mediation in Canada*. Waterloo, ON: The Network: Interaction of Conflict Resolution, Conrad Grebel College.

Shang, E.H., R.M.K. Yu, and R.S. Wu. 2006. 'Hypoxia affects sex differentiation and development, leading to a male-dominated population in zebrafish'. *Environmental Science and Technology* 40: 3,118–22.

Shaw, R.W., and CCAF A041 Project Team. 2001. *Coastal Impacts of Climate Change and Sea-Level Rise on Prince Edward Island: Synthesis Report*. Climate Change Action Fund Project CCAF A041. Ottawa: Ministry of Supply and Services.

Shiklomanov, I.A. 2000. 'Appraisal and assessment of world water resources'. *Water International* 25: 11–32.

Shrubsole, D. 2001. 'The cultures of flood management in Canada: Insights from the 1997 Red River experience'. *Canadian Water Resources Journal* 26: 461–79.

Sierra Club of Canada. 2006. *National Forest Strategy 2003–2008: An Assessment in 2006—Is It Making a Difference?* Toronto: Sierra Club of Canada.

Simpson, C.G. 1964. 'Species density of North American recent mammals'. *Systematic Zoology* 13: 15–73.

Slocombe, S.D. 1993. 'Implementing ecosystem-based management: Development of theory, practice and research for planning and managing a region'. *BioScience* 43: 612–22.

———. 2004. 'Applying an ecosystem approach'. In B. Mitchell, Ed., *Resource and Environmental Management in Canada: Addressing Conflict and Uncertainty*, 3rd edn, 420–41. Toronto: Oxford University Press.

Smith, A., and J.B. MacKinnon. 2007. *The 100-Mile Diet: A Year of Local Eating.* Toronto: Random House Canada.

Smith, J.N.M., et al. 'Population biology of snowshoe hares II: Interactions with winter food plants'. *Journal of Animal Ecology* 57: 269–86.

Sprague, J.B. 2007. 'Great Wet North: Canada's myth of water abundance'. In K. Bakker, Ed., *Eau Canada*, 23–35. Vancouver: University of British Columbia Press.

Standing Committee on Environment and Sustainable Development. 2000. *Pesticides: Making the Right Choice for the Protection of Health and the Environment.* Ottawa: House of Commons.

Starr, J. 1991. 'Water wars'. *Foreign Policy* 82: 17–36.

Statistics Canada. 2001. '2001 Census of agriculture'. www.statcan.ca/english/Pgdb/census.htm#far.

———. 2003a. *Farming Facts 2002.* Ottawa: Ministry of Industry.

———. 2003b. *Human Activity and the Environment: A Statistical Compendium.* Ottawa: Minister of Supply and Services.

———. 2006. *2006 Census of Agriculture 2006.* Ottawa: Statistics Canada.

———. 2007a. *2006 Census of Agriculture.* Ottawa: Statistics Canada.

———. 2007b. 'Households and the environment survey 2006'. http://www.statcan.ca/Daily/English/070711/d070711b.html.

———. 2007c. 'Recycling in Canada'. *EnviroStats* 1 (1): 3–7.

Statistics Canada, Agriculture Division. 2003. 'Food statistics, 2003'. www.statcan.ca/english/freepub/21-020-XIE/21-020-XIE02002.pdf.

Steele, D.H., R. Andersen, and J.M. Green. 1992. 'The managed commercial annihilation of northern cod'. *Newfoundland Studies* 8 (1): 34–68.

Stewart, E.J., et al. 2007. 'Sea ice in Canada's Arctic: Implications for cruise tourism'. *Arctic* 60: 370–80.

Stinchcombe, K., and R.B. Gibson. 2001. 'Strategic environmental assessment as a means of pursuing sustainability: Ten advantages and ten challenges'. *Journal of Environmental Assessment Policy and Management* 3: 343–72.

Stirling, A., et al. 2008 'Unusual predation attempts of polar bears on ringed seals in the southern Beaufort Sea: Possible significance of changing spring ice conditions'. *Arctic* 61: 14–22.

Stockton, A.S., et al. 2005. 'A natural experiment on the impacts of high deer densities on the native flora of coastal temperate rain forests'. *Biological Conservation* 126: 118–28.

Stockwell, C.A., P. Hendry, and M.T. Kinnison. 2003. 'Contemporary evolution meets conservation biology'. *Trends in Ecology and Evolution* 18: 94–101.

Stokes, K., and R. Law. 2000. 'Fishing as an evolutionary force'. *Marine Ecology Progress Series* 208: 307–9.

Stonehouse, D.P. 1994. 'Canadian experiences with the adoption and use of soil conservation practices'. In T.L. Napier, S.M. Camboni, and S.A. El-Swaify, Eds, *Adopting Conservation on the Farm*, 369–95. Ankeny, IA: Soil and Water Conservation Society.

Stuart-Smith, J., et al. 2002. 'Conserving whitebark pine in the Canadian Rockies'. *Research Links* 10: 11–14.

Suffling, R., and D. Scott. 2002. 'Assessment of climate change effects on Canada's national park system'. *Environmental Monitoring and Assessment* 74: 117–39.

Sumaila, U.R., et al. 2007. 'Potential costs and benefits of marine reserves on the high seas'. *Marine Ecology Progress Series* 345: 305–10.

Sumaila, U.R., and D. Pauly, Eds. 2007. 'Catching more bait: A bottom-up re-estimation of global fisheries subsidies'. 2nd version. *Fisheries Centre Research Reports* 14 (6): 49–53. Vancouver: Fisheries Centre, University of British Columbia.

Swinnerton, G.S. 2002. 'The Banff–Bow Valley: Balancing human use and ecological integrity within Banff National Park'. In P. Dearden and R. Rollins, Eds, *Parks and Protected Areas in Canada: Planning and Management*, 2nd edn, 240–64. Toronto: Oxford University Press.

Tahera Diamond Corporation. 2008. 'Mining—Jericho Diamond Mine'. http://www.tahera.com/Operations/Mining/JerichoDiamondMine/default.aspx.

Theberge, J.C. 2004. 'Lessons learned during interagency negotiation regarding wolf conservation and forestry'. In N. Munro, et al., Eds, *Making Ecosystem Based Management Work: Connecting Researchers and Managers.* Proceedings of the 5th International Conference of Science and Management of Protected Areas, Victoria, BC, May.

Thérival, R. 1993. 'Systems of strategic environmental assessment'. *Environmental Impact Assessment Review* 13 (3): 145–68.

Thompson, I.D., J.A. Baker, and M. Ter-Mikaelian. 2003. 'A review of the long-term effects of post-harvest silviculture on vertebrate wildlife, and predictive models, with an emphasis on boreal forests in Ontario, Canada'. *Forest Ecology and Management* 177: 441–69.

Tinker, J. 1996. 'Introduction'. In J.B. Robinson et al., Eds, *Life in 2030: Exploring a Sustainable Future for Canada*, ix–xv. Vancouver: University of British Columbia Press.

Todhunter, P.E. 2001. 'A hydroclimatological analysis of the Red River of the North snowmelt flood catastrophe of 1997'. *Journal of the American Water Resources Association* 37: 1263–78.

Toronto and Region Conservation Authority. 2006. *Moving toward the Living City.* Toronto: Toronto and Region Conservation Authority.

Trainor, S.F., et al. 2007. 'Arctic climate impacts: Environmental injustice in Canada and the United States'. *Local Environment* 12: 627–43.

Travers, J. 2007. 'Harper takes big risk by ignoring climate change issue'. *The Record* 25 September: A6.

Trewartha, G.T. 1954. *An Introduction to Climate.* New York: McGraw-Hill.

Trites, A.W., et al. 2007. 'Killer whales, whaling, and sequential megafaunal collapse in the North Pacific: A comparative analysis of the dynamics of marine mammals in Alaska and British Columbia following commercial whaling'. *Marine Mammal Science* 23 (4): 751–65.

Trostel, K., et al. 2007. 'Can predation cause the 10-year hare cycle?' *Oecologia* 74: 185–92.

Tynan, C.T., and J.L. Russell. 2008. 'Assessing the impacts of future 2°C global warming on Southern Ocean cetaceans'. Paper presented to the International Whaling Commission Scientific Committee, Chile, June.

UNAIDS. 2007. *AIDS Epidemic Update 2007.* Geneva: UNAIDS.

UNDP (United Nations Development Programme). 2005. *Human Development Report 2005: International Cooperation at a Crossroads: Aid, Trade and Security in an Unequal World.* New York: UNDP. Reproduced with permission of Palgrave Macmillan.

———. 2007. *Human Development Report 2007/8: Fighting Climate Change: Human Solidarity in a Divided World.* New York: UNDP.

UNEP (United Nations Environment Programme). 2002. 'Global environmental outlook 3 (GEO 3)'. www.unep.org.geo/geo3.

———. 2007. *Global Environment Outlook: Environment for Development (GEO-4).* Nairobi: UNEP.

———. 2008. 'Meltdown in the Mountains'. http://www.unep.org/ Documents.Multilingual/Default.asp?DocumentID=530&ArticleID.

United Nations Population Fund. 2007a. *State of the World Population 2007.* New York: United Nations Population Fund.

———. 2007b. *World Population Prospects: The 2006 Revision.* Geneva: United Nations.

Vancouver Sun. 2003. 'Spotted owl survival threatened'. 11 July.

Venter, O., et al. 'Threats to endangered species in Canada'. *BioScience* 56 (11) 903–10.

VETAC (Vegetation Enhancement Advisory Committee). 2005. *Annual Report.* Greater Sudbury: Land Reclamation Program.

———. 2007. *Annual Report.* Greater Sudbury: Land Reclamation Program.

Volpe, J.P., et al. 2000. 'Evidence of natural reproduction of aquaculture escaped Atlantic salmon (*Salmo salar*) in a coastal British Columbia river'. *Conservation Biology* 14 (3): 899–903.

von Braun, J. 2007. *The World Food Situation: New Driving Forces and Required Actions.* Washington: International Food Policy Research Institute.

von Oech, R. 1990. *A Whack on the Side of the Head.* New York: Warner Books.

Voynick, S. 1999. 'Diamonds on ice'. *Compressed Air* 104: 60–8.

Walker, I.J., and R. Sydneysmith. 2008. 'British Columbia'. In D.S. Lemmen, et al., Eds, *From Impacts to Adaptation: Canada in a Changing Climate 2007*, 329–86. Ottawa: Government of Canada.

Walters, C.J. 2007. 'Is adaptive management helping to solve fisheries problems?' *Ambio* 36 (4): 304–7.

Warner, J.F., and C.L. Johnston. 2007. 'Virtual water—real people: Useful concept or prescriptive tool?' *Water International* 32: 63–77.

Warnock, R.G., and M.A. Skeel. 2004. 'Effectiveness of voluntary habitat stewardship in conserving grassland: Case of Operation Burrowing Owl in Saskatchewan'. *Environmental Management* 25 March. http:// wwwspringerlink.com.

Watson, R., and D. Pauly. 2001. 'Systematic distortion in world fisheries catch trends'. *Nature* 414: 534–6.

Watt-Cloutier, S. 2000. *Wake-up Call.* www.ourplanet.com/imgversn/124/watt.html.

WCED (World Commission on Environment and Development). 1987. *Our Common Future.* Oxford: Oxford University Press.

Weiss, W., I. Bergmann, and G. Faninger. 2007. *Solar Heat Worldwide: Markets and Contribution to the Energy Supply 2005.* Paris: International Energy Agency.

Werniuk, J. 1998a. 'Great Canadian diamonds'. *Canadian Mining Journal* (October): 8–22.

———. 1998b. 'Where the smart mining is going'. *Canadian Mining Journal* (December): 14–18.

Whitelaw, G.S., and P.E.J. Eagles. 2007. 'Planning for long, wide conservation corridors on private lands in the Oak Ridges Moraine, Ontario, Canada'. *Conservation Biology* 21: 675–83.

Wielgus, R.B., and P.R. Vernier. 2003. 'Grizzly bear selection of managed and un-managed forests in the Selkirk Mountains'. *Canadian Journal of Forest Research* 33: 822–9.

Wiken, E. 1986. *Terrestrial Ecozones of Canada*. Ottawa: Environment Canada.

Wiley-Blackwell. 2008. 'Birds migrate earlier, but some may be left behind as climate warms rapidly'. *ScienceDaily* 22 June. http://www.sciencedaily.com/releases/2008/06/080620115925.htm.

Wilkinson, C.J.A. 2008. 'Protected areas law in Ontario, Canada: Maintenance of eco-logical integrity as the management priority'. *Natural Areas Journal* 28: 180–6.

Williams, R., D. Lusseau, and P.S. Hammond. 2006. 'Estimating relative energetic costs of human disturbance to killer whales (*Orcinus orca*)'. *Biological Conservation* 133: 301–11.

Winfield, M.S., and G. Jenish. 1998. 'Ontario's environment and the "common sense revolution"'. *Studies in Political Economy* 57 (autumn): 129–45.

Wolf, A.T. 1995. *Hydropolitics along the Jordan River: Scarce Water and its Impact on the Arab-Israeli Conflict*. Tokyo: United Nations University Press.

Wolfe, B.C., et al. 2005. 'Impacts of climate and river flooding on the hydro-ecol-ogy of a floodplain basin, Peace-Athabasca Delta, Canada, since A.D. 1700'. *Quaternary Research* 64: 147–62.

———. 2006. 'Reconstruction of multi-century flood histories from oxbow lake sed-iments, Peace-Athabasca Delta, Canada'. *Hydrological Processes* 20 (4): 131–53.

Wong, P.Y., and I. Brodo. 1992. 'The lichens of southern Ontario'. *Syllogeus* 69: 1–79.

World Bank. 2008. *World Development Indicators 2008*. Washington: World Bank.

World Conservation Union (IUCN—International Union for the Conservation of Nature and Natural Resources). 2002. '2002 IUCN Red List of Threatened Species'. http://www.redlist.org.

———. 2003. Press release: 'Release of the 2003 IUCN Red List of Threatened Species'. http://www.iucn.org.

———. 2007. '2007 IUCN Red List of Threatened Species'. http://www.redlist.org.

World Wildlife Fund, Zoological Society of London, and Global Footprint Net-work. 2006. *Living Planet Report 2006*. Gland, Switzerland: World Wildlife Fund.

Worldwatch Institute. 2003a. 'Earth trends: Agricultural inputs 2003'. http://earth-trends.wri.org/datatables/index.cfm?theme=8&CFID=504446&CFTOKEN=74873335.

———. 2003b. *Vital Signs*. New York: W.W. Norton.

———. 2004a. *State of the World 2004*. New York: W.W. Norton.

———. 2004b. *Vital Signs: The Trends That Are Shaping Our World*. New York: W.W. Norton

———. 2007a. *Biofuels for Transport: Global Potential and Implications for Energy and Agriculture*. London: Earthscan.

———. 2007b. *Vital Signs: The Trends That Are Shaping Our Future, 2007*. New York: W.W. Norton.

Worm, B., et al. 2006. 'Impacts of biodiversity loss on ocean ecosystem services'. *Science* 314 (7): 87–90.

Woynillowicz, D., and C. Severson-Baker. 2006. *Down to the Last Drop: The Atha-basca River and Oil Sands*. Drayton Valley, AB: Pembina Institute.

Woynillowicz, D., C. Severson-Baker, and M. Raynalds. 2005. *Oils Sands Fever: The Environmental Implications of Canada's Oil Sands Rush*. Drayton Valley, AB: Pembina Institute.

Index